Infrared Applications
of Semiconductors II

MATERIALS RESEARCH SOCIETY
SYMPOSIUM PROCEEDINGS VOLUME 484

Infrared Applications of Semiconductors II

Symposium held December 1–4, 1997, Boston, Massachusetts, U.S.A.

EDITORS:

Donald L. McDaniel, Jr.
Air Force Research Laboratory
Kirtland AFB, New Mexico, U.S.A.
and
University of New Mexico
Albuquerque, New Mexico, U.S.A.

M. Omar Manasreh
Air Force Research Laboratory
Kirtland AFB, New Mexico, U.S.A.

Richard H. Miles
SDL, Inc.
San Jose, California, U.S.A.

Sivalingam Sivananthan
University of Illinois at Chicago
Chicago, Illinois, U.S.A.

Materials Research Society
Warrendale, Pennsylvania

CAMBRIDGE UNIVERSITY PRESS
Cambridge, New York, Melbourne, Madrid, Cape Town,
Singapore, São Paulo, Delhi, Mexico City

Cambridge University Press
32 Avenue of the Americas, New York NY 10013-2473, USA

Published in the United States of America by Cambridge University Press, New York

www.cambridge.org
Information on this title: www.cambridge.org/9781107413429

Materials Research Society
506 Keystone Drive, Warrendale, PA 15086
http://www.mrs.org

First published 1998
First paperback edition 2013

Single article reprints from this publication are available through
University Microfilms Inc., 300 North Zeeb Road, Ann Arbor, MI 48106

CODEN: MRSPDH

ISBN 978-1-107-41342-9 Paperback

Effort sponsored by the Air Force Office of Scientific Research, Air Force Material
Command, USAF, under F49620-98-1-0055. The U.S. Government is authorized to
reproduce and distribute reprints for Governmental purposes notwithstanding any
copyright notation thereon. The views and conclusions herein are those of the
authors and should not be interpreted as necessarily representing the official
policies or endorsements, either expressed or implied, of the Air Force Office of
Scientific Research or the U.S. Government.

CONTENTS

*Invited Paper

*Invited Paper

PART IV: IR DETECTORS

*Invited Paper

PART V: GROWTH AND DOPING OF II-VI MATERIALS

*Invited Paper

*Invited Paper

PREFACE

The symposium titled "Infrared Applications of Semiconductors II" was held as Symposium F of the 1997 MRS Fall Meeting in Boston, Massachusetts. The sessions filled the four days of December 1–4. This very successful symposium was the second annual event covering this subject area, establishing the symposium topic as a regular part of the annual Fall Meeting. The session covered recent progress in III-V and II-VI interband and intersubband transitions in semiconductor materials and devices, and in semiconductor nonlinear optical and OPO materials. Participation was truly international featuring participants from five continents. A total of 74 papers and 59 posters were presented, most of which are detailed in this volume. These presentations chronicle the extensive progress being made in the modeling, design, fabrication and characterization of a diverse array of infrared semiconductor structures.

Donald L. McDaniel, Jr.
M. Omar Manasreh
Richard H. Miles
Sivalingam Sivananthan

January 1998

ACKNOWLEDGMENTS

We wish to thank the following organizations for their generous financial support of this symposium.

Air Force Office of Scientific Research (AFOSR)

Air Force Research Laboratory
Semiconductor Laser Branch (AFRL/DELS)

Air Force Research Laboratory
Space Sensing and Vehicle Control Branch (AFRL/VSSS)

In addition, we wish to thank the invited speakers for setting the tone of the sessions, the session chairs for assistance in organizing and managing their sessions, the many people who assisted in refereeing the papers in this volume and the staff of Materials Research Society. Neither the symposium nor this volume would have been possible without the selfless participation of these people.

xvii

MATERIALS RESEARCH SOCIETY SYMPOSIUM PROCEEDINGS

MATERIALS RESEARCH SOCIETY SYMPOSIUM PROCEEDINGS

Prior Materials Research Society Symposium Proceedings available by contacting Materials Research Society

Part I

Antimonide Related Materials —
Growth, Characterization, and Analysis

MATERIALS FOR MID-INFRARED SEMICONDUCTOR LASERS

A. R. KOST
Hughes Research Laboratories, Malibu, CA 90265, arkost@hrl.com

ABSTRACT

A variety of semiconductor materials have been used to fabricate diode lasers for the mid-infrared. Lasers using the lead salts (e.g. PbSnTe) have been commercially available for some time. Mid-infrared emitting III-V semiconductors (e.g. InGaAsSb) have superior thermal conductivity, and diode lasers fabricated from these materials offer higher powers. Of particular interest are the III-V semiconductor lasers based on type-II superlattices (e.g. InAs/GaInSb). Among the many unique properties attributed to type-II superlattices are small hole mass, reduced Auger recombination, and less inter-valence band absorption - all important for better lasers. Recent results with Quantum Cascade-type lasers are also very encouraging. This paper summarizes the important semiconductor materials for mid-infrared lasers with emphasis on the type-II superlattices.

INTRODUCTION

Interest in diode lasers operating in the 2-5 µm (MID-IR) wavelength range is driven by applications in spectroscopy, trace gas sensing, and infrared counter measures. In many cases, diode lasers with high output power are desirable as are diode lasers that operate at or near room temperature but not necessarily both. Generally, application to spectroscopy or trace gas sensing requires low to moderate power and high operating temperature. For infrared counter measures, high power (~ 1 Watt) is most important. For all applications, device reliability is important.

Among the most promising devices are those that have been fabricated from III-V semiconductor alloys containing antimony [1-7]. In particular, it has been proposed that GaInSb/InAs type-II superlattices offer improved laser performance in the MID-IR as a result of reduced Auger recombination [8,9]. Here we survey the semiconductors for MID-IR lasers and summarize the work with type-II superlattices. Although most work on MID-IR semiconductor lasers has emphasized the development of improved active regions, it is important to note that it is not the way that lasers can be improved. Optical cladding layers with low electrical resistance are also important for good performance.

LEAD SALT MATERIALS

Semiconductors based on lead salts (e.g. PbSnTe for the active layer) have been commercially available for some time. They are characterized by relatively small Auger coefficients and a very broad range of operating wavelengths (~ 3 µm to ~ 30 µm). Perhaps the most impressive MID-IR lead salt laser was a PbSe-PbSrSe multiple quantum well (MQW) device that operated at 4.2 µm up to 282K under pulsed current injection [10]. The lead salt semiconductors have low thermal conductivity (~ 0.1 W/cm•K at 77K), and lead salt lasers have correspondingly small output power, on the order of 1 mW or less. All lead salt lasers operate below room temperature. Difficulty in growing high quality hetero-junctions with these materials translates into poor device reliability. Finally, lead salt technology is mature, so we don't expect big improvements in the performance of the lead salt lasers in the near future.

II-VI SEMICONDUCTORS

There has been relatively little work on lasers based on the II-VI semiconductors. Perhaps the lack of interest can be explained by modest device performance, e.g. pulsed operation of a 2.9 µm HgCdTe device at 90K [11], and problems which are also associated with the lead salt materials. Like the lead salts, the II-VI semiconductors have relatively small thermal

3

conductivities, and high quality epitaxial growth is difficult.

III-V ALLOYS

In order to achieve good lattice matching, much of the MID-IR laser work with III-V semiconductors has concentrated on the GaInAsSb alloys (or the InAsSb subset) grown on InAs or GaSb substrates. The III-V alloys have good thermal conductivity, and lasers based on these materials have moderate output, e.g. 215 mW/facet cw at 3.5 μm for an InAsSb-AlInAsSb MQW device at 80K [12]. Like the lead salt and II-VI semiconductor lasers, MID-IR lasers employing III-V alloys operate below room temperature, e.g. maximum operating temperature of 175K for cw operation of the InAsSb-AlInAsSb MQW device [12]. Recently, a differential efficiency of 90% was reported for an InAsSbP-InAsSb-InAs double heterostructure laser operating at 3.2 μm at 78K [13]. The technology for III-V alloy lasers is moderately mature, and we don't expect large improvements in device performance soon.

AUGER RECOMBINATION

Auger recombination in semiconductors (Figure 1) is a nonradiative process which increases rapidly with increasing temperature and emission wavelength. It is believed to be the most important loss mechanism for inter-band MID-IR diode lasers. The temperature and band gap dependence for Auger processes are a result of energy and momentum conservation rules that require participation of electrons (holes) that are above (below) the k=0 point of the energy band. Electrons and holes are distributed in a band of energies within about kT of k=0, and for small band gap semiconductors, electrons and holes may be closer to k=0 and still satisfy the conservation rules.

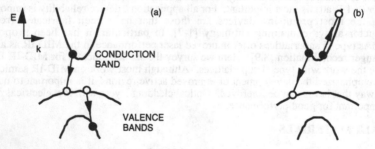

Figure 1. (a) Hole-hole Auger recombination and (b) electron-electron Auger recombination.

TYPE-II SUPERLATTICES

Two semiconductors in contact have a "type-II" band alignment the lowest energy conduction band states and the highest energy valence band states are in different semiconductors as pictured in Figure 2. We call a superlattice type-II if the bulk constituents have a type-II band alignment. If the bulk energy bands line up as in Figure 2a, it is called a staggered alignment superlattice. An example is InAsSb/InAs on a InAs substrate. Spatially indirect recombination occurs with an emission wavelength that can be longer than for the individual constituents. If the valence band in one constituent overlaps the conduction band in another, as illustrated in Figure 2b, it is called a broken gap superlattice. An example is GaInSb/InAs on a GaSb substrate.

Figure 3 provides a more precise description of the GaInSb/InAs broken gap superlattice. The straight solid lines represent bulk energy bands. The dashed lines labeled C1, HH1, and HH2 indicate the lowest energy conduction band state and first and second lowest heavy hole states for the superlattice. Optical transitions are between electrons with wavefunctions (curved lines) weakly localized in the InAs layers and holes with wavefunctions localized in GaInSb.

The energy gap of a type-II superlattice is adjusted by varying the composition of the constituents as well as layer thickness. For the GaInSb/InAs superlattice, the gap is readily adjusted from ~ 3 μm to the far infrared.

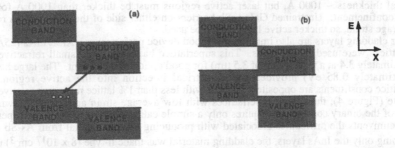

Figure 2. The two kinds of type-II superlattices: (a) staggered alignment and (b) broken gap.

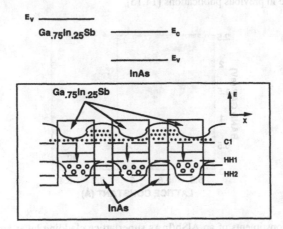

Figure 3. Electron and hole wavefunctions are weakly localized in a GaInSb/InAs superlattice.

Minimizing Auger Recombination

Auger recombination rates for GaInSb/InAs broken gap superlattices can be made small by varying one of the two degrees of freedom - layer thickness and ternary content [8,9]. Choices are subject to the following constraints: i) strain must not be so large as to introduce dislocations, ii) the emission wavelength is specified at the start, and iii) optical transitions must have sufficient oscillator strength for lasing to occur. Relatively thick (35 Å) GaInSb layers keep the growth-axis width of the conduction mini-band less than the energy gap to reduce electron-electron processes. A relatively high In content (25%) in the ternary (to increase the strain splitting of the valence bands) and a judicious choice of superlattice period reduce hole-hole recombination.

Laser Active And Cladding Regions

The type-II superlattice diode lasers discussed here were fabricated at the Hughes Research

5

Laboratories [14,15]. The active regions were MQW structures with $Ga_{0.75}In_{0.25}Sb/InAs$ superlattices for the wells. The most important reason for using MQWs is to reduce the average strain in the active region. GaInSb/InAs superlattices are compressively strained on GaSb, with a critical thickness ~ 1000 Å, but laser active regions must be thicker than 1000 Å for good optical confinement. Unstrained GaInAsSb barriers on either side of the superlattices reduced the average strain, so thicker active layers could be used.

Laser cladding layers are also important for good device performance. AlSb/InAs (25Å/25Å) superlattices were used for the clads. This superlattice has a relatively small refractive index (approximately 3.4 at a wavelength of 3.5 μm) for good optical confinement. The large band gap (approximately 0.85 eV) provides good electrical injection into the active region. The superlattice constituents are oppositely strained with less than 1% lattice mismatch relative to the substrate (Figure 4); thus, thick superlattices with low average strain are readily grown. The growth of the binary constituents requires only a simple calibration procedure. The superlattice also circumvents the difficulties associated with producing n-type material from As-Sb alloys.

By doping only the InAs layers, the cladding material was made n-type (8×10^{17} cm^{-3}) using a conventional dopant (Si) for molecular beam epitaxy. Current-voltage measurements on an AlSb(Be)/InAs - AlSb/InAs(Si) pn junction device showed that the junctions were rectifying and had relatively low resistance under forward bias.

More details of the device structure as well as fabrication methods and procedures for device testing can be found in previous publications [14,15].

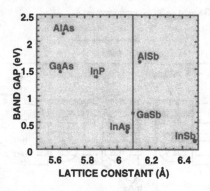

Figure 4. The constituents of an AlSb/InAs superlattice cladding layer are nearly lattice matched to the GaSb substrate.

Laser Performance

Electrically pumped GaInSb/InAs lasers have been demonstrated from 2.8 to 4.3 μm (Figure 5). All devices operated at liquid nitrogen temperature or above. Most of the lasers with a wavelength of 3.4 μm or less operated at temperatures that are accessible by thermoelectric cooling. A 3.2 μm laser operated up to 255 K.

With an 85 μsec current pulse (4.25% duty cycle), a 3.0 μm laser produced 75 mW per facet for the duration of the pulse (Figure 6). The differential quantum efficiency was approximately 4%.

These results are similar to the performance of devices that use the III-V alloys. Perhaps the lack of spectacular results with a type-II superlattice can be explained by differences between the optimized structures proposed by theoreticians (e.g. lasers with infinite superlattice active regions) and those that were actually fabricated. In any case, it is useful to investigate the mechanisms that limited performance.

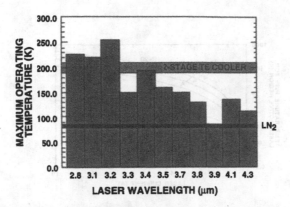

Figure 5. GaInSb/InAs superlattice lasers have been fabricated for a large portion of the mid-infrared.

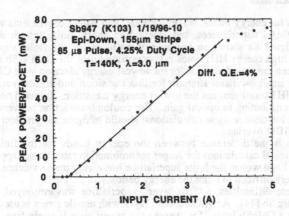

Figure 6. Light versus current for a GaInSb/InAs laser that was approximately 1 mm long.

Prospect For cw Power ≥ 1 Watt

Figure 7 shows calculations for the temperature rise from non-radiative recombination in an active region and from resistive heating in laser cladding layers. The results are for typical lasers with forward bias resistance between 0.5 and 2 ohm. Note that the calculations indicate that resistive heating dominates for all but the smallest input currents. For a 1 ohm laser, we expect an input current of 5 amps to heat the device by 70 K. Given the corresponding decrease in laser efficiency, it is no surprise that we observe saturating output as in Figure 6.

These results suggest that current injection levels should be kept to 5 A or less. Higher currents could be used only with devices that have less resistive cladding layers or heat sinks with higher thermal conductivity (e.g. diamond). To obtain a MID-IR device with cw power equal to or greater than one watt, with a current level of 5 amps, we need at least 20% efficiency. These results also explain why optically pumped lasers, with no resistive heating in the clads, can achieve higher output [16].

7

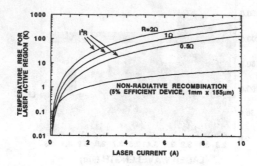

Figure 7. Calculations show that resistive heating is the dominate mechanism for laser heating for all but the smallest input current.

<u>Strain-Balanced Superlattices</u>

Initial calculations for energy bands, wavefunctions, and Auger coefficients were performed for infinite GaInSb/InAs superlattices, but lasers had short, finite length superlattices as described above. Figure 8a shows an unforeseen problem for the finite superlattice. The wavefunction for the high energy HH1 holes is concentrated near the GaInAsSb barriers. There is little overlap between the wavefunctions for the lowest energy electron state C1 and HH1. As a result, we expect a small oscillator strength for this transition and an unfavorable situation - lasing on the C1 to HH3 transition, not the lowest energy transition. Injected holes can occupy HH1 states without contributing to optical gain. The coulomb attraction between electrons and holes, which was not included in these calculations, should mitigate the problem to some extent by increasing the C1-HH1 overlap.

A bigger problem is the difference between the energy bands for the finite and infinite superlattices. The original calculations for Auger recombination rates do not apply to the actual laser active regions. Even worse, the finite superlattice has a very dense valence band structure, so it turns out to be very difficult to tailor the energy bands.

To circumvent these difficulties, a "four layer" superlattice was employed, with the basic period shown in Figure 8b [14]. An AlGaInAsSb layer with tensile strain is used to balance the compressive strain in a GaInSb layer. On average the superlattice is strain free, so it can fill an entire laser active region. The four-layer, strain-balanced superlattice also has a simplified valence band structure which is easier to tailor for reduced Auger recombination.

To date, we have had only limited success fabricating electrically pumped MID-IR lasers using the four-layer superlattice. Optical pumping experiments suggest that the difficulties are related to carrier transport into the active region.

QUANTUM CASCADE LASERS

Recent results with long wavelength, quantum cascade lasers are very impressive [17,18]. Pulsed, single mode, room temperature lasers have been demonstrated at 5.4 and 8 μm. By eliminating valence bands from the lasing process (the "type-I" quantum cascade laser), the quantum cascade design appears to have solved the Auger problem. Another plus is that the quantum cascade lasers use conventional InGaAs and InAlAs layers, lattice matched to InP substrates. On the other hand, the InGaAs/InAlAs conduction band offset limits operation to wavelengths longer than about 4 μm. This is not a fundamental difficulty, and it should be possible to use other semiconductor pairs for shorter wavelength emission.

A more fundamental problem, at least for the first quantum cascade lasers, has been low efficiency. Initially, it appeared that rapid, non-radiative, electron transitions by phonon

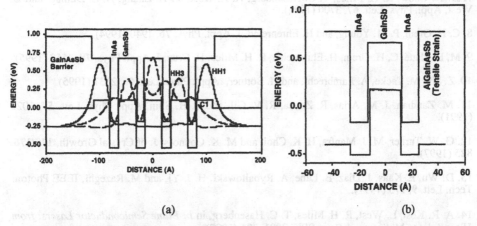

(a)   (b)

Figure 8. (a) Calculations predict small wavefunction overlap and weak oscillator strength for the transition from the lowest energy electron state C1 and the highest energy hole state HH1. (b) This four layer superlattice period balances compressive strain in GaInSb with tensile strain in AlGaInAsSb.

emission would make quantum cascade lasers unsuitable for applications that require very high optical power. More recently, dramatic improvements in device efficiency have been reported. A "type-II" (inter-band) quantum cascade laser developed by the University of Houston and the Naval Research Laboratories exhibited an internal efficiency exceeding 100% [19].

ACKNOWLEDGMENTS

I would like to acknowledge Linda West, Tom Hasenberg, Richard Miles, David Chow, Tom Boggess, and Michael Flatté for their work on type-II superlattice lasers. Work on MID-IR lasers at the Hughes Research Laboratories was supported in part by the Air Force Phillips Laboratory under Contract No. F29601-93-C-0037.

REFERENCES

1. R. J. Menna, D. R. Capewell, Ramon U. Martinelli, P. K. York, and R. E. Enstrom, Appl. Phys. Lett. **59**, 2127 (1991).

2. S. R. Kurtz, R. M. Biefeld, L. R. Dawson, K. C. Baucom, and A. J. Howard, Appl. Phys. Lett. **64**, 812 (1994).

3. A. N. Baranov, A. N. Imenkov, V. V. Sherstnev, and Yu. P. Yakovlev, Appl. Phys. Lett. **64**, 2480 (1994).

4. H. Q. Le, G. W. Turner, S. J. Eglash, H. K. Choi, D. A. Coppeta, Appl. Phys. Lett. **64**, 152 (1994).

5. Yong-Hang Zhang, Richard H. Miles, and David H. Chow, IEEE J. Of Selected Topics in Quantum Electron. **1**, 749 (1995).

6. H. K. Choi and G. W. Turner, Appl. Phys. Lett. **67**, 332 (1995).

7. D. H. Chow, R. H. Miles, T. C. Hasenberg, A. R. Kost, Y.-H. Zhang, H. L. Dunlap, and L. West, Appl. Phys. Lett. **67**, 3700 (1995).

8. C. H. Grein, P. M. Young, and H. Ehrenreich, J. Appl. Phys. **76**, 1940 (1994).

9. M. E. Flatté, C. H. Grein, H. Ehrenreich, R. H. Miles, H. Cruz, J. Appl. Phys. **78**, 4552 (1995).

10. Z. Shi, M. Tacke, A. Lambrecht, and H. Böttner, Appl. Phys. Lett. **66**, 2537 (1995).

11. M. Zandian, J. M. Arias, R. Zucca, R. V. Gil, and S. H. Shin, Appl. Phys. Lett. **59**, 1022 (1991).

12. G. W. Turner, M. J. Manfra, H. K. Choi, and M. K. Connors, J. of Crystal Growth, **175/176**, 825 (1997).

13. D. Wu, E. Kaas, J. Diaz, B. Lane, A. Rybaltowski, H. J. Yi, and M. Razeghi, IEEE Photon. Tech. Lett. **9**, 175 (1997).

14. A. R. Kost, L. West, R. H. Miles, T. C. Hasenberg, in *In-Plane Semiconductor Lasers: from Ultraviolet to Midinfrared*, Proc. SPIE **3001**, 321 (1997).

15. T. C. Hasenberg, R. H. Miles, A. R. Kost, and L. West, IEEE J. Quantum Electron. **QE-33**, 1403 (1997).

16. Y. H. Zhang, H. Q. Le, D. H. Chow, and R. H. Miles, in Proc. of the 7th Intl. Conference on Narrow Gap Semiconductors, (IOP, Brystol, 1995) pp. 36-40.

17. Jérome Faist, Federico Capasso, Carlo Sitori, Deborah L. Sivco, James N. Baillargeon, Albert L. Hutchinson, Sung-Nee G. Chu, and Albert Y. Cho, Appl. Phys. Lett. **70**, 2670 (1997).

18. S. Slivken, C. Jelen, A. Rybaltowski, L. Diaz, and M. Razeghi, Appl. Phys. Lett. **71**, 2593 (1997).

19. C. L. Felix, W. W. Bewley, I. Vurgaftman, J. R. Meyer, D. Zhang, C.-H. Lin, R. Q. Yang, and S. S. Pei, IEEE Photon. Technol. Lett. **9**, 1433 (1997).

THE GROWTH OF TYPE-II INFRARED LASER STRUCTURES

M. J. YANG, W. J. MOORE, B. R. BENNETT, B. V. SHANABROOK, AND J. O. CROSS
Naval Research Laboratory, Washington, D. C. 20375, yang@bloch.nrl.navy.mil

ABSTRACT

The MBE growth temperature for InAs/InGaSb/InAs/AlSb mid-infrared lasers has been studied. It is found that the best growth temperature is between 370°C and 420°C. A growth temperature above this range will result in excessive interlayer mixing, which degrades the radiative efficiency.

INTRODUCTION

The matched lattice constant and the broken-gap band alignment of InAs/GaSb heterostructures has encouraged considerable research on the 6.1Å lattice constant family, including InAs, InGaSb and AlSb materials. This material system provides tremendous potential for device applications, e.g., tunneling devices [1], microwave field-effect transistors [2], infrared detectors [3], and mid-infrared lasers [4,5].

EXPERIMENT

Recently, we have reported that the quality of type-II mid-infrared laser heterostructures is highly sensitive to the growth temperature [6]. In order to determine the optimum growth temperature, we have grown a series of infrared laser samples on GaSb substrates by molecular beam epitaxy at different growth temperatures within a two-week period. The samples consist of 20 periods of 5.5 ML InAs/10 ML $In_{0.28}Ga_{0.72}Sb$/5.5 ML InAs/14 ML AlSb 'W'-structure [5], clad with a 1 μm AlSb buffer and a 0.2 μm AlSb cap layer. The layer thickness was calibrated during the buffer growth and monitored during the 'W'-structure growth by RHEED oscillations, as shown in Fig 1. However, we do not see any correlation between the strength of the RHEED oscillation amplitude and the sample quality. The interfacial bonds are forced to be InSb-like by controlling the shutter sequence. For example, after 14ML of AlSb, the surface is soaked with Sb for 2 sec, followed by one monolayer of In. Subsequently, 4.5ML of InAs is grown, followed by one monolayer of In and then 2 sec of Sb soak. The total InAs layer thickness is 5.5ML by our convention which includes half an interfacial bond on each side.

The samples were characterized by photoluminescence (PL) with a Bomem Fourier transform infrared spectrometer in the temperature range from 300K to 5K. Fig. 2 shows the integrated PL intensity at 300K and 5K as a function of the substrate temperature employed during the growth of 'W'-structure. At 300K, the PL intensity for samples grown

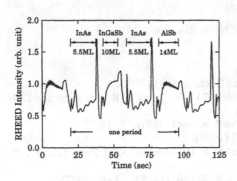

Fig. 1 RHEED intensity of the specular diffraction beam as a function of time.

between 370°C and 420°C is at least twice that of the other samples. This contrast is enhanced at low temperature. Samples with high luminescence efficiency have a PL intensity which increases monotonically as the temperature is decreased. The PL spectrum shows a single peak with a fast turn-on and a thermal tail when $T \geq 80K$. At 5K, the spectrum width is dominated by inhomogeous broadening with a full-width-at-half-maximum (FWHM) of 20 meV to 30 meV. In contrast, for poorer samples, as temperature decreases from 300K to 5K, the PL intensity first increases and then decreases. In some cases, the PL intensity at 5K is even smaller than that at 300K. At 5K, the PL spectrum has either a broad width with an FWHM around 50 meV, or a slow turn-on with a low-energy shoulder.

CONCLUSION

Our results show that the growth window for type-II mid-infrared laser structures is narrow, ranging at most from 370°C to 420°C. This suggests that it is important to minimize uncertainty in the substrate temperature. Our current transmission spectrum thermometry setup, designed for GaAs substrates, is not usable with GaSb substrates. As a result, growth temperature is always referred to the GaSb (1×5) to (1×3) RHEED phase transition. It has been reported [7] that the temperature of this transition is dependent on the type of Sb, i.e., cracked or uncracked, as well as the Sb beam equivalent pressure. We have

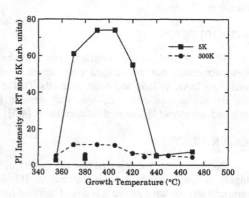

Fig. 2 The integrated photoluminescence intensity at 300K and 5K for samples grown at different temperatures.

standardized the method by determining the RHEED phase transition temperature under a carefully measured Sb to Ga flux ratio. The results are to be published [8].

ACKNOWLEDGEMENT

The authors acknowledge the support of Office of Naval Research.

REFERENCES

1. T. C. McGill and D. A. Collins, Semicond. Sci. Technol. 8, S1 (1993).
2. J. B. Boos, W. Kruppa, D. Park, Elec. Lett. 32, 1624 (1996); and C. R. Bolognesi and D. H. Chow, IEEE Elec. Dev. Lett. 17, 534 (1996).
3. C. Mailhiot and D. L. Smith, J. Vac. Sci. Technol. A7, 445 (1989).
4. R. Q. Yang, Superlatt. and Microstruct. 17, 77 (1995).
5. J. R. Meyer, C. A. Hoffman, F. J. Bartoli, and L. R. Ram-Mohan, Appl. Phys. Lett. 67, 757 (1995).
6. M. J. Yang, W. J. Moore, B. R. Bennett, and B. V. Shanabrook, Elec. Lett. (1997).
7. T. H. Chiu and W. T. Tsang, J. Appl. Phys. 57, 4572 (1985).
8. M. J. Yang, W. J. Moore, B. R. Bennett, B. V. Shanabrook, and J. O. Cross, to be published.

NON-DESTRUCTIVE AND WHOLE WAFER CHARACTERIZATION OF III-V INFRARED EPITAXIAL MATERIALS PREPARED BY TURBO DISK METALORGANIC CHEMICAL VAPOR DEPOSITION

Z. C. Feng, M. Pelczynski, C. Beckham, P. Cooke, I. Ferguson, R. A. Stall.
EMCORE Corporation, 394 Elizabeth Avenue, Somerset, NJ 08873

ABSTRACT

Multiple wafer growth of infrared III-V semiconductor materials of InSb and InGaAsP have been produced by metalorganic chemical vapor deposition technology employing a vertical reactor growth configuration with a high speed rotating disk. Three measurement techniques of sheet resistivity, Fourier transform infrared (FTIR) reflectance and photoluminescence have been used to characterize epitaxial films on wafers up to 4" diameter. Mapping distributions of the film thickness, sheet resistivity, surface morphology, and PL peak wavelength with uniformities better than 1% are illustrated. Data from our 2900 runs are produced. Variations of the characteristic features of the film with the growth conditions are discussed. These whole wafer and non-destructive material characterization techniques tightly coupled with the epitaxial processes are necessary to realize the high quality and high uniformity growth of state-of-art materials in a production environment.

INTRODUCTION

Infrared (IR) semiconductor materials and devices, in the 1-6 μm spectral range, have many important applications in telecommunications, detectors, automobiles, infrared radar, laser-guided weapons, laser rangefinders, remote sensors, environmental monitoring of pollutant gases, tracing of missiles and hydrocarbon fuel emissions, wavelength specific medical applications and spectroscopy [1,2]. InSb (5-6 μm) and InGaAsP (1.3 μm) are two of these important materials. Both are direct gap semiconductors. Several growth technologies, including molecular beam epitaxy (MBE) [3], metalorganic chemical vapor deposition (MOCVD) [4], liquid phase epitaxy (LPE) [5] etc., have been employed to grow these materials. With the rapid development of semiconductor IR devices, the requirement for the industrial production of these III-V materials is in great demand. Developments in modern electronics and optoelectronics require the production of different types of III-V IR material and microstructure wafers with high uniformity over the entire wafer area, coupled with the ability to maintain a wafer-to-wafer repeatability within a run and run-to-run, the ability to maximize the yield per wafer and the minimization of the costs of mass production. To achieve these goals, an advanced MOCVD turbo disk technology, which utilizes a vertical growth configuration and a high speed rotating disk reactor (RDR), has been developed for the epitaxial deposition of multiple wafers of various compound semiconductor materials [6-10].

High throughput production also demands the large area deposition of epitaxial compound materials and this raises a new challenge for whole wafer non-destructive material characterization. This is quite different from single point and destructive measurements. In the case of III-V compound semiconductors, such characterization is more complicated than for the case of single silicon semiconductors. In the mass production environment, there is also the requirements of convenience, rapid-turn around and reliability of these whole wafer non-destructive characterization techniques. As a flexible technique, whole wafer characterization has become an important part of advanced growth technology development.

In this study, three measurement techniques of sheet resistivity, Fourier transform infrared (FTIR) reflectance and photoluminescence (PL) are reported. Non-destructive whole wafer mapping of InSb and InGaAsP grown on GaAs or InP are produced for wafers up to 4" diameters. Single layer heterostructures are emphasized because they form the foundation of multi-layer microstructures, such as quantum wells, superlattices, Braggs and other device structures. Numerous wafers were examined and typical examples for each technique were shown. In practice, each wafer has to undergo various whole wafer characterizations before they are qualified for the end products.

EXPERIMENT

Two types of III-V IR epitaxial materials, including binary InSb films (with a band gap close to 6 µm) grown on 100 mm (4") GaAs substrates and quaternary InGaAsP (with a band gap at 1.3 µm) grown on 50 mm (2") InP with a lattice-matched heterostructure, are reported in this study. All materials were grown in an EMCORE MOCVD multiwafer system, which is capable of holding up to 42 x 2", or 17 x 3", or 9 x 4" wafers per run. It is equipped with a double walled water-cooled stainless-steel chamber, a two zone filament heater, and a UHV loadlock. This system offers a uniform temperature and reactant flow across a large deposition area covering one or multiple wafers [6-10]. *In-situ* optical reflectance is also installed in the growth chamber.

Trimethylindium (TMI) and Tris-dimethylaminoantimony (TDMASb) were used as In and Sb sources for the InSb growth. Hydrogen acted as carrier gas [7,9]. For the growth of InGaAsP, high purity TMI and trimethygallium (TMG) metalorganic sources were used to supply In and Ga, and hydrides AsH_3 and PH_3 were used for the As and P sources, respectively [6,10]. High purity H_2 was used as the carrier gas. Details can be found in [6-10].

After growth, the samples were characterized by a series of in-house characterization techniques, such as Hall, C-V, sheet resistivity, Nomarski microscope, scanning electron microscopy (SEM) and double crystal x-ray diffraction (DCXRD), PL etc., to be confirmed with mirror-like surface, good electrical properties, and high crystalline quality. This study, moreover, is concentrated on three mapping techniques: sheet resistivity/conductivity mapper using a contactless eddy current technique, thin film thickness mapper using FTIR spectroscopy technique, and multi-wavelength laser PL mapper. Examples are shown to demonstrate these capabilities and their applications with IR III-V materials.

RESULTS AND DISCUSSION

Sheet Resistivity Map of InSb on GaAs

EMCORE has been producing InSb materials by MOCVD for applications in the automotive and consumer industry. It had proved difficult to grow these materials in a production environment due to the tight temperature control required to maintain high mobilities using standard antimony sources such as triethylantimony (TESb) and Trimethylantimony (TMSb). To overcome this limitation EMCORE has used TDMASb as an antimony precursor. Typical values of mobility and carrier concentration for intrinsic grown InSb (1.2 µm thick) on 4" GaAs substrates using TDMASb and TMI to be >48,000 cm^2/v-sec and 1.8-2.2 x 10^{16}/cm^3, respectively. X-Ray data has a Full Width at Half Maximum (FWHM) of 260 - 320 aresec, almost the highest quality material produced to date. This is an order of magnitude improvement over previously published data which showed mobilities >40,000 cm^2/v-sec for layers 4 µm thick [11].

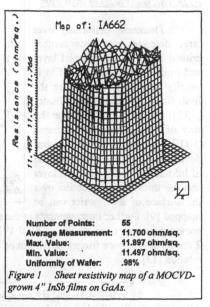

Number of Points:	55
Average Measurement:	11.700 ohm/sq.
Max. Value:	11.897 ohm/sq.
Min. Value:	11.497 ohm/sq.
Uniformity of Wafer:	.98%

Figure 1   Sheet resistivity map of a MOCVD-grown 4" InSb films on GaAs.

Over 2,000 growth runs have been completed. Early in the manufacturing cycle it was found necessary to closely relate destructive and non-destructive characterization techniques to reflect on the properties of the epitaxial material. It is obvious that there is a need to calibrate the material without destroying every wafer. We have relied on sheet resistance measurements and x-ray FWHM data to serve this purpose. The thickness and mobility are sensitive to the x-ray data. It is then possible to perform periodic destructive tests to verify that mobility and thickness data are consistent, using sheet resistance and x-ray measurements on all other wafers to sustain product needs. Figure 1 shows a 55-point map of the sheet resistivity of a typical InSb epilayer grown on a 4" (100 mm) diameter GaAs substrate with an excellent uniformity <1% over the entire wafer. The sheet resitance and mobility data are taken from over 2900 wafers while the thickness and x-ray data are from about 500 wafers. Table I shows the summarized characterization data.

Table I - Characterization data for 4" InSb/GaAs with nondestructive data
for >2900 wafers and destructive data for >500 wafers.

TEST	UNIFORMITY	STANDARD DEVIATION
SHEET RESISTANCE	1.94 %	.93 %
THICKNESS	1.63 %	.38 %
MOBILITY	39,817 cm^2/v sec	1,872 cm^2/v sec
X-RAY FWHM	151.2 arcsec	19.4 arcsec

FTIR Thickness Map of InSb Film

Thickness uniformity over large diameter wafers is another critical issue on IR epitaxial layer and multi-layer structures. SEM is usually employed to measure the InSb film thickness in the industry [7,12]. Recently a technique by way of Fourier transform infrared (FTIR) spectroscopy was developed to measure the thickness of InSb and other IR epitaxial layer and the thickness distribution over the surface of a 4" wafer can be

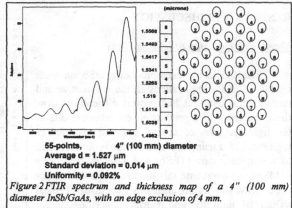

55-points,   4" (100 mm) diameter
Average d = 1.527 µm
Standard deviation = 0.014 µm
Uniformity = 0.092%

Figure 2 FTIR spectrum and thickness map of a 4" (100 mm) diameter InSb/GaAs, with an edge exclusion of 4 mm.

mapped [9]. Further improvements in the measurements and materials have been made. Figure 2 provides such an example for a typical 4" InSb/GaAs. The FTIR spectrum in the left side of Fig. 2 shows interference fringes which can be used to calculate the film thickness from the following relation [13]:

$$d = \frac{1}{2} \times \frac{\Delta m \lambda_1 \lambda_2}{\lambda_1 (n^2(\lambda_2) - \sin^2 \theta)^{\frac{1}{2}} - \lambda_2 (n^2(\lambda_1) - \sin^2 \theta)^{\frac{1}{2}}}, \tag{1}$$

where d is the film thickness to be measured, λ_1 and λ_2 represent the wavelengths at two maxima or minima of the interference fringes respectively, Δm is the number of fringes between the two maxima or minima corresponding to λ_1 and λ_2 respectively, $n(\lambda_1)$ and $n(\lambda_2)$ are the refractive indices of the materials being measured, and θ is the incident angle with the surface normal. The computer controlled mapper moves the wafer to measure the 55 points FTIR spectra. The film thickness is calculated by Eq. (1) and mapped automatically. A thickness map distribution is shown in the right part of Fig. 2. For this MOCVD-grown InSb/GaAs, we obtained an average InSb film thickness of 1.53 µm and a uniformity of 0.92%, better than 1%, indicating a high thickness uniformity of our InSb films grown on 4" GaAs substrates by Turbo Disk MOCVD. Both this thickness and sheet resistivity (in last sub-section) uniformity are superior than previously reported results in the literature for 2" InSb/GaAs in an early MOCVD system [7] and 3" InSb/GaAs produced by a planetary system [11].

InGaAsP epitaxial materials

Quaternary alloy InGaAsP, in the 1.3-1.5 µm spectral region, has become the key material for the fabrication of laser diodes in the NIR communication systems and quantum well infrared photo-detector (QWIP) structures. The major characterization technique for these materials is photoluminescence (PL) and mapping. PL spectrum provides information on alloy composition, crystalline properties, and impurities. PL maps directly measure the distribution and uniformity of the composition and crystalline quality of direct bandgap and the ability to produce a laser structure, compound semiconductor materials. Efforts have been made continually to control the composition distribution and improve the PL wavelength uniformity of

InGaAsP/InP within the whole wafer, from wafer to wafer in a batch reactor, and from run to run [14-17].

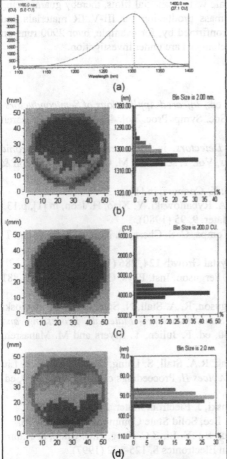

Figure 3   RT PL of an InGaAsP/InP, 2", D083, (a) single spectrum with λ(peak) at 1.30 μm and a FWHM of 90.9 nm, and PL maps over 50 mm wafer (with 3 mm edge exclusion) of (b) λ(peak) of an InGaAsP/InP, with a average λ(peak) of 1302.5 nm, a standard deviation (S-D) of 3.23 nm and a uniformity of 0.25%, (c) PL peak intensity, with an average I(peak) of 3840 count units (CU), a S-D of 190 CU and a uniformity of 5.0%, and (d) PL FWHM, with a average FWHM of 90.7 nm, a S-D of 2.25 nm and a uniformity of 2.5%.

Earlier work on 1.3 μm compositions showed uniformities of ±6 nm for a single (2") wafer system [14], ±4 nm for three [15] and six [16] wafer system. The main cause of compositional uniformity for InGaAsP/InP is the temperature sensitivity, which has been studied theoretically using the thermodynamic modeling [17] and experimentally [18]. In addition, a leading effect due to In being carried from the wafer carrier surface and changing the composition of the gas stream near the wafer edge has been studied by us recently [8]. A new design of a susceptor/wafer carrier combination has been presented to reduce the temperature difference between the wafer and its carrier, leading to better PL uniformities [8]. Figure 3 provides such an improved result of room temperature (RT) PL data for a 2" NIR InGaAsP/InP film, showing a single point spectrum in (a) and PL map (3 mm edge excluded) distributions of peak wavelength in (b), peak intensity in (c) and FWHM in (d). Excellent results have been obtained, showing in the figure caption, with uniformities of 0.25% for peak wavelength, 2.5% for FWHM, and 5.0% for PL peak intensity, over the entire wafer area.

SUMMARY

Non-destructive and whole wafer characterization on MOCVD-grown III-V infrared materials of InSb and InGaAsP was performed by mapping film sheet resistivity, thickness and major PL band. The obtained data show that the grown materials are of high crystalline quality and uniformity. For example, uniformities of our epitaxial film sheet resistivity, thickness and major PL band peak wavelength are typically better than 1%. Other mapping parameters such as the PL band intensity and band width possess also good uniformities among 2-5%. These wafer scale material characterizations were tightly combined with the epitaxial growth processes and helped to greatly improve the quality and uniformity of the large scale wafer epitaxial films, thereby guaranteeing the success, high yield and high efficiency of mass production for III-V IR materials and structures using MOCVD technology. These are confirmed by, for example, over 2900 runs on 4" InSb/GaAs wafers. Other full wafer mapping techniques are under investigation.

REFERENCES

1. M. O. Manasreh, T. H. Myers and F. H. Julien ed., *Infrared Applications of Semiconductors-Materials, Processing and Devices*, Mat. Res. Soc. Symp. Proc. Vol. **450**, Materials Research Society (1997).
2. M. Razeghi ed., *Long Wavelength Infrared Detectors*, in book series of *Optoelectronic Properties of Semiconductors and Superlattices*, Vol. 1, series ed. M. O. Mansreh, Gordon & Breach Press (1996).
3. G. W. Turner, H. K. Choi, M. J. Manfra and M. K. Connors, in [1], p. 3.
4. M. Razeghi, J. Diaz, H.J. Yi, D. Wu, B. Lane, A. Rybaltowski, Y. Xiao, H. Jeon, in [1], p.13.
5. D. E. Holmes and G. S. Kamath, J. Electron. Mater. **9**, 95 (1980).
6. M. A. McKee, P. E. Norris, R. A. Stall, G. S. Tompa, C. S. Chern, N. Noh, S. S. Kang and T. J. Jasinski, J. Crystal Growth **107**, 445 (1991).
7. M. A. McKee, B.-S. Yoo and R. A. Stall, J. Crystal Growth **124**, 286 (1992).
8. A.G. Thompson, R.A. Stall, A.I. Gurary and I. Ferguson, Inst. Phys. Conf. Ser. No. 155, 881 (1996).
9. Z. C. Feng, C. Beckham, P. Schumaker, I. Ferguson, R. A. Stall, N. Schumaker, M. Povloski and A. Whitley, in *Infrared Applications of Semiconductors - Materials, Processing, and Devices*, Mat. Res. Soc. Symp. Proc. Vol. **450**, ed. F. Julien, T. Myers and M. Manasreh, MRS, Pittsburgh, pp. 61-66 (1997).
10. I. Ferguson, C.A. Tran, R.F. Karlicek, Z.C.Feng, R.A. Stall, S. Laing, W. Cai, Y. Li, Y. Liu and Y. Lu, in *Photodetectors: Materials and Devices II*, Proceedings of SPIE, Vol. **2999**, ed. G.J. Brown & M. Razeghi, pp. 298-305 (1997).
11. E. Woelk, H. Jurgensen, R. Rolph and T. Zielinski, J. Electron. Mater. **24**, 1715 (1995).
12. B.-S. Yoo, M. A. McKee, S.-G. Kim and E.-H. Lee, Solid State Commun. **88**, 447 (1993).
13. K. Li, A.T.S. Wee, J. Lin, K.L. Tan, L. Zhou, S.F.Y. Li, *Z.C. Feng*, H.C. Chou, S. Kamra & A. Rohatgi, J. Materials Science: Materials in Electronics **8**, 125-132 (1997).
14. A.W. Nelson, P.C. Spurdens, S. Cole, R.H. Walling, S. Wong, M.J. Harding, D.M. Cooper, W.J. Devlin and M. J. Robertson, J. Crystal Growth 93, 792 (1988).
15. M.A. McKee, B.-S. Yoo and R.A. Stall, Proc. 4th Intl. Conf. On InP and Related Materials, p.151 (1992).
16. K. Mori, M. Takemi, T. Takiguchi, K. Goto, T. Nishimura, T. Kimura, Y. Mihashi and T. Murotani, Proc. 5th Intl. Conf. On InP and Related Materials, p.235 (1993).
17. A.S. Jordan, J. Electron. Mat. 24, 1649 (1995).
18. R.M. Lum, M.L. McDonald, E.M. Mack, M.D. Williams, F.G. Storz and J. Levkoff, J. Electron. Mat. 24, 1577 (1995).

RECENT PROGRESS IN THE GROWTH OF MID-ir EMITTERS BY METALORGANIC CHEMICAL VAPOR DEPOSITION

R. M. Biefeld, A. A. Allerman, S. R. Kurtz, and K. C. Baucom
Sandia National Laboratory, Albuquerque, New Mexico, 87185, USA

ABSTRACT

We report on recent progress and improvements in the metal-organic chemical vapor deposition (MOCVD) growth of mid-infrared lasers and using a high speed rotating disk reactor (RDR). The devices contain AlAsSb claddings and strained InAsSb active regions. These lasers have multi-stage, type I InAsSb/InAsP quantum well active regions. A semi-metal GaAsSb/InAs layer acts as an internal electron source for the multi-stage injection lasers and AlAsSb is an electron confinement layer. These structures are the first MOCVD multi-stage devices. Growth in an RDR was necessary to avoid the previously observed Al memory effects found in conventional horizontal reactors. A single stage, optically pumped laser yielded improved power (> 650 mW/facet) at 80 K and 3.8 μm. A multi-stage 3.8-3.9 μm laser structure operated up to T=170 K. At 80 K, peak power > 100 mW and a high slope-efficiency were observed in gain guided lasers.

INTRODUCTION

Mid-infrared (3-6 μm) lasers and LEDs are being developed for use in chemical sensor systems and infrared countermeasure technologies. These applications require relatively high power, mid-infrared lasers and LEDs operating near room temperature. The radiative performance of mid-infrared emitters has been limited by nonradiative recombination processes (usually Auger recombination) in narrow bandgap semiconductors. Potentially, Auger recombination can be suppressed in "band-structure engineered", strained Sb-based heterostructures. We have demonstrated improved performance for midwave infrared emitters in strained InAsSb heterostructures due to their unique electronic properties that are beneficial to the performance of these devices.[1-3] To further improve laser and LED performance, we are exploring the MOCVD-growth of novel multi-stage (or "cascaded") active regions in InAsSb-based devices.

Multi-stage, mid-infrared gain regions have been proposed for several material systems.[1-6] Ideally, a laser with an N-stage active region could produce N photons for each carrier injected from the external power supply. Thereby, multi-staging of the laser active region may increase gain, lower threshold current, and finally increase the operating temperature of gain-limited, mid-infrared lasers. The success of the unipolar, quantum cascade laser demonstrates the benefit of multi-stage gain regions.[7] However, the nonradiative (optical phonon) lifetimes of the unipolar devices are orders of magnitude shorter than the Auger-limited lifetimes for interband devices, and with multi-staging, mid-infrared interband Sb-based lasers may have lower threshold currents than unipolar quantum cascade lasers.

Gain regions with multiple electron-hole recombination stages have been proposed for Sb-based lasers.[1,5,6] These devices utilize a semimetal layer, formed by an InAs (n) / GaAsSb (p) heterojunction, as an internal electron-hole source between stages. Recently, cascaded lasers with type II InAs/GaInSb active regions have been demonstrated.[8] The type II lasers were grown by molecular beam epitaxy, and characteristic of multi-stage lasers, large differential quantum efficiencies (> 1) are now reported.[9] In this work we demonstrate 10-stage lasers composed of InAsSb quantum wells with type I band offsets.

A band diagram of our multi-stage active region, under forward bias, is shown in Figure 1. Electron-hole recombination occurs in compressively strained InAsSb quantum wells separated by tensile strained InAsP barriers. The type I, InAsSb/InAsP quantum wells have a large light-heavy hole splitting required for suppression of Auger recombination.[2] Electron-hole pairs for each stage are generated at a semi-metal, GaAsSb (p)/ InAs (n) heterojunction. An AlAsSb layer prevents electrons from escaping; nominal hole confinement is provided by the InAsSb quantum well valence band offset relative to the InAsP barrier layer in this initial device. Ideally, electron-hole generation replenishes the carriers which recombine in each stage, and for each carrier injected

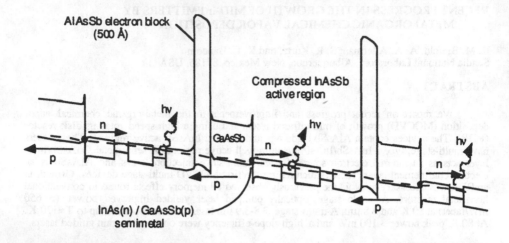

AlAsSb electron block
(500 Å)

Compressed InAsSb
active region

hv

hv

n

n

p

GaAsSb

hv

n

p

p

InAs(n) / GaAsSb(p)
semimetal

Figure 1 - Band diagram of a multi-stage laser active region with compressed, type I InAsSb quantum wells separated by InAsP barriers, electron-hole generation by an InAs(n)/GaAsSb(p) semimetal heterojunction, and an AlAsSb electron block

from the external circuit, the multi-stage active region can emit several photons resulting in an overall quantum efficiency greater than unity. In practice, charge can accumulate in the active region, shift the Fermi-level, and turn-off electron generation at the semi-metal. Previously, we have found that a AlGaAsSb graded layer between the AlAsSb and GaAsSb layers reduces hole trapping, thus increasing laser duty cycles and lowering turn-on voltages.

Unlike previous cascaded lasers, our device was grown by metal-organic chemical vapor deposition (MOCVD). Our devices are among the most complex structures ever grown by MOCVD, and it can be difficult to alternate Al, In, Ga, P, As, or Sb bearing materials while maintaining sharp interfaces and not having chemical carry-over into other layers. In particular, we have found that the combination of novel Al organometallic sources and an MOCVD, vertical, high speed rotating-disk reactor (RDR) is necessary to grow these structures to avoid the chemical carry-over previously observed with a horizontal reactor.[3] Even with an RDR, it is necessary to optimize the growth conditions and the reactor configuration as discussed below to minimize the effects of chemical carry over.

EXPERIMENTAL

This work was carried out in a previously described vertical, high-speed, rotating-disk reactor (RDR).[10] Ethyldimethylamine alane (EDMAA), trimethylindium (TMIn), triethylgallium (TEGa), triethylantimony (TESb), diethylzinc (DEZn), phosphine, and tertiarybutylarsine (TBAs) or arsine were used as sources. P-type doping was accomplished using diethylzinc (DEZn) in a dilution system. The structures were grown at 500 °C and 70 torr. The V/III ratios were optimized separately for each material and the InAsSb/InAsP strained-layer superlattice. The growth was performed on (001) InAs substrates. Hydrogen was used as the carrier gas at a flow of 18.5 slpm with a substrate rotation speed of 1500 rpm to retain matched flow conditions.[10] Semi-insulating, epi-ready GaAs or n-type InAs substrates were used for each growth.

Both InAsSb/InAs and InAsP/InAs multiple quantum wells (MQWs) were grown by MOCVD on n-type InAs substrates for calibration purposes to determine the solid-vapor distribution coefficients separately for Sb in InAsSb and P in InAsP. The InAsSb/InAsP SLSs were lattice matched to InAs with $\Delta a/a < 0.0004$. The MQW and SLS composition and strain were determined by double crystal x-ray diffraction.

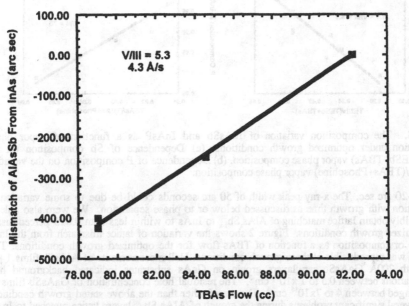

Figure 2. Effect of change in TBAs flow on the lattice match to the InAs substrate of an AlAsSb layer grown at 500 °C, 70 torr, a V/III ratio of 5.3, and a growth rate of 4.3 Å/s.

Double crystal x-ray diffraction (DCXRD) of the (004) reflection was used to determine alloy composition and layer thicknesses. Layer thicknesses were also determined using a groove technique and these were checked in several instances by cross sectional SEM. These techniques usually agreed within a few percent.

Infrared photoluminescence (PL) was measured on all samples from 14 K up to 300 K using a double-modulation, Fourier-transform infrared (FTIR) technique which provides high sensitivity, reduces sample heating, and eliminates the blackbody background from infrared emission spectra. Injection devices also were characterized with double modulation FTIR.

RESULTS AND DISCUSSION

MOCVD Growth

The optimum growth conditions for AlAs$_x$Sb$_{1-x}$ occurred at 500 °C and 70 torr at a growth rate of 4.3 Å/s using a V/III ratio of 5.3 assuming a vapor pressure of 0.75 torr for EDMAA at 19.8 °C and an [TESb]/([TBAs]+[TESb]) ratio of 0.83. The growth rate was found to be dependent on the EDMAA flow and independent of the group V flows for the conditions examined in this work. The best surface morphologies with the lowest number of defects were obtained by using a buffer layer grown before the AlAs$_x$Sb$_{1-x}$ layer. The defects consisted primarily of square pyramidal hillocks 10 to 20 μm on a side. Lattice matched AlAs$_x$Sb$_{1-x}$ films of high crystalline quality, as evidenced by double crystal x-ray diffraction (DCXRD) where full widths at half of the maximum intensity (FWHM) of less than 50 arc sec were obtained. Typical InAs substrate peaks

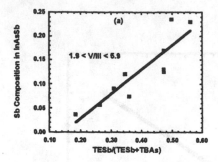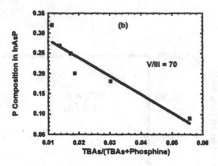

Figure 3. The composition variation of InAsSb and InAsP as a function of vapor phase composition under optimized growth conditions. (a) Dependence of Sb composition on the TESb/(TESb+TBAs) vapor phase composition. (b) Dependence of P composition on the variation of TBAs/(TBAs+Phosphine) vapor phase composition.

were 10-20 arc sec. The x-ray peak width of 50 arc seconds could be due to some variation in composition with growth time as discussed below or to phase separation. We were also able to reproducibly obtain lattice matching of $AlAs_xSb_{1-x}$ to InAs to within less than 0.015 percent using the optimized growth conditions. Figure 2 shows the variation of lattice mismatch from the InAs substrate or composition as a function of TBAs flow for the optimized growth conditions. This variation was reproducible at these growth conditions. Hall measurements of AlAsSb films 1 μm thick with 200Å GaAsSb cap layers grown on GaAs substrates indicated background hole concentrations between 0.5 to 1 x10^{17} cm^{-3}. The residual hole concentration of GaAsSb films on GaAs ranged between 4 to 7x10^{16} cm^{-3}. The use of other than the above stated growth conditions led to several significant problems during the growth of $AlAs_xSb_{1-x}$ layers lattice matched to InAs. These included composition control and reproducibility. For instance, growth at higher V/III ratios resulted in a large drift in composition from run to run. Composition variations have also been observed due to excess cooling of the chamber walls, probably due to an Sb memory effect on the inlet to the reactor.

We have successfully doped the $AlAs_xSb_{1-x}$ layers p-type using DEZn. We achieved p-type levels of 1 x 10^{16} to 6 x 10^{17} cm^{-3}. The mobilities for the $AlAs_xSb_{1-x}$ layers ranged from 200 to 50 cm^2/Vs with no clear trend that could be associated with the carrier concentration or type. SIMS measurements on the Zn doped samples indicated a similar level of Zn compared to the p-type carriers indicating complete activation of the Zn.

The growth of InAsSb/InAs and InAsP/InAs MQW structures were examined to optimize the growth of the ternaries. The growth conditions for both of these MQW's were examined at 500 °C and 70 torr using a rotation speed of 1500 rpm and a total H$_2$ flow of 18.5 SLM to retain matched flow conditions.[10] Growth rates of 2.8 Å/s were used for both systems. The growth rate was found to be proportional to the TMIn flow into the reaction chamber and independent of the group V flows. A purge time of 15 to 20 seconds with arsine flowing during the purge was used between InAs and the ternary layer growth to allow for source flow changes during the growth of the MQWs. Figure 3(a) illustrates the dependence of Sb composition on the TESb/(TESb+TBAs) vapor phase composition for optimized growth conditions. In this set of experiments the flow of TESb was held constant while the TBAs flow was varied to change the Sb solid composition. The V/III ratio varied from 1.9 to 5.9. A low V/III ratio is necessary for the growth of high quality InAsSb due to the low vapor pressure of Sb; excess Sb tends to cause surface morphology defects. In Figure 3(b) the dependence of P composition in InAsP on the variation of TBAs/(TBAs+Phosphine) vapor phase composition is illustrated. The flow of phosphine was held constant and the flow of TBAs was varied. In this case the V/III ratio is dominated by the excess phosphine flow and was approximately constant at 70. The high V/III ratio and excess phosphine,

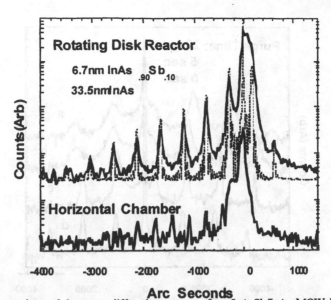

Figure 4. Comparison of the x-ray diffraction pattern of an InAsSb/InAs MQW laser structure grown in an MOCVD RDR(top pattern) versus the pattern for a MQW laser grown using a standard horizontal chamber configuration (bottom). The dotted line is a simulated x-ray spectrum for the RDR grown structure.

flow are necessary because of the high decomposition temperature of phosphine. In both cases, InAsSb/InAs and InAsP/InAs, the composition dependence was reproducible and approximately linear over the composition range that was examined.

Several test laser structures were prepared similar to the one previously reported using InAsSb/InAs MQW active regions.[3] These structures consisted of 2 µm AlAsSb top and bottom cladding layers with InAsSb/InAs 10 period MQW active regions and a (p)-GaAsSb/(n)-InAs semi-metal heterojunction for charge transfer.[1-3] The difference in the quality of the x-ray diffraction pattern of the continuously grown structure from the RDR and the re-grown active region and top cladding is illustrated in Figure 4. The continuously grown structures can be seen to be of much better quality from the x-ray diffraction pattern (top pattern). Similar lasing characteristics were obtained for both the RDR and re-grown structures. The advantage of the RDR growths is that no Al carry-over was observed as was found for the re-grown structures using the horizontal reactor.[3] The multi-staged laser structures described in this paper could not be grown using a conventional horizontal MOCVD system.

The growths of the InAsSb/InAsP ternary strained-layer superlattices (SLSs) used similar conditions as those for the MQW growths. The growth conditions for a given composition of the SLS was easily predicted from the compositions of the MQW's. However, very rough surface morphologies and poor x-ray diffraction patterns and photoluminescence characteristics were found for the 15-20 second purge times used between layers. The purge times were optimized using both x-ray diffraction patterns, as illustrated in Figure 5 for the x-ray diffraction patterns, as well as photoluminescence. As shown in Figure 5(a) and (b) for purge times of 20 and 5 seconds, very broad x-ray diffraction patterns were observed. The x-ray diffraction patterns shown in Figure 5(c) and (d) differ only in composition; both were grown using no purges between layers. The sample in (d) was grown with a slightly different composition to achieve lattice matching with the InAs substrate. The optimized time for the ternary SLS's was found to be less than 5 seconds. Similar characteristics were found for purge times of 0 or 1 second with arsine continuing to flow during the purge times as well as during the growth of the layers.

23

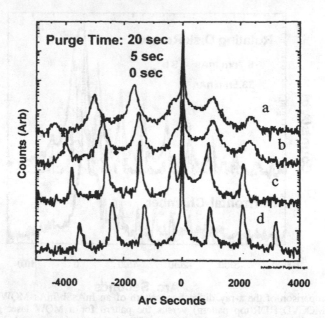

Figure 5. Comparison of the x-ray diffraction patterns for InAsSb/InAsP SLS's grown using identical growth conditions in (a), (b), and (c) but with the indicated purges between layers [(a) = 20 sec., (b) = 5 sec., and (c) = 0 sec.]. The sample in (d) was grown with a slightly different composition than (c) with no purge time to achieve lattice matching with the InAs substrate.

Optically Pumped Laser Design and Characteristics

Several laser structures have been examined by optical pumping in our previous work.[1] These structures consisted of an InAs substrate with a 2.5 μm thick AlAsSb lower cladding, a 1.0 μm thick, $InAs_{0.89}Sb_{0.11}$ /$InAs_{0.77}P_{0.33}$ (83Å / 87Å) SLS active region and several different top terminations. For the limited number of devices studied, neither a top cladding nor a semi-metal injection layer seemed to significantly affect laser performance under optical pumping. The SLS laser was pumped with a Q-switched Nd:YAG (1.06 μm, 20 Hz, 10 nsec pulse, focused to a 200 μm wide line), and emission was detected with an FTIR operated in a step-scan mode. Laser emission was observed from cleaved bars, 1000 μm wide, with uncoated facets. A lasing threshold and spectrally narrowed, laser emission was seen from 80 K through 240 K, the maximum temperature where lasing occurred. At 80 K, peak powers >100 mW could be obtained. The temperature dependence of the SLS laser threshold is described by a characteristic temperature, T_0 = 33 K, over the entire range. Similar experiments on structures containing 2000 Å InAs top and bottom wave-guiding regions resulted in improved power (>650 mW) at 80 K and 3.8 μm.[11] This result was obtained using an 808 nm diode stack pump laser, a 2 % duty cycle with 50 μsec pulses. The temperature behavior for this laser is illustrated in Figure 6(a) and (b). The lower T_0 (17 K) is most likely due to the increased duty cycle and pump laser characteristics used in this experiment.

We have recently demonstrated InAsSb/InAsP SLS injection lasers at 3.4 μm and 180 K. Under pulsed operation, peak power levels of 100 mW/facet (average power of 0.5 mW/facet) could be obtained at 80 K. A characteristic temperature (T_0) of 39 K was observed. We speculate that injection and transport of carriers in the SLS is presently limiting the performance of these devices.

24

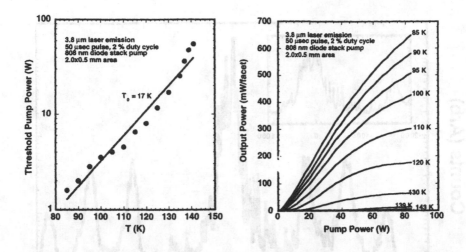

Figure 6. (a) Threshold optical pump power versus temperature. (b) Pulsed laser emission output power versus optical pump power for various temperatures for a 50 period $InAs_{0.84}Sb_{0.16}$ /$InAs_{0.68}P_{0.32}$ (83 Å / 87 Å) SLS active region with 2000 Å InAs waveguides and 2 μm AlAsSb claddings.

10-Stage, Cascaded InAsSb Quantum Well Laser at 3.9 μm

We have prepared multi-stage, cascaded InAsSb quantum well lasers that lase at 3.9 μm at 170 K. The gain region of our laser was composed of 10 stages. Each stage consisted of a $GaAs_{0.09}Sb_{0.91}$ (p, 300 Å) /InAs (n, 500 Å) semi-metal, 3 $InAs_{0.85}Sb_{0.15}$ (n, 94 Å) quantum wells separated by 4 $InAs_{0.67}P_{0.33}$ (n, 95 Å) barriers, an $AlAs_{0.16}Sb_{0.84}$ (p, 50 Å) electron block, a compositionally graded, Zn-doped AlGaAsSb (p = 5 x 10^{17} cm^{-3}, 300 Å) layer and a final 50 Å, Zn-doped $GaAs_{0.09}Sb_{0.91}$ layer. The total thickness of the gain region is 1.9 μm. The excellent crystalline quality of a similar laser structure (5 quantum wells instead of 3) is demonstrated in the x-ray diffraction spectrum where groups of satellite peaks corresponding to a gain-stage period of 2100 Å and the InAsSb/InAsP period of 190 Å are observed as illustrated in Figure 7. The composition and layer thickness for the InAsSb/InAsP superlattice can be determined from the satellites with the large repeat distance where the smaller repeat distance as shown in the inset is determined by the 10-stage structure.

Optical confinement is provided by 2 μm thick, Zn-doped $AlAs_{0.16}Sb_{0.84}$ (p=1 x 10^{17}cm^{-3}) claddings on both sides of the active region. A top 1500 Å Zn-doped $GaAs_{0.09}Sb_{0.91}$ layer (p=2 x 10^{18} cm^{-3}) is used as a contact and protective layer. The Zn doping levels were determined from Van der Pauw/Hall measurements and confirmed by secondary ion mass spectroscopy (SIMS) on thick calibration samples. Two orders of magnitude higher DEZn levels were required to obtain equivalent dopant levels in $AlAs_{0.16}Sb_{0.84}$ compared to $GaAs_{0.09}Sb_{0.91}$. This indicates a possible depletion reaction between DEZn and EDMAA. The structure is lattice matched to the InAs substrate. An InAs (n, 500 Å) /$GaAs_{0.09}Sb_{0.91}$ (p, 1000 Å) semi-metal is used to enable carrier transport from the n-type InAs substrate into the p-type $AlAs_{0.16}Sb_{0.84}$ cladding layer.

Lasing was observed from gain-guided stripe lasers. The facets were uncoated, and stripes were indium soldered to the heat sink with the epitaxial side up. Under pulsed operation with 100 nsec pulses at 1 kHz (10^{-4} duty-cycle), stimulated emission was observed from 80-170 K. Laser

25

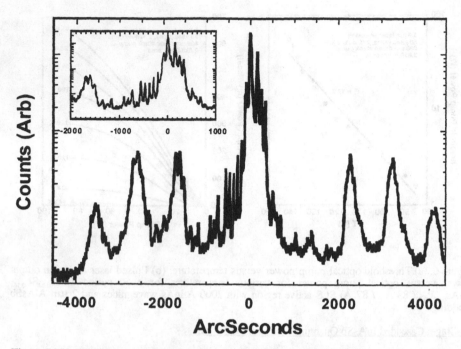

Figure 7. X-ray diffraction pattern for a 10-stage, 5-well InAsSb/InAsP laser structure with 2 μm AlAsSb claddings and p-GaAsSb/n-InAs semi-metal heterojunction injection layers.

emission spectra at 80 K and 160 K are shown in Figure 8. Emission occurred at 3.8-3.9 μm. For 750 μm long stripes with 80 μm wide metallizations, several longitudinal modes are observed with a mode spacing of 1.7 cm^{-1}. The longest laser pulses were 1 μsec. These initial devices were easily damaged by increased heating associated with higher temperature operation or longer duty cycles. Threshold current densities for these 10-stage devices were ≥ 1 kA/cm^2. Although these values are over-estimates due to current spreading in our gain-guided structures, reduction of threshold current density has not yet been demonstrated in type I or type II cascaded interband lasers.[9] Lower threshold current densities (0.1 kA/cm^2) have been demonstrated in single-stage, type I mid-infrared lasers at 80 K.[12] The characteristic temperature of a similar laser was about 34 K similar to our previously reported optically pumped InAsSb/InAsP lasers.[2]

The laser output was collected and focused directly onto an InSb detector to obtain power-current data shown in Figure 9 for the 750 μm stripe device. At 80 K, peak power values > 100 mW were obtained. At 80 K, the threshold current density was 1 kA/cm^2. The maximum slope-efficiency was 93 mW/A, corresponding to a differential external quantum efficiency of 29 % (2.9 % per stage). These slope-efficiencies are under-estimates due to current spreading in the gain-guided structures. This initial result is promising when compared to the value obtained for a second generation, 23-stage type II cascaded laser (3.9 μm) with a differential quantum efficiency of 131% (5.7 % per stage).[9]

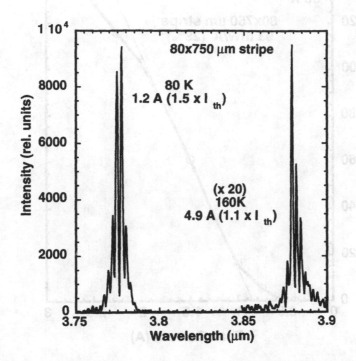

Figure 8. Laser emission spectra at 80 and 160 K for a 10-stage, 3-well laser with a 750 μm long stripe.

SUMMARY

In conclusion, we have demonstrated the first cascaded lasers and LEDs with type I InAsSb quantum well active regions. Also, these are the first cascaded devices grown by MOCVD. The 10-stage, 3.8-3.9 μm laser operated up to 170 K. At 80 K, peak laser power > 100 mW and a slope-efficiency of 29% (2.9% per stage) were observed. We are optimistic that advances in material quality and device design will improve carrier confinement and reduce loss, leading to higher efficiencies and higher temperature operation of cascaded InAsSb lasers.

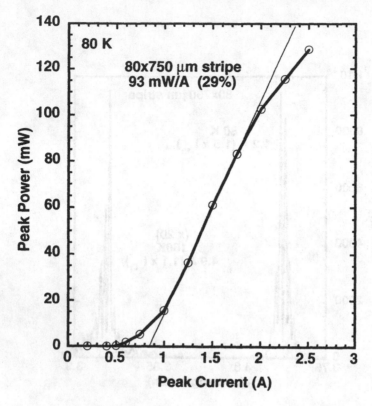

Figure 9. Peak laser power versus current for a 10-stage InAsSb/InAsP/AlAsSb/GaAsSb/InAs laser with a 750 μm long stripe at 80 K.

ACKNOWLEDGMENTS

We thank D. McDaniel of the USAF Phillips Laboratory for the results illustrated in Figure 6 and J. A. Bur and J. Burkhart for technical assistance. This work was supported by the U.S. Dept. of Energy under contract No. DE-AC04-94AL85000. Sandia is a multiprogram laboratory operated by Sandia Corporation, a Lockheed Martin Company, for the United States Department of Energy.

REFERENCES

[1] A. A. Allerman, R. M. Biefeld, and S. R. Kurtz, Appl. Phys. Lett. **69**, 465 (1996).
[2] S. R. Kurtz, A. A. Allerman, and R. M. Biefeld, Appl. Phys. Lett. **70**, 3188 (1997).
[3] R. M. Biefeld, S. R. Kurtz, and A. A. Allerman, J. Electronic Mater., **26**, 903 (1997).

[4] J. Faist, F. Capasso, D. L. Sivco, C. Sirtori, A. L. Hutchinson, and A. Y. Cho, Science **264**, 553 (1994).
[5] R. Q. Yang, Superlatt. Microstruct. **17**, 77 (1995).
[6] J. R. Meyer, I. Vurgaftman, R. Q. Yang, and L. R. Ram-Mohan, Elect. Lett. **32**, 45 (1996).
[7] J. Faist, F. Capasso, C. Sirtori, D. L. Sivco, J. N. Baillargeon, A. L. Hutchinson, S. N. G. Chu, and A. Y. Cho., Appl. Phys. Lett. **68**, 3680 (1996).
[8] C. H. Lin, R. Q. Yang, D. Zhang, S. J. Murry, S. S. Pei, A. A. Allerman, and S. R. Kurtz, Elect. Lett. **33**, 598 (1997).
[9] R. Q. Yang, B. H. Yang, D. Zhang, C. H. Lin, S. J. Murry, H. Wu, and S. S. Pei, Appl. Phys. Lett. **71**, 2409 (1997).
[10] W. G. Breiland and G. H. Evans, J. Electrochem. Soc., **138**, 1806 (1991).
[11] Private Communication from D. McDaniel, USAF Phillips Laboratory, Albuquerque, NM.
[12] H.K. Choi and G.W. Turner, Appl. Phys. Lett. **67**, 332 (1995).

TEM INVESTIGATION OF $Al_{0.5}Ga_{0.5}As_{1-y}Sb_y$ BUFFER LAYER SYSTEMS

E. CHEN*, J. S. AHEARN**, K. NICHOLS**, P. UPPAL** and D. C. PAINE*

*Brown University, Division of Engineering Providence RI 02912
**Sanders Lockeed-Martin, Nashua, NH 03061

ABSTRACT

We report on a TEM study of Sb-adjusted quaternary $Al_{0.5}Ga_{0.5}As_{1-y}Sb_y$ buffer-layers grown on <001> GaAs substrates. A series of structures were grown by MBE at 470°C that utilize a multilayer grading scheme in which the Sb content of $Al_{0.5}Ga_{0.5}As_{1-y}Sb_y$ is successively increased in a series of eight 125 nm thick layers. Post growth analysis using conventional bright field and weak beam dark field imaging of these buffer layers in cross-section reveals that the interface misfit dislocations are primarily of the 60° type and are distributed through out the interfaces of the buffer layer. Plan view studies show that the threading dislocation density in the active regions of the structure (approximately 2 μm from the GaAs substrate) is $10^{5-6}/cm^2$ which is comparable to equivalent $In_xGa_{1-x}As$ buffers. Weak Sb-As compositional modulations with a period of 1.8 nm were observed that provide a marker for establishing the planarity of the growth process. These features reveal that the growth surface remains planar through out the buffer layer growth sequence.

INTRODUCTION

The use of buffer layer structures for the creation of lattice-matched surfaces has been explored in the III-V system primarily by using indium-graded $In_xGa_{1-x}As$ alloys to form either continuously graded or step graded structures. Buffer layer structures fabricated using graded $In_xGa_{1-x}As$ steps have shown (e.g. [1]) at least a three orders of magnitude reduction in threading dislocation density (to $10^5/cm^2$) when compared to equivalent compositionally-uniform (i.e. single layer) films. An alternative approach to $In_xGa_{1-x}As$ step-based buffer layers that has not been widely explored is the use of Sb-graded alloys. From a lattice parameter perspective, ternary In-grading of the GaAs -InAs (a=6.0584 Å) pseudobinary is nearly identical to the alloying of GaAs with GaSb (a=6.095Å) since they both produce the same final relaxed lattice parameter. The mechanism of relaxation and the resultant threading dislocation density in Sb-graded layers is not, to our knowledge, available in the literature. The Sb-graded approach allows greater flexibility in the selection of the group III flux which might be desirable if, for example, In and/or Ga were replaced with Al to decrease the surface mobility of the metal species to effect an improvement in the growth surface

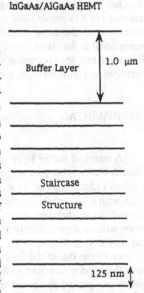

Fig. 1: Schematic view of the eight-step stair case buffer where each step represents a 0.02Å increment in lattice parameter.

31

morphoplogy. For this reason, quaternary alloys of $Al_{0.5}Ga_{0.5}As_{1-y}Sb_y$ were grown on GaAs substrates to produce buffer layer structures suitable for the growth of InGaAs HEMT structures.

The goal in fabricating buffer layer systems is to produce a heterostructure which is fully relaxed via the presence of interface dislocations but which contains a minimum number of threading defects. The glide of threading dislocations is necessary for the introduction of interface misfit but those threading segments that remain in the active regions of the structure after the growth is completed will degrade device performance and hence must be minimized. To reduce the number of threading defects that remain in the film after growth, their nucleation must be impeded but not eliminated. Those threading dislocations that nucleate must lay down sufficient interface misfit segments to fully relax the structure then, ideally, the threads would be removed either by glide out of the crystal (to the sample edges) or by reaction with other dislocations in the crystal. The rate of glide of a threading dislocation segment is increased by increasing the effective shear stress acting on it and by minimizing the blocking [3] effect caused by dislocation-dislocation strain-field interaction.

The features of graded-layer growth that reduce the density of threading defects (compared to the growth of a compositionally uniform single layer) have been summarized [2]. First, in compositionally graded structures, interface dislocations are not confined to a single interface but instead find a minimum energy position. As the graded-layer structure is grown, the minimum energy position for new dislocations moves further from the substrate/film interface. This is important since threading dislocations gliding in one <110> direction can be impeded by the strain field of interface misfits lying in the orthogonal direction [3]. This blocking effect is reduced by moving the minimum energy position of interface misfit dislocations to positions higher in the film as the buffer layer is grown so that the gliding thread is further removed from misfits that were formed earlier in the growth process. Second, the residual elastic strain that provides the driving force for the glide of threading dislocations is greatest near the free surface of the buffer layer structure (as it is grown) and is greatly enhanced compared to uniform layers. This aids in the movement of threading dislocations to the sample edges since the dislocation velocity is proportional to the elastic stress felt by the defect. Third, threading dislocation nucleation through dislocation reactions occurring near the substrate/film interface (e.g. [4]) are minimized because, compared to a uniform layer, the residual elastic strain deep in the buffer layer is reduced.

EXPERIMENTAL

A series of buffer layer structures were grown in a Varian MOD GEN II MBE on <001>-oriented GaAs substrates at a temperature of 450-500°C. The Al, Ga, and Sb sources were standard 125 cc cells and As was generated in a valved cracker as As_2. The buffer layer fabrication began with a 100 nm thick homoepitaxial GaAs layer and continued with the growth of eight 125-nm-thick layers that were grown with fixed (and equal) Al:Ga fluxes. Each subsequent layer was grown with an approximately 5% increment in the Sb flux to form a concentration stair-case shown schematically in Fig. 1. The buffer structures were designed to increase the relaxed lattice parameter from that of the GaAs substrate (5.65Å) to match the lattice parameter of the materials in the active device structure (5.84Å corresponding to $In_{0.46}Ga_{0.54}As$). Between each layer, the growth was paused by shuttering the metal sources and the Sb cell temperature was ramped. When close to the Sb-cell set-point temperature, the metal cell shutters were opened and the next layer was grown. A total of eight layers were grown all with equal Al:Ga fluxes and, at each step, an

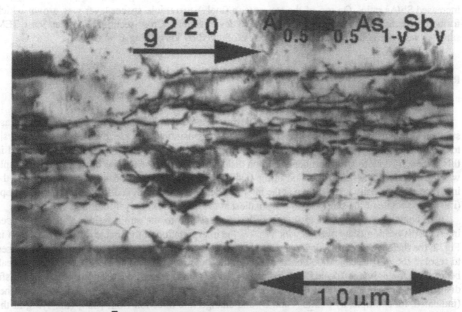

Fig. 2: Bright field $2\bar{2}0$ image showing the eight-step buffer layer.

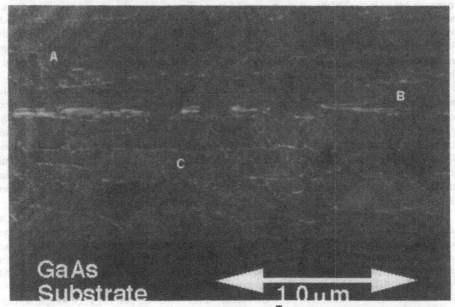

Fig. 3: Weak beam dark field image using g-3g and g=$2\bar{2}0$. Features A, B, and C represent examples of dislocations, respectively, threading from one interface to another, changing line direction, and interacting.

increased Sb/As ratio. On top of the staircase, a final 1 μm thick spacer layer was grown which formed the template for the growth of the active device layers. Cross-sectional and plan view TEM samples were prepared by mechanical thinning followed by ion milling at 5 keV.

RESULTS

Figure 2 shows a bright field 004 2-beam cross-sectional image of the 8 step stair case structure. The dislocations at the interfaces of each compositional step are revealed by g•b-analysis to be primarily 60°-type. The interface misfit dislocation density at each interface was estimated in this sample by counting the number of end-on dislocations visible when the cross-sectional sample was viewed along the [220] direction in the 004 2-beam condition. As can be seen in Fig. 2, the interface misfit density is maximum at the interface between the fourth and fifth compositional steps (i.e. in the middle of the staircase) and is smallest at the GaAs/buffer layer interface.

In conventional bright field imaging, the contrast due to the strain fields of closely spaced interface dislocations may overlap which makes the interpretation of the image difficult. As seen in Fig. 2, the dislocation density is large and, for this reason, weak beam dark field imaging was used to resolve the misfit dislocation configuration in the buffer layer. Figure 3 shows a g-3g weak beam dark field image formed using the 004 reflection. This image shows that the misfit dislocations are not confined to one interface but instead thread from one interface to another (indicated, for example, by points A and B) and change line direction from [220] (i.e. lying in the plane of the TEM sample) to $\overline{2}20$ (i.e. lying in the thickness direction). The dislocation line segments that lie in the 220 direction are revealed by oscillatory contrast such as that indicated at the point labeled C in Fig. 3. The possibility of changing interface position as the equilibrium misfit position changes during growth is one of the requirements for a successful buffer layer structure. Further, it is clear that the dislocations in this buffer layer scheme thread to other interfaces in the buffer where dislocation reactions which eliminate the threading portion may result. This reduces the number of dislocations that thread up through the spacer layer.

The effectiveness of the buffer layer in minimizing the number of threading dislocations that pass from the 8-step staircase buffer through the 1 μm spacer layer and into the active device regions is the critical measure of the success of a buffer layer scheme. In this work, this measurement was made using plan view TEM to estimate the threading defect density at the top surface of the structure. A set of samples were thinned from the backside so that only the top-most 300 nm of the structure remained. The number of threading dislocations that could be seen while crystal was tilted through a range of diffracting conditions were counted. The resulting threading dislocation density was estimated at approximately 10^{5-6} /cm^2 which is equivalent to the dislocation density measured for similar indium-graded In$_x$Ga$_{1-x}$As structures.

Figure 4 shows, in part (a), a 004 2-beam bright field image and, in part (b), an accompanying selected area diffraction (SAD) pattern taken from one of the steps of the buffer layer compositional stair-case. Compositional modulations that have formed parallel to the (004) growth surface are visible as uniformly spaced bright/dark intensity oscillations in the image. The period of these compositional modulations was determined from the satellite reflections seen in the SAD pattern of Fig. 4(b) and is 1.8 nm. The contrast in Fig. 3(a) is greatly enhanced by the use of the 004 2-beam imaging conditions which allows elastic strain relaxation near the TEM sample surfaces to artificially enhance [5] the image contrast; the same area imaged using the 220 reflection does not reveal the compositional modulations. An analysis of the SAD satellite reflection intensity and

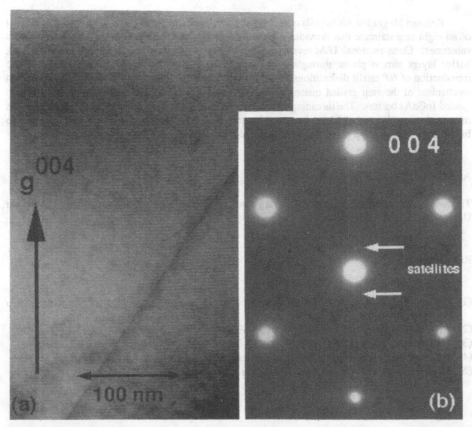

Fig. 4: (a) Bright-field 004 2-beam image showing compositional modulations in the spacer layer of the buffer structure. (b) Selected area diffraction pattern showing the satellite reflections oriented in the 004 direction around each of the primary reflections in the 110 zone.

investigation of the effect of the growth flux on the satellite period reveals that the modulations consist of small-scale fluctuations in the As:Sb ratio. While the origin of these composition modulations is presently obscure, they provide insight into the planarity of the growth surface. Observation of the compositional modulations through the thickness of the buffer layer and in the spacer layer reveal that the growth surface remains planar over a lateral scale of 100's of nms though out the growth. Although not yet clearly established experimentally, it appears that the use of Al in the metal flux, with its lower surface mobility compared to either In or Ga, plays an important role in preserving the growth surface planarity.

SUMMARY

Relaxed Sb-graded AlGaAsSb quaternary buffer layer structures were fabricated that consist of an eight-step staircase that provides a growth surface with a 5.85Å lattice parameter on GaAs substrates. Cross-sectional TEM reveals that both the 1 μm thick spacer layer and the underlying buffer layers remain planar throughout the growth and that the structure is relaxed via the introduction of 60° misfit dislocations at the interfaces of the buffer layer. Thus, the relaxation mechanism of the step graded quaternary AlGaAsSb buffer is similar to that observed for step graded InGaAs buffers. The threading defect density in the active regions of these structures was measured using plan view TEM to be approximately 10^{5-6} /cm^2 which suggests that this approach to buffer layer synthesis is a promising alternative to the use of InGaAs graded layer structures.

ACKNOWLEDGMENTS

This work was supported, in part, by a grant from Lockeed-Martin and by the MRSEC at Brown University.

REFERENCES

(1) D. Gonzalez, D. Araujo, S.I. Molina, A. Sacdon, E. Calleja, R. Garcia, Mat. Sci. Eng B28, p 497-501(1994).
(2) J. Tersoff, Appl. Phys. Lett., 62(7), p693-695(1993).
(3) L.B. Freund, J. Appl. Phys., 68(5), p2073-80(1990).
(4) F.K. LeGoues, B.S. Meyerson, and J.F. Morar, Phys. Rev. Lett. 66, p2903(1991).
(5) M.M.J. Treacy and J.M. Gibson, J. Vac. Sci. Technol. B4(6), p1458-1466(1986).

SUBSTRATE MISORIENTATION EFFECTS ON EPITAXIAL GaInAsSb

C.A. WANG*, H.K. CHOI*, D.C. OAKLEY*, and G.W. CHARACHE**
*Lincoln Laboratory, Massachusetts Institute of Technology, Lexington, MA 02173-9108
**Lockheed Martin Corporation, Schenectady, NY 12301

ABSTRACT

The effect of substrate misorientation on the growth of GaInAsSb was studied for epilayers grown lattice matched to GaSb substrates by low-pressure organometallic vapor phase epitaxy. The substrates were (100) misoriented 2 or 6° toward (110), (111)A, or (111)B. The surface is mirror-like and featureless for layers grown with a 6° toward (111)B misorientation, while a slight texture was observed for layers grown on all other misorientations. The optical quality of layers, as determined by the full width at half-maximum of photoluminescence spectra measured at 4K, is significantly better for layers grown on substrates with a 6° toward (111)B misorientation. The incorporation of Zn as a p-type dopant in GaInAsSb is about 1.5 times more efficient on substrates with 6° toward (111)B misorientation compared to 2° toward (110) misorientation. The external quantum efficiency of thermophotovoltaic devices is not, however, significantly affected by substrate misorientation.

INTRODUCTION

$Ga_{1-x}In_xAs_ySb_{1-y}$ is an important material for optoelectronic devices that operate in the mid-infrared. This alloy can be lattice matched to GaSb or InAs susbtrates and has a direct energy gap adjustable in the wavelength range from 1.7 (0.726 eV) to 4.2 μm (0.296 eV). Although most quaternary alloy compositions are predicted to exhibit thermodynamic immiscibility at typical growth temperatures [1,2], stable alloys with a cutoff wavelength of 2.39 μm were grown by liquid phase epitaxy (LPE) [3], and metastable alloys were grown by organometallic vapor phase epitaxy (OMVPE) [4] and molecular beam epitaxy (MBE) [5]. Devices that include lasers [6,7], photodetectors [8], and thermophotovoltaic devices [9,10] have been reported, and the technological interest of GaInAsSb continues to increase.

The growth of GaInAsSb has been performed, in general, on nominally (100) oriented GaSb substrates. The use of vicinal substrates for the growth of III-V semiconductors, however, can play an important role in the resulting material quality. For example, the surface morphology of GaAs layers was reported to be smoother for OMVPE growth on (100) substrates with a 2° misorientation toward (110) [11], and higher optical quality of MBE-grown GaAs/AlGaAs quantum wells was observed on vicinal substrates [12]. Furthermore, the device performance of GaAs/AlGaAs quantum-well lasers was significantly improved on tilted substrates [13]. In this paper, we report the growth and material characteristics of GaInAsSb alloys lattice matched to vicinal GaSb substrates with misorientations of 2 and 6° off (100) substrates toward (110), (111)A, and (111)B. The best surface morphology and highest optical quality are obtained for substrates with a 6° toward (111)B misorientation. The performance of thermophotovoltaic devices grown on (100) GaSb substrates with either a 2° toward (110) or 6° toward (111)B misorientation are compared.

EPITAXIAL GROWTH AND CHARACTERIZATION

$Ga_{1-x}In_xAs_ySb_{1-y}$ epilayers were grown in a vertical rotating-disk reactor with H_2 carrier gas at a flow rate of 10 slpm and reactor pressure of 150 Torr [14]. Solution trimethylindium, triethylgallium, tertiarybutylarsine, and trimethylantimony were used as organometallic sources. The growth rate was typically 2.5 μm/h. The V/III ratio ranged from 1.1 to 1.3 and the growth temperature ranged from 525 to 575°C. For doping studies, diethyltellurium (DETe) (10 ppm in H_2) and dimethylzinc (DMZn) (1000 ppm in H_2) were used as n- and p-type doping sources, respectively.

GaInAsSb was grown on (100) Te-doped GaSb substrates with misorientation angles of either 2 or 6° toward (110), (111)A, or (111)B. For direct comparison, epilayers were grown side by side on substrates of various misorientation angles to minimize the effects of run-to-run variability. The surface morphology was examined using Nomarski contrast microscopy. Double-crystal x-ray diffraction (DCXD) was used to measure the degree of lattice mismatch to GaSb substrates. Photoluminescence (PL) was measured at 4 and 300K using a PbS detector. The composition of epilayers was determined from DCXD splitting, the peak emission in PL spectra, and the energy gap dependence on composition based on the binary bandgaps as described in a previous reference [14]. For electrical characterization, GaInAsSb was grown on semi-insulating (SI) (100) GaAs substrates misoriented 2° toward (110) or 6° toward (111)B. GaAs substrates were used since SI GaSb substrates are not available.Carrier concentration and mobility were obtained from Hall measurements based on the van der Pauw method.

RESULTS

The surface morphology of $Ga_{1-x}In_xAs_ySb_{1-y}$ layers grown at 550°C on vicinal GaSb substrates is shown in Figs. 1a-f. The Nomarski interference micrographs correspond to layers grown on (100) substrates with a 2 or 6° misorientation toward (110), (111)A, and (111)B, respectively. The composition of these layers is x=0.18 and y=0.15 and corresponds to 300K PL peak emission at ~2.4 μm. The layers grown on substrates with a 2° misorientation exhibit considerable texture. A smoother surface morphology is observed for layers grown on substrates with a 6° misorientation, and the smoothest surface is observed for 6° toward (111)B misorientation.

On the other hand, the dependence of GaInAsSb surface morphology on substrate misorientation is less sensitive as the In concentration is decreased (i.e., composition moving away from the miscibility gap). For x~0.1, y~0.08 (~2.1 μm cutoff at room temperature), smooth $Ga_{1-x}In_xAs_ySb_{1-y}$ surfaces could be obtained for epilayers grown at a growth temperature of 550°C and with 2° misorientation toward (110). Other factors that affect the surface morphology are the growth temperature and V/III ratio. In general, for layers with similar composition, a smoother morphology is observed for those layers grown at lower temperatures. We also observed that the surface texture of layers with a 2° toward (110) misorientation increased when the V/III ratio was increased above the minimum value for stoichiometric GaInAsSb. In contrast, the 6° toward (111)B misorientation was less sensitive. These observations suggest that the 6° toward (111)B misorientation provides a wider operating range of growth parameters for which a smooth morphology can be obtained.

Since the (100) 2° toward (110) substrate is widely used for OMVPE growth, subsequent comparisons are made for layers grown on either (100) 2° toward (110) or (100) 6° toward (111)B substrates. The 4K PL spectra also show striking differences that depend on substrate misorientation. Figure 2 shows the PL spectra for GaInAsSb grown at 550°C. The full width at half-maximum (FWHM) for the (100) 2° toward (110) misorientation is 15.4 meV compared to a FWHM value of 7.5 meV for the (100) 6° toward (111)B misorientation. In addition, the peak emission is longer at 2045 nm for the 2° misorientation compared to 1995 nm for the 6° misorientation. Data for 4K FWHM of layers of different In and As composition and grown at 550 or 525°C are shown in Fig. 3, and plotted versus the PL peak position. Over the whole range, the FWHM value is consistently lower for the 6° misorientation. In addition, the PL peak position is at higher energy indicating a lower In incorporation. A dependence of In incorporation on substrate misorientation has also been reported for InGaAs [15].

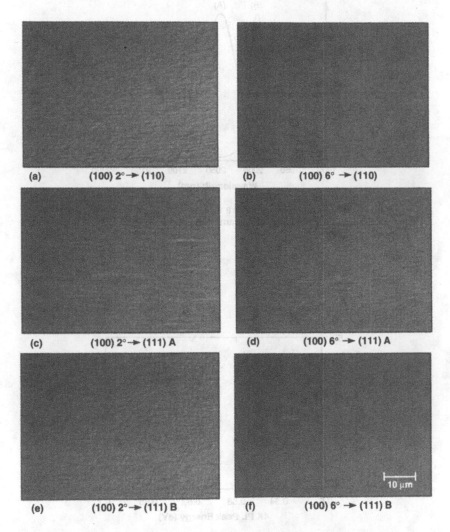

Figure 1. Surface morphology of $Ga_{0.82}In_{0.18}As_{0.15}Sb_{0.85}$ epilayers grown on GaSb substrates at 550°C and various substrate misorientations.

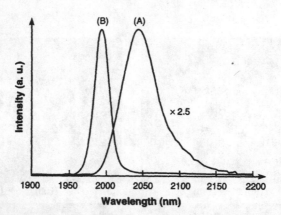

Figure 2. Photoluminescence spectra measured at 4K of GaInAsSb grown on
(100) GaSb with 2° toward (110) (sample A) and 6° toward (111)B (sample B)
misorientations.

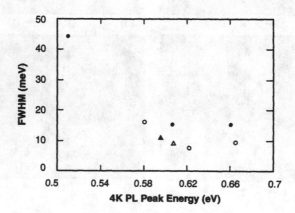

Figure 3. Photoluminescence FWHM measured at 4K of GaInAsSb grown on
(100) GaSb with 2° toward (110) (closed symbols) and 6° toward (111)B
misorientations (open symbols) at 525° C (triangles) and 550°C (circles).

The data in Fig. 3 also indicate that the PL FWHM is dependent on the growth temperature. A smaller difference in FWHM is observed for layers grown at 525°C compared to those at 550°C.

Figure 4 summarizes our best FWHM data for GaInAsSb epilayers. These samples were grown at 550°C on (100) 6° toward (111)B substrates or at 525°C on either (100) 2° toward (110) or 6° toward (111)B substrates. Also shown for comparison are data for layers grown by OMVPE on (100) substrates [16,17]. The FWHM values for samples in this study are significantly smaller than those reported previously, especially at the lower PL peak energy. The smallest FWHM value measured is 7.1 meV at 0.606 eV (2047 nm). The corresponding peak energy at 300K for this sample is 0.547 eV (2267 nm). A slight increase in FWHM is observed at lower PL peak energy. Our FWHM values are comparable to those reported for layers grown by MBE [18]. For GaInAsSb grown by LPE, a FWHM value of 9 meV at 0.587 eV (2112 nm) was reported [19].

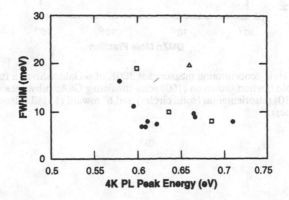

Figure 4. Photoluminescence FWHM measured at 4K of GaInAsSb layers grown on GaSb substrates. Solid circles are this work, open squares from reference 16, and open triangles from reference 17.

The p- and n-doping of GaInAsSb is also dependent on the substrate misorientation. Figures 5 and 6 show the hole and electron concentration versus the dopant mole fraction in the gas phase, respectively, for the 2° toward (110) and 6° toward (111)B misorientations. These layers were grown at 550°C on a 0.4-μm-thick undoped GaSb buffer layer, which has been shown to reduce the contribution of electrically active defects due to the lattice mismatch between the GaInAsSb and the SI GaAs substrate [18]. For p-GaInAsSb, the hole concentration ranges from 6.3×10^{16} to 1.7×10^{18} cm^{-3} with corresponding mobility values between 326 and 180 cm^2/V-s, respectively. The hole concentration is 1.5 to 1.6 times greater for the layers grown with a 6° toward (111)B misorientation compared to the 2° toward (110) misorientation. These results suggest that there is a preferential incorporation of Zn for the 6° toward (111)B misorientation. For n-GaInAsSb, the electron concentration ranges from 2.3×10^{17} to 2.3×10^{18} cm^{-3} with corresponding mobility values between 5208 and 2084 cm^2/V-s, respectively, and is 0.86 to 0.9 times lower for the 6° toward (111)B misorientation compared to the 2° toward (110) misorientation.

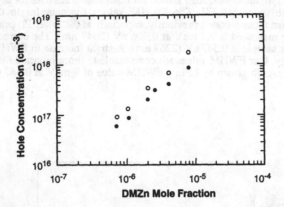

Figure 5. Hole concentration measured at 300K of p-GaInAsSb as a function of DMZn mole fraction grown on (100) semi-insulating GaAs substrates with 2° toward (110) misorientation (solid circles) and 6° toward (111)B misorientation (open circles).

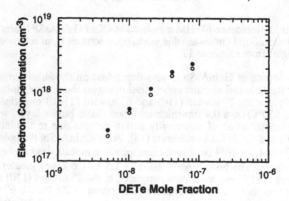

Figure 6. Electron concentration measured at 300K of n-GaInAsSb as a function of DETe mole fraction grown on (100) semi-insulating GaAs substrates with 2° toward (110) misorientation (solid circles) and 6° toward (111)B misorientation (open circles).

To investigate the effect of substrate misorientation on the performance of thermophotovoltaic devices, structures were grown on (100) n-GaSb substrates with a 2° toward (110) or 6° toward (111)B misorientation. The structure consists of 1-μm-thick n-GaInAsSb, 5-μm-thick p-GaInAsSb, 0.1-μm-thick p-AlGaAsSb, and 0.05-μm-thick p-GaSb. For this comparison, the structures were grown in two separate growth runs. The growth temperature was 550°C. Mesa diodes, 1 cm^2, were fabricated by a conventional photolithographic process. Ohmic contacts to p- and n-GaSb were formed by depositing Ti/Pt/Au and Au/Sn/Ti/Pt/Au, respectively, and alloying at 300°C. A single 1-mm-wide central busbar connected to the 10-μm-wide grid lines spaced 100 μm apart was used to make electrical contact to the front surface. No anti-reflection coatings were deposited on the devices.

The external quantum efficiency (QE) versus wavelength for the two devices is shown in Fig. 7. Surprisingly, the QE of the devices grown on the two misorientations are comparable, with only a slightly higher QE measured for the device grown with the 6° misorientation. Since the QE is highly dependent on the minority carrier lifetime, this result suggests that the lifetime is less affected by substrate misorientation. The cutoff wavelength for the device grown on (100) 6° toward (111)B is longer. This is a result of the gas flows used to grow the structure, and not a consequence of the misorientation angle.

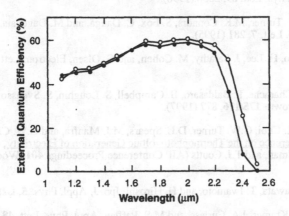

Figure 7. External quantum efficiency of GaInAsSb thermophotovoltaic devices grown on (100) GaSb substrates with 2° toward (110) misorientation (solid circles) and 6° toward (111)B misorientation (open circles).

CONCLUSIONS

The quality of GaInAsSb epilayers grown by OMVPE is dependent on the substrate misorientation, especially for alloy compositions approaching the miscibility gap. A smoother surface morphology and narrower PL FWHM was observed for layers grown on (100) substrates with a 6° toward (111)B misorientation compared to that grown on (100) 2° toward (110) substrates. These effects of substrate misorientation can be minimized by growing the layers at lower temperatures. However, for the growth of device structures that include other Sb-based alloys such as AlGaAsSb, the higher growth temperatures may be more desirable to avoid growth interruptions. Consequently, the use of (100) substrates with a 6° toward (111)B misorientation will allow a wider growth operating window, and subsequently will provide more consistent results. It was also found that the performance of thermophotovoltaic devices is less sensitive to the substrate misorientation.

ACKNOWLEDGMENTS

The authors gratefully acknowledge D.R. Calawa, J.W. Chludzinski, M.K. Connors, and V. Todman-Bams for technical assistance, K.J. Challberg for manuscript editing, and D.L. Spears for continued support and encouragement.

REFERENCES

1. K. Onabe, Jpn. J. Appl. Phys. **21**, 964 (1982).

2. G.B. Stringfellow, J. Cryst. Growth **58**, 194 (1982).

3. E. Tournie, F. Pitard, and A. Joullie, J. Cryst. Growth **104**, 683 (1990).

4. M.J. Cherng, H.R. Jen, C.A. Larsen, G.B. Stringfellow, H. Lundt, and P.C. Taylor, J. Cryst. Growth **77**, 408 (1986).

5. H.K. Choi, S.J. Eglash, and G.W. Turner, Appl. Phys. Lett. **64**, 2474 (1994).

6. D.Z. Garbuzov, R.U. Martinelli, H. Lee, P.K. York, R.J. Menna, J.C. Connolly and S.Y. Narayan, Appl. Phys. Lett. **69**, 2006 (1996).

7. H.K. Choi, G.W. Turner, M.K. Connors, S. Fox, C. Dauga, and M. Dagenais, IEEE Photon. Technol. Lett. **7**, 281 (1995).

8. Y. Shi, J.H. Zhao, H. Lee, J. Sarathy, M. Cohen, and G. Olsen, Electron. Lett. **32**, 2268 (1996).

9. P.N. Uppal, G. Charache, P. Baldasaro, B. Campbell, S. Loughin, S. Svensson, and D. Gill, J. Cryst. Growth **175/176**, 877 (1997).

10. C.A. Wang, H.K. Choi, G.W. Turner, D.L. Spears, M.J. Manfra, and G.W. Charache, in 3rd NREL Conference on the Thermophotovoltaic Generation of Electricity, edited by J.P. Benner, C.S. Allman, and T.J. Coutts (AIP Conference Proceedings **401**, Woodbury, NY, 1997) pp. 75-87.

11. M. Mizuta, S. Kawata, T. Iwamoto, and H. Hiroshi, Jpn. J. Appl. Phys. **5**, L283 (1984).

12. R.K. Tsui, G.D. Kramer, J.A. Curless, and M.S. Peffley, Appl. Phys. Lett. **48**, 940 (1986).

13. H.Z. Chen, A. Ghaffari, H. Morkoc, and A. Yariv, Appl. Phys. Lett. **51**, 2094 (1987).

14. C.A. Wang and H.K. Choi, J. Electron. Mater. **26**, 1231 (1997).

15. J. te Nijenhuis, P.R. Hageman, and L.J. Giling, J. Cryst. Growth **167**, 397 (1996).

16. M. Sopanen, T. Koljonen, H. Lipsanen, and T. Tuomi, J. Cryst. Growth **145**, 492 (1994).

17. J. Shin, T.C. Hsu, Y. Hsu, and G.B. Stringfellow, J. Cryst. Growth **179**, 1 (1997).

18. C.A. Wang, G.W. Turner, M.J. Manfra, H.K. Choi, and D.L. Spears, in Infrared Applications of Semiconductors - Materials, Processing and Devices, edited by M.O. Manasreh, T.H. Myers, and F.H. Julien (Mater. Res. Soc. Symp. Proc. **450**, Pittsburgh, PA, 1997) pp. 55-60.

19. E. Tournie, J.-L. Lazzari, F. Pitard, C. Alibert, A. Joullie, and B. Lambert, J. Appl. Phys. **68**, 5936 (1990).

BRIDGMAN GROWTH AND CHARACTERIZATION OF BULK SINGLE CRYSTALS OF Ga$_{1-x}$In$_x$Sb FOR THERMOPHOTOVOLTAIC APPLICATIONS

J.R. BOYER, W.T. HAINES
Lockheed Martin Corporation, Schenectady, NY 12301

ABSTRACT

Thermophotovoltaic generation of electricity is attracting renewed attention due to recent advances in low bandgap (0.5-0.7 eV) III-V semiconductors. The use of mixed pseudo-binary compounds allows for the tailoring of the lattice parameter and the bandgap of the material. Conventional deposition techniques (i.e., epitaxy) for producing such ternary or quaternary materials are typically slow and expensive. Production of bulk single crystals of ternary materials, for example Ga$_{1-x}$In$_x$Sb, is expected to dramatically reduce such material costs. Bulk single crystals of Ga$_{1-x}$In$_x$Sb have been prepared using a Bridgman technique in a two-zone furnace. These crystals are 19 mm in diameter by approximately 50 mm long and were produced using seeds of the same diameter. The effects of growth rate and starting materials on the composition and quality of these crystals will be discussed and compared with other attempts to produce single crystals of this material.

INTRODUCTION

III-V compound semiconductors are important materials for optoelectronic applications, in particular for thermophotovoltaic generation of electricity [1]. By alloying different compounds, the bandgap and the lattice parameter of the resulting combination can be adjusted. Tailoring these properties allows for the fabrication of a wider array of devices than is possible using only the simple binary materials. One of the many possible combinations currently receiving significant attention is the GaSb-InSb system. Combinations of these compounds can have lattice constants in the range of 0.6096 to 0.6479 nm and bandgaps between 0.17 and 0.726 eV [2].

Typical deposition techniques, such as liquid phase epitaxy, are expensive for fabricating these materials. The ability to grow large, bulk single crystals of these materials is expected to reduce their cost. For this to be effective, however, the composition of the alloy material must be uniform over a significant portion of the resulting boule. This is difficult to achieve for most compound semiconductor materials because of the wide and non-uniform liquidus-solidus separation of the pseudo-binary phase diagram. Additionally, as with the growth of crystals of most materials, solute transport and temperature variation within the melt also have important effects on the composition and quality of the final material produced. Previous work to produce GaSb-InSb alloys via different techniques and conditions is reported in [3] - [7] and [10].

In this paper, results of some experiments conducted to prepare bulk alloys of Ga$_{1-x}$In$_x$Sb (x~0.03 and 0.2) and to explore the effects of growth parameters on the composition of the resulting alloy are presented. A Bridgman technique was used to solidify the melt; single crystal and polycrystalline boules were produced. Seeds of the same diameter as the crucible were used to grow single crystals of the alloy to reduce the likelihood of twinning and other defects.

45

EXPERIMENT

Preparation of Binary Starting Compounds

The binary compounds, GaSb and InSb, were prepared in the laboratory. The starting materials were elements with 6N purity. In both cases, the Sb-III atom ratio was ~1.001 to allow for volatilization of Sb during processing. A fused silica crucible, with an inner diameter of 19 mm and an overall length of approximately 125 mm, was used to contain the mixture. The crucible was cleaned with HF, followed by rinses of deionized water and isopropanol, then dried with filtered argon. The starting materials, without any etching or cleaning, were placed into the cleaned crucible, then inserted into the furnace.

A two-zone transparent furnace (Trans-Temp, Chelsea, MA) was used for all experiments (Figure 1). The crucible was held in a graphite support (Figure 2a). A quartz rod, prepared in the same way as the crucible, was inserted from the top for use in mixing the melt.

After the crucible was inserted, the furnace was evacuated to ~100 millitorr, then flushed with argon (three cycles). An argon flow of ~1 slpm was maintained during furnace operation.

For GaSb synthesis, the hot zone and cold zone temperatures were 780 C and 600 C, respectively; for InSb synthesis, the hot zone and cold zone temperatures were 750 C and 450 C, respectively. The Type S control thermocouples were located in each zone such that a thermal gradient zone of $G \sim 25$ K/cm was established in the middle portion of the furnace. The crucible

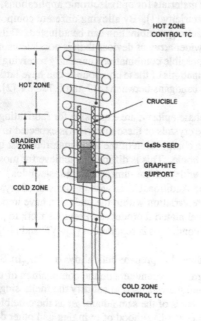

Fig. 1. Schematic of two-zone transparent furnace. The inner diameter of the inner quartz tube is 32 mm; the inner diameter of the gold-coated outer tube is 80 mm. Overall heated length is 60 cm.

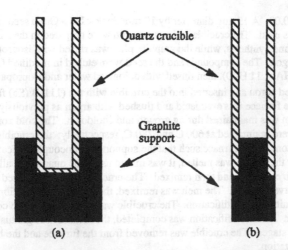

Quartz crucible

Graphite support

(a)   (b)

Fig. 2. Schematic of crucible and graphite supports used for synthesis and crystal growth. (a) Flat-bottomed crucible for binary compound synthesis and polycrystalline alloy fabrication. (b) Open tube crucible for single crystal growth.

was positioned within the furnace such that the support temperature was above the melting point of the compound being synthesized (712 C for GaSb and 535 C for InSb). After the charge material was completely melted, the quartz rod was used to mix the melt. Mixing was done every 45 to 60 minutes during the synthesis. After 4 to 5 hours, the crucible was pulled into the cold zone at a rate of 7 mm/hr, then the furnace was slowly cooled to room temperature in stages.

The polycrystalline boules were removed from the crucible, sectioned using a diamond saw, then characterized by electron microprobe and X-ray diffraction.

Synthesis of Polycrystalline Ga$_{1-x}$In$_x$Sb Boules

Polycrystalline Ga$_{1-x}$In$_x$Sb boules were made using flat bottomed crucibles without seeds (Fig. 2a). Starting materials were either the binary compounds or the three elements (6N purity) in appropriate ratios to achieve initial melt compositions of x~0.03 or x~0.20. When the elements were used, the Sb-III ratio was ~1.001; material and crucible preparation, melt mixing, and homogenization were performed in the same manner as described for the binary synthesis experiments. When the starting materials were the binary compounds, the same processes as described below for single crystal growth were used. The crucible translation rate, v, for these experiments ranged from 3 to 19 mm/hr. Furnace preparation and operation were the same as described for single crystal growth experiments.

Growth of Single Crystals of Ga$_{1-x}$In$_x$Sb

Single crystals of Ga$_{1-x}$In$_x$Sb were grown in a quartz crucible supported by a graphite plug (Fig. 2b). Starting materials were GaSb and InSb in amounts to achieve an initial melt composition,

x, of approximately 0.03. A 19 mm diameter by 10 mm thick <111> GaSb seed (Atramet, Inc., Hempstead, NY) was used. The crucible and a quartz rod were prepared in the same manner as that used for compound synthesis, while the graphite plug was rinsed with isopropanol and dried with filtered argon. The compounds and the seed were etched in modified CP-4 etchant (5 HNO_3 : 3 HF : 3 HAc : 11 H_2O), then rinsed with deionized water and isopropanol. The seed was dried with filtered argon and inserted into the crucible with the (111)B (Sb) face oriented toward the melt. The furnace was evacuated and flushed with argon as previously described; argon flow at ~1 slpm was maintained during growth and cooldown. The cold zone and hot zone temperatures were maintained at 600 C and 780 C, respectively; the crucible was located within the gradient zone of the furnace such that the support thermocouple indicated approximately 690 C. After the charge was melted, it was mixed using the quartz rod, allowed to homogenize for up to one hour, and then remixed. The crucible was repositioned until the top of the seed was observed to melt. The melt was remixed, then allowed to equilibrate for about 20 minutes prior to initiating solidification. The crucible was pulled toward the cold zone at rates of 3 to 7 mm/hr. After solidification was completed, the furnace was gradually cooled to room temperature in stages. The crucible was removed from the furnace and the boule was sectioned for characterization.

RESULTS AND DISCUSSION

All of the GaSb and InSb boules were sectioned longitudinally and etched using modified CP-4 etchant. The boules were polycrystalline, with elongated grains oriented with their major axis parallel to the solidification direction. Electron microprobe analysis (EMPA) indicated that each boule had a III-Sb ratio of 1:1 within analytical error. X-ray analysis produced diffraction patterns consistent with standard patterns for the respective compounds. Figure 3 shows a photograph of a typical GaSb boule.

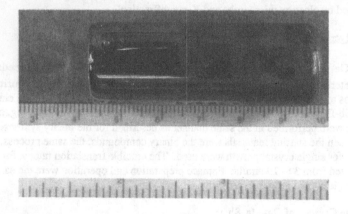

Fig. 3. Typical GaSb boule as removed from the crucible.

Figures 4 and 5 illustrate the longitudinal variation in composition for two polycrystalline boules grown at 19 mm/hr with elemental starting materials in different initial compositions. The former had an initial composition of $x=0.03$, while the latter was initially at $x=0.2$. Figure 6 shows the axial chemical composition of a polycrystalline boule grown from a melt of the binary compounds at 19 mm/hr with an initial melt composition of $x=0.03$. In all cases, the boules grown from a melt with lower x produced a more uniform composition along the boule compared with that produced from the higher melt fraction. Some constitutional supercooling

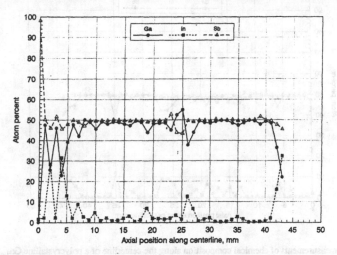

Fig. 4. EMPA measurements of chemical composition along the centerline of a polycrystalline $Ga_{1-x}In_xSb$ boule. The starting materials were elements with an initial composition of $x=0.03$; the solidification rate was 19 mm/hr.

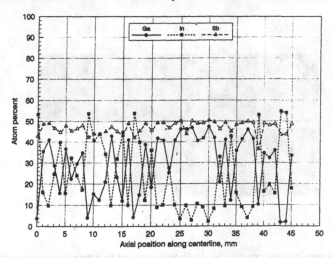

Fig. 5. EMPA measurements of chemical composition along the centerline of a polycrystalline $Ga_{1-x}In_xSb$ boule. The starting materials were elements with an initial composition of $x=0.2$; the solidification rate was 19 mm/hr.

49

may have occurred in the former boules, although it was not as significant as that observed in the boule produced from the less dilute melt. Additionally, the boule grown from the binary compounds did not exhibit high indium levels in the first-to-freeze portion, unlike those found in the boule grown from a melt of equivalent x produced from the elements.

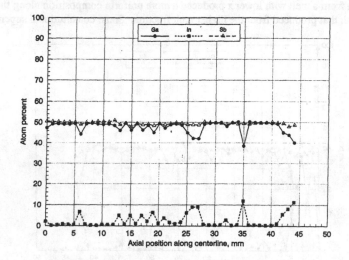

Fig. 6. EMPA measurements of chemical composition along the centerline of a polycrystalline $Ga_{1-x}In_xSb$ boule. Starting materials were GaSb and InSb with an initial composition of $x=0.03$; solidification rate was 19 mm/hr.

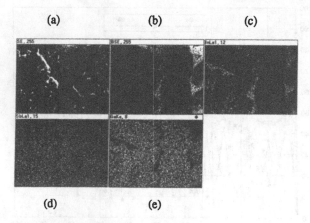

Fig. 7. Elemental maps of a polycrystalline $Ga_{1-x}In_xSb$ boule produced from the elements with an initial composition of $x=0.03$ at a solidification rate of 19 mm/hr. 30X magnification. (a) Secondary electron image. (b) Backscattered electron image. (c) Indium map. (d) Antimony map. (e) Gallium map.

50

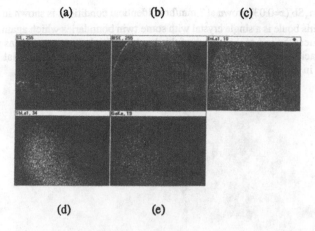

(d)   (e)

Fig. 8. Elemental maps of a polycrystalline $Ga_{1-x}In_xSb$ boule produced from GaSb and InSb with an initial composition of $x=0.03$ at a solidification rate of 19 mm/hr. 30X magnification. (a) Secondary electron image. (b) Backscattered electron image. (c) Indium map. (d) Antimony map. (e) Gallium map.

Elemental maps of the polished surfaces of each boule were made using a scanning electron microscope. Boules grown from elemental starting materials were found to have localized segregation of the elements throughout the boule, as illustrated in Figure 7. No sharp compositional changes were noted across any of the cracks in these boules. When the binary compounds were used as the starting materials, no elemental segregation was observed (Fig. 8) except in the last-to-freeze region.

The formation of elemental clusters in the alloy boules grown rapidly from elemental starting materials is in contrast to the uniform local elemental distributions found in the binary compounds grown under similar conditions. The presence of the third atom appears to have a pronounced effect on the diffusion and reaction rates within the melt. The rapid solidification rate exacerbated this effect by producing a locally varying composition before equalization of the composition could occur. This is similar to results reported in [5], in which it was determined that zone melting would produce equilibrium solid solutions only at zone translation rates of less than 3 mm/hr.

Figure 9 shows a single crystal of $Ga_{1-x}In_xSb$ ($x=0.03$) grown at 7 mm/hr. Although no grain boundaries are present, several twins were formed; all of these appear to have initiated at the crucible wall. Cracks were observed throughout the sectioned boule. X-ray diffraction confirmed uniform crystal orientation along the length of the boule. The results of EMPA measurements are shown in Figure 10. The predicted composition variation based on the Pfann distribution for normal freezing [8] is also shown in Figure 10. This curve was calculated using a constant segregation coefficient, k, of 0.22, which was obtained from the pseudo-binary phase diagram [9] at the initial melt composition. This growth resulted in significant compositional variation due to constitutional supercooling.

51

A boule of Ga$_{1-x}$In$_x$Sb (x=0.03) grown at 3 mm/hr at identical conditions is shown in Figure 11. The first 75% of this boule is a single crystal with some twin boundaries which again initiated at the cruible wall; near the end of the growth of this boule, multiple grains and twins were formed. Small cracks were found only in the upper quarter of this boule. The axial InSb composition is shown in Figure 12.

Fig. 9. Single crystal of Ga$_{1-x}$In$_x$Sb grown at 7 mm/hr. The initial melt composition was x=0.03. A (111) GaSb single crystal seed with the B face toward the melt was used.

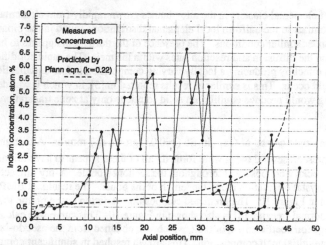

Fig. 10. Axial indium distribution in the single crystal shown in Fig. 9 as measured by EMPA. The indium composition predicted by the Pfann equation [8] for k=0.22 is shown for comparison.

The change in InSb composition in the more slowly grown boule is consistent with stable directional crystal growth from a well-mixed melt with a variable k. In these experiments, mixing of the melt resulted from natural convection. The variation in k as solidification progressed is consistent with that measured and tabulated for Czochralski growth of this alloy [7], and is due to

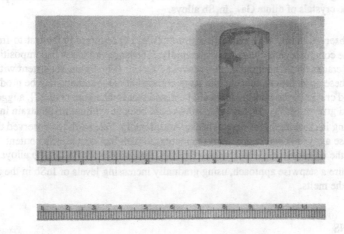

Fig. 11. Single crystal of $Ga_{1-x}In_xSb$ grown at 3 mm/hr. The starting materials were GaSb and InSb with an initial melt composition of $x=0.03$. A (111) GaSb single crystal seed with the B face toward the melt was used.

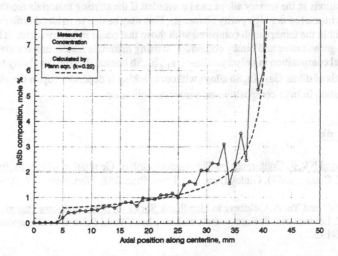

Fig. 12. Axial indium distribution in the single crystal shown in Fig. 11 as measured by EMPA. The indium composition predicted by the Pfann equation [8] for $k=0.22$ is shown for comparison.

the increasing separation between the liquidus and solidus as the fraction of InSb increases in the melt. These results show that constitutional supercooling of this alloy can be avoided if the ratio G/v is greater than about 8 K-hr/cm^2. This is well below the value of 260 K-hr/cm^2 determined by Yee, et al. [6], for $Ga_{1-x}In_xSb$ ($0.1 \leq x \leq 0.5$) grown in space by a gradient freeze technique. The lower critical value of G/v indicates that significantly lower thermal gradients and faster growth rates than those expected from previous works can be used to produce large bulk single crystals of dilute $Ga_{1-x}In_xSb$ alloys.

The cracking observed in the boule grown at 3 mm/hr (Fig. 11) occurred in the last-to-freeze region where the composition is changing most rapidly. The large variations in composition and the presence of cracks throughout the boule grown at 7 mm/hr (Fig. 9) are consistent with this observation. These observations suggest that large compositional gradients may be producing lattice strain and cracks in the boules. It should be noted that Bachmann, et al. [7], suggested that Czochralski growth of $Ga_{1-x}In_xSb$ up to $x\sim0.4$ could be done without misfit strain in the neck resulting in cracking in the crystal neck. Additionally, Yee, et al. [6] observed that the tendency of these alloys to develop microcracks increased with increasing InSb content. These results support the conclusions of Bachmann, et al. [7], that growth of $Ga_{1-x}In_xSb$ alloys with high x will require a stepwise approach, using gradually increasing levels of InSb in the seed crystals and in the melts.

CONCLUSIONS

Bridgman growth of single crystals of dilute $Ga_{1-x}In_xSb$ alloys was shown to be feasible for G/v ratios in excess of ~8 K-hr/cm^2. The use of seed crystals of the same diameter as the growth crucible has also been shown to be an effective means of growing these crystals. Local compositional variations in the ternary alloys can be avoided if the starting materials are the binary constituents instead of the high purity elements. This may be due to reduced diffusion and reaction rates within the ternary melt compared with those that occur in binary melts. The combination of high growth rates and binary compound starting materials results in a reasonably uniform axial composition in polycrystalline $Ga_{1-x}In_xSb$ boules. Lastly, the ability to grow bulk single crystals of dilute $Ga_{1-x}In_xSb$ alloys without cracking is dependent upon the ability to control the variation in InSb composition on a macroscopic scale.

REFERENCES

1. The Second NREL Conference on Thermophotovoltaic Generation of Electricity, edited by J. Benner, T. Coutts, and D. Ginley (Am. Inst. Phys. Proc. 358, Woodbury, NY, 1996).

2. A. Ya. Vul' and Yu. A. Goldberg in Handbook Series on Semiconductor Parameters, Vol. 1, edited by M. Levinshtein, S. Rumyantsev, and M. Shur (World Scientific, Singapore, 1996), p. 125 and 191-192.

3. A. Tanaka, A. Watanabe, M. Kimura, and T. Sukegawa, J. Crystal Growth 135, p. 269 (1994).

4. R. Hamaker and W. White, J. Electrochem. Soc. **116**, p. 478 (1969).

5. V. Ivanov-Omskii and B. Kolomiets, J. Sov. Phys. Solid State **1**, p 834 (1959).

6. J. Yee, M. Lin, K. Sarmi, and W. Wilcox, J. Crystal Growth **30**, p. 185 (1975).

7. K. Bachmann, T. Thiel, H. Schreiber, and J. Rubin, J. Electronic Mat. **9**, p. 445 (1980).

8. W. Pfann, Zone Melting, 2nd Ed., John Wiley & Sons, Inc., New York, 1966, p. 11.

9. P. Dutta, H. Bhat, and V. Kumar, J. Appl. Phys. **81**, p. 5821 (1997).

10. T. Ashley, J.A. Beswick, B. Cockayne, and C.T. Elliott, 7th Int. Conf. on Narrow Gap Semicond., edited by J. L. Reno (Inst. Phys. Conf. Ser. No. 144, Bristol, 1995).

Photoreflectance Study of MBE Grown Te-doped GaSb at the $E_0+\Delta_0$ Transition

S. Iyer, S. Mulugeta, J. Li, B. Mangalam, and S. Venkatraman
Department of Electrical Engineering, North Carolina A & T
State University, Greensboro, NC 27411

K. K. Bajaj
Department of Physics, Emory University, Atlanta, GA 30322

ABSTRACT

Photoreflectance (PR) measurements have been performed on MBE grown GaSb samples at the $E_0+\Delta_0$ transition. The transition energy has been measured from 4 to 300 K in the 1.35 to 1.8 eV energy region. Transition energy of 1.614 eV and broadening of 53 meV have been obtained at 4 K from a representative Te-doped sample while no PR signals were obtained from undoped samples. The logarithmic dependence of the PR amplitude and invariance of the line shape on the modulating pump intensity, and absence of Franz-Keldysh oscillations are typical of the PR spectra in the low-field regime. Good fits were obtained using line shape analysis of the PR spectra and the temperature dependence of the transition energy and broadening have been determined.

INTRODUCTION

Photoreflectance is a contactless and nondestructive characterization technique for optical studies of the critical point energies and assessment of the quality of semiconductors. The dominant mechanism in the PR signal has been identified to arise from the modulation of the built-in surface field and hence the application of this technique to narrow band gap III-V semiconductor materials has been somewhat limited[1,2]. Beaulieu et al.[1] and Hwang et al.[2] have reported PR studies of undoped GaSb at room temperature and InSb at 8 K, respectively. The latter, in their study on InSb found that the PR spectra and the mechanism involved were different for n- and p-type InSb. In this paper we report on the investigation of the PR of MBE grown GaSb epilayers at the $E_0+\Delta_0$ transition.

EXPERIMENT

The samples used in this study were MBE grown GaSb at a growth temperature of 350 °C on semi-insulating GaAs substrate. The epilayer thicknesses ranged from 4 to 5 μm. The Te-doped sample concentration was estimated to be $4\text{-}5\times10^{17}$ cm^{-3} at 300 K.

A standard PR system was setup. A probe beam from a 100 W tungsten halogen lamp source dispersed through a 0.25 m monochromator was used to scan the samples. Modulation was achieved by using a 10 mW He-Ne laser operating at 632.8 nm, chopped at 250 Hz. The laser intensities were varied between 0.08 to 0.8 mW/mm^2 by placing neutral density filters in front of the laser output to investigate the dependence of PR line shape and signal amplitude on laser beam intensity. Sample cooling for low-temperature measurements was provided by a Joule-Thompson cryogenic refrigerator system.

Mat. Res. Soc. Symp. Proc. Vol. 484 © 1998 Materials Research Society

RESULTS

Figure 1 shows PR signal at 4 K of a Te-doped GaSb sample. The line shape exhibits no Franz-Keldysh oscillations and is fitted very well using the first derivative of a Lorentzian functional form[3] which is appropriate for an excitonic transition:

$$\Delta R/R = Re[Ce^{i\theta} (E - E_{cp} + i\Gamma)^{-2}] \tag{1}$$

where $E = \hbar\omega$ is the energy of the probe beam, C and θ are an amplitude and phase factor that slowly vary with E, E_{cp} and Γ are the critical-point energy and the broadening parameter, respectively.

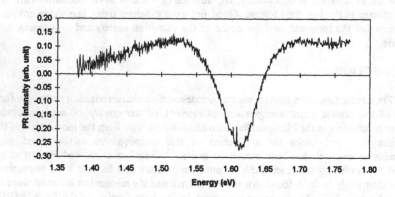

Figure 1. PR signal of MBE sample at 4 K

The PR line shapes measured in the 4 to 250 K range were found to be independent of the modulating pump intensity, as illustrated in Fig. 2 in the limited intensity range investigated, while the amplitude of the PR signal varied logarithmically with the intensity. These observations are indicative of the PR in the low-field regime.

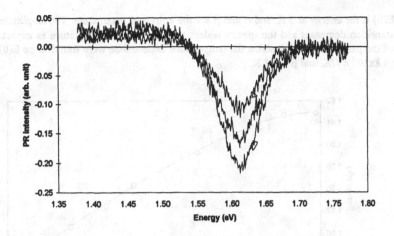

Figure 2. PR amplitude as a function of modulating laser intensity, top curve: 0.08, middle: 0.4, and bottom: 0.8 mW/mm²

Figure 3 shows the theoretical fit to the experimental data using Eq. (1). The best fit results obtained correspond to $E_0+\Delta_0$= 1.614 eV and Γ= 53 meV.

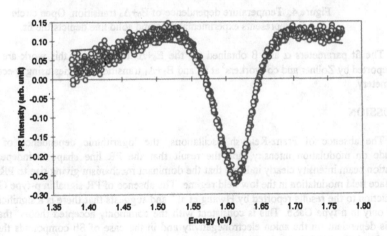

Figure 3. Theoretical simulation of the PR of MBE sample at 4 K. Open circle represents experimental data and solid line represents theoretical simulation.

The temperature dependence of the $E_0+\Delta_0$ transition energy (see Fig. 4) has been investigated using the semi-empirical Varshni's[4] relation:

$$E(T)=E(0)- \alpha T^2/ (\beta+T) \qquad (2)$$

59

where E(0) is the energy at 0 K, and α and β are the empirical parameters. The amplitude of the $E_0+\Delta_0$ transition decreases and the spectra widens with increasing temperature as expected. The values of the parameters that describe the temperature dependence were found to be E(0)=1.615 eV, $\alpha = 4.8 \times 10^{-4}$ eV/K, and $\beta = 188$ K.

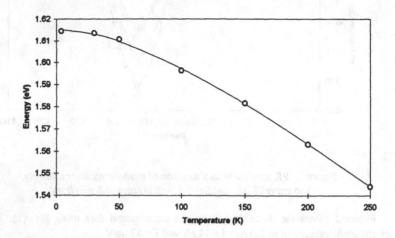

Figure 4. Temperature dependence of $E_0+\Delta_0$ transition. Open circle represents experimental data and solid line depicts the fit.

The fit parameters α and β obtained for the $E_0+\Delta_0$ transition in this work are close to those reported by Zollner and co-workers[5] at E_1 and $E_1+\Delta_1$ transition energies using spectroscopic ellipsometery.

DISCUSSION

The absence of Franz-Keldysh oscillations, the logarithmic dependence of the PR amplitude on modulation intensity, and the result that the PR line shape is independent of modulation beam intensity clearly indicate that the dominant mechanism giving rise to PR signal is the surface field modulation in the low field regime. The absence of PR signal in p-type GaSb is in contradiction to the results reported by Hwang et al.[2] and suggests that there is a significant built-in field only in n-type GaSb. This is consistent with the commonly accepted theory[6] that barrier height is dependent on the anion electronegativity and in the case of Sb compounds there is no barrier for holes and large barrier for electrons. Earlier studies[7,8] on GaSb using x-ray photoemission spectroscopy indicate that the presence of only submonolayer of oxygen/metal at the surface is sufficient to pin the Fermi level close to the valence band. We believe that this explains the PR signal dependence on the conductivity type of GaSb.

In conclusion, a detailed study of Te-doped MBE grown GaSb at the $E_0+\Delta_0$ transition has been made by analyzing the temperature dependence of the PR spectra. The absence of PR signal from undoped GaSb material confirms the pinning of Fermi level close to the valance band in GaSb.

ACKNOWLEDGMENT

This work has been supported by the Air Force Office of Scientific Research (AFOSR) Grant No. F49620-93-1-0111DEF and F49620-95-1-05.

REFERENCES

1. Y. Beaulieu, J. B. Webb, and J. L. Brebner, Solid State commun., 90, 683 (1994)

2. J. S. Hwang, S. L. Tyan, M. J. Lin, and Y. K. Su, Solid State Commun., 80, 891 (1991)

3. D. E. Aspnes, Surf. Sci. 37, 418 (1973)

4. Y. P. Varshni, Phys. 34, 149 (1967)

5. S. Zollner, M.Garriga, J. Humlicek, S. Gopalan, and M. Cardona, Phys. Rev. B43, 4349 (1991)

6. S. Tiwari and D. J. Frank, Appl. Phys. Lett. 60, 630 (1992)

7. W. E. Spicer, P. W. Chye, P. R. Skeath, C. Y. Su, and I. Lindau, J. Vac. Sci. Technol., 16, 1422 (1979)

8. P. W. Chye, I. Lindau, P. Pianetta, C. M. Garner, C. Y. Su, and W. E. Spicer, Phys. Rev. B18, 5545 (1978)

SURFACE AND INTERFACE PROPERTIES OF InSb EPITAXIAL THIN FILMS GROWN ON GaAs BY LOW PRESSURE METALORGANIC CHEMICAL VAPOR DEPOSITION

K. Li#, K.L. Tan*, M. Pelczynski+, Z.C. Feng+, A.T.S. Wee*, J.Y. Lin*, I. Ferguson+, R.A. Stall+
#Institute of Materials Research and Engineering, Singapore 119260
*Department of Physics, National University of Singapore, Singapore 119260
+EMCORE Corporation,394 Elizabeth Avenue, Somerset, NJ 08873

ABSTRACT

There is increasing interest in the epitaxial growth of high quality InSb thin films on GaAs substrates for many device applications such as infrared optoelectronics. The large lattice mismatch (14.6%) between InSb and GaAs has meant that both growth techniques and conditions have a large influence on the interface properties and consequently the film quality. A surface science study, by X-ray photoelectron spectroscopy (XPS) and Auger electron spectroscopy (AES) together with Nomarski microscopy, on the surface and interface properties of InSb/GaAs by metalorganic chemical vapor deposition is presented. It is found from the XPS data that the ambient surface is composed of InSb, In_2O_3, Sb_2O_3 and Sb_2O_5. The interdiffusion phenomena are studied by AES depth profiling; the width of interdiffusion region is determined to be 50 ± 10 nm for all the samples grown at different V/III ratios. This is narrower than the data previously obtained for InSb/GaAs interfaces produced by metalorganic magnetron sputtering. The results also demonstrate that uniform and stoichiometric InSb films have been obtained, and that the reproducibility of the MOCVD technique is excellent.

INTRODUCTION

Indium antimonide (InSb) exhibits the smallest bandgap (0.17 eV), highest intrinsic electron mobility (7×10^5 cm²/Vs at 77 K), highest maximum electron drift velocity, and lowest electron effective mass among all the binary III-V semiconductor compounds [1-4]. These properties make it a promising candidate for the application of infrared devices, high-speed transistors and magnetic sensors [3-9]. Numerous efforts have been completed to grow high quality, epitaxial InSb on silicon (Si) and gallium arsenide (GaAs) to take advantage of the advanced, mature and relatively low-cost Si and GaAs integration technology. GaAs provides a semi-insulating substrate that facilitates electrical isolation for device application as well as simpler analysis of the electrical properties of the lower bandgap InSb [3]. The epitaxial growth of InSb on GaAs has been achieved through several techniques such as molecular beam epitaxy (MBE) [2-4,6,10], metalorganic chemical vapor deposition (MOCVD) [7,11], and metalorganic/r.f. magnetron sputtering [8,12]. To commercially fabricate InSb-based devices, large-scale production is essential. MOCVD has been proved to be the most promising technique for this purpose [7].

In this paper a vertically configured MOCVD system has been employed to grow InSb films on GaAs substrates for production applications. The aim is to investigate the surface morphology, surface chemical states, bulk chemical composition depth distribution and their relevance to the III/V precursor input ratio. The emphasis is to examine the reproducibility of the MOCVD technique in terms of both surface and bulk chemical compositions.

EXPERIMENT

InSb films, 1.4-1.6 µm thick, were grown on GaAs (100) substrates using a low pressure MOCVD system of vertical configuration. This system was equipped with a high-speed rotating disk (180 mm in diameter) capable of growing 4-inch wafer size films. Trimethylindium (TMIn) and tris(dimethylamino)antimony (TDMASb) were used as In and Sb precursors respectively, while hydrogen was used as carrier gas. The growth temperature wastypically 395 °C.

XPS and AES studies were carried out in a VG ESCALAB MkII system. The XPS system employed a Mg Kα X-ray (120 W) as the radiation source. The spectra were collected under the constant analyzer energy (CAE) mode with the pass energy set at 50 eV for wide scan and 20 eV for narrow scan. All XPS binding energies have been referenced to the C 1s peak of adventitious carbon at 284.8 eV [13]. AES was done employing a 5 keV electron beam with the electron energy analyzer set at a constant retard ratio of 4. Ar$^+$ was used as the sputtering ion during AES depth profiling at a current of 0.31 µA, corresponding to a sputtering current density of 41.4 µA/cm^2. The sputtering rates under the conditions used in this study were calibrated with a Form Talysurf 120 profilometer, and found to be 350 ±25 Å/minute. A Nomarski microscope was also used to study the surface morphology.

RESULLTS AND DISCUSSION

Surface morphologies

Two different sets InSb samples were grown under V/III input ratios of 4.4 and 6.2 respectively. InSb films grown at the V/III ratio of 4.4 typically show specular surface with small indium droplets on top. Fig. 1 (a) shows such an example. When grown at a higher V/III input ratio of 6.2, the InSb surface is of better quality, typically free of indium droplets, as shown in Fig. 1(b). These observations suggest that a higher V/III ratio can give a smoother surface.

(a)   (b)

Fig. 1 Nomarski Micrographs, under a magnification factor of 100, of two MOCVD-grown InSb/GaAs, (a) IS17 with In droplets observed and (b) IS52 with a good surface morphology and without In droplets.

Surface chemical states

A typical XPS survey scan is shown in Fig. 2, which indicates the presence of carbon and oxygen due to ambient exposure. To obtain detailed chemical states information, Sb3d and In 3d narrow scan spectra were obtained at two photoelectron take-off angles of 75° and 20° with respect to the sample surface. Fig. 3 shows typical 75° and 20° Sb $3d_{3/2}$ XPS narrow scan spectra. For simplicity, the Sb $3d_{5/2}$ peak is not shown here, as it overlaps with the O 1s peak. Peak fitting is conducted using pure Gaussian line shape. Both of the peaks are fitted with three components, of which InSb component is located at 536.8±0.1 eV, Sb_2O_3 component at 539.4±0.1 eV, and Sb_2O_5

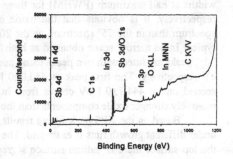

Fig. 2 A typical XPS survey scan obtained from sample IS17.

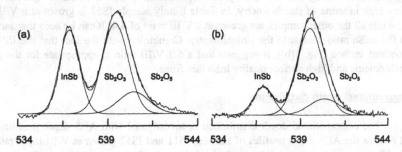

Fig. 3 Sb $3d_{3/2}$ narrow scans of sample IS17 obtained at the photoelectron take-off angles of (a) 75° and (b) 20°.

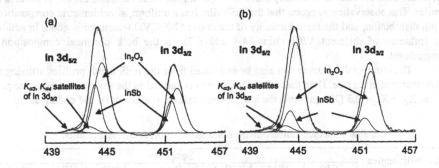

Fig. 4 In 3d narrow scans of sample IS17 obtained at the photoelectron take-off angles of (a) 75° and (b) 20°.

65

component at 540.3±0.1 eV [13]. Correspondingly the binding energies for the three Sb $3d_{5/2}$ components are 527.5±0.1 eV, 530.0±0.1 eV and 530.9±0.1 eV, respectively. The peak full widths at half maximum (FWHM) for these three components are 1.3 eV, 1.7 eV and 2.0 eV, respectively. It is obvious that the oxide component is much more pronounced in the 20° spectrum than in the 75° spectrum, as the 20° spectrum is more surface sensitive. Fig. 4 shows typical In 3d narrow scans obtained at both 75° and 20° photoelectron take-off angles. Both In $3d_{5/2}$ peaks are fitted with two peaks using pure Gaussian fitting with a FWHM of 1.3 eV and 1.8 eV respectively. The first peak at 440.0±0.1 eV binding energy corresponds to InSb, and the second one at 444.7±0.1 eV comes from In_2O_3 [13]. Similarly, the 20° spectrum is seen with relatively stronger oxide components than the 75° spectrum.

Based on the XPS peak fitting results, the surface stoichiometry of seven samples grown under different growth runs is examined. The results are summarized in Table I. It is clear that at the top surface there is indium surface segregation; most of the samples have an In/Sb ratio of about 1.4. The scattering of the data is also more severe as compared with that from the subsurface. This should be related to the more complicated environment at the top surface than at the subsurface. At the subsurface, the In/Sb ratio is stoichiometric, and the data scattering is also very small, indicating the film quality and reproducibility of our MOCVD growth technique isvery high in terms of stoichiometry. In Table I, only sample IS52 is grown at a V/III ratio of 6.2, while all the other samples are grown at a V/III ratio of 4.4. It can be seen that sample IS52 has the In/Sb ratio closest to the stoichiometry. Combining with the fact that the IS52 has the smoothest surface (Fig. 1(b)), it suggests that a 6.2 V/III ratio is appropriate for the growth of stoichiometric and high surface quality InSb thin films.

Compositional depth distribution

The compositional depth distribution is investigated with AES depth profiling. Figs. 5 and 6 show the AES depth profiles of samples IS11 and IS52 grown at V/III input ratios of 4.4 and 6.2 respectively. As in XPS analysis, indium surface segregation can be seen in both samples. In the bulk, the In/Sb ratio is uniform throughout the whole thickness in sample IS11, being around 1.04±0.05 from the AES data in Fig. 5, very close to the value of 1.10±0.05 obtained from XPS. For IS52, there is very small variation in In/Sb ratio in the surface region, but in the bulk the ratio is extremely close to 1. For all the other samples, the depth profiles are similar. This observation suggests that the InSb film has a uniform, stoichiometric compositional depth distribution, and the reproducibility of the current MOCVD technique is good. In addition the influence of different V/III ratios (4.4 and 6.2) on the bulk chemical composition is insignificant.

The interfacial diffusion can also be estimated from the AES depth profiles utilizing the calibrated sputtering rate. Here the interdiffusion area is defined as the region starting at the point where $X_{Ga} + X_{As} = 0.3$ (X_i represents the atomic composition of element i), and ending at the

Table I Surface stoichiometry obtained from XPS analysis results.

Take-off angles	Samples						
	IS10	IS11	IS14	IS15	IS17	IS26	IS52
75°	1.1	1.1	1.2	1.1	1.0	1.1	1.0
20°	1.4	1.4	1.5	1.4	1.2	1.4	1.2

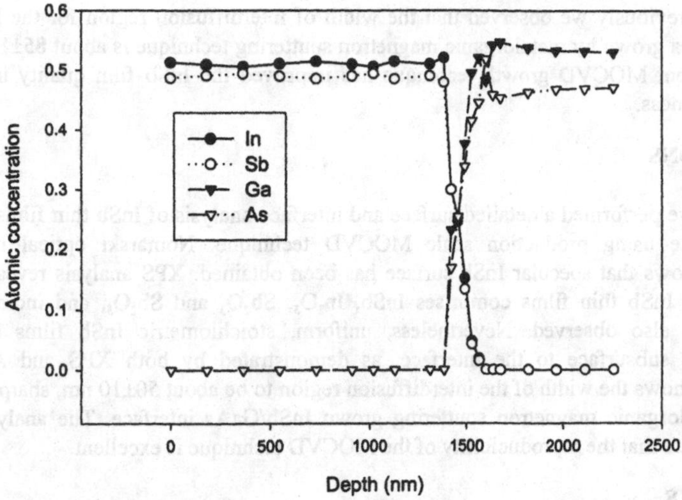

Fig. 5 AES depth profile of sample IS11grown under a V/III input ratio of 4.4.

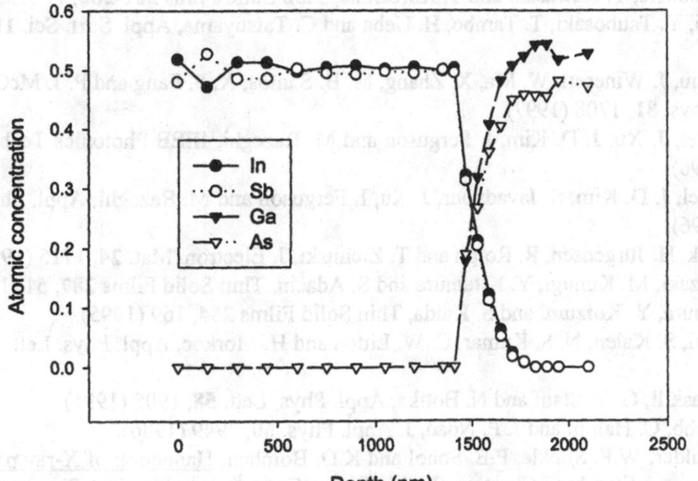

Fig. 6 AES depth profile of sample IS52 grown under a V/III input ratio of 6.2.

point where $X_{Ga} + X_{As} = 0.7$ without corrections for instrumental effects. Based on this definition, the width of the interdiffusion region is found to be 50 ± 10 nm. For all the samples studied, the width of interdiffusion falls in this region. This is reasonable, since the width of interdiffusion is dominated by the lattice mismatch between the epilayer and the substrate, as we demonstrated before [14]. Previously we observed that the width of interdiffusion region for the InSb/GaAs heterostructures grown by metalorganic magnetron sputtering technique is about 85 ± 10 nm [14]. Thus the present MOCVD growth technique has improved the InSb film quality in terms of interface sharpness.

CONCLUSIONS

We have performed a detailed surface and interface analysis of InSb thin films grown on GaAs substrate using production scale MOCVD technique. Nomarski optical microscope observation shows that specular InSb surface has been obtained. XPS analysis reveals that the surface of the InSb thin films comprises InSb, In_2O_3, Sb_2O_3 and Sb_2O_5, and indium surface segregation is also observed. Nevertheless, uniform, stoichiometric InSb films have been obtained from subsurface to the interface, as demonstrated by both XPS and AES. AES investigation shows the width of the interdiffusion region to be about 50 ± 10 nm, sharper than the previous metalorganic magnetron sputtering grown InSb/GaAs interface. The analysis results also demonstrate that the reproducibility of the MOCVD technique is excellent.

REFERENCES

1. D.L. Partin, J. Heremans and C.M. Thrush, J. Appl. Phys. **71** 2328 (1992).
2. M. Kitabatake, T. Kawasaki and T. Korechika, Thin Solid Films **281-282**, 17 (1996).
3. M. Mori, Y. Tsubosaki, T. Tambo, H. Ueba and C. Tatsuyama, Appl. Surf. Sci. **117-118**, 512 (1997).
4. W. K. Liu, J. Winesett, W. Ma, X. Zhang, M. B. Santos, X. F. Fang and P. J. McCann, J. Appl. Phys. **81**, 1708 (1997).
1. E. Michel, J. Xu, J. D. Kim, I. Ferguson and M. Razeghi, IEEE Photonics Technol. Lett. **8**, 673 (1996).
2. E. Michel, J. D. Kim, S. Javadpour, J. Xu, I. Ferguson and M. Razeghi, Appl. Phys. Lett. **69**, 215 (1996).
5. E. Woelk, H. Jürgensen, R. Rolph and T. Zielinski, J. Electron. Mat. **24**, 1715 (1995).
6. T. Miyazaki, M. Kunugi, Y. Kitamura and S. Adachi, Thin Solid Films **287**, 51 (1996).
7. H. Okimura, Y. Koizumi and S. Kaida, Thin Solid Films **254**, 169 (1995).
8. J.-I. Chyi, S. Kalen, N. S. Kumar, C. W. Litton and H. Morkoc, Appl. Phys. Lett. **53**, 1092 (1988).
9. D. K. Gaskill, G. T. Stauf and N. Bottka, Appl. Phys. Lett. **58**, 1905 (1991).
10. J. B. Webb, C. Halpin and J. P. Noad, J. Appl. Phys. **60**, 2949 (1986).
11. J.F. Moulder, W.F. Stickle, P.E. Sobol and K.D. Bomben, Handbook of X-ray photoelectron microscopy, edited by J.Chastain (Perkin-Elmer Corporation, Physical Electronic Division, USA, 1992).
12. K. Li, A. T. S. Wee, J. lin, K. K. Lee, F.Watt, K. L. Tan, Z. C. Feng and J. B. Webb, Thin Solid Films **302**, 111 (1997).

Part II
Antimonide Related Devices

THEORETICAL PERFORMANCE OF MID-INFRARED BROKEN-GAP MULTILAYER SUPERLATTICE LASERS

MICHAEL E. FLATTÉ*, J.T. OLESBERG*, AND C.H. GREIN**
* Department of Physics and Astronomy, University of Iowa, Iowa City, IA 52242
** Department of Physics, University of Illinois, Chicago, IL 60607

ABSTRACT

We present calculations of the intersubband absorption and Auger recombination rate of superlattices based on the InAs/GaInSb material system involving more than two layers in the repeating unit cell and strain balanced to match the GaSb substrate. We demonstrate theoretically the presence of final-state optimization in a 4.0 μm strain-balanced broken-gap superlattice. This system's band structure is optimized not only at the band edge, where the valence density of states has been reduced, but also at resonance energies, where reside final states for Auger and intersubband processes. The spectral structure of the intersubband absorption, which for some wavelengths near the lasing wavelength can exceed 500 cm^{-1} at lasing threshold, has been considered when designing this active region. Fortunately, final-state optimized designs which minimize Auger recombination tend to minimize intersubband absorption as well. The effectiveness of final-state optimization is evaluated by considering band structures with identical band edge structure, but different final-state structure.

INTRODUCTION

Recent work on mid-infrared lasers can be separated into two rough categories according to whether the light is generated in an interband[1-7] or intersubband[8] transition. The dominant non-radiative transition for the second type occurs via electron-phonon scattering, and has a typical timescale of picoseconds. The dominant non-radiative transition for interband lasers, by contrast, has a timescale of nanoseconds, and is due to carrier-carrier (Auger) scattering. We will focus here on the issues associated with improving interband lasers in the mid-infrared through band-structure engineering of a strain-balanced broken-gap superlattice active region.

Optimizing band structures is a key challenge in the design of mid-infrared lasers. Considerable optimization is possible through changing the band structure at the band edges. By introducing strain, the valence density of states can be reduced, thus decreasing the threshold carrier density n_{th}. Reduction of n_{th} leads to a reduced intervalence absorption (which is proportional to n_{th}) and Auger rate (proportional to n_{th}^3 at low density). We will refer to this form of band structure optimization as *band-edge* optimization.

Our focus will be on *final-state* optimization, a term which refers to reducing the availability of final states for Auger processes and intervalence absorption. For a fixed density these designs reduce intervalence absorption and Auger rates. Reduction of intervalence absorption indirectly reduces n_{th}, but reduction of the Auger rate does not. Reduction of the Auger rate does, however, reduce the more important physical quantity J_{th}, the threshold current density. Since a bulk system has neither band-edge nor final-state optimization we emphasize that demonstrating a suppressed Auger rate relative to a bulk system does not necessarily mean there is final-state optimization.

71

A particular example of a system exhibiting final-state optimization will be presented. Its 300K Auger rate is a factor of two smaller than another optimized type II system with the same band gap but without final-state optimization at 300K[3]. We then explore the sensitivity of this type of band structure to changes which affect the final states while leaving the band-edge electronic structure unchanged.

We continue by addressing possible problems associated with final-state optimization, such as vertical transport in a system with a highly quantized band structure. We will conclude with a system which should have acceptable transport properties, but which maintains the high degree of structure in its final states for Auger processes and intersubband transitions.

THEORY

Calculations of the band structure including non-parabolicity, and the **K**-dependent optical matrix elements, were performed with a superlattice **K · p** technique (see references in Ref. 9). Our confidence in this type of calculation is bolstered by exceptional agreement with measurements on a type II superlattice multiple quantum well[10]. The radiative and non-radiative recombination rates were then calculated with the procedure described in Ref. 11.

RESULTS

Figure 1 shows a structure which exhibits final-state optimization, 15Å InAs, 25Å $In_{0.40}Ga_{0.60}Sb$, 15Å InAs, and 39Å $Al_{0.30}In_{0.28}Ga_{0.42}As_{0.50}Sb_{0.50}$. The band edges are shown in Fig. 1(a) and the band structure in Fig. 1(b). Final-state optimization is possible in

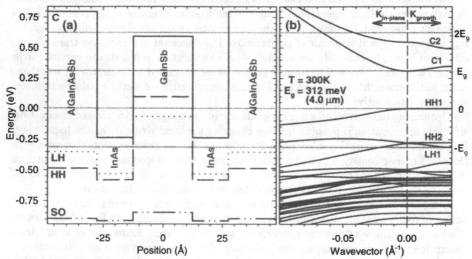

Figure 1: (a) Band offsets for the strain-balance broken-gap superlattice described above. (b) Band structure in the growth direction $(K_\perp)$ and in-plane direction $(K_\parallel)$. The resonance energies in the superlattice are indicated, as are the valence and conduction edges.

this structure because of the large range in band edges, from an energy gap E_g above the superlattice conduction band to E_g below the superlattice valence band edge. This range of band edges produces structure in the band structure at E_g above the superlattice conduction minimum and at one E_g below the superlattice valence band. Resonance energies are indicated in gray. Final states for Auger processes must lie above the $2E_g$ line in the conduction band (for electron Auger processes) or below the $-E_g$ line in the valence band (for hole Auger processes). In electron Auger processes the energy from the recombining electron-hole pair is transferred to an electron; in hole Auger processes it is transferred to a hole.

These structures in the conduction and valence band produce gaps in the intersubband absorption as shown in Fig. 2(a). The gap is somewhat smaller than either the zone-center or growth-direction zone-edge LH1—LH2 gap because transitions elsewhere in the Brillouin zone also contribute. Transitions which narrow the gap occur at the growth-axis zone boundary at finite in-plane momentum. This gap is placed around the peak of the gain region (the gain is shown in Fig. 2(b)). Thus n_{th} is lower, for the intersubband absorption does not need to be overcome. In contrast, a type I system[7] (whose intersubband

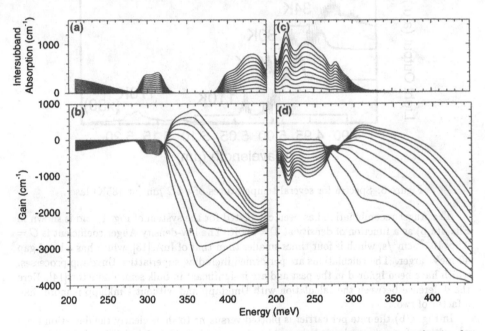

Figure 2: (a) Intersubband absorption for the strain-balanced broken-gap superlattice for several carrier densities, showing a pronounced gap between transitions associated with the LH1 and LH2 subbands. (b) Gain spectrum for the same structure and the same densities. The peak of the gain is in this gap in the intersubband absorption. (c) Intersubband absorption for a type I system[7], 80Å $InP_{0.24}As_{0.76}/80Å$ $InAs_{0.88}Sb_{0.12}$. The densities plotted are the same as in (a) and (b). (d) Gain spectrum for the type I system, showing a smaller peak gain than the strain-balanced broken-gap superlattice.

absorption is shown in Fig. 2(c) and whose gain is shown in Fig. 2(d)) has hundreds of cm^{-1} of intersubband absorption at the gain region, which increases n_{th}. In addition, since the intersubband absorption increases with n, the differential gain is much larger with the superlattice of Fig. 1 than that associated with Fig. 2(c,d).

Lasers with this type of active region[9] were grown and lased under optical pumping at room temperature. Possible problems with vertical transport will be addressed later in this article. A device with a strain-balanced broken-gap superlattice active region similar to that described above, but with a band gap of 5.2 μm, also lased under optical pumping up to 185K, making it the longest-wavelength III-V interband laser to date[12].

The spectrum of this 5.2 μm laser was of some interest, and is shown below in Fig. 3 for several temperatures. There is a "hole" in the spectrum at intermediate temperatures. It does not originate with an atmospheric absorption line, for the hole moves with temperature. The movement of the line with temperature roughly corresponds to the expected movement of an intervalence absorption feature. This hole, therefore, may be due to an intervalence absorption peak which has unfortunately ended up interfering with the gain region. The characteristic temperature (T_o) of this laser was 37K from 40K to 140K.

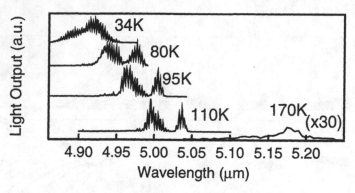

Figure 3: Spectra for several temperatures of a 5.2 μm (at 185K) laser.

The Auger recombination has been calculated for the system of Fig. 1, and is shown in Fig. 4(a,b) as a function of density at $T = 300$K. The low-density Auger coefficient is $C = 2.5 \times 10^{-27}$ cm^6/s, which is four times smaller than that of InAs[13], which has a band gap 40 meV larger. The calculations are performed including superlattice Umklapp processes, which have been ignored in the past and are insignificant in bulk semiconductors[14]. Here the difference between the calculation with Umklapp and without Umklapp is more than a factor of two.

In Fig. 4(b) the rate per carrier is plotted versus n^2 to show clearly the deviation from n^2. Although a sub-quadratic behavior has been predicted at high densities[15,16], this is the first realistic calculation (i.e., which does not assume $T = 0$K[15] or rely on the density dependence of the "most-probable transition"[16]). The deviation occurs at approximately the density where the valence band becomes degenerate (the conduction band is degenerate for the entire range of densities considered here). This correlation implies that the rate is dominated by hole Auger. The agreement with experimental measurements (see article by J. T. Olesberg, *et al.*, in this volume) is exceptional.

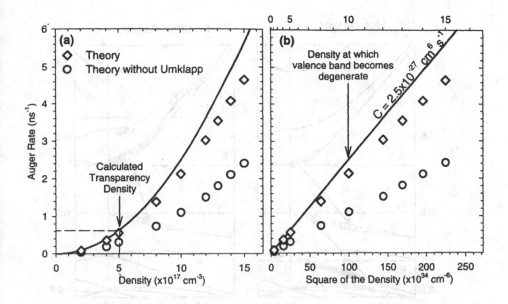

Figure 4: (a) Non-radiative recombination rate for the strain-balanced broken-gap super-lattice, calculated without (circles) and with (diamonds) Umklapp processes. The solid line fits the calculations at low densities, and is an n^2 extrapolation to higher densities. (b) Same calculations plotted versus n^2 to better indicate the deviation of the calculations from n^2.

The dominant hole and electron processes are shown below in Figs. 5(a,b) respectively. The most probable holes are indicated with open circles, and the most probable electrons are indicated with filled circles projected onto the in-plane direction of the band structure of the strain-balanced broken-gap superlattice. The band edges and the resonance energy are indicated by dashed lines. A guide to the importance of Umklapp processes is the relative size of the typical in-plane momentum transfer to the Brillouin zone width in the growth direction. For this system it is clear that the Brillouin zone width is smaller than this typical momentum transfer.

As mentioned before, comparison with a bulk system's Auger rate is not a good test of final-state optimization for the bulk system does not have band-edge optimization. Fig. 6, below, shows a comparison of the Auger rate per carrier for the strain-balanced broken-gap superlattice versus a T2QW structure[3] at the same wavelength. The difference in rate is about a factor of two. Since the T2QW structure is expected to have approximately the same band-edge optimization as the strain-balanced broken-gap superlattice, the Auger rate difference must originate from final-state optimization.

We now explore the importance of final-state optimization in influencing Auger rates. Shown in Fig. 7(a) are the most probable transitions in a 3.7 μm strain-balanced broken-gap superlattice[9] at 77K. Our focus is entirely on the dominant Auger process: hole Auger. Momentum conservation constraints require the electrons at 77K to have excess

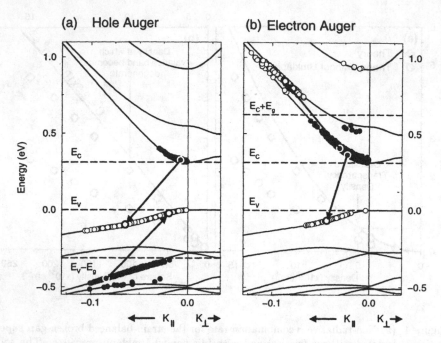

Figure 5: (a) Most probable transitions for hole Auger in the strain-balanced broken-gap superlattice. The arrows indicate the single most probable transition. (b) Most probable transitions for electron Auger in the strain-balanced broken-gap superlattice.

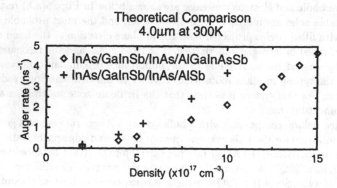

Figure 6: Comparison of the Auger rate per carrier between the strain-balanced broken-gap superlattice and a T2QW structure at the same wavelength.

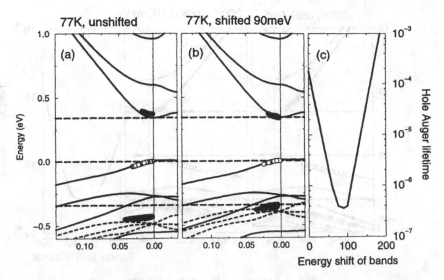

Figure 7: (a) Most probable transitions for hole Auger at 77K indicated on the band structure of a 3.7 μm strain-balance broken-gap superlattice. (b) Most probable transitions when the fourth, fifth, and sixth valence subbands have been moved up by 90 meV. (c) Carrier lifetime due to hole Auger as a function of the band shifts.

energy relative to the conduction minimum of approximately 30 meV, corresponding to $\sim 4k_BT$. Since the occupation factors are proportional to $\exp(-E/k_BT)$, this energy difference produces a reduction in the occupation factor, and thus in the hole Auger rate, of two orders of magnitude.

Our procedure is to take the band structure shown in Fig. 7(a) and manually shift the fourth, fifth, and sixth valence subbands from the valence edge (shown as dashed lines) and observe the effect on the carrier lifetimes. As the bands are shifted up more states at zero in-plane momentum become accessible as final states for Auger processes. We show in Fig. 7(b) the most probable transitions for the same situation as Fig. 7(a), but with the band shifted 90 meV up. It is evident that the electrons in the conduction band which were 30meV above the band edge are now at the band edge. Hence the final-state optimization has been removed. Fig. 7(c) shows the carrier lifetime as a function of the energy shift of the bands towards the band edge. The difference in the lifetime between the optimized case of Fig. 7(a) and the unoptimized case of Fig. 7(b) is over two orders of magnitude.

In Fig. 8(a,b,c) the situation at 300K is shown. Due to the marked sensitivity of the Auger rate on band gap the differences in the band structure at 77K and 300K are ignored; we use the 300K band structure for both carrier temperatures. Whereas the non-equilibrium carriers were quite concentrated at the zone center at 77K, here they are substantially more spread out. In addition, the non-equilibrium electrons involved in the most probable transitions are located at the conduction band minimum. Hence the contribution of final-state optimization to reducing the 300K Auger rate in this structure appears small. This is supported by Fig. 8(c), which shows the lifetime as a function of the band shift. We note that the 4.0 μm structure mentioned above as having final-

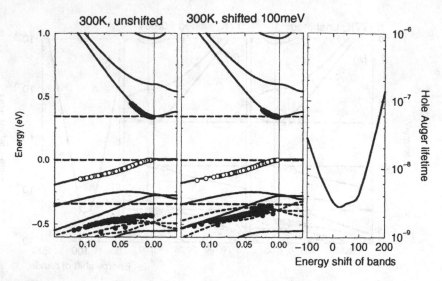

Figure 8: Same as Fig. 7, but for 300K.

state optimization roughly corresponds on Fig. 8(c) to an energy shift of −30 meV, which indicates an Auger rate suppression of a factor of 2.

We now address the issue of vertical transport in these structures. Since the valence band structure has little dispersion up to energies greater than the band gap, one might be concerned about vertical hole transport. This issue may be avoided by designing a system which allows transport near the band edge, but maintains the quantized structure at the resonance energies. Such a structure is shown in Fig. 9. Fig. 9(a) shows the band edge diagram for a superlattice injector into a superlattice well. The superlattice well has a band structure (shown in Fig. 9(b)) similar to the strain-balanced broken-gap superlattices described above. The barrier, however, is designed to allow for electron and hole transport at certain energies (indicated on the band structure diagram in Fig. 9(c) with gray shading), but to prevent such transport at the resonance energies. Thus a minigap of the superlattice injector is positioned near the resonance energies. The combination of the two pieces of the structure yields the band structure shown in Fig. 9(d). Transport is allowed for the highest valence subbands, except for the very top one, but is difficult for subbands near the resonance region. With such a design we can have quantized electronic structure at the resonance energies, but dispersive structure closer to the band edges.

CONCLUSIONS

The intersubband absorption and Auger rates have been calculated for strain-balanced broken-gap superlattices in the mid-infrared. Final state optimization of a 4.0 μm structure has been demonstrated by comparison with a structure with band-edge optimization. The sensitivity of final-state optimization to band location and temperature has been explored. A new design is proposed which allows for optimized intersubband absorption and Auger rates, and is expected to have good vertical transport.

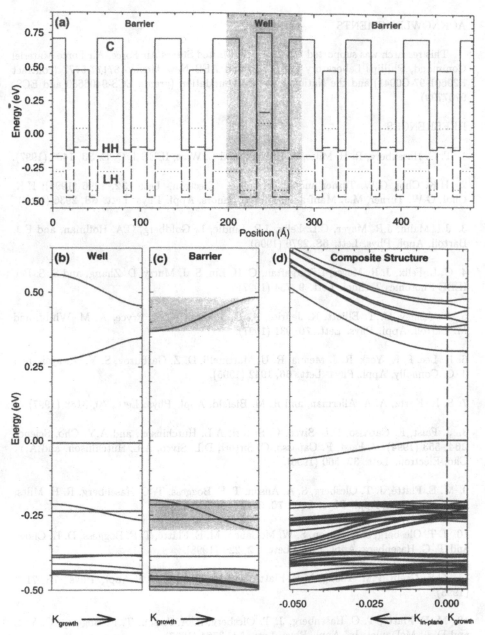

Fig. 9: (a) Band edges for a superlattice injection region and well (gray). (b) Growth-direction band structure of repeating well units. (c) Same as (b) but for the injector/barrier. Transport is allowed for those energies with dispersive subbands (indicated by gray). (d) Band structure of combination, showing dense states right below the valence maximum and gaps at the resonance energies in the valence and conduction bands.

ACKNOWLEDGMENTS

This research was supported in part by the United States Air Force, Air Force Materiel Command, Phillips Laboratory (PL), Kirtland AFB New Mexico 87117-5777 (contract F29601-97-C0041) and the National Science Foundation (grants ECS-9406680 and ECS-9707799).

REFERENCES

1. T. C. Hasenberg, R. H. Miles, A. R. Kost, and L. West, IEEE J. Q. E. **33**, 1403 (1997).

2. H.K. Choi, G.W. Turner and M.J. Manfra, Electron. Lett. **32**, 1296 (1996); H.K. Choi, G.W. Turner, M.J. Manfra and M.K. Connors, Appl. Phys. Lett. **68**, 2936 (1996).

3. J.I. Malin, J.R. Meyer, C.L. Felix, J.R. Lindle, L. Goldberg, C.A. Hoffman, and F.J. Bartoli, Appl. Phys. Lett. **68**, 2976 (1996).

4. C. L. Felix, J. R. Meyer, I. Vurgaftan, C. H. Lin, S. J. Murry, D. Zhang, and S. S. Pei, IEEE Photonics Technol. Lett. **9**, 734 (1997).

5. T. Ashley, C. T. Elliott, R. Jeffries, A. D. Johnson, G. J. Pryce, A. M. White, and M. Carroll, Appl. Phys. Lett. **70**, 931 (1997).

6. H. Lee, P. K. York, R. J. Menna, R. U. Martinelli, D. Z. Garbuzov, S. Y. Narayan, and J. C. Connolly, Appl. Phys. Lett. **66**, 1942 (1995).

7. S. R. Kurtz, A. A. Allerman, and R. M. Biefeld, Appl. Phys. Lett. **70**, 3188 (1997).

8. J. Faist, F. Capasso, D.L. Sivco, C. Sirtori, A.L. Hutchinson, and A.Y. Cho, Science **264**, 553 (1994). J. Faist, F. Capasso, C. Sirtori, D.L. Sivco, A.L. Hutchinson, and A.Y. Cho, Electron. Lett. **32**, 560 (1996).

9. M. E. Flatté, J. T. Olesberg, S. A. Anson, T. F. Boggess, T. C. Hasenberg, R. H. Miles, and C. H. Grein, Appl. Phys. Lett. **70**, 3212 (1997).

10. J. T. Olesberg, S. A. Anson, S. W. McCahon, M. E. Flatté, T. F. Boggess, D. H. Chow, and T. C. Hasenberg, Appl. Phys. Lett. **72**, 229 (1998).

11. C.H. Grein, P.M. Young, M.E. Flatté, and H. Ehrenreich, J. Appl. Phys. **78**, 7143 (1995).

12. M. E. Flatté, T. C. Hasenberg, J. T. Olesberg, S. A. Anson, T. F. Boggess, C. Yan, and D. L. McDaniel, Jr., Appl. Phys. Lett. **71**, 3764 (1997).

13. K. L. Vodopyanov, H. Graener, C. C. Phillips, and T. J. Tate, Phys. Rev. B **46**, 13194 (1992).

14. S. Brand and R. A. Abram, J. Phys. C **17**, L571 (1984).

15. A. Haug, J. Phys. C **16**, 4159 (1983).

16. V. Chazapis, H. A. Blom, K. L. Vodopyanov, A. G. Norman, and C. C. Phillips, Phys. Rev. B **52**, 2516 (1995).

AUGER RECOMBINATION IN ANTIMONY-BASED, STRAIN-BALANCED, NARROW-BAND-GAP SUPERLATTICES

J.T. OLESBERG*, THOMAS F. BOGGESS*, S.A. ANSON*, D.-J. JANG*, M.E. FLATTÉ*, T.C. HASENBERG*, C.H. GREIN**
* Department of Physics and Astronomy, University of Iowa, Iowa City, IA 52242
** Department of Physics, University of Illinois, Chicago, IL 60607

ABSTRACT

Time-resolved all-optical techniques are used to measure the density and temperature dependence of electron-hole recombination in an InAs/GaInSb/InAs/AlGaInAsSb strain-balanced superlattice grown by molecular beam expitaxy on GaSb. This 4 μm bandgap structure, which has been designed for suppressed Auger recombination, is a candidate material for the active region of mid-infrared lasers. While carrier lifetime measurements at room temperature show unambiguous evidence of Auger recombination, the extracted Auger recombination rates are considerably lower than those reported for bulk materials of comparable bandgap energy. We find that the Auger rate saturates at carrier densities comparable to those required for degeneracy of the valence band, illustrating the impact of Fermi statistics on the Auger process. The measured results are compared with theoretical Auger rates computed using a band structure obtained from a semi-empirical 8-band K·p model. We find excellent agreement between theoretical and experimental results when Umklapp processes in the growth direction are included in the calculation. Measured recombination rates from 50 to 300 K are combined with calculated threshold carrier densities to determine a material T_0 value for the superlattice.

INTRODUCTION

Auger recombination is one of the main factors limiting operation of mid-infrared semiconductor lasers. One strategy for minimizing Auger recombination is to utilize semiconductors with large type-II band offsets in the design of materials with a small band-edge density of states and minigaps in the band structure one band gap from the band edges. It is important to experimentally verify the extent to which Auger suppression is achieved in these designs. Comparison of measured and calculated Auger rates also provides a stringent test of the accuracy of band structure and matrix element calculations used in theoretical optimization of material properties.

EXPERIMENT

The structure under investigation was grown for optical characterization measurements using a Perkin-Elmer 430P MBE machine on a nominally undoped GaSb substrate. A schematic of the epitaxial structure is shown in Figure 1(a). Forty periods of a four-layer superlattice are grown between AlGaInAsSb barriers. The barriers serve to increase the absorption of the optical excitation source and keep the photogenerated carriers in the superlattice. Figure 1(b) shows one period of the superlattice. Room temperature photoluminescence from this structure indicates a band gap near 4.0 μm and good structural quality. A four-layer superlattice using the same alloys but with slightly different layer thicknesses has been described earlier.[1] An optically-pumped 5.2 μm laser using a similar superlattice has operated at 185 K.[2]

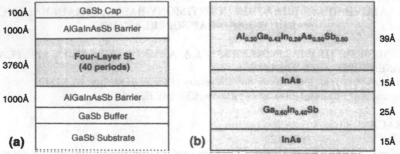

Figure 1: (a) Schematic of the epitaxial structure of the sample and (b) schematic of one period of the four-layer superlattice.

Two optical methods have been used to measure recombination rates in this sample: time-resolved differential transmission[3] and time-resolved photoluminescence upconversion.[4] A schematic of the differential transmission measurement is shown in Figure 2(a). A pulse from a Ti:sapphire laser is used to create a dense distribution of nonequilibrium carriers in the sample. The transmission of the sample is then probed with the mid-infrared output of an optical parametric oscillator.[5] The wavelength of the mid-infrared probe is tunable from 2.6 - 4.4 μm.[6] After passing through the sample, the probe is directed through a monochromator which increases the spectral resolution of the measurement. The transmitted probe intensity is measured with a liquid nitrogen cooled InSb detector. By chopping the pump and probe separately, we can extract a signal proportional to the change in transmission of the probe induced by the non-equilibrium carriers and a signal proportional to the equilibrium transmission of the probe. These values are used to calculate the differential transmission, which is the change in transmission induced by the presence of the non-equilibrium carriers normalized by the equilibrium transmission.

A schematic of the time-resolved photoluminescence upconversion experiment is shown in Figure 2(b). Each pulse from the Ti:sapphire laser is split into two. One is used to excite the sample, which is mounted in a variable temperature dewar. The mid-infrared photoluminescence

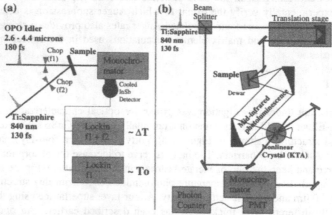

Figure 2: (a) Schematic of the differential transmission experiment. (b) Schematic of the time-resolved photoluminescence upconversion experiment.

from the sample is collected and imaged onto a nonlinear crystal. The remaining portion of the Ti:sapphire pulse is focused onto the same spot on the nonlinear crystal. A portion of the mid-infrared photoluminescence is upconverted to visible wavelengths and directed into a monochromator and photon-counted using a photomultiplier tube. The measured signal is the intensity of the upconverted photoluminescence, which is proportional to the intensity of the original mid-infrared photoluminescence. The temporal resolution in this experiment comes from the sum frequency generation: the Ti:sapphire pulse involved in the upconversion process acts as a temporal gate. In both experiments, the temporal resolution is approximately 200 fs.

Recombination rates (the inverse of the instantaneous lifetime) are extracted from the data using

$$R = \frac{1}{N}\frac{dN}{dt} = \frac{1}{N}\frac{\partial N}{\partial(Signal)}\frac{d(Signal)}{dt} \tag{1}$$

where R is the recombination rate per carrier, N is the carrier density, and "$Signal$" is either differential transmission or upconverted photoluminescence intensity. We assume that the initial carrier density in our superlattice is uniform over the length of the structure and that it is proportional to the incident photon density. The first assumption is reasonable since the carriers are injected with almost 1 eV of excess energy and will be able to distribute themselves throughout the superlattice before cooling to the band edge. The second assumption is necessary to avoid the very difficult process of modeling the density and excess energy dependent capture properties of the superlattice.

Since the photoexcited carriers are created with a large excess energy, it is expected that some will escape into the substrate before cooling to the band edge of the superlattice. Thus the photoexcited carrier density calculated using the pump photon density, pump absorption coefficient, and sample thickness represents an upper limit of the true carrier density in the superlattice. In the pump-probe experiment, we can also calculate the carrier density in the superlattice by comparing the saturation of the experimental differential-transmission with the calculated saturation due to the $(1 - f_e - f_h)$ term in the density-dependence of the absorption coefficient at the probe wavelength. In the photoluminescence-upconversion experiment, we can calculate the carrier density by comparing the widths of the measured and calculated photoluminescence spectra. All three methods are in reasonable agreement and we estimate the uncertainty in the density to be on the order of 25%.

RESULTS

Recombination rates at 300K are shown as a function of density in Figure 3(a). From the zero density intercept, we can extract a Shockley-Read-Hall rate corresponding to a lifetime of 2.7 ns. This is a characteristic lifetime for good quality structures grown in this MBE machine. The calculated transparency density for the superlattice is indicated in the figure. This gives a lower bound on the densities of interest for laser operation. At material transparency, the total recombination rate is just over 1 ns^{-1} (a lifetime of 0.95 ns). We expect radiative recombination to be insignificant in this structure at all but the lowest densities due to the type-II nature of the superlattice. Calculations of the radiative recombination rate show it to be more than a factor of 30 smaller than the total recombination rate for all densities considered at 300 K.

The total recombination rate increases nearly quadratically ($R \approx A + CN^2$ for $BN \ll A + CN^2$, where A, B, and C are the Shockley-Read-Hall, Radiative, and Auger

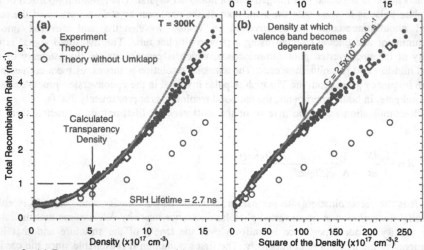

Figure 3: (a) Room temperature recombination rate as a function of density. (b) Recombination rate as a function of the square of the density.

coefficients, respectively) with density for densities below 9×10^{17} cm^{-3}, but becomes sub-quadratic at higher densities. The sub-quadratic dependence can be seen more clearly in Figure 3(b) where the recombination rate is shown as a function of the square of the density. On this figure, quadratic density-dependence appears as a straight line. The recombination rates lie on a straight line at low densities (up to approximately 9×10^{17} cm^{-3}), from which an Auger coefficient of 2.5×10^{-27} cm^{-3} is extracted (the solid line in Figure 3(a) is the same as that in Figure 3(b)). This value of the Auger coefficient is approximately four times smaller than the value of 11×10^{-27} cm^6 s^{-1} measured in InAs.[7] In light of the fact that the superlattice being studied has a bandgap 40 meV smaller than that of InAs, the measured Auger coefficient indicates that the Auger optimizations performed on this sample have been successful.

Sub-quadratic density-dependence of the Auger rate has previously been experimentally observed in bulk InSb.[8] Quadratic density-dependence is appropriate only for nondegenerate bands. The transition to sub-quadratic behavior is a consequence of Fermi statistics on the Auger process. The fact that we observe the transition to sub-quadratic behavior at densities very close to those at which the valence band becomes degenerate implies that Auger recombination in our system is dominated by hole-hole processes.

Auger rates for this structure have been calculated using band structures and momentum-dependent matrix elements calculated using 8-band superlattice **K·p** theory.[9] The accuracy of the band structure and matrix element calculation has been verified experimentally with a different type-II structure.[10] The results of the theoretical calculations are shown in Figure 3 along with the experimental data. There are no adjustable parameters used in the theoretical calculation. The experimental Shockley-Read-Hall rates have been added to the theoretical Auger rates to simplify comparison. The agreement between the experimental measurements and theoretical calculations is excellent given the difficulty of both the experiment and calculation. Theory and experiment are in agreement in both the magnitude and the density dependence of the Auger rate.

Figure 4(a) shows the measured and calculated Auger rates as a function of lattice temperature for three densities. The experimental Auger rate is extracted from the data by

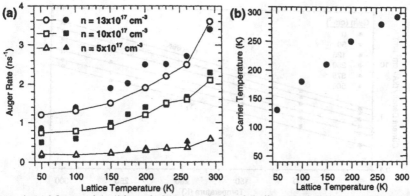

Figure 4: (a) Measured (solid symbols) and calculated (hollow symbols) Auger rates as a function of temperature. The lines connecting the calculated points are merely a guide to the eye. (b) Average carrier temperature as a function of lattice temperature.

subtracting the experimental Shockley-Read-Hall rate and the calculated radiative rate from the data. The calculated radiative rate is less than 20% of the total rate at all densities and temperatures. Both theory and experiment show increasing Auger rates at increasing temperature. Again, the agreement between theory and experiment is very good.

Figure 4(b) shows the average carrier temperature corresponding to the different lattice temperatures in the measurement. The carrier temperature was determined by fitting the high-energy tail of the upconverted photoluminescence spectra to a Boltzmann distribution. Since the Auger rate is strongly dependent on the carrier temperature, the experimental carrier temperatures were used in the calculation of the Auger rates. The shown temperature is the average carrier temperature over the interval from 10 ps - 1 ns.

The measured recombination rates can be used to calculate threshold current densities for this material. In Figure 5, threshold current density is shown as a function lattice temperature for a range of material gains. The density required for a given amount of gain is taken from absorption calculations. However, the gain has been scaled with experimental data from the differential transmission experiment and is thus semi-empirical. From the temperature dependence of the injected current densities, we calculate a material T_0 of 78 K for transparency and 69 K for 500 cm^{-1} of gain. These values of T_0 are for the superlattice material. The T_0 for a device which utilizes this superlattice as the active region will likely be smaller since there are additional temperature-dependent effects in a real device which are not present in our optical tests (e.g., leakage current).

CONCLUSIONS

We have measured recombination rates as a function of density and temperature in a narrow-bandgap superlattice designed to suppress Auger recombination. At low densities, the room temperature Auger coefficient is $C = 2.5 \times 10^{-27}$ cm^6 s^{-1}. Excellent agreement is demonstrated between measured and calculated Auger recombination rates for all temperatures and densities considered. Temperature-dependent current densities have been determined. For 500 cm^{-1} of material gain, the superlattice has a characteristic temperature of 69 K. The small

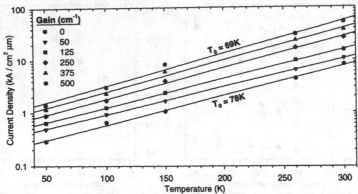

Figure 5: Injected carrier densities as a function of lattice temperature for different amounts of gain.

room temperature Auger coefficient and the large characteristic temperature are a direct consequence of the optimization of the band structure of this superlattice.

ACKNOWLEGMENTS

This research was supported in part by the U.S. Air Force, Air Force Materiel Command, Phillips Laboratory (PL), Kirtland AFB, NM 87117-5777 (Contract No. F29601-97-C0041) and the National Science Foundation (Grant Nos. ECS-9406680 and ECS-97-07799).

REFERENCES

1. Michael E. Flatté, J. T. Olesberg, S. A. Anson, Thomas F. Boggess, T. C. Hasenberg. R. H. Miles, and C. H. Grein, Appl. Phys. Lett. **70**, 3212 (1997).
2. Michael E. Flatté, T. C. Hasenberg. J. T. Olesberg, S. A. Anson, Thomas F. Boggess, Chi Yan, and D. L. McDaniel, Jr., Appl. Phys. Lett. **71**, 3764 (1997).
3. S. W. McCahon, S. A. Anson, D.-J. Jang, M. E. Flatté, Thomas F. Boggess, D. H. Chow, T. C. Hasenberg, and C. H. Grein, Appl. Phys. Lett. **68**, 2135 (1996).
4. D.-J. Jang, J. T. Olesberg, M. E. Flatté, Thomas F. Boggess, and T. C. Hasenberg, Appl. Phys. Lett. **70**, 1125 (1997).
5. S. W. McCahon, S. A. Anson, D.-J. Jang, and Thomas F. Boggess, Opt. Lett. **20**, 2309 (1995).
6. The additional wavelength tuning range as compared to Ref. 5 is made possible in part by the use of a potassium-titanyl-arsenate (KTA) rather than potassium-titanyl-phosphate (KTP) for the nonlinear crystal in the optical parametric oscillator cavity.
7. K. L. Vodopyanov, H. Graener, C. C. Phillips, and T. J. Tate, Phys. Rev. B **46**, 13194 (1992).
8. V. Chazapis, H. A. Blom, K. L. Vodopyanov, A. G. Norman, and C. C. Phillips, Phys. Rev. B **52**, 2516 (1995).
9. See references contained in Ref. 1.
10. J. T. Olesberg, S. A. Anson, S. W. McCahon, Michael E. Flatté, Thomas F. Boggess, D. H. Chow, and T. C. Hasenberg, Appl. Phys. Lett. **72**, 229 (1998).

HIGH POWER MID-INFRARED INTERBAND CASCADE LASERS

B. H. YANG, D. ZHANG, RUI. Q. YANG, C.-H. LIN, S. J. MURRY, H. WU, S. S. PEI
Space Vacuum Epitaxy Center, University of Houston, Houston, TX 77204-5507
BYang@space.SVEC.UH.edu

ABSTRACT

We have demonstrated 4-μm InAs/InGaSb/AlSb interband cascade lasers with optical output power close to 0.5 W per facet with 1-μs pulses at 1 kHz repetition rate. At 10% duty cycle, an average output power ~20 mW was realized. External and internal quantum efficiencies exceeding 200% have been achieved at 80 K.

INTRODUCTION

High power near-infrared semiconductor lasers have been extensively investigated for applications in data storage, solid-state laser pumping, medical therapy, etc. However, high power semiconductor lasers at mid-infrared atmospheric transmission window are urgently demanded for a number of military and commercial applications such as IR countermeasures, high resolution molecular spectroscopy, industrial process control, free space communications, etc. We have proposed [1-3] and demonstrated [4] a new type of mid-IR quantum cascade (QC) lasers based on interband transitions in type-II quantum wells. This interband cascade laser takes advantage of the broken-gap band alignment in the InAs/Ga(In)Sb heterostructure to recycle carriers from the valence band back to the conduction band, thus enabling sequential photon emission from active regions stacked in series. Recently, we demonstrated high-power and high-efficiency operation of the interband cascade laser. A peak optical output power of ~ 0.5 W/facet with a slope efficiency of 211 mW/A per facet at ~3.9 μm was obtained with 1 μs current pulses at 80 K [5]. The threshold current density at 80K was 290 A/cm^2 which is substantially lower than those reported for the QC lasers. The characteristic temperature was 80 K at temperatures up to 165 K. The slope efficiency was 211 mW/A per facet, corresponding to a differential external quantum efficiency of 130%, or 1.3 emitted photons per injected electron. Comparable external quantum efficiency, with a peak output power of 430 mW/facet and a slope efficiency of 274 mW/A per facet were also reported with a "W" configuration 2.9 μm cascade laser when pumped with 100 ns current pulses at 100 K [6]. At 180 K, the maximum output power was 252 mW/facet with a slope efficiency of 188 mW/A per facet. We report here 4-μm interband cascade lasers with quantum efficiencies > 200% and average optical output powers ~ 20 mW for the first time.

EXPERIMENT

The laser structure was grown with a Riber 32 MBE system on a rotating p-type GaSb substrate. The substrate temperature was kept at about 440°C during growth and the wafer was annealed at higher temperatures to improve the material quality. High-resolution double crystal X-ray diffraction was used to characterize the crystalline quality. The active region of the laser structure comprises 23 periods of coupled InAs/GaInSb/AlSb QWs which are connected by digitally graded n-type InAs/AlInSb injection layers as reported previously [5].

89

Gain-guided lasers were processed using standard photolithographic and lift-off techniques. To reduce current spreading, ridge waveguide structures were formed by chemical etching. The laser bars were cleaved into various cavity lengths ranging from 0.5 to 2 mm with the facets left uncoated. The samples were indium heat sunk to a copper submount with epilayer-side up, wire-bonded, and mounted onto the temperature controlled cold finger of a cryostat.

The lasers were pumped using a pulsed current source with variable pulse lengths and duty cycles. The emission spectra were measured by focusing the laser beam into a 0.5-m monochromator using f/4 optics. The lasing powers were measured using either a thermopile power meter if the average power is high enough, or an InSb detector. Neutral density filters were used to avoid the saturation of the detector when needed. The collection efficiency is calculated by integrating the measured far-field pattern in the perpendicular direction.

RESULTS

Figure 1 shows the optical output power versus injection current from a 900 µm long broad area laser. At 80 K, a peak output power as high as ~0.5 W has been obtained under 1-µs pulses at 1-kHz repetition rate. The maximum output power is limited only by the current source which has a peak current output of 3A. The optical output power drops quickly as the temperature rises. This device has lased up to 170 K as shown in the inset. The lasing spectra of this device show single mode at injection level just above threshold over the whole range of operating temperatures. At higher injection current, the multi-longitudinal modes appeared.

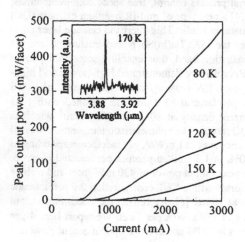

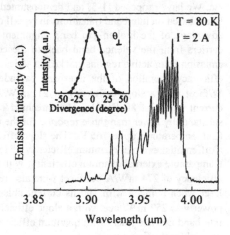

Figure 1. Peak optical output power per facet versus injection current from a 900 µm long laser. Inset shows the lasing spectrum at 170 K.

Figure 2. A lasing spectrum with a corresponding peak output power of 300 mW per facet. Inset shows the far-field pattern perpendicular to the junction.

Figure 2 represents a lasing spectrum measured at 80 K with an injection current of 2 A. The spectra are much broader at higher injection levels. This broadening in multi-mode interband cascade laser is caused by the gain broadening at high injection current densities. The separation between adjacent modes is about 25 Å as expected from a 900 µm cavity. The far-field pattern as shown in Figure 2 revealed that there is a single lobe in the direction perpendicular to the junction plan with divergence angle of ~ 36-40°. This pattern is very stable even at the highest injection level. However, the far-field pattern at lateral direction is much narrower and displays multiple lobes.

To achieve higher average power, we have operated several lasers at duty cycles up to 10%. Figure 3 shows L-I curves from a 0.9 mm-long gain-guided laser with 5 μs long pulses at 10% duty cycle and different heatsink temperatures. Average powers up to 15-20 mW have been achieved at 60-80 K. However, the output power and the external quantum efficiency degrade rapidly when the heatsink temperature rises. For example, the external quantum efficiency decreased from 64% at 60 K to about 2% at 140 K, correspondingly, the peak optical power dropped from 200 mW to about 1.6 mW.

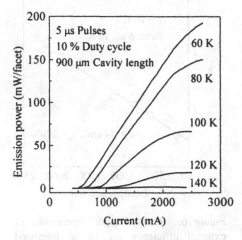
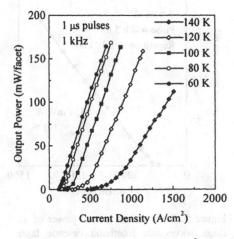

Figure 3. L-I curves of a 900×200 μm² broad area laser pumped with 5 μs pulses at 10% duty cycle.

Figure 4. L-I curves of a 430×200 μm² ridge waveguide interband cascade laser at various heat sink temperatures with 1 μs pulses at 0.1% duty cycle.

For lasers with the ridge structure, both power efficiency and threshold current density improved considerably. Figure 4 shows the peak output power per facet versus injection current measured with 1-μs long pulses at 1-kHz repetition rate. A slope efficiency of 334 mW/A is achieved at 80 K, corresponding to an external quantum efficiency of 214 % assuming identical emissions from both facets. To our knowledge, this is the highest external quantum efficiency ever reported at this wavelength. The external quantum efficiency was still higher than 100 % at temperatures up to 140 K.

The threshold current density, as shown in Fig. 5, is as low as 90A/cm² at 60 K which is considerably lower than any reported cascade laser. A peak output power higher than 180 mW per facet has also been obtained, corresponding to an average power >18 mW, which is comparable to that of the broad area laser as shown in Figure 3, but at half the injection current. In addition, it seems that the peak power was limited by a sudden damage with unknown origin. Similar failure mode was also observed in the 3-μm interband cascade laser reported previously [6]. From this curve we have determined a slope efficiency of 158 mW/A and external quantum

91

efficiency of 101%. The slope remains constant with injection current density until damage occurs, no saturation of the output power was observed up to the failure point suggesting that global heating may not be the cause of the failure. Material imperfections is more likely to be the failure mechanism.

The internal loss and internal quantum efficiency were obtained from the cavity length dependence of the external quantum efficiency as shown in Fig. 6. From the intersection and the slope, the internal quantum efficiency and the cavity loss were determined to be 220 % and 14 cm^{-1}, respectively. To the best of our knowledge, this is the first demonstration of an interband laser diode with an internal quantum efficiency > 200%.

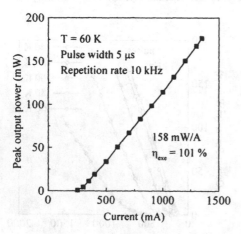

Figure 5. Peak optical output power of an ridge waveguide interband cascade laser with cavity length of 810 μm.

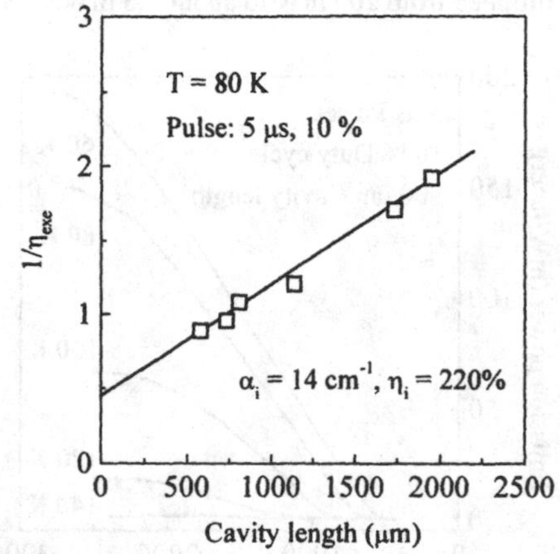

Figure 6. Cavity length dependence of external efficiency η_{ext} of an interband cascade laser measured with 5 μs pulses at 10% duty cycle.

DISCUSSIONS

The optical output power of our interband cascade laser is limited by a sudden failure with unknown origin. Conventionally, two common failure modes that limit the laser power are the catastrophic optical damage (COD) and the overheating. The former may arise from surface recombination of carriers at the facet that causes more absorption, eventually causing the thermal runaway. The junction overheating results in the saturation of the output power at higher injection levels. It seems that neither factors explains our experimental observations. Examining of our failed lasers with an optical microscope, we did not find any indication of the mirror facet damaged by COD. On the other hand, the output power at short pulses doesn't seem to be limited by junction heating because no saturation of the L-I curves was observed when the damage occurred. We are not yet sure the exact mechanism that caused the damage. We suspect that crystalline defect induced local breakdown and heating under high electrical field may be the cause of the failure.

The optical output power is determined by a number of factors including internal quantum efficiency, internal loss, and the leakage current. At low temperature, they all contributed to the relatively high optical output power in these type-II interband cascade lasers. However, these characteristics degraded rapidly as the temperature was raised. It is well known that the high temperature operation of the interband QW laser based on type-I narrow bandgap semiconductors is limited by the Auger recombination. It have been predicted theoretically and demonstrated experimentally that the Auger recombination could be suppressed in type-II superlattices [7]. It is clear from cross sectional scanning tunneling microscopy studies that the quality of the InAs/GaInSb/AlSb heterostructure is not optimum [8]. However, detailed analysis of optically-pumped type-II QW lasers showed that the internal loss may also play an important role in limiting the performance of type-II QW lasers at high temperatures [9-10]. The interband cascade lasers has similar active region. The low temperature internal loss is also comparable to those of the type-II QW lasers. We speculate that the intra-valence band absorption may have also contributed to the low external quantum efficiency of the interband cascade lasers at high temperatures. We anticipate significantly improvement of the optical output power with optimized device design and better material quality.

CONCLUSIONS

We demonstrated high efficiency and high power operation of 4-µm InAs/InGaSb interband cascade lasers. An internal quantum efficiency of 220% with 5-µs pulses at 10% duty cycle, and an external quantum efficiency of 214% with 1-µs pulses and 0.1% duty cycle have been achieved. At 80 K, an average output power of 16 mW has been achieved. These results demonstrated the potential of the interband cascade configuration. The optical output power of the interband cascade lasers is currently limited by failures probably due to crystalline imperfections in the material which causes local breakdown in the injection region under high electrical field. Further improvements in performance is expected with better device design and crystal quality.

ACKNOWLEDGMENTS

This work is supported in part by NASA under Cooperative Agreement -NCC8-127 and TcSUH. The research at QET Inc. (Rui Q. Yang) is partially supported by Ballistic Missile Defense/Innovative Science and Technology and managed by the Office of Naval Research.

REFERENCES

[1] R.Q. Yang, Superlattices and Microstructures 17, 77 (1995).

[2] J.R. Meyer, I. Vurgaftman, R. Q. Yang, and L. R. Ram-Mohan, Electron. Lett. 32, 45 (1996).

[3] R.Q. Yang and S.S. Pei, J. Appl. Phys. 79, 8197 (1996).

[4] C.-H. Lin, R.Q. Yang, D. Zhang, S.J. Murry, S.S. Pei, A.A.Allerman, and S.R. Kurtz, Electron Lett. 33, 598 (1997).

[5] R.Q. Yang, B.H. Yang, D. Zhang, C.-H. Lin, S. Murry, H. Wu, and S.S. Pei, Appl. Phys. Lett. 71, 2409 (1997).

[6] C.L. Felix, W.W. Bewley, I. Vurgaftman, J.R. Meyer, D. Zhang, C.-H. Lin, R.Q. Yang, and S.S. Pei, IEEE Photonics Technology Letters 9, 1433 (1997).

[7] E. R. Youngdale, J.R. Meyer, C.A. Hoffman, F.J. Bartoli, C.H. Grein, P.M. Young, H. Ehrenreich, R.H. Miles, and D.H. Chow, Appl. Phys. Lett., 64, 3160 (1994).

[8] J. Harper and M.Weimer, D. Zhang, C.-H. Lin and S. S. Pei, submitted to J. Vac. Sci. Technol.

[9] H. Q. Le, G. W. Turner, C.-H. Lin, S. Murry, R. Q. Yang, and S. S. Pei, (to be published).

[10] W. W. Bewley, I. Vurgaftman, C. L. Felix, J. R. Meyer. C.-H. Lin, D. Zhang, S. J. Murry, and S. S. Pei, submitted to J. Quan. Electron..

MID-IR VERTICAL CAVITY SURFACE-EMITTING LASERS

I. Vurgaftman*, W. W. Bewley*, C. L. Felix*, E. H. Aifer*, J. R. Meyer*, L. Goldberg*, D. H. Chow**, E. Selvig**
*Code 5600, Naval Research Laboratory, Washington, DC 20375
**Hughes Research Laboratory, MS RL63, Malibu, CA 90265

ABSTRACT

An optically pumped mid-infrared vertical-cavity surface-emitting laser based on an active region with a "W" configuration of type-II antimonide quantum wells is reported. The emission wavelength of 2.9 μm has a weak temperature variation ($d\lambda/dT \approx 0.07 - 0.09$ nm/K), and the multimode linewidth is quite narrow (2.5-4 nm). Lasing is observed up to $T = 280$ K in pulsed mode and up to 160 K cw. Under cw excitation at $T = 78$ K, the threshold pump power is as low as 4 mW for a 6 μm spot, and the differential power conversion efficiency is 4.5%.

INTRODUCTION

Vertical-cavity surface-emitting lasers (VCSELs) have such key advantages over edge-emitters as intrinsic single-longitudinal-mode operation, circular low-divergence output beam profiles, and the potential for fabricating monolithic 2D laser arrays. While near-IR surface emitters have advanced significantly in recent years, there has been only one previous report of a VCSEL operating in the mid-IR (2.5-5 μm), and lasing in that device was confined to very low temperatures and high pump intensities [1].

THEORY

A recent theoretical study [2] of antimonide type-II mid-IR VCSELs has predicted that optimized optically pumped devices employing the type-II quantum-well "W" configuration in the active region [3] should operate cw up to 250 K and be capable of producing single-mode output powers in the 1 W range. The comprehensive model of the VCSEL operation included full bandstructure calculations using the 8-band $\mathbf{k} \cdot \mathbf{p}$ finite-element algorithm, photon propagation, nonradiative, spontaneous, and stimulated recombination, free-carrier absorption, carrier heating and a 3D finite-difference treatment of lattice heating. Furthermore, the potential for ambient-temperature cw operation was identified for injection VCSELs based on interband cascade active regions [4].

GROWTH AND FABRICATION

First, a 12.5-period GaSb/AlAs$_{0.08}$Sb$_{0.92}$ distributed Bragg reflector (DBR) bottom mirror was grown by molecular beam epitaxy (MBE). From an analysis of transmission measurements, the DBR reflectivity plateau had a magnitude corresponding to an internal (from inside the cavity) reflectivity of $\approx$ 95% and was centered at $\lambda \approx 3.0$ μm. This was followed by a 79-period InAs/GaSb/InAs/AlSb (18.5 Å/ 28 Å/ 18.5 Å/ 40 Å) "W" quantum-well active region that was grown under conditions producing predominantly InSb-like interface bonds. The active region was designed to yield a gain maximum at $\lambda = 3.0$ μm for temperatures

Mat. Res. Soc. Symp. Proc. Vol. 484 © 1998 Materials Research Society

close to 200 K. The 0.83 μm net thickness of the active region was intended to realize a λ cavity with resonant wavelength coinciding with the mirror reflectivity and gain maxima at 3.0 μm. The active region was capped with 160 Å of AlSb to block the escape of holes, followed by 150-200 Å of GaSb. To complete the VCSEL cavity, a dielectric multilayer coating with a nominal refractive index contrast ratio of 2.25:1.48 and $\approx$ 99% reflectivity in the vicinity of 3.0 μm was deposited on top of the sample. The substrate was mechanically polished to a thickness of 50-70 μm and attached to the cold finger of an Air Products Heli-Tran dewar.

PULSED PUMPING RESULTS

The pulsed optical excitation was by 85 ns pulses from a Q-switched 2.1 μm Ho:YAG laser with a 1 Hz repetition rate. The TEM$_{00}$ mode was incident on the sample at an angle of 20° with respect to normal. At $T = 86$ K, the lasing spectrum had a peak wavelength of 2.893 μm with a full width at half maximum (FWHM) of 3.5 nm, which was typical of the data for all temperatures and spot sizes. Even though multiple transverse modes were present, the FWHM was considerably smaller than the 10-35 nm values characteristic of multi-longitudinal-mode mid-IR edge emitters pumped well above threshold. The average wavelength shift with temperature was $d\lambda/dT = 0.07$ nm/K, which is roughly a factor of 20 smaller than the usual rate of 1-2 nm/K for mid-IR edge emitters. This is because the wavelength shift in the VCSEL is governed by the change in the resonant wavelength as the refractive indices in the active region and the mirrors change with T rather than by the temperature shift of the energy gap.

The peak output power from the VCSEL is shown in Figure 1 as a function of pump intensity at several temperatures. The FWHM diameter of the pump beam was $\approx$ 600 μm. Note that over 2 W of peak power is generated even at $T = 260$ K. Lasing was observed up to 280 K, and analogous light-light characteristics were obtained for a much smaller spot (FWHM $\approx$ 30 μm). The large-spot incident-intensity threshold at $T = 86$ K is 1.2 kW/cm^2, which is only about an order of magnitude higher than the best results reported to date for optically pumped type-I and type-II edge-emitting mid-IR lasers [5]. The modeling shows that this I_{th} implies a net carrier lifetime of $\approx$ 7 ns at threshold, which may contain Shockley-Read, Auger, and spontaneous emission contributions. For the 30 μm spot at the same temperature, the threshold pump power is as low as 22 mW. Except at the very highest operating temperatures, the increases of I_{th} with T for both spot sizes yield characteristic temperatures of $T_0 = 34$ K. A far-above-threshold differential power conversion efficiency of 4% at $T < 120$ K is comparable to the best results reported to date for edge emitters in this wavelength range.

RESULTS FOR CW LASING

Optical pumping in the cw mode was accomplished by a 1.06 μm Nd:YAG laser operating in a TEM$_{00}$ mode and incident on the sample at an angle of 28° from normal. Figure 2 shows the VCSEL output spectra at $T = 78$ K and 120 K for a pump spot size of $D = 75$ μm, which yield peak emission wavelengths of 2.897 μm and 2.901 μm respectively. Both FWHM multimode linewidths of 2.9 nm are again considerably narrower than the typical values for mid-IR edge emitters, and $d\lambda/dT$ of 0.09 nm/K is consistent with the value for pulsed operation.

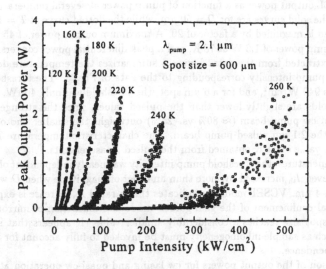

Figure 1. Peak power vs. pump intensity for the pulsed-mode VCSEL at several temperatures. The maximum lasing temperature was 280 K.

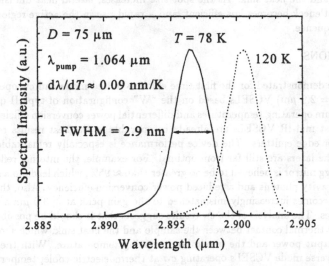

Figure 2. Spectral output vs. wavelength at T = 78 K and 120 K. The spectrometer resolution was 0.5 nm. The spectral width (FWHM) at 78 K is 2.9 nm, which is typical for emission at all temperatures. The average shift of wavelength with temperature ($d\lambda/dT$) is $\approx$0.09 nm/K.

The VCSEL output power as a function of pump power at several temperatures is shown in Figure 3. The solid curves are for $D = 50$ μm, while the dotted curve at $T = 160$ K is for a 10 μm spot and is magnified by a factor of 20. A maximum output power of 45 mW was obtained for a pump power of 1.3 W at 78 K, and a peak differential power conversion efficiency of 4.5% was extracted from the data. Figure 4 summarizes the temperature dependence of the threshold pump intensity corresponding to the data of Fig. 3. The threshold intensity at $T = 78$ K is 940 W/cm^2, and for a 6 μm spot, the threshold was only 4 mW. For $T \leq 150$ K, the thresholds are slightly lower than the pulsed values, since the stronger absorption of the 1.06 μm cw pump beam ($\approx$ 80% vs. 25%) outweighs the smaller photon-conversion decrement of the 2.1 μm pulsed pump beam. The characteristic temperature T_0 of 44 K is also larger the value of 34 K obtained from the pulsed measurements.

At low temperatures, the threshold pump intensity was nearly independent of spot size for large D. However, I_{th} increased by more than an order of magnitude when D was decreased from 20 μm to 4 μm. VCSEL modeling indicates that a rapid I_{th} increase is expected owing to the marginal confinement of the gain-guided mode and diffraction of mirror stacks, but not until the spot size reaches a considerably smaller value. It appears that some further mechanism such as sample inhomogeneity must be invoked to fully account for the observed pump size dependence.

A comparison of the output powers for cw lasing and quasi-cw operation at a 27% duty cycle and 90 μs pulse length is given in Figure 5. The cw output power as a function of pump spot size has a maximum near 100 μm ($\approx$ 60 mW), while the quasi-cw peak power increases with the spot diameter over a much wider range of spot sizes. Numerical simulations indicate that this functional dependence is consistent with a relatively poor contact between the sample and the heat sink. As the spot size increases, lateral heat diffusion around the thermal bottleneck becomes less efficient, and a rapid rise in the active region temperature is the consequence.

CONCLUSIONS

We have demonstrated for the first time efficient pulsed-mode and cw operation of mid-infrared ($\lambda = 2.9$ μm) VCSELs based on the "W" configuration of type-II quantum wells. The maximum operating temperatures and differential power conversion efficiencies obtained for these first mid-IR VCSELs are already comparable to the best results reported in the literature for edge emitters. The device performance is especially remarkable since several aspects of the lasers are still far from optimal. For example, the internal reflectivity of the bottom Bragg mirror is believed to be no greater than $\approx$ 95%, which leads to a significant loss rate of the cavity photons and a reduced power conversion efficiency. Also, the cavity mode at 2.9 μm becomes increasingly mismatched to the gain peak at $\geq$ 3.0 μm as the temperature increases. The analysis of the cw results suggests that it should be possible to achieve a much better thermal contact between the sample and the heat sink, which should raise both the peak output power and the maximum operating temperature. With the realization of single transverse mode VCSELs operating cw at thermoelectric-cooler temperatures, important applications in chemical sensing and other areas will become open to these devices.

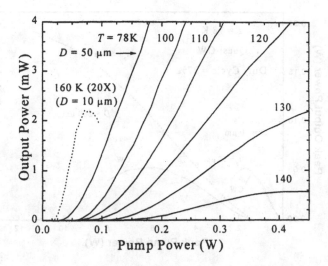

Figure 3. CW output power vs. pump power at a series of temperatures. Solid curves are for a spot size D of 50 μm. The dashed curve is for D = 10 μm at the maximum lasing temperature of 160 K. It is magnified by a factor of 20 to show on the same scale as the other data.

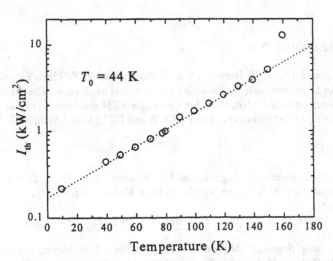

Figure 4. Threshold pump intensity vs. temperature. The characteristic temperature T_0 = 44 K is valid except at the highest operating temperature. These threshold are extracted from L-L data for various spot sizes.

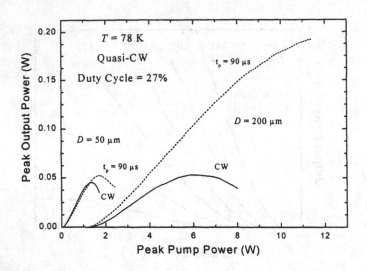

Figure 5. Peak output power vs. peak pump power for quasi-cw operation with a 27% duty cycle and a pulse width of 90 μm and for true cw operation. The results are shown for spot sizes $D = 50$ and 200 μm.

ACKNOWLEDGMENTS

We are grateful to F. J. Bartoli, C. A. Hoffman, and J. F. Pinto for valuable interactions, and to L. Warren and C. Haeussler for technical assistance. The authors also thank Quantum Semiconductor Algorithms for use of the FEM software. Work at NRL was supported by ONR, while two of the authors (WWB and EHA) held ASEE/NRL Associateships.

REFERENCES

1. E. Hadji, J. Bleuse, N. Magnea, and J. L. Pautrat, *Appl. Phys. Lett.* **68**, 2480 (1996).

2. I. Vurgaftman, J. R. Meyer, and L. R. Ram-Mohan, *IEEE J. Quantum Electron.* **34**, (1998), in press.

3. J. R. Meyer, C. A. Hoffman, F. J. Bartoli, and L. R. Ram-Mohan, *Appl. Phys. Lett.* **67**, 757 (1995).

4. R. Q. Yang, *Superlatt. Microstruct.* **17**, 77 (1995); J. R. Meyer, I. Vurgaftman, R. Q. Yang, and L. R. Ram-Mohan, *Electron. Lett.* **32**, 45 (1996).

5. C.-H. Lin, S. S. Pei, H. Q. Le, J. R. Meyer, and C. L. Felix, *Appl. Phys. Lett.* **71**, (1997), in press.

3.2 AND 3.8 μm EMISSION AND LASING IN AlGaAsSb/InGaAsSb DOUBLE HETEROSTRUCTURES WITH ASYMMETRIC BAND OFFSET CONFINEMENTS

M.P.MIKHAILOVA, B.E.ZHURTANOV, K.D.MOISEEV, A.N.IMENKOV, O.G.ERSHOV, YU.P.YAKOVLEV
Ioffe Physico-Technical Institute, RAS, 194021, St.Petersburg, Russia

ABSTRACT

We report the first observations of electroluminescence (EL) and lasing in laser structures with high Al-content (x=0.64, Eg=1.474 eV) cladding layers and a narrow-gap InGaAsSb active layer (Eg=0.326 eV at T=77K). The structures are LPE-grown lattice-matched to GaSb substrate. Band energy diagrams of the laser structures had strongly asymmetric band offsets. The heterojunction between high Al-content layer and InGaAsSb narrow-gap active layer has a type II broken-gap alignment at 300K. In this laser structure spontaneous emission was obtained at λ=3.8μm at T=77K and λ=4.25 μm at T=300K. Full width at half maximum (FWHM) of emission band was 34 meV. Emission intensity decreased by a factor of 30 from T=77K to 300K. Lasing with single dominant mode was achieved at λ=3.774 μm (T=80K) in pulsed mode. Threshold current as low as 60 mA and characteristic temperature T_0=26K were obtained at T=80-120K.

INTRODUCTION

Lately there has been intensive research in mid-infrared III-V (MIR) semiconductor diode lasers emitting from 3 to 5 μm. An important application of these lasers is ecological monitoring and tunable diode laser spectroscopy. Sb-based lasers operating at up to 180-200K in pulsed mode and 110-120K in cw mode were realized [1-4]. Novel type II laser structures using an intersubband transition were demonstrated [3,4]. Main physical processes limiting operation temperature of the longwavelength lasers are non-radiative Auger recombination, intervalence band absorption, carrier heating, as well as current leakage, due to poor electron and hole confinement. Attempts were made to improve electron and hole confinement by using MBE grown laser structures with high Al-content cladding layers [5]. Further progress in improving MIR laser performances is connected with new physical approaches to laser structure desing optimization.

Recently in Ref.[6] it was proposed to use high-barrier stopper confined layers for electron and holes in laser structures as a way of reducing thermoionic emission of carriers out of the active layer and decreasing the hole leakage. We try to apply this method to mid-infrared Sb-based laser structures grown by liquid phase epitaxy (LPE). New technical approach was proposed to create narrow-gap laser heterostructures with highly asymmetric band off-set confinements.

We report here the first observations of electroluminescence and lasing in MIR laser structures with high Al-content (64%) cladding layers (Eg=1.474 eV) and a narrow-gap $In_{0.94}$GaAsSb active layer (Eg=0.326 eV at T=77K) grown by LPE and lattice-matched to GaSb substrate.

EXPERIMENT

Two kinds of double heterostructure (DH) diode lasers were fabricated on N- and P-GaSb substrates, below we will be referring to them as structure A and structure B respectively (Fig.1).

101

The N-GaSb substrates were doped by Te to $5*10^{17}cm^{-3}$ and the P-GaSb ones were either undoped or Ge doped to $2*10^{19}cm^{-3}$. The laser structures consisted of the following layers: structure A-N-GaSb/N-Al$_{0.64}$Ga$_{0.36}$AsSb/n-In$_{0.94}$Ga$_{0.06}$As$_{0.82}$Sb$_{0.18}$/P-Al$_{0.64}$Ga$_{0.36}$AsSb/P-GaSb and structure **B** had inverted sequence of the layers of the same compositions of quaternary solid solutions, P-GaSb/P-Al$_{0.64}$Ga$_{0.36}$AsSb/p-In$_{0.94}$Ga$_{0.06}$As$_{0.82}$Sb$_{0.18}$/N-Al$_{0.64}$Ga$_{0.36}$AsSb/N-GaSb. The N- and P-type layers of the AlGaAsSb solid solutions were obtained by Te and Ge doping, respectively. The narrow-gap active layer of the n-InGaAsSb was undoped and the p-InGaAsSb layer was doped by Zn to $1*10^{17}cm^{-3}$. The thickness of the confined layers was as high as 2 μm, and the active layer thickness was in the range 0.4-1.4 μm.

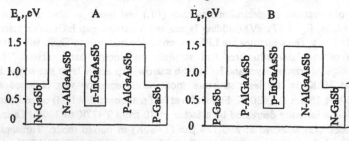

The main problem the growing (by LPE) the InAs-rich narrow-gap solid solutions lattice-matched to GaSb and the wide-gap AlGaAsSb is the big difference in the values of thermoconductivity and thermal expansion coefficients of these materials.

Fig.1. Energy band profiles of laser structures A and B.

To solve this problem we used a special thermodynamic calculation of equilibrium phase diagrams of the quaternary solid solutions and an original growth technique. High quality lattice-matched epilayers were grown onto GaSb (100) substrate by liquid phase epitaxy using a horizontal graphite multiwell sliding boat. The temperature of epitaxy was about 600°C. The epilayers were grown using melts supersaturated by 5-8°C and a cooling rate of 0.6°C/min. The lattice mismatch of the Al$_{0.64}$GaAsSb epitaxial layers as low as 0.05% was obtained. The lattice mismatch of the In$_{0.94}$GaAsSb epitaxial layers was about of 0.3% at room temperature. To obtain high quality multilayer laser structures with minimal strain at the heteroboundary between InAs-rich and GaSb-rich solid solutions we used InGaAsSb active layers having lattice constants equal or slightly higher than the lattice constant of AlGaAsSb emitter layers at growth temperature.

Mesa-stripe laser structures with stripe widths 11-45 μm and the cavity length ~300μm were fabricated by standard photolithography. EL spectra were measured using MDR-4 grating monochromator and a lock-in amplifier. The emission signal was registered by liquid N$_2$-cooled InSb photodetector.

We studied current-voltage characteristics, spectra of spontaneous emission and emission intensity versus drive current at 77 and 300K. The investigations of spontaneous EL spectra were carried out under quasi steady-state conditions with pulse duration of τ=2.5ms and pulse period-to-pulse duration ratio equal to 1. Spectra of coherent emission in pulsed mode were studied at T=80-150K, as well as temperature dependence of the threshold current.

RESULTS AND DISCUSSION

In laser structures **A** spontaneous emission was obtained at λ=3.8 μm (hν=326 meV) at T=77K and λ=4.25 μm (hν=291 meV) at room temperature which corresponds to energy gap of the InGaAsSb active layer. The emission band had a Gaussian symmetric shape. Full width at half maximum (FWHM) of the emission band was 34 meV (77K) and increased to 90-115 meV at 300K (fig.2,a). The emission intensity varied linearly with drive current and decreased by a factor of 30 as temperature was raised from T=77K to 300K. Lasing with single dominant mode at

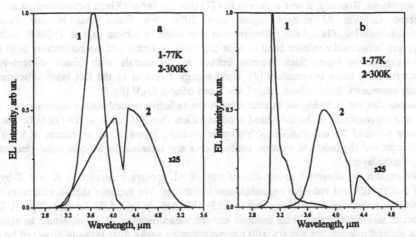

Fig.2. Electroluminescence spectra of the N-AlGaAsSb/n-InGaAsSb/P-AlGaAsSb (a)
and P-AlGaAsSb/p-InGaAsSb/N-AlGaAsSb (b)

λ=3.776 µm (T=80K) was achieved (See fig.3). Threshold current as low as ~60 mA and the characteristic temperature T_o=26K in the temperature range 80-120K were observed.

On the other hand, in structures **B** an intensive spontaneous emission and superluminescence were only obtained. Electroluminescence with very narrow asymmetric bands (FWHM~10 meV at 77K and 30 meV at 300K) was observed (fig.2,b). A blue shift of the emission band maximum of up to 60-70 meV was found, relative the emission band maximum in structures A. EL was peaking at photon energies hv=380-402 meV (λ=3.08-3.26 µm), depending on the drive current.

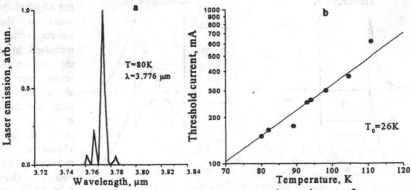

Fig.3. Coherent emission (a) and temperature dependence of
threshold current (b) in N-AlGaAsSb/n-InGaAsSb/P-AlGaAsSb
double heterostructure lasers

103

To explain the obtained experimental results we examined the energy band diagrams of both laser structures. Recently, it was shown in Ref.[7] that the InAs/AlGaSb heterojunction is type II staggered for the Al content higher than 30%. We found that in our case the $Al_{0.64}Ga_{0.36}AsSb/In_{0.94}Ga_{0.06}AsSb$ heterojunction was close to broken gap at T=300K with zero energy gap between the valence band of wide-gap semiconductor and the conduction band of the narrow-gap active layer. Such systems behave as semimetals with ohmic current-voltage characteristics at room temperature [8]. Band energy diagrams of the DH laser structures had strongly asymmetric band offsets: ΔE_C=1.46eV and ΔE_V=0.31eV (fig.4).

As one can see from Fig.4a, in structures A the radiative recombination occurs in the active layer and corresponds to band-to-band recombination (hv=326 meV at T=77K). In this laser structure we used N- and P-AlGaAsSb stopper cladding layers doped in excess of $8*10^{17}cm^{-3}$ which improved the hole and electron confinements and reduced the built-in serial resistance of the confining layers.

We explain the observed strong dissimilarity of EL spectra in structures A and B by their being due to different radiative recombination transitions. We suppose that in structures B the radiative recombination transitions occur at the P-AlGaAsSb/p-InGaAsSb interface(fig.4b). This is supported also by electron beam induced current measurements. It is important to note that similar electroluminescence spectra with narrow emission peaks were recently observed by us on type II broken-gap p-GaInAsSb/p-InAs heterojunctions [9]. The observed blue shift of the emission band maximum can be satisfactorily explained by strong electron confinement and carrier accumulation in the quantum well near the p-p interface [10].

In structures B there were used lightly doped $(1-5*10^{17}cm^{-3})$ P-AlGaAsSb confined layers. In this case a two-dimensional electron gas can form in the quantum wells near the interface on the side of narrow-gap semiconductor due to electron transfer from deep acceptor levels of the lightly doped wide-gap AlGaAsSb to the conduction band of the narrow-gap InGaAsSb. So, at the thermodynamic equilibrium we have an electron channel near isotype P-AlGaAsSb/p-InGaAsSb heteroboundary. The external bias applied to the laser structure is a reverse bias for the type II P-AlGaAsSb/p-InGaAsSb heterojunction. With increasing applied bias the band bending at the interface is changes on the two sides: the valence band edge of the P-AlGaAsSb solid solution goes down and the conduction band edge of the p-InGaAsSb solid solution goes up (See fig.5).

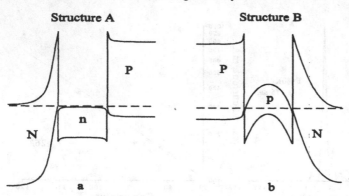

Thus, the depth of the adjacent quantum wells at the P-p heterointerface is increased. In this case the electron and hole levels are raised above the quantum wells and the energy separation between them is increased. Therefore, the radiative transition energy (hv~0.37–0.40eV) will be different from the band gap of the

Fig.4. Energy band diagrams of laser structures A and B with high asymmetric band offset confinement.

narrow-gap semiconductor (Eg~0.326eV) and exceeding it. Thus, the nature of type II

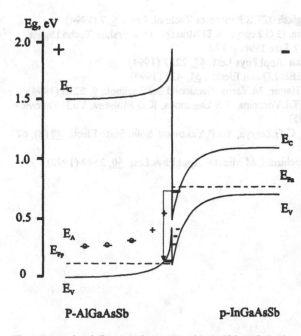

heterointerface between lightly doped P-AlGaAsSb solid solution layer and the layer of p-InGaAsSb can explain the observed blue shift of emission.

Intensive radiative recombination occurs due to indirect (tunnel) optical transitions of localized electrons and holes from the quantum well levels across the p-p heterointerface. It is necessary to note that in this case the radiative recombination channel can be as follows: electrons from the quantum well on the narrow-gap InGaAsSb solid solution side tunnel through the potential barrier in the wide-gap P-AlGaAsSb material and recombine with the holes in the quantum well on the side of the P-AlGaAsSb solid solution.

Fig.5. Energy band diagram of type II P-AlGaAsSb/p-InGaAsSb heterojunction under applied bias.

CONCLUSIONS

In conclusion, two kinds of diode laser structures with high (64%) Al-content asymmetric band offset confined layers and narrow-gap InGaAsSb active layers lattice-matched to GaSb substrates were grown for the first time by LPE method. Single mode lasing (λ=3.776 μm) was achieved at T=80-120K in pulsed mode in structures contained by heavily doped AlGaAsSb stopper layers. Intensive spontaneous emission was obtained in the temperature range 77-300K in the 3.5-4.4 μm spectral region in DH light emitting diodes grown on N-GaSb substrates and 3.0-4.0 μm in structures grown on P-GaSb substrates demonstrating good prospect of producing as new light-emitting diodes for gas analysis in atmospheric pollution control applications. Substantial performance improvement is expected from further optimization of the design of the proposed laser structures and the growth conditions.

ACKNOWLEDGMENTS

Authors thank Prof.G.Zegrya for the valuable discussion.
This work was supported in part by the Russian project #1-074 "Solid-State nanostructures physics"and RBRF project #96-0217841a.

REFERENCES

1. H.K.Choi, G.W.Turner Appl.Phys.Lett. **67**, 332 (1995)
2. T.N.Danilova, A.N.Imenkov, O.G.Ershov, M.V.Stepanov, V.V.Sherstnev, Yu.P.Yakovlev Semiconductors, **29** (7), 667 (1995)

3. H.K.Choi, G.W.Turner, S.Y.Eglash IEEE Photonics Technol. Lett., **6**, 7 (1994)
4. Yu.P.Yakovlev, M.P.Mikhailova, G.G.Zegrya, K.D.Moiseev, O.G.Ershov Techn.Digest CLEO-96, USA, Anaheim CA, 2-7 June 1996, p.170
5. H.K.Choi, G.W.Turner, Z.L.Liau Appl.Phys.Lett. **65**, 2251 (1994)
6. R.I.Kazarinov, G.L.Belenky IEEE J.Quant.Electr., **31**, 423 (1995)
7. K.B.Wong, G.K.A.Gopir, Y.P.Hagon, M.Yaros Semicond.Sci.Technol., **9**, 2210 (1994)
8. M.P.Mikhailova, I.A.Andreev, T.I.Voronina, T.S.Lagunova, K.D.Moiseev, Yu.P.Yakovlev Semiconductors, **29** (4), 353 (1995)
9. M.P.Mikhailova, K.D.Moiseev, G.G.Zegrya, Yu.P.Yakovlev Solid State Electr. **40** (8), 673 (1996)
10. S.Ideshita, A.Furukawa, Y.Mochizuki, M.Mizuta Appl.Phys.Lett. **60**, 2549 (1992)

High-Power Low-Threshold Optically Pumped Type-II Quantum-Well Lasers

Chih-Hsiang Lin*, S. J. Murry, Rui Q. Yang, and S. S. Pei
Space Vacuum Epitaxy Center, University of Houston, Texas 77204-5507

H. Q. Le
MIT Lincoln Laboratory, Lexington, Massachusetts 02173

Chi Yan and D. M. Gianardi, Jr.
Rocketdyne Technical Services, Boeing Defense & Space Group, Kirtland AFB, NM 87117-5776

D. L. McDaniel, Jr. and M. Falcon
Semiconductor Laser Branch, Air Force Phillips Lab., Kirtland AFB, NM 87117-5776

. Abstract

Stimulated emission in InAs/InGaSb/InAs/AlSb type-II quantum-well (QW) lasers was observed up to room temperature at 4.5 μm, optically pumped by a pulsed 2-μm Tm:YAG laser. The absorbed threshold peak pump intensity was only 1.1 kW/cm^2 at 300 K, with a characteristic temperature T_0 of 61.6 K for temperatures up to 300 K. We will also study the effects of internal loss on the efficiency and output power for type-II QW lasers via optical pumping. Using a 0.98-μm InGaAs linear diode array, the devices exhibited an internal quantum efficiency of 67% at temperatures up to 190 K, and was capable of > 1.1-W peak output power per facet in 6-μs pulses at 85 K. The internal loss of the devices exhibited an increase from 18 cm^{-1} near 70 K to ~ 60-100 cm^{-1} near 180 K, which was possibly due to inter-valence band free carrier absorption.

* also with Applied Optoelectronics Inc., Houston, Texas 77081

107

I. INTRODUCTION

High-power high-temperature mid-infrared (MIR) lasers at 2 to 5 μm are highly desirable for a variety of applications such as infrared (IR) countermeasures, medical surgery and diagnosis, covert illumination for night vision, and molecular spectroscopy, etc. In addition, many important atmospheric molecules including industrial pollutants and green house gases (H_2O, CO, HF, CH_4, C_2H_6, HCl, CO_2, N_2O, O_3, etc.) have strong, fundamental vibrational transitions in the MIR wavelength range. Even though they also have overtones at shorter, near IR wavelengths, the absorption there are much weaker. For example, the absorption of CH_4 is two orders of magnitude stronger at 3.3 μm than at 1.65 μm. Both military and commercial MIR systems are limited mainly due to a lack of adequate sources. Currently, MIR solid state lasers tend to be bulky with little wavelength agility, while optical parametric oscillators, which use a nonlinear crystal to down convert or up convert the output wavelength of pump sources, are complex and expensive. Semiconductor diode lasers have significant advantages in terms of cost, volume, weight, reliability, and power dissipation. The availability of compact high-power MIR semiconductor lasers operating at thermo-electrically (TE) cooled temperatures > 205 K would significantly enhance the capability of current MIR technology.

Recently, MIR semiconductor lasers have improved significantly in output power and operating temperature [1-15]. The Sb-based lasers [1-13] demonstrated the best power and temperature performance for 2 μm < λ < 4.5 μm, and the InP-based intersubband quantum cascade (QC) lasers [14, 15] performed best in the long-wavelength range > 5 μm. Both types have the potential to cover a wider wavelength range with more advanced material development. While both technologies have demonstrated high powers at low temperatures, or higher temperature but with substantially lower powers, a challenging problem for both is the net power efficiency at any temperature.

Room-temperature operation has been realized in Sb-based type-I interband lasers at wavelengths shorter than 2.8 μm [1, 2]. However, due to the intrinsic nature of the type-I interband transitions, the lasing wavelength is limited to the material bandgap, and the device performance suffers from non-radiative Auger recombination [3]. For wavelengths longer than 4 μm, MIR lasers based on type-I interband structure have been proven to be quite challenging. Electrically pumped MIR lasers based on InAs/InGaSb type-II superlattices (SLs) [9, 10] have achieved similar performance to those based on Sb-based type-I structures. A maximum operating temperature T_{max} = 260 K was achieved at 3 μm with a characteristic temperature T_0 of 33 K [10]. At 100 K, the maximum peak output power per facet was about 800 mW with a pulse length of 200 nsec and a repetition rate of 2.5 kHz. The corresponding differential external quantum efficiency was 31.4 %.

However, the two-constituent InAs/InGaSb type-II SL is non-optimal, since the strong penetration of the electron wavefunctions into the thin InGaSb barriers leads to a nearly isotropic electron mass, i. e., the electrons have strong energy dispersion along all three coordinate axes. It is well known from the wide-bandgap diode lasers that quantum-well (QW) devices with a step-like 2D density of states have a much higher gain per injection carrier. Figure 1 shows the band profiles along with four constitutes in each period of the active region (InAs/InGaSb/InAs/AlSb). Even though the electron wavefunction has its maximum in the InAs layers and the hole wavefunction is centered in the InGaSb layers, their overlap is sufficient to yield an interband optical matrix element ≈ 70% as large as those in typical type-I heterostructures. Note also that the electrons and holes both have 2D dispersion due to the lack of penetration through the AlSb barriers. Because of the coupled-well nature of the conduction band profile, the electron split into symmetric (E1S) and anti-symmetric (E1A) levels.

By simply changing the InAs and InGaSb layer thicknesses, the lasing wavelength of type-II QW lasers could vary from MIR to long IR. This is an important advantage of InAs/InGaSb type-II lasers compared to the Sb-based type-I devices in terms of epitaxial growth, since the composition control, composition uniformity, and growth reproducibility of InGaAsSb, InAlAsSb, and AlGaAsSb are very difficult [3]. Additionally, due to the unique feature of the type-II structure, the Auger recombination could be significantly suppressed through careful bandgap engineering [16-18]. It was recently demonstrated experimentally in Ref. 18 that at 77 K, the Auger lifetimes in InAs/InGaSb SLs can be two orders of magnitude longer than those in HgCdTe alloys with the same energy gap.

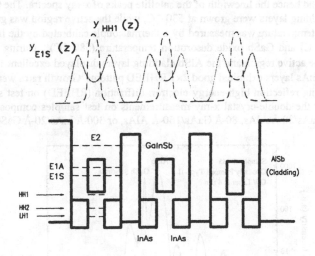

Figure 1. Conduction and valence band profiles for a four-constituent InAs/InGaSb/InAs/AlSb type-II QW lasers, and the electron and hole wavefunctions.

Recently, type-II QW lasers have been demonstrated at temperatures > 300 K at wavelengths of 2.9 to 4.5 µm optically pumped by a 2.06 µm Q-switched Ho:YAG laser [6, 7]. For the 3.2-µm lasers, the maximum lasing temperature was 350 K with a characteristic temperature T_0 of 68 K, the peak power per facet was 270 mW at 300 K [6]. For the 4.5-µm devices, lasing was observed up to 310 K, and the peak output power per facet was more than 3.65 W at 160 K and about 360 mW at 280 K [7]. However, these devices were operated at short pulses < 100 nsec and low duty cycles. Here, we report a 4.5 µm type-II QW laser with significantly improved performance over our previously reported devices in the differential external quantum efficiency, threshold pump intensity, and average output power, because of the improved material quality and device design in suppressing Auger recombination. Additionally, we will discuss the factors limiting the maximum average output power.

II. EXPERIMENT

The type-II QW lasers were grown on p-type GaSb substrates in a Riber 32 molecular beam epitaxy (MBE) system, equipped with EPI valved cracker As and P cells, an EPI Sb cracker cell, EPI In and Ga SUMO cells, an EPI Be dopant cell, and Riber Al and Si effusion cells. The active region was composed of 50 periods of un-intentionally doped InAs/In$_{0.35}$Ga$_{0.65}$Sb/InAs/AlSb (21.5 Å/37 Å/21.5 Å/42 Å) type-II QWs lattice matched to the AlSb cladding layers, 1.7 μm for the top cladding layer and 2.6 μm for the bottom cladding layer. Figure 2 shows the x-ray spectrum of the type-II QW lasers. Even the type-II QWs were lattice matched to the AlSb cladding layers instead of the GaSb substrate, the x-ray spectrum clearly shows many satellite peaks. Further improvement in the material quality is required to reduce the defect density and hence the linewidth of the satellite peaks of x-ray spectra. The GaSb buffer layer and AlSb cladding layers were grown at 530 °C, while the active region was grown at 440 °C. The substrate temperature was measured by a thermalcouple calibrated by the InSb melting temperature (525 °C) and GaSb oxide desorption temperature (580 °C). During growth, the InGaSb layers in the active region and the AlSb cladding layers displayed excellent 1×3 RHEED patterns while the InAs layers exhibited good 2×1 RHEED patterns. Growth rates were calibrated to within ± 2% using reflection high-energy electron diffraction (RHEED) on test samples and were confirmed by the double-crystal x-ray measurements on test samples composed of many periods of 300-Å GaAs/30-Å AlAs, 80-Å GaAs/160-Å AlAs, or 300-Å InAs/20-Å GaSb SLs.

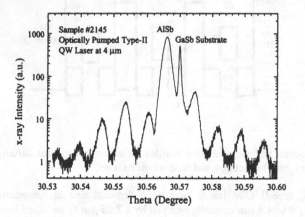

Figure 2. x-ray spectrum of the type-II QW laser.

For the 2-μm pumping measurements, the laser wafer was cleaved into laser bars with various cavity lengths without lapping. The lasers with uncoated facets were indium heat sunk to a copper cold finger, epitaxial-side-up. The devices were optically pumped by 650 ns pulses from a 2.016 μm Tm:YAG laser at a repetition rate of 2 kHz. The TEM$_{00}$ YAG output was line-focused with a single cylindrical lens to form a gain-guided pump stripe roughly 200 μm wide on the laser. The mid-IR output was relayed to the input slit of a 50-cm spectrometer with a pair of off-axis parabolic mirrors. Under pulsed operation, lasing was observed up to 300 K limited by the available cryostat temperature. The lasing peak wavelength displayed a red shift with increasing temperatures, from 3.84 μm at 44 K to 4.48 μm at 300 K, yielding a red shift of 2.47

nm/K. Figure 3a shows the peak output power vs. absorbed pump intensity at 300 K with an absorbed threshold pump intensity of 1.1 kW/cm². Here, we have assumed only 14% of incident power absorbed in the active region due to a small absorption coefficient of the type-II QWs at 2 μm calculated using a $k \bullet p$ program. The major contributions in the absorption coefficient at 2 μm are due to the type-II interband transitions from the valence band states of InGaSb to the conduction band states of InAs, and the type-I interband transitions in the InGaSb layers.

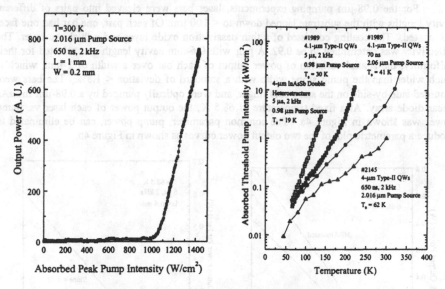

Figure 3 (a). Peak output power vs. absorbed peak pump intensity at 300 K. L denotes the device length and W denotes the pump beam width. (b). Comparison of the absorbed threshold pump intensity of two type-II QW lasers and one InAsSb type-I interband laser, optically pumped either by a 2-μm YAG laser or 0.98-μm InGaAs diode laser arrays.

Figure 3b compares the absorbed threshold pump intensity vs. temperature of the presented laser (sample #2145) with one InAsSb double heterostructure laser [5] and our previous type-II QW laser (sample #1989) [7], optically pumped either by a 2-μm YAG laser or 0.98-μm InGaAs diode laser arrays. Comparing sample #1989 with the InAsSb heterostructure laser operated under identical conditions, both lasers exhibited comparable thresholds at 70 K, indicating the good material quality of the type-II QW lasers since the thresholds at low temperatures are dominated by defect-related non-radiative recombination. But, the type-II QW laser (sample #1989) showed a T_0 of 30 K instead of 19 K, and thus operated up to 226 K. The difference in T_0 indicates that Auger recombination has been significantly suppressed in the type-II QW lasers. Figure 3b also clearly shows the advantage of a 2-μm pump source over a 1-μm pump source in the threshold pump intensity and T_0. With the same external quantum efficiency, the power efficiency using a 2-μm pump source will be two times of that using a 1-μm pump source, and hence the heat load will be much smaller. The significant improvements of the current device over our previously reported type-II QW laser in the threshold pump intensities and T_0 are mainly due to the improved material quality and device design in suppressing Auger recombination. The absorbed threshold pump intensity of sample #2145 was only 0.27 kW/cm²

111

at 220 K and 1.1 kW/cm² at 300 K, with a characteristic temperature T_0 of 61.6 K for temperatures up to 300 K. The observed thresholds are much lower than any previous data reported in the literature for the whole temperature range up to 300 K because of such high T_0. The devices have also been evaluated under continuous-wave (CW) operation. At 20 K, the CW lasing spectrum showed a full width at half maximum of 1.35 meV. The observed maximum CW operation temperature was 100 K limited by the 68 mW of absorbed pump power.

For the 0.98-μm pumping experiments, laser bars were cleaved into pairs of different cavity lengths with the substrate lapped down to < 100 μm. Of each pair, one bar had one facet HR coated. The coating consisted of a thin passivation oxide layer and a thick Au layer. The reflectivity was determined to be 0.92. A pair with 0.6-mm cavity length was selected for their uniformity; the spatial variation of power output of each bar over a width of 1 mm, which is much wider than the pump stripe width with a standard of deviation < 1.8%. The bars were mounted side-by-side on the same cold finger, and were optically pumped by a 0.98-μm InGaAs linear diode array. At a fixed temperature of 68.5 K, the output power of each laser vs. pump power was shown in Figure 4a. The common parameter, pump power, can be eliminated to produce a parametric plot of the two output power curves as shown in Figure 4b.

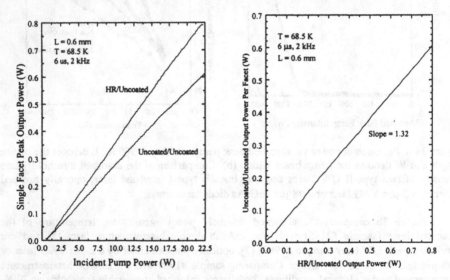

Figure 4 (a). Single facet output power vs. pump power for two lasers with facet conditions as indicated. (b). Output power correlation of the two lasers.

Let ρ_1 and ρ_2 be the single-facet external quantum efficiencies of two different lasers, one with two uncoated facets, and one with an uncoated facet and the other HR-coated, they can be expressed as [19]:

$$\rho_1 = \eta_i \frac{\ln\left[\frac{1}{R_1}\right]}{2\ln\left[\frac{1}{R_1}\right] + 2\alpha L}; \rho_2 = \eta_i \frac{\ln\left[\frac{1}{R_1}\right]}{\ln\left[\frac{1}{R_1 R_2}\right] + 2\alpha(1-\delta)L}, \qquad (1)$$

where η_i is the internal quantum efficiency, R_1 is the reflectivity of the cleaved facet, R_2 is the reflectivity of the HR facet, L is the laser length, and α is the internal loss coefficient. A factor

(1-δ), with $0 \leq \delta \ll 1$, is used as a small adjustment for ρ_2 to account for the possibility that the internal loss can be slightly less for ρ_2 [19].

From Eq. 1 and Figure 4, the internal loss was determined to be 18 cm^{-1} at 68.5 K. Figure 5a. show the measured single-facet external quantum efficiency of each laser at various temperatures. Here, we have assumed that the absorption of pump power in the active region was about 70%. From Figure 5a, the ratio ρ_2/ρ_1 was calculated, and the behavior of α vs. temperature is shown in Figure 5b. The lower curve assuming δ=0 in Eq. 1 serves as a lower-bound estimate of internal loss. The upper curve indicating the upper-bound estimate of α was obtained by using an estimate of δ from the threshold measurement [19].

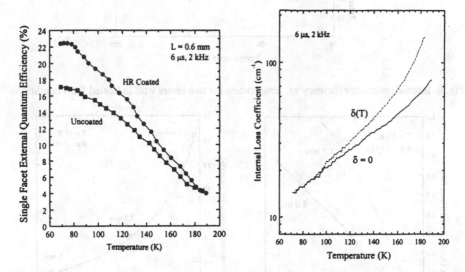

Figure 5 (a). Single-facet external quantum efficiency vs. temperature for the two lasers with different facet conditions as indicated. (b). Internal loss coefficient as a function of temperature. Solid curve is a lower bound estimate, and dashed curve is an upper bound estimate.

With the temperature dependent external quantum efficiency shown in Fig. 5a and the internal loss shown in Fig. 5b, Eq. 1 can be used to derive the internal quantum efficiency as a function of temperature. The results using the upper bound internal loss are shown in Figure 6. The internal quantum efficiency is approximately constant (~67%) at temperatures up to 190 K. The experimental results on the internal loss coefficient in Fig. 5b are further supported by the data on the external quantum efficiency vs. cavity length. Figure 7a shows the output power from 0.6, 1.7, and 3 mm long lasers at 85 K. Their near threshold external quantum efficiencies, in the regime where active region heating is insignificant, are depicted in Fig. 7b, and the solid curve is the calculation based on Eq. 1 using the upper-bound value of α in Fig. 5b. The agreement between the 0.6-mm laser data and the solid curve is not meaningful since the curve was derived from the data itself. But the agreement between the 1.7-mm and 3-mm laser data with the curve supports the internal loss measurement.

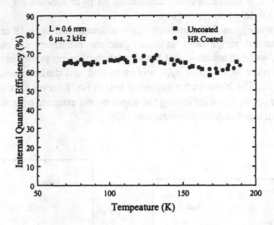

Fig. 6. Internal quantum efficiency vs. temperature for two lasers with indicated facet conditions.

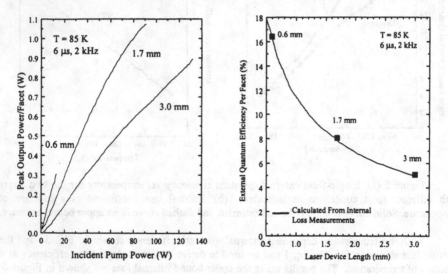

Figure 7 (a). Output power per facet vs. pump power for devices with various cavity lengths. (b). Symbols are the measured external quantum efficiencies from Fig. 7a, and the solid curve is the calculated efficiency using the upper-bound internal loss value in Fig. 5b.

The main result with direct consequence on the high power capability is the internal loss shown in Fig. 5b, which is ~18 cm⁻¹ at low temperature and steadily increases up to ~60 to 100 cm⁻¹ at high temperature. With such a high internal loss, power scaling by increasing device length is not possible as evidenced in Fig. 7b. At high temperatures, while lasing was possible under intensive and short pulses [6, 7], the lasing efficiency and average output power was not sufficient for practical use. Understanding the origin of the internal loss and reducing it are crucial to practical devices. The increase of internal loss vs. temperature in Fig. 5b is consistent with the increase in threshold carrier density. Therefore, the high internal loss is possibly due to the free carrier absorption, especial the inter-valence band absorption. Figure 8 shows the band structure of the type-II QWs calculated using a 8×8 $k \cdot p$ program. It is clear that the free carrier absorption due to the LH2 to HH1 transitions should be the dominant factor. Further improvement in the active region design would be necessary to reduce the internal loss and increase the power efficiency.

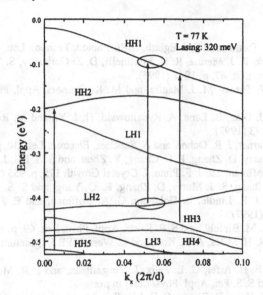

Figure 8. Band structure of the type-II QWs calculated using a 8×8 $k \cdot p$ program.

III. CONCLUSIONS

We have presented the room-temperature operation of type-II QW lasers at 4.5 μm optically pumped by a 2-μm source. The absorbed threshold peak pump intensity was only 1.1 kW/cm² at 300 K, with a characteristic temperature T_0 of 61.6 K for temperatures up to 300 K. Because of such high T_0, the observed thresholds are much lower than any previous data reported in the literature for the whole temperature range up to 300 K. The effects of internal loss on the efficiency and output power for type-II QW lasers have also been studied experimentally via 0.98-μm optical pumping. The devices exhibited an internal quantum efficiency of 67%, which is relatively high for MIR materials, and was capable of > 1.1-W peak output power per facet in 6-μs pulses with a 2 kHz repetition rate at 85 K, which is among the highest near 4 μm. The internal loss of the QWs exhibited an increase from 18 cm⁻¹ near 70 K to ~ 60-100 cm⁻¹ near 180 K, which was possibly due to inter-valence band free carrier absorption. Further improvements in the active region design and material quality should reduce the internal loss significantly, and hence high power efficiency and high average output power should be achievable.

ACKNOWLEDGMENT

The authors at University of Houston wish to thank Jun Zheng for the detailed band structure calculations. The work at Univ. of Houston is partially supported by NASA contract No. NAGW-977 and TcSCH. The work at MIT Lincoln Laboratory is supported by the US Department of the Air Force. The work at Applied Optoelectronics Inc. is partially supported by Air Force Phillips Lab. under the contract No. F29601-97-C-0072.

REFERENCES

1. H. K. Choi, G. W. Turner, and S. J. Eglash, IEEE Photon. Technol. Lett. **6**, p. 7 (1994).
2. H. Lee, P. K. York, R. J. Menna, R. U. Martinelli, D. Z. Garbuzov, S. Y. Narayan, and J. C. Connolly, Appl. Phys. Lett. **67**, p. 1942 (1995).
3. H. K. Choi, G. W. Turner, M. J. Manfra, and M. K. Connors, Appl. Phys. Lett. **68**, p. 2936 (1996).
4. D. Wu, E. Kaas, J. Diaz, B. Lane, A. Rybaltowski, H. J. Yi, and M. Razeghi, IEEE Photon. Technol. Lett. **9**, p. 173 (1997).
5. H. Q. Le, G. W. Turner, J. R. Ochoa, and A. Sanchez, Electron. Lett. **30**, p. 1944 (1994).
6. C.-H. Lin, S. J. Murry, D. Zhang, P. C. Chang, Y. Zhou, and S. S. Pei, J. I. Malin, C. L. Felix, J. R. Meyer, C. A. Hoffman, and J. F. Pinto, J. Crystal Growth **175**, p. 955 (1997).
7. C.-H. Lin, P. C. Chang, S. J. Murry, D. Zhang, R. Q. Yang, and S. S. Pei, J. I. Malin, J. R. Meyer, C. L. Felix, J. R. Lindle, L. Goldberg, C. A. Hoffman, and E. J. Bartoli, J. Electron. Materials **26**, p. 440 (1997).
8. A. A. Allerman, R. M. Biefeld, and S. R. Kurtz, Appl. Phys. Lett. **69**, p. 465 (1997).
9. T. C. Hasenberg, R. H. Miles, A. R. Kost, and L. West, IEEE J. Quantum Electron. **33**, p. 1403 (1997).
10. W. W. Bewley, E. H. Aifer, C. L. Felix, I. Vurgaftman, and J. R. Meyer, C. H. Lin, S. J. Murry, D. Zhang, and S. S. Pei, Appl. Phys. Lett., in press.
11. R. Q. Yang, B. H. Yang, D. Zhang, C. H. Lin, S. J. Murry, H. Wu, and S. S. Pei, Appl. Phys. Lett. **71**, p. 2409 (1997).
12. C. L. Felix, W. W. Bewley, I. Vurgaftman, and J. R. Meyer, D. Zhang, C. H. Lin, R. Q. Yang, and S. S. Pei, IEEE Photon. Technol. Lett. **9**, p. 1433 (1997).
13. A. N. Baranov, N. Bertru, Y. Cuminal, G. Boissier, C. Alibert, and A. Joullie, Appl. Phys. Lett. **71**, p. 735 (1997).
14. C. Sirtori, J. Faist, F. Capasso, D. L. Sivco, A. L. Hutchinson, and A. Y. Cho, IEEE Photon. Technol. Lett. **9**, p. 294 (1997).
15. C. Sirtori, J. Faist, F. Capasso, D. L. Sivco, A. L. Hutchinson, and A. Y. Cho, IEEE J. Quantum Electron. **33**, p. 89 (1997).
16. C. H. Grein, P. M. Young, and H. Ehrenreich, J. Appl. Phys. **76**, p. 1940 (1994).
17. L. R. Ram-Mohan and J. R. Meyer, J. Nonlinear Opt. Phys. Mat. **4**, p. 191 (1995).
18. E. R. Youngdale, J. R. Meyer, C. A. Hoffman, F. J. Bartoli, C. H. Grein, P. M. Young, H. Ehrenreich, R. H. Moles, and D. H. Chow, Appl. Phys. Lett. **64**, p. 3160 (1994).
19. H. Q. Le, and G. W. Turner, C.-H. Lin, S. J. Murry, R. Q. Yang, and S. S. Pei, submitted to IEEE J. Quantum Electron.

MODELING OF MID-INFRARED MULTI-QUANTUM WELL LASERS

A.D. ANDREEV
A.F. Ioffe Physico-Technical Institute of Russian Academy of Sciences,
Polytechnicheskaya 26, St.-Petersburg 194021, Russia
FAX: +7-812-247-1017; E-mail: andreev@theory.ioffe.rssi.ru

ABSTRACT

Threshold characteristics of mid-infrared MQW lasers have been studied theoretically. Auger recombination rates in strained quantum wells have been calculated in the framework of 8x8 Kane model taking account of spin-orbit interaction. It is demonstrated that the Auger coefficients of both CHCC and CHHS processes essentially depends on the QW parameters (strain, QW width, barrier bandgap), but have relatively weak (non-exponential) temperature dependence. It is shown that the laser characteristics can be improved for optimized laser structures, when the Auger rate is decreased.

INTRODUCTION

Semiconductor diode lasers operating at mid-infrared wavelengths (2-5 μm) are of considerable interest for a number of application such as remote sensing of pollution and gases, molecular spectroscopy and laser radar. Multi-quantum well (MQW) structures based on ternary and quaternary III-V alloys are promising candidates for the creation of high-performance mid-infrared lasers [1-3]. At high temperatures the laser threshold current is mainly controlled by non-radiative Auger recombination processes. The Auger process in QWs is very sensitive to the variation of the QW parameters [4]. Choosing optimal structure parameters it is possible to decrease the Auger rate and, hence, improve laser performance. Therefore modeling of the threshold characteristics of these lasers with the aim of structure optimization is important.

In this paper the author presents the calculation of the Auger recombination rates in strained QWs in the framework of 8x8 Kane model with the aim of modeling of the laser threshold characteristics and structure optimization.

AUGER RECOMBINATION PROCESSES IN QWs

During the Auger process the energy of the recombining electron-hole pair is transferred to another carrier (electron or hole), which is excited to a higher energy band. There are two dominant Auger processes in QWs: CHHS (when a hole is excited to the so-band) and CHCC (excitation of the electron). The Auger processes can also be classified by type of the excited carrier state, whether it is unbound or localized in the QW. In the first case, when the excited carrier is unbound, the Auger process is thresholdless and its rate has a weak (non-exponential) temperature dependence [4]. In the other case, the excited localized carrier has relatively large momentum component in the QW plane, and the Auger process is bulk-like and threshold. Thresholdless mechanism of Auger recombination is the dominant one for the structures with QWs.

Previously, we have reported the results of the calculations of CHCC Auger process rates in InAlAsSb QWs in the framework of 4x4 Kane model without taking account of spin-orbit interaction. The Auger rate was shown to be very sensitive to the variation of the structures

117

parameters (strain, QW width, emission wavelength) [4]. Therefore more realistic multi-band model is required as a basis for calculations. Also, in some case CHHS process is the dominant and should be calculated.

In this paper for band-structure and Auger rate calculation we have used 8x8 Kane model which includes the effects associated with strain and spin-orbit interaction [6]. This model was proved to describe properly the carrier states in strained QWs [7]. The 8x8 model takes account of the main band-structure features which considerably influences the Auger rate: (i) non-parabolicity of the electron and hole energy dispersion in the QW; (2) mixing of heavy, light and so-hole states in QWs (3) non-parabolicity of the energy dispersion of highly excited electron and so-hole. The electron and hole energy spectra in the QW were calculated by solving numerically the corresponding dispersion equation, which is the determinant of 4x4 matrix for electrons and 12x12 for holes.

The key point in calculation of the Auger recombination rate is derivation of the Auger transition matrix element, which should be calculated using wave functions obtained from 8x8 multiband model. To considerably reduce the time of numerical calculation of the Auger rate, an exact analytical formula for the Auger matrix element has been obtained. Matrix element is expressed through the overlap integrals: $M = M_I - M_{II}$,

$$M_I = \int \frac{dq}{(2\pi)} \frac{I_{13}(q)I_{24}(-q)}{q^2 + (k_{1\parallel} - k_{3\parallel})^2 + \lambda_D^2}, \quad I_{\alpha\beta}(q) = \int \Psi_\alpha^*(x)\Psi_\beta(x)e^{iqx}dx, \qquad (1)$$

where $k_{\alpha\parallel}$ is in-plane momentum component of the carrier in state α, λ_D is inverse screening length of the Coulomb potential, Ψ_α is eight-component carrier wave function derived from 8x8 Kane model (the product $\Psi_\alpha^*\Psi_\beta$ denotes the scalar product of two eight-component vectors), $I_{\alpha\beta}$ is the overlap integral between the states α and β. Overlap integrals have essentially non-linear dependence on hole momentum (see Fig.1) and are strongly sensitive to the variation of the structure parameters. Therefore, previously used phenomenological approach for calculation of the overlap integrals [8] is not valid. According to this approach, the electron-hole overlap integral had a linear dependence on momentum. In this paper, all overlap integrals have been calculated from first principles employing eight-component wave functions (see Eq.1). Fig.1 demonstrates that both hole - so-hole and electron-hole overlap integrals decrease with strain $\xi = (a_W - a_B)/a_B$ (here a_W and a_B are the lattice constants of the unstrained well and barrier materials respectively).

To derive analytical formula for the Auger transition matrix element, we perform the integration over q using the residue theorem [9]. The calculation of the Auger rate was carried out numerically employing Monte-Carlo method for computation of 5D integral.

For numerical calculations we consider structures based on $In_xAl_{1-x}As_ySb_{1-y}$, $In_xGa_{1-x}As_ySb_{1-y}$ and $InAs_xSb_{1-y}P_{1-x-y}$. For all series of structures, the structure composition is determined from the condition that the emission wavelength is constant at given strain and QW width. The bandgap dependence on composition was calculated using bowing parameters [10], other band parameters are assumed to have linear dependence on composition [10].

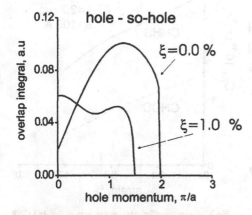

Fig.1a. Hole - so-hole overlap integral dependence on the transferred momentum for two values of strain in InAsSbP QW (a=100 A, λ=2.7 μm).

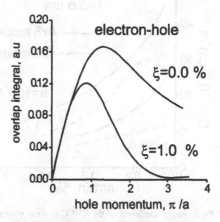

Fig.1b. Electron-hole overlap integral dependence on the momentum transferred for two values of strain in InAlAsSb QW (a=80 A, λ=3.5 μm).

Is is convenient to characterize the Auger rate by the Auger coefficient, $C_A = G / n^2 p$ for CHCC process and $C_A = G / np^2$ for the CHHS process (here n and p are 2D concentration of electron and holes in QW). Figures 2 and 3 shows the Auger coefficient versus strain, in this case the interval of possible variation of ξ depends on the QW parameters [4]. For InAlAsSb and InGaAsSb structures the dominant Auger process is the CHCC process since for these structures $(\Delta_{so} - E_g^{eff}) >> T$. However, for InAsSbP-based structures the rate of the CHHS process is more that one order larger than that of CHCC process (see Fig. 3). The Auger coefficient decreases with strain which mainly due to decrease of the overlap integrals. It should be also noted that previously used 4x4 model for the Auger rate calculations underestimates the result (see Figure 1), this reflect the fact that the Auger coefficient C_A is extremely sensitive the band structure variation. Figure 4 shows the depedence of C_A on the barrier bandgap for InGaAsSb QW with AlGaAsSb barriers. The Auger coefficient increases with the barrier bandgap, which characteristic for thresholdless Auger recombination mechanism [11]. It is well known that in bulk materials the Auger rate usually has an exponential dependence on temperature [12]. In QWs due to the lack of the conservation of the momentum component perpendicular to the QW plane, thresholdless Auger process with weak (non-exponential) temperature dependence is possible. This mechanism is the dominant one in QWs of "classical" widths (a<150 A). The coefficient of this process increase only by factor of 2-3 in the temperature range 50-250 K (see Fig. 5).

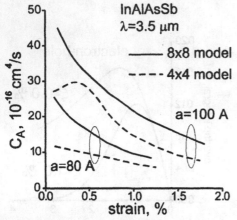

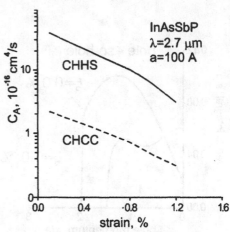

Fig.2. Auger coefficient of the CHCC process versus strain for InAlAsSb QWs, dashed lines - 4x4 Kane model without spin-orbit interaction [4,9], solid lines - present work (8x8 Kane model).

Fig.3. Auger coefficients versus strain for InAsSbP QWs.

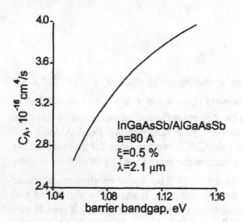

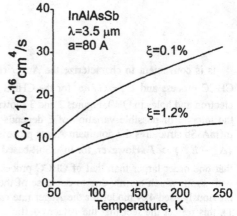

Fig. 4. Auger coefficient versus the barrier band gap for InGaAsSb/AlGaAsSb QW

Fig. 5. Auger coefficient versus temperature for InAlAsSb QW. Temperature dependence of the bandgap is taken into account.

LASER THRESHOLD CHARACTERISTICS

The threshold current density consists of three contribution:

$$J_{th} = J_A + J_R + J_L,$$
(2)

where J_A, J_R, J_L is the Auger, radiative and leakage current components, respectively. The Auger current is mainly determined by the thresholdless Auger recombination in QWs, but in structures with small barrier height for electrons or hole, it is also important to take account of Auger recombination in the barrier [5]. The threshold concentration was calculated from the threshold condition taking account 3D carriers in the barrier region. Internal absorption was calculated using 8x8 Kane model. In agreement with previous calculation [5] in 4x4 model, the internal absorption coefficient decreases with strain. Due to strong absorption, the threshold concentration has non-linear temperature dependence (see Fig. 6), which causes strong (exponential-like) temperature dependence of the threshold current. Thus, the temperature sensitivity of the laser threshold current is mainly controlled by temperature dependence of threshold concentration.

Since the main contribution to the threshold current is given by the Auger components, it is possible to considerably improve laser performance if the Auger recombination rate is reduced. It is shown in the previous section of this paper that the Auger rate is extremely sensitive to the variation of the band structure. In particular, Auger rate decreases with strain. Therefore optimal structure parameters, at which laser threshold current has a minimum and internal quantum efficiency has a maximum exists. This fact is illustrated for InAlAsSb laser (Fig. 7). When the strain increases, the Auger rate and, consequently, threshold current decrease. At large strain barrier height for electrons becomes relatively small which results in large leakage currents and the Auger current in the barrier. Thus, in define value of strain, laser performance is optimized. Such kind of laser structure optimization is possible also for lasers based on $In_xGa_{1-x}As_ySb_{1-y}$ and $InAs_xSb_{1-y}P_{1-x-y}$. However, it should be noted that for real lasers it is necessary to pay special attention to the details of the laser design in order to take account of such processes as lateral diffusive leakage, carrier heating and others.

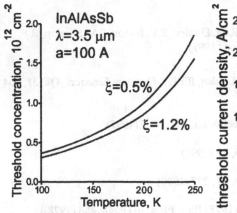

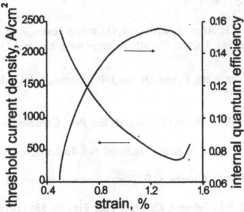

Fig. 6. Threhsold concentration versus temoerature for InAlAsSb MQW laser (4 QWs).

Fig. 7. Threhsold current density and internal quantum efficiency versus strain for InAlAsSb MQW laser, T=150 K, 4 QWs, λ=3.5 μm

121

CONCLUSION

In this paper we have performed the theoretical study of Auger recombination processes in strained QWs in the framework of 8x8 Kane model. It is demonstrated that the Auger rate is extremely sensitive to the variation of the band structure and QW parameters (strain, barrier bandgap, QW width).

Modeling of the MQW mid-infrared laser threshold characteristics has been carried out with paying special attention to the optimization of the laser structure. It is demonstrated that threshold current has a minimum and internal quantum efficiency has a maximum at definite value of strain in InAlAsSb structures. In summary, laser performance can be considerably improved for the optimized laser structures.

REFERENCES

1. C.A. Wang and H.K. Choi, Appl. Phys. Lett., **70**, 802 (1997)

2. B. Lane, D. Wu, A. Rybaltowski, H. Yi, J. Diaz, M. Razeghi, Appl. Phys. Lett., **70**, 443 (1997)

3. D.Z. Garbuzov, R.U. Martimelli, R.J. Menna, P.K. York, H. Lee, S.Y. Narayan, Appl. Phys. Lett., **67**, 1346 (1995)

4. A.D. Andreev, G.G. Zegrya, Appl. Phys. Lett., **70**, 603 (1997)

5. A.D. Andreev, G.G. Zegrya in *"In-Plane Semiconductor Lasers: from Ultraviolet to Mid-Infrared"*, ed. by H.K. Choi, P.S. Zory, Proceeding SPIE, **3001**, 364-376 (1997)

6. M.S. Hyberten, R.F. Kazarinov, G.A. Baraff, D.A. Akerman, G.E. Shtengel, in *"Physics and Simulation of Optoelectronic Devices III"*, ed. by M. Osinski, W.Chow, Proceedings SPIE, **2399**, 132 (1995)

7. G.N. Aliev, A.D.Andreev, O.Coshug-Toates, R.M. Datsiev, S.V. Ivanov, S.V. Sorokin, R.P. Seisyan, J. Cryst. Growth, accepted for publication (1997)

8. Jin Wang, P. von Allmen, J.-P Leburton, K.J. Linden, IEEE J. Quantum Electron., **QE-31**, 864 (1995)

9 A.D. Andreev, G.G. Zegrya, IEE Proc. Optoelectronics, **144**, No 5 (1997)

10. M.P.C.M. Krijn, Semicond. Sci. Technol., **6**, 27 (1991)

11. A.D. Andreev, G.G. Zegrya, Semiconductors, **31**, 297 (1997)

12. B.L. Gelmont, Zh. Eksp. Teor. Fiz., **75**, 536 (1978) [Sov. Phys. JETP, **48**, 268 (1978)]

Tunneling effects in InAs/GaInSb superlattice infrared photodiodes

U. Weimar, F. Fuchs, E. Ahlswede, J. Schmitz, W. Pletschen, N. Herres, and M. Walther
Fraunhofer–Institut für Angewandte Festkörperphysik, Tullastrasse 72, D-79108 Freiburg, Germany

ABSTRACT

The optical and electrical properties of InAs/GaInSb superlattice mesa photodiodes with a cutoff wavelength around 8 μm are investigated. The influence of the surface potential at the mesa sidewalls on the device properties was studied by fabricating gate-controlled diodes. At least two mechanisms determining the dark current in the reverse bias region can be identified. At high reverse biases bulk band-to-band tunneling dominates while the current at low reverse biases is most likely governed by surface effects. Bulk interband tunneling is further investigated by applying magnetic fields B up to 7 T parallel and perpendicular to the electric field E across the p-n junction.

INTRODUCTION

The InAs/GaInSb short-period superlattice (SPSL) system is suitable for long-wavelength detector applications due to the broken gap type II band alignment. An effective band gap between minibands in the superlattice can be opened due to quantization effects. After the first proposal as a possible alternative to HgCdTe in 1987 [1] the epitaxial growth has matured and the successful preparation of diodes has been reported by several groups [2–4]. The performance of a photodiode is generally limited by the dark current, which determines its dynamic impedance and its thermally generated noise. However, little work has been reported on dark current mechanisms in InAs/GaInSb SPSL photodiodes.

It is known from other low gap materials that tunneling currents may dominate the current-voltage characteristic in reverse bias [5, 6]. Tunneling may occur either across the metallurgical junction or across field induced junctions which are due to surface accumulation or inversion at the mesa sidewalls. Gate electrodes can be placed on the mesa sidewalls in order to control the surface potential and the surface leakage currents. Thus bulk properties may be distinguished from surface effects.

DEVICE PREPARATION AND DETECTOR CHARACTERIZATION

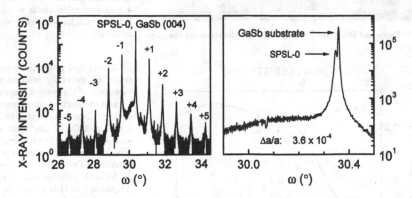

Fig. 1: (004) HRXRD rocking curves of a 100 periods 14 ML InAs/8 ML Ga$_{0.8}$In$_{0.2}$Sb SPSL.

The samples were grown by solid source MBE on undoped, but residual p-type, (100) GaSb at substrate temperatures around 410°C. Alternating InSb-like and GaAs-like interfaces were employed, terminating each individual layer of the superlattice by a group V soak and starting the following layer by a mono-layer (ML) of the respective group III material. This growth sequence was found to result in a superior quality in comparison to the inverted interface alternation [7, 8]. Growth parameters such as substrate temperature and III/V beam equivalent pressure ratios were optimized using Fourier-transform IR photoluminescence spectroscopy and high-resolution X-ray diffraction (HRXRD) yielding bandgap, average strain and superlattice period, respectively. Meanwhile, a residual mismatch below 5×10^{-4} is routinely achieved. The HRXRD profile of a recent sample, displayed in Fig. 1, shows besides the (004) reflection of the GaSb substrate narrow superlattice diffraction peaks up to $\pm 5^{th}$ order. The superlattice stack is under small biaxial compressive strain with a residual lattice mismatch of 3.6×10^{-4} with respect to the substrate (Fig. 1, right panel). The interference fringes seen in the diffraction profile give direct evidence for the excellent structural quality of the SPSL. However, a larger lattice mismatch is tolerable for good detector performance. The epitaxial layer structure used for the present study is under biaxial tension with a residual lattice mismatch of $\Delta a/a = 2\times10^{-3}$.

A series of detector structures consisting of a 13 MLs InAs/8 MLs $Ga_{0.85}In_{0.15}Sb$ SPSL sandwiched between 10 nm InAs:Si (2×10^{18} cm^{-3}) and 500 nm GaSb:Be (1×10^{18} cm^{-3}) contact layers was grown [9]. The superlattice forms a p-n-n$^+$ homojunction diode structure with a 85-period p-type region, a 15-period not intentionally doped but residual n-type region and 50-period n$^+$-type region doped to a concentration of 1×10^{18} cm^{-3}. The p-doping concentration was chosen to be just above the n-background concentration of about 5×10^{16} cm^{-3}. By varying the p-doping concentration the reduced carrier concentrations $N_{red} = (1/N_A + 1/N_D)^{-1}$ could be adjusted between 1.4×10^{16} cm^{-3} and 4.9×10^{16} cm^{-3} as determined by capacitance-voltage analysis. By proper choice of the doping level interband tunneling currents across the p-n junction could be suppressed below a critical limit and the resulting R_0A product of these diodes has been shown to be close to the diffusion limit down to 77 K [9].

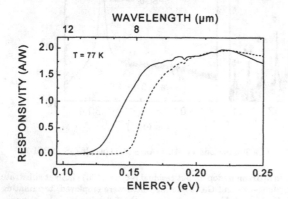

Fig. 2: Processed 50x50 μm^2 device with gate electrode on the mesa sides.

Photodiodes were processed employing standard optical lithography. Following the mesa definition by wet chemical etching using citric acid and the contact metallization by conventional lift-off technology, a dielectric was deposited. Then the contact metallization was opened by reactive ion etching. In the final processing step, a gate metallization was deposited on the sidewalls of part of the mesa diodes. An example of a gate controlled diode is seen in Fig. 2 which has an device area of 50x50 μm^2.

The optical properties were assessed by spectrally resolved measurements of the current responsivity and the electroluminescence. A series of detectors varying in area down to 110x160 μm^2 was used with a window in the top metallization for simple access by top illumination. Two response curves of diodes with different cut-off wavelengths at T = 77 K are seen in Fig. 3. The observed current responsivity of about 2 A/W was found to be independent of the diode size. Electroluminescence was observed up to a temperature of 240 K, decreasing in intensity relative to the 10 K value by only a factor of 20.

Fig. 3: Responsivity of two SPSL homojunction diodes with different cut-off wavelengths at T = 77 K.

GATE CONTROLLED DIODES

The results presented in this and the following section were obtained on diodes prepared from a layer structure with a reduced carrier concentration $N_{red} = 3.2 \times 10^{16}$ cm^{-3} and a cutoff wavelength of 7.8 μm. Data were taken on 120x120 μm^2 devices where the gate ring was contacted by wire bonding onto a separate pad.

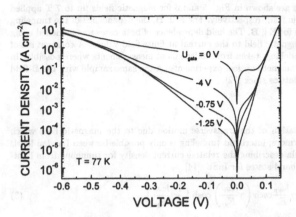

Fig. 4: Current-voltage characteristic for different gate voltages at T = 77 K.

The effect of applying a voltage between the gate and the common p-contact on the current-voltage characteristic of a diode operated at 77 K is shown in Fig. 4. By applying an optimum gate voltage of -1.25 V the dark current was reduced to a minimum. The resulting R_0A product of 720 Ωcm^2 is of the same order for gate-controlled and reference diodes without gate metallization. We therefore attribute the shift of the optimum gate voltage from $U_{gate} = 0$ V to -1.25 V to the work function difference of the gate metallization and the semiconductor. Applying a gate voltage larger than -1.25 V results in an inversion of the p-type region at the surface and the formation of an n-type channel. This field-induced lateral p-n junction is biased by applying a voltage between the p-contact and the top contact which is in ohmic contact with the inversion layer. Esaki tunneling is observed in the forward biased lateral junction resulting in a negative differential resistance in the I-V characteristic of the diode for gate voltages ≥ 0 V. Therefore the p-region of the superlattice as well as the n-type inversion layer must be degenerate. Applying a voltage < -1.25 V, the sidewalls of the n-region of the superlattice are inverted and holes are accumulated. Again, tunneling across this gate-induced lateral junction leads to an increase of the dark current.

From the inspection of the gate voltage dependence of the I-V curves it is possible to separate bulk from surface leakage currents. At large reverse bias voltages all curves merge into one, which is determined by bulk properties. In that region, the current-voltage characteristic is dominated by interband tunneling across the p-n junction which will be studied in detail in the following section.

MAGNETOTUNNELING

To get further insight into the tunneling mechanism we studied the effect of a magnetic field on the tunneling current. The particular device under study has a mesa size of 320x220 μm^2 and was fabricated without gate metallization.

The tunneling current density at zero magnetic field was modelled using the following relationship valid for a simple triangular barrier [10]

$$J_t = \frac{e^3 EV}{4\pi^2 \hbar^2} \sqrt{\frac{2m_t}{E_g}} exp\left(-\frac{4\sqrt{2m_t E_g^3}}{3eE\hbar}\right),\tag{1}$$

where E is the electric field associated with the barrier, E_g the band gap, V the applied voltage, and m_t the tunneling mass which is defined as the reduced mass $1/m_t = 1/m_e + 1/m_h$ of electron mass m_e and hole mass m_h in the direction of the electric field [11].

Assuming an abrupt p-n junction, the dependence of electric field on bias voltage is given by

$$E(V) = \sqrt{1 - \frac{V}{V_{bi}}} E_0,\tag{2}$$

where V_{bi} is the built-in potential and E_0 is the electric field at zero applied voltage. The tunneling contribution at zero magnetic field was modelled using $m_t^* = 0.041$ as determined by magneto-optical experiments [12] and $E_g = 145.8$ meV, measured by photoluminescence. V_{bi} was set to the value of the band gap. The experimental I-V curves can be fitted using $E_0 = 2.6 \times 10^4$ V/cm (see Figs. 5 and 6). Thus interband tunneling in the superlattice diodes can be described using models derived for bulk p-n diodes. The such determined field E_0 is consistent with the value calculated for an abrupt p-n junction, which yields is $E_0 = 3.2 \times 10^4$ V/cm for $N_{red} = 3.2 \times 10^{16}$ cm^{-3} [13].

Magnetic field dependent I-V curves are shown in Fig. 5 and 6 for magnetic fields up to 7 T applied parallel and perpendicular to the electric field, respectively. For $\mathbf{E} \perp \mathbf{B}$, the dependence of the tunneling current on magnetic field is weaker than for $\mathbf{E} \parallel \mathbf{B}$. The field dependence of both cases was modelled using formulas relating the current at zero magnetic field to the current at finite field values. A constant set of input data was used throughout the modelling, taken from independent measurements where possible. In particular, the masses were taken from magneto-optical experiments on the same sample with the B-field parallel and perpendicular to the superlattice stack [12].

Longitudinal magnetic field

$\mathbf{E} \parallel \mathbf{B}$ results in a Landau quantization of the transverse motion due to the magnetic field which increases the effective bandgap. Furthermore, interband tunneling is only possible between Landau levels of same index. Argyres derived a formula describing the relative current density for tunneling from light-hole valence band states to conduction band states for InSb [14]

$$\frac{J_t(B)}{J_t(0)} = \gamma \frac{\hbar \omega_r}{E_g} \mathrm{csch}\left(\gamma \frac{\hbar \omega_r}{E_g}\right) \cosh\left(\frac{1}{2}\gamma \frac{\hbar \omega_r}{E_g}\right), \tag{3}$$

where

$$\gamma = \frac{\pi}{2}\frac{\sqrt{m_\parallel E_g^3}}{\hbar e E}, \quad \omega_r = \frac{eB}{m_\parallel}$$

and $m_\parallel$ is the in-plane mass describing the Landau quantization.

The B-field dependence of the tunneling current has been modelled using Eq. 3. For the reduced mass we take the value determined by magneto-optical measurements [12] which results in $m_\parallel^* = 0.017$. Apart from $m_\parallel$ the same parameters were used as for zero magnetic field, which are given above. Resulting curves are shown in Fig. 5. Even though theory slightly underestimates the measured tunneling currents the experimentally observed effect is well reproduced. The assumptions made while deriving Eq. 3, which is to neglect the tunneling contribution from the spin-orbit split-off band and the heavy-hole valence band, are well justified for the InAs/GaInSb SPSL system. An 8x8 $k\cdot p$-calculation [15] shows that the energy difference between the two uppermost valence bands exceeds the superlattice band gap. Therefore only the uppermost valence band should contribute to the tunneling current.

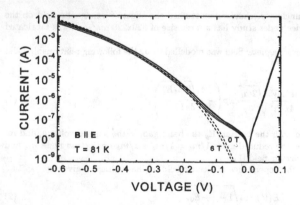

Fig. 5: Experimental current-voltage characteristic at 81 K for magnetic fields of 0, 3, 4.7, and 6 T parallel to the electric field (full lines). Dashed curves are the modelled tunneling contributions using Eq. 1 for the case of zero magnetic field and Eq. 3 for the field dependence. For clarity only curves for 0 and 6 T are plotted.

Transverse magnetic field

For $\mathbf{E} \perp \mathbf{B}$, the influence of both the electric and magnetic field are coupled leading to a much stronger suppression of the tunneling current than for $\mathbf{E} \parallel \mathbf{B}$. This holds in spite of the fact that the B-field induced widening of the SPSL energy gap is smaller for $\mathbf{E} \perp \mathbf{B}$ than for $\mathbf{E} \parallel \mathbf{B}$. Following [16], the electric field in Eq. 1 is transformed according to $E \to E(1 - \gamma^2)^{1/2}$ resulting in the following expression describing the relative current density

$$\frac{J_t(B)}{J_t(0)} = \sqrt{1 - \gamma^2} \exp\left[\lambda\left(1 - \frac{1}{\sqrt{1 - \gamma^2}}\right)\right] \tag{4}$$

where

$$\gamma = \frac{B}{E}\sqrt{\frac{E_g}{2m_\perp}}, \quad \lambda = \frac{4\sqrt{2m_t E_g^3}}{3eE\hbar}.$$

For the reduced effective mass $m_\perp^*$ we take the value of 0.041 as determined by magneto-optical experiments [12]. For $\gamma < 1$, a Lorentz transformation can be performed into a coordinate system where the magnetic field vanishes and only an effective electric field governs the motion of electrons and holes [17]. Likewise for $\gamma > 1$, a transformation can be carried out resulting in a zero electric field for which tunneling is expected to vanish. In this case no real solution of Eq. 4 exists.

The dependence of the tunneling current on magnetic field was modelled using Eq. 4. The experimentally observed strong effect of applied magnetic field on the tunneling contribution is reproduced semi-quantitatively as can be seen in Fig. 6. The discrepancy between experiment and theory results primarily from the predicted sharp decrease of the tunneling current for voltages approaching a critical maximum value, for which $\gamma > 1$ and thus tunneling is not allowed. Zav'ialov and Radantsev [16] observed similarly an unexpectedly high tunneling current density in their experiments on Pb-p-HgCdTe Schottky barriers, which they attributed to non-resonant scattering during the tunneling process. As a result, tunneling should be possible at any magnetic field strength, even at fields resulting in $\gamma > 1$. However, it is beyond the scope of the present paper to include defect assisted scattering in the tunneling model.

An alternative explanation to be considered is that interband tunneling may indeed be fully suppressed by the magnetic field as predicted by theory and a different current mechanism is dominant. In the above section we were able to distinguish at least two mechanisms contributing to the reverse bias current. A close inspection of Fig. 6 reveals that the low bias current component extends further and further towards larger reverse bias voltages with increasing magnetic field. This contribution may be due to surface induced tunneling which was discussed above. Further investigations are required to decide which explanation is correct.

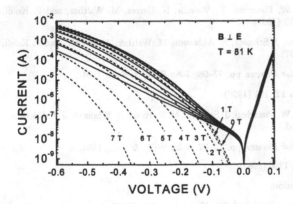

Fig. 6: Experimental current-voltage characteristic at 81 K for magnetic fields of (0–7) T perpendicular to the electric field (full lines). Dashed curves are the modelled tunneling contributions using Eq. 1 for the case of zero magnetic field and Eq. 4 for the field dependence.

127

SUMMARY

The I-V characteristics of InAs/GaInSb SPSL IR photodiodes were studied using mesa diodes with an additional gate electrode on the mesa sidewalls. Gate voltage dependent measurements show additional surface induced tunneling currents, which can be varied over several orders of magnitude by changing the gate voltage. At large reverse bias voltages the leakage currents are dominated by band-to-band tunneling currents. Magnetic fields up to 7 T were applied leading to a reduction of the tunneling currents. This effect was found to depend strongly on the orientation of the magnetic field. For B parallel to the electric field across the p-n junction, the reduction in tunneling current is mainly caused by the widening of the band gap due to the Landau quantization of electrons and holes with cyclotron orbits lying in the layer plane. For $E \perp B$ the suppression of the tunneling currents is stronger because of the B-field induced change of the momentum of the carriers during the tunneling process.

ACKNOWLEDGEMENTS

The technical assistance of H. Güllich, M. Kaufmann, J. Linsenmeier, J. Schleife, and K. Schwarz is gratefully acknowledged. The authors would further like to thank J. Wagner for his valuable contributions. The work was financially supported by the Bundesministerium für Verteidigung.

REFERENCES

1. D. L. Smith and C. Mailhoit, J. Appl. Phys. **62**(6), 2545 (1987).

2. D. H. Chow, R. H. Miles, and C. W. Nieh, J. Cryst. Growth **111**, 883 (1991).

3. J. L. Johnson, L. A. Samoska, A. C. Gossard, J. L. Merz, M. D. Jack, G. R. Chapman, B. A. Baumgratz, K. Kosai, and S. M. Johnson, J. Appl. Phys. **80**(2), 1116 (1996).

4. F. Fuchs, W. Pletschen, U. Weimar, J. Schmitz, M. Walther, J. Wagner, and P. Koidl, Proc. 8th Int. Conf. On Narrow Gap Semiconductors, Shanghai (1997), in press.

5. M. B. Reine, A. K. Sood, and T. J. Tredwell, vol. 18 of *Semiconductors and Semimetals*, pp. 201–311, Academic Press, 1981.

6. R. Adar, Y. Nemirovsky, and I. Kidron, Solid-State Electron. **30**(12), 1289 (1987).

7. G. Tuttle, H. Kroemer, and J. H. English, J. Appl. Phys. **65**(12), 5239 (1989).

8. J. Wagner, J. Fuchs, J. Schmitz, W. Pletschen, U. Weimar, N. Herres, M. Walther, and P. Koidl, Electrochem. Soc. Proc. **97**(21), 171 (1997).

9. F. Fuchs, U. Weimar, W. Pletschen, J. Schmitz, E. Ahlswede, M. Walther, J. Wagner, and P. Koidl, Appl. Phys. Lett. **71**(2), 3251 (1997).

10. S. M. Sze, *Physics of Semiconductor Devices*, pp. 97–98, John Wiley & Sons, 1981.

11. E. O. Kane, J. Phys. Chem. Solids **12**, 181 (1959).

12. F. Fuchs, E. Ahlswede, U. Weimar, W. Pletschen, J. Schmitz, M. Hartung, A. Wixforth, J. P. Kotthaus, and F. Szmulowicz, to be published.

13. S. M. Sze, *Physics of Semiconductor Devices*, pp. 74–77, John Wiley & Sons, 1981.

14. P. N. Argyres, Phys. Rev. **126**(4), 1386 (1962).

15. F. Szmulowicz, private communication.

16. V. V. Zav'ialov and V. F. Radantsev, Semicond. Sci. Technol. **9**, 281 (1994).

17. A. G. Aronov and G. E. Pikus, Sov. Phys. JEPT **24**(1), 188 (1967).

MID-IR PHOTODETECTORS BASED ON InAs/InGaSb TYPE-II QUANTUM WELLS

G. J. BROWN*, M. AHOUJJA*, F. SZMULOWICZ*, W. C. MITCHEL * and C. H. LIN**
*Air Force Research Laboratory, Materials Directorate, Wright Patterson AFB, OH 45433-7707
**Space Vacuum Epitaxy Center, University of Houston, Houston, TX

ABSTRACT

We report on the growth and characterization of $InAs/In_xGa_{1-x}Sb$ strained-layer superlattices (SLS) designed with a photoresponse cut-off wavelength of $10\mu m$. The structural parameters, layer thicknesses and compositions, were chosen to optimize the infrared absorption for a superlattice with an energy band gap of 120 meV. The energy band structure and optimized absorption coefficient were determined with an 8x8 envelope function approximation model. The superlattices were grown by molecular beam epitaxy and were comprised of 100 periods of 43.6Å InAs and 17.2Å $In_{.23}Ga_{.77}Sb$ lattice-matched to the GaSb substrates. In order to reduce the background carrier concentrations in this material, superlattices grown with different substrate temperatures were compared before and after annealing. This set of superlattice materials was characterized using x-ray diffraction, photoresponse and Hall measurements. The measured photoresponse cut-off energies of 116 ± 6 meV is in good agreement with the predicted energy band gap for the superlattice as designed. The intensity of the measured mid-infrared photoresponse was found to improve by an order of magnitude for the superlattice grown at the lower substrate temperature and then annealed at 520 °C for 10 minutes. However, the x-ray diffraction spectra were very similar before and after annealing. The temperature dependent Hall measurements at low temperatures (<25K) were dominated by holes with quasi two-dimensional behavior. An admirably low background carrier concentration of 1×10^{12} cm^{-2} was measured at low temperature.

INTRODUCTION

Since the first calculations of Smith and Mailhiot[1], type-II superlattices of $InAs/In_xGa_{1-x}Sb$ have shown great promise as a III-V alternative to the conventional HgCdTe alloys for infrared imaging applications.[2] These superlattices have absorption coefficients comparable to HgCdTe alloys, but are predicted to have other superior properties such as long carrier lifetimes.[3] The wavelength tunability of these superlattices combined with the suppression of band to band Auger recombination are particularly advantageous for infrared detection at long to very long wavelengths. In addition, the molecular beam epitaxy (MBE) technology used to grow these structures promises to yield uniform and structurally robust detector arrays.

In last the five years, significant progress has been made in the growth of these superlattices by MBE.[2,4,5] One of the materials issues in these SLS materials has been the high background carrier concentration in these nominally undoped materials. Identifying the source of these charge carriers and eliminating them through improved growth techniques is a necessary step toward increasing the photodetection in these materials. It is easiest to examine the effects of these background carriers on the photoresponse by studying the SLSs as photoconductors rather than

129

as photodiodes. Hall measurements can also be directly made on these simple photoconductive superlattices, rather than on the more complex photodiode heterostructures. The initial results of our efforts to design, grow, and characterize these basic superlattices are covered in this paper.

THEORY

Because of the number of design parameters (layer widths, indium composition, substrate orientation, interface type, etc.) available to tailor the properties of these superlattices (SLs), theoretical models provide essential guidance for the experimental effort. In order to design optimized SLs responding in different wavelength windows, a quick but accurate EFA model for calculating the band gaps of InAs/In$_x$Ga$_{1-x}$Sb SLs and the oscillator strengths at the center of Brillouin zone was implemented.[6] This model employs a 3x3 linear in k model for the C, LH, and SO bands and a quadratic in k 1x1 model for the HH band. As input into the model, the band offsets are obtained from experiment[7] and the model-solid theory of van de Walle[8] , and other parameters are from tables. Strain effects were taken into account via the Bir-Picus deformation potential theory.

This model was then used to design structures with optimum absorption for a given detection threshold.[6] For example, Figure 1 shows the results of such a calculation for a 10 micron structure, where the solid curves give the values of (InAs, InGaSb) widths for which the band gap is constant at 10 microns. The dashed curves show the corresponding oscillator strengths, which can be used to identify optimized structures. Circles indicate when the SL is lattice-matched to the GaSb substrate. In order to maintain a minimum net strain and to keep the InGaSb layers greater than a few monolayers, the best choice is for an indium content near 20%.

The calculation of the complete optical response of SLs requires the knowledge of their electronic structure throughout the Brillouin zone. To this end, a new coupled-band envelope function approximation (EFA) formalism was developed.[9] This model was used to calculate the linear absorption coefficient spectrum of superlattices of interest, including the contributions from both the heavy-hole and light-hole bands.

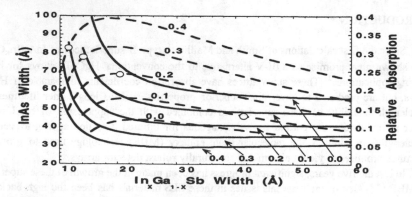

Figure 1. Solid curves trace pairs of (In$_x$Ga$_{1-x}$Sb/InAs) widths yielding a 10 μm bandgap for different x. Dashed curves are the corresponding relative absorption (right scale) versus InGaSb widths. The circle indicates lattice match to GaSb substrate.

EXPERIMENT

The InAs/InGaSb type-II superlattice photoconductors were grown on p-type (001) GaSb substrates in a Riber 32 molecular beam epitaxy (MBE) system, equipped with EPI valved cracker As and P cells, an EPI Sb cracker cell, EPI In and Ga SUMO cells, an EPI Be dopant cell, and Riber Al and Si 35 c.c. effusion cells. The active regions of samples #2218 and #2219 (#2228) were composed of 100 (150) periods of unintentionally doped 43.6Å InAs/17.2Å $In_{0.23}Ga_{0.77}Sb$ type-II SLs. The GaSb buffer layer was grown at 500 °C, while the active region was grown at 450 °C for sample #2219 and at 390 °C for samples #2218 and #2228. The annealing was performed under As overpressure at 510 ± 10 °C for 10 minutes in the MBE chamber. The substrate temperature was calibrated by the InSb melting temperature (525 °C) and GaSb oxide desorption temperature (580 °C). The shutter sequences to control the SL interfaces have been discussed in detail in Ref. 10. During growth, the GaSb and InGaSb layers displayed excellent 1x3 reflection high-energy electron diffraction (RHEED) patterns while the InAs layers exhibited good 2x1 RHEED patterns. The growth rates were calibrated to within ± 2% using RHEED on test samples and were confirmed by the double-crystal x-ray (DXRD) measurements on test samples composed of many periods of 300Å GaAs/30Å AlAs, 80Å GaAs/160Å AlAs, or 300Å InAs/20Å GaSb SLs. From the DXRD spectra on bulk materials, the Sb background in InAs was about 0.7 % and the As background in GaSb was about 1.3 % with the As valve open, while the growth rates for the InAs and GaSb were the same, < 0.7 monolayer per second.

Electron transport properties of this set of InAs/InGaSb SLs were studied using temperature dependent as well as variable magnetic field Hall measurements. Indium solder was used for contacts to the van der Pauw samples. The sample temperature was estimated at 200 °C during contact application. The temperature dependent Hall system measured the horizontal areal resistivity, carrier concentration and mobility of the superlattices from 10K to 300K. The variable magnetic field Hall measurements were performed at fixed temperatures to determine the mobility spectrum of the material as a function of magnetic field.

For spectral photoresponse measurements, a BIO-RAD Fourier Transform spectrometer is used, covering the mid-infrared range from 2 to 50 micrometers. The sample temperature is varied from 8K to room temperature with the use of a helium gas closed-cycle coldhead. The test samples have a cleaved surface area of 6 mm by 6 mm. Indium strip metallization is used to attach gold wires near the edges of the top surface. The samples are run in a constant current-bias mode with a typical bias current of 0.5 mA. The samples are mounted at normal incidence in the infrared beam and the light enters through the non-metallized portion of the top superlattice surface.

RESULTS

Photoresponse spectra were obtained on five different samples in this study: #2218 (as grown and annealed), #2219 (as grown and annealed), and #2228. Although the growth conditions were varied between these samples, the same 120 meV SL bandgap design was maintained. Only minor shifts in the measured band edge were noted as seen in Figure 2. The average measured cut-off energy, where the photoresponse intensity drops to 50%, was 116 ± 6 meV. This demonstrates good repeatability of the photoresponse cut-off wavelength in these samples, and good agreement with the designed SL band gap. The SL period of these samples, determined by

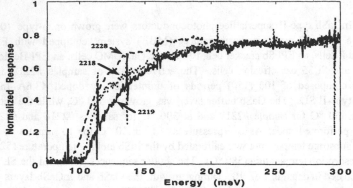

Figure 2. Comparison of the normalized photoresponse (T = 10K) for three samples with the same design parameters: 43.6Å InAs, 17.2Å $In_{0.23}Ga_{0.77}Sb$ but different MBE growth conditions.

DXRD, was on the average 59.6Å, versus an "ideal" period of 60.8Å. The measured superlattice periods ranged from 58.72 to 61.18 Å, which is a variation of less than one monolayer.

Between samples #2218 and #2219, the main growth change was the growth temperature of the SLs, i.e. 390 °C versus 450 °C. The photoresponse of these samples was nearly identical in intensity, but the band edge was much sharper in energy for #2218. This may indicate sharper interfaces were formed in this sample. The most remarkable difference between these samples occurred after a short post-growth anneal. After annealing, the photoresponse for #2218 increased by nearly an order of magnitude. However, the photoresponse of #2219 remained the same before and after annealing. Clearly, the annealing was very beneficial for the SL grown at lower temperature. The x-ray diffraction spectra were very similar before and after annealing. And, the SL band edge remained sharp and only shifted by about 5 meV toward higher energy after annealing.

We performed Hall measurements to see if the increased photoresponse could be correlated with the measured background carrier concentrations in these samples. The Hall concentration and mobility curves are shown in Figure 3. For all the samples, the low temperature (T<25 K) carrier concentrations were around 1×10^{12} cm^{-2}. The Hall concentration is constant at low temperatures and increases sharply at about 25 K, indicating the onset of the intrinsic conduction in GaSb substrates. This is one of the drawbacks in making Hall measurements on samples with conducting substrates. The results for a GaSb substrate are therefore included for comparison. It should be noted that the mobility of the GaSb substrate is much lower than those of the SLs at low temperatures. So, in this temperature range, the electrical conduction should be dominated by the SLs rather than the GaSb where the conduction is low.

For sample #2218, before and after annealing, the hole sheet carrier concentration remained nearly the same. After annealing, the in-plane hole mobility decreased slightly. This slightly lower mobility is probably a factor in boosting the sample's resistance, and hence photoresponse intensity. However, the differences aren't large enough to account for a factor of eight increase in photodetected signal. The expected increase in resistivity based on the 10% reduction in the mobility and hole concentration is about 50%.

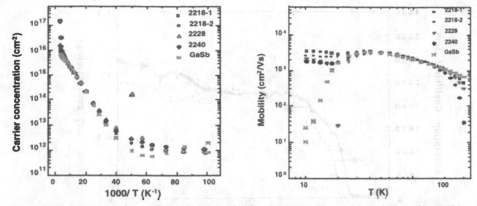

Figure 3. Temperature dependent Hall results for the charge carrier concentrations and mobilities. The designation 2218-1 is for the as-grown superlattice, and 2218-2 is the annealed sample.

For the next SL (#2228), the As valve was reduced to 5 mils after InAs growth to reduce the As background in the InGaSb layer. The cracker temperature of the Sb was slightly lowered to reduce impurities. This lower Sb cracker temperature may also reduce the InSb-like interface bonds. The number of SL periods was increased to 150 to enhance the infrared absorption. Despite these changes, the photoresponse intensity was slightly lower in #2228 compared to #2218 after annealing. This reflects the slightly lower sample impedance of #2228. However, these growth changes did create a significant difference in the temperature dependence of the photoresponse signal. For #2228, the signal dropped by an order of magnitude between 10 and 60 K. For #2218, an order of magnitude drop in signal had occurred by only 30K.

There was no similar difference in the temperature dependence between these two samples noted in the Hall results (see Fig. 3). However, #2228 was unique in that it changed to n-type at low temeperature while #2218 remained p-type. Figure 4 shows the mobility spectrum results of sample 2228 at 10 K. In the mobility spectrum we see two peaks, one is due to the InAs layers and the other due to the InGaSb layers. The overall conductivity due to InAs layers is slightly higher than that due to InGaSb layers and hence the n-type nature observed at low temperatures.

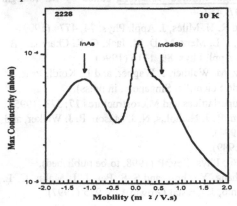

Figure 4. Maximum conductivity as a function of discrete mobility in sample 2228. The negative mobility axis corresponds to electron mobility and positive x-axis corresponds to hole mobility. The arrows indicate the peaks that correspond to the respective conducting carriers in the SLS.

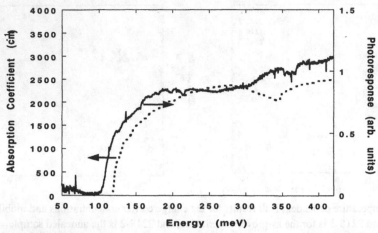

Figure 5. The normal-incidence linear absorption coefficient for a (43.6 Å/17.3 Å) InAs/InGaSb, x = 0.23, SL on GaSb (dashed curve) and the measured photoresponse of sample #2218.

CONCLUSIONS

We have designed, grown and tested a set of high quality InAs/InGaSb superlattices. The SL structural parameters were very repeatable between samples as evidenced by the consistency of the SL periods and the long wavelength photoresponse cut-off. The measured photoresponse spectra were in excellent agreement with the calculated absorption spectrum, as seen in Figure 5. Very low background carrier concentrations were achieved in this sample set, and clear separation of the electrons and holes in their respective SL layers was observed in the mobility spectrum. A short anneal at 510 °C improved the SL response without smearing out the infrared band edge.

REFERENCES

1. D. L. Smith and C. Mailhiot, J. Appl. Phys. **62**, 2545 (1987).
2. R. H. Miles and D. H. Chow, Long Wavelength Infrared Detectors, edited by M. Razeghi (Gordon and Breach, Philadelphia, 1996), pp. 397- 452.
3. P. M. Young, C. H. Grein, H. Ehrenreich, and R. H. Miles, J. Appl. Phys. **74**, 4774 (1993).
4. J. L. Johnson, L. A. Samoska, A. C. Gossard, J. L. Merz, M. D. M. Jack, G. R. Chapman, B. A. Baumgratza, K. Kosai, and S. M. Johnson, J. Appl. Phys. **80**, 1116 (1996).
5. F. Fuchs. W. Pletschen, U. Weimar, J. Schmitz, M. Walther, J. Wagner, and P. Koidl, Proc. 8th Int. Conf. on Narrow Gap Semiconductors, (World Scientific, Singapore, in press).
6. F. Szmulowicz, E. R. Heller, and K. Fisher, Superlattices and Microstructures **17**, 373 (1995).
7. D. M. Symmons. M. Lakrimi, R. J. Warburton, R. J. Nicholas, N. J. Mason, P. J. Walker, and M. I. Eremets, Semicond. Sci. Technol. **9**, 118 (1994).
8. C. G. Van de Walle, Phys. Rev. **B 39**, 1871 (1989).
9. F. Szmulowicz, Phys. Rev. **B 54**, 11539 (1996); Phys. Rev. B (1998, to be published).
10. C.-H. Lin, S. J. Murry, D. Zhang, P. C. Chang, Y. Zhou, and S. S. Pei, J. I. Malin, C. L. Felix, J. R. Meyer, C. A. Hoffman, and J. F. Pinto, J. Crystal Growth **175**, 955 (1997).

PROGRESS ON GaInAsSb AND InAsSbP PHOTODETECTORS FOR MID-INFRARED WAVELENGTHS

Z.A. SHELLENBARGER, M.G. MAUK, P.E. SIMS, J.A. COX, J.D. LESKO, J.R. BOWER, J.D. SOUTH*, and L.C. DINETTA
AstroPower Inc., Solar Park, Newark, DE 19716-2000
*Materials Science Department, University of Delaware, Newark, DE 19717

ABSTRACT

Progress on mid-infrared photodetectors fabricated by the liquid phase epitaxial growth of GaInAsSb, InAsSbP, and AlGaAsSb on GaSb and InAs substrates is reported. GaInAsSb p/n and p-i-n detectors, InAsSbP p/n detectors and AlGaAsSb/GaInAsSb avalanche photodiode (APD) structures were fabricated. Preliminary results indicate that these devices can have higher detectivity with lower cooling requirements than commercially available detectors in the same wavelength range. Infrared p/n junction detectors made from GaInAsSb and InAsSbP showed cut-off wavelengths of 2.3 μm and 2.8 μm respectively. Room temperature background noise-limited detectivity (D*$_{BLIP}$) of 4×10^{10} cmHz$^{1/2}$/W for GaInAsSb detectors and 4×10^{8} cmHz$^{1/2}$/W for InAsSbP was measured. Room-temperature avalanche multiplication gain of 20 was measured on AlGaAsSb/GaInAsSb avalanche photodiodes.

INTRODUCTION

The main objective of this work is to further develop liquid-phase heteroepitaxy of lattice-matched GaInAsSb[1-4] and InAsSbP[4-6] on GaSb and InAs substrates in developing high performance, room temperature, mid-infrared detectors. These quaternary alloy systems will provide room-temperature IR detectors adjustable over the 1.7 to 4.5 μm wavelength range. In this work, detectors with response out to 2.8 μm were demonstrated. With proper device design, both the cut-on and cut-off wavelengths can be tailored. A significant advantage of using quaternary systems over binary and ternary compounds is the ability to "tune" the bandgap while still providing lattice-matched growth to the substrate material. Another expected advantage is that this detector structure will have good detectivity at room temperature, while most commercially available detectors need cooling to liquid nitrogen temperatures. Photodetectors operating in the mid-IR range will have a variety of commercial applications in air pollution monitoring, industrial process control, automobile emission monitoring, and future lightwave communication systems using novel fiber materials.

EPITAXIAL GROWTH AND DEVICE FABRICATION

Device layers of GaInAsSb, InAsSbP, and AlGaAsSb were grown by liquid phase epitaxy (LPE) to fabricate devices. A standard sliding graphite boat system was used. The different detector structures are detailed in the next section. Several solid-compositions for these detector materials were explored. The compositions used in this work and their growth parameters are given in Table I. The GaInAsSb and AlGaAsSb layers were grown on 500 μm thick, chemically polished, (100) oriented, n-type GaSb wafers doped to 3-5 x 10^{17} cm^{-3} with tellurium, while the InAsSbP layers were grown on 500 μm thick, chemically polished, (100) oriented, n-type InAs wafers doped to 2-3 x 10^{18} cm^{-3} with sulfur. InAsSbP layers were also grown on GaSb substrates, but dissolution of the substrate caused poor quality layers not suitable for device fabrication. Starting melt compositions and temperatures for the growth of the epitaxial layers were determined from published liquidus phase equilibria data[7-9] and experimentation. The metal components of the melt were added as high purity (99.9999%) gallium, indium, and antimony shot pieces and aluminum wire. The arsenic was added as either undoped GaAs or InAs polycrystalline material. Phosphorous was added as undoped InP polycrystalline material. Prior to growth, the metal components of the melts were heated to 700 °C for fifteen hours under flowing hydrogen to de-oxidize the metallic melt components and outgas residual impurities. After cooling the melt 7 to 15 °C below the equilibrium temperature, the substrate was contacted with the melts at a cooling rate of 1 °C/min. to grow the device layers. Growth rates were typically 2-3 µm/min.

Table I. Growth parameters for epitaxial layers.

solid composition	liquid composition						growth temp (°C)	bandgap (eV)
	X^L_{Al}	X^L_{Ga}	X^L_{In}	X^L_P	X^L_{As}	X^L_{Sb}		
$Ga_{0.85}In_{0.15}As_{0.17}Sb_{0.83}$		0.19	0.59		0.01	0.21	515	0.53
$InAs_{0.61}Sb_{0.18}P_{0.21}$			0.589	0.00088	0.01	0.40	565	0.44
$Al_{0.28}Ga_{0.72}As_{0.015}Sb_{0.985}$	0.015	0.957			0.001	0.028	515	1.10

The compositions of the epitaxial layers given in Table I were determined by electron microprobe analysis using wavelength dispersive spectroscopy (WDS). The compositions are very close to what was expected except for the InAsSbP which had a higher As concentration than expected. This may be due to the layer being so thin that the electron beam penetrated into the InAs substrate skewing the measurement towards higher As concentration. The quality of the device layers was assessed using optical and scanning electron microscopy. Good quality layers exhibited no cross-hatched pattern associated with lattice-mismatch of the substrate and epitaxial layers. A 5 K photoluminescence (PL) spectrum for the AlGaAsSb material is shown in Figure 1. FWHM is 53 meV. A longer wavelength detector was not available on this system for PL of the GaInAsSb and InAsSbP materials.

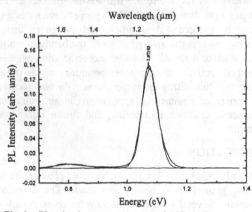

Wavelength (µm)

Energy (eV)

Fig. 1. Photoluminescence spectrum of AlGaAsSb.

For doping of the GaInAsSb and AlGaAsSb, tellurium (Te) was used as an n-type dopant and germanium (Ge) was used as a p-type dopant. For InAsSbP, zinc (Zn) was used as a p-type dopant. Absorbing layers for detectors were not intentionally doped. Some absorbing layer melts had small amounts of the rare earth elements gadolinium (Gd) or ytterbium (Yb) added in order to reduce the background carrier concentration by gettering impurities. Doping levels for GaInAsSb layers using Gd were reduced from 7-8 x 10^{15} cm^{-3} down to 1-3 x 10^{15} cm^{-3} as determined by capacitance-voltage measurements. Secondary ion mass spectroscopy (SIMS) measurements were also used to measure the Te and Ge concentrations of epitaxial layers.

Mesa photodiodes with 200 µm diameter active area were formed using photolithography and chemical etching. Metallization for back n-type contacts was planar Au/Sn while front p-type contacts were annular Au/Zn. A spin-on, photosensitive polyimide layer was deposited and patterned before front contact deposition. This had several functions including planarizing the surface for the front contact deposition, providing insulation of the junction, and passivating the edges of the device area. After the photodiodes were formed, the substrate was diced into 1 mm² pieces with a single device in the middle of each square. These were mounted to TO-18 headers using silver conducting epoxy and wirebonded from the bonding pad of the front contact to the header post. The devices were encapsulated with epoxy (Epo-Tek 301) to protect the device surface and wirebond connection as well as to passivate and reduce reflection from the device surface. As explained later, the transmission of this epoxy was not optimal for the wavelengths of interest.

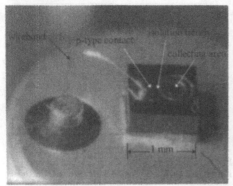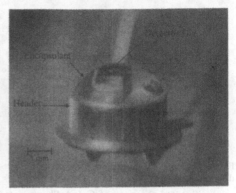

Fig. 2. Packaged detector structures.

DEVICE RESULTS

A computer controlled spectrometer was used to measure spectral response versus wavelength. The light source is a 1000 W tungsten halogen lamp. A calibrated pyroelectric detector with a calcium fluoride window is used as a reference. This detector has flat response in the wavelength range from 200 nm to over 10 μm. An x-y stage with positional control in the 5 micron range is used to ensure that response of the test detector and reference detector are measured in the same position. Capacitance-voltage and dark current-voltage measurements were also taken. Background noise limited detectivity was calculated using the equation[10]

$$D^*_{BLIP} = \frac{S(\lambda) \cdot (R_0 A)^{1/2}}{2(kT)^{1/2}}$$

where $S(\lambda)$ is the peak spectral response, R_0 is the zero bias resistance, A is the area of the diode, k is the Boltzmann constant and T is the temperature. The zero-bias resistance is obtained from the I-V measurement with $R_0 = dv/di$ at 10 mV.

GaInAsSb and InAsSbP p/n and p-i-n Detectors

The structure of GaInAsSb p/n and p-i-n detectors is shown in Figure 3. Structures fabricated as p/n detectors consisted of an undoped, 3 μm thick, n-GaInAsSb layer followed by a Ge doped, 0.5 μm, p-GaInAsSb layer. Structures fabricated into p-i-n detectors consisted of a Te doped, 0.5 μm, n-GaInAsSb layer, an undoped, 3 μm, i-GaInAsSb layer, and a Ge doped, 0.5 μm, p-GaInAsSb layer. Doping levels were approximately 1×10^{18} cm⁻³ for Ge and Te doped layers and 5×10^{15} cm⁻³ for undoped layers. The GaInAsSb/GaSb structure can be fabricated as either frontside or backside illuminated detectors. In this way, the device can be designed as either a broad or narrow band detector. For the narrow band detector, the cut-on wavelength of

1.7 μm is determined by the GaSb substrate while the cut-off wavelength is determined by the GaInAsSb active layer. One possible advantage of the backside illuminated design is that surface recombination effects caused by illumination through the top of the device are avoided, causing more light to be absorbed within the depletion region. A possible disadvantage of the backside-illuminated design is that the efficiency of the device can be lowered by free carrier absorption in the substrate and filter layers. Reductions in the carrier concentration and thickness of the substrate should be able to alleviate this problem. The spectral response of frontside and backside illuminated detectors fabricated from the same GaInAsSb/GaSb growth is shown in Figure 4.

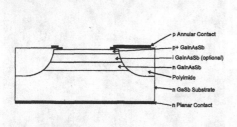

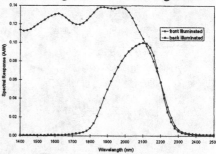

Fig. 3. Structure of GaInAsSb/GaSb p/n and p-i-n detectors.

Fig. 4. Spectral response of frontside and backside illuminated GaInAsSb on GaSb p/n detector structures.

Spectral response of GaInAsSb p/n detectors with and without encapsulant is shown in Figure 5. This spectral response indicates a quantum efficiency near 100% for encapsulated devices near 2 μm. The large increase in response seen for devices with encapsulant is likely due to anti-reflection properties of the encapsulant as well as possible light concentration due to the domed shape of the encapsulant. Absorption of the encapsulant at wavelengths longer than 2 μm causes the peak response to shift from 2.06 μm for non-encapsulated to 1.94 μm for encapsulated devices. The large dip in response of the encapsulated device may also be due to absorption of the encapsulant. Figure 6 shows spectral response for a p-i-n detector with the encapsulant.

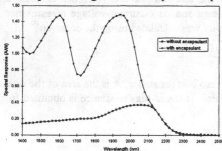

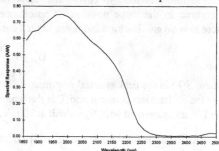

Fig. 5. Spectral response of GaInAsSb p/n detectors.

Fig. 6. Spectral response of GaInAsSb p-i-n detector.

Detector structures were fabricated from InAsSbP on InAs growths in a manner similar to that described for GaInAsSb. Structures fabricated into p/n detectors consisted of a undoped, 3 μm thick, n-InAsSbP layer followed by a zinc doped, 0.5 μm thick, p-InAsSbP layer. The structure of these devices is shown in Figure 7 while the spectral responses for encapsulated and non-encapsulated devices are shown in Figure 8. The absorption of the encapsulant at wavelengths longer than 2 μm is obvious in these devices. The results of measurements on

GaInAsSb and InAsSbP detectors is summarized in Table II. Reverse saturation current is measured at -0.1 V reverse bias.

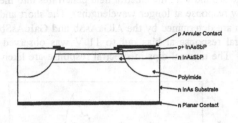

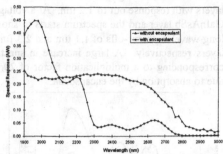

Fig. 7. Structure of InAsSbP/InAs p/n detectors.   Fig. 8. Spectral response of InAsSbP detectors.

Table II. Measurement results on GaInAsSb and InAsSbP detectors.

material	structure	encapsulant	λ_{peak} (µm)	λ_{cutoff} (µm)	SR (A/W)	I_{sat} (µA)	R_0A (Ωcm^2)	D^*_{BLIP} (cmHz$^{1/2}$/W)
GaInAsSb	p/n	no	2.06	2.27	0.37	0.8	40.8	1.8×10^{10}
GaInAsSb	p/n	yes	1.94	2.20	1.49	2.0	14.1	4.3×10^{10}
GaInAsSb	p-i-n	yes	1.98	2.24	0.75	1.0	28.6	3.1×10^{10}
InAsSbP	p/n	no	2.4	2.8	.024	865	0.045	4.0×10^{8}

GaInAsSb Avalanche Photodiodes

More recently, the fabrication of GaInAsSb avalanche photodiodes (APDs) has been investigated. The structure is a separate absorption and multiplication (SAM) APD. The SAM-APD has proven to be an effective means of producing gain in the absence of large dark current[11]. Figure 9 shows the design of the IR SAM-APD. For the IR SAM-APD shown, photo-absorption occurs in the narrow bandgap GaInAsSb layer. The photogenerated carriers are subsequently swept into the wide bandgap AlGaAsSb layer where multiplication occurs. In this structure, the dark current that undergoes avalanche gain should be small. For avalanche breakdown, the breakdown voltage increases with increasing temperature[12]. Therefore, temperature dependent dark I-V measurements are usually used to determine which breakdown mechanism is dominant. The temperature dependent I-V curves showed a definite increase in breakdown voltage with increased temperature as shown in Figure 10, indicating that the devices were definitely avalanche photodiodes. Breakdown voltages as high as 15 V were measured.

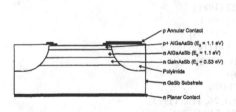

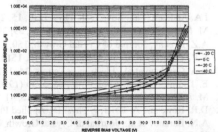

Fig. 9. Structure of AlGaAsSb/GaInAsSb SAM-APD.   Fig. 10. Temperature dependent I-V for SAM-APD.

139

The spectral response under different values of reverse bias voltage of the SAM-APD device is shown in Figures 11. At zero bias, the photosensitivity is determined by the AlGaAsSb layers with response out to 1.2 μm. At a voltage around 3 V, the electric field penetrates into the GaInAsSb layer and the spectrum starts to show response at longer wavelengths. The short and long wavelength cut-offs of 1.1 μm and 2.3 μm are determined by the AlGaAsSb and GaInAsSb layers respectively. A large increase in spectral response at bias out to 11 V was observed, corresponding to a multiplication factor of 20. The unusual dips in spectral response are likely due to absorption by the encapsulant.

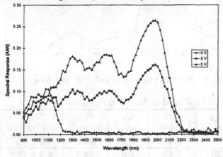 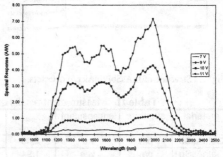

Fig. 11. Spectral response under different values of reverse bias voltage for the SAM-APD.

CONCLUSION

Progress on mid-infrared photodetectors fabricated by the liquid phase epitaxial growth of GaInAsSb and InAsSbP on GaSb substrates was reported. Both p/n junction and avalanche photodiode (APD) structures were fabricated. Preliminary results indicate that these devices can have higher detectivity with lower cooling requirements than commercially available detectors in the same wavelength range. Future plans for these devices include better encapsulation, optimization of device layer doping levels and thicknesses, addition of window layers and other surface passivation methods, and incorporation of anti-reflection coatings.

ACKNOWLEDGMENTS

We gratefully acknowledge the support of the National Aeronautics and Space Administration under contract number NAS5-97005 and the Department of the Air Force under contract number F09630-96-C-0051.

REFERENCES

1. I.A. Andreev et al., Sov. Tech. Phys. Lett., **15**, 253 (1989).
2. M. Mébarki et al., Solid-State Electronics, **39**, 39 (1996).
3. A.Z. Li et al., Journal of Crystal Growth, **150**, 1375 (1995).
4. Y.P. Yakovlev et al., Proceedings of SPIE, **1510**, 170 (1991).
5. R.A. Garnham et al., Electronics Letters, **24**, 1416 (1988).
6. E. Tournie, et al., J. Appl. Phys., **68**, 5936 (1990).
7. M. Astles, et al., J. Electronic Materials, **15**, 41 (1986).
8. D.R. Rowe and A. Krier, J. Phys. D: Applied Phys, **26**, 1103 (1993).
9. A.N. Baranov et al., Journal of Crystal Growth, **66**, 547 (1984).
10. X.Y. Gong et al., Jpn. J. Appl. Phys., **36**, 2614 (1997).
11. K. Taguchi, Y. Matsumoto, and K. Nishida, Electronics Letters, **15**, 453 (1979).
12. S.M. Sze, Physics of Semiconductor Devices, (NY: John Wiley & Sons 1981) p.766-774.

Part III

Innovative IR Devices

NON-EQUILIBRIUM InSb/InAlSb DIODES GROWN BY MBE

A D JOHNSON, A B J SMOUT, J W CAIRNS, G J PRYCE, A J PIDDUCK, R JEFFERIES,
T ASHLEY & C T ELLIOTT

DERA, St. Andrews Road, Great Malvern, WR14 3PS, UK
adjohnson@dera.gov.uk

ABSTRACT

The application of non-equilibrium transport techniques to Molecular Beam Epitaxy (MBE) grown InSb/InAlSb heterostructure diodes has produced practical devices such as midinfrared LED's and negative luminescent sources that operate at room temperature. By extending the epitaxial growth to vicinal InSb substrates it has been demonstrated that the temperature window for high quality epitaxy can be lowered by ~120°C, giving greatly improved epilayer morphology. The degree of misorientation needed for given growth temperatures is shown from Atomic Force Microscope (AFM) measurements to be only ~2°. In addition, the lower growth temperature gives improved dopant activation, lower trap densities and lower reverse bias leakage currents, with consequent benefits to device performance.

INTRODUCTION

Indium Antimonide (InSb) has the narrowest forbidden energy gap of the binary III-V semiconductors with a cut-off wavelength corresponding to 7µm at room temperature and 5.5µm at 77K. It has therefore traditionally found uses in infrared photon detection. More recently, InSb has attracted a lot of attention due to its unique material properties which are a result of its narrow energy gap. A low electron effective mass and very high saturation velocity offer the opportunity of very high speed, low power and low noise electronic devices and its narrow bandgap offers the possibility of emitters in the mid infrared waveband that have applications in such areas as gas sensing and spectroscopy. However, the large intrinsic carrier density generated by thermal excitation of electrons into the conduction band has, until recently, limited practical devices to being operated at 77K.

Recently, we have reported on growth and fabrication of InSb/In$_{1-x}$Al$_x$Sb heterostructure diodes that make use of a barrier in the conduction band of the structure to allow extraction and exclusion of the intrinsic carriers in the active region of the device to allow room temperature operation [1]. We have used similar structures operated in forward bias to demonstrate infrared LEDs which show emission with a peak spectral output at ~5.8µm [2]. Operated in reverse bias, such structures act as negative luminescent devices, which contravene Kirchhoff's Law because they will absorb radiation without emitting it [3] and have important applications in gas sensing, radiometric reference sources and dynamic infrared scene projectors. We have also reported the first InSb diode lasers with emission wavelengths of 5.1µm and pulsed operation up to 90K [4].

143

Although there are several reports of MBE growth of InSb on (100) surfaces in the literature [5-7], complete optimisation of all the growth conditions remains to be demonstrated. Conditions necessary for Reflection High Energy Electron Diffraction (RHEED) oscillations [8] and maps of surface reconstructions versus substrate temperature and surface stoichiometry [5,6] have also been reported. There have been several studies of doping of MBE grown InSb, both n-type with Si and p-type with Be [9,10]. In reference [10] it was reported that anomalous migration of Be towards the growing surface was seen when the substrate temperature exceeded 340°C although no mention was made as to how the substrate temperature was calibrated, nor as to the quality of the epilayer morphology observed. However, by maintaining a low growth temperature (<340°C), a Be doping level of $2 \times 10^{19} \text{cm}^{-3}$ was achieved, which is the highest value ever reported. In the case of Si, it was claimed that at a substrate temperature of 300°C an n-type doping level of $8 \times 10^{18} \text{cm}^{-3}$ was possible, again the highest value ever reported, corresponding to 100% electrical activation of the Si as measured by Hall and Secondary Ion Mass Spectrometry (SIMS). On increasing the substrate temperature it was found that the Si electrical activation decreased, which was explained by aggregation of the Si in the epilayers as indicated by increased X-ray rocking curve widths.

In reference [9] Be and Si doping of InSb epilayers grown by MBE onto GaAs (100) were studied. Again, as in reference [10], it was claimed that for increasing substrate temperatures there is a decrease in Si dopant activation. Growth temperatures below 340°C again indicated 100% Si activation but the authors were only able to dope up to $3 \times 10^{18} \text{cm}^{-3}$ in this case. The cause of the reduced Si activation was inferred to be due to the amphoteric behaviour of the Si at higher growth temperatures. In the case of Be, it was claimed that doping up to a level of $2 \times 10^{19} \text{cm}^{-3}$ was possible but no comment was made about possible surface segregation of the Be. Also no comment regarding the layer morphology was made despite the large lattice mismatch (~14%) in the InSb-GaAs system. It would therefore be preferable to maintain a substrate temperature of <340°C to maintain optimum doping conditions for both p and n-type doping.

EXPERIMENTAL

All the layers were grown in a modified VG V80H MBE reactor equipped with elemental Sb, In and Al effusion sources and Si and Be sources for n and p-type doping respectively. InSb(100) wafers were supplied by Wafer Technology Ltd, Milton Keynes, UK, doped n-type with Te at $\sim 2 \times 10^{15} \text{cm}^{-3}$. Prior to insertion into the MBE reactor the wafers were degreased in acetone, trichloroethylene and iso propyl alcohol and etched in a modified CP4 etch described previously [11]. Native surface oxide was removed using atomic hydrogen produced by an Astex AX4300 Electron Cyclotron Resonance (ECR) plasma source, using conditions which have been shown to provide an oxide free, atomically flat surface [11]. InSb growth rates were measured by In induced RHEED oscillations from the (100) surface and nominally set to 0.5μm/hr. The Sb:In ratio was measured by using Sb induced RHEED oscillations from the same sample and set for a Sb:In ratio of 1.5:1. For InAlSb barrier layers the Al composition was calibrated by use of AlSb RHEED oscillations from a GaSb(100) surface cleaned with the ECR plasma source.

Substrate temperature during growth was calibrated with reference to the c(4x4)→1x3 surface phase transition (designated T_t) from the (100) InSb surface. With a growth rate of 0.5μm/hr and V:III ratio of 1.5:1, T_t is at a temperature of ~385°C and all subsequent substrate temperatures stated are relative to it. All dopant levels were measured by comparison with calibrated implants using SIMS and by Hall measurement.

Figure 1. Nomarski micrograph of InSb epilayer grown at T_t-50°C and magnification of x200. The surface is characterized by a series of hillocks of height of tens of nm and diameter of tens of μm.

RESULTS

MBE growth of InSb/InAlSb diodes has indicated a narrow substrate temperature window in which acceptable epitaxy can occur. The upper limit on growth temperature is defined by the required doping level both p and n-type. In the case of Be, for a doping level of $3x10^{18}cm^{-3}$, a maximum temperature of $T_t+35°C$ is permissible before surface segregation of the Be occurs with the consequence of diffuse junctions. For Si doping, we find that at $T_t+35°C$ there is ~10-20% electrical activation as seen by Hall and SIMS measurement. The maximum chemical Si level that can be incorporated at that temperature is ~$4x10^{19}cm^{-3}$ which gives an electrical concentration of ~$2x10^{18}cm^{-3}$.

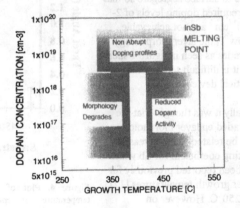

Figure 2. Schematic diagram showing narrow temperature window for MBE growth of InSb onto on-axis InSb(100) substrates.

As has been stated previously [9,10], it is possible to increase the Si incorporation by reducing the substrate temperature during growth. However, we have found that on reducing the substrate temperature to T_t and below, there is a degradation of the layer morphology as seen in figure 1. At growth temperatures of T_t and below the layer surface consists of a series of 'hillocks' with dimensions of tens of nm high and tens of μm across and with a density of $\sim10^6\,\mathrm{cm}^{-2}$. The epitaxy is thus defined by a temperature window of $\sim35°C$ as shown schematically in figure 2. Increasing the substrate temperature above $T_t+35°C$ ($\sim420°C$) means that Si dopant activation drops further and Be starts to surface segregate so that our required doping levels of 2-$3\times10^{18}\,\mathrm{cm}^{-3}$ are not achievable. Below T_t the layer morphology degrades as seen in figure 1 such that it is difficult to fabricate practical devices.

Initially it was thought that the degraded epilayer morphology may be related to an increased sticking coefficient for Sb which has been reported previously [12] of $\sim250°C$. However on inspection of layers grown below T_t on cleaved pieces of InSb(100) wafer, it could be seen that

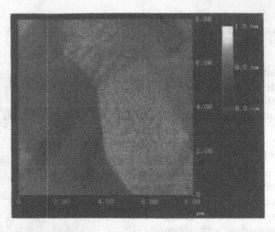

Figure 3. AFM image of hillock feature on epilayer grown at T_t-50°C on an 'on axis' InSb(100) wafer. The step edges that can be seen are of monolayer height.

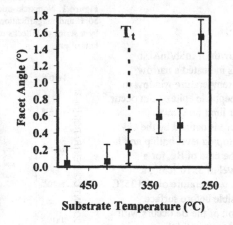

Figure 4. Plot of average hillock gradient versus growth temperature for epitaxy onto 'on axis' InSb(100). With decreasing growth temperature it can be seen that the measured slope increases.

although the hillock morphology continued right up to the cleaved edge it disappeared close to the rounded edge of the wafer in all cases. The only explanation that could be found to explain this was that the areas near the rounded wafer edge had an orientation that is not (100) but oriented away from it and that this misorientation was the cause of the improved morphology. On growth onto substrates that had tighter tolerance on its orientation ((100) ±

0.1° compared with ± 0.25° previously) we found that the hillock morphology persisted even at growth temperatures of Tt+35°C and was somewhat smoothed out as we increased the growth temperature further. To further study this effect we undertook a series of experiments at various substrate temperatures.

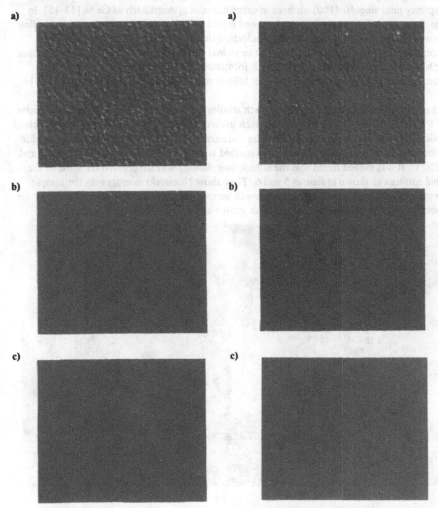

Figure 5. Nomarski micrographs of epilayers grown at T_t-50°C on vicinal surfaces with miscuts of a) 'on axis', b) 0.5° and c) 1.0°.

Figure 6. Nomarski micrographs of epilayers grown at T_t-85°C on vicinal surfaces with miscuts of a) 'on axis', b) 0.5° and c) 2.0°.

Initially, we grew a series of layers grown at temperatures ranging from T_t+35°C to T_t-120°C and compared the gradients of the resultant hillocks studied with AFM. In each case the AFM

147

showed high quality epitaxy with monolayer height steps between terraces along the side of the hillocks as can be seen in Figure 3. Figure 4 shows the gradients of the sides of the hillocks plotted versus growth temperature and it can clearly be seen that as the growth temperature is reduced the slope of the features increases with consequent increased step density and reduced terrace length. Pyramidal features such as those we are reporting have been seen previously for epitaxy onto singular (100) surfaces in other material systems such as GaAs [13-15]. In those cases, the origin of the features has been stated to be due to the step spacing exceeding adatom diffusion lengths, causing island nucleation that then act as step edges for further adatoms to attach to. By extending epitaxy to vicinal surfaces, such that the adatom diffusion length exceeds the step spacing such hillock formation has been shown to be suppressed although under certain conditions rippled or hillock morphologies can still be attained [15].

We have therefore extended our InSb growth studies to vicinal surfaces, with misorientations of 0.5°, 1°, 2°, 4° and 10° from (100). At each growth temperature studied, a piece of each of the vicinal wafers was mounted alongside an 'on axis' InSb(100) sample which was used for T_t measurement. The growth temperatures studied were Tt+35°C, Tt, Tt-50°C, Tt-85°C and Tt-120°C. It was indeed found that the hillock morphology was suppressed on some of the vicinal surfaces as shown in figures 5 and 6. They show Nomarski micrographs for sample with various angles of misorientation at growth temperatures of Tt-50°C and Tt-85°C respectively. It can be seen that for the layer grown at Tt-50°C, the hillock features are

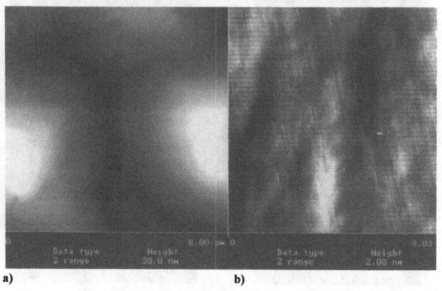

a)   b)

Figure 7. AFM images of InSb epilayers grown at T_t-50°C
a) 'on axis' substrate and b) 2° miscut towards (111)B.
Note the difference in vertical scales.

evident on the 'on axis' sample (5.a.) but that for the layers grown on surfaces miscut by 0.5°

and above, no such morphology is evident with apparently smooth epilayer surfaces. For the Tt-85°C growth temperature it can be seen that hillocks are still visible at a miscut of even 1.0° and that a miscut of 2° is required before a smooth morphology is seen (figure 6.c.).

AFM images of the layers grown at T_t-50°C are shown in figure 7; 7.a) for the 'on axis' and 7.b) the 2° miscut sample. Again the contrast between the samples is very marked with 7.b) having a much smoother surface as shown by the two images having greatly different vertical scales, 30nm in the case of the 'on axis' sample and 2nm in the other. The effect of the vicinal surface is therefore to introduce a greater step edge density that permits two-dimensional growth to occur whereas the 'on axis' sample requires the nucleation of the hillocks to generate the required step edge density.

The complete set of layer morphology data as seen under Nomarski is summarized in table 1.

	T_t +35°C	T_t	T_t -50°C	T_t -85°C	T_t -120°C
On Axis (0°)	Hillocks	Hillocks	Hillocks	Hillocks	Hillocks
0.5°	Smooth	Smooth	Smooth	Hillocks	Hillocks
1.0°	Smooth	Smooth	Smooth	Hillocks	Hillocks
2.0°	Smooth	Smooth	Smooth	Smooth	Hillocks
4.0°	Smooth	Smooth	Smooth	Smooth	Smooth + Defects
10.0°	Smooth	Smooth	Smooth	Smooth	Smooth + Defects

Table 1. Comparison of surface morphology of layers grown on vicinal surfaces at various substrate temperatures. 'Hillocks' defines morphology as shown in figure 5.a. and 'Smooth' defines morphology as shown in figure 5.c.

The data in table 1. follows the same trend as seen in figure 4. That is, as the substrate temperature is lowered, a higher step density is required for epitaxy to proceed, in line with the kinetic model proposed for GaAs [13-15]. The defects seen in some of the layers grown at T_t-120°C (see table 1.) may be related to the change in sticking coefficient of Sb which has previously been seen at around the same substrate temperatures for InSb MBE [12]. This remains to be confirmed experimentally.

Having found a way to achieve high quality epitaxy at growth temperatures <340°C, an InSb/InAlSb heterostructure diode was grown with the following structure; 3μm of n$^+$ InSb (3×10^{19}cm^{-3}) / 1μm of p$^+$ InSb (3×10^{18}cm^{-3}) / 200Å In$_{0.85}$Al$_{0.15}$Sb p$^+$ barrier (3×10^{18}cm^{-3}) / 3μm undoped InSb / 1μm n$^+$ InSb (3×10^{19}cm^{-3}). The layers were grown at T_t-85°C (~300°C) on an InSb(100) wafer miscut 2° towards (111)B. The resultant morphology was excellent and the structure was fabricated into 100μm square mesas using techniques described previously [1]. The resultant room temperature I-V characteristic is shown in figure 8 along with the resultant differential resistance-voltage curve.

There are two important results from this curve. Firstly, the reverse bias leakage current is greatly reduced from what we would expect from a similar structure grown at T_t+35°C. At a bias of -150mV the measured current is ~800µA, compared with ~2mA for a similar structure grown under normal growth conditions ($T_t+35°C$); i.e.a reduction by a factor of ~2.5. Secondly, it can be seen that as the diode is driven into reverse bias there is a slight reduction in the leakage current with consequent demonstration of negative differential resistance (NDR). This has been seen previously in Cadmium Mercury Telluride (CMT) heterostructures [16] but is the first demonstration at room temperature in InSb devices. It has been shown to be associated with suppression of Auger generation mechanisms due to the non-equilibrium operation of the diode. In reverse bias, the intrinsic carriers (electrons) are swept out of the

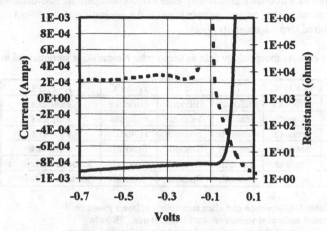

Figure 8. Current-Voltage and differential resistance - voltage curves for InSb/InAlSb diode grown at T_t-85°C. The reverse bias leakage current drops slightly at ~-130mV, giving negative resistance indicative of Auger generation suppression. The I-V curve is the solid line corresponding to the left hand axis.

active region of the device to the n^+ contact and the barrier in the conduction band due to the wide gap InAlSb prevents electrons from being re-injected. The carrier density in the undoped region then falls to the extrinsic level, which in this case is ~10^{15}cm^{-3} p-type, due to background doping of the MBE grown material.. The consequence of this is that the dominant Auger generation process, which requires conduction band electrons, is suppressed with the consequent reduction in reverse bias leakage current. In the undoped region, the dominant generation mechanism is then expected to be due to the Shockley Read traps, and so the further reduced reverse bias leakage current measured here is indicative of a reduced trap density in the undoped region of the diode. This has important implications for future devices; it offers the possibility of reduced currents in negative luminescent diodes, higher dynamic range in InSb transistors and improved detectivities in InSb based infrared photon detectors.

CONCLUSIONS

Using conventional 'on axis' InSb(100) substrates for InSb MBE there is a narrow temperature window of ~35°C within which high quality epitaxy can take place. By growing onto vicinal InSb(100) substrates, we have demonstrated that the growth temperature can be reduced by as much as 120°C with consequent advantages in dopant activation and reduced dopant segregation. The angle of misorientation required is only 2-3° to allow high quality epitaxial growth even at the lowest temperatures we have studied here. Growth at even lower temperatures will be hampered by the increased sticking coefficient of Sb, which eventually leads to the three-dimensional growth if the surface is at a low enough temperature [12].

MBE growth of an InSb/InAlSb diode at a substrate temperature of T_t-85°C (300°C) has demonstrated negative differential resistance in such a structure for the first time at room temperature. This reduced current is indicative of Auger suppression in reverse bias and may be due to a reduced Shockley Read trap density in the undoped part of the diode. This is very important results since it offers a route to reducing the current in negative luminescent devices and improved detectivity (D^*) in minimally cooled IR photon detectors.

ACKNOWLEDGEMENTS

We would like to thank staff at Wafer Technology (Milton Keynes, UK) for their help in preparing vicinal InSb substrates. We would also like to thank D E J Soley, L M Smith, V D Stimson and P K Moores for their invaluable technical assistance in enabling these measurements to be made.

REFERENCES

[1]   T Ashley, A B Dean, C T Elliott, G J Pryce, A D Johnson & H Wills, Appl. Phys. Lett. **66**, No.4, pp481-3, (1995).

[2]   T Ashley, C T Elliott, N T Gordon, R S Hall, A D Johnson & G J Pryce, Appl. Phys. Lett. **64**, No.18, pp2433-5, 1994.

[3]   T Ashley, C T Elliott, N T Gordon, R S Hall, A D Johnson & G J Pryce, Infrared Physics & Technology, **36**, pp1037-44, 1995.

[4]   T Ashley, C T Elliott, R Jefferies, A D Johnson, G J Pryce & A M White, Appl. Phys. Lett. **70**, (8), pp. 931-33, 1997.

[5]   A J Norieka & M H Francombe, J. Appl. Phys. **52**, (12), pp. 7416-7420, 1981.

[6]   Kunishige Oe, Seigo Ando & Koichi Sugiyama, Jap. Journ. Appl. Phys. 19, No. 7, pp. L417-20, 1980.

[7]   G M Williams, C R Whitehouse, T Martin, N G Chew, A G Cullis, T Ashley, D E Sykes, K Mackay & R H Williams, J. Appl. Phys. 63, p1526, 1988.

[8]   R Droopad, R L Williams & S D Parker, Semicond. Sci. & Technol. 4, p111, 1989.

[9]   S D Parker, R L Williams, R Droopad, R A Stradling, K W J Barnham, S N Holmes, J Laverty, C C Phillips, E Skuras, R Thomas, X Zhang, A Staton- Bevan and D W Pashley, Semicond. Sci. Technol. 4, pp. 663-676 (1989).

[10]   P E Thompson, J L Davis, M-J Yang, D S Simons and P H Chi, J. Appl. Phys. 74, (11), pp. 6686-6690 (1993).

[11]   A D Johnson, G M Williams, A J Pidduck, C R Whitehouse, T Martin, C T Elliott & T Ashley, Inst. Phys. Conf. Ser. No. 144, Section 4, pp204-8, 1995.

[12]   M B Santos & W K Liu, Inst. Phys. Conf. Ser. No. 144, Section 4, pp199-203, 1995.

[13]   G W Smith, A J Pidduck, C R Whitehouse, J L Glasper & J Spowart, J. Cryst. Growth 127, p996, 1993.

[14]   C Orme, M D Johnson, K-T Leung, B G Orr, P Smilauer & D Vvedensky, J. Cryst. Growth 150, p128, 1995.

[15]   M Rost, P Smilauer & J Krug, Surf. Sci. 369, pp. 393-402, 1996.

[16]   C.T. Elliott, N.T. Gordon, T.J. Phillips, A.M. White, C.L. Jones, C.D. Maxey and N.E. Metcalfe, Journal of Electronic Materials, 25, (8), p1139, 1996

POSITIVE AND NEGATIVE LUMINESCENT INFRARED SOURCES AND THEIR APPLICATIONS

T. ASHLEY
Defence Evaluation and Research Agency
St. Andrews Road, Malvern, Worcestershire, WR14 3PS, UK.
email: tashley@dera.gov.uk

ABSTRACT

We describe uncooled mid-infrared light emitting and negative luminescent diodes made from indium antimonide based III-V compounds, and long wavelength devices made from mercury cadmium telluride. The techniques to produce large area positive and negative luminescent sources, and the application of these devices to gas sensing, improved thermal imagers and imager testing is discussed.

1. INTRODUCTION

There is increasing interest in mid and long wavelength infrared semiconductor sources for a number of applications including sensing of atmospheric or pollutant gases; eye-safe range finding; and secure communications. In addition, diodes displaying negative luminescence with a good efficiency extend the potential applications to include components for improved thermal imaging and for imager testing. In all cases it is desirable that the infrared diodes should operate with little or no cooling in order to minimise system cost, size and power consumption.

In section 2, we describe the structures for uncooled mid and long wavelength infrared light emitting and negative luminescent diodes. The mid-IR devices are made from indium antimonide based III-V compounds and comprise strained $In_{1-x}Al_xSb/InSb$ on InSb substrates, or, for greater wavelength flexibility, $In_{1-x}Al_xSb/In_{1-y}Al_ySb$ heterostructures on $In_{1-z}Ga_zSb$ substrates. The growth of these structures is described by Johnson in an associated paper at this meeting [1]. The long wavelength IR devices are made from mercury cadmium telluride (MCT), and comprise double heterostructures. Techniques to fabricate large area InSb devices, up to 1cm^2, which include the use of degenerately doped substrates to provide transparency and the integration of optical concentrators in the substrate material to improve optical efficiency are described in section 3. Emission data from both material systems are presented in section 4. The application of positive and negative luminescent devices to gas sensing, improved thermal imagers and imager testing is discussed in section 5.

2. DEVICE STRUCTURES

The structures, for each material system, used to make the luminescent devices have been described previously [2-4], and are shown in figure 1, together with the associated equilibrium band diagrams. Briefly, the devices have an 'active', narrow-gap region between n- and p-type contact regions which have a very high doping level and, ideally, a wider energy-gap. For optical devices, the active region is doped p-type, normally to a low level such that it is intrinsic at room temperature and so is referred to as π-type. The structure, therefore, contains one pn junction and one isotype junction. The diffusion lengths in the contact regions are sufficiently short that ohmic metal contacts do not influence the carrier densities at the junctions to the active region. The products of generation/recombination rates and diffusion lengths are also low in the contact

Mat. Res. Soc. Symp. Proc. Vol. 484 © 1998 Materials Research Society

regions, so the thermally mediated carrier flows to and from the active region are small and the majority of the device current arises only from processes in the active region. This has benefit for conventional detectors, which are operated at zero bias, by constraining the volume of material which generates thermal noise to be no more than is necessary to obtain the required quantum efficiency.

Under reverse bias, minority carrier extraction occurs at the pn junction and exclusion takes place at the isotype junction. At ambient temperature, this leads to a reduction in the minority carrier density in the active region by several orders of magnitude. In order to maintain approximately zero space charge, there is also a large reduction of the majority carrier density by, typically, two orders of magnitude down to the background doping level. The consequence is that the Auger carrier generation processes, in particular Auger-1 and Auger-7, and their associated noise, are suppressed leading to an improvement in performance of uncooled detectors and minimisation of the current necessary to drive negative luminescent devices.

Under forward bias, the extracting and excluding contacts become injecting and accumulating, respectively, and serve to confine the electrons and holes in the active region, with little leakage into the contact regions, so maximising the radiative recombination rate and giving efficient emission.

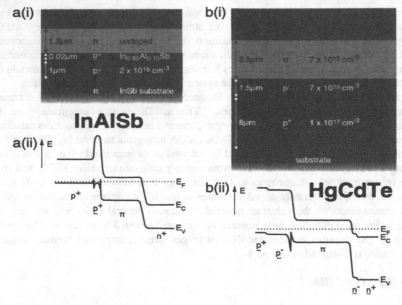

Figure 1: Schematic cross-sections showing the structure of InAlSb and HgCdTe devices (a(i) and b(i) respectively) and their respective equilibrium energy band diagrams (a(ii) and b(ii))

Referring now to figure 1a, the InSb/In$_{1-x}$Al$_x$Sb structures are grown by molecular beam epitaxy onto (001) InSb substrates, as described by Johnson [1]. This material system is not lattice matched and the thickness of the wide-gap region is limited to values of order tens of nanometres to avoid strain relaxation. A four layer p$^+$p$^+$πn$^+$ structure is used therefore, the purpose of the p$^+$ layer being to provide a low resistance, low injection contact to the p$^+$ region [5]. The p$^+$ and p$^+$ regions are doped using beryllium to a level of 2×10^{18}cm^{-3} and have thicknesses of 1-3μm and 20nm respectively. The composition, x, of the $In_{1-x}Al_xSb$ barrier is 0.15 giving a conduction band barrier height which is estimated as 0.26eV. The n^+ region is typically 1μm thick, though thicknesses as little as 0.15μm have been found to be satisfactory, and is doped using silicon to an electrical level of $2 \times 10^{18} cm^{-3}$. At this doping the InSb is degenerate, and the Fermi energy is about 0.19eV above the conduction band minimum giving a valence band step of about 0.22eV between the n^+ and active regions.

In order to facilitate positive luminescent and negative luminescent sources covering a variety of wavelengths in the mid-IR region, we have developed the growth of ternary $In_{1-z}Ga_zSb$ material, to provide lattice matched substrates for the growth of $In_{1-x}Al_xSb$ active region devices. The growth of the $In_{1-z}Ga_zSb$ is performed on the [111] orientation by a double crucible Czochralski technique [6], which provides replenishment of the melt from the outer crucible to ensure that the melt composition in the inner crucible is maintained at the correct value to permit a uniform ingot composition. This is important both to maximise the amount of usable material from the ingot and to ensure minimum strain and hence high quality. The ingot is pulled under hydrogen at atmospheric pressure to minimise carbon incorporation. A typical gallium proportion of 0.03 to 0.07 is required, to provide a lattice constant suitable for subsequent epitaxial growth. There is still some variation in composition along the ingots, therefore they are cut normal to the growth direction, to provide [111] wafers rather than the usual [001] used for MBE growth. X-ray measurements have determined that the wafers generally comprise large areas which are single crystal, but with a lateral variation of up to 1% at present. The MBE growth has been performed under similar conditions to those used for the conventional [001] material, and found to be satisfactory [7].

$Hg_{1-x}Cd_xTe$ heterostructure devices of the type shown in figure 1b are supplied by GEC-Marconi Infrared Limited. This material system is more closely lattice matched than the $In_{1-x}Al_xSb$ one, so full heterostructures can be used. The structure is $p^+ p^- \pi n^- n^+$, with the p^- and n^- layers designed to minimise generation in the grades between regions of different composition [8]. The material is grown by metal organic vapour phase epitaxy using the inter-diffused multilayer process [9,10] on a GaAs substrate using arsenic and iodine for p- and n-type doping respectively. $Hg_{1-x}Cd_xTe$ devices with a variety of compositions have been investigated. The values given here apply to the devices with the longest wavelength luminescence, which have x = 0.184 in the active π region, x = 0.35 in the p^+ and p^- regions and x = 0.23 in the n^+ and n^- regions.

3. LARGE AREA DEVICES

The applications of the positive and negative luminescent diodes require device areas ranging from $0.01 cm^2$ to several cm^2. The smaller devices are simple, chemically etched mesas, with concentric annular top contacts. The larger devices, $>0.1 cm^2$, have only been implemented in the InSb material to date and benefit from a more complex arrangement designed to minimise the overall current supply requirement. This involves inversion of the structures so that light is transmitted through the substrates, which must therefore be transparent, and fabrication of optical concentrators into the substrate material to improve optical efficiency.

3.1 Transparent Substrates

A good transmission can be achieved, in principle, by the use of degenerately doped n-type InSb which has a Moss-Burstein shift [11], instead of the normal low p-type substrates. Modelling has indicated that a doping of approximately $3 \times 10^{18} cm^{-3}$ is necessary to achieve adequate transmission for wavelengths in excess of 4μm. The doping levels which are normally available are insufficient, so substrates have been specially grown by a commercial supplier. Free electron absorption necessitates thinning of the substrates, by a bromine ethanediol polish, to maintain good transmission. The resulting transmission is within 10% of the maximum value determined by reflections from the un-coated surfaces. An example of the transmission through such a substrate is shown in figure 2, where it is compared with that through one of the usual substrates. Further iterations are required to obtain acceptable uniformity, however we believe the viability of the technique has been proven.

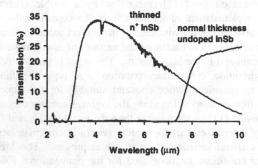

Figure 2: Transmission of un-coated low p-type InSb 500μm thick and n$^+$ InSb 100μm thick.

3.2 Optical Concentrators

A problem common to all light emitting diodes, but particularly acute for devices made from the narrow-gap materials such as InSb owing to its very high refractive index of 4, is that of total internal reflection at the semiconductor to air interface which means that only a few percent of light generated within the semiconductor escapes. One solution is the use of immersion lenses, see for example Gooch [12], which are made ideally from material of the same refractive index as the source. These require intimate contact with the source (gap < λ/10) to minimise attenuation at the interface. Separate immersion lenses are ill-suited to volume production techniques, and for large area sources the immersion lens becomes impractically large. The use of individual immersion lenses has been described for infrared detector arrays [13], however, whether hybridised or integrated, they can be difficult to align and are not efficient for emission (or detection) over large apertures.

We have chosen a solution involving Winston cone [14] concentrators, integrated into the substrate on which the epitaxial material is grown, see figure 3. The active region is distributed about the focus of the paraboloid cone, so that light emitted at large angles, which would normally suffer total internal reflection at the opposite, radiating surface, is reflected by the surface of the cone so that it impinges on the radiating surface at a low angle and so is emitted from the semiconductor. The cone structure defines the area of the active region, and the maximum ratio of cone emission area to active region area is a factor of n^2, where n is the refractive index. Hence, in the absence of other losses, a factor of n^2 (i.e. 16 for InSb) lower current is required to obtain the same light output. At present, the area ratio is conservatively designed to be 6. The cones have a hexagonal distribution on a pitch of 55μm, and have a height of typically 35μm.

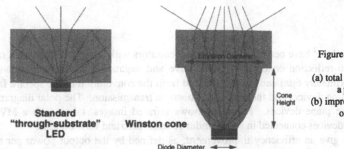

Figure 3: Schematic diagram
indicating
(a) total internal reflection loss at
a plane surface, and
(b) improved emission properties
of a Winston cone.

**Standard
"through-substrate"
LED**

Winston cone

The concentrators are fabricated by reactive ion etching using a 3-d photoresist image. The resist structures are made by conventional lithography to form cylinders, which are then re-flowed to a spherical section. An iterative reactive ion etching schedule is then used, alternating between a $CH_4/H_2/N_2$ gas mixture which etches the InSb and O_2 which ashes the resist, as illustrated in figure 4.

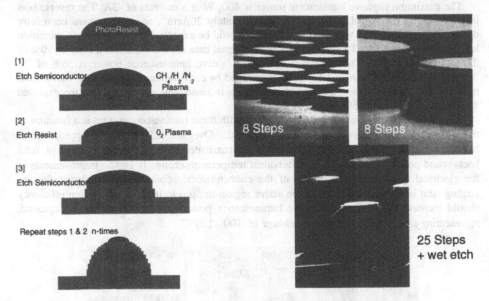

Figure 4: Etching procedure for cones   Figure 5: SEM images of partial and completed cones

The material is given a short wet chemical etch which removes approximately 1μm of InSb to eliminate the minor surface damage resulting from the RIE and to smooth the surfaces. Scanning electron micrographs of some concentrators at various stages of etching are shown in figure 5. The resulting structure is planarised and contact windows opened up for metal contacts. The whole device is then inverted, mounted with conducting epoxy onto a suitable substrate and, in some cases, anti-reflection coated with tin oxide.

4. DEVICE RESULTS

4.1 InSb Devices

InSb devices up to $1cm^2$ have been made, using concentrators with a nominal cone area ratio of 6, but without anti-reflection coatings. The positive and negative luminescence emission spectra peak at approximately $6\mu m$ and are as expected from the convolution of the spectra from the front surface emitting devices and the measured substrate transmission. The polar diagram is Lambertian, as for the plane devices. Figure 6 shows infrared images, taken with a MWIR camera, of four $1cm^2$ devices connected in series under reverse, zero and forward bias.

The concentrators give an efficiency improvement, as defined by the output power per unit input current, of a factor of three compared with plane devices, which is approximately half of their area ratio. Two factors determine this figure - the electrical efficiency compared with plane material, where for example additional surface leakage currents around the cones may occur; and the optical efficiency compared with plane material, where radiation might fall between the cones or when within a cone might not be collected in the diode active region. Both factors are the subject of further research to increase their efficiencies

The maximum negative luminescent power is $400\mu W$ at a currents of -3A. The reverse bias leakage current density of the diodes is approximately $20Acm^{-2}$, so a one square centimetre device with concentrators with an area ratio of 6 would be expected to have a reverse saturation leakage current of 3.3A. The reverse bias experimental data are therefore taken at up to 90% of the saturation current. The maximum observed negative luminescence power is 36% of the theoretical maximum figure of 1.1mW, as determined by convolution of the thermal background radiation level and the absorption spectrum. This is in reasonable agreement with the expected value of approximately 45%.

The equivalent temperature change associated with these luminescent powers is a function of the wavelength interval over which it is measured. Over the $3.5-5\mu m$ band, appropriate to MWIR cameras, the experimental power is approximately $90\mu W$, compared with a total background power of $440\mu W$, so the equivalent temperature change is 16°C. Improvements in the electrical and optical efficiencies of the concentrators; application of an anti-reflection coating; and increase in thickness of the active region to $5\mu m$ to improve absorption efficiency should increase the maximum negative luminescence power to 80 - 90% of the background, representing an equivalent temperature change of 100 - 120°C.

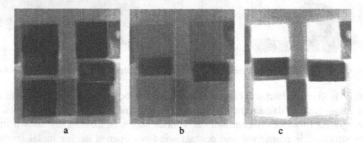

Figure 6: Infrared images of four $1cm^2$ devices (a) under reverse bias, showing negative luminescence, (b) at zero bias, and (c) under forward bias showing positive luminescence

Under forward bias, the maximum positive luminescent power is also 400µW. Unlike the negative case, however, there is no fundamental upper limit to the power, which will be determined solely by the current supply capability and any consequent heating effects. Equivalent positive temperature changes in excess of 100°C are anticipated.

4.2 InAlSb Devices

The room temperature characteristics of diodes, of area $10^{-4}cm^2$, made from one of the ternary layers are compared with a conventional, InSb active region device in table 1. The maximum resistance in reverse bias and the R_o at lower temperatures for the ternary devices fall below the InSb ones, suggesting a higher density of defects at present. Notwithstanding this, the ambient temperature behaviour is broadly as expected from the increase in energy-gap in terms of the increase in R_o, the increase in optical efficiency and the shift of the peak emission to shorter wavelengths.

Table 1: Electrical and optical characteristics of InSb and $In_{0.933}Al_{0.067}Sb$ active region diodes at 294K

Active region material	I_{sat} (mA)	R_o (Ω)	Peak wavelength (µm)	Power at peak at 100mA (µW/µm)
InSb	0.47	55	6.2	4
$In_{0.933}Al_{0.067}Sb$	0.10	550	4.1	11

4.3 MCT Devices

MCT devices 150µm in diameter have been made [15]. These devices display very pronounced negative conductance under reverse bias, as shown in figure 7a. This is due to the suppression of Auger processes following the removal of carriers from the active region. The relative efficiency of radiative recombination processes rises dramatically, as is seen in figure 7b by the increase in quantum efficiency. Devices operating in a negative luminescent mode, therefore, provide a much more efficient source of modulated radiation at long IR wavelengths than the normal positive luminescent case [16]. The negative luminescent spectrum of the above device is shown in figure 8. It has a peak at 9µm, with a response extending to 12µm. The negative luminescent power of the device is 0.7µW, equivalent to an intensity of $4.2mWcm^{-2}$, at a current of 1.7mA,

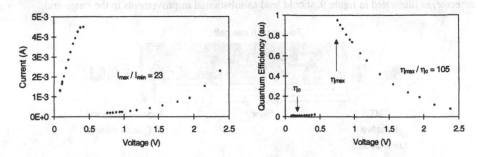

Figure 7. (a) Current vs. voltage and (b) relative quantum efficiency vs. voltage characteristics of a MCT diode under reverse bias at 295K.

159

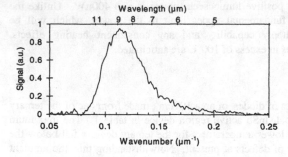

Figure 8. Spectrum of an MCT diode under
reverse bias at 295K

giving an internal efficiency of 10%. A barrier to larger area devices is that of high series resistance in the devices, which would lead to a large voltage requirement and inefficient operation. Research to grow the MCT on conducting GaAs substrates, so that the substrate can be used to provide a low resistance current path is in progress. Contact resistances as low as 10Ω have been obtained by growth on n^{++} capped GaAs using an initial $Hd_{0.4}Cd_{0.6}Te$ layer on top of a HgTe nucleation layer.

5. APPLICATIONS OF POSITIVE AND NEGATIVE LUMINESCENT DIODES

The potential for relatively efficient sources of infrared radiation which can be modulated at high frequencies both above and below the background level opens up a number of applications [17], which will be reviewed briefly here.

5.1 IR Sources for Gas Sensing

Conventional IR absorption based local gas sensors for gases such as CH_4 or CO_2 use a filament lamp as the thermal source and a thermopile or pyroelectric device for the detector. The glass envelope in the lamp absorbs radiation at wavelengths longer than 4.2μm, so these sources are not applicable to the sensing of carbon monoxide, nitrogen oxides, sulphur oxides, ammonia etc. Furthermore, the hot filament lamps present a safety hazard in explosive environments which necessitates bulky, expensive enclosures. Replacement of the lamp with an IR LED and, ideally, also replacement of the detector with an uncooled high performance Auger suppressed photon detector, as illustrated in figure 9, should lead to substantial improvements in the range and

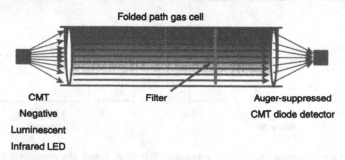

Figure 9. Gas sensor using an LED as the IR source and an Auger suppressed diode detector.

sensitivity of gases which can be detected. This example is a sensor operating at long wavelengths (> 6µm), where use of an MCT source is necessary and where improved efficiency is obtained by operation in a negative luminescence mode. A prototype sensor has been constructed for NO_2 detection using an uncooled InSb LED in positive luminescence mode [18]. This has demonstrated viability of 10ppm sensitivity but presently uses a cooled detector. Further improvements in the epitaxial InSb detector technology are expected to yield a similar sensitivity without the requirement for cryogenic cooling.

5.2 Radiometric Reference Planes for Thermal Imagers

Thermal imagers are required to resolve very small temperature differences in scenes with a mean temperature of, typically, 300K. Current generation imagers can resolve between 20mK and 100mK, and future generations of equipment will reduce this to a few milli-Kelvin. This sensitivity imposes severe demands on the uniformity of detectors within an array, and, even then, necessitates non-uniformity correction to remove fixed pattern noise. The correction is performed conventionally using Peltier devices. These have a slow response time and therefore can not follow rapid changes in mean flux level and can require several minutes to perform a correction sequence. Positive and negative luminescent sources can provide any arbitrary flux, within the range of scene temperature normally encountered, very rapidly and so overcome many of the deficiencies of the Peltier devices.

DC restoration in scanned thermal imagers

Scanned imagers are normally ac coupled, as illustrated in figure 10, which removes offsets and enhances contrast. This introduces other image defects - droop will occur in a step function signal from the detector, because of discharge of the coupling capacitor, and undershoot will follow as the step function returns to zero, see figure 11a. An additional problem associated with

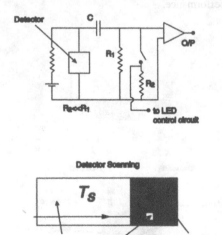

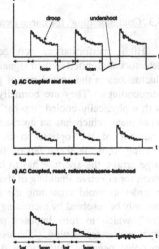

Figure 10. Schematic of the detector bias circuit and ac coupling of the output with a reset switch. The lower diagram shows the detector scanning across the scene then onto the reference plane

Figure 11. Detector output waveforms for: a) ac coupling, b) ac coupling with reset, c) ac coupling, reset and reference balancing

161

ac coupling is that the removal of the dc component removes the signal relating to the absolute temperature of the scene. This is particularly undesirable when performing radiometric monitoring or, for example, when imaging scenes with a prominent horizon.

The solution adopted is illustrated in figure 10b. At the end of each line-scan the detector views a uniform reference scene and the channels are reset by short-circuiting the coupling capacitor to ground. Offset and droop problems still arise if the reference temperature is not close the mean scene temperature, as shown in figure 11b, which are countered by automatically adjusting the reference temperature to correspond to that of the scene, as shown in figure 11c. The time constant of this can be very long if the reference plane is a Peltier device, whereas the positive/negative LEDs can respond essentially instantaneously.

Non-uniformity correction in 'staring' arrays

'Staring' array imagers are dc coupled and so are particularly sensitive to non-uniformities in the responses of individual detectors and their associated read-out circuits. The detector outputs are related to the flux by non-linear polynomial equations, so the non-uniformities can be corrected by allowing the array to see a number of uniform reference temperatures, in order that the constants of the polynomial can be calculated and applied to the output signal. In practical systems, however, this correction has to be approximated by a linear relationship based on only two reference temperatures ('gain and offset' correction) owing to problems of having multiple Peltier reference sources or the time for a single source to stabilise at a number of temperatures. This means that the correction in less effective for scene temperatures not equal to one of the reference temperatures, and can lead to increases in the minimum resolvable temperature difference (MRTD) of several tens of milli-Kelvin. The non-uniformities often change with time, so the calibration must be performed periodically - from perhaps every hour to as often as every frame in particularly demanding environments. In a similar manner to the scanned systems, the use of positive/negative luminescent devices will allow a larger number of effective temperatures, both above and below the physical scene temperature, to be produced virtually instantaneously so leading to substantially improved performance.

5.3 'Cold Shielding' for Large Detector Arrays

High performance, cryogenically cooled detectors are limited in their signal to noise ratio by fluctuations in the arrival rate of photons from the surroundings. They are normally 'cold shielded' with a physically cooled 'top-hat' mounted on the focal plane which has an aperture matched to the f/number of the optics, so that only photons arriving from the scene fall on the detector. For large arrays, the aperture has to be opened up too much for effective shielding of the central elements in order to avoid vignetting the edge ones. This can only be resolved by increasing the height of the 'hat', which in turn imposes problems for the cryogenic encapsulation. The use of an external, uncooled negative luminescent device overcomes this problem as illustrated in figure 12.

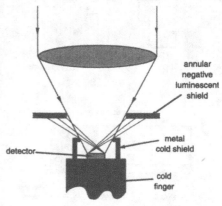

Figure 12. Negative luminescent 'cold-shield'

5.4 Dynamic Infrared Scene Projection

Infrared detector arrays in systems such as missile seekers require dynamic testing, so-called 'hard wired in the loop'. This is performed with the use of a dynamic infrared scene projector (DIRSP) to simulate a variety of IR scenes. Such test systems often employ 2d arrays of resistors whose temperature is changed by passing current through them which leads to Joule heating. The positive/negative luminescent diodes provide an alternative, as illustrated in figure 13, which has the advantage of being able to simulate cold objects as well as warm ones. This is important as a typical scene may include the cold sky or reflections of it in metallic or wet objects. As in previous applications, the fast response of the LEDs, compared with devices which have to undergo a physical temperature change, should enable high frame rates to be simulated.

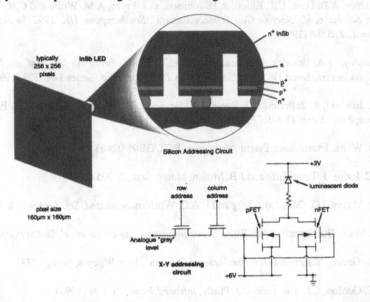

Figure 13. DIRSP using positive/negative luminescent diodes and a simplified example of a drive circuit.

6. CONCLUSIONS

We have demonstrated positive and negative luminescence from $InSb/In_{1-x}Al_xSb$ and MCT diodes, covering wavelengths from 4µm to 12µm. The techniques to achieve large area devices have been demonstrated for the InSb case, which indicates that a number of potential applications of these devices will be viable.

ACKNOWLEDGEMENTS

The author gratefully acknowledges the co-operation of colleagues at DERA in providing results for inclusion in this summary. The work was supported by the UK MOD, Corporate Research Programme.

REFERENCES

1. A. D. Johnson, this conference, paper F3.5

2. T. Ashley, C.T. Elliott and A.T. Harker, *Infrared Phys.*, **26**, 303 (1986)

3. C.T. Elliott, *Semicond. Sci. Technol.*, **55**, S30 (1990)

4. T. Ashley and C.T. Elliott, *Proc. NATO workshop 'Narrow Gap Semiconductors'*, Oslo, Norway, 1991, in Semicond. Sci. Technol., **6**, C99 (1991)

5. T. Ashley, A.B. Dean, C.T. Elliott, A.D. Johnson, G.J. Pryce, A.M. White and C.R. Whitehouse, *Proc. 6th Int. conf. 'Narrow Gap Semiconductors', Southampton, UK, 1992, in Semicond. Sci. Technol.*, **8**, S386 (1993)

6. T. Ashley, J.A. Beswick, B. Cockayne and C.T. Elliott, *Proc. 7th Int. conf. on Narrow-Gap Semiconductors, Santa Fe, NM, USA, 9-12 Jan 1995*, IoP Conf. Series **144**, 209 (1995)

7. A.D. Johnson, R. Jefferies, G.J. Pryce, J.A. Beswick, T. Ashley, J. Newey, C.T. Elliott and T. Martin, *Proc. Symp. O, MRS Fall Meeting 1996*

8. A.M. White, International Patent Application PCT/GB96/02403

9. S.J.C. Irvine, J. Tunnicliffe and J.B. Mullin, *Mater. Lett.*, **2**, 305 (1984)

10. C.D. Maxey, I.G. Gale, J.B. Clegg and P.A.C. Whiffen, *Semicond. Sci. Technol.*, **8**, S183 (1993)

11. T.S. Moss, G.J. Burrell and B. Ellis, *'Semiconductor Opto-Electronics'*, Butterworths, 1973

12. C.H. Gooch, *'Injection Electroluminescent Devices'*, John Wiley & Sons, 1973

13. N. T. Gordon, C.L. Jones and D.J. Purdy, *Infrared Phys.*, **31**, 599 (1991)

14. W.T. Welford and R. Winston, *'The Optics of Nonimaging Concentrators'*, Academic Press, 1978

15. C.T. Elliott, N.T. Gordon et. al., *J. Elect. Mater.*, **25**, 1139 (1996)

16. N.T. Gordon, private communiation

17. T. Ashley, C.T. Elliott, N.T. Gordon, T.J. Phillips and R.S. Hall, *Infrared Phys. Technol.*, **38**, 145 (1997)

18. C.H. Wang, J.G. Crowder, V. Mannheim, T. Ashley, D.T. Dutton, A.D. Johnson, G. J. Pryce and S.D. Smith, *to be published*

GaAs/AlGaAs INTERSUBBAND MID-INFRARED EMITTER

G. STRASSER, S. GIANORDOLI, L. HVOZDARA, H. BICHL, K. UNTERRAINER
E. GORNIK, TU Wien, Solid State Electronics, Vienna, AUSTRIA
P. KRUCK, M. HELM, Univ. Linz, Semiconductor Physics, Linz, AUSTRIA
J.N. HEYMAN, Macalester College, St. Paul, MN, USA

ABSTRACT

We report here the growth and characterization of intersubband quantum well structures based on the GaAs/AlGaAs material system, designed to emit radiation at approximately 7 micrometers. We present transport behavior, infrared photocurrent spectra, and electroluminescence data. First attempts to fabricate a laser structure from this devices encountered difficulties with the electrical properties of the AlGaAs waveguide cladding layers. Thus, we present measurements with different waveguide concepts as doped AlAs cladding layers and doped superlattice cladding structures.

INTRODUCTION

Many designs have been proposed for infrared lasers based on intersubband transitions. Spontaneous emission from intersubband transition at photon energies well below the optical phonons (36 meV) was demonstrated via resonant tunneling through quantum wells[1] and electron transport in inversion layers[2]. The first room temperature electroluminescence based on intersubband transitions in quantum wells at energies above the optical phonon energy showing population inversion was demonstrated by Faist et al.[3] and was the final step to an intersubband laser. The advent of this unipolar semiconductor laser in 1994[4] marked a breakthrough in the application of band-structure engineering.

The emission wavelength of these devices is determined by the quantum well thickness in contrast to usual diode lasers, where it is dictated by the energy gap of the semiconductor material. Great progress has been achieved in the performance and operation characteristics of quantum cascade lasers[5] based on InGaAs/InAlAs, lattice matched to InP, lasers operating between 3.5 and 11 μm have been demonstrated. High-power room-temperature operation has been realized in pulsed mode up to a wavelength of 8.5 μm. Only recently single-mode operation in distributed feedback QC lasers has been observed.

Competing efforts for the development of mid-infrared sources, which have many potential applications related to the detection of trace gases, are based on narrow-gap semiconductors such as lead salts[6] and, more recently, III-V compounds containing Sb. In this latter system lasing between 2 and 5 μm has been demonstrated[7] and also a novel type-II QCL has been implemented by applying the quantum cascade concept to interband transitions in a staggered-line up heterosystem in combination with interband tunneling. On this basis, a 3.8 μm laser[8] and a light emitting diode in the 5-8 μm region[9] have been fabricated.

In the GaAs/AlGaAs material system intersubband photoluminescence[10], stimulated emission[11] and finally lasing[12] in the long wavelength range was achieved by optical pumping with a free electron or a CO_2 laser. Electrically pumped GaAs/AlGaAs intersubband luminescence at cryogenic temperatures was demonstrated by Li et al.[13] and by our group[14], room temperature electroluminescence of a QCL-structure was reported recently[15].

For electrical pumping the realization of the waveguide cladding layers has to be considered carefully, since, in contrast to the InGaAs/InAlAs/InP system, the substrate material has the smallest energy gap and largest refractive index and, thus, cannot be used as a lower cladding layer. In addition, AlGaAs with high Al-content has deep-lying donor states (DX centers), which give rise to inferior transport properties and could therefore induce a large series resistance. An alternative cladding concept is based on a short-period GaAs/AlAs superlattice.

SAMPLES

We report the growth and characterization of three different light emitting diodes described in table 1: sample #1 without any cladding layers, sample #2 with a highly doped superlattice cladding and sample #3 with highly doped thick AlAs-layers.

sample number	substrate	lower cladding	active zone	top cladding
sample #1	S.I.-GaAs	none	25 periods	none
sample #2	S.I.-GaAs	n^+-GaAs/AlAs superlattice	25 periods	n^+-GaAs/AlAs superlattice
sample #3	n^+-GaAs	n^+-AlAs	25 periods	n^+-AlAs

Table1: sample description

The MBE grown diode structures essentially follow the design considerations for a QC emitter given by Faist et al. in Ref.[3] with the modifications necessary due to the other material system and consist of 25 periods of GaAs/$Al_{0.45}Ga_{0.55}As$ coupled quantum well layers.

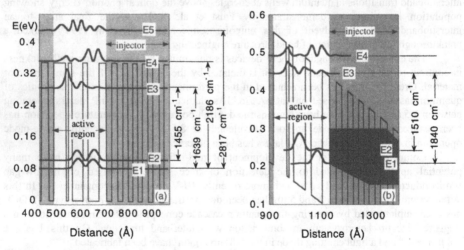

Figure 1a: *Calculated band structure of the active cell with injector (sample #1)*

Figure 1b: *Calculated band structure under bias: 70 kV/cm corresponds to V_b= 7.5 V*

The growth sequence of the active cell of samples #2 & #3 (#1) is: 10 Å GaAs, 15 Å AlGaAs, 47 (45) Å GaAs, 22 (20) Å AlGaAs, and 40 (45) Å GaAs (see Fig. 1). After characterization of sample #1 the growth sequence of the active layers was slightly changed to suppress the direct 3-1 transition. The active cells are separated by doped (n ≈ 10^{17} cm^{-3}) miniband funnel injectors, which provide effective injection through the tunnel barrier into the n=3 state and prevent the escape of electrons into the continuum. Tunnel barriers and active cells are left undoped in order to suppress the influence of doping impurities on the linewidth of the intersubband luminescence[16]. The whole structure is embedded between two highly doped cladding or n+GaAs contact layers (table 1).

When an electric field of 70 kV/cm is applied to the structure, which corresponds to the calculated operating bias of V_b=7.5 V, the separation of the n=1 and n=2 subbands is around 35-40 meV. This energy difference ensures fast depopulation of the n=2 level due to efficient optical-phonon emission, which is required for population inversion and potential lasing. The separation between the n=2 and n=3 levels, which determines the emission wavelength, is calculated to 1510 cm^{-1}, (corresponding to a wavelength of 6.6 μm). The present choice was dictated by two considerations: on the one hand, shorter wavelength emission (or lasing) is generally easier to achieve due to the larger spontaneous photon emission rate and the smaller optical-phonon emission rate, which increases the internal quantum efficiency. On the other hand it was aimed to keep the Al content in the barriers below 45% in order to avoid effects related to the X-point, which becomes the lowest conduction-band valley in AlGaAs above this value.

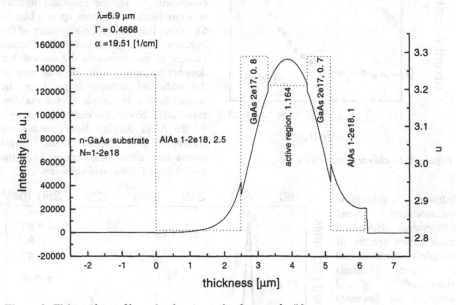

Figure 2: TM$_0$ mode profile and refractive index for sample #3

Mode profile calculations of sample #1 and sample #2 (doped superlattice cladding) revealed, that the surface plasmon on the top metal contact could not be suppressed enough so that the waveguide loss are ca. α_1=200 cm^{-1}. Fig. 2 illustrates the calculated mode profiles (TM$_0$) and refractive index (calculated values for highly doped and literature values for low doped material) for sample #3 (wavelength: 6.9 μm). To enhance the optical confinement in the active region pure AlAs was taken. AlAs is the material with the lowest refractive index lattice matched on GaAs. On the upper 1 μm thick AlAs-cladding a 50 nm thick highly doped (5.10^{18} cm^{-3}) is added to decrease the surface plasmon and to guarantee a good ohmic top contact. The calculated confinement factor is $\Gamma = 0.46$. The internal losses are estimated to be $\alpha_I = 20$ cm^{-1}. It is a combination of free carrier absorption, lateral losses and plasmon losses on the top metal contact. For a 3 mm ridge the mirror losses are about $\alpha_M = 9$ cm^{-1}.

EXPERIMENTS

For characterization, samples were processed into 40 x 70 μm^2 mesa structures. As expected,

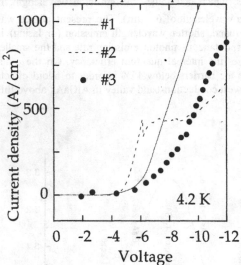

Figure 3: *IV-characteristics at 4.2K*

the current-voltage characteristic is very asymmetric, with much higher current in forward direction. As shown in Fig. 3 forward current-voltage (IV) characteristics (here: towards negative bias) were recorded at 4K. Pulsed-IV techniques are used (f=21kHz duty cycle=2%) to prevent sample heating and damage at high bias conditions. At 4K the miniband injector structure blocks the current up to a bias of 6V, above that voltage the alignment of the injectors leads to a current. Further increase in bias increases the current but does not detune the levels, because most of the additional voltage drops over the contact layers. In sample #2 the cladding superlattice blocks the current, in sample #3 the AlAs cladding layers reduce the current. Highly doped contact layers causes free carrier losses and reduces the probability of stimulated emission.

Infrared absorption measurements and temperature dependend photocurrent spectra of sample #1 are discussed in great details elsewhere[14,17] and therefore not discussed here.

A comparison of the photocurrent at low temperatures of the three samples is shown in Fig. 4. The transitions (indicated in Fig.1a) for samples #2 and #3 are

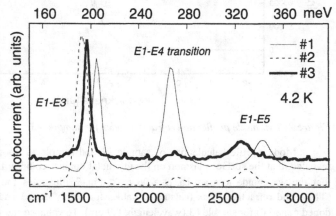

Figure 4: *Photocurrent spectra of samples #1-#3 at 4.2K*

slightly shifted to lower energies due to the different quantum well thicknesses and show the sensitivity and accurancy of the device (and the performance of the growth equipment). The parameters of the active wells were changed to reduce the probability for the 3-1 transition, this effects also the photocurrent probability as it can be seen by direct comparison of the peak intensities (1 -3 and 1-4 transition) within the same sample.

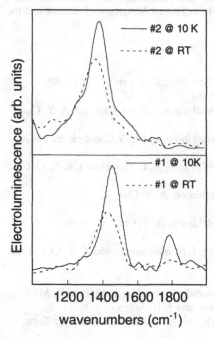

Figure 5: Electroluminescence spectra at 10K and room temperature(RT) for sample #1 & #2

Emission measurements were performed with a Fourier transform infrared spectrometer operating in the step-scan mode and using a Lock-In amplifier. During the measurement the samples were mounted in a cold-finger cryostat kept at 10 K or room temperature, respectively. The emission from $100 \times 100 \ \mu m^2$ mesas was coupled out of the sample through a polished 45° facet and was detected with a LN_2 cooled HgCdTe detector. Electrical pulses of typically 15 ms length with a 3 kHz repetition rate were applied to the samples (5% duty cycle). For the present structures strong injection occurs for bias voltages above V_b=7.5 V (sample #1 and #2), corresponding to a current of the order of I=100 mA (current density j=1 kA/cm^2). Typical emission spectra are shown in Fig. 5. The main emission peak (sample #1: 1450 cm^{-1} or 6.9 μm; sample #2: 1380 cm^{-1} or 7.25 μm) is due to the 3-2 transition. In contrast to bandgap transitions there is almost no shift in the transition energy between 10K and room temperature. The second emission peak of sample #1 (3-1) is suppressed in sample #2 due to the better overlap of the wavefunctions of level E1and E2 (see fig. 1b). The width of the luminescence line is found to be as low as 14 (17) meV for sample #1(#2) at 10 K and broadens to 20 meV at room temperature for both samples. The narrow linewidth of both the photocurrent and emission is due to the lack of doping impurities in the active region and promises lower threshold currents in laser application, since the peak gain is inversely proportional to the linewidth. The electroluminescence of sample #3 at cryogenic temperatures consists mainly of unpolarized black body emission from the heated AlAs layers. Although the waveguide losses of sample #3 (Fig. 2) are low enough to provide guided modes, the heating of the contact layers due to the inferior transport properties of AlAs prevent measurable stimulated emission.

CONCLUSIONS

In conclusion, we have demonstrated the operation of quantum cascade light emitting diodes based on the GaAs/AlGaAs material system up to room temperature. Different cladding concepts have been tested to prepare the ground for the fabrication of a quantum cascade laser.

Although further investigations are neccessary, we are confident to get stimulated emission from an electrically pumped QCL structure on the basis of the present results.

ACKNOWLEDGMENTS

We gratefully acknowledge support by the Fonds zur Förderung zur wissenschaftlichen Forschung (FWF, project P 10211) and by the Austrian Federal Ministery of Science and the Gesellschaft für Mikroelektronik (GMe).

REFERENCES

[1] M. Helm, P. England, E. Colas, F. De Rosa, S. J. Allen, Phys. Rev. Lett. **63**, 74 (1989)

[2] E. Gornik, D.C. Tsui, Phys. Rev. Lett. **37**, 1425 (1976)

[3] J. Faist, F. Capasso, C. Sirtori, D. L. Sivco, A. L. Hutchinson, S.-N. G. Chu, and A. Y. Cho, Appl. Phys. Lett. **64**, 1144 (1994)

[4] J. Faist, F. Capasso, D. L. Sivco, C. Sirtori, A. L. Hutchinson, and A. Y Cho, Science **264**, 553 (1994)

[5] J. Faist, F. Capasso, C. Sirtori, D. L. Sivco, J. N. Baillargeon, A. L. Hutchinson, S.-N. G. Chu, and A. Y. Cho, Appl. Phys. Lett. **68**, 3680 (1996)

C. Sirtori, J. Faist, F. Capasso, D.L. Sivco, A. L. Hutchinson, A. Y. Cho, Appl. Phys. Lett. **69**, 2810 (1996)

C. Sirtori, J. Faist, F. Capasso, D. L. Sivco, A. L. Hutchinson, A. Y. Cho IEEE Photonics Technol. Lett. **9**, 294 (1997)

J. Faist, C. Gmachl, F. Capasso, C. Sirtori, D. L. Sivco, J. N. Baillargeon, and A. Y. Cho, Appl. Phys. Lett. **70**, 2670 (1997)

[6] M. Tacke, Infrared Phys. Technol. **36**, 447 (1996)

[7] e.g. "Infrared Applications of Semiconductors - Materials, Processing, and Devices", MRS Proc. **450** (1997), ed by M. O. Manasreh, T. H. Myers, and F. H. Julien.

[8] C.-H. Lin, R. Q. YXang, D. Zhang, S. J. Murry, S.-S. Pei, A. A. Allerman, and S. R. Kurtz, Electron. Lett. **33**, 598 (1997)

[9] R. Q. Yang, C.-H. Lin, S. J. Murry, S. S. Pei, H. C. Liu, M. Buchanan, and E. Dupont, Appl. Phys. Lett. **70**, 2013 (1997)

[10] S. Sauvage, Z. Moussa, P. Boucard, F.H. Julien, V. Berger and J. Nagle, Appl. Phys. Lett. **70**, 1345 (1997)

[11] O. Gauthier-Lafaye, P. Boucaud, F.H. Julien, R. Prazeres, F. Glotin, J.-M. Ortega, V. Thierry-Mieg, R. Planel, J.-P. Leburton, V. Berger, Appl. Phys. Lett. **70** (24), 3197 (1997)

[12] O. Gauthier-Lafaye, P. Boucaud, F.H. Julien, S. Sauvage, S. Cabaret, J.-M. Lourtioz, V. Thierry-Mieg, R. Planel, Appl. Phys. Lett., in print

[13] Y.B. Li, J W Cockburn, M. S. Skolnick, M.J. Birkett, J. P. Duck, R. Grey, G. Hill, Electron. Lett. **33**, 1874 (1997)

[14] G. Strasser, P. Kruck, M. Helm, J.N. Heyman, L. Hvozdara, E. Gornik Appl. Phys. Lett. **71** (20), 2892 (1997)

[15] P. Kruck, G.Strasser, M. Helm, L. Hvozdara, E.Gornik, Physica B, in print

[16] J. Faist, F. Capasso, C. Sirtori, D. L.Sivco, A. L. Hutchinson, S.-N. G. Chu, and A. Y. Cho, Appl. Phys. Lett. **65**, 94 (1994)

[17] L. Hvozdara, J.N. Heyman, G. Strasser, K. Unterrainer, P. Kruck, M. Helm, E. Gornik, Proc. ISCS24, San Diego (1997), to be published

NOVEL PIEZOELECTRIC HETEROSTRUCTURE FOR ALL-OPTICAL INFRARED LIGHT MODULATION

V. ORTIZ*, N. T. PELEKANOS*, GUIDO MULA*,**
*Département de Recherche Fondamentale sur la Matière Condensée, CEA/Grenoble, SP2M/PSC, 17 rue des Martyrs, 38054 Grenoble, France
** INFM and Dipartimento di Scienze Fisiche, Università degli Studi di Cagliari, 09124 Cagliari, Italy

ABSTRACT

We present an innovative-design heterostructure based on the exploitation of in-the-barrier piezoelectric field for all-optical light modulation. The novel layout allows an efficient light modulation with low power densities (few tens of W/cm^2), easily attainable with standard laser diodes. The modulation mechanism relies upon drastic photocarrier separation by the piezoelectric field in the barrier layers. We present room temperature results showing that an optical "control" power of 70 W/cm^2 creates in the heart of the structure a space-charge field of about 30 kV/cm, inducing large spectral shifts (~100 nm) in the photoluminescence spectra of a CdHgTe quantum well in the 1.5 μm range.

INTRODUCTION

Fast parallel image processing applications require optical modulators that are sufficiently fast (MHz modulation rates [1]) and that can operate with the optical power of a standard laser diode (W/cm^2). In this context, the aim of this paper is to demonstrate an innovative-design piezoelectric heterostructure suitable for realising all-optical light modulators that matches these requirements.

Previous work on piezoelectric heterostructures used the in-well screening of the piezoelectric field by photocarriers as a mechanism for enhanced all-optical nonlinearity [2,3,4]. However, the continuous depletion of the screening photocarriers on the nsec time scale by radiative recombination, makes such all-optical modulators very demanding in optical "control" powers, typically of the order of kW/cm² [3]. We redesigned the structure of standard piezoelectric devices by "moving" the piezoelectric field from the QW to the barrier layers. This new modulator device is all-optical in the sense that a "control" laser beam modulates a "read" one without the use of an external bias. We further require that it operates at room temperature with a "control" power of the order of 10 W/cm² .

A piezoelectric field can be generated in compound semiconductors by growing strained layers on a high-index surface, such as for example (111) or (211) [5,6]. This field, directed along the growth axis, exceeds typically 100 kV/cm for rather moderate lattice mismatch strains of $\varepsilon = 0.7\%$ [7,8]. We took advantage of the possibilities of Molecular Beam Epitaxy (MBE) for growing coherent heterostructures with carefully controlled amount of strain selectively inserted in the barrier layers.

The use of the piezoelectric field inside the barriers emphasises efficient spatial separation of the screening photocarriers lengthening by orders of magnitude their effective lifetime, which is essential in order to reduce the optical power requirements. We present here a proof-of-principle demonstration of such an all-optical modulator, operating at room temperature and with optical powers of a few tens of W/cm².

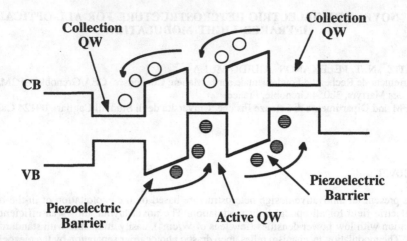

Figure 1. Principle of operation of an all-optical light modulator device utilising photocarrier separation at the piezoelectric barriers. The photocarriers generated in the barriers are separated via the built-in piezoelectric field and then accumulated in the *collection* QWs. The ground state of the *active* QW is then modulated by the electric field generated by the accumulated carriers.

EXPERIMENT

Samples' Structure

In Fig.1 we show a schematic of one period of the heterostructure we propose. It is grown along a polar axis and its period consists of three QWs and two piezoelectric barriers. The outer QWs are the carrier collection QWs, while the central (active) QW is the one whose optical properties are to be modulated by the presence of stored carriers in the collection QWs. It is important to note that the lowest exciton resonance of the active QW is the ground state of the whole heterostructure.

Principle of Operation

The principle of operation is as follows: a "control" optical beam with photon energy larger than the barrier bandgap photogenerates electrons and holes in the piezoelectric barrier layers. There, the piezoelectric field separates the carriers on either side of each barrier. The collected carriers in the active QW deplete radiatively on the nsec time scale, and the end result of the optical excitation is an accumulation of electron and hole populations in the outer QWs. This creates a space-charge electric field which shifts the excitonic resonance of the active QW according to the quantum-confined Stark effect. An optical "read" beam resonant to the exciton of the active QW is thus modulated.

The switch-on time of the modulating electric field is of the order of nsec depending mainly on the depletion by radiative recombination of the carriers in the active QW. The switch-off time, however, is directly related to the escape rates from the collection QWs due to tunneling and thermionic emission and then depends largely on the heterostructure parameters, such as layer thicknesses, composition and band-offsets. We estimate for various heterostructure choices that

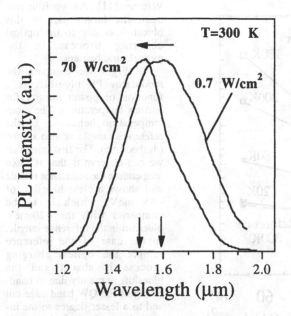

Figure 2. Room temperature PL spectra with increasing optical power density of excitation from the active QW of a modulator sample designed to operate at 1.5μm. A clear blue-shift (~100 nm) of the luminescence peak is observed from low (0.7 W / cm²) to high (70 W / cm²) pump power. The intensities are normalised.

the switch-off times can be easily tuned between 0.1-1 μsec allowing large design flexibility depending on the device response requirements. Assuming 100 nsec switch-off time and based on a simple rate equation model, we estimate that a few tens of W/cm² of CW optical excitation is sufficient to generate 50 kV/cm across the active QW.

Samples' Preparation

The samples discussed here are grown coherently on (211) $Cd_{0.96}Zn_{0.04}Te$ substrates by MBE and consist of a single heterostructure period, as shown in Fig.1, embedded between a 2000 Å-thick $Cd_{0.75}Mg_{0.25}Te$ buffer layer (lattice-matched with the substrate) and a 2000 Å-thick $Cd_{0.75}Mg_{0.25}Te$ cap layer to minimise surface-field effects. The collection QWs are 400 Å of $Cd_{0.70}Hg_{0.30}Te$. For a 1.5 μm-modulator sample, the active QW is 200 Å of $Cd_{0.60}Hg_{0.40}Te$ with its bandgap around 1.5 μm. Please note the large flexibility inherent to the CdHgTe system in choosing the modulation wavelength throughout the entire infrared spectrum by simply varying the Hg-composition in the active QW [9]. The piezoelectric barriers are 400Å-thick $Cd_{0.68}Zn_{0.12}Mg_{0.20}Te$. They are under dilation (ε = -0.75%) and the strain induced piezoelectric field is over 100 kV/cm [10]. It is important to note that both the active and the outer QWs are under a small compression (ε = 0.10-0.15%). Hence there exists also in these layers a piezoelectric field, which we estimate by extrapolating the available data of ref.[10] to be around 30-40 kV/cm. Finally, a reference sample is grown identical to the modulator with the exception that, instead of the piezoelectric barriers, lattice-matched $Cd_{0.75}Mg_{0.25}Te$ barriers are used having no piezoelectric field. It should be noted that the bandgap difference between $Cd_{0.68}Zn_{0.12}Mg_{0.20}Te$ and $Cd_{0.75}Mg_{0.25}Te$ is negligible, of the order of 10 meV.

RESULTS

We monitor the photogenerated electric field by measuring the shifts of the photoluminescence (PL) peak of the active QW, as a function of incident power density and for various temperatures. The Argon blue laser lines having photon energies larger than the piezoelectric barrier bandgap served simultaneously for the PL excitation and the optical charging of the structure. In Fig.2, we show room temperature PL spectra as a function of power density for the modulator sample. A pronounced PL blueshift of about 100 nm is observed as we increase the power from 0.7 to 70

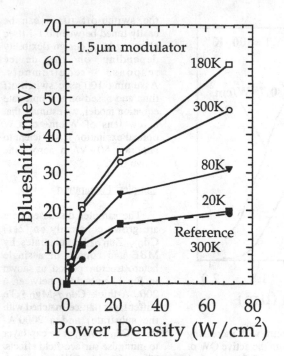

Figure 3. PL blueshift versus optical power density
from 20K to room temperature for the modulator (solid
lines). The room temperature modulation from the
reference sample is also shown (dashed line).

W/cm² [11]. As we illustrate next, the larger part of this blueshift is due to the optical charging process in the piezoelectric barriers.

In Fig.3 we compare the modulator PL blueshift as a function of power density for various temperatures. The room temperature behaviour of the reference sample is also shown (dashed line). The first thing that we can observe is that at room temperature the modulator sample still shows a clear blueshift of ~45 meV, which is to be compared with the ~20 meV blueshift in the reference sample. In the case of the reference sample the optical charging process is absent and the blueshift is mainly due to band-filling of the QW band edge tail and to a lesser degree to the in-well screening of the small piezoelectric field present in the QW. We should also mention that for the reference sample the blueshift does not depend on temperature, remaining always less than 20 meV [12]. The modulator's blueshift is for all power densities larger than the reference's one due to exactly the optical charging process. This is true for all temperatures except at T = 20K, where the two curves are almost superposed. A maximum blueshift of ~60 meV for the modulator is obtained at T = 180K. At room temperature the comparison between the two samples allows us to conclude that the net blueshift due to the optical charging process when the modulator is excited by 70 W/cm² is ~25 meV. This blueshift corresponds for a 200 Å QW to a photogenerated electric field of about 30 kV/cm. Thus, the efficiency of the optical charging process in our sample compares well with the estimates of the model described above, assuming that the electron and hole escape times from the collection QWs are of the order of 100 nsec.

The temperature dependence of the modulator's blueshift shown in Fig. 3 is worth a more detailed analysis. Contrary to what is usually the case, the optical charging process is more effective at higher temperatures. To understand this result we compared the temperature dependence of the integrated piezoelectric barrier PL peak intensity and of the maximum blueshift, defined as the PL blueshift between 0.7 and 70 W/cm², for the modulator sample.

The temperature evolution of the maximum blueshift is shown in Fig. 4 (right-hand axis). At T = 20K the blueshift is limited to 20 meV and coincides to that of the reference sample. With increasing temperature the blueshift increases and reaches its maximum value at T = 180K. A further increase in the temperature gives a small decrease in the blueshift and at room temperature, as we saw before, it remains quite large, ~45 meV. The dotted line in Fig.4 represents the

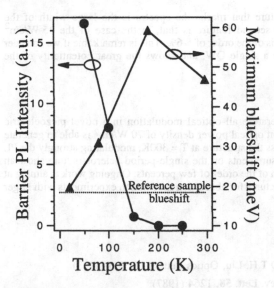

Figure 4. Spectrally integrated barrier PL intensity (left-hand side) and maximum PL blueshift of the active QW (right-hand side) in the modulator sample as a function of temperature. The dotted line indicates the modulation of the reference sample.

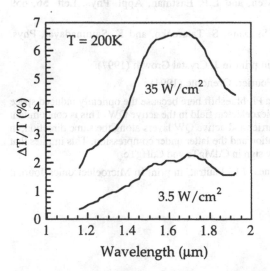

Figure 5. Percentage variation of the transmission of a single period modulator sample at T = 200K for two pump power densities (3.5 and 35 W/cm²).

maximum blueshift for the reference sample, which is constant for all temperatures. The temperature dependence of the integrated barrier PL peak intensity is also shown in Fig. 4 (left-hand axis). At T = 20K the PL intensity is maximum, and decreases with increasing the temperature. For temperatures above 150K the barrier PL almost disappears.

We can then explain qualitatively the temperature behaviour of the modulator as follows: the optical charging process is inefficient at low temperatures due to carrier trapping in shallow levels in the barriers, screening the piezoelectric field and hindering the photocarrier separation. This is supported by the observation of intense barrier PL at T = 20K. Hence, at T = 20K the observed blueshift is essentially due to band-filling of the band-edge tail and in-well screening. Increasing the temperature gives a reduction of the barrier PL intensity due to thermal excitation of the trapped carriers and enables the optical charging process. For T ≥ 80K, the optical charging process becomes more and more efficient inducing larger blueshifts, allowing for the observation of 60 meV blueshift at T = 180K. Finally, at room temperature the thermionic emission of the carriers (mainly the holes) out of the collection QWs becomes important and results in the small decrease of the optical charging effect.

For the realisation of a real light modulator the crucial parameter is the variation of the transmission as a function of the optical control power. We present here preliminary measurements of transmission modulation of our single period modulator sample, with and without the pump laser. In Fig. 5 we show the percentage variation of the transmission for two pump power densities of 3.5 W/cm² and

35 W/cm^2 at T = 200K. The first feature that marks the spectra is the large width of the modulation region (~150nm), and the second feature is that, in the case of the 35 W/cm^2 measurement, the maximum modulation is of the order of 5-6%. This is remarkable if we consider that this is the modulation obtained by a single QW, and shows the great potentiality of the heterostructure we propose.

CONCLUSIONS

We have demonstrated room temperature all-optical modulation in a novel piezoelectric heterostructure. We show that an incident optical power density of 70 W/cm^2 is able to generate an electric field of about 30 kV/cm across the structure at T = 300K, modulating strongly the PL spectrum. Moreover, transmission measurements on the single-period heterostructure show an encouraging variation of the transmission of the order of few percents. Ongoing work is aiming at obtaining multiple period modulator structures for absorption modulation experiments with larger modulation contrast.

REFERENCES

1. A. Partovi, A.M. Glass, T.H. Chiu, D.T.H. Liu, Optics Letters **18**, 906 (1993).

2. D.L. Smith and C. Mailhiot, Phys. Rev. Lett. **58**, 1264 (1987).

3. T.S. Moise, L.J. Guido, R.C. Barker, J.O. White, A.R. Kost, Appl. Phys. Lett. **60**, 2637 (1992).

4. M. Livingstone, I. Galbraith, B.S. Wherrett, Appl. Phys. Lett. **65**, 2771 (1994).

5. D.L. Smith and C. Mailhiot, Rev. Mod. Phys. **62**, 173 (1990).

6. E. Anastassakis, Phys. Rev. B **46**, 4744 (1992).

7. E.A. Caridi, T.Y. Chang, K.W. Goossen, and L.F. Eastman, Appl. Phys. Lett. **56**, 659 (1990).

8. R. André, C. Deshayes, J. Cibert, Le Si Dang, S. Tatarenko, and K. Saminadayar, Phys. Rev. B **42**, 11392 (1990).

9. V. Ortiz, N.T. Pelekanos, Guido Mula, in print in J. Crystal Growth (1997).

10. R. André, Ph.D. thesis, Université J. Fourier, Grenoble (1994).

11. The optical charging process results in a PL blueshift here because the optically induced space charge field compensates the pre-existing piezoelectric field in the active QW. This is concomitant with having the piezoelectric field in the barrier and active QW layers along the same direction, in spite of the fact that the former is under dilation and the latter under compression. This implies that the piezoelectric coefficient e_{14} has opposite sign in CdMgTe and CdHgTe.

12. N.T. Pelekanos, Guido Mula, N. Magnea, J.L. Pautrat, in print in Microelectronics Journal (1997).

MULTIVALENT ACCEPTOR-DOPED GERMANIUM LASERS:
a solid-state tunable source from 75 to 300 μm

D. R. Chamberlin[1], O. D. Dubon[1,†], E. Bründermann[1], E. E. Haller[1], L. A. Reichertz[1,‡],
G. Sirmain[2], A. M. Linhart[2], H. P. Röser[2]
1 - Lawrence Berkeley National Laboratory and UC Berkeley, Berkeley, CA
2 - DLR, Institute of Space Sensor Technology, Berlin GERMANY

ABSTRACT

We report on the performance of far-infrared hole inversion lasers made from germanium doped with the multivalent acceptors beryllium and copper. Commonly used hole inversion lasers are made from Czochralski-grown Ga-doped Ge single crystals and show emission from 75 to 125 and 170 to 300 μm. The emission gap between 125 and 170 μm, originating from absorption of the far-infrared light due to internal hole transitions in the neutral Ga acceptor, is absent in the new Be and Cu-doped lasers. We also find a mechanism for inversion depopulation through neutral Ga which hinders lasing at low electric fields. This same mechanism is shown to cause population inversion in the Be-doped laser and allows lasing at lower fields. This reduces the power input into the germanium crystal and has allowed us to increase the duty cycle up to 2.5% which is one order of magnitude higher than the maximum duty cycle reported for Ga-doped Ge lasers. These new lasers may offer an opportunity for achieving continuous-wave operation.

In addition we have performed preliminary studies on the effect of uniaxial stress on the lasing in these new materials. We demonstrate that small uniaxial stress increases laser action in Ge:Cu. We propose that this is due to an increased population inversion because under these conditions two separate mechanisms cause heavy holes to enter the light hole band.

INTRODUCTION

There are a number of important applications for a compact, tunable, continuous-wave far-infrared (FIR) laser. For instance, astronomers need such a device for high-resolution heterodyne spectroscopy of FIR light from star-forming regions to determine their chemical constituents. The most promising device for this application is the p-type germanium laser based on a hole population inversion. However, so far this device is only operable in a pulsed mode because of excessive power dissipation. The objective of our research is to achieve continuous-wave operation so that such a laser can be operated aboard satellites or high-altitude airborne observatory.

The p-Ge hole laser was invented in 1984 independently by Andronov et al [1] and Komiyama et al [2]. The laser operates near liquid helium temperatures under crossed electric (E) and magnetic fields (B). Under these conditions the maximum kinetic energy $E_{kinetic}$ of a charge carrier with effective mass m* is given by the equation:

$$E_{kinetic} = 2 \ m^* \ (E/B)^2 \quad (1)$$

as determined from the Lorentz force. Equation (1) shows that for any set value of electric and magnetic field, heavy holes in the germanium valence band will have a higher kinetic energy than light holes. Under certain E and B conditions heavy holes will accelerate with little energy loss until they reach the optical phonon energy (37 meV). At this energy the cross-section for optical phonon emission rises by several orders of magnitude. An optical phonon is emitted and the hole returns to the top of the valence band (k=0). In this process there is a finite probability of heavy holes entering the light hole band. If the ratio of E and B is chosen properly the light holes do not gain enough energy to emit an optical phonon and the light hole lifetime will be relatively large. In this way a population inversion is created between the light and heavy hole sub-bands within the valence band. Far-infrared stimulated emission originates from transitions of holes between the

† - current address: Harvard University, Gordon McKay Lab, 8 Oxford St., Cambridge, MA
‡ - current address: Max Planck Institute for Radio Astronomy, Bonn, GERMANY

light and heavy hole band or between light hole Landau levels. Because of the hole current necessary for operation, a large amount of resistive heating occurs during operation. This increases the acoustic phonon density (temperature) and destroys the population inversion necessary for lasing. As a consequence the lasers must be operated in a pulsed mode.

From 1984 until 1996 all germanium hole inversion lasers were fabricated from germanium doped with gallium or other shallow hydrogenic acceptors. These lasers showed an emission gap in the critical range between 60 and 80 cm^{-1}. This emission gap was shown to originate from absorption of the emitted photons by the dopant acceptors.[3] Multivalent dopants have much larger ionization energies than hydrogenic acceptors, well above the range of laser emission. We have since demonstrated operation of lasers doped with the double acceptors beryllium and zinc [4] and the triple acceptor copper [5]. These lasers show emission throughout the range from 40 to 130 cm^{-1} without an emission gap [6]. We present in this paper for the first time data on all three lasers and compare their inversion mechanisms. We show preliminary measurements of the effect of uniaxial stress on the lasing region in the standard E-B plot for all three materials. In addition, we show that multivalent dopants allow operation at reduced electric fields and currents and demonstrate higher duty cycles than are achievable with single-acceptor doped germanium.

EXPERIMENTS

For this study, germanium single crystals were Czochralski-grown with the dopants Ga or Be. Copper doping was achieved by sputtering a 1000 Å layer of copper on both sides of ultra-pure Ge wafers with residual shallow impurities of $2x10^{11}$ cm^{-3}. The copper was then diffused at 700°C for 24 hours and the wafers were quenched in ethylene glycol to achieve uniform doping. The germanium crystals were characterized with variable temperature Hall effect and photothermal ionization spectroscopy to determine the dopant concentrations and types. The laser crystals were cut, etched, and contacted as described elsewhere [5] to fabricate the laser crystals.

The laser emission was measured under liquid helium by a fast, highly compensated Ge:Ga photoconductor. All the crystals used in this study had a 10 mm^2 cross-section parallel to the electric field direction and had a total crystal volume of 45 mm^3. These crystals are small relative to those commonly found in the germanium laser literature. Because the refractive index of germanium is high, the lasers can operate without external resonators using only the polished crystal surfaces as mirrors for internal reflections. The doping concentrations were $9x10^{13}$, $1x10^{14}$, and $3x10^{15}$ cm^{-3} for the gallium, beryllium, and copper-doped lasers, respectively. The increased doping in the Ge:Cu laser leads to the same hole current as the lower doped Ge:Be and Ge:Ga, since ionization of the deeper copper acceptor is not complete [5]. The electric field was oriented parallel to the [0 0 1] direction and the magnetic field was pointed in the [1 1 0] direction. The light was measured in the same [1 1 0] direction as the magnetic field. In the stress experiments, pressure was applied in the [0 0 1] direction by using a piston as the top electrode. Force was applied to the piston by a spring located outside of the liquid helium cryostat.

RESULTS

The magnitudes of electric and magnetic fields were measured at the onset of lasing. The heavy hole energy under these conditions was calculated from equation (1). The results for the gallium-doped laser are shown in figure 1. Stimulated emission is expected to occur as soon as the heavy holes reach enough energy to emit an optical phonon, i.e., 37 meV. However, laser action is not observed in the Ge:Ga laser until the heavy holes reach much higher energies. This can be understood by calculating both the heavy and light hole energies as a function of electric and magnetic fields, shown in figure 2. This graph shows that as the heavy holes reach the energy to emit an optical phonon and enter the light hole band, the light holes have the same energies as ground to excited state transitions if holes in neutral Ga acceptors. This provides a mechanism for non-radiative transitions between the sub-bands. When a heavy hole enters the light hole band upon emission of an optical phonon, the conditions of E and B are such that it will have an energy of about 7 meV as a light hole. The energy of the lowest, optically active gallium excited state lies at 6.7 meV. If a light hole with this energy encounters a neutral gallium impurity, it can transfer its

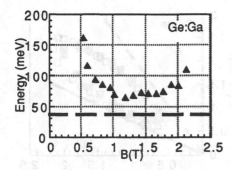
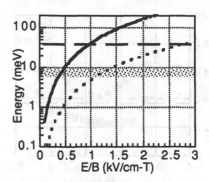

Figure 1: Heavy hole energies at observed conditions for the onset of stimulated emission. ▲ - heavy hole energy, dashed line - optical phonon energy.

Figure 2: Calculated hole and state energies as a function of applied fields. solid line - heavy holes, dotted line - light holes, dashed line - optical phonon energy, dotted area - region of Ga excited states.

energy to a bound hole in the ground state and raise it to a bound excited state. The bound, excited state hole returns to the ground state by emission of phonons. We will term this energy exchange between free holes and neutral acceptors "hole-neutral acceptor energy transfer." At this point the light hole has a very good probability to reenter the heavy hole band, shortening the light hole lifetime. Lasing is not observed until the probability for a radiative transition is greater than the probability of energy transfer from a light hole to a neutral acceptor. A similar analysis of hole energies has been performed for the Ge:Tl laser, and light hole-neutral Tl energy transfer occurs in this laser where the light hole energies coincide with the energies of the neutral Tl excited states.[7]

Since the energies of the excited states for the multivalent acceptors are much higher, we have measured the heavy hole energy at the onset of lasing in the Ge:Be and, for the first time, in the Ge:Cu lasers. Indeed, the Ge:Cu laser does show the onset of lasing at heavy hole energies directly above the optical phonon energy (figure 3). However, the Ge:Be laser displays population inversion at heavy hole energies as low as 21 meV, well below the optical phonon level. This energy coincides with the lowest, optically active beryllium ground to excited state transition. The inversion at these low energies has therefore been attributed to transfer of energy from the heavy holes to neutral beryllium acceptors [4]. The energy of the lowest copper ground to excited state transition lies at 38 meV, very near the optical phonon energy. Since these energies lie so close together, it is not possible to determine whether the inversion in the Ge:Cu laser occurs because of emission of optical phonons, transfer of heavy hole energy to neutral copper acceptors, or a combination of both mechanisms.

To gain more insight into the population inversion in the Ge:Cu laser, preliminary experiments with uniaxial stress have been performed. This is the first time that the effect of stress on multivalent acceptor-doped lasers has been investigated. Uniaxial stress destroys the symmetry of the lattice and splits the heavy and light hole bands at the Brioullin zone center. In doing this it also warps the bands and changes their effective mass. In particular, the heavy hole mass decreases [8]. Because the effective mass is reduced, the ionization energy and the energy of the bound excited states are also reduced. The results from the application of small uniaxial stress in the [0 0 1] direction on the lasing are shown in figures 4 and 5. Figure 4 shows the settings of electric and magnetic field for which lasing is observed in the Ge:Be laser. Upon the application of 0.3 kbar uniaxial stress, the minimum E/B ratio necessary for lasing to be observed increases. This can be understood because the reduced heavy hole effective mass increases the E/B ratio necessary to achieve the same energy. However, a quantitative, more detailed analysis relating the pressure-induced bandstructure changes to the observed changes in laser operation to occur has yet to be performed.

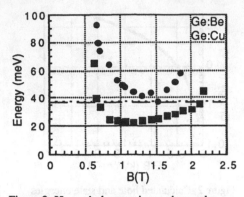

Figure 3: Heavy hole energies at observed conditions for the onset of stimulated emission. ■ - Ge:Be, ● - Ge:Cu, solid line - Be excited state, dashed line - optical phonon energy, dotted line - Cu excited state

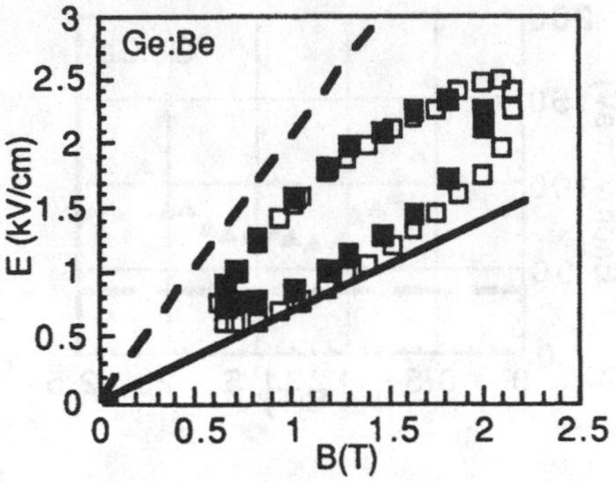

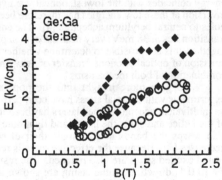

Figure 4: Electric and magnetic field conditions for lasing in Ge:Be under the application of 300 bar [001] stress. □ - zero stress, ■ - 300 bar, solid line - E_{heavy} = 21 meV, dashed line - E_{light} = 21 meV.

The effect of [001] stress on the conditions for lasing of Ge:Cu is shown in figure 5. With the same 0.3 kbar stress applied to the Ge:Be laser, the area of E and B field under which lasing is observed *increases*. This can be tentatively attributed to the decrease in the energy of the bound excited state of the neutral copper impurity. Since the energy decreases upon application of uniaxial stress, more heavy holes will transfer to the light hole band through heavy hole-neutral Cu energy transfer. This will increase the population inversion. Upon further increase of the pressure to 0.6 kbar the area of E and B under which lasing occurs decreases. This can be attributed to the decrease in the heavy hole mass, just as in the Ge:Be laser. Again, a more quantitative assessment of the relative importance of these factors in enhancing or reducing the population inversion is necessary.

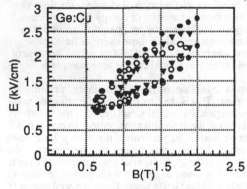

Figure 5: Electric and magnetic field conditions for lasing in Ge:Cu under the application of [001] stress. o - zero stress, ● - 0.3 kbar, ▼ - 0.6 kbar.

Figure 6: Electric and magnetic field conditions for lasing in Ge:Ga and Ge:Be. Lasing is observed inside the boundaries given by the data points. o - Ge:Be, ◆ - Ge:Ga.

Figure 6 shows the conditions of electric and magnetic fields for which lasing is observed at zero stress in the Ga and Be-doped lasers. The Ge:Be laser demonstrates stimulated emission at much lower electric fields than in Ge:Ga, and therefore uses smaller currents. Because the currents needed for lasing in Ge:Be are significantly lower, we have tested lasers made from this material for high duty cycle. The best duty cycle achieved has been 2.5% in a Be-doped laser doped 2×10^{13} cm^{-3} with the dimensions 2 x 4 x 8 mm oriented along [0 0 1] x [1 1 0] x [1 -1 0], respectively. The electric field was applied in the [0 0 1] direction, and the 4 x 8 mm crystal faces were attached to large copper electrodes for maximum cooling. This laser reached a maximum pulse length of 25 μs at 1 kHz repetition rate and emitted power in the milliwatt range. This corresponds to an external efficiency of about 10^{-4} percent.

CONCLUSIONS

We have demonstrated two mechanisms for population inversion in germanium hole inversion lasers doped with three different types of acceptors. In gallium-doped lasers, the mechanism for inversion is dominated by emission of optical phonons. Holes switch between the heavy and light-hole bands in beryllium-doped lasers through transfer of energy to neutral Be atoms. In both lasers, we show that the inversion is weakened or destroyed when light hole-neutral acceptor energy transfer occurs. In the copper-doped lasers, the excited state energy lies very close to the optical phonon level so it is not known whether phonon emission or heavy hole-neutral copper energy transfer dominates the inversion process at zero pressure. However, with the application of small uniaxial stress it appears that the increased contribution of heavy hole-neutral Cu energy transfer enhances the population inversion.

Because the inversion in beryllium-doped germanium occurs at low heavy hole energies, lasers fabricated from this material allow operation at lower electric fields and currents. This has allowed us to achieve a record duty cycle of 2.5% for a germanium hole inversion laser, which is one order of magnitude better than the best duty cycle reported for gallium-doped lasers [9]. With further improvements such as the use of external resonators and further miniaturization, continuous-wave operation should be achievable in the foreseeable future.

ACKNOWLEDGMENTS

We acknowledge partial support for this work by the German DLR and the use of facilities at the Lawrence Berkeley National Laboratory operated under U. S. Department of Energy Contract DE-AC03-76SF00098. D. R. C. acknowledges the support of the NASA Office of Space Science for her support.

REFERENCES

1. A. A. Andronov, I. V. Zverev, V. A. Kozlov, Yu. N. Nozdrin, S. A. Pavlov, and V. N. Shastin, JETP Lett. 40, pp.804-807 (1984).

2. S. Komiyama, N. Iizuka, and Y. Akasaka, Appl. Phys. Lett. 47, pp. 958-960 (1985).

3. W. Heiss, K. Unterrainer, E. Gornik, W. L. Hansen, and E. E. Haller, Semicond. Sci. Techonol. 9, pp.638-640 (1994).

4. E. Bründermann, A. M. Linhart, L. Reichertz, H. P. Röser, O. D. Dubon, W. L. Hansen, G. Sirmain, and E. E. Haller, Appl. Phys. Lett. 68, pp.3075-3077 (1996).

5. G. Sirmain, L. A. Reichertz, O. D. Dubon, E. E. Haller, W. L. Hansen, E. Bründermann, A. M. Linhart, and H. P. Röser, Appl. Phys. Lett. 70, pp.1659-1661 (1997).

6. L. A. Reichertz, O. D. Dubon, G. Sirmain, E. Bründermann, W. L. Hansen, D. R. Chamberlin, A. M. Linhart, H. P. Röser, and E. E. Haller, to be published in Phys. Rev. B.

7. A. M. Linhart, E. Bründermann, L. A. Reichertz, H. P. Röser, O. D. Dubon, G. Sirmain, W. L. Hansen, and E. E. Haller, Conference Proceedings of the 21st International Conference of Infrared and Millimeter Waves, Berlin, July 1996.

8. J. C. Hensel and K. Suzuki, Phys. Rev. B **9**, 4219 (1974).

9. E. Bründermann, A. M. Linhart, H. P. Röser, O. D. Dubon, W. L. Hansen, and E. E. Haller, Proceedings of the 7th International Symposium on Space Terahertz Technology, Charlottesville, VA, March 1996.

Band-reject infrared metallic photonic band gap filters on flexible polyimide substrate

SANDHYA GUPTA*, GARY TUTTLE*, MIHAIL SIGALAS** AND KAI-MING HO**
*Department of Electrical and Computer Engineering and the Microelectronics Research Center
Iowa State University, Ames IA 50011, sgupta@iastate.edu.
**Department of Physics and Astronomy and the Ames Laboratory, U. S. Department of Energy
Iowa State University, Ames IA 50011

ABSTRACT

Metallic photonic band gap (MPBG) structures are multi-layer metallic meshes imbedded in a dielectric medium. We report the successful design, fabrication, and characterization of infrared band-reject filters using MPBG structures in a flexible polyimide substrate. The metal layers of the MPBG have square grid patterns with short perpendicular cross-arm defects added halfway between each intersection. The transmission characteristics of these filters show a higher order band-reject region in addition to a lower order band gap that extends from zero to particular cut off frequency. The critical frequencies of the filters depend on the spatial periodicity of the metal grids and length of the cross-arm defects. Optical transmission measurements of the band-reject filters show lower edge cutoff frequency of about 2 THz and the higher order bandgap region centered around 4.5THz with attenuation of more than 35 dB in the bandgap region. This is in good agreement with the theoretical calculations. The filters maintain their optical characteristics after repeated bending, demonstrating mechanical robustness of the MPBG structure and have minimal dependence on angle of incidence.

INTRODUCTION

Photonic band gap (PBG) structures are periodic dielectric structures exhibiting a region of frequency in which electromagnetic waves cannot propagate. PBG structures have generated a lot of interest in scientific community because of their possible application as electromagnetic filters, reflectors, or waveguides [1] operating over a wide range of frequencies. Most of the research in the field of PBG has focused on dielectric structures.

Metallic photonic band gap(MPBG) structures have recently garnered more attention because of advantages over dielectric PBG structures[2],[3]. MPBG structures offer the potential of lighter weight, reduced size and lower cost as compared to dielectric structures. The use of metal can also lead to fundamentally different PBG characteristic. For an interconnected mesh structure, the stop band extends from zero frequency up to some cutoff frequency, determined by the periodicity of the structure[4]. On the other hand, isolated metal patches show a bandstop behavior very similar to a dielectric PBG characteristic[5].

In this paper, we describe an MPBG structure which shows both these behaviors, a lower order bandgap extending from zero to cutoff frequency and another bandstop region at higher frequency. The fabrication technique and the transmission behavior of this mechanically flexible MPBG structure in the far infrared are discussed in detail.

The MPBG structure described in this paper are related to frequency selective surfaces (FSSs), which are two dimensional arrays of metallic patches or aperture elements with frequency filtering properties. Frequency selective surfaces have been studied in great detail because of their application in the microwave region.[6] Most of the FSS work has focused on single layer metal structures.

THEORETICAL CALCULATION

We calculated the expected transmission spectrum of the three-layer MPBG structure before fabricating it. The transfer matrix method (TMM) originally introduced by Pendry and MacKinnon has been used for the theoretical calculations.[7] In the TMM technique, the total volume is divided into small cells, and the fields associated with each cell are coupled to neighboring cells. The final transfer matrix relates the incident wave on the PBG from one side to the outgoing wave on the other side. The TMM can be used to calculate the band structure of an infinite periodic system. However, for these studies, we used the TMM to determine the electromagnetic transmission and reflection coefficients as functions for frequency for waves incident on a PBG of finite thickness. The transfer matrix method has previously been applied in studies of defects in 2D PBG structures, of 3D layer-by-layer PBG structure, and of 2D and 3D metallic structure. In all those previous investigations, the theoretical results matched very well with experimental measurements.[8]-[10]

EXPERIMENT

The metallic photonic bandgap structures were fabricated in a layer-by-layer fashion using alternating layers of polyimide for the dielectric and aluminum for metal grids, as depicted in figure 1(a). The MPBG structures covers an area of 2cm x 2cm. The lattice constant of the metal grid is 32μm and the width of the metal lines is 2.5μm as shown in figure 1(b). Halfway between each intersection, a short perpendicular cross-arm defect has been added. The cross-arm is 2.5μm wide and has lengths(L) of 10.5μm, 18.5μm and 22.5μm for three different samples. The thickness of the dielectric separation layer is 11µm in all the samples. Standard spin-on polyimide supplied by DuPont Pyralin® SP series PI-1111 with a dielectric constant $\varepsilon_r = 2.8$ was used for dielectric.

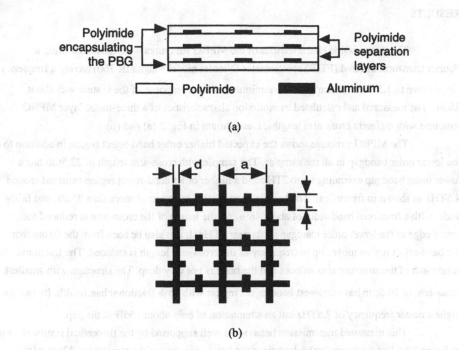

(a)

(b)

Figure 1. (a) Cross-sectional view of the three-layer metallic photonic bandgap structure, and (b) the grid pattern of each metal layer. The lattice constant of the grid is $a = 32$µm, the line width is $d = 2.5$µm, inter-layer separation is $w = 11$µm and additional metal bar length L is 10.5µm, 18.5µm and 22.5µm for three different samples.

The sequence of layers starts by spinning a 5µm thick layer of polyimide on a GaAs substrate. A 2000Å thick aluminum film is deposited on top and patterned using standard lift-off techniques. Another layer of polyimide is spun on over the metal grid and cured. The desired thickness of the polyimide layer is achieved by controlling the spin speed of the polyimide. Another metal layer is deposited and patterned, with this second metal grid pattern aligned directly over the first. The sequencing of layers continues with another thick polyimide layer followed by a third metal grid aligned to the previous layers, and a final cap layer of polyimide. After the structure is completed, the GaAs substrate is etched away in a citric acid/hydrogen peroxide solution, leaving a thin, polyimide-encapsulated MPBG structure.

RESULTS

The transmission characteristics of the MPBG structures were measured using a Fourier transform infrared (FTIR) spectrometer (Nicolet model Magna IR 760) having a frequency range down to 1.5 THz (50 cm^{-1}). The minimum relative response of the system was about 35dB. The measured and calculated transmission characteristics of a three-metal-layer MPBG structure with different cross-arm lengths, L are shown in Fig. 2 (a) and (b).

The MPBG structure shows the expected higher order band-reject region in addition to the lower order bandgap in all the samples. The sample with cross-arm length of 22.5µm has a lower order bandgap extending up to 2THz and a higher order band-reject region centered around 4.5THz as shown in figure 2(a). The band is deep, with a rejection of more than 35dB, and fairly wide, with a fractional bandwidth of about 48%. As the length of the cross-arm is reduced the cutoff edge of the lower order bandgap shifts near 3THz. It can also be seen from the figure that the band-reject region moves up in frequency as the cross-arm length is reduced. The fractional bandwidth of the structure also reduces and the band is not very deep. The structure with smallest cross-arm of 10.5µm has narrowest band-reject region with 25% fractional bandwidth. Its has the highest center frequency of 7.5THz but an attenuation of only about 15dB in the gap.

This measured transmission behavior is well supported by the theoretical results shown in figure 2(b). For the theoretical calculations the lattice constant of the structure is 32µm with a line width of 3.2µm. The two different sets of cross-arm lengths chosen for the calculations are 9.6µm and 22.4µm. The structure with cross-arm of length 22.4µm shown a lower bandgap cutoff edge near 1.8THz and a higher order bandgap centered around 3.5THz with a fractional bandwidth of 51% and rejection of about 50dB. By decreasing the cross-arm length to 9.6µm, the fractional bandwidth of the structure reduces and the bandgap moves up in frequency which is also seen in measured samples. The calculations do not include the effects of absorption in polyimide.

The filter characteristics change very little as the incidence angle varies from normal to 45°. The cut-off frequency was nearly independent of incidence angle in that range, and the transmission level in the band-reject region changes from 35dB at normal incidence to about 30dB at 45°. All the measurements are done using an unpolarized beam. However, theoretical studies show that for normal incidence, the cutoff frequency is independent of the two polarizations. As the incidence angle increases, there are small differences between the two polarizations. Even for

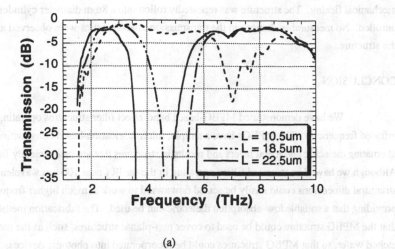

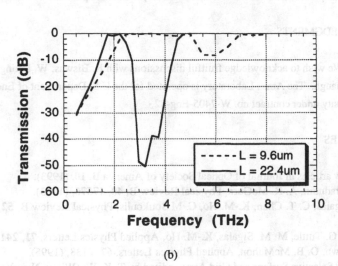

Figure 2: (a) Measured and (b) Calculated transmission characteristics of the band-reject filter as a function of the length of the cross-arm. The structure has a lattice constant of 32μm. Measured samples have metal line width of 2.5μm while the calculation results have 3.2μm line width.

angles as high as 80°, the difference in the cutoff frequencies of the two polarizations is less than

10%. Also, the MPBG structure maintains its transmission properties after considerable mechanical flexing. The structure was repeatedly rolled into a 8mm diameter cylinder and then unrolled. No measurable changes in the transmission characteristics were observed after flexing the structure.

CONCLUSION

We have demonstrated MPBG-based band-reject filter structures operating at far infrared frequencies. The MPBGs have a simple, microfabrication-based construction that uses alternating dielectric and metal layers and results in structures that are mechanically flexible. Although we have demonstrated filters operating in the far IR (near 100 μm wavelength), the structural dimensions could easily be scaled downward to work at much higher frequencies, providing that a suitable low-absorption dielectric can be used. The fabrication method suggests that the MPBG structure could be used to cover non-planar structures, such as the mesas on an etched wafer, so that MPBG structures could be incorporated into photonic devices.

ACKNOWLEDGMENTS

We wish to acknowledge fruitful discussions with R. Biswas, W. Leung, S.D. Cheng, and J. Kavanaugh. The Ames Laboratory is operated for the U.S. Department of Energy by Iowa State University under contract no. W-7405-Eng-82.

REFERENCES

1. See review articles in Journal of Optical Society of America B, 10, (1993).
2. A. A. Maradudin, A. R. McGurn, Physical Review B, 48, 17576, (1993).
3. M. M. Sigalas, C. T. Chan, K.-M. Ho, C. M.Soukoulis, Physical Review B, 52, 11744, (1995).
4. S. Gupta, G. Tuttle, M. M. Sigalas, K.-M. Ho, Applied Physics Letters, 71, 2412, (1997)
5. E. R. Brown, O. B. McMahon, Applied Physics Letters, 67, 2138, (1995).
6. Frequency Selective Surface and Grid Array, edited by T. K. Wu(Wiley, New York, 1995).
7. J. B. Pendry, A. MacKinnon, Physical Review Letters, 69, 2772, (1992).
8. M. M. Sigalas, C. M. Soukoulis, E. N. Economou, C. T. Chan and K.-M. Ho, Physical Rev. B, 48, 14121, (1993).
9. E. Ozbay, G. Tuttle, M. M. Sigalas, C. M. Soukoulis, K.-M.Ho, Physical Review B, 51, 13961, (1995).
10. J. S. McCalmont, M. M. Sigalas, G. Tuttle, K.-M. Ho, C. M. Soukoulis, Applied Physics Letters, 68, 2759, (1996).

Part IV

IR Detectors

High Temperature Infrared Photon Detector Performance

C.H. GREIN*, H. EHRENREICH**
*Department of Physics, M/C 273, University of Illinois at Chicago, 845 W. Taylor St. #2236, Chicago IL 60607-7059
**Department of Physics and Division of Engineering and Applied Sciences, Harvard University, Cambridge, MA 02138

ABSTRACT

The combined effects of suppressing Auger recombination by band structure engineering strained-layer superlattices (SL), suppressing radiative recombination by photon recycling, and suppressing both Auger and radiative recombination with carrier depletion are calculated quantitatively for long-wavelength and mid-wavelength InAs/ $In_xGa_{1-x}Sb$ SL-based infrared detectors operating at temperatures between 200 and 300K. The results are compared to their HgCdTe counterparts. The SL performance is better in all cases. However, the carrier concentrations required for typical background limited performance (300K, 2π field of view), ranging between about 1×10^{13} and 4×10^{13} cm^{-3} at 300K in long-wavelength to mid-wavelength SLs, are seen to be impractically low. The carrier concentration in a 11 μm photon detector yielding equivalent performance to an ideal 300K thermal detector is about 10^{14} cm^{-3}. Large performance enhancement using carrier depletion therefore appears impractical even in optimized SLs.

INTRODUCTION

The cost, weight, and reliability of cryogenic coolers severely inhibits the widespread use of mid-wavelength (3-5μm) and long-wavelength (8-12μm) infrared photon detectors. Photon detector detectivities are increased by enhancing carrier recombination lifetimes. Hence a strategy for reducing and possibly eliminating cooling requirements is to exploit a variety of techniques for increasing carrier lifetimes. In this paper, we explore three effects: (i) the suppression of Auger recombination using band structure engineering in strained layer superlattices[1]; (ii) the suppression of radiative recombination using photon recycling[2, 3]; and (iii) the suppression of both Auger and radiative recombination with carrier depletion[4, 5]. These effects are all based on reducing the device noise by increasing carrier lifetimes without additional cooling, hence improving the device detectivity. Effects (ii) and (iii) may be employed in both bulk and superlattice-based infrared detectors, whereas (i) is unique to heterostructure-based designs. While promising, we find the combination of all three effects on device performance to be insufficient for typical background limited performance at room temperature in such detectors because of unrealistically low required carrier concentrations. However, the gains are substantial for realistic carrier concentrations, which may lead to reduced cooling requirements in a variety of infrared systems.

The present work considers photoconductive detectors with active regions consisting of InAs/InGaSb strained layer superlattices. For reference, comparisons are made with bulk HgCdTe-based detectors. The strained layer superlattices have been extensively explored in recent years due to the possibility of significantly suppressing Auger recombination rates with band structure engineering methods in comparison with unstrained bulk materials[6]. The extent to which Auger recombination is suppressed is mainly determined by the energy splitting between the light and heavy holes, associated with the strain between the layers. The more this splitting exceeds the energy gap, the greater the suppression[1]. Suppression of Auger rates of up to two orders of magnitude has been observed in long-wavelength InAs/InGaSb superlattices[6].

Radiative recombination rates may be substantially suppressed by exploiting photon recycling. This involves the reabsorption of photons emitted within a sample before they have the opportunity to escape from it, and can result in order of magnitude enhancements[7] in radiative lifetimes over those obtained from the standard van Roosbroeck and Shockley expression[8], which assumes photons immediately leave the sample after emission. Photon recycling was treated in the recombination rate calculations reported here employing the methods described in ref. [3].

The third effect, carrier depletion, suppresses both radiative and Auger recombination rates by reducing carrier concentrations below equilibrium. Specific designs of photoconductive detectors that permit carrier depletion, and associated limits on electric fields within the detectors, are discussed in ref. [4]. The present work on photoconductive detectors treats carrier depletion by performing recombination rate calculations involving carriers at concentrations below thermal equilibrium. The nonequilibrium carrier distribution functions are obtained by assigning electron and hole quasi-chemical potentials to produce the chosen carrier concentrations at the chosen lattice temperature.

METHODS

The photoconductor detectivity is calculated following the assumptions of Kinch and Borrello[10] and Piotrowski et al.[5]. The specific detectivity is given by

$$D^* = \mathcal{R} \frac{\sqrt{A\Delta f}}{I_n} \tag{1}$$

where $\mathcal{R}$ is the responsivity, A is the detector area (optical and electrical areas are assumed to be equal), Δf is the bandwidth, and I_n is the noise current. Under conditions where the noise is dominated by generation and recombination processes, the noise current is given by[11]

$$I_n = e\sqrt{2d(G + R)A\Delta f}, \tag{2}$$

where G and R are the carrier generation and recombination rates respectively, and d is the detector thickness. The responsivity for photons of energy $h\nu$ is defined as

$$\mathcal{R} = \frac{I_s}{P_I} = \frac{I_s}{\phi_s Ah\nu}, \tag{3}$$

where I_s is the signal current, P_I is the signal power, and ϕ_s is the signal flux. If η electron-hole pairs are produced and contribute to I_s for each incident photon and assuming unity photoelectric gain, then $I_s = \eta e \phi_s A$, and

$$\mathcal{R} = \frac{\eta e}{h\nu}. \tag{4}$$

Employing (2) and (4) in (1) gives the device detectivity

$$D^* = \frac{\eta}{h\nu} \frac{1}{\sqrt{2d(G + R)}}. \tag{5}$$

In the present work, two generation-recombination mechanisms are incorporated into the calculations: (i) the net recombination rate due to carrier-carrier scattering, namely the Auger recombination rate R_A minus the impact ionization rate G_A[1]; and (ii) the net recombination rate due to radiative transitions, namely the radiative recombination rate R_r minus the photon absorption rate G_r, including the effects of photon recycling[3]. The net recombination rate is $R - G = (R_A - G_A) + (R_r - G_r)$. The calculations of $G+R$ may be simplified by noting that in the limit of strong carrier depletion $G_A >> R_A$ so $G_A + R_A \approx G_A - R_A$, whereas for near equilibrium carrier distributions $G_A \approx R_A$. The prediced detectivities are theoretical upper bounds which can be achieved in ideal, defect free materials because Shockley-Read-Hall recombination is neglected in the present calculations.

Superlattice and bulk band structures were calculated employing 8 band K·p formalisms using an envelope function approach which is described in detail in ref. [12]. The input parameters for the band structure calculations are given in ref. [1]. The temperature dependence of the band structure was neglected.

RESULTS

Figure 1 plots calculated device detectivities at temperatures ranging from 200K to 300K for nonequilibrium electron concentrations of 10^{12} to 10^{16} cm^{-3}. The background radiation was not incorporated in the evaluation of the detectivity, hence the plots display limits based solely on internal device noise. Fig. 1(a) corresponds to bulk HgCdTe with an 11μm energy gap. Some controversy exists regarding discrepancies between theoretical and experimental p-type lifetimes in HgCdTe[13, 14]. The Auger lifetimes calculated and employed here for HgCdTe are dominated by n-type recombination and are hence unaffected by p-type rates. Fig. 1(b) corresponds to a 35.9Å InAs/15.7Å $In_{0.225}Ga_{0.775}Sb$ superlattice, which, according to previous calculations, has an 11μm energy gap. The superlattice, having the same average lattice constant as that of a GaSb substrate, was designed according to the band structure engineering methods of ref. [1] to provide a large reduction in Auger recombination rates[15]. The detector thickness in both cases was chosen to be 1×10^{-3} cm, the quantum efficiency equal to be unity, and the donor concentration equal to 10^{12} cm^{-3}. Such a low donor concentration is probably not possible to achieve experimentally, but is considered here to permit the examination of nonequilibrium electron concentrations down to this value. Plotted also is the background limited detectivity for an assumed 300K background and a 2π field of view. Note that the plotted device detectivities can exceed this limit because the background radiation and associated noise are not included in the evaluation of the device detectivity. The device detectivities would asymptotically approach the background limit if the background radiation were included.

The 300K equilibrium intrinsic carrier concentration for 11μm bulk HgCdTe is found to be 6.0×10^{16} cm^{-3}, and that for the 11μm superlattice is 3.8×10^{16} cm^{-3}. For 11μm bulk HgCdTe, the electron concentrations required to achieve background limited performance (BLIP) are 4.8×10^{13}, 1.4×10^{13}, 4.9×10^{12}, and 2.6×10^{12} cm^{-3} at the temperatures of 200, 250, 275, and 300K respectively. The corresponding electron concentrations for the 11μm superlattice are 3.2×10^{14}, 3.8×10^{13}, 2.3×10^{13}, and 1.4×10^{13} cm^{-3}. The higher carrier concentrations in the superlattice are a direct consequence of the suppression of Auger recombination in this system.

These results illustrate the need for very strong carrier depletion in order to achieve background limited performance, even in the Auger suppressed superlattices. The best experimental results to our knowledge are for bulk 9μm HgCdTe, in which 295K carrier concentration reductions of about one order of magnitude have been reported.[9]. Photon recycling is found to provide no significant performance enhancements for both the SL- and bulk-based long wavelength detectors because Auger recombination is the dominant recombination mechanism. For equal carrier concentrations and temperatures, the SL detectivities are approximately a factor of two greater than those of the bulk system due to band structure engineered Auger suppression in the SL system.

Results for two 3.5μm systems are plotted in Figure 2. Fig 2(a) corresponds to 3.5μm bulk HgCdTe, and Fig. 2(b) corresponds to a 3.5μm 16.7Å InAs/35Å $In_{0.25}Ga_{0.75}Sb$ superlattice. The superlattice is optimized to provide suppression in

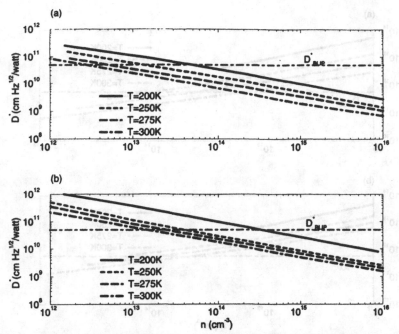

Figure 1: Calculated device detectivities at temperatures ranging from 200K to 300K as a function of nonequilibrium carrier concentration. Fig. 1(a) corresponds to bulk HgCdTe with an 11μm energy gap. Fig. 1(b) corresponds to a 35.9Å InAs/15.7Å In$_{0.225}$Ga$_{0.775}$Sb superlattice with an 11μm energy gap.

Auger recombination[16]. For both systems, the detector thickness was chosen to be 3×10^{-3} cm, the donor concentration equal to 10^{12} cm^{-3}, and the quantum efficiency equal to 0.7. The BLIP conditions are chosen to be the same as for Fig. 1. Note that the detectivity is plotted only up to the equilibrium carrier concentration at the various temperatures, resulting in the curves ending within the plots. For 3.5μm bulk HgCdTe, BLIP performance is achieved under equilibrium conditions at 200K, and the nonequilibrium electron concentrations of 1.1×10^{14}, 4.5×10^{13}, and 1.6×10^{13} at the temperatures of 250, 275, and 300K respectively. For the 3.5μm superlattice, BLIP performance is achieved under equilibrium conditions at 200 and 250K, whereas the nonequilibrium concentrations of 9.5×10^{13} and 4.3×10^{13} cm^{-3} are needed at 275 and 300K respectively. All three carrier lifetime enhancement methods play a role in the mid-wavelength detectors. Photon recycling enhances radiative lifetimes to the point that the detectors are Auger limited; band structure engineering of Auger lifetimes results in superior performance of the SL-based detectors; and carrier depletion produces substantial performance enhancements in both bulk and SL systems.

It is interesting to compare the predicted performance of photon detectors with that of thermal detectors. Present day thermal detectors operate at close to their theoretical limit. This corresponds to having their performance limited by noise due to fluc-

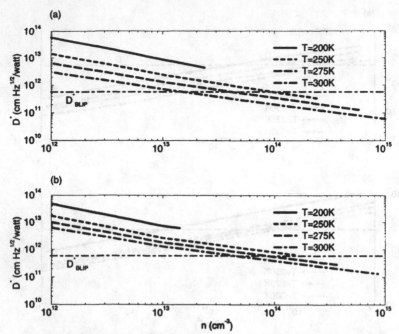

Figure 2: Calculated device detectivities at temperatures ranging from 200K to 300K as a function of carrier concentration. Fig. 2(a) corresponds to bulk HgCdTe with a 3.5μm energy gap. Fig. 2(b) corresponds to a 16.7Å InAs/35Å In$_{0.25}$Ga$_{0.75}$Sb superlattice with a 3.5μm energy gap.

tuations in the number of photons from the background and emitted by the detector. Assuming that the background and detector are at the common temperature of 300K, the detector couples only radiatively to its environment, does not employ lenses, is limited only by fluctuations in power flowing to and from the responsive element, and acts as a perfect blackbody radiator, results in a 300K ideal thermal detector detectivity of 1.8×10^{10} cm Hz$^{1/2}$/watt[17]. The 11μm 35.9Å InAs/15.7Å In$_{0.225}$Ga$_{0.775}$Sb superlattice would require a carrier concentration suppression to 9.3×10^{13} cm^{-3} to achieve comparable performance at 300K.

ACKNOWLEDGEMENTS

Support was provided by DARPA under ONR Contract N00014-96-1-0887.

References

[1] C.H. Grein, P.M. Young, M.E. Flatté, and H. Ehrenreich, J. Appl. Phys. **78**, p. 7143 (1995).

[2] W.P. Dumke, Phys. Rev. **105**, p. 139 (1957).

[3] C.H. Grein, H. Ehrenreich, and E. Runge, SPIE Proceedings **2999**, p. 18 (1997); C.H. Grein, E. Runge, and H. Ehrenreich, in preparation.

[4] T. Ashley, C.T. Elliott, and A.T. Harker, Infrared Physics **26**, p. 303 (1986).

[5] J. Piotrowski, W. Gawron, and Z. Djuric, Optical Engineering **33**, p. 1413 (1994).

[6] E.R. Youngdale, J.R. Meyer, C.A. Hoffman, F.J. Bartoli, C.H. Grein, P.M. Young, H. Ehrenreich, R.H. Miles, and D.H. Chow, Appl. Phys. Lett. **64**, p. 3160 (1994).

[7] R.G. Humphreys, Infrared Phys. **23**, p. 171 (1983).

[8] W. van Roosbroeck and W. Shockley, Phys. Rev. **94**, p. 1558 (1954).

[9] C.T. Elliott, N.T. Gordon, R.S. Hall, T.J. Phillips, A.M. White, C.L. Jones, C.D. Maxey, and N.E. Metcalfe, J. Elect. Mater. **25** , p. 1139 (1996).

[10] M.A. Kinch and S.R. Borrello, Infrared Physics **15**, p. 111 (1975).

[11] A. Rose, Concepts in Photoconductivity and Allied Problems (Interscience, New York, 1963).

[12] N.F. Johnson, H. Ehrenreich, P.M. Hui, and P.M. Young, Phys. Rev. B **41**, p. 3655 (1990).

[13] G.M. Williams, J. Appl. Phys. **77**, p. 4153 (1995).

[14] C.H. Grein, M.E. Flatté, H. Ehrenreich, and R.H. Miles, J. Appl. Phys. **77**, p. 4156 (1995).

[15] Y. Juan, private communication.

[16] P. Young, private communication.

[17] The Infrared Handbook, edited by W.L. Wolfe and G.J. Zissls, p. 11-29 (Office of Naval Research, Washington, 1978).

FREE CARRIER ABSORPTION IN P-TYPE EPITAXIAL Si AND GaAs FILMS FOR FAR-INFRARED DETECTION

A. G. U. PERERA*, W. Z. SHEN*, M. O. TANNER**, K. L. WANG**, W. SCHAFF***
* Department of Physics and Astronomy, Georgia State University, Atlanta, GA 30303,
uperera@gsu.edu
**Department of Electrical Engineering, University of California at Los Angeles, Los Angeles, CA 90095
*** School of Electrical Engineering, Cornell University, Ithaca, NY 14853

ABSTRACT

We report the investigation of free-carrier absorption characteristics for epitaxially grown p-type thin films in the far-infrared region (50 ~ 200 μm), where homojunction interfacial workfunction internal photoemission (HIWIP) detectors are employed. Five Si and three GaAs thin films were grown by MBE over a range of carrier concentrations, and the experimental absorption data were compared with calculated results. The free-hole absorption is found to be almost independent of the wavelength. A linear regression relationship between the absorption coefficient and the carrier concentration, in agreement with theory, has been obtained and employed to calculate the photon absorption probability in HIWIP detectors. The detector responsivity follows the quantum efficiency predicted by concentration dependence of the free carrier absorption coefficient.

INTRODUCTION

Recent interest in the development of internal photoemission homojunction far-infrared (FIR) detectors[1] has created a need for the FIR absorption results. These homojunction structures for FIR detection were classified as homojunction interfacial work-function internal photoemission (HIWIP) detectors[1,2], which are suitable for space astronomy applications at wavelengths greater than 50 μm[1]. Free carrier absorption in the highly doped thin emitter layers is followed by the internal photoemission occurring at the interface of the highly doped and intrinsic regions, with barrier penetration and collection form the basis of FIR detection mechanism of HIWIP detectors. Recent modeling and experimental studies[1,2] have shown that HIWIP FIR detectors could have a performance comparable to that of conventional Ge FIR photoconductors[3] or blocked-impurity-band (BIB) FIR detectors[4], with unique material advantages. Hence, it is important to understand the FIR free carrier absorption behavior in epitaxial thin films, both for fundamental as well as device performance reasons. However, previous studies of optical absorption in Si and GaAs were limited to relatively short wavelengths ($\leq$ 40 μm)[5,6]. No free carrier absorption data are available for the wavelength range $\geq$ 40 μm, where the HIWIP FIR detectors usually work.

EXPERIMENT

Five Si (doped with Boron) and three GaAs (doped with Beryllium) thin films were grown by molecular beam epitaxy (MBE) on pieces from Si and GaAs (100) wafer substrates. The carrier concentrations and mobilities in the films are extracted from Hall measurements, while the epilayer thicknesses are obtained by profilometer thickness measurements. Details of the samples are listed in Table I. The mobilities of the Si samples are

Table I: Characteristics of thin films. All samples are doped p type with Boron for Si and Beryllium for GaAs.

sample	thickness (μm)	concentration P (cm^{-3})	mobility μ (cm^2/Vs)
MT324F (Si)	0.68	5.5×10^{19}	29
MT325F (Si)	0.73	1.5×10^{19}	42
MT326F (Si)	1.37	7.1×10^{18}	55
MT327F (Si)	2.46	1.9×10^{18}	110
MT328F (Si)	2.42	8.1×10^{17}	126
GSA (GaAs)	0.10	1.0×10^{18}	–
GSB (GaAs)	0.10	5.0×10^{18}	–
GSC (GaAs)	0.10	2.0×10^{19}	–

similar to the high dopant level average values reported in the literature[7]. To determine the infrared absorption of the samples, the transmission and reflectance were measured using a Perkin-Elmer, system 2000, Fourier transform infrared spectrometer (FTIR) and a Si composite bolometer detector. The measurements were performed at room temperature with a resolution of 2 cm^{-1} and no changes are expected at low temperatures. Both the transmission and reflectance measurements were made using a normal incidence geometry with light incident on the doped layer surface. A 2.5 μm Mylar Pellicle film was used as a beamsplitter by placing it at 45° with the incident light in the reflectance measurements.

THEORY

The absorption (A) in thin films is determined from the transmission (T) and reflection (R) in conjunction with the expression

$$A = 1 - T - R \tag{1}$$

and further subtraction of the absorption of the substrates. The dielectric constant of the thin films is derived from the frequency–dependent conductivity for free carriers by

$$\sigma = \frac{\sigma_0}{1 - i\omega\tau} \tag{2}$$

where σ_0 is the dc conductivity and τ is the relaxation time, which is independent of frequency ω in the semiclassical transport theory. Since our main interest is in the FIR range (≥ 50 μm), the other contributions, e.g., intervalence band transitions and lattice vibrations, have been ignored. Using the measured values of mobility μ for Si samples in Table I, the relaxation time was determined by the relation $\mu = e\tau/m_p^*$, where $m_p^* = 0.37 m_0$[5] is the heavy-hole effective mass in Si, m_0 is the free electron mass, and e is the magnitude of the electron charge. For GaAs, the relaxation time 1.7×10^{-14} s is used based on the measured mobility of 60 cm^2/Vs for similar doping level of 2.5×10^{19} cm^{-3}[6]. This means that no free parameters are used to fit the experimental data with the modeling results.

The absorption depends not only on the real part of the refractive index, but also on its imaginary part, which is proportional to the absorption coefficient (α) defined by

$$\alpha = 2kq \tag{3}$$

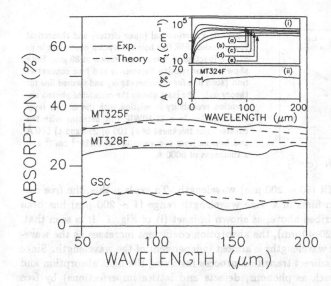

Figure 1: Experimental FIR free hole absorption in two p-Si (MT325F and MT328F) and one GaAs (GSC) thin films at room temperature (solid curves). The dashed curves are the theoretical results without any fitting parameters. The inset i) shows the calculated free carrier absorption coefficient (α_t) using the parameters in Table I of five samples a) MT324F b) MT325F c) MT326F d) MT327F and e) MT328F in a wavelength range from 1 to 200 μm, and the inset ii) shows the experimental absorption (A) of sample MT324F in the same range using two detectors; HgCdTe ($2 \sim 20 \mu$m) and Si bolometer ($20 \sim 200 \mu$m), displaying the wavelength square dependence at shorter wavelengths.

where k is the imaginary part of the refractive index, and q is the wave number of the incident radiation. In experiments, the absorption coefficient α of a thin film with thickness d can be obtained from

$$\alpha = \frac{1}{d} \ln\left(\frac{1-R}{T}\right) \tag{4}$$

RESULTS

The absorption results of two of the Si and one GaAs thin films over the wavelength range from 50 to 200 μm are shown in Fig. 1. The experimental curves have been smoothed using the FTIR software. The theoretical absorption curves shown in Fig. 1 were calculated from the complex dielectric constant of the Si and GaAs layers by matching electric and magnetic fields at the interfaces[8]. The reasonably good agreement between the experimental and theoretical results strongly demonstrates that the absorption is actually due to the contribution of free carriers. Further evidence for the identification of free carrier absorption in these thin films can be clearly seen from the absorption of their substrates, where our experiments show that their absorption can generally be neglected [the absorption coefficient is in the order of 10^{-2} cm^{-1}, in comparison with the order of $10^3 \sim 10^4$ cm^{-1} in thin films (see below)]. The measured values of absorption in the films were found to be almost independent of wavelength, which is similar to the results for Schottky barrier IR detector samples beyond the lattice bands[9] and SiGe layers above 15 μm[10]. The other four samples measured displayed the same absorption features. Measurements also show that the free carrier absorption increases with increasing doping concentration. This is expected as the free carrier absorption is proportional to both the carrier concentration and scattering rate (also increases with doping).

Silicon Films

In contrast to the wavelength squared dependence of the free carrier absorption coefficient for shorter wavelengths[5], the measured free carrier absorption coefficients are

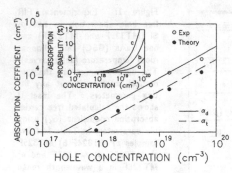

Figure 2: Experimental (open circles) and theoretical (solid circles) FIR free hole absorption coefficient in p-Si as a function of hole concentration at 80 μm. Both show a linear relation between α and hole concentration (solid line for experiments α_d and dashed line for theory α_t). The inset shows the calculated photon absorption probability in relation with the emitter layer concentration in a Si HIWIP FIR detector with the emitter layer thickness of a) 100 b) 250 and c) 500 Å and bottom layer concentration of 2×10^{19} cm^{-3} and a thickness of 6000 Å.

almost independent of the FIR (50 ~ 200 μm) wavelength. To explain this, the free carrier absorption of five Si thin films in a wide wavelength range (1 ~ 200 μm) has been calculated by the model described above, as shown in inset (i) of Fig. 1. It is seen that, at shorter wavelengths (1 ~ 20–40 μm), the absorption coefficient increases as the wavelength squared, and at longer wavelengths is almost independent of the wavelength. Since free carrier absorption is an indirect transition process involving the light absorption and quasi-particle interactions (such as phonons, defects and lattice imperfections) by free carriers, the weak energy of photons in the FIR region results in a reduced excitation of carriers to higher energy levels within the same energy valley[11], reducing the wavelength squared dependence. The samples were also measured in the 2 ~ 20 μm range by using a HgCdTe detector with the results for Si sample displayed in inset (ii) of Fig.1. This shows wavelength squared dependence of the free carrier absorption and the trend (12-20μm) coincidences with the longer wavelength results by Si bolometer detector, in good agreement with the theory shown in inset (i). The above arguments strongly demonstrate the reliability of both the modeling and experimental results.

The most important result in connection with the HIWIP detector is the relationship between the free carrier absorption coefficient and the carrier concentration. The strength of the free hole absorption at 80 μm, obtained from both the measurements and calculation, is shown in Fig. 2. Both show that the absorption can be well described by a linear relation between the absorption coefficient and the concentration of holes, in accordance with the case for shorter wavelengths[5]. The fitted regression formula as a function of hole concentration (P) below 10^{20} cm^{-3} gives the experimental absorption coefficient as

$$\alpha_d = 3.71 \times 10^{-16} \text{cm}^2 * P \tag{5}$$

which are close to the classical expressions[5] at wavelengths λ=20–40μm. From the modeling, we get $\alpha_t = 2.06 \times 10^{-16} cm^2 * P$, where we can see that the theoretical calculation is in reasonable agreement with the experimental results. This level of agreement is also observed in the shorter wavelength case, where it also is seen that the experimental absorption coefficient is larger by a factor of 1.7 than the theoretical value mainly due to the relative simplification of the theory[5].

Gallium Arsenide Films

The free hole absorption coefficient in three highly doped p-GaAs thin films is also almost independent of the FIR wavelength. The strength of the free hole absorption at a wavelength of 80 μm is shown in the inset of Fig. 3. The absorption can be well described

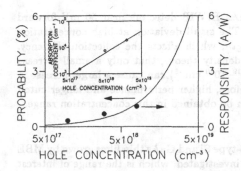

PROBABILITY (%)

RESPONSIVITY (A/W)

HOLE CONCENTRATION (cm⁻³)

Figure 3: Theoretical (solid curve) photon absorption probability as a function of hole concentration in p-GaAs by using the linear relationship [Eq. (6)] shown in the inset. Solid circles are the experimental p-GaAs detector responsivity near 50 mV forward bias. The inset shows the experimental free hole absorption coefficient at 80 μm (open circles) as a function of hole concentration at room temperature, together with its linear regression relation.

by a linear relation between the absorption coefficient and the concentration of holes, just as in the case of Si thin films. The fitted regression formula as a function of hole concentration (P) is found to be:

$$\alpha = 8.28 \times 10^{-16} \text{cm}^2 * P \qquad (6)$$

APPLICATION IN DETECTORS

A high absorption coefficient $(10^3$–10^4 cm$^{-1})$ in the FIR range is an important advantage for HIWIP detectors, since the absorption coefficient is almost independent of temperature due to the almost invariant carrier concentration and mobility with temperature. The total quantum efficiency of a HIWIP detector is the product of photon absorption probability, internal quantum efficiency, and barrier collection efficiency[1]. The photon absorption probability for HIWIP detectors can be calculated as[1]:

$$\eta_a = \{1 + \exp[-(\alpha_e W_e + 2\alpha_b W_b)]\}[1 - \exp(-\alpha_e W_e)] \qquad (7)$$

where α_e, α_b are the free carrier absorption coefficients in the emitter layer (thickness W_e) and the bottom contact layer (thickness W_b), respectively. By using the experimental relationship in Eqs. (5), (6) and the layer thicknesses, the photon absorption probability was calculated for HIWIP detectors as a function of carrier concentration, which is shown in the inset of Fig. 2 (for Si) and Fig. 3 (for GaAs). Apart from the increase of absorption probability due to the increase of hole absorption in thicker emitter layers, it can be seen that the photon absorption probability increases rapidly when the carrier concentration over $\sim 10^{19}$ cm^{-3}. The collection efficiency also depends on the diffusion length, however, the effect is much weaker compared with the effect of absorption probability as seen from the mobility data in Table I. Therefore, this strong enhancement of the photon absorption probability with the carrier concentration shows that highly doped emitter layers are more attractive for higher quantum efficiency in HIWIP detectors, which has been demonstrated in our recent GaAs HIWIP experimental results[2].

The responsivity of a HIWIP detector is proportional to its total quantum efficiency, Since the Fermi level E_F, work function Δ and hole mobility do not change very much with the hole concentration in the experimental region $(1 - 8 \times 10^{18}$ cm$^{-3})$, it is assumed that the internal quantum efficiency and barrier collection efficiency do not change with the doping concentration. Taking these into account, the variation of the detector responsivity with the doping concentration should follow that of the absorption probability. Fig. 3 shows that the experimental responsivity of the GaAs HIWIP detectors near 50 mV forward bias follows the absorption probability well. The small deviation at high concentration is due to the slight decrease in diffusion length, which affects the collection efficiency. In addition, it was shown[1], from the high density theory, that only a small increase in the emitter layer concentration (around 10^{19}–10^{20} cm^{-3}) can cause a large increase in the cutoff wavelength of the detectors. Therefore, higher performance and longer cutoff wavelength Si and GaAs HIWIP detectors can be obtained in this concentration range.

CONCLUSIONS

In summary, the free carrier absorption in p-type Si and GaAs thin films grown by MBE in the far-infrared region (50 ~ 200 μm) was investigated, which is the range of interest for HIWIP detectors. Both the calculations and experimental data reveal that the hole absorption is almost independent of wavelength. The theoretical results also confirm the wavelength squared dependence of the absorption coefficient at shorter wavelengths, which agrees with the previous experimental results. A linear regression relationship, essential for the performance of HIWIP detectors, between the absorption coefficient and carrier concentration was obtained and employed to calculate the photon absorption probability, which is found to follow the detector responsivity well.

ACKNOWLEDGMENTS

This work was supported in part by the NASA under contract #NAG5-4950 and the NSF under grant #DMR-95-20893.

REFERENCES

1. A. G. U. Perera, in Physics of Thin Films, Vol. 21, edited by M. H. Francombe and J. L. Vossen, Academic Press, NY, 1995, pp. 1-75.

2. A. G. U. Perera, H. X. Yuan, S. K. Gamage, W. Z. Shen, M. H. Francombe, H. C. Liu, M. Buchanan and W. J. Schaff, J. Appl. Phys. **81**, 3316 (1997).

3. E. E. Haller, Infrared Physics & Technology **35**, 127 (1994).

4. D. M. Watson, M. T. Guptill, J. E. Huffman, T. N. Krabach, S. N. Raines and S. Satyapal, J. Appl. Phys. **74**, 4199 (1993).

5. D. K. Schroder, R. N. Thomas, and J. C. Swartz, IEEE Transactions on Electron Devices **ED-25**, 254 (1978).

6. M. L. Huberman, A. Ksendzov, A. Larsson, R. Terhune and J. Maserjian, Phys. Rev. B **44**, 1128 (1991).

7. C. Jacoboni, C. Canali, G. Ottaviani, and A. A. Quaranta, Solid State Electron. **20**, 77 (1977).

8. O. S. Heavens, Optical properties of thin solid films, Dover Publications, Inc., New York, ch.4, 1965.

9. C. K. Chen, B. -Y. Tsaur, and M. C. Finn, Appl. Phys. Lett. **54**, 310 (1989).

10. T. L. Lin, J. S. Park, S. D. Gunapala, E.W. Jones, and H. M. Del Castillo, Optical Engineering **33**, 716 (1994).

11. J. I. Pankove, Optical process in semiconductors, Dover Publications, Inc., New York, ch.3, 1975.

MONOLITHICALLY INTEGRATED DUAL-BAND QUANTUM WELL INFRARED PHOTODETECTOR

D.K. Sengupta, S.D. Gunapala, S.V. Bandara, F. Pool, J.K. Liu, M. McKelvey, E. Luong,
J. Torezan, J. Mumulo, W. Hong, and J. Gill
Center for Space Microelectronics Technology, Jet Propulsion Laboratory,
California Institute of Technology, Pasadena, CA 91109

G.E. Stillman, A.P. Curtis, S. Kim, L.J. Chou, P.J. Mares, M. Feng, K.C. Hseih, S.L. Chuang,
S.G. Bishop, and Y.C. Chang
Microelectronics Laboratory, Department of Electrical Engineering,
Department of Physics & Materials Research Laboratory,
University of Illinois at Urbana-Champaign, Urbana, IL 61801

H.C. Liu
Institute of Microstructural Sciences, National Research Council,
Canada K1A 0R6

W.I. Wang
Department of Electrical Engineering, Columbia University,
New York, NY 11007

ABSTRACT

A monolithic quantum well infrared photodetector (QWIP) structure has been presented that is
suitable for dual bands in the two atmospheric transmission windows of 3 - 5.3μm and 7.5 - 14μm,
respectively. The proposed structure employs dual stacked, strain InGaAs/AlGaAs and lattice-
matched GaAs/AlGaAs quantum well infrared photodetector for mid wavelength and long wavelength detection. The response peak of the strain InGaAs/AlGaAs quantum well is at 4.9 μm and
the lattice-matched GaAs/AlGaAs is at 10.5μm; their peak sensitivities are in the spectral regions
of 3 - 5.3μm and 7.5 - 14μm. The peak responsivity when the dual-band QWIP is biased at 5 Volts
is ~0.065A/W at 4.9μm and ~0.006A/W at 10.5μm ; at this voltage the dual-band QWIP is more
sensitive at the shorter wavelengths due to its larger impedance thus exhibiting wavelength tunability
characteristics with bias. Additionally, single colored 4.9 and 10.5μm QWIPs were fabricated from
the dual-band QWIP structure to study the bias-dependent behavior and also to understand the
effects of growing the strain layer InGaAs/AlGaAs QWIP on top of the lattice-matched GaAs/
AlGaAs QWIP. In summary, two stack dual-band QWIPs using GaAs/AlGaAs and strained InGaAs/
AlGaAs multiquantum wells have been demonstrated with peak spectral sensitivities in the spectral region of 3 - 5.3μm and 7.5 - 14μm. Also, the voltage tunable dual-band detection have been
realized for this kind of QWIP structure.

INTRODUCTION

Multispectral infrared detectors would be highly beneficial for a variety of applications such as guidance, surveillance, detection, tracking, imaging and monitoring [1,2]. Dual-band detection in the mid-IR and far-IR atmosphere window have been performed using HgCdTe photodiodes or photoconductive devices [3,4]. Compared with HgCdTe, the quantum well infrared photodetector (QWIP) based on intersubband transitions in GaAs/AlGaAs, InGaAs/AlInAs, InGaAs/InP and InGaAs/InGaAsP superlattices offer greater potential as infrared detectors than HgCdTe because they are easier to grow with high quality and uniformity [5,6,7,8,9,10]. In all cases its detection wavelength and thus the relevant energy bands are chosen prior to growth, and the layer structures are designed accordingly. Two or multiple color QWIPs have recently been demonstrated [11,12]. The structures of the dual-color QWIPs include multi-stack structures [13], symmetric [14], and asymmetric [15,16] quantum well structures. However, the operation windows with a photoresponse in the 3 - 5.3μm and 7.5 - 14μm using GaAs/AlGaAs quantum wells is quite difficult to be realized and especially in designing 3 - 5.3μm QWIPs [17]. For the n-type detector operation in the 3-5.3μm wavelength range, a large conduction band discontinuity is required. Hence, conventional quantum well structures based on Al(x)Ga(1-x)As/GaAs material system are modified by inserting thin (~15A°) high aluminum content (y ~ 1.0) AlyGa(1-y)As barriers on either side of a GaAs quantum well [11]. These thin aluminum content barriers effectively shift the quasi-bound second subband to higher energies thus allowing for operation in the 3 - 5 - 3μm wavelength range. However, the growth of these structures requires the use of two aluminum cells in the growth chamber [18]. In order to achieve a shorter wavelength response, we have designed a QWIP based on the strained InGaAs/AlGaAs material system. Each period of the QWIP structure consists of a single InGaAs quantum well surrounded by AlGaAs barriers. To date, little work has been done using the smallest possible indium mole fraction that satisfies the requirement of operating in the 3-5.3μm wavelength range using n-type strained layer InGaAs/AlGaAs QWIP [19, 20]. More recently, development of the crystal growth technology has made strained quantum well structures with high optical qualities possible. The deformation of the band structure caused by the lattice mismatch in the strained quantum well not only changes the effective band gap, but also changes the effective mass and polarization selection rules [21, 22]. Here we report the realization of a dual-band QWIP by combining the conventional GaAs/AlGaAs QWIP with the strained InGaAs/AlGaAs QWIP.

EXPERIMENTAL PROCEDURES

The QWIP samples were grown by solid source molecular beam epitaxy (MBE) on semi-insulating GaAs (001) substrates. The dual-color QWIP structure used in this study is shown schematically in Fig. 1. The quantum wells between the n-type (2.0 x 10 18 cm -3) 1μm contact layers consist of two stacks of quantum wells each. The first stack was designed as a long wavelength QWIP, with each period of quantum wells in the stack consisting of a 40A° GaAs well (doped to n = 1 x 10 18 cm -3) and 300A° Al(0.23) Ga(0.77) As barriers. The second stack was designed as a middle wavelength QWIP with each period of quantum wells consisting of a 30A° In(0.25) Ga(0.75)As well (doped to n = 2 x 10 18 cm -3) and 400A° Al(0.38)Ga(0.62)As barriers. The 7.5 - 14μm absorbing region is implemented using the AlGaAs/GaAs material system while the 3 - 5 .3μm absorbing region is implemented using the InGaAs/AlGaAs material system. To model a 3 - 5.3μm QWIP based on the strained InGaAs/AlGaAs material system, we modified our absorption coefficient model based on the Kroning - Penney model approach to account for the compressive strain that exists in the AlGaAs/InGaAs material system.

1 μm	n = 2.0 x 10^{18} cm^{-3}	GaAs:Si
400Å	UNDOPED	AlGaAs(x = 0.38)
30Å	InGaAs(x = 0.25) DOPED 2 x 10^{18} cm^{-3} WITH Si OVER CENTER 20Å	AlGaAs(x = 0.38)
400Å	UNDOPED	AlGaAs(x = 0.38)
1 μm	n = 2.0 x 10^{18} cm^{-3}	GaAs:Si
300Å	UNDOPED	AlGaAs(x = 0.223)
40Å	GaAs DOPED 1 x 10^{18} cm^{-3} WITH Si OVER CENTER 20Å	
300Å	UNDOPED	AlGaAs(x = 0.23)
1 μm	n = 2.0 x 10^{18} cm^{-3}	GaAs:Si

x30 (for the upper group)

x30 (for the lower group)

(100) SEMI-INSULATING GaAs SUBSTRATE
THICKNESS: 625 μm

Fig. 1   Schematic layout of the dual-band QWIP structure designed and grown
for operation in both the 7.5μm - 14 μm and the 3 - 5.3μm wavelength bands.

One impact of the compressive strain is to cause a reduction in the conduction band edge discontinuity. Therefore, in order to be able to accurately predict the intersubband transition energy, and
thus the operating wavelength, it is necessary to properly model its reduction in conduction band
offset via the inclusion of strain in our model. The model is also modified to account for the
material parameters of the InGaAs material system as well as a band offset ratio of Qc/Qv = 65/35.
To design our 3 - 5.3μm QWIPs, we use the following criteria. First, the structure should, prefer-
ably, not contain wells less than 30A° in width. Second, the smallest possible indium mole fraction
that satisfies the requirement of operating in the 3 - 5.3μm wavelength range should be used. The
former criteria is desirable primarily for growth reproducibility and uniformity. The latter criteria
is desirable in order to limit the strain in the QWIP because very large strains could adversely
impact the QWIP performance. Keeping these criteria in mind, we chose A1(0.38) Ga(0.62)As for
the barriers which gives the maximum direct bandgaps in the A1GaAs material system. We then
investigate the various combinations of InGaAs well width and mole fraction that will satisfy QWIP
operations in the 3 - 5.3μm wavelength range.   Our design consists of 30 periods of 30A°
In(0.25)Ga(0.75)As quantum wells doped over the center 20A° to 2 x 10 18 cm -3 and surrounded
by 400A° undoped A1(0.38)Ga(0.62)As barriers.

The material quality and optical properties of the dual-color MQW samples were investigated using cross-sectional TEM, PL, and infrared absorption measurements. TEM was performed with a
120KV Phillips CM12 microscope. The intersubband absorption was measured at room temperature using a DA3 Bomem FTIR spectrophotometer. The MQW samples were polished into multipass
waveguides and sandwiched between two infrared transmitting KRS-5 slabs. IR measurements
were done using a glowbar source and a KBr beam splitter. Detection of the infrared beams was
acheived using a 77K photoconductive HgCdTe (MCT) detector.

Quantum well infrared detectors were fabricated from the dual-band samples into 240 mm2 square
mesas by etching through the upper contact layer and the multiple -quantum well structure down to
the bottom contact layer. Ohmic contacts to the n-doped contact layers were subsequently formed

by evaporating and alloying AuGe/Ni/Au metallization. All dark current, and spectral response on these QWIPs were performed with the detectors mounted on a stage which is in thermal contact with the cold end of a continous flow helium cryostat. The cryostat has an infrared transmitting KRS-5 window which gives the detectors a 2π Sr field-of-view to a 300K background. In order to set the detector to the desired operating temperature, a small electrical heating coil is attatched to the cold end of the continous flow helium cryostat, which allows the temperature to be varied from 20K to 350K with a stability of 0.05C. The dark current (I-V) characteristics of the QWIPs were measured with a HP 4145 semiconductor parameter analyzer at variuos sample temperatures. In order to obtain the absolute spectral response of the as-grown QWIP, both blackbody and relative spectral response measurements were performed. The detectors were electrically connected to an external bias circuit, and bias currents were supplied from a battery in series with a large resistor which maintains a constant bias current. Light from an infared source was chopped at 200Hz by a mechanical chopper with a 50% duty cycle and shined on the detectors through the KRS-5 window of the continous flow helium cryostat. The incident radiation entered the detector structure through a 45° polished face on the substrate. The signal current produced by the detectors was measured by a lock-in-amplifier. In the blackbody measurement, a 1000K blackbody was used as the radiation source, and the photocurrent from the QWIP was detected for various bias currents and blackbody apertures. In the relative spectral response measurements, a glowbar source filtered by a SPEX 340E monochromator was used as the infrared source. Infrared light from the monochromator was collimated and then focussed on to the 45 polished input face using two ZnSe plano convex lenses. The spectral variation of the light intensity from the spectrometer was calibrated using a window-less pyroelectric detector. The blackbody response measurement of the QWIP was required to calculate their absolute response. The total absorbed optical intensity was determined by integrating over wavelength the product of the normalized spectral response $r(\lambda)$ and the blackbody irradiance spectrum $W(\lambda,T)$. The peak absolute response R_p(A/W) was calculated from the absorbed optical intensity and the peak blackbody photocurrent I_p as

$$ R_p = \frac{I_p}{M_F \dfrac{a^2}{a^2 + d^2} \, A \, \text{Cos} \, \theta * T \int W(\lambda,T)\, r(\lambda)\, d\lambda} \tag{1} $$

where M_F was the modulation factor of the chopped blackbody source, a was the aperture radius, d was the distance between the blackbody and QWIP, A was the QWIP area, t was the transmission through the KRS-5 window, T was the blackbody temp, and θ was the angle of incident radiation.

MATERIALS RESULTS

Cross-sectional transmission electron microscopy have been applied to examine the microstructures of the superlattices. Standard techniques have been used for the preparation of electron-transparent cross-sectional foils. First, two cleaved samples, each of 2mm x 2mm have been glued face to face using epoxy. The composite was then diced into 200nm wide slices. To make it electron transparent, each sample was then thinned down to 30nm by mechanical lapping before further thinning by ion-milling with Ar ions at 5KV while the sample was kept cold at the liquid nitrogen temperature. A Philips CM-12 electron microscope operated at 120KV was used to examine the microstructures. A (002) reflection imaging is shown in Fig. 2. The striations in both ternary layers can be attributed to sample heating during the growth. The radiation heat makes the substrate slightly warmer and the incorporation of In changes very sensitively with the substrate temperature. We have performed FTIR room temperature transmittance measurements on dual-color

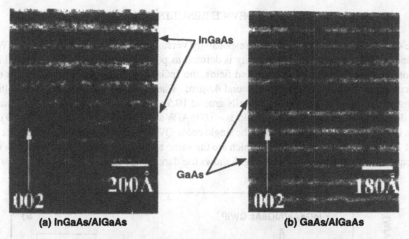

(a) InGaAs/AlGaAs   **(b) GaAs/AlGaAs**

Fig. 2. XTEM image of the dual-band QWIP structure

structure using Bomen DA3 0.0002 cm -1 system. The room temperature FTIR transmittance spectrum of the dual-band QWIP is shown in Fig. 3. The labeled dips in the FTIR transmittance spectrum (A through D in Fig. 3) correspond to absorption peaks that are located at approximately :900/cm [11.1μm] (A), 1400/cm [7.14μm] (B), 2150/cm [4.65μm] (C) and 3790/cm [2.64μm] (D). Of particular interest are the dips A and C of Fig. 3. Dip A of Fig. 3 corresponds to a relatively broad absorption peak centered around 11.1μm which responds to radiation in the 7.5 - 14μm wavelength band. Dip C of Fig. 3 corresponds to an absorption peak located at approximately 4.65μm and which responds to radiation in the 3 - 5.3μm wavelength band. The FTIR spectrum also indicates the presence of a small absorption peak near 2.64μm. Therefore, based on the FTIR data, the two-color structure that we have designed and grown holds great promise for dual wavelength operation in both the 7.5 - 14μm and 3 - 5.3μm wavelength bands.

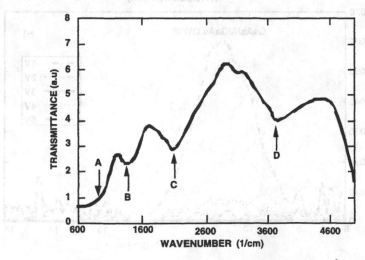

Fig. 3. Room Temp FTIR specturm of the dual-band QWIP structure

209

DEVICE RESULTS

Figure 4(a) & 4(b) displays the absolute responsivity versus applied bias of the dual-band QWIP at 25K under different voltages. The polarity is defined as positive when the higher potential is applied on the top of a mesa. At low applied fields, the InGaAs/AlGaAs wells provide most of the photocurrent with the peak response around 4.9μm; when the bias is increased to 5 Volts, the response of the GaAs/AlGaAs quantum wells around 10.5μm begins to occur. The 25K responsivity of the detector with a bias voltage of 5Volts is ~0.006A/W at 10.5 μm and ~0.065A/W at 4.9 μm as shown in Fig. 5. The response peaks of the single color QWIP devices shown in Figures 6 & 7 are also at 10.5 and 4.9 μm, respectively, which are the same as the peak response wavelengths of the dual-band QWIPs as shown in Fig. 5. Fig. 8 shows the dark current measured for both forward and

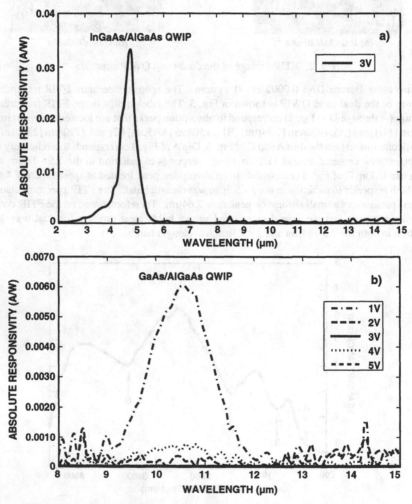

Fig. 4. Absolute responsivity versus bias of a dual-band QWIP at T = 25K.

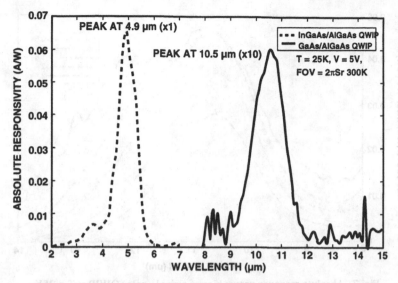

Fig. 5. Absolute response versus bias of a dual-band QWIP at T = 25K.

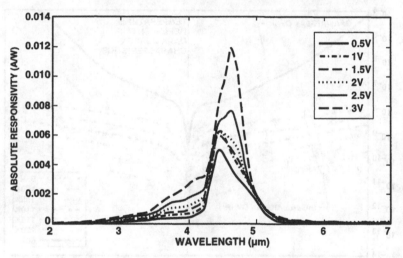

Fig. 6. Absolute response versus bias of a single- color QWIP at T = 25K.

reverse bias with a 2π Sr 300K background. The dark current throughout the entire biases for both devices, Strain layer single color InGaAs/AlGaAs and lattice-matched single color GaAs/AlGaAs appears to be thermionically assisted in nature rather than resonant tunneling. Comparision of dark current versus bias reveals that dark current of the single color strain layer InGaAs/AlGaAs QWIP is lower in comparision to the single color lattice-matched GaAs/AlGaAs QWIP and is attributed to higher thermal barriers for the strain layer InGaAs/AlGaAs shorter wavelength QWIP. Also, the 60K dark current characteristics of the single color lattice-matched GaAs/AlGaAs QWIP is higher and may be attributed to growth related effects such as migration of the dopants in the growth

211

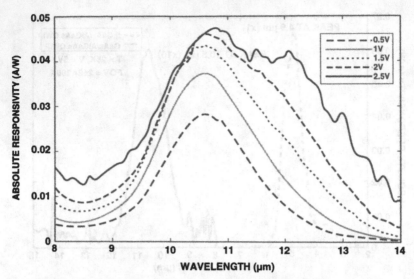

Fig. 7. Absolute response versus bias of a single-color QWIP at T = 25K.

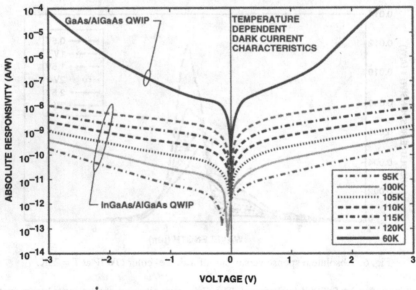

Fig. 8. Dark current versus bias for temperatures from 95 to 120K for the strain layer InGaAs/ AlGaAs QWIP with 2π 300K background. Also shown is the 60K dark current characteristics of the lattice matched GaAs/AlGaAs QWIP.

direction [23]. Recent advances in growth, complimented by innovative structures (random gratings and reflector layers) should offset any degradation in performance and make feasible dualband QWIPs.

In conclusion, based on the FTIR and device data, the dual-band structure that we have designed and grown holds great promise for dual wavelength operation in both the 3-5.3μm and the 7.5-14μm wavelength bands and dual-band QWIP focal plane arrays can be expected in the near future.

ACKNOWLEDGMENT

Research described in this paper was performed by the Center for Space Microelectronics Technology, Jet Propulsion Laboratory, California Institute of Technology, Pasadena, CA 91109 and by the Microelectronics Laboratory, University of Illinois at Urbana-Champaign, Urbana, IL 61801. The authors would also like to acknowledge Dr. T.N. Krabach and other members of the Microdevices Laboratory (JPL) for encouragement and support during the preparation of this work. The authors would also like to thank Prof. N. Holoynak (UIUC) and Dr. K.K. Choi of US Army Research Labs, New Jersey for many helpful discussions. We would also like to acknowledge J. Park (JPL), B. Payne and M. Day (UIUC), J. Washington, and D. Cuda (JPL) for help with the manuscript preparation. One of the authors (D.K.S) acknowledges the fellowship awarded by National Research Council.

REFERENCES

1. K. Konuma, Y. Asano, K. Masubuchi, H. Utsumi, S. Tohyama, T. Endo, H. Azuma, and N. Teranishi, IEEE Trans. Electron Devices, 43, 282 (1996).

2. E.R. Blazejewski, J.M. Arias, G.M. Williams, W. Mclevige, M. Zandian, and J. Pasko, J. Vac. Sci, Technol, B, 10, 1626 (1992).

3. M.B. Reine, P.W. Norton, R. Starr, M.H. Weiler, M. Kestigian, B.L. Musicant, P. Mitra, T. Schimert, F.C. Case, I.B. Bhat, H. Ehsani, and V. Rao, J. Electron. Mater, 24, 669 (1995).

4. B.F. Levine, J. Appl. Phys. 74, R1 (1993).

5. A. Zussman, B.F. Levine, J.M. Kuo, and J. de Jong, J. Appl. Phys. 70, 5101 (1991).

6. G. Sarusi, B.F. Levine, S.J. Pearton, S.V. Bandara, and R.E. Leibenguth J. Appl. Phys. 64, 960 (1994).

7. S.D. Gunapala, B.F. Levine, D. Ritter, R.A. Hamn, and M.B. Panish, Appl. Phys. Lett. 58, 2024 (1991).

8. D.K. Sengupta, S.L. Jackson, D. Ahmari, H.C. Kuo, J.I. Malin, S. Thomas, M. Feng, and G.E. Stillman, Appl. Phys. Lett. 69, 3209 (1996).

9. A. Hiromitou, and K. Yuich, Appl. Phys. Lett. 56, 746 (1990).

10. M.Z. Tidrow, and K. Bacher, Appl. Phys. Lett. 70, 859 (1997).

11. K.L. Tsai, K.H. Chang, C.P. Lee, K.F. Huang, J.S. Tsang, and H.R. Chen, Appl. Phys. Lett. 62, 3504 (1993).

12. I. Grave, A. Shakouri, N. Kuze, and A. Yariv, Appl. Phys. Lett. 60, 2362 (1992).

13. A. Kock, E. Gornick, G. Abstreiter, G. Bohm, M. Walker, and G. Weimaun, Appl. Phys. Lett. 60, 2011 (1992).

14. K. Kheng, M. Ramsteiner, H. Schneider, J.D. Ralston, F. Fuchs, and P. Koidl, Appl. Phys. Lett. 61 666 (1992).

15. Y.H. Wang, S.S. Li, and P. Ho, Appl. Phys. Lett. 62, 93 (1993).

16. E. Martinet, E. Rosencher, F. Luc, Ph. Bois, E. Constard, and S. Delaitre, Appl. Phys. Lett. 61, 246 (1992).

17. Y. Zhang, D.S. Jiang, J.B. Xia, L.Q. Song, Z.Q. Zhou, and W.K. Wu, Appl. Phys. Lett. 68, 2114 (1996).

18. Personal communication (W.I. Wang).

19. M.Z. Tidrow, K.K. Choi, A.J. DeAnni, W.H. Chang, and S.P. Svensson, Appl. Phys. Lett. 67, 1800 (1995).

20. M.Z. Tidrow, J.C. Chiang, S. li, K. Bacher, Appl. Phys. Lett. 70, 859 (1997).

21. H.C. Liu, J. Li, J.R. Thompson, Z.R. Wasilewski, M. Buchanan, and J.G. Simmons, IEEE Electron Device Letters, 14, 566 (1993).

22. S.L. Chuang, Physics of Optoelectronics Devices (New York: Wiley, 1995).

23. H.C. Liu, Z.R. Wasilewski, M. Buchanan, and H. Chu, Appl. Phys. Lett. 63, 761 (1993).

GERMANIUM FAR INFRARED BLOCKED IMPURITY BAND DETECTORS

C.S. Olsen,[a,b] J.W. Beeman,[a] W.L. Hansen,[a] and E.E. Haller[a,b]
[a] Lawrence Berkeley National Laboratory and [b] University of California at Berkeley
Berkeley, California 94720 USA

ABSTRACT

We report on the development of Germanium Blocked Impurity Band (BIB) photoconductors for long wavelength infrared detection in the 100 to 250 μm region. Liquid Phase Epitaxy (LPE) was used to grow the high purity blocking layer, and in some cases, the heavily doped infrared absorbing layer that comprise theses detectors. To achieve the stringent demands on purity and crystalline perfection we have developed a high purity LPE process which can be used for the growth of high purity as well as purely doped Ge epilayers. The low melting point, high purity metal, Pb, was used as a solvent. Pb has a negligible solubility $<10^{17}$ cm^{-3} in Ge at 650°C and is isoelectronic with Ge. We have identified the residual impurities Bi, P, and Sb in the Ge epilayers and have determined that the Pb solvent is the source. Experiments are in progress to purify the Pb. The first tests of BIB structures with the purely doped absorbing layer grown on high purity substrates look very promising. The detectors exhibit extended wavelength cutoff when compared to standard Ge:Ga photoconductors (155 μm vs. 120 μm) and show the expected asymmetric current-voltage dependencies. We are currently optimizing doping and layer thickness to achieve the optimum responsivity, Noise Equivalent Power (NEP), and dark current in our devices.

INTRODUCTION

There are few sensitive, low noise far infrared detectors which operate in the 100 to 250 μm (40 to 100 cm^{-1}) range and can reach low background limited operation in space borne far IR missions. Uniaxially stressed Ge:Ga detectors are current the only devices which have demonstrated excellent sensitivity and sufficiently low Noise Equivalent Power (NEP),[1,2] but the necessary mechanical stressing apparatus poses serious difficulties for large array fabrication. Properly operating Ge BIB detectors should be competitive with stressed photoconductors regarding the spectral onset, because the photoionization threshold of shallow dopants should drop to 6 meV (50 cm^{-1}) in the heavily doped absorbing layer (~10^{16} cm^{-3}). Furthermore, BIB detectors have additional advantages over conventional photoconductors which we will discuss later.

The fabrication of a Ge BIB requires the use of a high purity epitaxial technique. State-of-the-art epitaxial methods such as Molecular Beam Epitaxy (MBE) or Chemical Vapor Deposition (CVD), allow fabrication of sharply defined epilayer structures but have inherent limitations regarding achievable purity.[3,4] LPE has many advantages for growth of epitaxial Ge layers for BIB detectors. LPE uses low melting point metals as solvents to transport solute or growth materials to the substrate. During the LPE process most impurities are incorporated in the solid at a concentration which is lower than that in the liquid.[5] This segregation effect is significantly less pronounced in gas to solid interfaces as they exist in CVD and MBE. LPE also allows growth to occur at lower temperatures than MBE and CVD because of the solvent mediated transport. At these lower temperatures, less impurities are released from the growth chamber materials and incorporated in the epitaxial film. Despite these advantages, very limited research on Ge LPE and no systematic studies on high purity Ge LPE have been performed.[6,7] Earlier attempts have been made to fabricate Ge BIB detectors using CVD[8,9,10] and Boron implantation.[11] Si BIB detectors doped with As & Sb have been successfully developed using CVD.[12,13] Ge CVD has been less successful for fabricating Ge BIB detectors. This is due, in part, to inexperience with Ge CVD but also to the much more stringent dopant and purity requirements for Ge BIB detectors compared to Si BIBs. Here we report on our newest detectors produced with a high purity liquid phase epilayer growth process.

BIB DETECTOR THEORY

The BIB detector concept was invented by Petroff and Stapelbroek.[14] A BIB detector consists of a two layer structure: a heavily doped (IR absorbing) region ($\sim 10^{16}$ cm^{-3} majority shallow impurities, $\sim 10^{12}$ cm^{-3} minority impurities) and a thin, pure (blocking) layer ($< 10^{14}$ cm^{-3}) thus the need for a Ge epitaxial technique. The two regions are sandwiched between two degenerately doped contacts. Just like conventional photoconductors, BIB detectors are operated at cryogenic temperatures where the dopant carriers are frozen out. The IR absorbing region is typically 10 to 100 μm thick and the blocking is much thinner, 2 to 5 μm. Such epitaxial layer thicknesses are too large for MBE, but are achievable with CVD and LPE.

Properly operating BIB detectors have three advantages over conventional photoconductors. In the following we focus on n-type BIB detectors. First, the effective ionization energy from the donor band to conduction band is decreased by high dopant concentrations resulting in donor wave function overlap. These decrease in ionization energy provides a photoconductive response at wavelengths longer than regular photoconductors doped with the same donor species. Second, the high doping concentration in the IR absorbing layer provides increased absorbance requiring a thinner layer for efficient photon absorption. The resulting reduction in total detector volume of about 100 decreases the rate of cosmic ray interactions. A third advantage is that a BIB device has unity photoconductive gain, i.e., every photoinduced charge effectively crosses the device once. This reduces the noise caused by the photoconductive gain fluctuations in regular photoconductors.

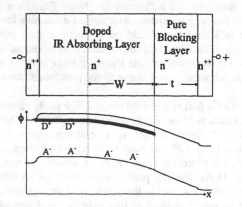

Fig. 1 (top) Schematic of components of n-type BIB detector
(bottom) energy band diagram of n-type BIB under reverse bias.

The energy band structure of a BIB detector is shown schematically in Fig. 1. When a photon excites an electron into the conduction band it is free to drift all the way to the positively biased electrode through the conduction band. The positive donor state D^+ created in this ionization event propagates via hopping of electrons from neighboring neutral donors in the impurity band all the way to the negative electrode. Independent of the location of the photoionization event, a total of one charge crosses the total distance of the depletion layer plus blocking layer.

The application of a voltage bias to a BIB detector creates a space charge layer W, devoid of ionized donors. The ionized donors are neutralized by electrons however minority acceptor remain negatively charged and represent the space charge. This space charge region is where collection of photon excited electrons occurs. The width of the space charge layer W depends on the minority concentration N_a, and the applied voltage V_a, and the thickness t of the blocking layer:[14]

$$W = \sqrt{\frac{2\varepsilon\varepsilon_o(V_a - V_{bi})}{eN_a} + t^2} - t \qquad (1)$$

$\varepsilon\varepsilon_o$ is the dielectric constant and V_{bi} the built in voltage. Depletion widths of ~100 μm are needed for responsive detectors. This in turn sets upper limits of the order of 10^{12} cm^{-3} for the minority dopant concentrations in the IR absorbing layer. More complete descriptions of device operation and advantages are given by other researchers.[8,14,15]

EPILAYER GROWTH AND CHARACTERIZATION

LPE growth was performed in a quartz tube furnace in a Pd-diffused hydrogen ambient. The growth chamber allows tipping of the solution without any rotating seals. The quartz tube and associated vacuum system all rotate together about the quartz tube axis. Pb solvents of 99.9999% purity were saturated with high-purity Ge under 1 atm of flowing H_2. Densified graphite crucibles, commercially purified and degassed in vacuum at 950°C, were used to contain the growth materials. Dissolution of the high-purity Ge into the Pb was performed at 665°C for 6 hours. Epitaxial growth was initiated by tipping the saturated solution onto the substrate and cooling to 340°C at 0.36°C per minute. Upon reaching 340°C the Pb solution was tipped off the substrate and cooled to room temperature. Any residual Pb present on the surface is removed with a 1:1 Acetic Acid: H_2O_2 mixture.

The Ge epilayers grown from Pb solvents on high-purity substrates have been characterized by X-ray diffraction rocking curves, variable temperature Hall effect, and PhotoThermal Ionization Spectroscopy (PTIS).[16] X-ray diffraction rocking curves of the (111) reflection of the epilayers had FWHM values of 21.5 arcseconds while the substrates gave widths of 13 arcseconds. The narrow peak width of the epilayer suggests high quality single crystal material. Variable temperature Hall effect measurements give residual donor impurity concentrations of $2x10^{14}$ cm^{-3} and minority acceptor impurity concentrations of $5x10^{11}$ cm^{-3}. From PTIS, the characteristic ground state to bound excited-state peaks identify the majority donors to include phosphorus, bismuth, and antimony. PTIS measurement give only the impurities in the n-type epilayer, because it is electrically separated from the p-type substrate, due to the formation of a depletion region at the epi-substrate interface. Through a series of carefully designed experiments using a number of crucibles and different Pb sources, the main source of the group V impurities has been determined to be the high purity Pb. Five different commercial sources of high purity Pb have been shown to all contain similar contaminants. Experiments are in progress to purify the Pb by distillation methods. Since it is the group V contaminants in the Pb that limit our ability to fabricate sufficiently pure blocking layers, we decided to only grow doped n-type IR absorbing layers with low acceptor concentrations. For the pure blocking layers, we resorted to using lightly doped n-type bulk Ge material.

DETECTOR FABRICATION AND OPERATION

For detector fabrication single crystal Ge substrates were chosen with (111) orientation and As doping of 3×10^{13} cm^{-3}. Sb was selected as the dopant for the IR absorbing layer because it has the lowest ionization energy of all group III and group V dopants in Ge. In order to dope the epilayer accurately, the segregation coefficient for Sb had to be measured and was found to be 0.015. The segregation coefficient is the ratio of Sb concentrations in the Ge epilayer and in the Pb solvent. A 100 μm Ge epilayer was doped to a concentration of approximately 10^{16} cm^{-3} with Sb. Only the residual acceptors and not the residual donors in the epilayer limit the IR absorbing depletion layer thickness. After epitaxial growth, the substrate was mechanically thinned to form the blocking layer. We stopped at 35 microns to prevent accidental removal of the blocking layer. This device would operate much more optimally with a 5 to 10 micron blocking layer. Ohmic contacts were fabricated by implanting P at 40 keV and 100 keV with doses of 2×10^{14} cm^{-2} and 4×10^{14} cm^{-2} respectively into the sample maintained at 105 K. Detectors were annealed to activate the P implant at 450°C for 2 hours in a quartz furnace under N$_2$ atmosphere. The contacts were metallized with 40 nm of Pd and 400 nm of Au. Devices were cut to an optical active area of 2.6 mm^2 and the sides were etched to remove residual saw damage. The detectors were mounted in an Infrared Labs dewar model HD-3 with black polyethelyene filter to remove above Ge bandgap energy photons.

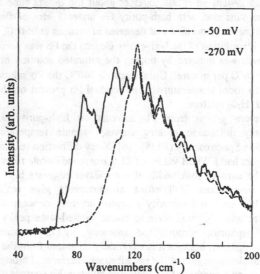

Fig. 2. Spectral response of Ge:Sb BIB detector as a function of bias.
Spectra were scaled to match high wavenumber response.

Spectral response measurements were performed on a far infrared Fourier transform spectrometer with a mercury arc lamp as the infrared source. At low bias, the detector response at 114 cm^{-1} originates mainly from the As doped blocking layer. At higher bias, the electric field increases in the blocking layer and penetrates into the IR absorbing layer doped with Sb. Sb has an ionization energy of 83 cm^{-1} at low concentration, but the higher doping concentration decreases the ionization energy to 65 cm^{-1}. Our earlier results showed response out to 60 cm^{-1} [15], but these devices suffered from low responsivity.

The dark current of the device shown in Fig. 3 displays the expected asymmetric current-voltage dependence. The dark current measurement at low biases is limited by the resolution of our amplifier.

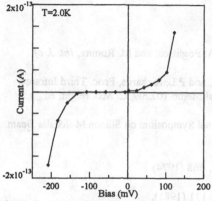

Fig. 3. Current voltage curve for Ge BIB detector.

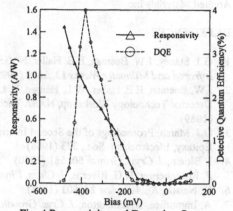

Fig. 4 Responsivity and Detective Quantum Efficiency of Ge:Sb BIB detector under low photon flux at 89.9 μm.

The responsivity shown in Fig. 4 was measured at 89.9 μm with a 11.9 μm bandwidth under a photon power of 2.3x10^{-12} W on the detector. The responsivity of this detector at the Detective Quantum Efficiency (DQE) peak is 0.9 A/W. The responsivity was measured for a single pass absorption without an integrating cavity. We conclude that the improvement in responsivity over previous Ge BIB detectors[15] was due to a reduction in the residual minority dopant concentration to 5x10^{11} cm^{-3} in the IR absorbing layer. The responsivity at the DQE peak occurred at a depletion width of approximately 16 μm and an absorption volume of 0.042 mm^3. In comparison a 1 mm^3 Ge:Ga detector has a responsivity of 4.8 A/W for single pass absorption geometry.[17] The NEP was found to be 4.0x10^{-16} W/Hz$^{1/2}$ at the peak DQE.

CONCLUSION

Ge epilayers have been grown by LPE from a Pb solution and have been characterized for structural perfection and residual electrical impurities. Group V impurities currently limit fabrication of sufficiently pure blocking layers. Ge BIB detectors have been fabricated that respond out to 65 cm^{-1} with responsivity comparable to Ge:Ga photoconductors for a comparably reduced volume, thus reducing the sensitivity to cosmic ray interaction. Reduction of the residual minority impurity concentration increased the space charge layer and the performance. Optimization of the Sb doping concentration should further decrease the photoionization energy of these Ge BIB detectors. Pb purification efforts are in progress and we expect to be able to grow both the blocking and IR absorbing layer with LPE.

ACKNOWLEDGMENTS

This work was supported by National Aeronautics and Space Administration under contracts W17605 and A59513C through interagency agreement with U.S. Department of Energy under contract DE-AC03-76SF00098. C.S. Olsen acknowledges fellowship support from Applied Materials Inc.

REFERENCES

1. G.J. Stacey, J.W. Beeman, E.E. Haller, N. Geis, A. Poglitsch, and M. Rumitz, *Int. J. of Infrared and Milimeter Waves* 13, 1689 (1992).
2. J.W. Beeman, E.E. Haller, W.L. Hansen, P. Luke, and P.L. Richards, Proc. Third Infrared Detector Technology Workshop, NASA Technical Memo 102209, C. McCreight, ed., p.5. (1989).
3. O.J. Marsh, Proceedings of the Second International Symposium on Silicon Molecular Beam Epitaxy, Electrochem. Soc., 333 (1988)
4. J. Bloem, *J. Cryst. Growth* 50, 581 (1980)
5. F. Rosenberger, H.G. Riveros, *J. Chem. Phys.* 60, 668 (1974).
6. H. Nelson, *RCA Review* 24, 603 (1963).
7. A. Immorlica, B. Ludington, *J. Crys. Growth.* 51, 131 (1981).
8. C.S.Rossington, "Germanium Blocked Impurity Band Far Infrared Detectors," (Ph.D. thesis UCB, 1988 unpublished, LBL report #25394).
9. M.P. Lutz, "Development of Ultra Pure Germanium Epi Layers for Blocked Impurity Band Far Infrared Detectors," (Ph.D. thesis UCB, 1991 unpublished, LBL report #30822).
10. D.M Watson, J.E. Huffman, *Appl. Phys. Lett.*, 52, 1602 (1988).
11. I.C. Wu, J.W. Beeman, P.N. Luke, W.L. Hansen and E.E. Haller, *Appl. Phys. Lett.* 58, 153 (1991).
12. D.B. Reynolds, D.H. Seib, S.B. Stetson, T. Herter, *IEEE Trans. Nucl. Sci.*, NS-36, 857 (1992).
13. J.E. Huffman, A.G. Crouse, B.L. Halleck, T.V. Downes, T.L. Herter, *J. Appl. Phys.* 72, 273 (1992).
14. M.D. Petroff and M.G. Stapelbroek, J.J. Speer, D.D. Arlington and C. Sayre, IRIS Specialty Group on IR Detectors, Boulder, CO, (1984).
15. C.S. Olsen, J.W. Beeman, E.E. Haller, Proc. SPIE, Vol. 3122, 1997
16. E.E. Haller, W.L. Hansen, *IEEE Trans. Nucl. Sci.*, NS-21, 279 (1974)
17. J.W. Beeman, E.E. Haller, Infrared Phys., 35, 827 (1994)

HgCdTe-- AN UNEXPECTEDLY GOOD CHOICE FOR (NEAR) ROOM TEMPERATURE FOCAL PLANE ARRAYS

W. E. TENNANT, C. CABELLI
Rockwell Science Center, LLC, 1049 Camino Dos Rios, Thousand Oaks, CA 91360,
wetennant@rsc.rockwell.com

ABSTRACT

Fueled by a broad range of government needs and funding, HgCdTe materials and device technology has matured significantly over the last two decades. Also in this same time period, we have come to understand better the phenomenology which limits imager performance. As a result of these developments, it appears that HgCdTe arrays may be tailored in wavelength to outperform GaAs-based image intensifier devices in sensitivity and to compete with bolometric and pyroelectric imaging arrays in NEDT at temperatures at or near room temperature (250K-295K). These benefits can be fully realized, however, only if HgCdTe can be brought to a level of maturity where the material and detectors made from it are limited by fundamental mechanisms. We will discuss the state of HgCdTe near room temperature performance and the practical and theoretical limits which constrain it.

INTRODUCTION

Over the last two decades government funding of HgCdTe for defense, earth resources, and scientific applications has totaled several billion dollars. The resulting technology has broad applicability to imaging throughout the infrared spectral region at the highest temperatures available to any quantum detector for any given wavelength. Traditional development of HgCdTe has focused on high-sensitivity (few mK noise equivalent temperature difference or NEDT) thermal detection applications which typically require cooling. To obtain high operability focal plane arrays, however, the technology has had to develop materials and device architectures which have relatively good high-temperature performance as well, for detecting both thermal radiation and reflected light.

During this same period, room temperature image intensifiers based on GaAs photocathodes have been improved to give background-limited performance (or "BLIP") at the extremely low light levels associated with overcast starlight. Some slightly-cooled Si CCD's also provide sensitivity near the single photon level. Uncooled thermal detector arrays based on pyroelectric or bolometric detection principles have attained moderate sensitivities of a few 10's of mK. However, more sensitivity is always desirable, as is the ability to see new spectral phenomena without sacrificing the convenience of room temperature performance. The near-IR (NIR) spectral range from ~1.0 μm to 2.5μm appears to offer additional illumination sources (both passive and active) as well as additional spectral information to enhance various imaging applications. Eye-safe lasers typically operate at wavelengths >1.5μm and are therefore invisible to the best current GaAs- or Si-based low-light-level (LLL) imagers. Even under adverse weather conditions the night sky has considerable emission in this spectral region as is seen in Figure 1. Here the radiant stearance in both clear and overcast starlight is at least 10X greater over the near-IR spectral band than it is in the visible-to-very-near-IR (VNIR-- to <900nm for GaAs and to <1100nm for Si) region accessible to current LLL imagers.

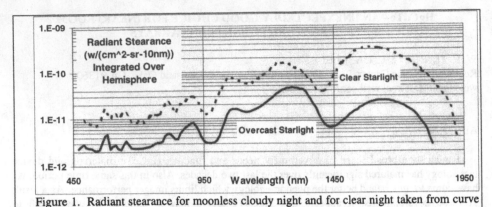

Figure 1. Radiant stearance for moonless cloudy night and for clear night taken from curve 1, figure 13 and curve 1, figure 20, respectively from reference [1].

In this paper we consider HgCdTe operating at or near room temperature, as a candidate to address imaging of thermal infrared, low-light-level (LLL) reflected light, or both.

THEORY

This paper analyzes p/n HgCdTe photodiodes whose architecture will be described below. A simplified version of the one-dimensional analytical model of Williams and DeWames [2] provides interpretation of the diode results. A spreadsheet version of the model runs easily on a personal computer and exploits the spreadsheet fit routines. This model assumes all diode currents are diffusion currents from the bulk material n-side of the junction (i. e. that the substrate interface is passive, and that the depletion region and the p-side of the junction contribute no additional currents). The model calculation includes radiative and Auger recombination mechanisms and allows a Shockley-Read-Hall lifetime to be introduced as an independent parameter. Also, the model assumes that the diffusion length is much greater than the layer thickness. The Appendix gives a listing of the program, references for (or explanations of) the formulae used, definitions of the parameters, and a sample calculation.

Note that the model is not self-consistent with regard to the radiative lifetime. The literature expressions for the lifetime derive from an approximate expression for the absorption coefficient rather than the actual empirical one used for the optical parameters. Calculations show that in the NIR-mid-wave infrared (MWIR) spectral and high operating temperature range (250K-295K) of this investigation the self-consistent lifetime would be about 2 times longer than the model lifetime used for ~2μm detector material and only a few percent longer for ~4μm material. This deviation does not alter the conclusions of the paper.

Holding substrate transmission and contact reflection to be unity, fixing the SRH lifetime to be so large as to be insignificant in determining performance, setting operating bias at -0.1V, and maximizing D* for a given wavelength with respect to Cd fraction (x) and active layer thickness gave upper performance bounds for the technology. Figure 2 shows the model calculations for 250K and 295K at a range of wavelengths (as well as representative data which we shall discuss in greater detail below). The calculations can be compared to a similar calculation by Piotrowski and Gawron [3]. The latter's results agree with these to within a factor of 2 for >4μm wavelengths, the differences possibly being due to different model assumptions about the radiative lifetime. The differences in the models increase markedly, however, at shorter

222

wavelengths. That Piotrowski and Gawron's NIR calculations at wavelengths below about 2.4 μm actually exceed BLIP for 2p steradians and unity quantum efficiency (QE) seems unphysical.

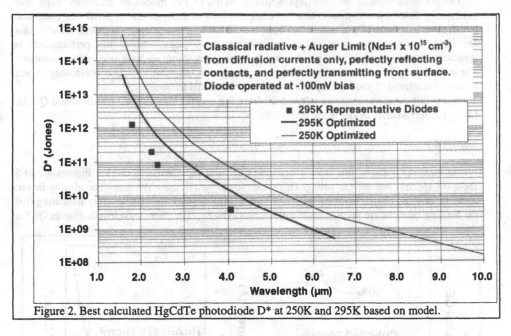

Figure 2. Best calculated HgCdTe photodiode D* at 250K and 295K based on model.

EXPERIMENT

The above discussion assumes that interfaces (and surfaces) are sufficiently passive that they need not be considered in the diode analysis. The last few years' advances in HgCdTe materials and device technology legitimize these assumptions. Molecular beam epitaxy (MBE) crystal growth provided in a single growth run the triple-layer architecture(wide-band-gap buffer, the active, and the wide-band-gap cap). This structure was indium-doped in-situ with ~1-2x1015cm-3 for n-type conduction [4-6]. Selective arsenic ion implantation followed by a high-temperature near-Hg-saturated anneal allowed planar junction formation to produce the double-layer planar heterostructure (DLPH) device geometry as has been described previously [7]. Figure 3 shows a schematic of this

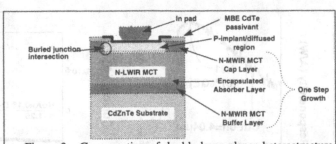

Figure 3. Cross-section of double-layer planar heterostructure (DLPH) device geometry.

architecture. The buffer and cap layers encapsulate the active region of the junction with high quality materials, thus removing potential sources of surface/interface currents. Almost all the devices discussed have this geometry.

Our analysis focuses on four representative MBE-DLPH diodes of relatively large size (~125µm x 125µm junction area) to avoid complications of analysis due to small size compared to diffusion length (9-17 µm as inferred from optical area measurement). Although these diodes are large compared to those found in imaging focal plane arrays, the performance is representative of measured focal plane arrays in those cases where we have been able to compare. These diodes represent not the absolute best we have seen but what we might reasonably expect to be characteristic of good focal plane arrays made with this technology.

Regarding measurements, we inferred D* from the reverse bias saturation current and QE, by fitting the spreadsheet model in the text.

RESULTS

As mentioned above, this work concentrates on four representative diodes. Figures 4 and 5 show the spectral and current-voltage characteristics of two of these devices. The ideality factor, n, of the devices shown brackets the range seen for all devices and is near unity indicating that the limiting currents are preponderantly diffusion currents. The short wavelength drop in QE for

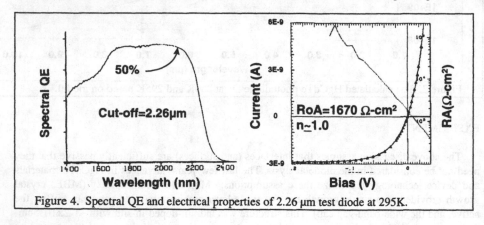

Figure 4. Spectral QE and electrical properties of 2.26 µm test diode at 295K.

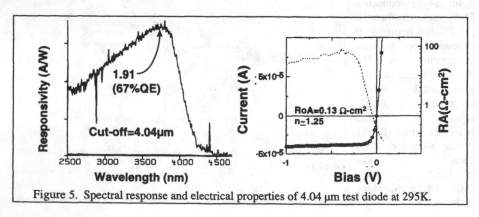

Figure 5. Spectral response and electrical properties of 4.04 µm test diode at 295K.

the 2.26μm device is likely due to some interfacial recombination causing recombination in the buffer layer. This drop in QE in the buffer layer's spectral region does not always appear. In this sample the buffer layer is evidently serving its purpose of protecting the active layer. Table I summarizes relevant diode properties. Two of these devices were reported by DeWames, et al. [8]. These devices were limited by SRH centers. Both layers had trap levels at 0.78 of the energy gap. The hole recombination lifetime (typically called tp0) was 500ns for the 1.81μm layer and 700ns for the 2.4μm layer.

Table I. Properties of four representative diodes used in this analysis

Cut-off Wavelength (μm)	QE (% at cutoff/% at measurement wavelength in μm])	RoA (Ohm-cm²)	D* at cutoff wavelength (Jones at 295K)	Inferred lifetime/ theoretical lifetime
1.81 (QE=1/2 meas QE)	28/ 57@1.6μm	80,000	1.3×10^{12}	.12
2.26 (QE=1/2 meas QE)	25/ 50@2.1μm	1670	2.0×10^{11}	.19
2.40 (QE=1/2 meas QE)	30/ 59@2.1μm	180	8.3×10^{10}	.12
4.04 (1/2 peak spectral response)	29/ 67@3.5μm	0.13 (inferred from saturation leakage current)	3.8×10^{9}	.31

It is clear from the D* data shown in the table and plotted in Figure 2 that these diodes fall below the theoretical maximum value by ~4-7X. A portion of this shortfall is accounted for by the imperfect contact reflectance and lack of front surface AR coating. Using the model analysis to correct these deficiencies results in a net shortfall of 2.5-3.5X. Some loss is also incurred from evaluating QE at the cutoff wavelength where it is below the optimum value, but this is a relatively small effect since a higher QE requires a shorter wavelength which corresponds to a higher theoretical D* value.

Most of the shortfall of the experiment from the theory appears due to recombination centers which shorten the minority carrier lifetime in the n-region. The D* should be depressed from theory by the square root of this lifetime ratio.

Recent data reported by Edwall, et al., [9] indicate that process modification may restore the lifetime to near-theoretical values. With conventional annealing typical NIR layer lifetimes were ~30% or less of theory-- roughly similar to our inferred results. With improved annealing four NIR layers showed 295K lifetimes to be ~90% of theory. If this improvement can be preserved through device processing, outstanding detector performance will result. MWIR diodes apparently receive a similar benefit from this anneal.

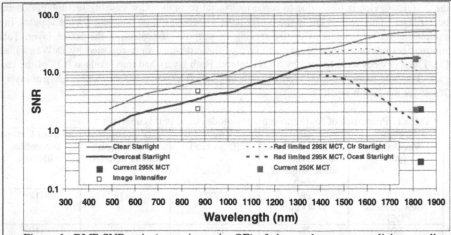

Figure 6. BLIP SNR ratio (assuming unity QE) of clear and overcast starlight as well as the expected SNR of actual and ideal detectors.

We have analyzed the potential for NIR devices to observe scenes in adverse conditions. Figure 6 shows as a function of the cut-off wavelength the signal-to-noise ratio expected for an imager with 25μm pixels integrating for 30ms and using f/1.3 optics. For comparison sake, an image intensifier is portrayed as a BLIP detector with 40% QE. Current room temperature HgCdTe imagers would have comparable SNR to the image intensifier for clear starlight conditions. At 250K the SNR is comparable in overcast starlight conditions and superior in clear starlight conditions. If materials improvements mentioned above can translate into detector performance, cooling will not be needed to match image intensifier performance even under overcast starlight conditions.

Of course the image intensifier is a very mature, high resolution device and will not likely be displaced soon by the NIR HgCdTe. On the other hand, the NIR imager may offer complementary information to that of the image intensifier. Figure 7 shows the NIR image of a dark-haired man wearing a black shirt and standing by a tree. Clearly colors are substantially different in the NIR. Note that this imager actually used an InGaAs detector array with a cutoff ~1.7μm, but whose detectors performed comparably to our best HgCdTe NIR devices.

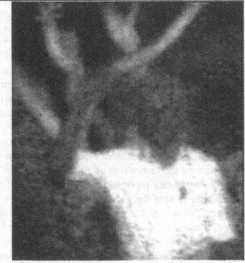

Figure 7. NIR image of a dark-haired, dark-shirted man standing near a tree showing the difference in color phenomenology between NIR and visible.

Thermal imaging is another potential application area for near-room-temperature HgCdTe. Figure 8 shows the projected NEDT for HgCdTe detectors performing at the theoretical limit at 295K and 250K. Also shown is the contribution to NEDT from various components of the leakage current. BLIP is the performance expected from a high quality cooled detector. Also shown are the current performance and goals [10] for conventional uncooled imagers (whose detectors use thermal isolation structures). Clearly no uncooled imager, either current or planned, offers performance compared to cooled devices. However, near-room-temperature MWIR HgCdTe appears to offer sensitivity comparable to conventional uncooled devices. Microlenses on the CdZnTe substrate [11] can decrease NEDT by 2X or more. Cooling slightly will allow MWIR HgCdTe to compete with future uncooled imagers. Suppressing Auger recombination along the lines of the techniques being pioneered in Britain's DRA laboratories [12] may allow this performance level at room temperature.

Another potential enhancement of devices which are limited only by radiative recombination is suggested by Humphries [13]. He points out that the primary source of leakage currents in such diodes is generation due to radiation originating within the semiconductor by the recombination of minority carriers. In reverse-biased junctions fabricated in epitaxial layers whose thickness is much less than a diffusion length, these minority carriers will diffuse to the junction and be collected before they recombine. Since they are not available to generate photons, the radiative background is reduced, along with the diode leakage. In the absence of non-radiative mechanisms, only photons generated outside the semiconductor (or at least outside the region of low minority carrier concentration) will contribute to dark currents. With proper device design, it should be possible to arrange that the only photons contributing to the current are those entering through the system optics-- namely the background photocurrent. Thus, a "Humphries" device will be BLIP, regardless of operating temperature. Figure 9 illustrates this concept in cross-section.

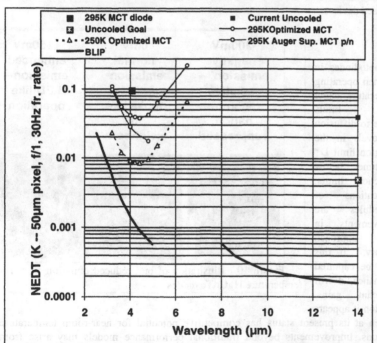

Figure 8. Noise equivalent temperature difference (NEDT) calculated for actual and theoretical MWIR HgCdTe with and without Auger recombination compared to BLIP limit and uncooled detectors

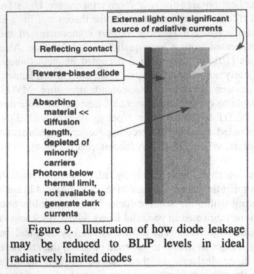

Figure 9. Illustration of how diode leakage may be reduced to BLIP levels in ideal radiatively limited diodes

No one has as yet demonstrated a HgCdTe device which operates in this limit, but it is clear that reverse biasing diodes can reduce thermally generated radiation or "negative luminescence," as reported by Ashley and co-workers [14]. Figure 10 shows a dramatic illustration of such an effect in an optically thin room temperature MWIR HgCdTe diode as seen by a cooled HgCdTe thermal imager. The room temperature diode was observed through the CdZnTe substrate. Because the measurement wavelength was well below the bandgap, the effect was small and needed enhancement for clarity by a manual frame-to-frame subtraction. A slight position shift between frames caused the metal pads of varying reflectance and emission to give a bipolar background structure to the picture. Nonetheless, the figure shows clearly that the reverse-biased diode has lower thermal emission than that of neighboring unbiased devices.

CONCLUSIONS

P/n DLPH photodiodes in MBE HgCdTe when operating near room temperature show performance limits predominantly due to diffusion currents. Deep levels in the charge neutral region limit D* to levels 4-7 times below traditional theoretical limits. Recent modifications of annealing schedules are showing improvements in lifetimes to approach fundamental theory. If this improvement can be captured in device performance, it will produce significant gains.

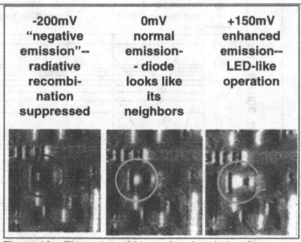

Figure 10. Illustration of bias-reduced emission from room temperature HgCdTe diodes

From this discussion it appears that HgCdTe, even at its present status has competitive potential for near-room temperature imaging applications. Improvements beyond traditional performance models may arise from suppressing Auger and even radiative generation, provided other recombination mechanisms can be removed through improved materials growth and processing.

APPENDIX

References, definitions, sample calculation, and listing for simplified 1-D HgCdTe analytical model

Physics			
Boltzmann's constant	k=	1.381E-23	
Electronic Charge	q=	1.602E-19	
Plank's Constant	h=	6.624E-34	
Permittivity of Free Space (F/cm)	eo=	8.850E-14	
Speed of Light (microns/sec)	cl=	2.998E+14	
Independent Parameters			
Composition	x=	3.052E-01	
Operating Temperature	T=	2.950E+02	
Donor Density	Nd=	8.000E+14	
Wavelength	lam=	4.040E+00	
Bias voltage	Vb=	-1.000E-01	
Layer Thickness	d=	4.898E-04	
Diode Area	Ad=	1.444E-05	
Diode Ideality Factor	nid=	1.000E+00	
Single Surface Substrate Transmission	Trans=	1.000E+00	
Backsurface Reflectance	Rb=	1.000E+00	
extrinsic lifetime limit selected by user to match experimental data or separate SRH calculation	taulow=	1.000E+00	
Dependent parameters			
Energy Gap--Hansen, Schmit, Casselman, J. Appl. Phys. 53, 7099 (1982)	Eg=	2.967E-01	=-0.302+1.93*x-0.81*x^2+0.832*x^3+0.000535*T*(1-2*x)
Gap wavelength--Hansen, Schmit, Casselman, J. Appl. Phys. 53, 7099 (1982)	lg=	4.180E+00	=1.24/Eg
Energy Gap-- Schacham and Finkman, J. Appl. Phys. 57, 2001, (1985)	Egsf=	2.967E-01	=-0.3424+1.838*x+0.148*x^4+(0.000768*T+0.0629)*((1-2.14*x)/(1+x))
Gap wavelength-- Schacham and Finkman, J. Appl. Phys. 57, 2001, (1985)	lgsf=	4.179E+00	=1.24/Egsf
Intrinsic Carrier Concentration-- Hansen and Schmit, J. Appl. Phys. 54, 1639 (1983)	ni=	2.871E+15	=(5.585-3.82*x+0.001753*T-0.001364*x*T)*(1000000000000000*Eg^0.75*T^1.5*EXP(-q*Eg/(2*k*T)))
minority carrier equilibrium concentration from n0=p0 and (Nd+n0)*p0=ni^2	p0=	2.498E+15	=((Nd^2+4*ni^2)^0.5-Nd)/2
Electron Mobility-- Rosbeck, Starr, Price, and Riley, J. Appl. Phys. 53, 6430 (1982)	mue=	5.553E+03	=900000000*(0.2/x)^7.5/T^(2*(0.2/x)^0.6)
Hole Mobility-- using ratio of effective masses, could also use Rosbeck, Starr, Price, and Riley, J. Appl. Phys. 53, 6430 (1982) : muh=mue/100	muh=	1.183E+02	=mue/mci
Electron Effective Mass (inverse)--Weiler, Semiconductors and Semimetals 16 (Willardson & Beer), 119 (1981)	(1/mc=)mci=	4.695E+01	=-0.6+6.33*((2/Eg)+1/(Eg+1))
Hole Effective Mass (inverse)-- Weiler, Semiconductors and Semimetals 16 (Willardson & Beer), 119 (1981)	(1/mv=)mvi=	2.000E+00	=2

Description	Symbol	Value	Formula
High Freq Dielectric Const-- Rogalski & Pietrowski, Prog. Quant. Elect. 12, 160 (1988) citing Dornhaus, Nimitz & Schlicht, Narrow Gap Semicond. Springer-Verlag, Berlin, 1983	einf=	1.162E+01	=15.2-13.7*x+6.4*x^2
Auger overlap integral--Schacham and Finkman, J. Appl. Phys. 57, 2001, (1985)	FF=	2.000E-01	=0.2 (values in literature range from 0.1-0.3)
effective mass ratio mc/mv	mcv=	4.259E-02	=mvi/mci
Auger lifetime-- Schacham and Finkman, J. Appl. Phys. 57, 2001, (1985)	tA=	4.304E-06	=1*2*(ni^2/((Nd+2*p0)*(Nd+p0)))*(0.000 0000000000000038*einf^2*mci*(1+mcv)^0.5*(1+2*mcv)*(q*Eg/(k*T))^1.5*EXP((1+2*mcv)*(q*Eg/(k*T))/(1+mcv))/FF^2)
Radiative Lifetime-- Rogalski & Pietrowski, Prog. Quant. Elect. 12, 160 (1988) mod by Schacham and Finkman, J. Appl. Phys. 57, 2001, (1985)	tr=	5.670E-06	=1/(0.00000000000058*(Nd+2*p0)*einf^ 0.5*(1/mci+1/mvi)^-1.5*(T/300)^- 1.5*(1+mci+mvi)*(Eg^2+3*k*T*Eg/q+3.7 5*(k*T/q)^2))
Total lifetime	tt=	2.447E-06	=1/(1/tr+1/tA+1/taulow)
Minority Carrier (hole) Diffusion Length	Lh=	2.712E-03	=(muh*k*T*tt/q)^0.5
Intermediate parameter--Schacham and Finkman, J. Appl. Phys. 57, 2001, (1985)	Ego=	2.198E-01	=Egsf-(0.0629+0.000768*T)*(1- 2.14*x)/(1+x)
Intermediate parameter--Schacham and Finkman, J. Appl. Phys. 57, 2001, (1985)--my approximate solution (pretty close) from their formulae	alf1=	1.209E+03	=IF(((h*cl/(q*lam))- Egsf)<=0,0,800*(32670*(1+x)/((LN(800) +18.88-53.61*x)*(T+81.9)- (0.0629+0.000768*T)*(32670)*(1- 2.14*x)))^0.5*((h*cl/(q*lam))-Egsf)^0.5)
Intermediate parameter--Schacham and Finkman, J. Appl. Phys. 57, 2001, (1985)-- their stated solution	alf1a=	1.246E+03	=IF(((h*cl/(q*lam))- Egsf)<=0,0,210900*((1+x)/(81.9+T))^0. 5*((h*cl/(q*lam))-Egsf)^0.5)
Intermediate parameter--Schacham and Finkman, J. Appl. Phys. 57, 2001, (1985)	alf2=	1.518E+03	=EXP(53.61*x- 18.88)*EXP(32670*(1+x)*((h*cl/(q*lam)) -Ego)/(81.9+T))
Intermediate parameter--Schacham and Finkman, J. Appl. Phys. 57, 2001, (1985)	alf=	1.209E+03	=IF(alf1>800,alf1,alf2)
Zero Bias Resistance-area product	RoA=	3.174E-01	=(Nd+p0)*Lh^2/(q*muh*d*ni^2)
Diode Current	I=	-1.134E-06	=(k*T*Ad/(q*RoA))*(EXP(q*Vb/(nid*k*T))-1)
Differential Resistance Diode Area Product at bias Vb	RA=	1.620E+01	=RoA*EXP(-q*Vb/(nid*k*T))
Intermediate parameter	alfL=	3.278E+00	=alf*Lh
Intermediate parameter	dL=	1.806E-01	=d/Lh
Intermediate parameter	alfd=	5.919E-01	=alf*d
Quantum Efficiency-- Adapted from Rogalski & Pietrowski, Prog. Quant. Elect. 12, 111 (1988)	QE=	6.863E-01	=Trans*(alfL/(alfL^2-1))*((((alfL-EXP(- alfd)*SINH(dL))/COSH(dL))-alfL*EXP(- alfd))+Rb*EXP(-alfd)*((alfL*EXP(-alfd)- SINH(dL))/COSH(dL)))
	D*=	1.371E+10	=0.8*lam*QE/((2*q*ABS(I)/Ad)+4*k*T/R A)^0.5

ACKNOWLEDGEMENTS

The authors thank Roger DeWames and Dennis Edwall for many useful technical discussions regarding basic limiting mechanisms and the specific mechanisms which limit our samples. We also recognize the substantial contribution of George Williams who researched, constructed, and verified the complete analytical model on which the above simplified version is based.

REFERENCES

[1] Mishri L. Vatsia, U. Karl Stich, Douglas Dunlap, Technical Report ECOM-7022, U. S. Army Electronics Command, Night Vision Laboratory, (Sep. 1972), obtained from Defense Technical Information Center (DTIC).

[2] G. M. Williams, and R. E. DeWames, J. Electron. Mat. **24**, 1239 (1995)

[3] J. Piotrowski, and W. Gawron, Infrared Phys. & Tech. **38**, 63 (1997)

[4] 4. J.M. Arias, "Growth of HgCdTe by molecular-beam epitaxy," in Properties of Narrow Gap Cadmium-based Compounds, EMIS Datareview Series No. 10, Inspec publication, edited by Peter Capper, 30-35 (1994).

[5] J.M. Arias, J.G. Pasko, M. Zandian, J. Bajaj, L.J. Kozlowski, R.E. DeWames and W.E. Tennant, Proceedings of SPIE Symposia on "Producibility of II–VI Materials and Devices", Volume 2228, 1994.

[6] J. Bajaj, J.M. Arias, M. Zandian, J.G. Pasko, L.J. Kozlowski, R.E. DeWames and W.E. Tennant, "Molecular Beam Epitaxial HgCdTe Materials Characteristics and Device Performance: Reproducibility Statistics," J. Electronic Materials 24, 1067 (1995).

[7] J.M. Arias, J.G. Pasko, M. Zandian, S.H. Shin, G.M. Williams, L.O. Bubulac, R.E. DeWames, and W.E. Tennant, Appl. Phys. Lett. 62, 976 (1993).

[8] R. E. DeWames, D. D. Edwall, M. Zandian, L. O. Bubulac, J. G. Pasko, W. E. Tennant, J. M. Arias, a. D'Souza, Proceedings of the 1997 Workshop on the Physics and Chemistry of II-VI Materials, to be published.

[9][6] D. D. Edwall, R. E. DeWames, W. V. McLevige, J. G. Pasko, J. M. Arias, Proceedings of the 1997 Workshop on the Physics and Chemistry of II-VI Materials, to be published.

[10] Defense Advanced Research Projects Agency BAA 96-32.

[11] M. Edward Motamedi, William E. Tennant, Haluk O. Sankur, Robert Melendes, Natalie S. Gluck, Sangtae Park, Jose M. Arias, Jagmohan Bajaj, John G. Pasko, William V. McLevige, Majid Zandian, Randolph L. Hall, Patricia D. Richardson, Opt. Eng. **36**(5), 1374 (1997)

[12] C. T. Elliott, N. T. Gordon, R. S. Hall, T. J. Phillips, and A. M. White, J. Electron. Mat. **38**, 1139 (1996)

[13] R. G. Humphries, Infrared Phys. **26**, 337 (1986)

[14] T. Ashley, C. T. Elliott, N. T. Gordon, T. J. Phillips, and R. S. Hall, Infrared Phys. & Tech. **38**, 145 (1997)

BANDGAP-ENGINEERING OF HgCdTe FOR TWO-COLOR IR DETECTOR ARRAYS BY MOVPE

P. MITRA[a], F.C. CASE[a], S.L. BARNES[a], M.B. REINE[b], P. O'DETTE[b] and S.P. TOBIN[b]

[a] Lockheed Martin Vought Systems, Dallas TX 75265-0003
[b] Lockheed Martin IR Imaging Systems, Lexington, MA 02173-7393

ABSTRACT

Recent results on MOVPE growth of multilayer two-color HgCdTe detectors, for simultaneous and independent detection of medium wavelength (MW, 3-5 μm) and long wavelength (LW, 8-12 μm) bands, are reported. The structures are grown *in situ* on lattice matched (100) CdZnTe in the double-heterojunction p-n-N-P configuration. A barrier layer is placed between the LW and MW absorber layers to prevent diffusion of MW photocarriers into the LW junction and thereby eliminate spectral crosstalk. X-ray double crystal rocking curve widths are ~ 45 arc-secs, indicating good epitaxial quality. SIMS depth profile measurements of these 28 μm thick structures show well-defined alloy compositions, and arsenic and iodine doping. SIMS data on a series of thirteen films show that good run-to-run repeatability is obtained on thicknesses, compositions, and dopant levels with values close to the device design targets. Depth profile of etch pits through the thickness of the films show etch pit densities in the range of $8 \times 10^5 - 5 \times 10^6$ cm^{-2}.

INTRODUCTION

Vapor phase epitaxy of $Hg_{1-x}Cd_xTe$ alloys has emerged as the materials technology of choice for growth of multilayer and multijunction infrared devices such as the independently accessed dual-band detectors.[1,2] To achieve detector performance comparable to that of state-of-the-art single-band devices, the dual-band detector structures require precise control of alloy composition, p-type and n-type doping levels, high dopant activation, as well as precise positioning of the p-n junctions. Both molecular beam epitaxy[2] (MBE) and metalorganic vapor phase epitaxy[1,3] (MOVPE) have been recently demonstrated to be capable of *in situ* growth of these devices.

At Lockheed Martin we are pursuing MOVPE technology for growth of dual-band IR detectors. Recent progress in MOVPE of HgCdTe now permits the growth of multilayer structures[3,4] *in situ*, with single or multiple p-n junctions, with complete flexibility in the control of alloy composition, iodine donor[5] and arsenic acceptor doping,[6-8] both with 100% dopant activation. The mobilities and minority carrier lifetimes are essentially the same as those observed in comparably doped HgCdTe grown by liquid phase epitaxy (LPE). These attributes of MOVPE enable the realization of IR detector devices that incorporate optimal device designs through bandgap engineering. Using MOVPE we have previously demonstrated *in situ* growth of single-band LW detector arrays in the p-on-n heterojunction configuration with 77 K average zero-bias resistance-area product (R_0A) values of 430 ohm-cm^2 and quantum efficiencies (QE) of ≥60% for a cutoff wavelength of 10.1 μm.[8] The trend in average R_0A values for a number of LW 64x64 arrays exhibits essentially the same trend as that achieved in state-of-the-art LPE or MBE based HgCdTe detector arrays. In the MW band a new N-on-P heterojunction[6] detector was shown with R_0A products above 2×10^5 ohm-cm^2 for cutoff wavelengths of 5.0 μm at 77 K and

QE in the range of 70-75%. In addition, independently accessed, p-n-N-P two-color HgCdTe detectors[1] and 64x64 focal plane arrays[3] (FPA) were demonstrated, that operate simultaneously in the MW and LW spectral bands. The QE and detectivities of these detectors approach those achieved in state-of-the-art single band detectors in the respective spectral bands.

In this paper new results are reported for MOVPE growth of stacked two-color IR detectors in the double-heterojunction p-n-N-P configuration that operate simultaneously in the MW (3-5 μm) and LW (8-12 μm) spectral bands. As compared to our previous demonstration[3] of two-color detectors, a barrier layer between the MW and LW absorber layers has been introduced to minimize spectral cross-talk between the two detectors. Repeatability results based on SIMS analysis are reported for a series of thirteen films grown under nominally identical conditions. X-ray double crystal rocking curve (DCRC) mapping data, and dislocations in the two-color detector films evaluated by etch pit density (EPD) depth profiles through the film thickness, are presented.

THE p-n-N-P TWO-COLOR DETECTOR

Lockheed Martin's independently accessed, backside illuminated, simultaneous, p-n-N-P two-color HgCdTe detector is based upon an LW p-on-n heterojunction on top of an N-on-P MW heterojunction grown *in situ* on nominally lattice-matched CdZnTe. We have shown[3] that this device architecture allows the fabrication of FPAs with high QE and detectivities at 78 K. The spectral response data showed well behaved sharp profiles corresponding to the compositions of the absorber layers. The LW → MW spectral crosstalk was measured to be at 0.4% but a more significant MW → LW spectral crosstalk was observed. The latter was attributed to spillover of photocarriers from the MW absorber layer to the LW junction. To eliminate the residual MW → LW crosstalk a small composition barrier has been added at the MW/LW absorber layer interface. A cross-section of the double-heterojunction p-n-N-P two-color IR detector design with the barrier layer and its energy band profile is shown in Figure 1. Two indium bump interconnects in each detector provide independent electrical access to the LW and MW photodiodes and allow the respective photocurrents to be separated. The wide gap SW p-type layer serves as the common contact.

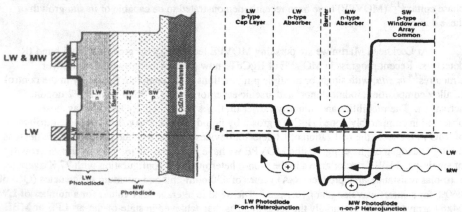

Figure 1. Cross section and energy band profile for the simultaneous p-n-N-P two-color HgCdTe detector with a composition barrier.

MOVPE growth of the multilayer HgCdTe films was carried out by the interdiffused multilayer process (IMP) which involves the growth of alternating thin layers of HgTe and CdTe that are subsequently allowed to interdiffuse at the growth temperature to form a homogeneous alloy.[9] Growth was carried out at 360° C and the precursors, growth system and the conditions used are identical to those described previously.[3,8] Iodine and arsenic doping were achieved with ethyl iodide (EI) and *tris*-dimethylamino-arsenic (DMAAs), respectively. The two-color films were grown on nominally lattice-matched CdZnTe, oriented (100) 4°→<111>B.

Secondary ion mass spectrometry (SIMS) depth profile data, taken at Charles Evans & Associates, on an LW/MW two-color film grown with a composition barrier between the MW and LW absorber layers are shown in Figure 2. The Cd mole fractions (HgCdTe x-values) were obtained from measurements of the yield of $CsCd^+$ molecular ions resulting from Cs^+ ion bombardment and recombination of the sputtered neutral Cd atoms.[10] These values were generally higher than that measured by FTIR transmission on the smallest bandgap layer and were therefore corrected for each of the layers in the entire profile. The data in Figure 2 however, are shown as obtained by SIMS without the correction. The x-value profile clearly shows well-defined alloy composition regions in the multilayer film structure. The interdiffused width between two x-values is ≤ 1.5 μm and is well within acceptable limits for these devices. The small composition barrier layer between the MW and LW absorber layers is also well-defined and demonstrates the degree of composition control that is achieved in MOVPE of HgCdTe.

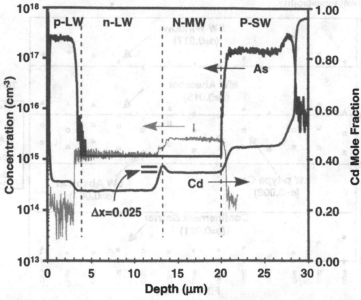

Figure 2. SIMS depth profile data of the Cd mole fraction, arsenic and iodine doping levels in a p-n-N-P two-color HgCdTe film grown with a small composition barrier between the I-doped LW and MW layers.

The dopant concentrations were also measured by SIMS with Cs^+ ion bombardment and were quantified against calibrated standards. The depth profile data for iodine show sharp onset and falloff due to its low diffusion rate in HgCdTe at the growth temperature. The iodine level is at $2x10^{15}$ cm^{-3} in the MW and at $1x10^{15}$ cm^{-3} in the LW absorber layers, in agreement with the respective target values of $(2-4)x10^{15}$ cm^{-3} and $(1-2)x10^{15}$ cm^{-3}. The intentionally graded reduction in the doping level through the barrier layer is also clearly evident. The As doping levels in the SW p-type layer and the LW p-type cap layer are also well within the targeted values of $(1-2)x10^{17}$ cm^{-3} and $(2-4)x10^{17}$ cm^{-3}, respectively.

Repeatability in the growth of the two-color films by MOVPE was investigated by SIMS depth profile analysis for a series of 13 films grown with the same parameters. The film parameters measured were thicknesses, x-values and doping levels. Figure 3 shows the x-value repeatability data for each of the four layers and the confinement barrier. The target values are included along with the standard deviation in x-values for each layer. The x-value of the LW n-type layer was determined by FTIR transmission measurements and was used to calibrate each of the SIMS x-profiles. The values for each of the layers are close to the target values and demonstrate good run-to-run repeatability.

Figures 4 and 5 show repeatability data for thicknesses and doping levels, respectively, for the same thirteen films along with their target values. Good repeatability is achieved in the thickness data. In Figure 5 the iodine doping results in the first four films show considerable variation. These early films were grown with a new flow control system that had to be adjusted to achieve the targeted values. Following the initial results both the iodine and arsenic doping show excellent repeatability.

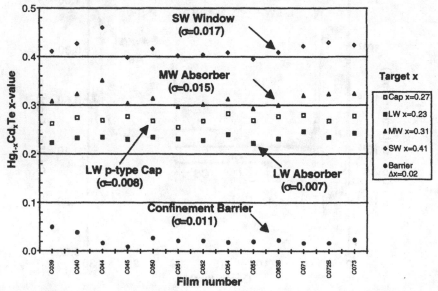

Figure 3. Repeatability results for x-values for each of the layers in a series of p-n-N-P two-color films. The target x-values for each layer are shown on the right. The values in parentheses are the standard deviations.

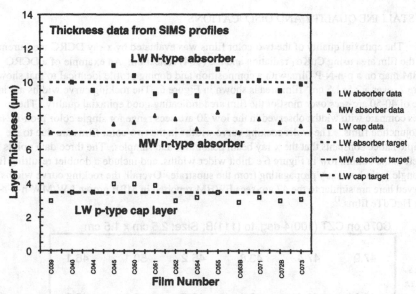

Figure 4. Repeatability data for thicknesses for each of the layers in p-n-N-P two-color films.

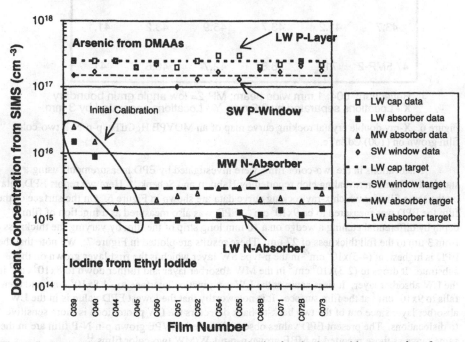

Figure 5. Repeatability data for arsenic and iodine doping levels along with target values for p-n-N-P two color HgCdTe films.

CRYSTALLINE QUALITY AND DISLOCATIONS

The epitaxial quality of the two-color films was evaluated by x-ray DCRC measurements over the film area using CuKα_1 radiation and the (004) reflection. An example of a DCRC FWHM map on a p-n-N-P film, with a composition and doping profile identical to that shown in Figure 2, over a 2.5x1.5 cm^2 film area is shown in Figure 6. The rocking curve widths are in the range of 40-50 arc-secs over most of the film area indicating good epitaxial quality. These values compare with widths observed in the low 30 arc-secs range for single color p-on-n heterojunction films. The wider rocking curve widths in the two-color films are due to the multiple lattice constants that the x-ray beam penetrates and samples. The three data points from the left in the bottom row in Figure 6 exhibit wider widths, and include a doublet resulting from a low angle grain boundary propagating from the substrate. Overall, the rocking curve widths observed here are similar to the 47 arc-secs FWHM reported in MBE grown LW/MW n-p-n two-color HgCdTe films.[11]

C075 on CZT (100)4 deg. to (111)B; Size: 2.5 cm x 1.5 cm

47.9	47.4	43.8	45.2	50.1	46.1
45.9	46.4	43.4	48.2	45.8	45.8
43.7	41.2	43.7	45.9	43.2	41.1
47.5MP-2	71	58.8	42.7	40.6	41

Reflection: 004; 1 mm wide beam; MP-2= low angle grain boundary
X - Locations separated by 3.5 mm; Y - Locations separated by 3 mm

Figure 6. X-ray double crystal rocking curve map of an MOVPE HgCdTe p-n-N-P two-color film grown on (100) CdZnTe.

Dislocations in the two-color films were investigated by EPD measurements using a standard dislocation revealing etch reported by Hahnert and Schenk.[12] Here we report EPD data on the same film for which x-ray rocking curve data are shown in Figure 6. On the surface of the film the EPD was measured to be 9x10^5 cm^{-2}. EPD was also measured as a function of film depth by differential etching a wedge on a 10 mm long strip of the film by varying the thickness from 3 μm to the full thickness of 27 μm. These results are plotted in Figure 7. We note that the EPD is highest, at (4-5)x10^6 cm^{-2} in the p-type SW layer which is the first layer grown on the substrate. It drops to (2-3)x10^6 cm^{-2} in the MW absorber layer and further down to 8x10^5 cm^{-2} in the LW absorber layer. It rises again at the LW p-n junction interface to ~1.8x10^6 cm^{-2} and then falls to 9x10^5 cm^{-2} at the film surface. It is noteworthy that the lowest EPD value is in the LW absorber layer since out of the two back-to-back detectors the LW photodiode is more sensitive to dislocations. The present EPD values observed in the MOVPE grown p-n-N-P film are in the same range as those reported in MBE grown n-p-n LW/MW two-color films.[11]

238

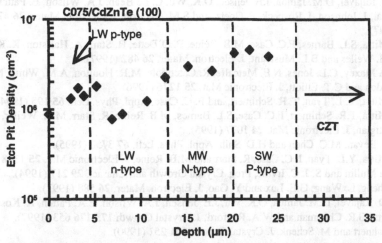

Figure 7. EPD depth profile in an MOVPE p-n-N-P LW/MW double-heterojunction two-color film with a confinement barrier.

CONCLUSIONS

The results presented in this paper demonstrate that MOVPE growth of HgCdTe has matured significantly to grow bandgap engineered films with complete flexibility in alloy composition and doping control. LW/MW two-color films have been grown in the double-heterojunction p-n-N-P configuration with a confinement barrier between the LW and MW absorber layers. SIMS depth profile data indicate that all of the targeted composition and doping levels have been achieved. SIMS data on a series of two-color film growth runs show good repeatability in the composition, thicknesses, iodine and arsenic doping in the individual layers of the multilayer films. These repeatability results are significant since the p-n-N-P two-color detector has an especially challenging architecture and no *in situ* monitors were used to correct unintended variations in growth parameters. X-ray rocking curve widths in the 40-50 arc-secs range indicate good epitaxial quality. An EPD depth profile through the thickness of a two-color film indicates surface EPD of ~$9x10^5$ cm^{-2} and a maximum EPD of $(4-5)x10^6$ cm^{-2} in the SW p-type layer. Further improvements in run-to-run repeatability and reductions in the dislocation density in the MOVPE two-color films are expected to be achieved through tighter controls by the utilization of *in situ* growth monitors.

ACKNOWLEDGEMENT

We gratefully acknowledge the support of the Air Force Phillips Laboratory Contract F29601-96-C-0008 monitored by Dr. Paul D. LeVan and the support of the BMDO -Advanced Sensor Technology Program managed by Dr. Walt Dyer. This work was also supported in part by Lockheed Martin Vought Systems internal research funds.

REFERENCES

1. M.B. Reine, P.W. Norton, R. Starr, M.H. Weiler, M. Kestigian, B.L. Musicant, P. Mitra, T.R. Schimert, F.C. Case, I.B. Bhat, H. Ehsani and V. Rao, J. Electron. Mater. **24** 669 (1995).

2. R.D. Rajavel, D.M. Jamba, J.E. Jensen, O.K. Wu, C. Le Beau, J.A. Wilson, E. Patten, K. Kosai, J. Johnson, J. Rosbeck, P. Goetz, and S.M. Johnson, J. Electron. Mater. **26** 476 (1997).

3. P. Mitra, S.L. Barnes, F.C. Case, M.B. Reine, P. O'Dette, R. Starr, A. Hairston, K. Kuhler, M.H. Weiler and B.L. Musicant, J. Electron. Mater. **26** 482 (1997).

4. C.D. Maxey, C.L. Jones, N.E. Metcalfe, R. Catchpole, M.R. Houlton, A.M. White, N.T. Gordon and C.T. Elliott, J. Electronic Mat. **25** 1276 (1996).

5. P. Mitra, Y. L. Tyan, T. R. Schimert, and F. C. Case, Appl. Phys. Lett. **65** 195 (1994).

6. P. Mitra, T.R. Schimert, F.C. Case, S.L. Barnes, M.B. Reine, R. Starr, M.H. Weiler, and M. Kestigian, J. Electronic Mat. **24** 1077 (1995).

7. M.J. Bevan, M.C. Chen and H.D. Shih, Appl. Phys. Lett. **67** 3750 (1995).

8. P. Mitra, Y.L. Tyan, F.C. Case, R. Starr and M.B. Reine, J. Electronic Mat. **25** 1328 (1996).

9. J. B. Mullin and S. J. C. Irvine, Prog. Crystal Growth and Charact. **29** 217 (1994).

10. J. Sheng, L. Wang, G.E. Lux and Y. Gao, J. Electron. Mater. **26** 588 (1997).

11. R. D. Rajavel, D.M. Jamba, O.K. Wu, J.E. Jensen, J.A. Wilson, E.A. Patten, K. Kosai, P. Goetz, G.R. Chapman and W.A. Radford, J. Crystal Growth **175/176** 653 (1997).

12. I. Hahnert and M. Schenk, J. Crystal Growth **101** 251 (1990).

INFLUENCE OF STRUCTURAL DEFECTS AND ZINC COMPOSITION VARIATION ON THE DEVICE RESPONSE OF $Cd_{1-x}Zn_xTe$ RADIATION DETECTORS

H. Yoon*, J.M. Van Scyoc*, T.S. Gilbert*, M.S. Goorsky*, B.A. Brunett**, J.C. Lund[†], H. Hermon[†], M. Schieber[†], R.B. James[†]
*Department of Materials Science and Engineering, University of California, Los Angeles, Los Angeles, CA 90095, yhojun@seas.ucla.edu
**Department of Electrical and Computer Engineering, Carnegie Mellon University, Pittsburgh, PA 15213
[†]Sandia National Laboratories, Livermore, CA 94551

ABSTRACT

Zinc composition variation and gross structural defects including grain and tilt boundaries, twins, and mechanical cracks in high pressure Bridgman $Cd_{1-x}Zn_xTe$ are characterized and correlated to various detector-related responses. Triple axis x-ray diffraction, double crystal x-ray topography, infrared microscopy, and etch pit density measurements are used to reveal and quantify the spatial distribution and the nature of the structural defects. Mechanical cracks in the material are found to act as conductive "shorting paths", indicated by excessive leakage currents and reduced charge (electron) collection measured along these cracks. Reduced charge collection is also obtained across grain boundaries and in regions with poor crystallinity, indicating that they serve as carrier recombination sites. Finally, the effects of the zinc composition variation on the measured leakage current and the amount of electrons collected are found to be masked by gross structural defects. These characterization techniques provide a wealth of information which can be used not only to study the relationship between the structural and device properties of CdZnTe but also to screen production material for subsequent device fabrication.

INTRODUCTION

There has been a rapid development of solid state radiation detectors fabricated from high pressure Bridgman cadmium zinc telluride (HPB CdZnTe) since its discovery in 1991.[1,2] Material uniformity is still an important issue that has received much attention,[3,4] since it has a strong influence on detector yield. Applications such as imaging devices or large volume spectrometers require uniform detector properties. Structural defects such as mechanical cracks, grain/tilt and twin boundaries, and variations in the zinc content can contribute to non-uniform detector response. Indeed, it has been reported that mechanical cracking of the ingot significantly reduces the material yield,[4] and that grain boundaries act as carrier recombination sites.[5] However, the relationship between these structural defects and device response is not well understood at this time and requires further investigation, especially on a systematic basis. In this paper, we report on the use of an automated spatial mapping apparatus to obtain various detector related properties from large area CdZnTe wafers. In particular, variations in the leakage current and maximum alpha particle pulse height are measured and related to various structural defects and zinc variation. Structural perfection (crystallinity) and zinc composition are determined from triple axis x-ray diffraction (TAD) techniques.[6,7] Infrared (IR) transmission microscopy and double crystal x-ray topography (DCXRT) are employed to image the structural defects.

EXPERIMENT

CdZnTe samples investigated in this study are grown by the vertical HPB process.[8] One wafer (polycrystalline) examined in this study was sliced along the growth direction with area ~100 cm^2 and thickness ~5 mm. Another sample is a single crystal material of ~5 cm^2 and thickness 2.5 mm. Figure 1 shows a schematic experimental setup of the detector mapping apparatus, which is a modified version of the setup used in Ref. [9]. Contact is achieved onto a bare crystal surface via a conductive rubber probe of 2.5 mm in diameter, and is equipped with a guard ring, which helps to provide uniform electric field and eliminate surface leakage current. The opposite face of the bare crystal is coated with electroless gold which serves as the grounding electrode. An alpha particle source (^{244}Cm) is used as the

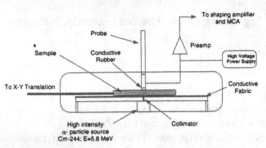

Figure 1. Schematic illustration of the automated spatial mapping apparatus utilizing a collimated alpha particle source.

source, and is collimated to about 1 mm spot size onto the cathode side of the detector such that electrons are collected. Spatial mapping is performed by translating the sample at 2 mm increments; at each location, leakage current and alpha particle pulse height spectrum are recorded. The leakage current is measured (averaged over 45 seconds), at a bias of 100 volts and the alpha particle spectrum is simultaneously collected using an amplifier shaping time of 5 μs. The entire process is fully automated via a LabVIEW program and all information collected by a computer. IR transmission imaging is obtained with a commercial CCD camera with its IR cut filter removed. Triple axis x-ray diffraction measurements[6,7] are conducted on a Bede D^3 diffractometer. TAD ω scans ("rocking curves") employ a 4-bounce (220) silicon analyzer crystal and are used to assess the crystallinity and to reveal the presence of tilt or low angle grain boundaries. Precise lattice parameter measurements[7] use the same analyzer crystal to determine the exact 2θ position of the sample. Vegard's law between the lattice parameter of CdTe and ZnTe is assumed to determine the zinc composition to an accuracy of 0.1%. Double crystal x-ray topography[10,11] is performed using a modified Bede 150 diffractometer, and utilizes a (422) silicon first crystal which is set to diffract the CuKα$_1$ line. Kodak SR1 or Kodak DEF 5 films are used for recording the topographs.

RESULTS

Figure 2 presents the detector response mapping data performed on a ~100 cm^2 CdZnTe. Spatial variations of (a) the leakage current and (b) the maximum alpha particle pulse height, along with (c) an IR transmission image are shown. The left side corresponds to the first-to-freeze end of the boule. The energy channel at which the peak number of counts occur in the alpha particle spectrum represents the maximum pulse height, which is plotted in Figure 2(b). The IR image clearly shows the grain boundaries, twins, and cracks. The cracks show up as dark (black) regions in the IR image, as they provide strong scattering centers for the incident light. The leakage current varies by 3 orders of magnitude within this wafer, mainly due to the

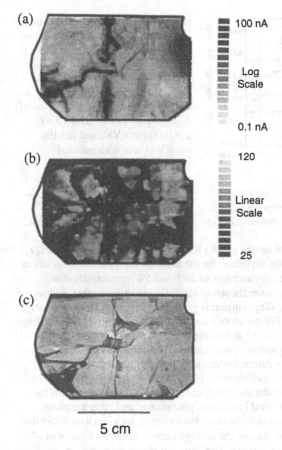

(a)

100 nA

Log
Scale

0.1 nA

(b)

120

Linear
Scale

25

(c)

5 cm

Figure 2. Mapping results of (a) leakage current, (b) maximum alpha particle pulse height, and (c) IR transmission image of a ~100 cm^2 longitudinal slice CdZnTe wafer.

excessively high leakage current along the cracks. The leakage current, I_L, is related to the resistivity of the material according to

$$I_L = \frac{V \cdot A}{\rho \cdot d} \qquad \text{Eq.(1)}$$

where V is the detector bias, A is the contact area, ρ is the resistivity, and d is the detector thickness. Under our experimental conditions of $V=100$ volts, $A=0.05$ cm^2, and $d=0.55$ cm, I_L of 0.1 nA [the lowest value on the scale in Figure 2(a)] corresponds to a resistivity of ~9×10^{10} Ω·cm.

To study the possible effects of the zinc composition on the leakage current (resistivity) we measured the lattice parameter at various locations across the wafer using triple axis x-ray diffraction,[7,12] and Figure 3 displays the zinc composition variation determined from these measurements. As seen on the graph, the zinc content decreases monotonically from approximately 11% from the first-to-freeze end down to about 4% towards the end of the boule. The solid line represents the normal freezing equation[13]

$$C_s = k \cdot C_a (1-x)^{k-1} \qquad \text{Eq.(2)}$$

using an effective segregation coefficient of k=1.15 where C_s is

the amount of zinc in the solid at the point where a fraction x of the original melt has solidified, C_a is the initial amount of zinc (10%), and k is the segregation coefficient. From this composition variation, the resistivity [and hence the leakage current from Eq.(1)] can be predicted based on the changes in the electron and hole concentrations, since the resistivity varies as

$$\rho = \frac{1}{n \cdot e \cdot \mu_e + p \cdot e \cdot \mu_h} \qquad \text{Eq.(3)}$$

where n and p are electron and hole concentrations, respectively, e is the charge, and μ_e and μ_h are the electron and hole mobility, respectively.

To calculate the values of n and p, we utilize Shockley's approach[14] of solving the equations governing the carrier concentrations for extrinsic semiconductors incorporated with a deep level defect. Here we assume the deep level to be an acceptor type, which compensates the shallow donor impurities to make the material semi-insulating by effectively pinning the Fermi

243

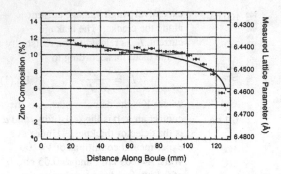

Figure 3. Experimentally determined lattice parameter and zinc composition variation across the CdZnTe wafer shown in Figure 2. The solid line represents the normal freezing equation using k=1.15.

energy near the center of the bandgap. We use the following values: donor concentration, $N_d = 10^{16}$ cm^{-3}, donor energy level, $E_d = 0.010$ eV, and deep level acceptor concentration, $N_a = 10^{17}$ cm^{-3}. Using commonly accepted values of $E_{g, CdTe}$=1.45 eV, $E_{g, ZnTe}$=2.3 eV, μ_e=1000 cm^2/(V·s), and μ_h=100 cm^2/(V·s), and with material parameters from Ref. [15], we consider two cases. First, we assume that the deep level is at the midgap independent of the zinc composition. For the second case,

we assume that the deep level is at a fixed distance (0.8 eV) away from the valence band edge, regardless of the zinc composition. For the first case, the calculations yield resistivity values of 5.4×10^9 Ω·cm and 2.3×10^9 Ω·cm for zinc compositions of 10% and 5%, respectively (for simplicity, the two zinc composition values are chosen as approximate variation that is observed along the boule). Correspondingly, the leakage current is predicted to increase (monotonically) by only a factor of 2.3 across the boule. For the second case, the results are similar: the leakage current is predicted to increase by a factor of 2.7 across the boule. By contrast the experimental leakage current varies by 3 orders of magnitude. These calculations show that the effects of the zinc composition variation on the leakage current is "masked" by other factors such as that induced by the cracks. Another important consideration is the local fluctuations of the carrier mobility (which was assumed constant for the above calculations), which may arise from the charge scattering centers provided by chemical impurities, precipitates, and other structural imperfections such as grain boundaries and dislocations. For example, by using a hole mobility value of μ_h=10 cm^2/(V·s) in the above calculation, the leakage current varies by a factor of ~7. These results and analysis indicate that the zinc variation does not play a major role in influencing the leakage current in this detector material.

The maximum pulse height mapping shown in Figure 2(b) also reveals the non-uniform detector response from this wafer. The arbitrary scale is converted to represent a relevant detector property, namely the electron mobility-lifetime product ($\mu\tau$), which defines the drift length, λ, of the carrier by the relationship $\lambda=\mu\tau E$ where E is the electric field. This is done following a similar procedure as in Ref. [1]. This is done by measuring the amount of charge collected as a function of bias and fitting the Hecht relation[16]

$$Q(V) = \frac{Q_o \cdot \mu\tau \cdot V}{d^2} \cdot \left(1 - \exp\left(\frac{-d^2}{\mu\tau \cdot V}\right)\right)$$ Eq.(4)

where $Q(V)$ is the amount of charge collected at a bias of V, Q_o is the amount of incident charge created, and d is the detector thickness. Using this procedure, the end points of the scale in Figure 2(b) corresponds to electron $\mu\tau$ values of 0.5×10^{-3} (low) and 6.4×10^{-3} cm^2/V (high), respectively. The greater part of the entire wafer area exhibits low $\mu\tau$, unacceptable for spectrometer grade detectors. By comparing this mapping to the IR image, it is generally observed that the electron transport property abruptly changes across a grain boundary (see for example, the upper left region of the wafer) but not necessarily across twin boundaries (see for example the upper right region of the wafer). Also, regions of high leakage current correspond to regions of low $\mu\tau$ values, indicating that above a certain threshold leakage current (in this case, $>\sim 10$ nA) detector property is significantly degraded. However, there are regions (in particular, lower right area of the wafer) where the leakage current is fairly low but display poor $\mu\tau$ values. This can be attributed to defects which can significantly shorten the electron lifetime, τ, or lower the mobility, μ. These include chemical defects such as impurities and precipitates and structural defects such as dislocations and grain boundaries. The effects of the structural defects have been observed and is described in the following paragraph.

The poor charge collection (such as the lower right area of the wafer as discussed above) in CdZnTe was correlated to the crystal quality of the material, as quantified by the full width at half maximum (FWHM) of TAD ω scans. "$\mu\tau$ mapping" was performed on single crystal CdZnTe samples of area ~ 5 cm^2, and compared to the TAD ω scan FWHM mapping.[7] Large FWHM values are obtained in regions where tilt boundaries and mosaic structure are prevalent, as observed by multiple peaks and broadened widths in the ω scans. Dislocation density, as determined by EPD measurements on (111)A surfaces using the Nakagawa etch,[17] is also higher ($\sim 10^5$ cm^2) in regions where large FWHM ($>\sim 40$ arcsecs) are obtained. These results confirm the effects of the structural defects on the detector performance, namely that they serve to lower the electron mobility and/or lifetime.

Analysis performed for the leakage current as a function of the zinc variation was similarly conducted for the variation in the measured maximum pulse height. Here, one would expect the zinc content to affect the number (N) of electron-hole pairs created per incident alpha particle ($N=E_{incident}/\varepsilon$), since the energy, ε, required to create one electron-hole pair is approximately constant ($\sim 3 \cdot E_g$) for most semiconductor detectors.[18] This would only have an effect in the Q_o in Eq.(4), and the predicted variation as a function of zinc content is only $\sim 2.5\%$ from one end of the wafer to the other. By contrast, the maximum pulse height measured (Q) varies by an order of magnitude and rather randomly throughout the wafer. This indicates that the effects of the zinc variation on the amount of charge collected is masked by other factors, namely the large $\mu\tau$ variation due to structural defects.

Additionally, structural features imaged by DCXRT are shown in Figure 4. This is an image of the upper right corner region of the wafer shown in Figure 2 (lighter regions are areas where diffraction is occurring). The image clearly reveals twin boundaries and misoriented regions, with the range of misorientation being approximately 200 arcsecs. The numerous small line features running $\sim$parallel to the growth direction are believed to be void "pipes" as reported in Ref. [8], and have been confirmed by IR microscopy. One key advantage to this technique compared to, for example, the IR imaging technique, is that DCXRT technique is able to provide a quantitative assessment of the probed region. For example, the amount of angular misorientation between tilted regions can be determined. In summary, DCXRT is demonstrated to be a very useful characterization tool for nondestructively imaging various structural defects in CdZnTe.

CONCLUSIONS

Various spatial mapping techniques were developed and utilized to characterize the structural defects, zinc composition variation, and detector-related parameters of HPB CdZnTe. Results showed that cracks in the material provide conductive "shorting paths" and hence lead to high leakage current and degraded detector performance. Other structural defects including grain and tilt boundaries are also shown to be responsible for effectively serving as carrier

Figure 4. Double crystal x-ray topograph of the upper right corner section of the CdZnTe sample shown in Figure 2. Lighter regions are areas where diffraction is occurring. The misorientation between the tilt regions is measured to be ~200 arcsecs.

recombination sites and thereby reducing the amount of charge collected. The effects of the zinc composition variation (~11% to 4%) that exists across the wafer on the leakage current and the maximum pulse height are masked by gross structural defects, confirming that the material is extrinsic in nature. A combination of TAD techniques, DCXRT, and IR imaging is demonstrated to provide a wealth of information on the structural properties of CdZnTe and hence enables one to study the relationship between the device performance and structural properties of semiconductor materials.

ACKNOWLEGMENTS

Funding for this research has been provided by the U.S. Department of Energy, Office of Nuclear Nonproliferation, and by Digirad Corporation, San Diego, CA and the UC MICRO program.

REFERENCES

[1] F.P. Doty, J.F. Butler, J.F. Schetzina, and K.A. Bowers, J. Vac. Sci. Technol. **B10** (1992) 1418.

[2] J.F. Butler, C.L. Lingren, and F.P. Doty, IEEE Trans. Nucl. Sci. **39** (1992) 605.

[3] J.M. Van Scyoc, J.C. Lund, D.H. Morse, A.J. Antolak, R.W. Olsen, R.B. James, M. Schieber, H. Yoon, M.S. Goorsky, J. Toney, and T.E. Schlesinger, J. Electron. Mat. **25** (1996) 1323-7.

[4] Scientific Workshop on Room Temperature Semiconductor Nuclear Radiation Detectors, March 1997, Sandia National Laboratory, Livermore, CA.

[5] P. N. Luke and E. E. Eissler, IEEE Trans. Nucl. Sci. **43** (1996) 1481.

[6] M.S. Goorsky, H. Yoon, M. Schieber, R.B. James, D.S. McGregor, and M. Natarajan, Nucl. Instr. and Meth. **A380** (1996) 6-9.

[7] H. Yoon, S.E. Lindo, and M.S. Goorsky, J. Crystal Growth **174** (1997) 775.

[8] K. B. Parnham, Nucl. Instr. Meth. **A377** (1996) 487.

[9] H. Yoon, J.M. Van Scyoc, M.S. Goorsky, H. Hermon, M. Schieber, J.C. Lund, and R.B. James, J. Electron. Mat. **26** (1997) 529.

[10] B. Jenichen, R. Köhler, and W. Möhling, J. Phys. E: Sci. Instrum. **21** (1988) 1062.

[11] M. Meshkinpour, M.S. Goorsky, B. Jenichen, D.C. Streit, and T.R. Block, J. Appl. Phys. **81** (1997) 3124.

[12] H. Yoon, J.M. Van Scyoc, T.S. Gilbert, T. McGrath, M.S. Goorsky, J.C. Lund, and R.B. James, "Triple axis x-ray diffraction and Laue back reflection techniques for characterization of polycrystalline CdZnTe wafers grown by the high pressure Bridgman technique," presented August 7, 1997 at the Denver X-ray Conference, August 4-8, 1997, Steamboat Springs, CO.

[13] W.G. Pfann, *Zone Melting*, John Wiley & Sons, Inc., New York, 1958, p.10.

[14] W. Shockley, *Electrons and Holes in Semiconductors*, Krieger, New York, 1981, Ch.16.

[15] T.H. Myers, S.W. Edwards, J. Liu, and J.F. Schetzina, Phy. Rev. B **25** (1982) 1113.

[16] K. Hecht, Zeits. Phys. **77** (1932) 2335.

[17] K. Nakagawa, K. Naeda, and S. Takeuchi, Appl. Phys. Letters **34** (1979) 574.

[18] M. Cuzin, Nucl. Instrum. Meth. **A253** (1987) 407.

ANALYSIS OF GRAIN BOUNDARIES, TWIN BOUNDARIES AND Te PRECIPITATES IN $Cd_{1-x}Zn_xTe$ GROWN BY HIGH-PRESSURE BRIDGMAN METHOD

J. R. HEFFELFINGER, D. L. MEDLIN AND R. B. JAMES
Sandia National Laboratories, Materials and Engineering Sciences Center
P. O. Box 969, Livermore, CA 94551

ABSTRACT

Grain boundaries and twin boundaries in commercial $Cd_{1-x}Zn_xTe$, which is prepared by a high-pressure Bridgeman technique, have been investigated with transmission electron microscopy, scanning electron microscopy, infrared-light microscopy and visible-light microscopy. Boundaries inside these materials were found to be decorated with Te precipitates. The shape and local density of the precipitates were found to depend on the particular boundary. For precipitates that decorate grain boundaries, their microstructure was found to consist of a single, saucer-shaped grain of hexagonal Te (space group P3,21). Analysis of a Te precipitate by selected-area diffraction revealed the Te to be aligned with the surrounding $Cd_{1-x}Zn_xTe$ grains. This alignment was found to match the (111) $Cd_{1-x}Zn_xTe$ planes with the $(0\bar{1}11)$ planes of hexagonal Te. Crystallographic alignments between the $Cd_{1-x}Zn_xTe$ grains were also observed for a high-angle grain boundary. The structures of the grain boundaries and the $Te/Cd_{1-x}Zn_xTe$ interface are discussed.

INTRODUCTION

$Cd_{1-x}Zn_xTe$ is a leading material for use in room-temperature gamma-ray spectrometers. $Cd_{1-x}Zn_xTe$ possesses the necessary material properties of high bulk resistivity, good photoconductivity and acceptable electron and hole transport, that are required to fabricate gamma-ray detectors. Production of large-volume $Cd_{1-x}Zn_xTe$ material is currently hindered by the presence of chemical and structural inhomogeneities, such as grain boundaries, twin boundaries, inclusions and precipitates.[1,2] Although the effect of these inhomogeneities on device performance is still poorly understood, mounting evidence suggests that defects, such as grain boundaries, affect the electrical transport properties inside the material.[3] The local effect of the grain boundaries, and the precipitates associated with them, may act to channel charge along certain paths inside the $Cd_{1-x}Zn_xTe$, and thus, create a device where the electrical transport properties are anisotropic. In order to gain insight on how these defect structures affect device performance, the structural and chemical nature of grain boundaries and twin boundaries in $Cd_{1-x}Zn_xTe$ are explored.

Commercially available $Cd_{1-x}Zn_xTe$ is typically manufactured using a high-pressure Bridgman (HPB) technique.[4] In this method, a large over-pressure of an inert gas (typically Ar) is used to inhibit the evaporation of the charge material. Material prepared by this method has a polycrystalline microstructure with grains on the order of centimeters in size. In order to grow material by this method, several complications must be addressed. Perhaps one of the most challenging growth dilemmas is that of the retrograde solubility in the Cd-Te phase diagram.[5-7] As the $Cd_{1-x}Zn_xTe$ boule is cooled, Te and/or Cd precipitate out of the $Cd_{1-x}Zn_xTe$ lattice. The chemistry of these precipitates is determined by whether the starting material is rich in Te or Cd. Precipitates tend to decorate internal boundaries, such as grain boundaries and twin boundaries inside $Cd_{1-x}Zn_xTe$. The composition and distribution of the precipitates have been previously studied for CdTe and $Cd_{1-x}Zn_xTe$ materials grown by either HPB or Bridgman techniques.[2,8,9] Post-growth annealing treatments of the $Cd_{1-x}Zn_xTe$ boule has been used to reduce the number and concentration of precipitates, but the effect of annealing on the electrical properties of the material has yet to be determined.[10-12] In this study, Te precipitates have been investigated for their microstructure and crystallographic alignment with the surrounding $Cd_{1-x}Zn_xTe$ matrix. By analyzing the shape, alignment and internal microstructure of the Te precipitates, valuable insights into the mechanisms of their formation and their effect on the surrounding $Cd_{1-x}Zn_xTe$ lattice can be made.

EXPERIMENTAL

Polycrystalline $Cd_{1-x}Zn_xTe$ material, manufactured using HPB techniques, was obtained from commercial suppliers for analysis. This material was grown from a Te-rich charge with a nominal zinc concentration of 10%. An Ar over pressure was used to suppress the loss of Cd from the melt. Parameters related to the growth and cooling rates were not disclosed by the manufacturer. $Cd_{1-x}Zn_xTe$ samples were characterized using visible-light microscopy (VLM), transmission infrared-light microscopy (IR), scanning electron microscopy (SEM), X-ray energy-dispersive spectroscopy (XEDS) and transmission electron microscopy (TEM). Prior to examining by VLM, IR and SEM, $Cd_{1-x}Zn_xTe$ samples were polished using one-micron alumina slurry and a felt cloth. Although this procedure produces a surface with noticeable scratches, no chemical alteration of the surface is expected to occur. TEM samples were prepared using conventional dimpling and ion milling techniques. A JEOL 840 SEM operating at 15 keV and equipped with a beryllium-window, Noran XEDS detector were used for SEM and XEDS analysis. A JEOL 1200EX operating at 120 keV and equipped with a Tracor Norton 5500 XEDS were used for TEM and XEDS analysis. A JEOL 4000EX operating at 400 keV was used for high-resolution TEM analysis.

RESULTS AND DISCUSSION

Precipitation of Te at grain boundaries and twin boundaries in $Cd_{1-x}Zn_xTe$

The surface of polished $Cd_{1-x}Zn_xTe$ shows visible evidence of grain boundaries, twin boundaries and precipitates. Fig. 1 is a pair of VLM images showing a grain boundary (Fig. 1a) and a twin boundary (Fig. 1b). Such boundaries can be seen in the VLM due to the anisotropic polishing of the alumina slurry. Here, grains of $Cd_{1-x}Zn_xTe$ polish at different rates depending on their relative orientation to the surface. Likewise, precipitates of a different phase and chemistry from that of the $Cd_{1-x}Zn_xTe$ will also be visible at the surface. Precipitates, which decorate the grain boundary and twin boundary in Fig. 1, are seen to protrude out the surface. This observation suggests that the Te polishes at a slower rate than the surrounding $Cd_{1-x}Zn_xTe$. A three dimensional perspective of the distribution of Te precipitates can be acquired using IR imaging. Fig. 2 shows a transmission IR image of a grain boundary, which is the same boundary as seen in Fig. 1a. A schematic representation of the viewing direction and distribution of Te precipitates is seen next to the image in Fig. 2. The fact that $Cd_{1-x}Zn_xTe$ is transparent to IR light allows for analysis of internal defects, such as precipitates, voids and cracks that act to either scatter or absorb the IR light. Fig. 3a is a secondary-electron SEM image of a Te precipitate that has intersected the $Cd_{1-x}Zn_xTe$ surface. The Te precipitate was located along a grain boundary (i.e. the same grain boundary that is seen in Fig. 1a and Fig. 2). The chemistry of the Te precipitate was probed with XEDS inside the SEM. Figs. 3b and 3c show XEDS spectra from the $Cd_{1-x}Zn_xTe$ matrix and the Te precipitate. Light elements, such as oxygen, were not detected in the XEDS spectra due to the use of a beryllium-window XEDS detector. Constituents, other than the three staring materials of Cd, Te and Zn, where not detected by XEDS in the SEM.

The precipitation of Te from the $Cd_{1-x}Zn_xTe$ matrix is due to a retrograde solubility in the phase diagram, as described earlier. Grain boundaries and twin boundaries act as low-energy nucleation sites for the precipitation of Te. For this reason, the grain boundaries and twin boundaries contain a higher density of precipitates than the bulk $Cd_{1-x}Zn_xTe$. In general, grain boundaries were observed to have a higher density of precipitates than the twin boundaries. This observation suggests that the barrier for nucleation of a Te precipitate at a grain boundary is generally less than that for the twin boundary. The distribution of Te precipitates at grain boundaries was also seen to vary between different grain boundaries. Thus, as the character of the grain boundary changes the barrier for nucleation also changes. Some of the grain boundaries were found to be completely free of Te precipitates, while regions where grain boundaries met (i.e. triple junction) were found to contain some the highest concentration of precipitates. Often, these triple junctions contained a continuous precipitate or a wire of Te. If the wire were to

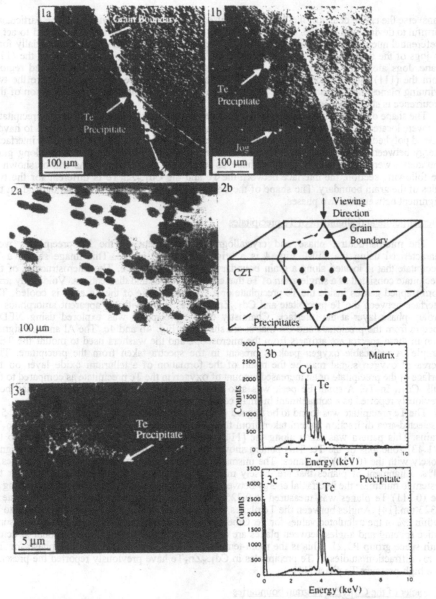

Fig. 1a) VLM image of a grain boundary decorated by Te precipitates. b) VLM image of
a twin boundary decorated by Te precipitates.

Fig. 2 a) IR transmission image showing the distribution of Te precipitates at a grain boundary.
b) Schematic illustration of the sample seen in Fig. 2a.

Fig. 3 a) Secondary-electron SEM image of the $Cd_{1-x}Zn_xTe$ surface along a grain boundary.
This image shows a Te precipitate that has intersected the surface. b) XEDS spectrum from the
$Cd_{1-x}Zn_xTe$ matrix. c) XEDS spectrum from the Te precipitate.

transverse the entire detector, the defect may cause a short circuit path and would be particularly harmful to device performance. Specific sites along twin boundaries were also observed to act as preferential nucleation sites. As seen in Fig. 1b, precipitates were observed to preferentially form on jogs of the twin boundary. In $Cd_{1-x}Zn_xTe$ the common twinning plane is that of the (111) plane. Jogs along the twin boundary would change the interface between the twinned regions from the (111) plane to some other plane, such as a (11$\bar{2}$) plane. The corner where the two twinning planes meet apparently acts as a low energy site for nucleation. An observation of this occurrence is shown in Fig. 1b.

The shape of the Te precipitates was found to depend on their location. In general, precipitates that were located either along twin boundaries or in the bulk of the $Cd_{1-x}Zn_xTe$ tended to have a faceted polyhedra shape. By faceting onto specific planes of the $Cd_{1-x}Zn_xTe$ lattice, the interfacial energy between the precipitate and matrix can be minimized. Precipitates that form along grain boundaries were found to have a saucer shape, as seen in Fig. 2 and Fig. 3. As will be shown in the following section, the interface between the Te and the $Cd_{1-x}Zn_xTe$ is different for the two sides of the grain boundary. The shape of the Te/ $Cd_{1-x}Zn_xTe$ boundary will thus depend on the alignment between the two phases.

Microstructure and alignment of Te precipitates

The microstructure, phase and crystallographic alignment of the Te precipitates were characterized using the TEM. Fig. 4 is a bright-field TEM image. This image shows a Te precipitate that is located along a grain boundary in the $Cd_{1-x}Zn_xTe$. The microstructure of the precipitate consisted of a single grain of Te that contained voids and dislocations. Voids may arise from trapped gasses inside the Te precipitate or a volume change of the Te as it is cooled. The interface between the Te precipitate and $Cd_{1-x}Zn_xTe$ was sharp with no apparent amorphous or tertiary phase layer at the interface. Chemistry of the precipitate was explored using XEDS. Spectra from the precipitate and the matrix are shown in Fig. 4b and 4c. The Al and Cu signals seen in these spectra are artifacts from the microscope and the washers used to mount the TEM sample. A noticeable oxygen peak is present in the spectra taken from the precipitate. This increased oxygen signal may be the result of the formation of a tellurium oxide layer on the surface of the precipitate or an increased amount of oxygen in the Te precipitate as compared to the bulk $Cd_{1-x}Zn_xTe$. A small Fe peak was also observed from the Te precipitate. Fe has been previously reported as a contaminant in small concentrations in $Cd_{1-x}Zn_xTe$ materials.[13]

The Te precipitate was found to be aligned with the surrounding $Cd_{1-x}Zn_xTe$ grains. Fig. 5 is a selected-area diffraction pattern taken from the precipitate and one of the adjacent $Cd_{1-x}Zn_xTe$ grains. This pattern was taken along the [11$\bar{2}$] zone axis of the $Cd_{1-x}Zn_xTe$ and close to the [5$\bar{1}$43] zone axis of the Te. This pattern shows the (111) planes of $Cd_{1-x}Zn_xTe$ to be aligned closely with the (0$\bar{1}$11) Te planes. The mismatch in lattice spacing between these two planes is 14%. Alignment of planes that are closely matched in lattice spacing would be a way for the system to minimize the interfacial energy between the precipitate and matrix. The lattice spacing of the (0$\bar{1}$11) Te planes was measured at 0.325 nm, while the experimentally reported value is 0.323 nm.[14] Angles between the Te planes, which are shown in Fig. 5, were measured to be within 2% of the calculated values for Te. Other zones of Te have been indexed, and all measured lattice spacing and angles between planes are consistent with Te having the hexagonal structure with space group P3$_2$21. This is the room-temperature and atmospheric pressure phase for Te. X-ray diffraction studies of Te precipitates in $Cd_{1-x}Zn_xTe$ have previously reported the presence of a high-pressure Te phase.[11]

Character of the $Cd_{1-x}Zn_xTe$ grain boundaries

An important question in the study of $Cd_{1-x}Zn_xTe$ is whether the high-angle grain boundaries contain a secondary phase at the interface. A secondary phase may have significant effects on the electrical properties of the boundary. Obviously, Te has been identified as a secondary phase, but these precipitates are assumed to be localized. To verify whether the interface between the $Cd_{1-x}Zn_xTe$ grains is sharp, high-resolution TEM was used to characterize a high-angle grain boundary. Fig. 6 is a high-resolution TEM image of a $Cd_{1-x}Zn_xTe$ grain boundary. This image shows lattice fringes from the planes of the $Cd_{1-x}Zn_xTe$ grains. Each grain has a different

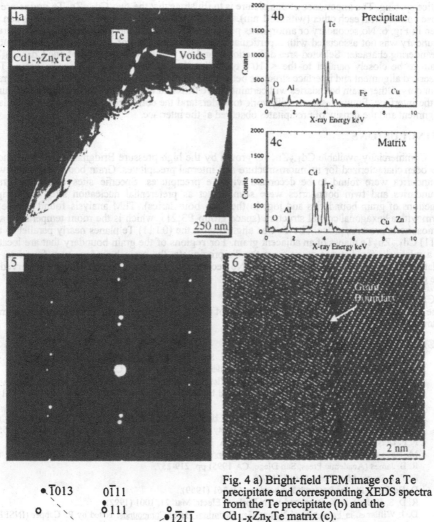

4a

Te

Cd$_{1-x}$Zn$_x$Te

Voids

250 nm

4b Precipitate

Counts

O Al Te Fe Cu

X-ray Energy keV

4c Matrix

Counts

O Al Cd Te Cu Zn

X-ray Energy keV

5

6

Grain Boundary

2 nm

$\bar{1}013$ $0\bar{1}11$

111 $1\bar{2}1\bar{1}$

46.4° 61.7°

$2\bar{2}0$ 40.0° $1\bar{1}0\bar{2}$

(o) Cd$_{1-x}$Zn$_x$Te
(•) Te

Fig. 4 a) Bright-field TEM image of a Te precipitate and corresponding XEDS spectra from the Te precipitate (b) and the Cd$_{1-x}$Zn$_x$Te matrix (c).

Fig. 5 Selected-area diffraction pattern showing the alignment between the Te and Cd$_{1-x}$Zn$_x$Te This pattern is seen along the [11$\bar{2}$] Cd$_{1-x}$Zn$_x$Te axis and the [5$\bar{1}\bar{4}$3] axis of the Te. This pattern shows the alignment of the (111) Cd$_{1-x}$Zn$_x$Te planes with the (0$\bar{1}$11) Te planes.

Fig. 6 High-resolution TEM image of the Cd$_{1-x}$Zn$_x$Te grain boundary.

orientation with respect to the electron beam and thus, the two grains appear to have quite different lattice spacing. The importance of this picture is to illustrate that the two $Cd_{1-x}Zn_xTe$ grains are in close contact with each other (within ~1 nm). A certain amount of overlap of the two grains can be seen in Fig. 6. No secondary or amorphous phase was observed at the grain boundary. The grain boundary was not associated with a particular plane of either grain of the $Cd_{1-x}Zn_xTe$ and had a meandering character. Selected-area diffraction found the <110> type direction of one $Cd_{1-x}Zn_xTe$ grain to be closely parallel to the <110> type direction of the other $Cd_{1-x}Zn_xTe$ grain. The observed alignment and interface structure between the grains was characterized for only one grain boundary. Other grain boundaries will certainly have different alignments and dissimilar structures at the interface. Future research will work to understand the correlation between grain boundary alignment and the density of precipitates observed at the interface.

CONCLUSIONS

Commercially available $Cd_{1-x}Zn_xTe$, grown by the high-pressure Bridgman (HPB) method, has been characterized for its microstructure and internal precipitates. Grain boundaries and twin boundaries were found to be decorated with Te precipitates. Specific sites along the grain boundaries and twin boundaries were seen to act as preferential nucleation sites (i.e. triple junctions of grain boundaries and jogs in the twin boundaries). TEM analysis found the Te to form with a hexagonal crystal structure (space group $P3_121$), which is the room temperature and atmospheric pressure phase. The Te was aligned with the $(0\bar{1}11)$ Te planes nearly parallel to the (111) $Cd_{1-x}Zn_xTe$ planes of an adjacent grain. For regions of the grain boundary that are located at a relatively far distance away from the Te precipitate, the grain boundary was found to be relatively sharp (within ~ 1 nm in width). No secondary or amorphous phases were observed at the grain boundary interface.

ACKNOWLEDGMENTS
The authors acknowledge the financial support of the U. S. Department of Energy, under contract number DE-AC04-94-AL85000.

REFERENCES
1. R. B. James, T. E. Schlesinger, J. Lund, and M. Schieber, in Semiconductors and Semimetals Vol. 43, edited by T. E. Schlesinger and R. B. James (Academic Press, San Diego, CA 1995) pp. 335-381.
2. J. R. Heffelfinger, D. L. Medlin, H. Yoon, H. Hermon and R. B. James, in Hard X-ray and Gamma-ray Detector Physics, Optics and Application, edited by R. B. Hoover and F. P. Doty (SPIE Proc. 3115, Bellingham, WA, 1997) pp. 40-50.
3. F. P. Doty, J. P. Cozzatti and J. P. Schomer, in Hard X-ray and Gamma-ray Detector Physics, Optics and Application, edited by R. B. Hoover and F. P. Doty (SPIE Proc. 3115, Bellingham, WA, 1997) pp. 51-55.
4. F. P. Doty, J. F. Butler, J. F. Schetzina, and K. A. Bowers, J. Vac. Sci. Technol. B10, 1418 (1992).
5. M. Hage-Ali and P. Siffert, in Semiconductors and Semimetals Vol. 43, edited by T. E. Schlesinger and R. B. James (Academic Press, San Diego, CA 1995) pp. 219-257.
6. F. T. J. Smith, Met. Trans 1, 617 (1970).
7. D. DeNobel, Philips Research Publications 14, 361 (1959).
8. R. D. S. Yadava, R. K. Bagai and W. N Borle, J. Electr. Mat. 21, 1001 (1992).
9. D. J. Williams, in Properties of Narrow Gap Cadmium Based Compounds, edited by P. Capper (INSPEC, Exeter, England 1994) pp. 510-515.
10. H. R. Vydyanath, J. Ellsworth, J. J. Kennedy, B. Dean, C. J. Johnson, G. T. Neugebauer, J. Sepich, and P.-K. Liao, J. Vac. Sci. Technol. B10, 1476 (1992).
11. T. S. Lee, J. W. Parks, Y. T. Jeoung, H. K. Kim, C. H. Chun, J. M. Kim, I. H. Park, J. M. Chang, S. U. Kim and M. J. Park, J. Electr. Mat. 24, 1053 (1995).
12. N. V Sochinskii, E. Diegues, U. Pal, J. Piqueras, P. Fernandez, A. F. Agullorueda, Semi. Sci. Tech. 10, 870 (1995).
13. S. Sen and J. E. Stannard, in Semiconductors for Room-Temperature Radiation Detector Applications, edited by R. B. James, T. E. Schlesinger, P. Siffert and L. Franks (Mater. Res. Soc. Proc. 320, Pittsburgh, PA 1993) p. 391.
14. CRC Handbook of Chemistry and Physics, edited by David R. Lide (CRC Press Inc. Boca Raton, FL 1997) p. 12-83.

MAPPING OF LARGE AREA CADMIUM ZINC TELLURIDE (CZT) WAFERS: APPARATUS AND METHODS

B. A. BRUNETT[a,b], J. M. VAN SCYOC[a,c], H. YOON[c], T. S. GILBERT[c],
T. E. SCHLESINGER[b], J. C. LUND[a], R. B. JAMES[a]
[a]Sandia National Laboratories, Livermore CA, 94550
[b]ECE Department, Carnegie Mellon University, Pittsburgh, PA 15213
[c]MSE Department, University of California, Los Angeles, CA 90095

ABSTRACT

Cadmium Zinc Telluride (CZT) shows great promise as a semiconductor radiation detector material. CZT possesses advantageous material properties over other radiation detector materials in use today, such as a high intrinsic resistivity and a high cross-section for x and γ-rays. However, presently available CZT is not without limitations. The hole transport properties severely limit the performance of these detectors, and the yield of material possessing adequate electron transport properties is currently much lower than desired. The result of these material deficiencies is a lack of inexpensive CZT crystals of large volume for several radiation detector applications. One approach to help alleviate this problem is to measure the spatial distribution (or map) the electrical properties of large area CZT wafers prior to device fabrication. This mapping can accomplish two goals: identify regions of the wafers suitable for detector fabrication and correlate the distribution of crystalline defects with the detector performance. The results of this characterization can then be used by the crystal manufacturers to optimize their growth processes. In this work, we discuss the design and performance of apparatus for measuring the electrical characteristics of entire CZT wafers (up to 10 cm x 10 cm). The data acquisition and manipulation will be discussed and some typical data will be presented.

INTRODUCTION

Numerous investigators have reported on the electrical characteristics of CZT crystals that have been fabricated into radiation detectors by commercial manufacturers. However, the manufacturers of detectors usually dice the raw boule of CZT into a large number of small pieces and select those with the best properties. Thus, the properties of CZT crystals represented in the literature do not necessarily represent the typical characteristics of material selected at random from a CZT boule. In this paper we describe an apparatus that measures the electrical properties of whole wafers of CZT. The apparatus we describe enables the rapid determination of regions of good quality material prior to dicing. In addition, the spatial mapping apparatus also enables the correlation of electrical characteristics over large length scales (many cm), enabling manufacturers to gain insight into the distribution of defects throughout the boule, and hopefully providing insight into methods to improve the crystal growth conditions.

The apparatus we describe in this paper measures the electrical properties of CZT boules that are relevant to ionizing radiation detector performance. In particular, the apparatus estimates the bulk resistivity of the CZT by measuring the leakage current at a large number of different locations uniformly sampled across the CZT wafer. The electron and hole transport characteristics (drift mobility and mean trapping time) are also measured. These characteristics are critical to the performance of radiation detectors operated as pulse height spectrometers. The design of the electrical and mechanical components that comprise the spatial mapping apparatus are described below, along with some representative data taken with the apparatus.

EXPERIMENTAL CONFIGURATION

The overall design of the spatial mapping apparatus is shown in **Figure 1**. The apparatus is designed to translate a wafer of CZT in two dimensions over a source of ionizing radiation. A computer controls the motion of the CZT wafer and raises and lowers a probe onto the surface of the CZT wafer. The probe is used to measure the leakage current at each sampled point. In addition, the transport properties of electrons are measured at each point using both pulse height analysis methods and transient charge techniques. The same computer that is used to translate the wafer and raise and lower the probe also controls a pulse height analyzer, the voltage-current source-measure unit, and a digital oscilloscope. Typically the computer provides real-time visualization of the electrical characteristics as they are being measured and saves all of the measurements for more detailed subsequent analysis. In the following sections we describe in more detail the design of the mechanical components, the electronic instrumentation, and some of the methods used to analyze and visualize the data.

Mechanical

The CZT wafer is mounted in a Teflon holder that clamps the periphery of the sample leaving both faces of the wafer accessible. The wafer and holder rest on a conductive table which has an aperture in the center for alpha particle irradiation. The bottom surface of the sample possesses an Au contact layer to ensure adequate electrical contact. Spatial translation is accomplished by sliding the sample on the conductive table. Contact to the upper surface of the wafer is achieved via a conductive rubber probe that can be translated perpendicular to the wafer surface. The probe is lifted for sample translation and then lowered for measurement. The actuating probe design allows slight pressure to be applied to the probe to ensure contact while eliminating any unnecessary friction between the sample and the table during translation. A diagram of this apparatus is shown in **Figure 1**.

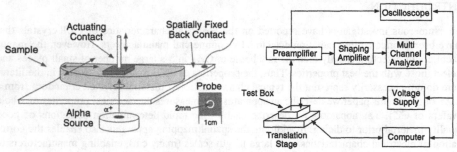

Figure 1: Diagram showing mechanical setup (left) and instrumentation.

The electrical probe consists of a 2 mm diameter center contact which is used to measure the signal. This inner contact is surrounded by a 1 cm square contact with approximately a 1 mm separation. Both of these contacts are maintained at equal potential. This guarded probe design was adopted to maximize the resolution while alleviating alignment constraints. This probe design allows electronic collimation of the carriers within the crystal. Any carriers generated outside the signal contact area will be collected by the guard probe and will not contribute to the measured signal. This allows excitation with a loosely collimated source thereby relaxing the alignment of the source with the probe.

The sample table is contained within a Faraday cage with ports that allow mechanical coupling of the probe and the sample holder to external translation stages.

Electronic Instrumentation

The electronic instrumentation used to measure the electrical characteristics is of three general types: a DC measurement component, pulse height analysis electronics, and a digital oscilloscope for capturing transients produced by radiation interaction with the sample.

The DC component of the system consists of a combination voltage/current source-measure unit (Keithley 237). The source/measure unit delivers the bias to the sample and allows measurement of the leakage current. Biases from -1200 to +1200 V can be provided by the unit, and currents as low as 10 pA can be sampled. The source/measure unit is controlled via an IEEE-488 (GPIB) parallel bus to the computer.

The pulse height analysis instrumentation consists of a typical nuclear spectrometer electronic chain: a charge sensitive preamplifier (Tennelec TC-170), a shaping amplifier (Tennelec TC-241), and a multichannel analyzer (Oxford PCA-3) mounted in the ISA bus of the host computer.

The output of the charge sensitive preamplifier is also monitored with a digital oscilloscope (LeCroy 9354). The digital oscilloscope is controlled via a GPIB interface.

Together, the three components of the electronic instrumentation allow simultaneous measurement of the localized leakage current, preamplifier risetime, pulse height spectrum, and transients from the preamplifier for each spatial point in the map. All translation, data acquisition, and data logging is accomplished via a computer. A block diagram showing the configuration of the instrumentation is shown in **Figure 1**.

DATA ANALYSIS

The data acquired by the mapping apparatus is processed in two ways: 1. a rapid analysis for real-time monitoring and diagnostics of the experiment as the mapping run proceeds, and 2. a more detailed off-line analysis and interactive visualization of the data acquired. In the subsequent sections we describe the theory and application behind the measurements performed.

Leakage Current

High efficiency spectroscopic applications require both large device area for maximum efficiency and low leakage current for noise considerations. This necessitates CZT material with high resistivity. The leakage current data obtained from the mapping apparatus yields a resistivity map of the wafer. First the equivalent resistance, R, of the current point is found by equation (1),

$$R = \frac{V}{I_l} - R_i,\tag{1}$$

where V is the applied bias, I_l is the leakage current, and R_i is the bias isolation resistor. Typically, the isolation resistor is small compared to the sample resistance and can be ignored. The local resistivity is then found by equation(2),

$$\rho = \frac{RA}{d},\tag{2}$$

in which A is the probe area and d is the sample thickness. A representative resistivity map of one half of a 10 cm diameter wafer is shown in **Figure 2**.

Pulse Height Spectrum (PHS)

An important parameter in semiconductor gamma-ray detector performance is the drift length of electrons. Two methods are used to measure electron drift length in the sample: pulse height analysis and a transient charge technique (described later). The pulse height analysis method is based on the theory of Schockley and Ramo [1]. In the Schockley-Ramo theorem, the induced charge on the contacts of a planar detector is a monotonic function of the ratio of the drift length of the carriers, λ, to the thickness of the detector, d,

$$Q(\lambda) = Q_o \frac{\lambda_e}{d}\left[1 - \exp\left(\frac{x-d}{\lambda_e}\right)\right] + \frac{\lambda_h}{d}\left[1 - \exp\left(\frac{-x}{\lambda_h}\right)\right], \tag{3}$$

where Q_o is the nominal charge generated and x is the depth of interaction measured from the negative contact (cathode).

If we assume a constant electric field in the sample and locally homogenous material, the drift length can be expressed as a function of the electron and hole transport parameters,

$$\lambda_e = \frac{(\mu\tau)_e V}{d} \qquad \lambda_h = \frac{(\mu\tau)_h V}{d}, \tag{4}$$

where μ and τ are the mobility and lifetime respectively.

In our experiments, we use alpha particles impinging on the negative contact of the device to excite carriers. Because alpha particles have a very short range in CZT (<100 μm), we can assume that all of the charge they generate in the crystal is formed very close to the negative contact ($x \approx 0$). Under these conditions equation (3) reduces to the familiar single carrier Hecht equation [2], shown as equation (5):

$$Q_m = Q_o \frac{\mu\tau V}{d}\left[1 - \exp\left(-\frac{d^2}{\mu\tau V}\right)\right], \tag{5}$$

where Q_m is the measured charge of the photopeak in the PHS, Q_o is the nominal charge generated for the system, V is the potential applied across the sample, and d is the sample thickness. The nominal charge generated for the system is found by measuring the bias dependent PHS at several points across the wafer prior to mapping and averaging the Q_o values. A map of the μτ values calculated from the PHS and the bias-dependent Hecht equation is given in **Figure 2**.

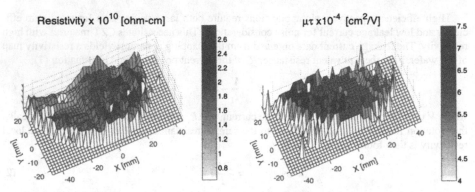

Figure 2: Processed data showing resistivity map (left) and mobility lifetime product map.

Transient Charge Method

An alternative method to pulse height analysis for determining the electron and hole transport properties is to analyze the shape of the signal at the output of the charge sensitive preamplifier as a function of time. Since the bandwidth of the charge sensitive amplifier is very large, the voltage output of the preamplifier is a close approximation to the induced charge on the contacts of the sample as a function of time (a typical transient is shown in **Figure 3**). Ignoring the effects of charge detrapping, the induced charge on the contacts of the detector as a function of time [3] can be represented by equation (6). In this case, the charge is represented as preamplifier output voltage for simplicity. A linear factor could be applied to transform the output signal of the preamplifier into units of charge,

$$
V(t) = \begin{cases}
V_o & t < t_o \\[2ex]
V_o + \dfrac{V_m \tau}{T_r}\left(1 - \exp\left(-\dfrac{t - t_o}{\tau}\right)\right) & t_o \leq t < T_r \\[2ex]
V_o + \dfrac{V_m \tau}{T_r}\left(1 - \exp\left(-\dfrac{T_r}{\tau}\right)\right) & T_r \leq t
\end{cases} \tag{6}
$$

where V_o represents the preamplifier offset voltage, V_m is equivalent to the nominal charge generated for the system, t_o is the time of interaction, τ gives the carrier lifetime, and T_r is the transit time of the carriers. As equation (6) shows, three time segments are necessary to represent the transient signal. The first is valid before the interaction occurs and is simply the offset voltage. The second segment represents the time during which the carriers traverse the crystal. And finally, the third segment corresponds to the time after all of the carriers have reached the opposite contact of the sample or have been trapped, and is equal to a constant value. All of the variables are

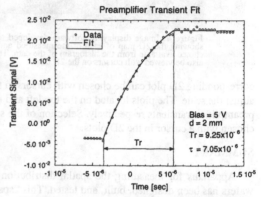

Figure 3: Plot showing fit of preamplifier transient signal.

employed in a fitting algorithm. The mobility and the lifetime of the carriers can be uniquely determined from the fit. The lifetime is given directly and the mobility can be found knowing the sample thickness and applied bias. A representative fit is shown in **Figure 3**, giving values for μ and τ of 865 cm^2/Vs and 7.05 μs respectively.

It should be noted that this method does not work well for high bias values and that there is a slight bias dependence of the obtained values. These factors are currently under investigation.

Data Visualization

A typical scan of a 100 cm diameter wafer sampled at uniform intervals of 2 mm produces 50 x 50 individual samples of information on the wafer. At each point on the wafer, a pulse height spectrum (1024 channels at 2 bytes per channel), 50 values of the transient signals (500 by 4 byte vectors), and the bias and leakage current information are acquired. Thus about 200 Mb of data are acquired in a typical wafer scan. Analysis and visualization of this large data set requires complex software. The graphical user interface (GUI) to the analysis and visualization software, developed by the author using LabVIEW®, is shown in **Figure 4**.

Shown in the bottom left corner of **Figure 4** are four 2D plots available for comparison of the various measured and calculated properties. The user can view properties such as leakage current and resistivity. The plot on the right displays a histogram of the selected data set. The range of the

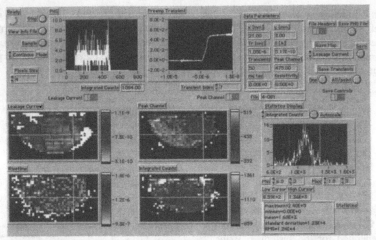

Figure 4: Image displaying software developed to analyze the data. Interactive display showing the 2D map of each of the measured and calculated properties with plot range choice available from the histogram of the data. The PHS and preamplifier transients can also be viewed with cursors on the 2D maps.

corresponding 2D plot can be chosen with cursors in the histogram plot allowing a simple way to adjust the scale. The plots located on the top left and top center are the pulse height spectrum and preamplifier transients respectively. Selection of the spatial point to be displayed in these plots is chosen with a cursor in the 2D plots.

SUMMARY

Apparatus for measuring the spatial distribution of electrical properties across entire CZT wafers has been designed, built, and tested. This "spatial mapping" apparatus is capable of measuring most of the CZT electrical properties important to radiation detectors. In particular, estimates of the spatial distribution of resistivity and electron transport characteristics have been made. Future improvements in software will allow additional electrical parameters to be measured. It is hoped that this apparatus can be used to visualize the large-scale distribution of electrical characteristics in CZT boules and, in so doing, provide insight into methods to improve the growth of CZT crystals for use in radiation detectors.

ACKNOWLEDGMENTS

We wish to acknowledge financial support from the DOE office of Research and Development, offices of Nuclear Nonproliferation.

REFERENCES

1. S. Ramo, Currents Induced by Electron Motion, Proc. IRE, 27, p. 584, (1939).

2. K. Hecht, Zum Mechanismus des lichtelektrischen Primärstromes in isolierenden Kristallen, Z. Physik, 77, p. 235, (1932).

3. G. Bertolini, Pulse Shape and Time Resolution, p. 243, in <u>Semiconductor Detectors</u>, ed. by G. Bertolini and A. Coche, Wiley Interscience, (1968).

RECENT IMPROVEMENTS IN DRY ETCHING OF $Hg_{1-x}Cd_xTe$
BY CH_4 BASED ELECTRON CYCLOTRON RESONANCE PLASMAS

M. Seelmann-Eggebert, A. Rar*, H. Zimmermann and P. Meisen
Fraunhofer Institut für Angewandte Festkörperphysik, 79108 Freiburg, Germany
*presently at National Research Institute for Metals, Tsukuba, Ibaraki 305, Japan.

ABSTRACT

Significant improvements of a previously reported etching process [1] for $Hg_{1-x}Cd_xTe$ have been achieved with respect to etch rate, surface morphology and surface stoichiometry by optimization of the process parameters. The gas phase and surface reactions driving the etching process have been analyzed by combined optical and electrical characterization of the plasma and surface analyses of the samples. A reaction scheme is suggested which allows to model and upscale the process in a consistent manner.

INTRODUCTION

Dry etching is an important processing step improving the performance of infrared focal plane arrays [2-6] and required to design novel nanostructure IR devices [7] based on $Hg_{1-x}Cd_xTe$. Electron cyclotron resonance (ECR) plasmas of CH_4/H_2 mixtures are commonly used for etching of HgCdTe which proceeds by reaction of the CH_3 radical to form volatile compounds. According to

$$CdTe + 4 CH_3 \Rightarrow Cd(CH_3)_2 \text{ (gas)} + Te(CH_3)_2 \text{ (gas)} \tag{1}$$

$$HgTe + 2 CH_3 \Rightarrow Te(CH_3)_2 \text{ (gas)} + Hg \text{ (gas)} \tag{2}$$

material is removed in the stoichiometric proportions of the $Hg_{1-x}Cd_xTe$ sample. Beside $Cd(CH_3)_2$ and $Te(CH_3)_2$, residual gas analysis showed TeH_2 as an additional etching product of $Hg_{1-x}Cd_xTe$ [8]. Produced in the plasma from H_2 or occuring as byproduct of methane dissociation atomic hydrogen etches HgTe according to

$$HgTe + 2H \Rightarrow TeH_2(\text{gas}) + Hg \text{ (gas)} \tag{3}$$

$$CdTe + 2H \Rightarrow TeH_2 \text{ (gas)} + Cd \text{ (solid)} \tag{4}$$

Since atomic hydrogen can break the bonds of HgTe, reaction (3) occurs at a rapid rate and leads to highly non-stoichiometric etching accompanied with considerable surface roughening. Moreover, hydrogen atoms diffuse rapidly in the $Hg_{1-x}Cd_xTe$ matrix and may cause a type conversion in p-material [9]. Hence, an optimization of the etching conditions should aim at supporting reactions (1) and (2) and inhibiting reactions (3) and (4).

An additonal boundary condition for the etching process exists as to polymer deposition has to be avoided. The addition of hydrogen to the plasma is usually required since atomic hydrogen has an inhibiting effect on polymer deposition. Polymer formation is initiated by the precursors CH and CH_2, which occur as additional byproducts of methane dissociation.

259

Regarding the detrimental effects of hydrogen and polymers considerable improvements had been already achieved [1] by addition of nitrogen to the etching gas, by which polymer formation could be suppressed and smooth surfaces were obtained. The objective of this investigation was to get insight into the microscopic reaction mechanisms and their control by the external process parameters.

EXPERIMENTAL

The plasma reactor chamber [8] is equipped with an ASTEX compact electron cyclotron resonance (ECR) source (250 W) and a temperature controlled sample stage ($-30°C \leq T_s \leq 150°C$) located at a distance of about 10 cm from the ECR exit. An aperture at the exit of the ECR liner acts as a pressure stage decreasing the pressure by a factor of ten as gas introduced into the ECR flows into the spherical reaction chamber (diameter 30 cm). No pressure gradient is present for CH_4 (or Hg) which was generally introduced in the downstream region to avoid contamination of the ECR liner. Under typical operating conditions the pressure inside the ECR source is 2-10 µbar and the elastic mean free path of molecules is comparable to the ECR dimensions. In this plasma stream [9] configuration the ECR exit can be crudely thought of as beam source of active species which move essentially collision free in the downstream region. Forming a divergent beam and acquiring relatively large kinetic energies, ions and electrons are guided and accelerated along the magnetic field lines (plasmapotential < 15 V). Radicals leave the source in a straight line motion and divergence losses of this reactive beam can be assumed to be still small at the sample stage. The mass balance between the afflux of methane into the open ECR end and the efflux of dissociation products links the supply rate of the methyle radical to the degree of dissociation in the ECR. Since gas phase recombination cross sections are small, reactive or charged particles generated in the ECR recombine predominantly at the sample surface or internal walls.

For the detection of reactive species in the downstream region optical emission spectroscopy (OES) was performed through a quartz port covering a region just above the sample stage. The electrical plasma parameters were determined by a single Langmuir probe (LP). A semiquantitative treatment of the OES data was facilitated by an actinometric approach (with nitrogen and argon transitions as actinometers) or by refering the OES data to the independently determined number density of the electrons [11].

RESULTS AND DISCUSSION

OES gave direct evidence for the presence of H, CH_3, CH, CN, NH, N_2^+, N, N_2^{ms} (N_2^{ms} is metastable nitrogen which is in the excited molecular $B^3\Pi_g$ state and carries the energy $E_a=6.2$ eV) in the reaction zone above the sample. The observation of radicals implies that in spite of the low pressure gas phase reactions occur and are likely to play a role in the etching process. On the basis of detailed actinometric analyses [11] of OES data in combination with Langmuir probe measurements and mass spectroscopy we propose that the following set of equations contains the most relevant reactions (preceding the etching reactions) in our setup for the $CH_4/H_2/Ar/N_2$ plasma. The parentheses indicate that the respective species participating in the reaction could not be unambiguously identified. Simple plasma reactions by electron impact such as the formation of N_2^{ms}, the ionization of N_2 and the dissociation of N_2 and H_2 were also substantiated by OES, but have been omitted.

Methane dissociation is verified to be exclusively driven by electron impact according to

$$CH_4 + e \quad \Rightarrow \quad CH_3 + H + e \tag{5}$$
$$(CH_3) + e \quad \Rightarrow \quad CH + (H_2) + e \tag{6}$$

With increasing electron density the two step process (6) for the CH production becomes increasingly important. For the CH_3 production mechanisms other than (5) are not effective in the accessable range of process parameters. In particular, experimental evidence was found that the following reactions are unimportant

$$CH_4 + H \quad \Rightarrow \quad CH_3 + H_2 \tag{7}$$
$$CH_4 + N_2^{ms} \quad \Rightarrow \quad CH_3 + H + N_2 \tag{8}$$
$$CH_4 + N \quad \Rightarrow \quad CH_3 + NH \tag{9}$$

The beneficial effect of the nitrogen addition was traced back to the reactions

$$(CH) + N_2^{ms} \quad \Rightarrow \quad CN + NH \tag{10}$$
$$(H) + N_2^{ms} \quad \Rightarrow \quad (NH) + N \tag{11}$$
$$N_{ad} + H_{ad} + (H_2) \Rightarrow \quad NH_3 \tag{12}$$

Reactions such as (10) - (12) between two radicals or metastable particles occur predominantly at the sample and internal surfaces of the chamber. (10) exemplifies the general mechanism of polymer inhibiting reactions. Atomic hydrogen is bound by nitrogen via the reactions (11) and (12). The primary guideline to optimization of the external process parameters is to increase the CH_3 supply by increasing the number density of high energetic electrons E>10 eV [12] (or the CH_4 flux) and to compensate for the increased production rate of CH by a respective amount of active nitrogen.

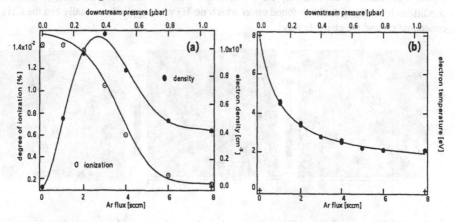

Fig. 1: Pressure dependence of (a) electron density, degree of ionization and (b) electron temperature as determined with a Langmuir probe in the downstream region for an Argon plasma (microwave power 200W).

TABLE I: Correlation between plasma conditions and etching properties (for $Hg_{0.78}Cd_{0.22}Te$ at $T_S=60°C$).OES signals $I(CH_3)$ and $I(H)$, etching rate E.R. and AFM surface roughness S.R. vs Ar, CH_4, N_2 and H_2 mass flow and microwave power settings ($v=I(CH_3)/I(H)$)

	Ar [sccm]	CH$_4$ [sccm]	N$_2$ [sccm]	H$_2$ [sccm]	W [W]	I(CH$_3$) [arb.u]	I(H) [arb.u]	v	E.R. [Å/min]	S.R. [Å]
a	1.75	1.75	0.45	0.6	200	880	780	1.1	140	50
b	1.75	1.75	0.6	0.6	200	790	710	1.1	120	50
c	1.75	1.75	0.6	0	200	1150	425	2.7	160	10
d	2	1	0.6	0	175	790	65	12	80	-
e	2	1	0.6	0	250	1300	210	6.2	270	-

A considerable acceleration [13] of the ions in the downstream region was confirmed by the LP measurements which indicated plasma potentials between 10 V and 15 V and a reduction of the plasma density by almost two orders of magnitude as compared to values typically present in ECR sources. Nevertheless the electron concentrations in downstream and upstream region should be correlated almost linearly. For all investigated plasma compositions the electron density n was found to increase with the microwave power W according to n ~ W^m where $0.6<m\leq1$. As exemplified by Fig. 1 for the case of a pure argon plasma, the electron density passes through a maximum as the pressure increases, since the degree of ionization decreases monotonicly. The electron temperature decreases with increasing pressure, but is found to be rather independent of the microwave power. Consequently, while the microwave power of the ECR source should be set at its maximum, there is an optimum operation pressure depending on the gas composition.

The optimization was carried out empirically with use of the OES signals for characteristic CH_3 and H transitions at 216 nm and 486 nm, respectively. The CH_3 emission signal was found to depend only on the methane flux and the electron density. In fact, conditions for etching could be found under which no H_2 was supplied externally and the CH_3 concentration and the etch rate was

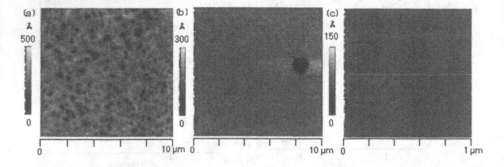

Fig.2: AFM images of $Hg_{0.78}Cd_{0.22}Te$ samples obtained after etching with Ar:1.75 sccm, CH_4:1.75 sccm, N_2:0.6sccm and W=200W under different conditions. RMS surface roughness is (a) 50 Å, (b) 10 Å and (c) 1 Å. Substrate temperatures were 60°C for (a) and (b) and -25°C for (c). Hydrogen was supplied externally only in case (a) (0.6 sccm).

even higher then in the previously used $Ar/N_2/CH_4/H_2$ based plasma (Tab. I). Under previously [1] reported conditions (denoted (a) in Tab.I) etch rates of 140 Å/min were obtained and a surface roughness (as measured by AFM) of about 50 Å resulted (Fig. 2a). If the plasma is operated under the similar condition denoted (b) etching proceeds even when the hydrogen flux is turned off (condition (c)). Surprisingly, the interruption of hydrogen flow increases the CH_3 concentration in the plasma and, consequently, a larger etch rate is obtained. At the same time a significantly smoother surface results (Fig. 2b), since the ratio of the H and CH_3 concentrations is reduced (by a factor of 2.5). The data of Tab. 1 show that the etch rate could be further increased (and the concentration of atomic hydrogen decreased) by reduction of the CH_4 flow and increase of the Ar flow. Surfaces which are smooth on an atomic scale were obtained by low temperature etching (Fig. 2c).

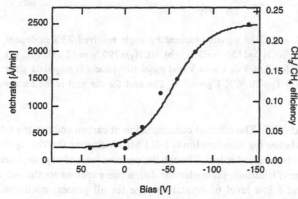

Fig. 3: Etch rate of $Hg_{0.78}Cd_{0.22}Te$ in dependence of sample bias. (Microwave power: 200W, CH_4: 1 sccm, N_2: 0.6 sccm, Ar: 2 sccm, $T_s= 60°C$)

The etch rate for $Hg_{0.78}Cd_{0.22}Te$ was found to be rather independent of the substrate temperature in the range -25°C to 80°C, but showed a strong dependence on the sample bias (Fig.3). A region of linear increase with bias voltage indicates that the energy provided by positive ions can substantially assist to overcome the reaction barriere. Metastable Ar and N_2 particles produced in the plasma may act in a similar way. The involvement of metastable particles is supported by the observed temperature independence and constancy of the etch rate in the positive bias region. Metastable excited Argon (which carries an energy of 11.6 eV) appears to be more important, since without argon in the plasma no etching could be achieved at all. At very high bias the etch rate shows a tendency to saturate. Using the etch rate by means of the reactions (1)-(4) as a measure of the CH_3 flux we can estimate the consumption efficiency on the basis of our process model and establish the right scale of Fig.1. Obviously, the saturation region has to be attributed to a complete consumption of the CH_3 flux emerging from the ECR. Our combined OES, LP and etchrate results imply that the order of the reaction rate with respect to the CH_3 radical concentration is larger than one but does not exceed two. All these findings and further results presented below suggest that the breaking of the first backbond of surface Te atoms (bound to CdTe) is the rate determining step of the etching process.

To analyse the surface composition the samples were transferred under UHV conditions into an analysis chamber equipped with a facility for angle resolved x-ray

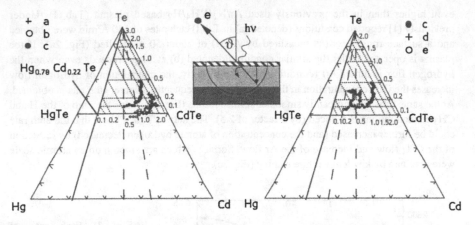

Fig. 4: Normalized Hg4f, Te3d, and Cd3d signals obtained by angle resolved XPS in dependence of the escape angle. (a) $I(CH_3)=1150$, $v=2.7$ (b) $I(CH_3)=790$, $v=12$ (c), (d), (e) $I(CH_3)=1300$, $v=6.2$. For (a),(b),(c) $T_S=60°C$, $U_B=0$ V and $p_{Hg}\leq 10^{-3}$ µbar. (d) $p_{Hg}=0.01$ µbar (e) $p_{Hg}=0.2$ µbar. For (d) and (e) $T_S=-20$ °C, , $U_B=-60$ V. The unit for the grid is the escape length λ.

photoelectron spectroscopy (ARXPS). The residual contamination of carbon and oxygen left on the surface was found to be below the detection limits (<0.1 ML) indicating that during the etching the coverage with adsorbed methyl is low. Though the surfaces were always depleted in HgTe and showed some excess tellurium, photoelectron diffraction experiments showed a good near range order [8] and a low level of crystal damage for all process conditions investigated in this study.

To obtain information on the extent of this depletion zone the XPS probing depth was reduced from its maximum value ($\vartheta=0°$) at normal electron escape by a factor 6 (grazing escape at $\vartheta=80°$). That the width of the depletion zone was not much larger than the escape length λ (≈ 20 Å) of the electrons was evident by the fact that the Hg/Cd ratio decreased by going from normal to grazing escape. For a quantitative assessment of the ARXPS data (Fig. 4) a grid has been constructed applicable to a uniform overlayer of CdTe and Te on a $Hg_{0.78}Cd_{0.22}Te$ substrate. The uniformity of this overlayer can be tested by reading a reduced thickness value at normal electron escape and comparing it with the respective value at grazing escape which should be approximately a factor 6 higher. The compositional triangle on the left of Fig.4 shows the influence of the hydrogen on the HgTe depletion zone. If the etching is performed without external hydrogen supply the width w of the depletion zone decreases (from w> 20 Å to w>10 Å) and the overlayer becomes more uniform. The presence of a tellurium surface layer is revealed by the trend at grazing angles. The vertical grid (applicable to a closed tellurium layer on $Hg_{1-x}Cd_xTe$ with arbitrary x) shows that this Te layer is relatively uniform and has a thickness of about 2 Å. Regarding the rate determining step the presence of surface tellurium implies that the tellurium backbonds are more difficult to break than those of Cd. If at low etch rates the supply of CH_3 and active nitrogen is unbalanced, a reaction of this surface Te with nitrogen is indicated by an appearance of a chemical shifted component of the Te3d electrons. Such a component was present for dataset (b) in Fig.4.

Another approach to improved surface stoichiometry was based on the attempt to shift the equilibrium of reaction (3) to the left side by external supply of mercury vapor. To hinder the desorption of Hg from the surface the sample was cooled to -25°C. No noticable influence on the etch rate was found upon the Hg addition and in the presence of Hg the surface composition was rather independent of the bias (for $U_B \geq -60V$). The data set (d) in Fig. 3 was obtained by maintaining a Hg pressure in the chamber close to the Hg vapore pressure above the sample. A further improvement in surface stoichiometry is seen when a Hg overpressure is used (dataset (e)). A detailed two-step model analysis [14] of the respective angle dependence confirms that the ARXPS data can be explained by a zone of altered x-value and a surface termination by Te atoms. For conditions (e) optimized to minimize Hg loss, the x-value is 0.45 at the surface and gradually approaches the bulk value within about 20 Å. For these conditions the etchrate is still 1300 Å/min.

SUMMARY

Recent improvements of a dry etching process for $Hg_{1-x}Cd_xTe$ have been presented. In this process Cd and Te are converted to volatile $Cd(CH_3)_2$ and $Te(CH_3)_2$ by reaction with CH_3 radicals created in an ECR plasma. Ar, N_2 and CH_4 are used as process gases. Since no supply of hydrogen is required, excellent surface morphologies are obtained. With this process etch rates of more than 2000 Å/min can be achieved if a bias of -100 V is applied. At minimum surface damage (ion energy $\approx$ plasma potential < 15 eV) etch rates are 200-300 Å/min. The addition of nitrogen is necessary as it binds atomic hydrogen and inhibits polymer formation by reaction with the polymer precursors CH and CH_2. These reactions occur mainly on the sample surface and are driven by nitrogen atoms and metastable nitrogen molecules produced by the plasma. Experimental evidence is found that the breaking of surface backbonds is the rate determining step. Since these bonds are preferentially broken by impact of high energetic particles, the etch rate is determined by the supply of active argon (as ions or metastables) from the plasma. At high sample bias, the supply with CH_3 becomes rate limiting. The detrimental side effect of HgTe depletion in the surface near sample region was reduced by adding Hg to the plasma in the downstream region and cooling of the sample.

ACKNOWLEDGEMENT

The samples have been kindly provided by T. Simon and H. Palm, AIM, Heilbronn, Germany.

REFERENCES

1. R.C. Keller, M. Seelmann-Eggebert, H.J.Richter, J. Electron. Mater. **25**, 1270 (1996); R.C. Keller, M. Seelmann-Eggebert, H.J.Richter, Appl. Phys. Lett. **67**, 3750 (1995)

2. J. E. Spencer, J. H. Dinan, P. R. Boyd, H. Wilson and S. E. Buttrill, J. Vac. Sci. Technol. **A7**, 676 (1989).

3. A. Semu, L. Montelius, P. Leech, D. Jamieson, P. Silverberg, Appl. Phys. Lett. **59**, 1752 (1992)

4. J. N. Johnson, J. H. Dinan, K. M.Singley, M. Martinka, Mat. Res. Soc. Symp. Proc. Vol. **450**, 293 (1997)

5. M. A. Foad, C.D.W. Wilkinson, C. Dunscomb, R.H. Williams, Appl. Phys. Lett. **60**, 2531 (1992)

6. G. J. Orloff, John A. Wollam, Ping He, William A. McGahan, J. R. McNeil, R. D. Jacobson and B. Johs, Thin Solid Films **233**, 46 (1993).

7. C. R. Eddy, C. A. Hoffmann, J. R. Meyer and E. A. Bobisz, J. Electron. Mat. **22** (8), 1055 (1993)

8. R.C. Keller, M. Seelmann-Eggebert, H.J.Richter, J. Electron. Mater. **24**, 1155 (1995)

9. J. W. Baars, R. C. Keller, H. J. Richter, M. Seelmann-Eggebert, SPIE Vol. **2816**, 98 (1996)

10. O. A. Popov, J. Vac. Sci. Technol. **A7**, 894 (1989)

11. M. Seelmann-Eggebert, A. Rar, to be published

12. A. Oumghar, J.C. Legrand, A.M. Diamy, N. Turillon, R.I. Ben-Aim, Plasma Chemistry and Plasma Processing **14**, 229 (1994)

13. N. Sadeghi, T. Nakano, D. J. Trevor, R.A. Gottscho, J. Appl. Phys. **70**, 2552 (1991)

14. M.Seelmann-Eggebert, R.C. Keller, Surf. and Interf. Anal. **23**, 589 (1995)

MODULAR 64x64 CdZnTe ARRAYS WITH MULTIPLEXER READOUT FOR HIGH-RESOLUTION NUCLEAR MEDICINE IMAGING

J.M. WOOLFENDEN*, H.B. BARBER*†, H.H. BARRETT*†, E.L. DERENIAK †, J.D. ESKIN †, D.G. MARKS †, K.J. MATHERSON †, E.T. YOUNG‡, F.L. AUGUSTINE**
*Department of Radiology, University of Arizona, Tucson, AZ 85724
† Optical Sciences Center, University of Arizona, Tucson, AZ 85721
‡ Steward Observatory, University of Arizona, Tucson, AZ 85721
**Augustine Engineering, 2115 Park Dale Lane, Encinitas, CA 92024

ABSTRACT

We are developing modular arrays of CdZnTe radiation detectors for high-resolution nuclear medicine imaging. Each detector is delineated into a 64x64 array of pixels; the pixel pitch is 380 µm. Each pixel is connected to a corresponding pad on a multiplexer readout circuit. The imaging system is controlled by a personal computer. We obtained images of standard nuclear medicine phantoms in which the spatial resolution of approximately 1.5 mm was limited by the collimator that was used. Significant improvements in spatial resolution should be possible with different collimator designs. These results are promising for high-resolution nuclear medicine imaging.

INTRODUCTION

Gamma-ray imaging in nuclear medicine demonstrates the biodistribution of radionuclides in the body. Spatial resolution of gamma-ray imaging devices has improved greatly in the nearly 50 years since rectilinear scanners were introduced. The intrinsic spatial resolution at present for gamma cameras with sodium iodide scintillation detectors is about 3-4 mm for planar images; the spatial resolution for tomographic images is about 10 mm [1]. Since spatial resolution is degraded by the collimator, by increasing distance from the collimator, and by scatter of gamma rays in the body, the resolution in clinical images seldom achieves these values. Better spatial resolution would be useful for many imaging applications in general and for single-photon emission computed tomography (SPECT) of the brain in particular. Radiopharmaceuticals are being developed that target specific receptors in brain, and the ability to image abnormalities in receptor binding that are in the range of 1-2 mm in size would be expected to have important clinical benefits. When such abnormalities do not have accompanying structural abnormalities, they are unlikely to be apparent on x-ray computed tomography (CT) or magnetic resonance imaging (MRI), and so the radionuclide images will contain unique information.

Semiconductor detectors for gamma-ray imaging are attractive alternatives to NaI(Tl) and other scintillators for imaging in nuclear medicine. The good energy resolution of semiconductors such as germanium (Ge) permits identification of photons that have undergone Compton scattering and that may therefore give misleading information about the site of radioactive decay in the body. Disadvantages of Ge include its relatively low atomic number, which results in poor absorption of medium and high-energy gamma rays, and its small band gap, which requires operation at cryogenic temperatures. Other semiconductors with higher effective atomic numbers and larger band gaps that permit room-temperature operation include mercuric iodide (HgI_2), cadmium telluride (CdTe), and cadmium zinc telluride (CdZnTe). An advantage of CdZnTe is its production in multi-kilogram boules by the high-pressure vertical Bridgman method, which may permit economies of scale and lower prices compared to other semiconductors. A disadvantage of HgI_2, CdTe and CdZnTe is lower

267

carrier mobilities than Ge, with an accordingly greater likelihood of carrier trapping at sites of impurities and defects. The use of small pixels in semiconductor arrays can make the effects of charge-carrier trapping much less significant [2].

We have previously developed a Ge 48x48 array [3, 4] and a CdZnTe 48x48 array [5] with multiplexer readout. We review here our progress in developing modular CdZnTe 64x64 arrays with multiplexer readout for high-resolution nuclear medicine imaging.

MATERIALS AND METHODS

Hybrid Detector Arrays

We are developing CdZnTe arrays with multiplexer readout for use in both a modular brain SPECT imager [6] and planar imaging devices. The detector is a 2.5 cm x 2.5 cm x 0.15 cm slab of $Cd_{0.9}Zn_{0.1}Te$ that has been polished flat to ± 2 μm on both sides. Gold is deposited on both sides; one side has a continuous Au electrode, and the other side has a 64x64 array of Au electrodes delineated by photolithography. Each pixel defined by the electrodes is square, 330 μm on a side; the interpixel space is 50 μm, which results in a pixel pitch of 380 μm. Pixel size was selected based upon predicted and measured charge-transport in pixels of different sizes [1].

Each pixel is connected by indium-bump bonds to a corresponding pad on a multiplexer readout circuit. The multiplexer is a custom-designed application-specific integrated circuit (ASIC) that includes a 64x64 array of unit cells as well as row and column shift registers to connect each unit cell in sequence to the readout line. Each pixel unit cell in the ASIC contains a capacitive-feedback transimpedance amplifier, which integrates and stores the leakage current as well as any charge resulting from a gamma-ray interaction in the pixel. The unit cell also incorporates correlated double-sample-and-hold circuitry. This unit-cell design was chosen for its established performance and low noise [7]. The pixel unit-cell schematic is shown in Figure 1.

Indium-bump bonding of detector and ASIC is commonly referred to as hybridization, and the resulting device is a hybrid. The indium-bump bonds are not mechanically strong enough to hold the hybrid together, so we use epoxy bonding between detector and ASIC for added strength. Prior to placement of indium bumps, a metallic layer is deposited to prevent chemical interactions between indium bumps and gold electrodes that might damage the electrode contact with the detector.

Image Acquisition System

Our current imaging system consists of a 2x2 array of hybrids, each mounted on a ceramic substrate or daughterboard that provides conditioning electronics. The four daughterboards are mounted on a larger ceramic

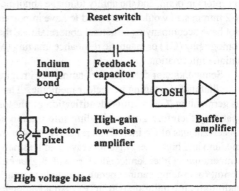

Pixel Unit Cell

Figure 1. Pixel unit cell. CDSH indicates correlated double-sample-and-hold.

motherboard, which in turn is mounted on a thermoelectric cooler. The cooler permits operating below room temperature, which reduces noise caused by leakage current. The cooler also minimizes drift in the electronics caused by changing temperature. The entire apparatus is housed in an aluminum container. Dry nitrogen is circulated through the container to prevent moisture condensation at low temperature.

The imaging system is controlled by a personal computer. A programmable data generator card provides the clock signals to drive the multiplexers. Software for data acquisition and control was written using LabVIEW™. Video signals coming from each detector are amplified and digitized at a rate of 4×10^6 samples per second; offset and gain corrections are applied on the fly [8]. The corresponding frame rate for each 64x64 array is 10^3 frames per second.

Imaging Methods

We used the imaging array to obtain emission images of two standard nuclear medicine phantoms: the Picker thyroid phantom and the Hoffman brain phantom. Both are hollow and designed to be filled with liquid containing a radionuclide. We filled each phantom with Tc-99m, which emits a 140 keV gamma ray and is the most commonly used radionuclide for nuclear medicine imaging. Image data were corrected for decay of Tc-99m, which has a 6-hour half-life. The detectors were operated at a bias voltage of -150 V. The integration time in the readout was 1 millisecond. The temperature of the array was approximately -10°C. Only one of the hybrids was used to acquire the images, which were obtained as a mosaic by mechanically stepping the phantoms over the array. The images were acquired using a parallel-hole lead collimator with 1.22 mm bore diameter and 31 mm length. The collimator was separated from the detector by about 5 mm, yielding a calculated spatial resolution of about 1.5 mm at 10 mm from the collimator face. The collimator efficiency was approximately 10^{-4}.

The Picker thyroid phantom simulates the shape of a human thyroid, although it is somewhat larger. Its overall dimensions are 10 cm x 10 cm. The right lobe has the same external dimensions as the left lobe, but it has a larger fillable volume than the left lobe. There are three simulated cold nodules of 6, 9 and 12-mm diameter, and one hot nodule of 12-mm diameter. The phantom contained 50 mCi of Tc-99m when imaging began. Data were acquired at each imaging position for 300 seconds.

The Hoffman brain phantom simulates a transverse slice through the human brain. There are three different levels of activity in the phantom. Structures in the phantom that simulate gray and white matter of the brain have features of about 1-2 mm in size, so the phantom can serve as a test of spatial resolution. The phantom contained 44 mCi of Tc-99m when imaging began. Data were acquired at each imaging position for 180 seconds.

RESULTS

An emission image of the Picker thyroid phantom is shown in Figure 2. One row and one column of nonfunctioning pixels have been filled in by nearest-neighbor averaging, and a median window filter has been applied. There are approximately 1.3×10^7 counts in the image. Detector efficiency was calculated to be about 40%. All three cold defects and the hot area are clearly seen.

An image of the Hoffman brain phantom is shown in Figure 3. As before, one row and one column of nonfunctioning pixels have been corrected by nearest-neighbor averaging, but the image is otherwise unprocessed. The image has about 4×10^7 counts. Detector efficiency was again about 40%. Fine structural details of the phantom are well demonstrated.

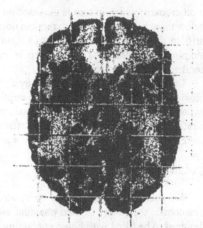

Figure 2. Picker thyroid phantom. The phantom contained 50 mCi Tc-99m when imaging began. The mosaic image was obtained for 300 seconds at each imaging position. The filling tubes are faintly visible at the bottom of the phantom.

Figure 3. Hoffman brain phantom. The phantom contained 44 mCi Tc-99m when imaging began. The mosaic image was obtained for 180 seconds at each imaging position.

CONCLUSIONS

The images that we obtained demonstrate the feasibility of high-resolution imaging using an array of hybrid CdZnTe detectors. Spatial resolution was limited by the parallel-hole collimator that was used, and it does not begin to approach the limit of the detector pixel size. We are currently working on collimator designs that will be a better match for the pixel size.

The activity levels that were used in the phantoms are much higher than would be achieved in the human body. The use of multiple detectors and more-efficient collimators should permit clinically reasonable imaging times with typical doses of radiopharmaceuticals.

Features in the detector crystals such as local changes in chemical composition and grain boundaries may affect charge transport. Our imaging system provides a method for measuring the effects of such features and assessing their impact on image formation [9].

ACKNOWLEDGMENT

This work was supported by the U.S. Department of Health and Human Services under National Cancer Institute Grants P01 CA 23417 and R01 CA 75288.

REFERENCES

1. R. J. Jaszczak and B. M. W. Tsui in Principles of Nuclear Medicine, 2nd ed., edited by H. N. Wagner, Jr., Z. Szabo, and J. W. Buchanan (Saunders, Philadelphia, 1995), p. 319.

2. H. H. Barrett, J. D. Eskin, H. B. Barber. Phys. Rev. Lett. **75**, 156 (1995).

3. H. B. Barber, F. L. Augustine, H. H. Barrett, E. L. Dereniak, K. J. Matherson, T. J. Meyers, D. L. Perry, J. E. Venzon, J. M. Woolfenden, E. T. Young. Nucl. Instr. and Meth. A **353**, 361 (1994).

4. H. B. Barber, F. L. Augustine, H. H. Barrett, E. L. Dereniak, J. D. Eskin, D. G. Marks, K. J. Matherson, J. E. Venzon, J. M. Woolfenden, and E. T. Young in 1995 IEEE Nuclear Science Symposium and Medical Imaging Conference Record, edited by P. A. Moonier (IEEE Service Center, Piscataway, NJ, 1995) pp. 113-117.

5. D. G. Marks, H. B. Barber, H. H. Barrett, E. L. Dereniak, J. D. Eskin, K. J. Matherson, J. M. Woolfenden, E. T. Young, F. L. Augustine, W. J. Hamilton, J. E. Venzon, B. A. Apotovsky, F. P. Doty. IEEE Trans. Nucl. Sci. **43**, 1253 (1996).

6. M. M. Rogulski, H. B. Barber, H. H. Barrett, R. L. Shoemaker, J. M. Woolfenden. IEEE Trans. Nucl. Sci. **40**, 1123 (1993).

7. F. L. Augustine, Nucl. Instr. and Meth. A **353**, 201 (1994).

8. K. J. Matherson, H. B. Barber, H. H. Barrett, J. D. Eskin, E. L. Dereniak, D. G. Marks, J. M. Woolfenden, E. T. Young, F. L. Augustine in 1997 IEEE Nuclear Science Symposium and Medical Imaging Conference Record (IEEE Service Center, Piscataway, NJ), in press.

9. H. B. Barber, E. L. Dereniak, J. D. Eskin, N. R. Hilton, D. G. Marks, K. J. Mathersòn, J. M. Woolfenden, and E. T. Young in 1997 IEEE Nuclear Science Symposium and Medical Imaging Conference Record (IEEE Service Center, Piscataway, NJ), in press.

PERFORMANCE OF P-I-N CdZnTe RADIATION DETECTORS AND THEIR UNIQUE ADVANTAGES

R.SUDHARSANAN*, C.C.STENSTROM*, P.BENNETT**, G.D.VAKERLIS*
* Spire Corporation, One Patriots Park, Bedford, MA, 01730-2396
** Radiation Monitoring Devices, Watertown, MA

ABSTRACT

We present the performance characteristics of CdZnTe radiation detectors with a new P-I-N design and their unique advantages over metal-semiconductor-metal (M-S-M) devices. In M-S-M CdZnTe detectors the bulk resistivity of the substrate largely determines the leakage current. High leakage current is a dominant noise factor for CdZnTe detector arrays, coplanar detectors, and detectors used for low X-ray energy applications. P-I-N devices provide low leakage currents. Early CdZnTe detectors exhibited polarization, were limited to small detection volumes, and some required high deposition temperatures. We have developed a new hetero-junction design which can be deposited at low temperatures so that even high-pressure Bridgman CdZnTe can be used. Using the P-I-N design, CdZnTe detectors with high detection volumes (>200 mm^3) were fabricated and exhibited low leakage current, good energy resolution, and no polarization. These detectors have significant advantages over M-S-M detectors in three specific areas. First, X-ray fluorescence studies require detectors with low leakage currents to provide less spectral broadening due to electronic noise. Second, less expensive vertical Bridgman CdZnTe material can be used for imaging applications since it normally possesses too low of a bulk resistivity to be useful as a M-S-M detector. Third, leakage currents across the anode grid in large volume coplanar detectors can be significantly reduced.

1   INTRODUCTION

Compound semiconductors such as CdTe and HgI$_2$ have been investigated for room temperature gamma-ray detection since they have a high average atomic number (resulting in large attenuation coefficient) and a wide bandgap (which enables room temperature operation). However, major issues such as poor charge collection efficiency and non-ideal ohmic contacts have limited their potential. In CdTe, deep traps reduce the mobility-lifetime product ($\mu\tau$) of holes causing long low-energy tails, leading to poor energy resolution. Poor charge collection also limits the detection volume. Increasing bias will improve the energy resolution; however, CdTe's low resistivity (10^9 ohm-cm) and M-S-M design prevent operation at high bias due to high leakage current [1,2].

A significant advancement in radiation detector technology has been the development of CdZnTe alloys by the high pressure Bridgman (HPB) technique [3,4,5]. CdTe with Zn (in the range of 4 to 20% Zn) increases the resistivity to mid 10^{10} ohm-cm. This improves the overall performance of the device, but there are still problems with low energy tailing and poor peak-to-valley ratios in large volume detectors. The electron $\mu\tau$ product does not improve, and the hole $\mu\tau$ product decreases when adding Zn; deep traps still limit the energy resolution and detection volume.

Several approaches such as thermoelectric cooling, improved pulse processing methods, and new device designs are being researched to improve detector energy resolution [6,7,8].

Cooling CdZnTe detectors to -30°C reduces leakage currents from nA to pA levels [7]. Such a reduction allows the use of low-noise amplifiers and operation at high bias, thus increasing the risetime and improving energy resolution. However, a major limitation to this approach is that the larger the detector volume, the more difficult it becomes to cool.

Pulse processing methods include pulse-shape discrimination and charge compensation. In pulse-shape discrimination techniques [6], signals arising from slow-moving holes are rejected improving the spectral shape. This method increases the energy resolution but results in large losses of photopeak efficiency requiring longer collection times for certain applications. In the charge compensation approach, appropriate charge is added to the signal to compensate for loss due to incomplete charge collection. The disadvantage is the need for sophisticated electronics to achieve high energy resolution.

Coplanar and P-I-N advanced designs are currently under investigation to improve energy resolution [8,9,10]. In the coplanar design, grids are formed on one side of the detector which collect carriers of one polarity efficiently, improving the energy resolution. This approach permits larger detection volumes. Recently, spectral energy resolution better than 2% at 662 keV was achieved using a 1 cm^3 CdZnTe detector [11].

The advantage of the P-I-N design is low leakage current, allowing use of higher bias voltages at room temperature. Increasing bias voltages improves the charge collection efficiency. Furthermore, it permits use of low-cost, low-resistivity CdZnTe such as that produced by the conventional vertical Bridgman (VB) technique. Several groups have developed P-I-N CdTe and CdZnTe detectors [8,10,12,13]. Two methods commonly used to fabricate P-I-N detectors are thermal diffusion and liquid phase epitaxy (LPE). In the thermal diffusion process, indium is thermally diffused on one side of the CdZnTe to form an N$^+$ layer, and a thin Au layer is deposited on the other side to provide a P$^+$ contact. The drawback is that indium, in addition to being a donor, forms defect complexes which could degrade the energy resolution. Also they exhibit polarization problems resulting in a degradation of performance when the detector is operated for an extended amount of time.

LPE-grown HgCdTe is used to form P and N-type layers on high resistivity CdZnTe substrates [8]. Even though this approach produces higher energy resolution P-I-N detectors compared to M-S-M detectors, the fabrication method is expensive and not practical for HPB CdZnTe because of the high temperatures involved. Several researchers have observed that high temperatures (greater than 150°C) severely degrade the resistivity and detector properties of HPB CdZnTe [14]. Since LPE growth of HgCdTe layers is typically done around 400 to 550°C, HPB CdZnTe is not suitable for this process.

We have developed a novel P-I-N design shown in Figure 1 which uses CdS (N+) and ZnTe (P+) layers to form a P-I-N structure on high resistivity CdZnTe. The advantage of the design is the reduced temperature of the fabrication process which is compatible with HPB CdZnTe. In addition, the process has the potential for large-volume manufacturing at low cost since inexpensive methods such as thermal evaporation can be used.

CdS and ZnTe have several attractive properties including wide bandgap (2.42 eV and 2.23 eV, respectively). CdS is always N-type and ZnTe is always P-type; hence, high doping can be easily achieved. Both materials can be deposited by inexpensive deposition methods. Similar structures have been used for thin film CdTe solar cells on SnO$_2$/glass substrates to achieve high efficiency [15,16]. In this paper we present the performance of our P-I-N design and the unique applications for which it is useful in overcoming present obstacles involved with CdZnTe material.

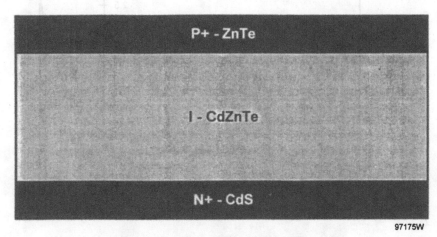

Figure 1   *Schematic of Spire's P-I-N design.*

2   EXPERIMENTAL METHODS AND RESULTS

We used HPB CdZnTe material of various sizes purchased from eV Products. Before fabrication, the CdZnTe was first characterized by applying standard M-S-M Au contacts. Later the Au contacts were stripped and the P-I-N structure was deposited on the same device. The details of the P-I-N fabrication method was described in a previous paper [19]. The devices are measured for I-V behavior. The detector performance was measured by using ^{241}Am, ^{57}Co, and ^{133}Ba sources to observe the resultant spectra, and measuring FWHM and energy resolution. Low temperature testing was performed by AMPTEK, Inc.

After optimizing the P-I-N fabrication process, we fabricated several P-I-N CdZnTe detectors varying detector volumes. In most cases the leakage current of the P-I-N detector decreased by a factor of 2-3 compared to the same detector fabricated as an M-S-M detector. Figure 2 shows the lowest leakage current observed for a 5x5x3 mm device after P-I-N fabrication. It measures only 868pA when biased at -200V. The largest detectors fabricated thus far are 6.5x6.5x5 mm (200 mm³). Figure 3 shows the best energy resolution for ^{57}Co measured at room temperature after P-I-N fabrication. The detector bias was 1000 V with 0.5µs shaping time, resulting in a FWHM of 5.4% at 122 keV. ^{133}Ba spectrum also taken with a 5 mm thick device appears in Figure 4. At 1000V and 1µs shaping time, the three major peaks of ^{133}Ba(356, 302.9, and 275.9 keV) are easily defined.

Several devices were sent to AMPTEK, Inc. for low temperature measurements. The detectors were tested at -23°C by mounting the crystals onto a thermoelectrically cooled platform. The response of a 6.5 x 6.5 x 5 mm device when irradiated with ^{241}Am is shown in Figure 5. The 60 keV photopeak shows an energy resolution of 1.2 keV FWHM.

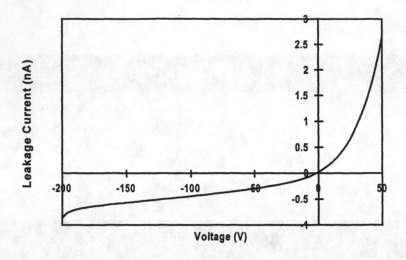

Figure 2   *I-V curve of a 5 x 5 x 3 mm P-I-N CdZnTe device.*

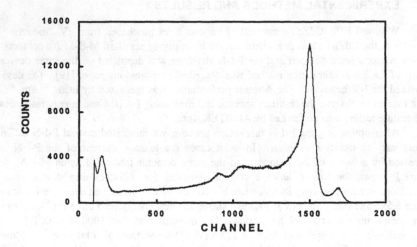

Figure 3   *Energy resolution of ^{57}Co using P-I-N CdZnTe device. Sample size is 6.5 x 6.5 x 5 mm.*

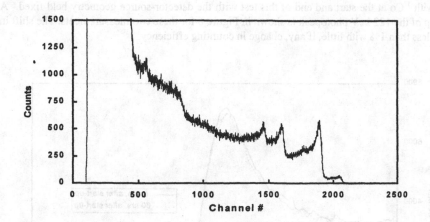

Figure 4 ^{133}Ba *energy spectrum measured using a 5 mm thick P-I-N CdZnTe device biased at 1000V with 1 μs shaping time.*

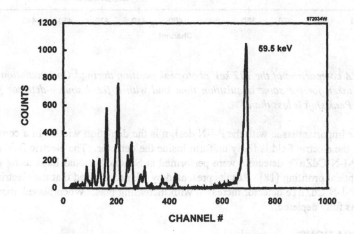

Figure 5 ^{241}Am *energy spectrum measured at -23°C for a 6.5 x 6.5 x 5 mm P-I-N CdZnTe device biased at 800V.*

277

A major advantage of our P-I-N design is the absence of degradation of spectral quality as a function of time as a result of polarization. Polarization has been observed in CdTe P-I-N detectors fabricated by indium diffusion. We tested our heterojunction P-I-N design by continuously irradiating a 7.5 x 7.5 x 2 mm detector under bias for a period of 60 hours. Spectra were taken with ^{57}Co at the start and end of this test with the detector-source geometry held fixed. A close-up of the 122 keV photopeak is shown in Figure 6 for these two measurements. The shift in gain is less than 1% with little, if any, change in counting efficiency.

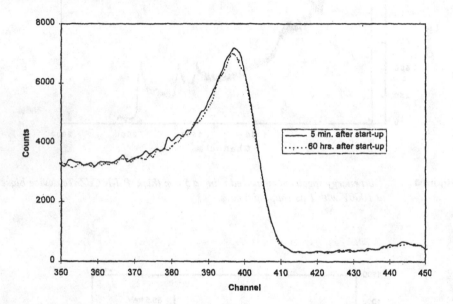

Figure 6   *A comparison of the 122 keV photopeak position during ^{57}Co irradiation. Spectra taken for the same acquisition time and with a fixed source-detector geometry Peak shift is less than 1%.*

Another important issue with the P-I-N design is the depletion width. In a conventional M-S-M design the electric field is fairly uniform inside the detector. The electric field measurements of the P-I-N CdZnTe detectors were performed at Sandia National Labs using polarized transmission optical profiling [18]. Results presented by W.Yao showed that the electric field of a 2mm thick P-I-N CdZnTe detector increased with increasing bias. When biased above 400V, the detector was fully depleted.

3   APPLICATIONS

3.1   Vertical Bridgman Applications

There are many advantages to using Vertical Bridgman (VB) CdZnTe material. VB substrates are low cost($100/cm^2), large area (up to 4 x 6 cm), and the manufacturing technology is more mature. VB substrates have been used for HgCdTe detectors for several decades and less expensive (three to four times) than HPB CdZnTe. However, VB substrates have not been explored extensively for radiation detectors because they have low resistivity (10^6 ohm cm) and produce high leakage current as M-S-M detectors. They also suffer from high trap density resulting in poor energy resolution.

We have shown earlier that our P-I-N structure on VB material helps to lower the leakage current of the device [19]. When comparing the I-V data from VB material fabricated with an M-S-M design and a P-I-N design, the leakage current of the M-S-M device was many orders of magnitude higher than the P-I-N device. Detector testing with ^{57}Co showed a clear 122 keV photopeak which was encouraging for VB CdZnTe. By further optimizing the P-I-N fabrication process and improving the quality of the substrates, detector performance comparable to HPB CdZnTe should easily be obtained.

3.2   Coplanar Applications

Another application for the P-I-N structure is to incorporate it into a coplanar CdZnTe detector. The coplanar detector offers an alternate geometry, based on the Frisch grid, which allows for single carrier sensing. With this geometry severely trapped carriers do not contribute to the detector's induced charge and only the electrons contribute to pulse formation. Position dependent charge collection effects are reduced. We have designed and fabricated coplanar CdZnTe detectors using a P-I-N grid on one face and a continuous ohmic contact on the other face. The use of the P-I-N structure reduces the dark current between the grids and allows the increase of the bias between the grids which results in improved charge collection. Figure 7 shows the design for our coplanar P-I-N grid detector. Using this coplanar design, room temperature CdZnTe radiation detectors with volumes as high as 1 cm^3 have been fabricated. Figure 8 shows a ^{137}Cs spectral comparison of a M-S-M and a P-I-N coplanar CdZnTe detector both tested at eV Products. A M-S-M coplanar detector was stripped of its contacts and refabricated as a P-I-N coplanar detector. The P-I-N detector's photopeak is more defined, with a significantly improved peak-to-valley ratio.

3.3   Imaging Arrays

The P-I-N structure can also be used for CdZnTe imaging arrays. The P-I-N design offers a significant advantage over M-S-M detector arrays through reduced leakage current. This will lead to improved performance with a variety of read-out schemes and particularly for gated integration techniques with long sampling periods. A 4x4 pixelated P-I-N CdZnTe array fabricated and packaged at Spire and tested at Harvard University is shown in Figure 9. It was fabricated on a 10 x 10 x 2 mm HPB CdZnTe substrate using 1.5 mm^2 pixels. The leakage current was measured in all 16 pixels. The array demonstrated exceptional uniformity with approximately 6 nA leakage current from each pixel. Figure 10 shows the ^{241}Am spectral uniformity of the 16 pixel detector array. The average FWHM taken over all the pixels is 4.4% while the best FWHM is 3.3% at 60 keV.

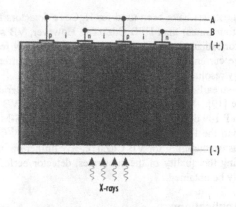

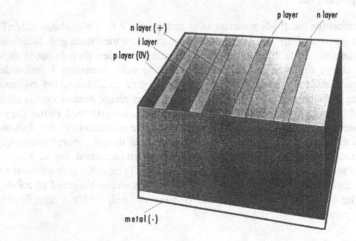

Figure 7   *Spire's coplanar grid design utilizing the P-I-N fabrication process.*

4   CONCLUSIONS

We have demonstrated the advantages of P-I-N CdZnTe radiation detectors useful for many applications. The fabrication process is low cost, low temperature, and compatible with HPB and VB CdZnTe. Spectral response data show improved energy resolution and peak-to-valley ratios for [241]Am and [57]Co at high bias voltages compared with M-S-M detectors. We also looked into the different applications for the P-I-N structure including VB CdZnTe to improve the resistivity and performance of this versatile material, coplanar CdZnTe detectors, and imaging arrays.

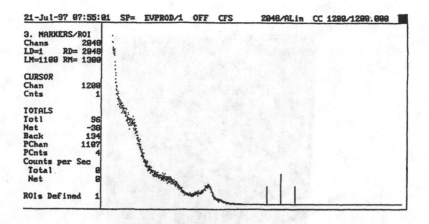

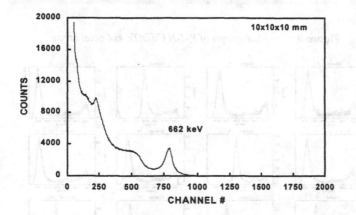

Figure 8   *^{137}Cs spectrum comparison of a) M-S-M and b) P-I-N coplanar CdZnTe detectors measured by eV Products.*

5   ACKNOWLEDGMENTS

This work was funded by DOE grant# DE-FG0294ER81869. We would like to thank Mr. K.B.Parnham from eV Products for his discussions and measurements on HPB CdZnTe detector, Dr. R.B.James from Sandia National Labs and Dr. H.W.Yao from the University of Nebraska for performing electric field measurements, J.Pantazis and R.Redus at AMPTEK for performing low temperature measurements on our P-I-N CdZnTe detectors, Prof. G.Grindlay's group at Harvard University for their work in characterizing our CdZnTe imaging arrays, and Dr. H.Glass at Johnson Matthey for supplying VB CdZnTe samples.

Figure 9   *Photograph of P-I-N CdZnTe 4x4 pixel array.*

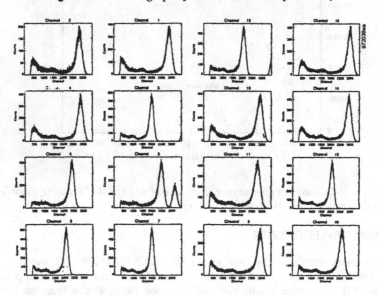

Figure 10   *Data demonstrating* ^{241}Am *uniformity among pixels.*

282

REFERENCES

1. Y.Eisen, *Nuclear Instruments and Methods in Physics Research*, **A322**, 596 (1992).
2. M.Richter and P.Siffert, *Nuclear Instruments and Methods in Physics Research*, **A322**, 596 (1992).
3. F.P.Doty, J.F.Butler, J.F.Schetzina, and K.A.Bowers, *Journal of Vacuum Science and Technology*, **B10**, 1418 (1992).
4. J.F.Butler, F.P.Doty, B.Apotovsky, S.F.Friesenhahn, *Materials Research Society Symposium Proceedings*, **V302**, 497 (1993).
5. K.B.Parnham, *Nuclear Instruments and Methods in Physics Research*, **A377**, 487 (1996).
6. A.Niemela, H.Sipila, and I.V.Ivanov, *Nuclear Instruments and Method in Physics Research*, **A377**, 484 (1996).
7. M. Richter and P.Siffert, *Nuclear Instruments and Method in Physics Research*, **A322**, 529 (1992).
8. W.J.Hamilton, D.R.Rhiger, S.Sen, M.H.Kalisher, K.James, C.P.Reid, V.Gerrish, and C.O.Baccash, *IEEE Trans. on Nuclear Science*, **41**, 989 (1994).
9. P.N.Luke, *IEEE Trans. on Nuclear Science*, **V42**, 207 (1995).
10. S.H.Shin, G.T.Niizawa, J.G.Pasko, G.L.Bostrup, F.J.Ryan, M.Zoshnevisan, C.T. Westmark, and C.Fueller, *IEEE Trans. on Nuclear Science*, **NS-32**, 487 (1985).
11. Z.He, G.F.Knoll, D.K.Wehe, and J.Miyamoto, Presented at the IEEE Nuclear Science Symposium and Medical Imaging Conference, Anaheim, CA.
12. A.Niemela, H.Sipila, and I.V.Ivanov, *IEEE Trans. on Nuclear Science*, **V43**, 1476 (1996).
13. W.J.Hamilton, D.R.Rhiger, S.Sen, M.H.Kalisher, G.R.Chapman, and R.E.Millis, *Journal of Electronic Materials*, **V25**, 1286 (1996).
14. Final Report to DOE, SBIR grant# DE-FG02-94ER81869, (1995).
15. P.V.Meyers, Solar Cells, 23, 59 (1988).
16. R.Sudharsanan and A.Rohatgi, Solar Cells, 31, 143 (1991).
17. Y.Nemirovsky, A.Ruzin, G.Asa, Y.Gorelik, and L.Li, Presented at the 1996 U.S. Workshop on the Physics and Chemistry of II-IV Materials, Las Vegas, NV, October 22-24, 1996.
18. H.W.Yao, R.J.Anderson, R.B.James, *SPIE Proceeding*, **V3115**, 62 (1997).
19. R.Sudharsanan, G.D.Vakerlis, and N.Karam, *Journal of Electronic Materials*, **V26**, 745 (1997).

FABRICATION OF CDZNTE STRIP DETECTORS FOR LARGE AREA ARRAYS

C. M. Stahle,* Z. Q. Shi,** K. Hu,** S. D. Barthelmy,*** S. J. Snodgrass,** S. J. Lehtonen,****
K. J. Mach,**** L. Barbier, N. Gehrels, J. F. Krizmanic,*** D. Palmer,*** A. M. Parsons, P. Shu

NASA Goddard Space Flight Center, Greenbelt, MD 20771

*Orbital Sciences Corp., NASA GSFC, Greenbelt, MD 20771, carl.stahle@gsfc.nasa.gov
**Hughes STX Corp., NASA GSFC, Greenbelt, MD 20771
***Universities Space Research Associates, NASA GSFC, Greenbelt, MD 20771
****Johns Hopkins University Applied Physics Laboratory, Johns Hopkins Road, MS 13-N273, Laurel, MD 20723

ABSTRACT

A CdZnTe strip detector large area array (~ 60 cm^2 with 36 detectors) with capabilities for high resolution imaging and spectroscopy has been built as a prototype for a space flight gamma ray burst instrument. The detector array also has applications in nuclear medical imaging. Two dimensional orthogonal strip detectors with 100 μm pitch have been fabricated and tested. Details for the array design, fabrication and evaluation of the detectors will be presented.

INTRODUCTION

A CdZnTe (CZT) strip detector array is being developed at NASA Goddard Space Flight Center for an instrument to accurately locate gamma ray bursts, determine their distance scale, and measure the physical characteristics of the emission region [1]. The energy range of interest is 10 - 150 keV. A CZT strip detector array with less than 100 μm spatial resolution combined with a coded aperture mask will allow a gamma ray burst to be positioned on the sky to a few arc second accuracy. This is an advance of more than an order of magnitude in angular resolution compared to the best previous instrumentation in hard x-ray and gamma ray astronomy. Previous work has shown the good imaging and spectral resolution for this CZT strip detector [2,3]. Recently, a 2 x 2 CZT strip detector array has been fabricated and tested to demonstrate the imaging properties of the array [4, 5].

A large area and compact CZT array would have an enormous impact on nuclear medical imaging. The prototype CZT strip detector array that is described in this paper has over 580,000 pixels at a spacial resolution of 100 μm. The number of pixels is over a factor of 100 greater than existing detector arrays. With this detector array, 3D tomographic images with 0.3 x 0.3 x 0.3 mm volume elements could be obtained. The rest of the imaging system could be designed such that 1 mm^3 details would be easily discernible which is a great improvement over the 1 cm^3 resolution that is available today [6].

EXPERIMENT

Design of Strip Detector Array

The 6 x 6 CZT strip detector array which we built is shown in Figure 1. Testing of this array is in progress. The array has 36 double sided CZT strip detectors. Each detector side has 127 strips of length 1.27 cm with 100 μm pitch plus a guard ring. A large area strip detector array is constructed by connecting the individual strips of 6 detectors in a row to make one long strip of 7.6 cm. Thus, the array will have 762 x 762 strips with 580,644 resolvable pixels. The size of the detectors was chosen to be 15 x 15 x 2 mm. This was the largest area detector of sufficient quality and quantity which was available. The specification for a noise threshold of 10 keV in the array limited the maximum length of a strip due to the interstrip capacitance. The strip detectors are biased through a monolithic bias resistor array (200 MΩ) and ac coupled through a capacitor array (1000 pF) to high density readout electronics (XA1.2 ASICs from Integrated Device Electronics,

285

IDE AS). The strips are connected to other detectors and to the electronics by wire bonding. Ceramic bridges are mounted between the detectors for the wire bonds. For this array, over 20,000 wire bonds were required. For fabrication of future arrays of this area, 27 x 27 x 2 mm CZT detectors may be available which would reduce the number of detectors and wire bond connections by a factor of four.

The configuration of the 6 x 6 strip detector array places challenging fabrication requirements on the detectors. Since 6 strips are connected in series to make 1 long (7.6 cm) strip, the leakage current in the individual strips of each detector needs to be small to have low leakage current noise from the long strip. Our goal for the fabrication run was to have an average strip leakage current less than 500 pA/detector strip. In order to achieve the science goal of > 90% active area in the array, greater than 99% [7] of the strips are required to be active after integration to the electronics. This yield of greater than 99 % includes strips which are not open or shorted due to fabrication defects or wire bond failures.

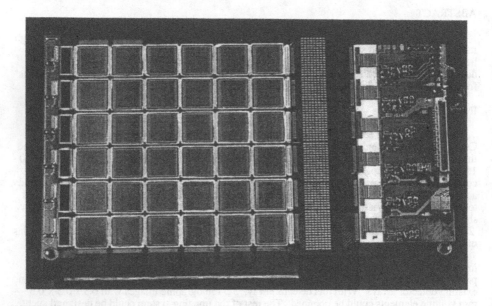

Figure 1: 6 x 6 CZT strip detector array. The 36 CZT detectors are in the center. The resistor arrays (6) are on the left of the detectors and the capacitor arrays (6) are on the right. To the right of the capacitor arrays are the readout electronics with the ASIC chips (6). A similar set of electronics is on the back side.

Fabrication of CZT Strip Detectors

The materials used in this work were discriminator grade 15 mm x 15 mm x 2 mm $Cd_{0.9}Zn_{0.1}Te$ purchased from eV Products. The two dimensional CZT strip detectors were produced by having strips on each side of the detector orthogonal to each other. The strip detector was patterned with 100 μm pitch, 50 μm wide metal strips. The fabrication process consisted of the following steps: mechanical polishing for producing a smooth starting surface, solvent cleaning to remove any debris from the polishing, chemical etching for removing damage from the polished surface and to ensure a fresh surface, photolithography to develop the strip pattern, metalization and lift-off. Platinum metal contacts were sputter deposited on the CZT surface followed by a sputtered gold film for wire bonding. For the second side process, the procedure was the same except that the strips on the finished side were protected with photoresist. After both sides of the detector were fabricated, the detectors were passivated to increase the interstrip resistance and annealed to improve the adhesion of the metal to the CZT. Figure 2 shows a closeup of a finished CZT strip detector. Additional details of the fabrication steps have been previously published [8].

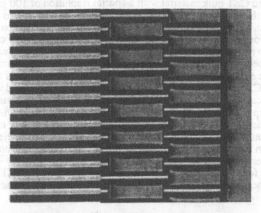

Figure 2: Close up view of the current CZT strip detector. The strip pitch is 100 μm with 50 μm wide strips. The bonding pads are staggered with a length of 400 μm and a width of 120 μm. The guard ring is 450 μm wide.

RESULTS

Evaluation of Strip Detectors

The three methods used to evaluate and qualify CZT strip detectors for the array were visual inspection, electrical testing, and wire bonding tests. The visual inspection detected open strips and shorted strips as well as cracks, voids, and other defects in the detector which might impair detector performance. The electrical tests measured the leakage current from individual strips through the bulk or volume of the detector and the interstrip leakage current. An interstrip resistance was calculated from the interstrip leakage current. The wire bonding tests were done on a small number of samples to determine the typical wire bonding yield on the metal bonding pads. In addition, pull tests were done on all the wire bonds of these samples to test the adhesion of the wire to the metal pad and the adhesion of the metal pads to the CZT.

Visual Inspection

In a visual inspection of 44 CZT strip detectors fabricated for the array, there were 38 open strips out of 11,176 total strips or 0.34%. The open or broken strips were caused by voids or pipes at the surface of the CZT. For a void across or along a strip, the photoresist was too thick in the void so the photoresist would be be completely exposed. Subsequently, the metal strip would be broken in this area during the lift-off process. For the same 44 strip detectors, there were 94 shorted strips out of a total of 11,264 strips and guard rings or 0.83%. The majority of shorts were between two strips or bonding pads. These strips were shorted either by voids, defect bumps or a metal liftoff problem where the metal did not break off between strips. A defect bump was caused by the different etching rate of the defect by the bromine-ethylene glycol solution. The defect bumps would break the photoresist between the strips when the detector was in contact with the mask which allowed metal to be deposited between the strips. The photoresist metal liftoff problem was solved for later detector samples by making sure the trenches for the metal films were deep enough so the metal lifted off easily and cleanly. For this set of 44 samples, the total yield of good strips (not open or shorted) was 98.8%. Although this is good for a first try at processing a large number of detectors, we need to improve. If CZT material without voids is used, we believe the fabrication yield for good strips will be close to 100%.

287

Electrical Tests

The electrical tests for the CZT strip detectors provided an indication of the quality of our metal contacts, surfaces, and CZT bulk material with a spatial resolution of 100 μm. We measured the strip leakage current though the detector and between strips for a few strips across the active area of the detector. Details of the test procedures have been published previously [9]. The results are shown in Figure 3. For 49 strip detectors, the average of the strip leakage current was 332 ± 232 pA, and the interstrip resistance was 5.4 ± 4 GΩ. The variation in strip leakage current and interstrip resistance between samples and across the strips for the same sample is large and not understood. Possible explanations include variations in our metal contacts and surfaces or material differences between CZT samples. Since all the detectors were fabricated with the same process and many were fabricated in the same batch, we can not identify an obvious processing variation. The average total strip leakage current for 6 strips would be 2 nA. For 2 nA of leakage current and a typical integration time of 2 μs, the expected leakage current or shot noise is 158 electrons rms. This is significantly less than the expected capacitance noise of 320 electrons (25 pF capacitance load from 6 strips) expected from the XA1.2 ASIC chip. From the geometry of the strip detector, our average interstrip resistance is calculated to be 1.4×10^{12} ohms/square. This is an important number for a variety of other electrode configurations for CZT detectors where interelectrode resistance is critical for a uniform electric field and low surface leakage current.

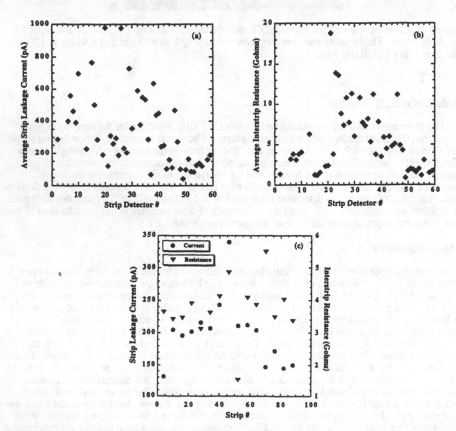

Figure 3: Measurement of (a) average strip leakage current, (b) average interstrip resistance for all of the detectors and (c) strip leakage current and interstrip resistance across a typical detector.

Wire Bonding Tests

One of the critical tests for evaluating the CZT detector fabrication process was the wire bond tests. Our goal was to achieve a wire bonding yield > 99.5%. From our final fabrication process, we prepared 4 strip detectors for wire bond tests. Two samples were used to test the "A" side or the gamma ray incident side of the detector, and two samples were tested on the other side or "B" side. For each side, 512 total wire bonds were attempted. An automatic ultrasonic wedge wire bonder was used with 25 μm (0.001 inch) diameter, 99.99% aluminum bonding wire. We have previously found that aluminum bonding wire with 1% silicon alloy (which is commonly used for wirebonding silicon chips) caused cratering in the CZT material under the bonding pads. We believe the cratering was due to the increased hardness of the wire and higher ultrasonic power required to achieve a satisfactory bond. To minimize stress in the CZT material, the mechanical layout of the CZT detectors and ceramic bridges was designed such that the first bond was placed on the CZT bonding pad and the second bond on the ceramic bridge. The second bond experiences a significant pull force immediately after the bond is made when the machine cycles through the wire tear-off portion of the cycle. Pull tests were done on the wire bonds to determine the strength of the bonds.

A histogram of the bond pull tests for the "A" and "B" sides for the 4 samples is shown in Figures 4. On the "A" side, we achieved 100% wire bonding yield and 0 bond peels on the pull tests. The wire broke before the bond peeled which is the best result. For the "B" side, we achieved 99.6% wire bonding yield (2 bonds peeled during bonding) and had 7 bonds peel during the pull test. However, all of the 7 bond peels resulted from a force between 2.5 and 4.6 grams which exceeds the MIL SPEC standard of 1.5 grams. For both sides with a total of 1024 wire bonds, we achieved a 99.8% wire bonding yield. We found that the "B" was more difficult to bond, and the ultrasonic power had to be increased by 20% to get the bonds to stick. This is probably the reason we had some bond peeling because the higher ultrasonic power caused greater damage to the CZT surface underneath the metal bonding pad. We have found in previous work that the CZT surface can be broken if the ultrasonic power is too high. Improved surface cleaning of the "B" side before wire bonding should alleviate this problem. The "B" side of the detector is fabricated first so it is susceptible to surface contamination while the "A" side of the detector is processed. For the 6 x 6 strip detector array, we achieved a 99.5% bonding yield. We believe the keys to successful wire bonding to CZT are to use an ultrasonic wedge bonder with soft aluminum wire, low ultrasonic power, excellent metal adhesion, thick metal pads (2-3 μm) to minimize damage to the CZT, and a large bonding pad area so the force and power to apply the bond are spread over a larger metal pad area.

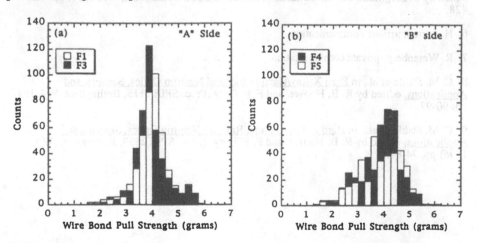

Figure 4: Wire bond pull tests for the (a) "A" side and the (b) "B" side for detector samples.

To summarize, the energy coupled into the CZT from the wire bonder should be minimized. Finally, the length of the bonding pad should be made long enough to have space for two wire bonds in case rebonding is necessary from a failed first attempt.

CONCLUSIONS

We have developed a robust fabrication process for CZT strip detectors to make the large number of detectors needed for a large area array. The strip detectors can be fabricated with a high yield of good strips, low leakage current, and high interstrip resistance. In addition, the strips can be connected to high density electronics with wire bonds with a yield close to 100% . A 6 x 6 strip CZT strtip detector array has been assembled and integrated to electronics.

ACKNOWLEDGMENTS

The authors acknowledge the work of Carol Sappington and Frank Peters in the earlier wire bonding tests, Andre Burgess in assisting with mechanical polishing of the CZT material, and Jim Odom with assembly of the electronics for the arrays.

REFERENCES

1. N. Gehrels et al., in Gamma-Ray and Cosmic Ray Detectors, Techniques, and Missions, edited by B. D. Ramsey and T. A. Parnell (Proc. SPIE 2806, Bellingham, WA 1996) pp. 12-19.

2. L. M. Bartlett et al., in Gamma-Ray and Cosmic Ray Detectors, Techniques, and Missions, edited by B. D. Ramsey and T. A. Parnell (Proc. SPIE 2806, Bellingham, WA 1996) pp. 616-628.

3. P. Kurczynski et al., IEEE Trans. Nucl. Sci. 43, 1011 (1997).

4. A. Parsons et al., in EUV, X-Ray, and Gamma-Ray Instrumentation for Astronomy VIII, edited by O. Siegmund and M. Gummin (Proc. SPIE 3114, Bellingham, WA 1997) pp. 341-348.

5. D. M. Palmer et al., in EUV, X-Ray, and Gamma-Ray Instrumentation for Astronomy VIII, edited by O. Siegmund and M. Gummin (Proc. SPIE 3114, Bellingham, WA 1997) pp. 422-428.

6. H. Barrett, private communication.

7. R. Wesenberg, private communication.

8. C. M. Stahle et al., in Hard X-Ray/Gamma-Ray and Neutron Optics, Sensors, and Applications, edited by R. B. Hoover and F. P. Doty (Proc. SPIE 3115, Bellingham, WA 1997) pp. 90-97.

9. C. M. Stahle et al., in Hard X-Ray/Gamma-Ray and Neutron Optics, Sensors, and Applications, edited by R. B. Hoover and F. P. Doty, (Proc. SPIE 2859, Bellingham, WA 1996) pp. 74-84.

MAPPING DETECTOR RESPONSE OVER THE AREA
OF A CdZnTe-MULTIPLE-ELECTRODE DETECTOR

C.L. LINGREN, B. APOTOVSKY, J.F. BUTLER, F.P. DOTY, S.J. FRIESENHAHN, A. OGANESYAN, B. PI, S. ZHAO
Digirad Corporation, 7408 Trade Street, San Diego, CA 92121-2410

ABSTRACT

Semiconductor multiple-electrode detectors have been developed for the purpose of reducing effects of hole trapping in room-temperature radiation detectors.[1,2] Some reported geometries maintain a nearly-uniform electric field inside the detector, but others generate an electric field that is very non-uniform and highly-concentrated at the anode. This paper reports the results of mapping such a detector (having a non-uniform electric field) with a finely collimated gamma-ray beam to determine the detector response as a function of position.

INTRODUCTION

The goal of this experiment was to determine the response of the detector as a function of position for a detector with non-uniform electric field to ensure that the entire volume provides useful signal and, therefore, to ensure that the detector sensitivity will be proportional to the total number of gamma rays absorbed in the volume of the detector.

Figure 1 shows the geometry of the detector that was used in this experiment and Figure 2 shows the shape of the electric field that would be created by the voltages imposed (typical for 3-mm-thick detector: anode=0, cathode=-300, and third electrode=-200 volts) on the electrodes if the detector were a homogeneous material with uniform dielectric constant.

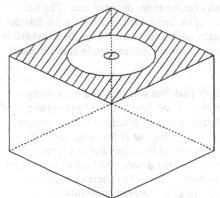

Figure 1 SpectrumPlus™ Detector
Pat. No. 5,677,539

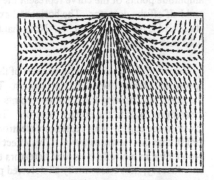

Figure 2 Electric Field in Detector

In this experiment, a spectrum was taken with the gamma-ray source flooding the entire cathode from a direction approximately normal to the cathode. The gamma-ray source was then finely collimated and spectra were taken at discrete positions as the source was scanned along the x and y coordinates of the cathode. The source to detector distance was maintained constant and the exposure time at each location was kept the same. The total number of counts were divided by the total exposure time at each position and plotted as counts per second as a function of x-y position along the cathode.

EXPERIMENT

The experiment was conducted by moving a collimated gamma-ray beam across the area of the detector under study in discrete steps. An energy spectrum was collected at each step and the data collecting time was the same for each measurement. The beam was kept approximately perpendicular to the surface of the detector and was formed by placing a concentrated source behind a single-hole collimator.

The detector was a single SpectrumPlus™ element in an array, with nominal dimensions of 3x3x3-mm. The anode is a small diameter pad in the center of the detector face opposite the cathode, the third electrode surrounds the anode, and the cathode covers the entire surface opposite the anode surface.

The beam had an effective diameter of approximately 1-mm. The collimator was about one inch thick and was formed by drilling a 1-mm diameter hole through the lead.

Figure 3 shows the magnitude of the count rate as a function of position in the detector for pulses in an energy window from 119 KeV to 153 KeV, using ^{99m}Tc as the gamma-ray source. Good uniformity of count rate sensitivity is achieved over the entire detector area. The half amplitude points of the curve represent the positions of the beam when it is half on the detector and half off the detector. The horizontal axis is source position as recorded for the x-translation stage and is measured in inches. Note that the distance between half-amplitude points is about 0.12 inch (3 mm).

Figure 4 shows the energy channel of the 140 KeV peak that was recorded in the energy spectrum taken at each detector position. The energy peak was near 140 KeV as long as part of the beam was striking the detector. Points were plotted at zero to indicate that there was no meaningful peak in the energy spectrum. The peak energy is highest at the center and slightly lower at the edges. This represents electron charge trapping at the edges due to a longer path length that the electrons travel. This effect is not seen in planar detector and is a source of peak broadening in multiple-electrode detectors that significantly reduce the effects of trapped holes and, thereby, move essentially all detected photons to the proper energy amplitude.

Figure 3. Detector Count Rate Scan

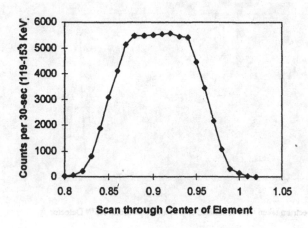

Figure 4. Detector Peak Energy Scan

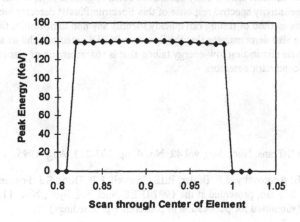

Figure 5 shows a typical spectrum for a collimated ^{57}Co source with a SpectrumPlus™ detector. Note the high photo-peak efficiency which results from measuring each photon detection event at full amplitude by eliminating effects of trapped holes.

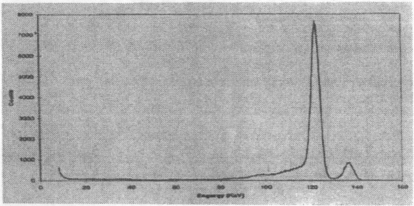

Figure 5. ^{57}Co Spectrum taken with collimated source and SpectrumPlus™ Detector

CONCLUSION

By scanning the detector with a finely collimated beam of gamma rays, it was demonstrated that the counting sensitivity spectral response of this SpectrumPlus™ detector were uniform over its entire area. This method of testing can quickly identify any non uniformities that may exist in a detector. The data also demonstrates that the SpectrumPlus™ detector exhibits single charge carrier response, thus eliminating low-energy tailing that is characteristic of planar, room-temperature, semiconductor detectors.

REFERENCES

1. P.N. Luke, IEEE Trans. Nucl. Sci., vol 42, No. 4, pp. 207-213, Aug. 1995.

2. C.L. Lingren, B. Apotovsky, J.F. Butler, R.L. Conwell, F.P. Doty, S.J. Friesenhahn, A. Oganesyan, B. Pi, S. Zhao, presented at the 1997 IEEE Nucl. Sci. Sym., Nov. 11, 1997, Albuquerue, NM (submitted for publication in conference proceedings).

NEW TYPE OF ir PHOTODECTORS BASED ON LEAD TELLURIDE AND RELATED ALLOYS

D.R. KHOKHLOV
Physics Department, Moscow State University, Moscow 119899, Russia,
khokhlov@mig.phys.msu.su

ABSTRACT

Doping of the lead telluride and related alloys with the group III impurities results in an appearance of the unique physical features of a material, such as such as persistent photoresponse, enhanced responsive quantum efficiency (up to 100 photoelectrons/incident photon), radiation hardness and many others. We review the physical principles of operation of the photodetecting devices based on the group III-doped IV-VI including the possibilities of a fast quenching of the persistent photoresponse, construction of the focal-plane array, new readout technique, and others. The advantages of infrared photodetecting systems based on the group III-doped IV-VI in comparison with the modern photodetectors are summarized.

INTRODUCTION

Most of the sensitive infrared photodetecting systems operating in the far infrared wavelength range (20-200) μm are based on germanium or silicon doped with the shallow impurities [1]. The highest cutoff wavelength reported $\lambda \approx 220$ μm corresponds to the uniaxially stressed Ge(Ga) [2]. The main advantage of germanium and silicon is a very well developed growth technology that allows to receive materials with extremely low uncontrolled impurity concentration.

Another opportunity arises if one tries to use a narrow-gap semiconductor system $Pb_{1-x}Sn_xTe$ heavily doped with the group III impurities.

INDIUM-DOPED LEAD-TIN TELLURIDES: THE MAIN PROPERTIES

Fermi level pinning

When the lead-tin tellurides are doped with indium in a concentration N_{In} exceeding the concentration of other impurities N_c, the Fermi level becomes pinned at some definite position E_0 that practically does not depend on N_{In} [3]. E_0 does not shift with respect to the middle of the bandgap under the action of external factors: additional doping [3], temperature variation [4], external pressure application [5].

The important consequence of the Fermi level pinning effect is the homogenization of material electrical properties. Indeed, in PbTe(In) with the pinned Fermi level up to 50 periods of the Shubnikov-de Haas (SdH) oscillations are resolved [6] indicating very high degree of the sample homogeneity. It should be noted that no more than 10 periods of the SdH oscillations have ever been observed in undoped PbTe with the comparable electron concentration n $\approx$ $7 \cdot 10^{18}$ cm^{-3}. So the electrical properties of the heavily doped $Pb_{1-x}Sn_xTe(In)$ system are much more reproducible and homogeneous than those of the undoped alloys.

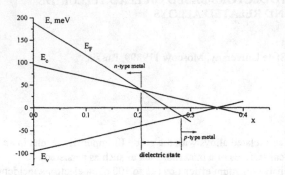

The position of the pinned Fermi level depends strongly on x. In PbTe the E_0 lies at ~ 70 meV above the bottom of the conduction band E_c. When the tin content x increases, the level approaches E_c, then crosses the bandgap at 0.22 < x < 0.28, and enters the valence band giving rise to the free hole concentration [7] (Fig.1). The conductivity of material being in the dielectric state (0.22 < x < 0.28) is defined by the activation from

Fig.1. Energy diagram of the $Pb_{1-x}Sn_xTe(In)$ alloys for different x. E_c - conduction band bottom, E_v - valence band edge, E_F - position of impurity level that pins the Fermi level [7].

E_0-level, and therefore the free carrier concentration is very low $n,p < 10^8$ cm^{-3} at the temperatures T < 10 K. It should be noted that the free carrier concentration in the undoped alloys is defined by the electrically active growth defects. Their concentration is never smaller then 10^{15} cm^{-3}. So one has a very unusual situation, when a heavily doped narrow-gap semiconductor with high number of the growth defects acts as an almost ideal one with practically zero background free carrier concentration and very high electrical homogeneity. This makes very attractive the idea to use this material as an infrared photodetector.

Persistent photoconductivity

The external infrared illumination leads to the substantial increment of material conductivity

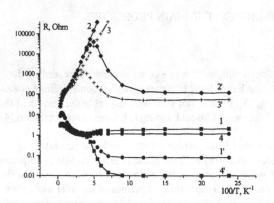

at the temperatures T < 25 K independently on the Fermi level pinning position [8] (Fig.2). If the E_0-level lies within the bandgap the relative conductivity change is much higher, because there is practically no free carriers in the initial "darkness" state.

A high amplitude of a photoresponse at the low temperatures is a consequence of the persistent photoconductivity effect: if the permanent radiation flux illuminates the sample the photoresponse increases linearly in time, and when

Fig.2. Temperature dependence of the sample resistance R measured in darkness (1-4) and under infrared illumination (1'-4') in alloys with x = 0.22 (1, 1'), 0.26 (2, 2'), 0.27 (3, 3') and 0.29 (4, 4') [8].

the light is switched off, the conductivity value remains stable (the photomemory). The effect is defined by the nature of impurity states.

According to the theory, the persistent photoconductivity effect in $Pb_{1-x}Sn_xTe(In)$ is due to the strong electron-lattice coupling in the system: when the electron is excited by the light quantum into the conduction band the impurity crystalline surrounding rearranges, and a barrier between the system states with the localized and the free electron appears [9]. This leads to the accumulation of the photoexcited free electrons in the conduction band, and therefore the photoresponse is linear in time.

PHOTODETECTORS BASED ON $Pb_{1-x}Sn_xTe(In)$

Quenching of the persistent photoconductivity

The maximal accumulation characteristic time is defined by the photoexcited electron lifetime τ In $Pb_{1-x}Sn_xTe(In)$ $\tau \sim 10^4$ s at 4.2 K < T < 10 K, then it sharply decreases with the temperature rising, and $\tau \sim 10^{-2}$ s at T ~ 20 K. So one may choose the proper temperature in order to provide the photodetector operation time required τ_o. The examples are known when the BLIP regime is realized for the $Pb_{1-x}Sn_xTe(In)$ photodetector operating at T = 25 K [10]. If the sample temperature is so that $\tau > \tau_o$, then the photoresistor may operate, only if there exists a way to resume the initial "darkness" state, i.e. to quench quickly the persistent photoconductivity. Moreover, periodical accumulation and successive fast quenching of the photosignal leads to the substantial gain in the S/N ratio with respect to the case of ordinary single photodetectors.

One of the most attractive techniques is quenching of persistent photoconductivity by the microwave pulses[11,12]. In this technique either a sample is installed in a microwave "resonator", or the microwave pulses are applied directly to the sample contacts. Experiments described in [11,12] were performed with the samples with x = 0.25, in which the Fermi level is pinned within the gap. The microwave frequency was ~ 400 MHz, power in the pulse - up to 0.9 W, the pulse length ~ 10 μs. The sample resistance after the end of each microwave pulse did not differ from the initial "darkness" resistance of the samples. Probably this is due to the fact that the electric field distribution along the sample is much more "soft" for the microwave pulse, so the high-state is formed throughout the sample volume. Besides that, whereas the strong electric field pulses induce nonequilibrium state of a sample, that slowly relaxes in time to the equilibrium low-ohmic one even without radiation, in the case of the microwave quenching the equilibrium long-living high-ohmic state is resumed. This is one more advantage of a microwave quenching. At last, application of the short microwave pulses affects the photoresponse after the pulse end. It was shown in [12], that if the persistent photoconductivity is quenched by the microwave pulses of minimal necessary length and amplitude, the quantum efficiency of a photoresistor increases up to ~ 10^2. Explanation of this effect of microwave stimulation of quantum efficiency was given in [12] in terms of localization of the photoexcited free carriers to the metastable local states after the short pulses. These states are strongly correlated, and the avalanche excitation from them to the conduction band is possible. This gives rise to the increment of the quantum efficiency.

Radiometric parameters

We have constructed the laboratory model of the IR-radiometer based on $Pb_{1-x}Sn_xTe(In)$ operating in the regime of periodical accumulation and successive fast quenching of the persistent photoconductivity. The blackbody and the sample were separated by the shutter, the diaphragm and the grid filter cooled by liquid helium. The blackbody temperature controlled by a thermocouple could be varied from 4.2 to 300 K by means of changing of the blackbody position with respect to the liquid helium level. The diaphragm restricted the field of view of a sample to the blackbody cavity independently on the blackbody position. The grid filter provided effective cutting of the blackbody radiation spectrum at the wavelengths $\lambda > 18$ μm. It is easy to calculate for this construction the incident intensity and photon fluxes with rather high accuracy of ~ 20%. Though the voltage measurement technique that we have used was quite poor (the oscilloscope with the voltage measurement accuracy of only 1 mV), we were able to detect the photon flux of $N \approx 2 \cdot 10^4 \, s^{-1}$ that corresponds to NEP $\approx 2 \cdot 10^{-16}$ W for the detector area 0.3·0.2 mm^2 and the operating rate 3 Hz.

Photoconductivity spectral characteristics

Measurements of the photoconductivity spectra of materials revealing the photomemory effect is a sophisticated problem. In fact the background radiation present in any spectrophotometer "pumps" the nonequilibrium free carriers, and one cannot begin the measurements from the ground "darkness" state. Moreover one can register only the spectral response of the processes with the characteristic time τ_p lower than the time of spectral scanning τ_s. In Pb$_{1-x}$Sn$_x$Te(In) $\tau_p \sim (10^4\text{-}10^5)$ s at T = 4.2 K, and the ordinary spectral measurements are practically impossible.

The indirect technique of the red cutoff energy determination based on the analysis of the photoresponse increase dynamics under the action of the black-body illumination has been proposed in [13]. The advantage of this technique is the complete screening of the background radiation. Though the method accuracy is low ~ (20 - 50)%, it allowed to estimate the cutoff energy $E_{co} \sim 20$ meV in Pb$_{0.75}$Sn$_{0.25}$Te(In). It corresponds to $\lambda_{co} \sim 100$ μm. The microwave stimulation effect should provide the red shift of λ_{co}.

Focal-plane "continuous" array. Readout technique

Specifics of impurity states may make very easy the construction of a focal-plane array on Pb$_{1-x}$Sn$_x$Te(In). The local infrared illumination leads to the local generation of the nonequilibrium free electrons, i.e. the persistent photoconductivity effect is observed only in the illuminated part of the sample, and the photoexcitation does not propagate into the darkened regions [14]. The characteristic time of the excitation propagation is at least more than 10^4 s at T = 4.2 K. The spatial characteristic scale is ~ 10 μm. The physical picture of the processes involved is the following. On one hand the photoexcited free electrons cannot diffuse far away from the region of generation due to the electrostatic attraction to the ionized impurity centers. On another hand these electrons cannot recombine because of the existence of a barrier between the local and extended states.

So the distribution of the radiation exposure on the sample surface reflects in a distribution of the concentration of the long-living free electrons. In other words, one may construct the focal plane "continuous" array, where the signal is internally integrated.

The readout technique is a special problem. The approach that we propose here seems to be the most promising. Let us consider a thin slice of $Pb_{1-x}Sn_xTe(In)$ with a semitransparent electrode deposited on one side (Fig.3). The investigated radiation flux illuminates the sample from this side. A buffer insulating fluoride layer is deposited on another side of a sample plate followed by a thin layer of silicon or some other semiconductor with the relatively wide gap, and a second semitransparent electrode. If one illuminates the plate by a shortwavelength laser from the right side (see Fig.3) one may create a local highly conductive region in the wide-gap semiconductor. If one applies a voltage between the electrodes, the current will be defined by the $Pb_{1-x}Sn_xTe(In)$ sample conductivity in the region of the laser spot, because the thickness of a semiinsulating wide-gap semiconductor layer will be much less in this point. Using of this idea allows also to reduce greatly the dark current. To some extent it is analogous to the ideas involved in BIB structures based on impurity Si and Ge. If the recombination rate in the wide-gap semiconductor is high enough, one may easily reconstruct the conductivity distribution over the $Pb_{1-x}Sn_xTe(In)$ sample simply by scanning the laser beam over the structure surface and by measuring the respective current. Unfortunately this idea is not realized in practice so far.

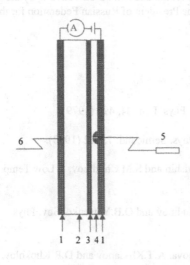

Fig.3. Device for readout of information from the "continuous" focal-plane array on $Pb_{1-x}Sn_xTe(In)$. 1 - semitransparent electrodes, 2 - active $Pb_{1-x}Sn_xTe(In)$ layer, 3 - fluoride buffer layer, 4 - layer of a silicon (or other semiconductor with a relatively wide gap), 5 - short-wavelength laser, 6 - incident infrared radiation flux.

Stability to the hard radiation action

High stability with respect to the action of hard radiation is one more advantage of the photodetectors based on $Pb_{1-x}Sn_xTe(In)$. This is a consequence of a very high density of impurity states ($\sim 10^{18} - 10^{19}$ cm^{-3}) that pin the Fermi level. The fast electron irradiation with fluencies up to 10^{17}-10^{18} cm^{-2} does not affect the photoresponse [15]. This value is at least by 4 orders of magnitude higher then for $Hg_{1-x}Cd_xTe$, doped Ge and Si.

SUMMARY

In summary, application of the lead-tin tellurides doped with the group III impurities as base elements for the infrared photodetectors gives a challenging opportunity to produce universal and sensitive systems. They have a lot of advantageous features that allow them to compete successfully with the existing analogs: internal accumulation of the incident radiation flux, possibility of effective fast quenching of an accumulated signal, microwave stimulation of the quantum efficiency up to 10^2, possibility of realization of a "continuous" focal-plane array, possibility of application of a new readout technique, high stability to the action of hard radiation. In our opinion, these features make the $Pb_{1-x}Sn_xTe(In)$-based photodetectors ideal for the space-borne applications, for example, for the infrared astronomy.

ACKNOWLEDGMENTS

The research described in this paper was supported in part by the grants of the Russian Foundation for the Basic Research (RFBR) No. 95-02-04658-a and No. 96-02-16275-a, by the INTAS-RFBR grant No. 95-1136 and by the Grant of the President of Russian Federation for the Young Scientists - Doctors of Science No. 96-15-96957.

REFERENCES

1. J.Wolf and D.Lemke, Infrared Phys. **25**, 327 (1985).

2. E.E.Haller, M.R.Hueschen and P.L.Richards, Appl. Phys. Lett. **34**, 495 (1979).

3. V.I.Kaidanov, R.B.Mel'nik and I.A.Chernik, Sov. Phys. Semicond. **7**, 522 (1973).

4. B.A.Akimov, N.B.Brandt, L.I.Ryabova, V.V.Sokovishin and S.M.Chudinov, J. Low Temp. Phys. **51**, 9 (1983).

5. B.A.Akimov, V.P.Zlomanov, L.I.Ryabova, S.M.Chudinov and O.B.Yatsenko, Sov. Phys. Semicond. **13**, 759 (1979).

6. B.A.Akimov, N.B.Brandt, K.R.Kurbanov, L.I.Ryabova, A.T.Khasanov and D.R.Khokhlov, Sov. Phys. Semicond. **17**, 1021 (1983).

7. B.A.Akimov, L.I.Ryabova, O.B.Yatsenko and S.M.Chudinov, Sov. Phys. Semicond. **13**, 441 (1979).

8. B.A.Akimov, N.B.Brandt, S.O.Klimonskiy, L.I.Ryabova and D.R.Khokhlov, Phys. Lett. A **88A**, 483 (1982).

9. B.A.Volkov and O.A.Pankratov, Sov. Phys. Doklady **25**, 922 (1980).

10. V.F.Chishko, V.T.Hryapov, I.L.Kasatkin, V.V.Osipov, E.I.Slinko, O.V.Smolin and V.V.Tretinik, Infrared Phys. **33**, 197 (1992).

11. B.A.Akimov, N.B.Brandt, D.R.Khokhlov and S.N.Chesnokov, Sov. Tech. Phys. Lett. **14**, 325 (1988).

12. B.A.Akimov and D.R.Khokhlov, Semicond. Sci. Technol. **8**, S349 (1993).

13. B.A.Akimov, N.B.Brandt, L.I.Ryabova and D.R.Khokhlov, Sov. Tech. Phys. Lett. **6**, 544 (1980).

14. B.A.Akimov, N.B.Brandt, S.N.Chesnokov, K.N.Egorov and D.R.Khokhlov, Solid State Commun. **66**, 811 (1988).

15. L.I.Ryabova, E.P.Skipetrov (private communication).

PbTe(Ga) - NEW MULTISPECTRAL IR PHOTODETECTOR

A.I. BELOGOROKHOV **, I.I. IVANCHIK *, D.R. KHOKHLOV*
*Physics Department, Moscow State University, Moscow 119899, Russia,
khokhlov@mig.phys.msu.su
**Institute of Rare Metals, Moscow, Russia

ABSTRACT

Doping of the lead telluride - narrow-gap semiconductor - with gallium results under certain conditions in the Fermi level pinning in the gap thus providing the semiinsulating state of material. Besides that, the persistent photoconductivity effect is observed at a temperatures $T < T_c = 80$ K. The photoresponse kinetics consists of two parts: the slow one with the characteristic time t_{char} going up to 10^4 s at $T = 4.2$ K, and the fast part with t_{char} of the order of 10 ms. We have measured the spectra of a fast part of the photoresponse using the Fourier-transform spectrometer "Bruker" IFS-113v. The photoconductivity is observed in two spectral regions: in the middle- and far - infrared. Response in the middle-infrared consists of the ordinary fundamental band and a strong superimposed resonance-like structure just at the bandgap energy. The position of this spectral line may be tuned in a wide range (3.5-5.5) μm by variation of temperature and/or composition of a lead telluride-based alloy. This middle-infrared photoresponse becomes considerable already at $T = 160$ K. The photoresponse in the far-infrared may be depending on the excitation conditions an analogous resonance-like structure at a wavelength 70 μm, or a broad band with the cutoff wavelength at least higher than 500 μm, which is the highest cutoff wavelength for the photon detectors observed up to date.

INTRODUCTION

IV-VI semiconductors are extensively used in the infrared optoelectronics. Some of their features, such as direct gap, high radiation recombination output, possibility of tuning of the gap value in a wide range - make them especially attractive for the construction of infrared lasers [1]. Application of IV-VI semiconductors as infrared photodetectors is restricted by a high free carrier concentration originating from a large number of growth defects in the as-grown materials.

Doping of the lead telluride-based alloys with some of the group III impurities results in an appearance of features that are not observed in the undoped material, such as Fermi level pinning and persistent photoconductivity effects [2].

Gallium is one of the dopants of the above-mentioned kind. Generally it acts as a donor in PbTe. However in some range of a gallium concentration the Fermi level becomes pinned at ~ 70 meV below the bottom of conduction band. In this regime the impurity centers reveal some of the features of DX-centers in III-V and II-VI materials, for instance, the persistent photoconductivity is observed at the temperatures $T < T_c \approx 80$ K [3].

At the same time there is a considerable difference from the case of DX-centers in III-V and II-VI. The photoconductivity kinetics is quite complicated and consists of two parts: the relatively fast and slow ones [3]. The characteristic time of the slow part of a photoresponse varies from more than 10^4 s at $T = 4.2$ K to ~ 10 ms at $T = 77$ K whereas for the fast part it varies from ~ 10 ms at $T = 4.2$ K down to ~ 10 μs at $T = 77$ K.

EXPERIMENTAL

Spectral measurements of the slow part of the photoconductivity are very uneasy to do. Indeed, since the characteristic energies of the energy spectrum of PbTe are quite small, the background infrared radiation present in any standard spectrophotometer would lead to the accumulation of electrons in the conduction band so that one would not be able to start the measurements from the initial «darkness» state.

We have measured the fast part of the photoconductivity spectra of PbTe(Ga) and related alloys in the wavenumber range $(10-5000)$ cm^{-1} at the temperatures $(45-160)$ K. At $T < 45$ K the changes in the photoconductivity that are due to the fast processes are not noticealbe on the background of a high concentration of free electrons accumulated in the conduction band due to slow processes. The spectra have been taken using a Fourier-transform spectrophotometer «Bruker IFS-113v».

The samples of PbTe(Ga), $Pb_{1-x}Mn_xTe(Ga)$, $Pb_{1-x}Ge_xTe(Ga)$ were grown by a range of techniques: Chokhralski method, Bridgeman technique and from the vapour phase. The amount of gallium introduced into the crystal varied in the range $(0.1-0.5)$ at. % in order to provide pinning of the Fermi level in the gap. The results practically did not depend on the growth method. The indium contacts used were ohmic at all temperatures. The photoconductivity measurements have been done using the 4-probe technique.

RESULTS

The photoresponse has been detected in two spectral regions. For the wavenumber range $(1000 - 5000)$ cm^{-1}, typical photoconductivity spectra of PbTe(Ga) at a different temperatures are shown in Fig.1. The spectra consist of a resonant-like structure at the energy that is very close to the gap value in the undoped PbTe superimposed on a rather flat broad band with a pronounced red cutoff wavelength. Position of the photoconductivity peak ω_{max} shifts with the temperature coefficient $\partial\omega_{max}/\partial T \approx$ 3 cm^{-1}/K that is equal to the temperature coefficient of the energy gap in PbTe. For some of the samples the photoresponse becomes noticealbe on the noise background already at $T =$

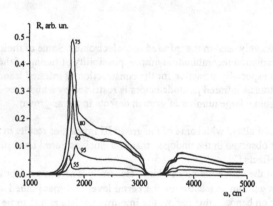

Fig.1. Photoconductivity spectra of PbTe(Ga) in the middle - infrared. Figures near the curves correspond to the sample temperature in K.

160 K, and the amplitude of a peak reaches its maximum at $T = (75-80)$ K. The ratio of the peak amplitude A_{max} to the broad background amplitude A_{bg} does not depend on the temperature, but changes from sample to sample. In most of the cases $A_{max}/A_{bg} \gg 1$.

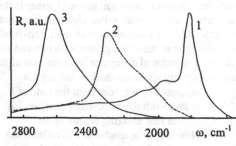

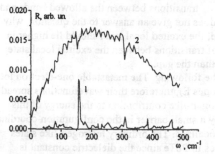

Fig.2. Photoconductivity spectra of PbTe(Ga) (curve 1), PbMnTe(Ga) (curve 2) and PbGeTe(Ga) (curve 3) taken at 90 K.

For the $Pb_{1-x}Mn_xTe(Ga)$, $Pb_{1-x}Ge_xTe(Ga)$ alloys, the energy gap increases with rising Mn and ge content. Consequently their photoconductivity spectrum, looking basically the same, shifts to a somewhat higher wavenumbers (Fig.2.). Again, the position of the photoconductivity peak is defined by energy gap and depends therefore both on the temperature and on the x value. It means that the position of the photoconductivity peak may be tuned by the temperature and the alloy composition.

The photoconductivity spectrum taken in the far-infrared region consists of a narrow resonant-like structure at $\omega \approx 155$ cm^{-1} (Fig.3). In most of the cases this line is resolved only in a narrow temperature range (57-63) K with the maximum at 60 K. Sometimes the photoconductivity resonance is not resolved at all, in some cases it exists in a more wide range of temperatures. In the latter situation not only the main peak is observed, but also some additional photoconductivity resonances at a multiple frequencies.

The main peak amplitude is maximal just after the sample is cooled down to 60 K. If then the temperature is fixed, the photoresponse gradually decreases and eventually disappears in ~ 40 min. Whenever the temperature is changed even for 10 K in any direction and then returned back, the main photoconductivity line at ~ 155 cm^{-1} appears again.

If the sample is cooled down to 60 K in the presence of an additional relatively short-wavelength radiation source with the emission spectrum corresponding to the fundamental absorption, then the evolution of the far-infrared spectrum first looks quite analogous: the photoconductivity line at ~

Fig.3. Photoconductivity spectra of PbTe(Ga) in the far-infrared just after the sample cooling to 60 K (curve1), and under the action of an additional short-wavelength radiation source (curve 2).

155 cm^{-1} disappears quite fast, but then the broad far-infrared photoresponse band with a high amplitude appears (see Fig.3). We were not able to detect the red cutoff wavelength of this band, it is at least higher than 500 μm. As long as the short-wavelength radiation source is switched on, the picture remains stable. When the source is off, this far-infrared photoresponse band disappears.

DISCUSSION

The impurity centers formed in the lead telluride-based alloys upon doping are quite analogous to the DX-centers observed in III-V's and II-VI's. It reveals, for instance, in an appearance of the persistent photoconductivity effect at the low temperatures in PbTe(Ga).

According to the commonly accepted point of view, gallium forms a negative-U center in the lead telluride [4]. Therefore the ground impurity state E_2 corresponds to two electrons localized on an impurity. At the same time there exists a metastable one-electron local state E_1, separated by a barrier in the configuration-coordinate space from the states of the system with two and zero localized electrons [5] (Fig. 4). The last point makes a substantial difference from the case of the DX-centers in III-V and II-VI where the metastable impurity states are shallow and are not separated by a barrier from the state of a system with the delocalized electrons [6].

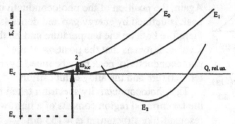

Fig.4. Configuration-coordinate diagram of PbTe(Ga). The curves E_n correspond to the states with n electrons localized on the center. 1 - transitions responsible for the middle-infrared photoresponse, 2 - possible optical transitions responsible for the far-infrared photoresponse, ω_{loc} - excitation of the local vibrational mode.

The first amazing feature of the photoresponse spectra in PbTe(Ga) is the absence of an impurity photoconductivity at the energies well below the gap even though the galvanomagnetic measurements show unambiguously that there exists an impurity level providing the Fermi level pinning at ~ 70 meV below the bottom of the conduction band [3]. It means that the transitions from the ground two-electron state to the conduction band most likely correspond to the slow part of the photoresponse, whereas the fast part we measure originates from the transitions between the allowed bands and the excited local state. This explanation however does not give an answer to the question, why there is no sub-bandgap photoconductivity. Indeed, the excited local level should lie higher in energy than the ground one, therefore the energy of transitions between the excited local state and the conduction band should be much smaller than the gap.

The possible resolution of this discrepancy is the following. The metastable one-electron local states E_1 are much less localized than the ground ones E_2, therefore their wavefunctions spread over many lattice periods. Consequently, the hydrogen-like contribution to the energy of this state may be predominant despite it is separated by a small barrier in the configuration-coordinate space from the conduction band and is therefore localized. The ordinary hydrogen-like states would be delocalized at all reasonable temperatures in PbTe since the dielectric constant is extremely high $\varepsilon \sim 1500$ [7] and the free carrier effective masses are only $\sim 0.01\ m_0$ [8], so the binding energy is very small < 0.1 meV.

The metastable E_1 - states keep many features of the hydrogen-like states. The near-bandgap resonant-like structure is most likely due to the transitions from the valence band to the E_1 state (see Fig. 4). Indeed, if this state is hydrogen-like, then its wavefunction originates mainly from the states of the conduction band bottom. Since the wavevector does not change in the optical transitions, their energy should be close to the energy gap value, and the photoresponse has a resonant-like structure. Optical transitions of electrons lying rather deep in the valence band to the E_1 - state are damped because of the momentum conservation law. The density of the E_1 - states is of the order of magnitude of overall gallium concentration and may be much higher than the density of states at the bottom of the conduction band that is quite small because of the small effective electron mass. As a result, the amplitude of photoconductivity due to the valence band - E_1 level transitions may be considerably higher than the photoresponse due to the fundamental absorption. Direct optical transitions from the E_1 - state to the conduction band are damped because the momentum conservation law mentioned above.

One could consider the far-infrared photoresponse at $\omega = 155$ cm^{-1} to be originating from the optical excitation from the E_1 - state to the conduction band. This possibility however seems to us to be not very likely. Indeed, as one can see from the photoconductivity spectra in the bandgap region, position of the E_1 - level is extremely close to the bottom of the conduction band, and the shift of 155 cm^{-1} would be quite noticeable. On the other hand one cannot completely exclude the above-mentioned possibility since the optical E_1 - conduction band transitions should occur without changing the impurity crystalline surrounding, i.e. they are direct in the configuration-coordinate space. (see Fig. 4).

In our opinion the far-infrared photoresponse has another origin. The frequency of $\omega = 155$ cm^{-1} directly corresponds to the frequency of an additional oscillator revealing itself in the reflection spectra of PbTe(Ga) at the temperatures $T < 80$ K [9]. An analogous oscillator at the frequency $\omega = 120$ cm^{-1} has been observed in the infrared reflection spectra of Pb$_{0.75}$Sn$_{0.25}$Te(In) alloy [10] and in the Raman spectra of PbTe(In) [11] at $T < 25$ K. These materials reveal the persistent photoconductivity effect at a temperatures $T < T_c = 25$ K. The frequencies of all these oscillators are quite close to, but somewhat higher than the frequency of the LO-phonon mode in PbTe $\omega_{LO} = 110$ cm^{-1} [12]. Therefore it is natural to assume that these oscillators originate from the local lattice vibrations corresponding to the E_1 - state. If one considers only the mass effect, the frequencies of the oscillators corresponding to In- and Ga- doped materials should satisfy a simple relation $\omega(Ga)/\omega(In) = (m_{In}/m_{Ga})^{1/2}$. It is easy to calculate that this relation is approximately valid for the oscillator frequencies on In- and Ga-doped alloys.

Coincidence of the frequency of the far-infrared photoconductivity with the oscillator frequency in the reflection spectra means that the far-infrared photoresponse is most likely due to the optical excitation of a local vibrational mode. The mechanism of the photoresponse could be the following. The background radiation excites one electron from the ground two-electron impurity state, and the impurity center is transferred to the metastable one-electron local state. If then the local vibrational mode is excited by the incident far-infrared radiation, the electron localized on a metastable impurity state may overcomes small barriers separating it from the conduction band and thus may take part in the photoconductivity. As the excitation of the local mode is resonant in frequency, the photoresponse should be resonant, too.

Disappearance of the far-infrared photoresponse in time may result from the interaction of E_1 local states when their population becomes considerable. At the same time one can see that this interaction does not lead to the delocalization of these states. Overlapping of the lattice deformation areas around the E_1 - centers, but not of the respective electronic wavefunctions is most likely responsible for such an interaction. An additional short-wavelength infrared irradiation further increases the population of the E_1 local states, and the intensity of a broad far-infrared photoconductivity band with a red cutoff wavelength at least higher than 500 µm appears. To our knowledge, it is the highest cutoff wavelength ever observed for the photon photodetectors.

SUMMARY

The resonance-like photoconductivity spectrum in the middle- and in the far-infrared has been observed in PbTe(Ga) and related alloys.

The resonant photoresponse structure in the middle infrared is observed at 45 K < T < 160 K in a wavelength range (3.5 - 5.5) µm depending on the temperature and the alloy composition. It corresponds to the optical transitions from the valence band to the «quasi-shallow» one-electron local state lying close to the conduction band bottom.

The selective photoconductivity observed in the far-infrared at 55 K < T < 65 K at the wavelength λ ≈ 70 μm originates from the optical excitation of a local vibrational mode providing the transfer of electrons from the metastable one-electron local state to the conduction band. An additional short-wavelength infrared irradiation transforms this selective photoresponse to an intense broad band with the red cutoff wavelength at least higher than 500 μm. Is the highest cutoff wavelength observed so far for the photon photodetectors.

ACKNOWLEDGMENTS

The research described in this paper was supported in part by the grants of the Russian Foundation for the Basic Research (RFBR) No. 96-02-18853 and No. 95-02-04658-a, No. 96-02-16275 by the INTAS-RFBR grant No. 95-1136 and by the Grant of the President of Russian Federation for the Young Scientists - Doctors of Science No. 96-15-96957.

REFERENCES

1. A. Lambrecht, H. Boettner, M. Ange, R. Kurbel, A. Fach, B. Halford, U. Sciessl and M. Tacke, Semicond. Sci. Technol. **8**, S334 (1993).

2. B.A. Akimov, A.V. Dmitriev, D.R. Khokhlov and L.I. Ryabova, Phys. Stat. Sol. (a) **137**, 9 (1993).

3. B.A. Akimov, N.B. Brandt, A.M. Gas'kov, V.P. Zlomanov, L.I. Ryabova and D.R. Khokhlov, Sov. Phys. Semicond. **17**, 53 (1983).

4. I.A. Drabkin and B.Ya. Moizhes, Sov. Phys. Semicond. **17**, 611 (1983).

5. I.I. Zasavitskiy, B.N. Matsonashvili, O.A. Pankratov and V.T.Trofimov, JETP Lett. **42**, 1 (1985).

6. P.M. Mooney, J. Appl. Phys. **67**, R1 (1990).

7. G. Nimtz and B. Schlicht in Narrow-Gap Semiconductors, edited by G. Hohler (Berlin, Springer, 1983).

8. B.A. Akimov, N.B. Brandt, L.I. Ryabova V.V. Sokovishin and S.M Chudinov, Journ. Low Temp. Phys. **51**, 9 (1983).

9. A.I. Belogorokhov, S.A. Belokon', I.I. Ivanchik and D.R. Khokhlov, Sov. Phys. Solid State **34**, 873 (1992).

10. N. Romcevic, Z.V. Popovic, D. Khokhlov, A.V. Nikorich and W. Koenig, Phys. Rev. B **43**, 6712 (1991).

11. N. Romcevic, Z.V. Popovic and D.R. Khokhlov, J. Phys.: Condens. Matt. **7**, 5105 (1995)

12. M.A. Kinch and D.D. Buss, Solid State Commun. **11**, 319 (1972).

Part V

Growth and Doping of II-VI Materials

CURRENT ISSUES OF HIGH-PRESSURE BRIDGMAN GROWTH OF
SEMI-INSULATING CdZnTe

CSABA SZELES, ELGIN E. EISSLER

eV Products a division of II-VI, Inc., Saxonburg, PA 16056

ABSTRACT

The availability of large-size, detector-grade CdZnTe crystals in large volume and at affordable cost is a key to the further development of radiation-detector applications based on this II-VI compound. The high pressure Bridgman technique that supplies the bulk of semi-insulating CdZnTe crystals used in X-ray, γ-ray detector and imaging devices at present is hampered by material issues that limit the yield of large-size and high-quality crystals. These include ingot cracking, formation of pipes, material homogeneity and the reproducibility of the material from growth to growth. The incorporation of macro defects in the material during crystal growth poses both material quality limitations and technological problems for detector fabrication. The effects of macro defects such as Te inclusions and pipes on the charge-transport properties of CdZnTe are discussed in this paper. Growth experiments designed to study the origin and formation of large defects are described. The importance of material-crucible interactions and control of thermodynamic parameters during crystal growth are also addressed. Opportunities for growth improvements and yield increases are identified.

INTRODUCTION

Semi-insulating (SI) CdZnTe grown by the high-pressure Bridgman (HPB) technique has been successfully used in a large variety of room-temperature X-ray and γ-ray radiation detector applications. The applications range from nuclear diagnostics, digital radiography, high-resolution astrophysical X-ray and γ-ray imaging, industrial web gauging and nuclear nonproliferation.[1,2]

These crystals offer a remarkable combination of electrical properties, notably sufficiently high electrical resistivity and acceptable carrier transport properties. In contrast to SI CdTe crystals grown by conventional Bridgman technique[3,4] or the traveling heater method[5], the HPB CdZnTe crystals also show improved long term stability and much better polarization properties.

Indeed, the introduction of Zn into the CdTe matrix strengthens the lattice due to the shorter bond length (0.2643 nm in ZnTe vs. 0.2794 nm in CdTe), less ionicity (0.49 for ZnTe vs. 0.56 for CdTe), and higher binding energy (-4.7 eV for ZnTe vs. -4.3 eV for CdTe). The stronger lattice and the randomness introduced by the Zn atoms slow down the long-range diffusion of native defects and impurities in the ternary compound, that is often associated with the slow deterioration of CdTe radiation detectors over long periods of time. The introduction of Zn also widens the band gap of the material increasing the intrinsic resistivity and shifts the characteristic temperature of polarization phenomena from around room temperature in CdTe to the -50°C to -70 °C temperature range for $Cd_{0.9}Zn_{0.1}Te$.

The typical electrical properties of HPB SI CdZnTe crystals used in radiation detector applications are listed in Table I. These crystals are commonly grown with 10% Zn content. Since the segregation coefficient of Zn is greater than 1, the first to solidify section of the ingot

309

(tip) contains Zn in excess of 10%. The Zn concentration gradually decreases along the growth axis of the ingot and drops well below 10% in the last to solidify section (heel). Because of this long-range Zn composition variation a range of characteristic physical properties exists within the same CdZnTe ingot as indicated in Table I.

Table I. Typical properties of HPB-grown SI CdZnTe crystals.

Property	Nominal	Actual
Zn composition (%)	10	5 –13
Band gap at room temperature (eV)	1.56	1.53 – 1.58
Intrinsic resistivity ($\times 10^{10}$ Ωcm)	3.0	1.7 – 4.0
$\mu\tau_e$ ($\times 10^{-3}$ cm^2/V)	2.5	0.5 – 5.0
$\mu\tau_h$ ($\times 10^{-5}$ cm^2/V)	2.0	0.2 – 5.0

The remarkable electrical properties of HPB SI CdZnTe crystals, however, are at present accompanied with rather poor crystalline perfection and unfavorable mechanical properties. The ingots are prone to macroscopic cracking and are often fragile. The crystals contain significant concentrations of large defects such as hollow pipes, grain boundaries, twins and Te rich inclusions and Te precipitates. In addition to the long-range Zn distribution, short-range Zn concentration variations are also often observed in the material. Most of the observed macro defects have significant influence on the electrical properties of the material and limit the yield of homogeneous detector crystals from the ingots. In addition, material inhomogeneities are potentially detrimental to the performance of monolithic detector arrays in imaging applications, further reducing the yield of useful material from the ingots.

In the following we will discuss the principles of the HPB technique, the structure of as-grown SI CdZnTe ingots, their typical physical properties, and the typical large defects occurring in these ingots. We will also describe crystal-growth experiments performed to suppress the formation of pipes and macroscopic cracking in the material. We will also discuss the origin of the brittleness of SI CdZnTe crystals and conclude with the future improvement potentials of the HPB technique.

THE PRINCIPLE OF THE HPB TECHNIQUE

Fig. 1a shows the schematic diagram of a HPB furnace that is variant of the classical vertical Bridgman furnace housed inside a high-pressure chamber. The technique allows for a wide choice of crucible materials including moderately porous materials such as graphite. Graphite crucibles with a tight lid are often the choice for growing SI CdZnTe with the HPB technique since the porous graphite can be well purified. The porosity allows the evacuation and high-temperature bake-out of the crucible and CdZnTe charge and the reduction of gaseous impurities, notably oxygen and nitrogen in the growth chamber. SI CdZnTe ingots with total impurity concentration better than 10^{16} cm^{-3} can be now grown with this technique. Since the CdZnTe melt consists of volatile components, there is a steady loss of the constituents from the vapor phase above the melt through the porous walls of the crucible during the growth (Fig. 1b). The loss of the constituents is suppressed by the application of an external inert gas pressure, typically 10-150 Atm of argon. Since Cd has the highest vapor pressure among the CdZnTe melt constituents, the vapor phase predominantly consists of Cd atoms. Although, the external Ar pressure greatly reduces evaporation, it does not completely eliminate Cd loss from the crucible. As the growth proceeds the melt is gradually enriched in Te. Due to this effect the composition of the initial charge is typically chosen at the Te rich side of the existence region of the Cd(Zn)Te phase diagram.[6]

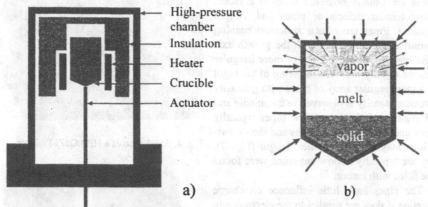

— High-pressure chamber
— Insulation
— Heater
— Crucible
— Actuator

vapor

melt

solid

a)   b)

Fig. 1. Typical high-pressure Bridgman furnace (a), and principle of HPB growth (b).

STRUCTURE OF THE CdZnTe INGOTS

A typical HPB CdZnTe ingot is shown at Fig. 2. The ingots are most often covered with a thin, porous Te-rich layer. Once this layer is removed and the surface of the material is revealed, usually macroscopic cracks are observed in the ingots (Fig. 3). These cracks are typically

Fig. 2 As-grown HPB SI CdZnTe ingot.

Fig. 3. As-grown HPB SI CdZnTe ingot with macroscopic cracking.

initiated perpendicular to the surface and run inward toward the center of the material as shown in Fig. 4. The material shows a characteristic columnar structure where the large grains grow parallel to the growth axis. The average size of monocrystal grains is typically 2-3 cm in the HPB SI CdZnTe ingots. The grains contain varying numbers of twin boundaries. The radiation detector crystals cut from these materials are typically polycrystalline or single crystals with twins (Fig. 5).

Fig. 6 shows an infra-red (IR) micrograph of a HPB SI CdZnTe ingot. Two types of macro defects are usually observed in these crystals: hollow tubular defects or pipes and Te-rich inclusions. Pipes are tubular structures running intermittently and parallel with the growth axis of the ingot. A higher density of more irregular pipes are usually observed in the tip of the ingot and a more regular array of pipes with gradually decreasing density is observed in the middle and heel regions of the ingot. The pipes typically appear and disappear in families and their cross-sections may vary along their length (Fig. 7). They are typically hollow but some were found to be filled with carbon. [7]

Fig. 4. Axial slice of a HPB CdZnTe ingot.

Fig. 5. Typical HPB SI CdZnTe radiation detector crystals.

The pipes have little influence on charge collection if they are parallel to the electrodes in radiation detectors. However, parallel pipes intersecting the surface may render the crystals useless in applications such as monolithic arrays where flat surfaces are a must for electrode patterning. Pipes perpendicular to the electrodes, on the other hand, have serious deteriorating effects on charge collection and detector performance. Perpendicular pipes intersecting the electrodes usually cause significantly increased leakage currents, distorted electric fields and voltage breakdown.

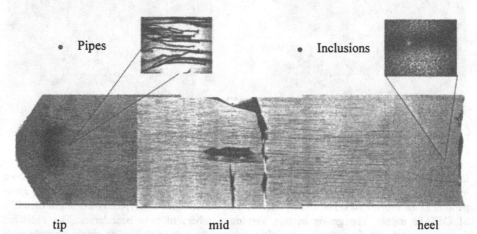

- Pipes

- Inclusions

tip   mid   heel

Fig. 6 IR micrograph showing the typical large defects in HPB CdZnTe ingots.

The formation of pipes parallel to the growth axis prohibits routine slicing of the ingots perpendicular to the growth axis. This poses a significant disadvantage in detector fabrication and inhibits the exploitation of the lower Zn composition variation in radially sliced detectors than in axially sliced detectors. Fig. 9 shows the ^{57}Co γ-ray response of two spectroscopy grade $10 \times 10 \times 2$ mm³ detectors cut radially and axially from the same monocrystal.

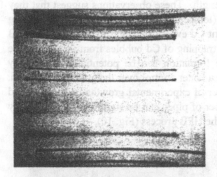

Fig. 7 Family of pipes in HPB SI CdZnTe. Fig. 8 SEM micrograph of pipes in SI CdZnTe.

The radially cut detector shows a 25% higher peak-to-valley ratio for the 122 keV photopeak. This behavior is in agreement with the theoretical prediction that the larger Zn concentration variation in axially sliced detectors leads to a larger band-gap variation and larger fluctuation in the number of electron-hole pairs created by the incident radiation. As a result, broadening of the photopeaks is predicted for detectors with larger Zn composition variation.[8] Further studies will be necessary to explore this effect and measure the magnitude of photopeak broadening due to Zn composition variation.

The origin and the formation mechanism of pipes are not yet resolved in HPB CdZnTe. The predominantly hollow nature of these defects indicates that they are formed by the trapping of gas bubbles at the growth interface. Both impurities and fluctuations in the growth conditions may have a significant role in the nucleation, stabilization and eventual collapse of the bubbles.

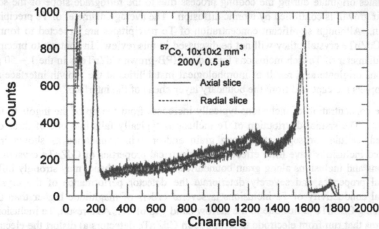

Fig. 9. Comparison of the response of axially and radially sliced detectors to ^{57}Co.

The SEM studies of Heffelfinger et al[7] indicated that some of the pipes are filled with C. It is not clear at this point, however, whether the C filled rod-like defects are of the same origin as the pipes or not. Our SEM studies showed the pipes to be hollow with some Te enrichment in the lattice at the ends of the pipes (Fig. 8). We generally find that the pipes appear and disappear in families and their diameter often varies along their length. These observations suggest that they are related to fluctuations in the growth conditions (temperature, pressure, growth rate) during the solidification process. Considering the significant Cd evaporation and loss during the HPB process, we suggest that the pipes are formed by the trapping of Cd bubbles from the melt at the growth interface. Based on this assumption, pipe formation can be potentially reduced by lowering the temperature gradient in the melt and by a better control over the growth conditions. To test this hypothesis, we have performed a number of experimental growths in an improved research furnace. The results indicate that the number of pipes can be significantly reduced by improving the control over the growth conditions in the HPB process (Fig. 10).

Fig. 10 IR micrograph of a 38 mm diameter experimental CdZnTe ingot showing a significant reduction in the number of pipes.

Te rich inclusions are the other macro defects usually abundant in HPB-grown CdZnTe ingots. We use here the definitions of Rudolph and Mühlberg for inclusions and precipitates.[9] Precipitates originate during the cooling process due to the retrograde slope of the solidus line and their growth is controlled by atomic diffusion. The average diameter of Te precipitates is 10 – 30 nm. Although significant concentration of Te precipitates are expected to form in HPB-grown CdZnTe crystals, they will not be discussed in this review. In contrast to precipitates, the typical diameter of Te-rich inclusions formed in HPB-grown CdZnTe is in the 1 – 50 μm range. Inclusions originate as a result of morphological instabilities at the growth interface as Te-rich melt droplets are captured from the boundary layer ahead of the interface.[9]

The concentration of inclusions typically increases from the tip of the ingots to the heel of the ingot. The spatial distribution of Te inclusions typically falls into one of three categories: dispersed, cellular or segregated along grain and/or twin boundaries as shown in Fig. 11. Dispersed inclusions have little effect on the electrical properties of CdZnTe crystals. Cellular inclusions and inclusions along grain boundaries, on the other hand, may strongly influence the electrical properties and severely deteriorate the detector performance of the crystals. The electrical conductivity of Te inclusions is several orders of magnitude higher than that of the surrounding CdZnTe lattice due to the narrow band gap of Te. As a result, Te inclusions aligned in patterns that run from electrode to electrode in CdZnTe detectors a) distort the electric field in the crystal reducing the active volume of the detector and causing voltage breakdown, b) increase the leakage current and deteriorate the energy resolution of the device. We often observe these effects if the Te inclusions run from electrode to electrode in a cellular pattern or along grain and/or twin boundaries.

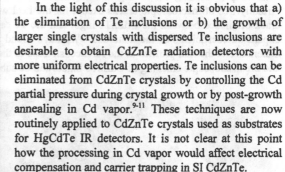

In the light of this discussion it is obvious that a) the elimination of Te inclusions or b) the growth of larger single crystals with dispersed Te inclusions are desirable to obtain CdZnTe radiation detectors with more uniform electrical properties. Te inclusions can be eliminated from CdZnTe crystals by controlling the Cd partial pressure during crystal growth or by post-growth annealing in Cd vapor.[9-11] These techniques are now routinely applied to CdZnTe crystals used as substrates for HgCdTe IR detectors. It is not clear at this point how the processing in Cd vapor would affect electrical compensation and carrier trapping in SI CdZnTe.

The mechanical properties of HPB-grown SI CdZnTe crystals are not ideal. The yield of high-quality crystals is limited by macroscopic cracking of the ingots (Figs. 3 and 4). In addition the crystals are sometimes brittle causing material loss during detector fabrication. The unfavorable mechanical properties are due to the relatively poor mechanical properties of CdTe compared to the properties of other materials used as radiation detectors (Table II). In addition, we usually observe more cracking in the ternary alloy than in the binary counterparts CdTe and ZnTe.

The macroscopic cracking of the CdZnTe ingots is due to excessive thermal and mechanical stresses accumulated in the material during crystal growth. Large temperature differences across the ingot can lead to the build up of thermal stresses. Improper matching of the thermal expansion coefficients of CdZnTe and the crucible materials can lead to thermal shocks during crystal growth. Unfavorable physical and chemical interactions between the CdZnTe melt and the crucible wall can lead to the accumulation of mechanical stresses at the periphery of the ingot.

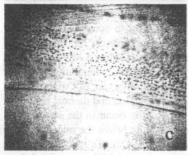

Fig. 11 The spatial distribution of Te inclusions in HPB CdZnTe: a) dispersed, b) cellular, and c) segregated along grain and twin boundaries.

Table II. The mechanical properties of radiation detector materials.

Material	Diamond	Ge	Si	GaAs	CdTe	ZnSe
Young modulus (10^{10} N/m²)	105	10	13.1	8.3	3.7	6.72
Rupture modulus (10^7 N/m²)	25	9.3	3.4	13.8	2.2	5.5
Knoop hardness (kg/mm²)	9000	692	1150	750	45	105-120

We have studied these effects in a series of growth experiments and established some correlations between the various sources of thermal and mechanical stresses and the degree of macroscopic cracking of the ingots. Fig. 12 shows an experimental CdZnTe ingot that is virtually free of macroscopic cracks except for the heel where the excess Te solidified. Here a crack originated between the CdZnTe ingot and the Te cap due to the difference in the thermal expansion coefficients of the materials. We have found that by choosing appropriate crucible materials and thermal profiles in the furnace, stresses in the material can be significantly reduced and the macroscopic cracking of CdZnTe ingots can be substantially suppressed.

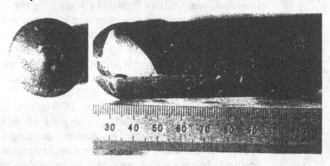

Fig. 12 Nearly crack-free experimental CdZnTe ingot.

SI CdZnTe crystals with good electrical properties and radiation detector performance are often quite brittle, posing significant problems for material handling during detector fabrication and leading to substantial fabrication yield losses. The origin of this brittleness was suggested to be related to the segregation of ZnTe

precipitates in the CdZnTe matrix. ZnTe segregation was suggested to occur as a result of a miscibility gap predicted to exist in CdZnTe below 700 K due to a strong repulsive mixing enthalpy in this compound.[12] ZnTe segregation was found to occur in strained MOCVD grown CdZnTe[12,13] layers, however, the existence of the miscibility gap and ZnTe precipitation is not yet confirmed in bulk CdZnTe.[14] It is to be noted that if ZnTe segregation occurs in CdZnTe due to the miscibility gap it should be strongly dependent on the Zn composition and thermal history of the material. Detectable ZnTe segregation would only be expected to occur in the 20% to 80% Zn composition range after prolonged annealing at temperatures right below the miscibility gap at 700 K. Negligible ZnTe precipitation is expected to occur in HPB CdZnTe ingots commonly containing 10% Zn, considering the relatively short cooling cycle used in the growth of these crystals.

Due to the fundamental and practical importance of this phenomenon, we have performed a systematic TEM and X-ray powder diffraction investigation of as-grown CdZnTe crystals of varying detector performance. We found no evidence for the existence of ZnTe particles in the CdZnTe matrix (Fig. 13). The observed satellite peaks in the electron diffraction patterns cannot be identified as ZnTe precipitates but instead suggest the formation of a superlattice. Independent TEM studies on HPB CdZnTe crystals with 50% Zn content and annealed below the miscibility gap found no evidence of ZnTe precipitation in this material.[15] These results indicate that ZnTe segregation is negligible in HPB CdZnTe and is very unlikely to be the cause of material brittleness. Further studies are necessary to understand the origin of the relative fragility of SI CdZnTe crystals and to improve the mechanical properties of these materials.

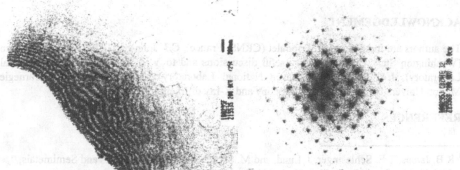

Fig. 13 TEM micrograph and electron diffraction pattern of an as-grown HPB CdZnTe crystal. The satellite peaks were associated with a superlattice.

YIELD OF HPB CdZnTe DETECTORS

Macroscopic cracking, brittleness, and macro defects such as pipes and grain and/or twin boundaries decorated with Te inclusions presently amounts to a majority of the material losses occurring during fabrication of radiation detector crystals and limit the yield of useful material to about 25% of the ingots. Improvements in furnace design, process control and crucible materials could significantly reduce the material losses due to these effects. We estimate that a useful detector yield of 45% should be achievable with the HPB growth process.

CONCLUSIONS

Macroscopic cracking and large defects such as pipes and Te inclusions affecting the electrical properties have a major role in limiting the yield of radiation detector crystals from HPB-grown SI CdZnTe materials. Growth experiments have shown that both macroscopic cracking and the formation of pipes can be significantly suppressed by improved growth techniques. The results suggest that the pipes are formed by the trapping of Cd bubbles at the growth interface from the melt, and their formation is related to fluctuations in the growth conditions. Macroscopic cracking is related to the thermal and mechanical stresses occurring during the growth and can be significantly reduced by appropriate furnace design and crucible materials. Te inclusions are formed in significant concentrations during HPB-growth of CdZnTe from a Te rich melt. Te inclusions dispersed inside the grains have no significant degradation effect on the electrical properties. Inclusions segregated along extended defects such as grain boundaries and twin boundaries on the other hand significantly degrade the electrical properties and detector performance of these crystals. It is not yet understood how the elimination of Te inclusions by Cd vapor growth or annealing would affect electrical compensation and carrier transport properties of HPB-grown SI CdZnTe. Growth of larger single crystals with dispersed Te inclusions is a desirable pathway for improving the homogeneity and useful yield of HPB CdZnTe material for radiation detector applications.

ACKNOWLEDGEMENT

The authors are indebted to R. Triboulet (CRNS, France), C.J. Johnson (II-VI, Inc.), K.G. Lynn (Washington State University) for useful discussions and to Y.M. Zhu (Brookhaven National Laboratory), J.A. Hriljac (Brookhaven National Laboratory) and V. Balakrishna (Carnegie Mellon University) for electron microscopy and X-ray diffraction results.

REFERENCES

[1] R.B. James, T.E. Schlesinger, J. Lund, and M. Schieber in Semiconductors and Semimetals, Vol.43, (Academic Press, New York, 1997), p. 335.

[2] K.B. Parnham, Nucl. Instrum. Meth. **A377**, 487 (1996).

[3] K. Zanio in Semiconductors and Semimetals, Vol. 13, ed. by R.K. Willardson and A.C. Beer (Academic Press, New York, 1978), p. 103.

[4] P. Sieffert, J. Berger, C. Scharager, A. Cornet, R. Stuck, R.O. Bell, H.B. Serreze, and F. V. Wald, IEEE Trans. Nucl. Science **NS-23**, 159 (1976).

[5] R. Triboulet, Y. Marfaring, A. Cornet and P. Siffert, J. Appl. Phys. **45**, 2759 (1974).

[6] J.H. Greenberg, V.N. Guskov, V.B. Lazarev, and O.V. Shebershneva, J. Solid State Chem. **102**, 382 (1993).

[7] J.R. Heffelfinger, D.L. Medlin, H. Yoon, H. Hermon and R.B. James, SPIE Proceedings Series, Vol. 3115, 40 (1997).

[8] J.M. Toney, talk at the 5th Symposium on Room-Temperature Semiconductor X-ray, Gamma-ray and Neutron Detectors, Sandia, March 1997.

[9] P. Rudolph and M. Mühlberg, Mater. Sci. Eng. **B16**, 8 (1993).

[10] H.R. Vydyanath, J. Ellsworth, J.J. Kennedy, B. Dean, C.J. Johnson, G.T. Neugebauer, J. Sepich, P.-K. Liao, J. Vac. Sci. Technol. **B10**, 1476 (1992).

[11] S. Sen, C.S. Liang, D.R. Rhiger, J.E. Stennard, H.F. Arlinghaus, J. Electron. Mater. **25**, 1188 (1996).

[12] A. Marbeuf, R. Druilhe, R. Triboulet, and G. Patriarche, J. Crystal Growth, **117**, 10 (1992).

[13] M.-O. Ruault, O. Kaitasov, R. Triboulet, J. Crestou, and M. Gasgnier, J. Crystal Growth, **143**, 40 (1994).

[14] P. Fougeres, M. Hage-Ali, J.M. Koebel, P. Sieffert, S. Hassan, A. Lusson, R. Triboulet, G. Marrakchi, A. Zerrai, K. Cherkaoui, R. Adhiri, G. Bremond, O. Kaitasov, M.O. Ruault, and J. Crestou, 6th Int. Conf. on II-VI Compounds, Grenoble, France 1997.

[15] F.P. Doty, Digirad, Inc., private communication.

Improved CdZnTe detectors grown by vertical Bridgman process

K.G. Lynn [*], M. Weber [*], H.L. Glass [**], J.P. Flint [**], Cs. Szeles [***]
[*]Dept. of Physics, Washington State University, Pullman, WA 99164-2814
[**]Johnson Matthey Electronics, Spokane, WA 99216
[***]eV Products division of II-VI, Inc., Saxonburg, PA 16056

ABSTRACT

The γ ray ([57]Co) and α particle ([241]Am) detector response of $Cd_{1-x}Zn_xTe$ crystals grown by vertical Bridgman technique was studied under both positive and negative bias conditions. Post-growth processing was utilized to produce a high-resistivity material with improved charge-collection properties. Samples of various Zn concentrations were investigated by I-V measurements and thermally stimulated spectroscopies to determine the ionization energies of deep levels in the band gap. When the post-processing conditions were optimized the low-energy tailing of the γ-ray photopeaks was significantly reduced and an energy resolution of under 5% was achieved for the 122 keV γ-photon line in crystals with $x=0.2$ Zn content at room temperature. A peak to background ratio of 14:1 for the 122 keV photopeak from [57]Co was observed on the best sample, using a standard planar detection geometry. The low-energy 14.4 keV X-ray line could also be observed and distinguished from the noise.

INTRODUCTION

Semi-insulating (SI) CdZnTe crystals have shown great potential for room-temperature radiation-detector and imaging applications [1,2]. The combination of the high efficiency and good energy resolution of CdZnTe detectors make these compounds attractive in a number of applications such as simple, low-cost, efficient, rugged, portable industrial sensors, high-sensitivity radiation detectors for nuclear safeguards and imaging devices for X-ray digital radiography and γ diagnostics. So far CdZnTe crystals grown by the high-pressure Bridgman (HPB) technique containing typically 10% Zn have shown the best detector properties and long-term stability [3,4]. The relatively poor hole transport properties of the HPB-grown CdZnTe crystals, however, result in a low-energy tailing of the photopeaks and lower energy resolution than desired. To compensate for the poor hole collection and achieve a better energy resolution single-carrier devices [5] and pulse-processing techniques [6] were developed. It would be however a tremendous advantage if the hole transport properties of SI CdZnTe could be improved and the low-energy tailing eliminated in CdZnTe radiation detectors.

In this report we discuss the radiation detector performance of $Cd_{1-x}Zn_xTe$ (x=0.2, 0.1, and 0.05) crystals grown by the classical vertical Bridgman technique at Johnson Matthey Electronics (JME) and subsequently processed to obtain SI crystals with good carrier transport properties. The response of the detectors to radiation from [57]Co (14.4, 122, and 136 keV γ rays) (shown in figure 1), from [133]Ba as well as α and γ radiation from a [241]Am source (shown in figure 2) under both positive and negative bias conditions was studied. In addition to the detector performance the carrier transport properties and deep electronic levels were also studied using thermoelectric effect spectroscopy (TEES) [7,8], thermally stimulated current (TSC) [9] and I-V measurements. The performances of the detectors were compared to the performance of planar and coplanar grid detectors made from commercial HPB CdZnTe crystals (see figure 1).

The results indicate that a detector performance superior to the HPB single-carrier and pulse-processing devices can be achieved by post-growth processing of $Cd_{1-x}Zn_xTe$ (x=0.2) crystals grown by the classical Bridgman technique. The low-energy hole tailing was significantly reduced for the 122 keV γ photopeak from [57]Co. The results also indicate that only a fraction of

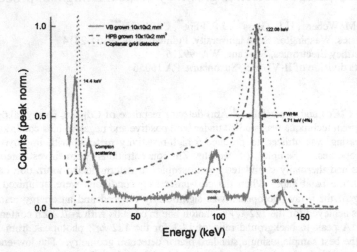

Figure 1: Detector spectra from this study (sample (B), VB grown) compared to a commercially available detector (HPB grown) and to a detector with a coplanar grid anode.

the total volume of the processed crystals shows the improved detector performance and the efficiency of the detectors is less than 15% of the same size HPB crystals. The thermally stimulated spectroscopy measurements show both electron and hole traps in the crystals.

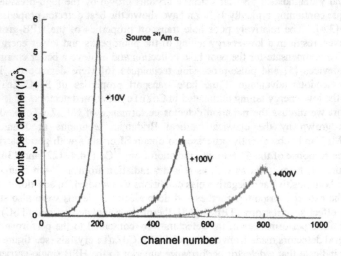

Figure 2: Detector response to α radiation emanating from a ^{241}Am source for different detector biases. (Positive bias corresponds to electron collection). Data were taken with sample (B).

EXPERIMENTAL

The $10\times10\times2$ mm^3 samples were cut form CdZnTe ingots grown by the classical vertical Bridgman technique at JME [10-12]. The samples were polished to mirror finish at the opposing 10×10 mm^2 surfaces. Three samples (B), (C), and (D) from the same bulk $Cd_{1-x}Zn_xTe$ crystal containing $x=0.20$ Zn were subjected to post-growth processing at temperatures ranging from 600 to 850 °C. Sample (A) was not treated. Several other samples with x=0.05 and 0.10 Zn were also investigated. After processing the samples were etched in a bromine solution, cleaned in methanol, and electroless gold contacts were applied to the 10×10 mm^2 surfaces from a gold chloride solution to form planar detectors. The samples were repeatedly etched and mounted with fresh electrodes to study if their detector performance could be improved. Sputtered contacts were also used on a second set of samples.

The samples were mounted in a light-tight fixture with a thin Be window and connected to a charge sensitive preamplifier, research amplifier and a PC-based multi-channel analyzer to study their radiation-detector performance as a function of applied bias and polarity. Reversing the polarity permits the collection of either holes (negative bias) or electrons (positive bias) and study of their transport properties. Most spectra were accumulated at a count rate of 2×10^4 cps.

The same samples were investigated with TEES, TSC, and I-V measurements in a separate apparatus. The I-V measurements were performed at room temperature. For the TEES and TSC experiments the samples were mounted in light-tight closed-cycle cryostat and cooled down to about 10 K in vacuum. Photoexcitation of the samples was performed with 1.33 eV photons from an infra-red light-emitting diode. TEES and TSC experiments were performed in the 10 - 350 K temperature range.

RESULTS

Detector performance

The ^{57}Co spectrum of sample (B) having the best detector performance is shown in figure 1. No rise-time correction or any other signal processing was applied to obtain this spectrum. Figure 1 also shows spectra from a standard planar detector and one with coplanar grid electrodes [13] fabricated from commercially available high-pressure Bridgman CdZnTe material for comparison. The best resolution was obtained for 200 V or greater in the case of electron collection. Clearly resolved are the photopeaks corresponding to the 14.4, 122, and 136 keV γ-radiation of the ^{57}Co source. The peak observable on the high-energy side of the 14.4 keV is associated with a backscatter peak associated with the ^{57}Co source. A rather large escape peak is observed near 100 keV suggesting a small active volume of the detector. The ratio of the 122 keV line to the escape peak diminishes with increasing bias. The maximum bias applied was 400 V in either direction. The area under the escape peak is approximately a factor of 8 greater than 3% prediction of a theoretical simulation. This is consistent with the lower detection efficiency compared to HPB grown detectors. No low energy tailing, a common feature of HPB-grown CdZnTe, was observed in the spectrum. The FWHM of the 122 keV line was less than 4% and 6% for the negative bias. Generally the detector performed well for either polarity but slightly better with positive bias. The response of the detector to α radiation was tested using a ^{241}Am source. The results for various bias voltages are shown in figure 2. Under positive bias conditions a broad distribution is visible and the channel of the peak maximum increases with applied bias.

Fitting the Hecht relationship to the peak to voltage curve for α-particle radiation gives $\mu\tau_e=$ $(1.1\pm0.14)\times10^4$ cm^2/V for electrons which is lower than the best values found for high quality HPB material. Hole transport was observed under negative bias but $\mu\tau_h$ of holes was found to be

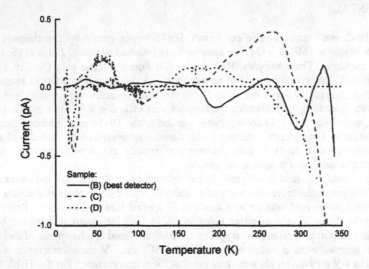

Figure 3: TEES results for the samples (B), (C), and (D). They originate from the same
boule but were post-processed to an increasing extent, respectively. Note that (B) has a
deep electron trap around 290 K which is consistent with the $\mu\tau_e$ results.

within a factor of two of the $\mu\tau_e$ for electrons. The two sides of the sample were different for the
hole $\mu\tau$. It is noted that the $\mu\tau$ measurements with the alpha particles were not performed on the
best detector sample but on samples that had been post processed in the same manner as sample
(B) and were cut from the same ingot which did not have as good as detector response. This ob-

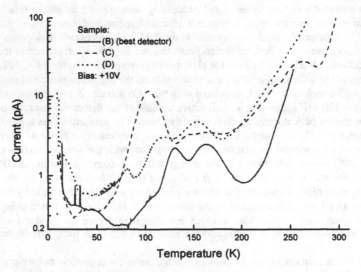

Figure 4: TSC results for the samples as in figure 3.

servation of comparable $\mu\tau$ for holes and electrons is probably the reason for the lack of hole tailing in the sample, as the carrier extraction factors for holes and electrons [14] are more closely matched than is observed in HPB or the THM grown samples. As pointed out by Knoll and McGregor the resolution is best when the collection efficiencies are equal [14]. This has been achieved by increasing $\mu\tau_h$ for holes and decreasing $\mu\tau_e$ for electrons thus making them similar in magnitude. Sample (B) yielded a $\mu\tau_e=(5.5\pm0.14)\times10^{-4}$ cm^2/V when irradiated with alphas and operated under positive bias for electron collection however hole collection (neg. bias) could not be measured on this sample owing to space charge effects. Further, the Hecht relationship fits the data poorly in the 30 to 100 V region suggesting potential problems due to a nonuniform electric field and voltage-dependent depletion.

In sample (B) it was also discovered that the detector continued to operate for extended periods of time after the positive bias was shut down. The α particle measurements also showed polarization during the runs, which limited the acquisition time. This did not occur for the gamma measurements as runs were made for one to two days with no loss of resolution. In some cases a series of runs were accumulated to determine the peak channel. These effects indicate the formation of a space charge region in the crystal by carrier trapping in deep levels near the contact region. Once the external bias voltage is removed the internal field due to the trapped charge is strong enough to sustain carrier drift for a prolonged period of time. As the deep traps gradually empty the detector slowly ceases to operate. Owing the poor fits and polarization effects one should not put to much credence in the accuracy of these $\mu\tau$ values.

None of the other samples cut from the x=0.2 Zn boule and subjected to different post-growth processing performed as well as sample (B). The unprocessed sample (A) did not work as a detector at all while samples (C) and (D) showed poorer detector performance. The samples containing 5% and 10% Zn also performed more poorly as detectors. Most of the samples exhibited different behavior depending on the mounting orientation in the detector fixture suggesting electrode asymmetry or possibly inhomogeneous material. This was confirmed in sample (B) when after repeated etching and electrode deposition the sample was polished again.

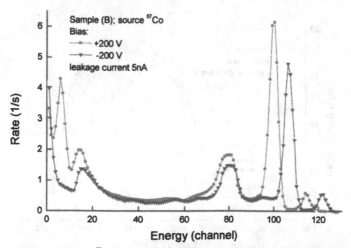

Figure 5: Spectra from a ^{57}Co source with sample (B) as the detector for positive and negative 200 V bias. The data are smoothed to eliminate the statistical fluctuations resulting from 200 s measurement times. It should be noted that the gain of the preamplifier is different for each polarity by approximately the shift in the 122 keV peak channel.

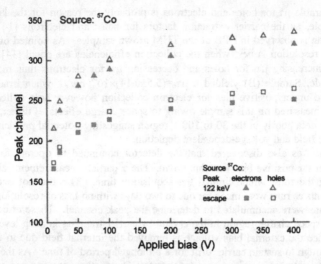

Figure 6: Amplitude of the 122 keV peak and the escape peak signal of sample (B)
(positive for electron collection and negative for hole collection).

I-V, TSC, and TEES measurements

The contacts of CdZnTe radiation detectors are crucial components of the devices. The barriers formed at the electrodes influence charge collection. Ohmic contacts are desirable for high-resistivity materials. Here electroless gold from gold-chloride solution was applied to the polished and etched surface of the CdZnTe crystals. Any remaining solution was immediately rinsed and the electrode dried to prevent diffusion of Cl into the samples.

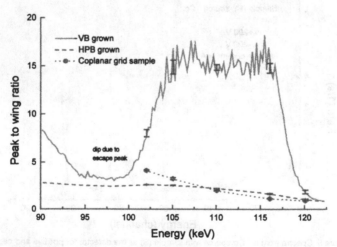

Figure 7: 122 keV peak value to low energy wing ratio as a function of the energy of
the wing value. The vertical Bridgman (VB) grown sample is sample (B).

324

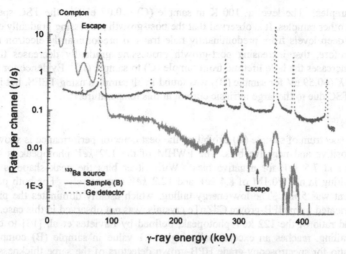

Figure 8: Spectrum of a ^{133}Ba source as measured with sample (B) and a high purity Ge detector. Sample (B) detector has a resolution of about 9.4 keV FWHM at 356 keV.

The I-V curves show small steps of 0.35±0.05 nA in a ±1 V region around 0 V for sample (B), indicating the formation of a small Schottky barrier at the contacts. Sample (B) showed the highest resistivity of (2.15±0.03)×10^{11} Ωcm at large bias voltages. Sample (C) displayed the highest resistivity at low bias but deteriorated at higher voltages. Sample (D) behaved like a diode. The unprocessed sample (A) had a resistivity of (5.15±0.05)×10^{8} Ωcm.

In radiation detectors localized levels capture the electrons or holes created by the incident radiation. The charge from shallow levels may be released during the transit time of the carriers in the devices. Capture of charge carriers causes incomplete charge collection in the detectors and severe low-energy tailing in the measured photopeaks. Thermally stimulated current measurements (TSC) are well suited to investigate carrier traps in SI materials [9]. Unfortunately, the method is insensitive to the polarity of the charge carriers. In addition, it is difficult to use for studying deep traps since at temperatures where carrier emission from deep levels is significant the current from intrinsic conduction dominates. In thermoelectric effect spectroscopy (TEES) one relies on a small temperature gradient across the sample to drive the mobile carriers instead of an external bias [7,8]. Thus electrons and holes drift in the same direction and the conduction type of the material, as well as the nature of the traps, can be determined from the polarity of the measured current. The background current is in general far lower in TEES experiments than in TSC measurements permitting the detection of deep levels close to the middle of the band gap with this technique [15].

The TEES and TSC measurements performed on samples (B) – (D) are shown in figures 3 and 4, respectively. The negative current above room temperature in figure 3 indicates that all of the post-growth processed samples are n-type. The increasing TSC current measured at 300 K from sample (C) to sample (D) indicates that the post-growth processing raises the Fermi level and decreases the resistivity of the CdZnTe samples. Electron and a significantly lower hole trap concentration were found in all samples compared to HPB samples. The results in figures 3 and 4 indicate that the distribution and concentration of traps observed in the 50 to 330 K temperature range is sensitive to the post-growth processing applied to the samples. The deep traps, with thermal ionization energies in the 0.35 – 0.75 eV range, observed in the 150 – 330 K temperature range play an important role in carrier transport and trapping in the detectors made from these samples. The level at 100 K in sample (C) ($\approx$0.07 eV) in the TSC spectra is not present in the other samples. It is observed that the post-growth processing gradually changes the nature of the deep levels from predominantly hole traps to predominantly electron traps. It is expected therefore, that increasing post-growth processing gradually decreases the electron lifetime and increases the hole lifetime from sample (C) to sample (D). Evidence for an electron trap near 290 K ($\approx$0.59 eV for sample (B)) was found in all samples using TEES, but it cannot be observed by TSC due to the large intrinsic current at this temperature.

DISCUSSION

The [57]Co spectrum of sample (B) exhibiting the best detector performance is shown in figure 5 for both positive and negative bias. The FWHM of the 122 keV photopeak is 4.7 keV for positive bias and 7.5 keV for negative bias. With either bias the peak shape is very much gaussian resulting in a FW0.1M of 8.4 keV and 12.2 keV respectively. For both polarities the leakage current was 5 nA. The low-energy tailing, which usually dominates the photopeak in detectors fabricated from HPB-grown CdZnTe crystals, was not observed in this case. The peak-to-background ratio of the 122 keV photopeak, defined by Lavietes et al. [13] to quantify the low-energy tailing, reaches an exceptionally high 14:1 value in sample (B) compared to the typical 3:1 ratio for spectroscopy-grade HPB-grown detectors of the same thickness. Figure 7 compares the peak-to-background ratio as a function of the mean energy of the low energy region for the spectra and detectors shown in figure 1. The performance and resolution of the detector depends on the applied bias. In figure 6 the position of the 122 keV photopeak is displayed as a function of the applied positive and negative bias.

A large escape peak was found in all the spectra suggesting that the active volume of the detector is less than the volume of the crystal ($10 \times 10 \times 2$ mm^3). The ratio of the 122 keV photopeak area and the total area of the escape peak and the 122 keV photopeak peak increases with the applied bias for both bias polarities. Near 400 V the ratio saturates, indicating that the active volume of the detector is fully depleted at this bias voltage. The efficiency of the classical vertical Bridgman-grown and processed sample (B) is less than 15% of a same size HPB-grown CdZnTe crystal for [57]Co. Interestingly the efficiency for positive bias is only 30% of the efficiency for negative bias.

The behavior of sample (B) was investigated in the −50 °C to 30 °C temperature range. The leakage current increases roughly exponentially from a 0.66 nA at −50 °C to 14 nA at 30 °C. Interestingly the peak amplitude also increased with decreasing temperature. While the peak channel in the spectrum remained constant above 0 °C, it increased by about 1% at the lowest temperature. These data were collected after the detector was etched several times and contacts were applied again which caused some degradation of the sample occurred. The energy resolution improved from about 7% at 30 °C to 5% at −20 °C and remained nearly constant at lower temperatures, in good correlation with the behavior of the leakage current.

A number of other samples with x = 0.05 and x = 0.10 Zn concentrations and post-growth processed in the same fashion as samples (B) – (D) were also investigated. None of them yielded more than counter grade detectors.

Figure 8 shows the spectrum of sample (B) exposed to radiation from a [133]Ba source compared to the same spectrum recorded with a high-purity Ge detector. All major peaks were clearly resolved below 400 keV energy. This result again indicates the superior peak-to-background ratio and energy resolution of sample (B).

The response of the detector is sensitive to the repeated etching-contacting of the crystal. It was observed that as the thickness of the detector decreases with repeated etching the negative bias performance of the detector gradually deteriorated. When the crystal was thinned to less than 1 mm the [57]Co spectrum could not be obtained with negative bias condition. After repolishing we found that the detector was still functional and improved from the repeated etching and recontacting. These results suggest that in post-growth processed SI CdZnTe detectors only a section of the whole crystal possesses the superior charge collection properties and detector performance observed in figures 1 and 5. It is to be noted that at present only one CdZnTe ingot has shown the superior detector properties discussed here. Further post-growth processing studies of vertical Bridgman grown CdZnTe ingots are underway to determine if the efficiency of the detectors can be improved.

CONCLUSIONS

We have shown the potential of CdZnTe crystals grown by the vertical Bridgman technique as radiation detectors for low-energy gamma, X-ray, and α radiation. The quality of the detectors depends on the post-growth processing conditions. The processing changes the nature of the dominant deep levels from hole traps to electron traps as observed by TEES experiments. The sample with the highest resistivity and low hole trap concentration showed the best detector performance. An exceptionally high peak-to-background ratio, 14:1, was found for the 122 keV photopeak for ^{57}Co γ radiation. The same high peak-to-background ratio was found for both positive and negative bias conditions in sharp contrast to HPB-grown CdZnTe radiation detectors. From the intensity of the large escape peak and the low detector efficiency compared to HPB-grown CdZnTe crystals it was estimated that the active volume of the detector is less than 15 % of same size HPB CdZnTe detectors. The observed exceptional detector performance of post-growth processed vertical Bridgman grown $Cd_{1-0.2}Zn_{0.2}Te$ crystals is very encouraging for high-resolution radiation detector applications. This is understood in closer matching of the transport properties of the holes and electrons by appropriate post processing. More studies are, however, needed to fully understand the structure of these detector crystals especially the efficiency and the reproducibility of the post-growth processing technique. Further correlation of TEES with GDMS will also be undertaken to understand the impurity associated with deep levels.

ACKNOWLEDGEMENTS

The authors would like to thank E. Eissler for valuable insights and Y. Dardenne for his theoretical simulations of CdZnTe detectors and Richard Olsen for some useful references.

REFERENCES

[1] CdTe and Related Cd Rich Alloys, edited by R. Triboulet, W.R. Wilcox, and O. Oda (North-Holland, Amsterdam ,1993).

[2] Semiconductors for Room Temperature Nuclear Detector Applications, edited by T.E. Schlesinger and R.B. James (Academic Press, San Diego, 1995).

[3] E. Raiskin and J.F. Butler, IEEE Trans. Nucl. Sci. NS-35, 81-84 (1988).

[4] F.P. Doty, J.F. Butler, J.F. Schetzina, and K.A. Bowers, J. Vac. Sci. Technol. B10, 1418-1422 (1992).

[5] P.N. Luke, Appl. Phys. Lett. 65, 2884-2886 (1994).

[6] Y. Eisen and Y. Horowitz, Nucl. Instr. Meth. A353, 60-66 (1994).

[7] B. Santic and U.V. Desnica, Appl. Phys. Lett., 56, 2636-2638 (1990).

[8] Z.C. Huang, K. Xie, and C.R. Wie, Rev. Sci. Instrum. 62, 1951-1954 (1991).

[9] R.H. Bube, Photoconductivity of Solids, (Wiley, New York, 1960), p. 292.

[10] H.L. Glass, A.J. Socha, D.W. Bakken, V.M. Speziale, and J.P. Flint, Fall 1997 MRS meeting, elsewhere in these proceedings (1997) to be published.

[11] P. Rudolph, U. Rinas, and K. Jacobs, J. Crystal Growth, **138**, 249-254 (1994).

[12] P. Rudolph, S. Kawasaki, S. Yamashita, S. Yamamoto, Y. Usuki, Y. Konagaya, S. Matada, and T. Fukuda, J. Crystal Growth, **161**, 28-33 (1996).

[13] A.D. Lavietes, J.H. McQuaid, W.D. Ruhter, T.J. Paulus, LLNL Preprint (1994).

[14] G.F. Knoll, and D.S. McGregor, Mat. Res. Symp. Proc. **302**, 17 (1993).

[15] Cs. Szeles, Y.Y. Shan, K.G. Lynn, and A.R. Moodenbaugh, Phys. Rev. B **55**, 6945-6949 (1997).

The Reduction of The Defect Density in CdTe Buffer Layers for The Growth of HgCdTe Infrared Photodiodes on Si (211) Substrates

H.-Y.Wei, L. Salamanca-Riba, Department of Materials and Nuclear Engineering, University of Maryland, College Park , MD; N. K. Dhar, U.S. Army Research Laboratory, Fort Belvior, VA

CdTe epilayers were grown by molecular beam epitaxy on As-passivated nominal (211) Si substrates using thin interfacial ZnTe layers. By using thin recrystallized (initially amorphous) ZnTe buffer layers, we utilized migration enhanced epitaxy (MEE) in the ZnTe layer and overcome the tendency toward three dimensional nucleation. The threading dislocation densities in 8-9 μm thick CdTe films deposited on the recrystallized amorphous ZnTe films were in the range of 2 to 5 x10^5 cm^{-2}. In addition to the reduction of threading dislocation density, the interface between the ZnTe layers and the Si substrate is much smoother and the microtwin density is an order of magnitude lower than in regular MEE growth. In order to understand the initial nucleation mechanism of the ZnTe on the As precursor Si surface, we also grew ZnTe epilayers on Te precursor treated Si substrates. The growth mode, microtwin density, and threading dislocation density are compared for films grown on Si substrates with different surface precursors and grown by different growth methods.

I. INTRODUCTION

CdTe has a wide range of applications including x-ray and neutron detectors, Gamma ray detector, and optical modulators and high efficiency solar cells. In addition, the study of CdTe/Si heteroepitaxial system is motivated by the need for alternative substrates with high crystalline quality to produce very large HgCdTe detector arrays. Currently, the IR detector array is fabricated using the HgCdTe/CdZnTe structure and the signal readout integrated circuit is fabricated on Si. Bulk CdZnTe is brittle, expensive, and not available in large-area wafer size. In addition, the growth of CdZnTe on Si has serious reliability problems because the thermal expansion coefficients between Si and CdZnTe are very different. Since 1989, ZnTe buffer layers have successfully been used to preserve the single-orientation of CdTe(111)B when grown on Si(001) substrates by molecular beam epitaxy[1].

In order to obtain device quality CdTe layers on Si wafers, there are several issues that have to be addressed. The lattice mismatches between CdTe and Si and ZnTe is 19.3% , and 12%,respectively. These large lattice mismatches usually will induce three dimensional island nucleation and large defect density. Moreover, the subsequent HgCdTe laye inherits the defects and rough faceted surface morphology of ZnTe layer. The defects in turn reduce the minority carrier lifetime. If CdTe/Si is to replace the bulk CdZnTe substrates, the defect density in the CdTe layers must be reduced to 10^5 cm^{-2} range. Besides, the CdTe surface crystallographic polarity must be Te terminated (B-type). To overcome this problem, we have previously introduced the concept of crystallized amorphous deposits for relaxed epitaxy (CADRE)[2]. During this growth process, the amorphous ZnTe template crystallizes in the 2-D growth mode on the As-treated Si substrates. The subsequent CdTe layers deposited on these films were very smooth and mirror like. In contrast, when the ZnTe template was prepared by MEE growth procedure on As-passivated Si substrate, it exhibited a Stranski-Krastanow mode (SK-mode). In this paper, we present a detailed study of the structure of the CdTe/ZnTe/Si (211) films grown by the CADRE technique compared to that of similar films grown by the standard MEE technique.

II. EXPERIMENTAL DETAILS

CdTe films of ~9μm were deposited on ZnTe/Si (211) substrates using molecular beam epitaxy (MBE). The thin interfacial ZnTe layers were deposited using two different techniques. With one technique the ZnTe layer was deposited by migration enhanced epitaxy(MEE), and the other technique was CADRE[2]. In brief, hydrogen terminated Si substrates were heated to 550°C under an As_4 flux.[3,4] Reflection high energy electron diffraction (RHEED) exhibited a bulk-like (1x1) As-passivated RHEED patterns. To study the ZnTe nucleation, Zn or Te_2 were also ad-sorbed.on clean or As -passivated Si surfaces. It was found that Zn atoms did not bind to either atomically clean or arsenic treated Si surfaces. Te_2 not only adsorbed readily on the atomically clean Si surface but it also adsorbed on the As-treated surfaces. The Te coverage was less in the latter case as monitored by Auger Electron Spectroscopy(AES). For the initial nucleation studies, Te- and As- terminated Si (211) surfaces were used.[5]

During MEE deposition, the Si substrate temperature was held at 250°C, while 400 ~ 500Å of a ZnTe layer was deposited. In the CADRE case, first a thin (50~80Å) amorphous ZnTe film was deposited at a temperature of 40°C. The amorphous ZnTe template was then annealed at ~ 390°C under a stabilizing Zn flux until recrystallized. The Si substrates were then cooled to 250°C, and 400 ~ 500Å of ZnTe was further deposited by the standard MEE method. In all cases, the CdTe overgrowth layers were deposited at 290°C from a CdTe compound source over the ZnTe layer. The grown structures were then characterized by double crystal x-ray rocking curve full width at falf-maxium (DCRC FWHM), x-ray θ-2θ diffraction scans, chemical etch pit density (EPD) and transmission electron microscopy (TEM). The cross-sectional TEM samples were prepared by mechanical thinning with a tripod polisher. After polishing down to ~ 10μm thick, the specimens were ion-milled using 3KV Ar^+ ion cooled with a liquid-N holder. The observa-tions were made with a JEOL 4000-FX high resolution electron microscope operated at 300 KV.

III. RESULTS and DISCUSSION

Table 1 summarizes the results for samples grown by these two types of growth techniques and different surface precursors. The crystalline quality of the CdTe layers deposited on Te pre-cursor surfaces was significantly inferior to that of the CdTe layers deposited on the As precursor surfaces, regardless of the growth process used. Besides, a set of CdTe layers grown on Te pre-cursor surfaces prepared at different temperatures showed different surface polarity. In contrast, the surface polarity of the CdTe layer on As precursor surface was always B-type.[5]

Figure 1(a) and (b) are the cross-sectional TEM [220] bright field images for CdTe/ZnTe/Si grown on As precursor Si substrates with regular MEE and CADRE techniques, respectively. The CdTe/ZnTe interface in Fig 1(b) is much smoother than that in Fig 1(a) for MEE growth. That is because the CADRE technique can promote 2-D growth for the initial ZnTe nucleation. For the MEE grown sample, the 3-D island growth of the ZnTe produces randomly distributed islands with a relative orientation of 180°. When these islands coalesce, twin boundaries and interfacial defects form. The twin densities in the ZnTe films grown by MEE and CADRE are $2x10^9$ cm^{-2} and $5x10^8$ cm^{-2}, respectively. The threading dislocation density near the CdTe/ZnTe interface is also lower in the CADRE (10^8 cm^{-2}) than in MEE grown sample (10^{10}cm^{-2}). Figure 2 shows the [110] high resolution image from the CdTe/ZnTe/Si on As precursor substrate grown by CADRE technique. In this figure, the ZnTe/Si interface shows a lighter contrast, and very high quality. The (111) lattice planes of ZnTe show a 3° tilt with respect to the (111) lattice planes of Si.

Table I. Film quality and crystal structure analysis for films grown under different growth process.

Characteristics	Te precursor MEE	Te precursor CADRE	As precursor MEE	As precursor CADRE	As precursor CADRE on vicinal Si (211) 5° toward <111>
Growth mode	island mode	poly-crystalline	S-K mode	layer-by-layer mode	layer-by-layer mode
Surface morphology	faceted, rough	—	smooth	mirror-like	mirror-like
Polarity	temperature dependence		B face (Te)	B face (Te)	B face (Te)
Twins (near surface)	high		very low	none	none
Twin density(near ZnTe/Si)	~10^{11} cm^{-2}		~10^{9} cm^{-2}	~10^{8} cm^{-2}	very low
Threading dislocation density(near CdTe/ZnTe)	~10^{12} cm^{-2}		~10^{10} cm^{-2}	~10^{8} cm^{-2}	~10^{8} cm^{-2}
Surface EPD	10^{7} to 10^{8} cm^{-2}		2.5 to 4x10^{6} cm^{-2}	5x10^{5} to 2.2x10^{6} cm^{-2}	2 to 8x10^{5} cm^{-2}
Tilt angle	-5°		+3°	+3°	+1.5°
FWHM(best)	>500 arc sec		120 arc sec	84 arc sec	72 arc sec

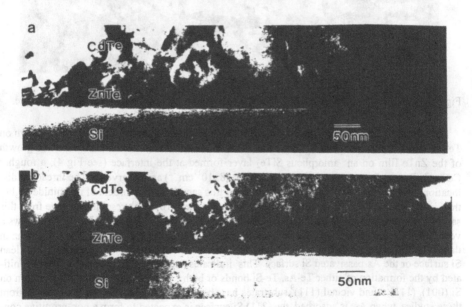

Figure 1 . [220] bright field images of CdTe/ZnTe/Si grown on As precursor. (a) MEE technique, (b)CADRE technique.

Figure 2. [011] high-resolution image of CdTe/ZnTe/Si grown on As precursor with CADRE method. Twin is marked by arrow head, and inclined ~ 20° with respect to interface

Figure 3. [220] bright field image of CdTe/ZnTe/Si grown on Te precursor by MEE technique.

Figure 3 shows the cross-sectional TEM [220] bright field image for CdTe/ZnTe grown on Te precursor Si substrate with MEE growth. Due to the three dimensional nucleation and growth of the ZnTe film on an amorphous $SiTe_2$ layer formed at the interface (see Fig 4), a rougher CdTe/ZnTe interface and a higher twin density ($8 \times 10^{11} cm^{-2}$) are observed. The threading dislocation density near the CdTe/ZnTe interface in this sample is two orders of magnitude higher than in the sample with As precursor. Figure 4, shows the [110] high resolution image from this sample. The amorphous $SiTe_2$ interfacial layer is marked by arrow heads. This figure shows a very high twin formation in the ZnTe layer indicating that the crystalline quality is worse than in the CADRE sample. As mentioned above, Zn atoms do not adsorb on either the atomically clean Si surface or the As-passivated Si surface. This indicates that the ZnTe nucleation must be initiated by the formation of either Te-As, Te-Si bonds or both. The mechanism of As adsorption on Si (001), (111)[7,8] and vicinal (111)[9] substrates has been reported . If we use the concept from these studies, when As is adsorbed, the (211) Si surface is expected to form a non-primitive configuration with the Si atoms on (111) terraces being replaced by As atoms. The Si atoms at the step edges are now having two free bonds per Si atom.

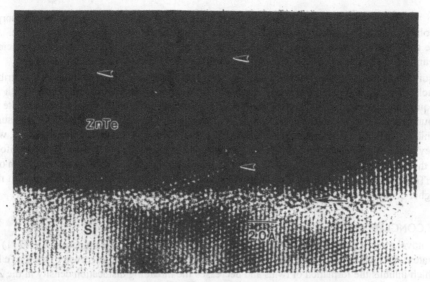

Figure 4. [011] high-resolution image of CdTe/ZnTe/Si grown on Te precursor by MEE technique. Amorphous SiTe₂ layer is marked by arrowhead. Twins in ZnTe/Si interface are also marked.

When Te₂ adsorbs on the As precursor surfaces, the Te sticking coefficient is significantly smaller than for the clean Si surface[5]. Also the lateral mobility of Te can be enhanced. This is because the bond strength of As-Si on (111) Si is 4eV, while that of Te-Si is 3.46eV. Therefore, displacement of As atoms bonded to Si by Te atoms is less favorable at low temperature. The Te-As bond strength is expected to be much smaller than both the As-Si and the Te-Si bond strengths. Therefore, any Te atoms bonded to As on the terraces will readily evaporate during growth or migrate to the step edges.

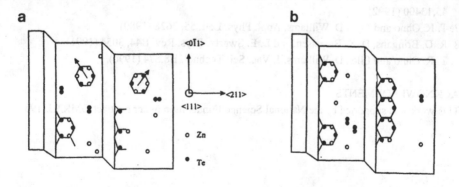

Figure 5. Schematic model for ZnTe nucleation mechanism on different surface precursor.(a) Te precursor. (b) As precursor

Figure 5(a) shows a schematic model of ZnTe nucleation on Te precursor[5]. The adsorption probabilities of Te on clean Si (211) are approximately the same for step and terrace sites. Also, the adsorbed Te atoms have very limited surface mobility because the Te sticking coefficient is near unity. The adsorbed Te on the terraces could possibly have two crystallographically equivalent configurations. When Zn exposure begins, during MEE growth randomly distributed nuclei form which could have a 180° relative rotational relationship between them. With subsequent deposition the nuclei grow into small islands and eventually coalesce and create twin boundaries. For As precursor Si surface, as shown schematically in figure 5(b), this surface inhibits Te bonding on terraces, and enhance the migration of Te atoms to the step edges, where they attach to the two fold coordinated Si. Therefore ,the growth is expected to proceed uniformly in the lateral direction and in a step-flow mode,which promotes a two dimensional growth of the ZnTe layer. This growth mode of the ZnTe layer in turn gives rise to a CdTe film with much higher quality.

IV.CONCLUSION

A novel growth procedure CADRE can provide epitaxial CdTe layers on nominal Si (211) substrates with dislocation of HgCdTe. This growth process promotes 2-D growth of the ZnTe layer which inhibits the formation of twins and decreases the threading dislocation density in the ZnTe layer and the CdTe layer grows over it.

V. REFERENCES

1. R. Sporken, S. Sivananthan, K. K. Mahavadi, G. Monfroy, M. Boukerche, and J. P. Faurie, Appl. Phys. Lett. 55,1879 (1989).
2. N. K. Dhar, C. E.C Wood, A. Gray, H.Y. Wei, L. Salamanca-Riba, and J. H. Dinan, J. Vac. Sci.Technol. B 14(3) (1996).
3. D. B. Fenner, D. K. Biegelsen, and R. D. Bringans, J. Appl. Phys. 66,419 (1989).
4. A. Million, N. K. Dhar, and J.H. Dinan, J. Cryst. Growth 159 (1996).
5. N. K. Dhar Ph. D thesis at Univ. of Maryland (unpublished).
6. R. D. Bringans, D. K. Biegelsen,L. E. Swartz, F. A. Ponce, and J. C. Tramontana, Phys. Rev.B 45,13400 (1992).
7. T. R. Ohno and Ellen D. Williams, Appl. Phys. Lett. 55, 2628 (1989).
8. R. D. Bringans, D. K. Biegelsen, and L. E. Swartz, Phys. Rev. B44, 3054 (1991).
9. T. R. Ohno and Ellen D. Williams, J. Vac. Sci. Technol. B8,874 (1990).

ACKNOWLEDEMENTS

This work was supported by the National Science Funderation under contract DMR 9321957

CONTROL OF DEFECTS AND IMPURITIES IN PRODUCTION OF CdZnTe CRYSTALS BY THE BRIDGMAN METHOD

H. L. GLASS, A. J. SOCHA, D. W. BAKKEN, V. M. SPEZIALE AND J. P. FLINT
Johnson Matthey Electronics, 15128 E. Euclid Ave., Spokane, WA 99216, hglass@eznet.com

ABSTRACT

Cadmium zinc telluride crystals were grown by vertical Bridgman processes using *in situ* compounding from high purity elements into pyrolytic boron nitride crucibles within sealed fused quartz ampoules containing cadmium vapor at a pressure of roughly one atmosphere. These conditions produce material having the low etch pit density, low precipitate density, high infrared transmission and high purity required for use as substrates for infrared focal plane detector arrays fabricated in epitaxial mercury cadmium telluride. Similar processes should be satisfactory for producing cadmium zinc telluride for gamma ray detectors.

INTRODUCTION

$Cd_{(1-x)}Zn_xTe$ (abbreviated CdZnTe) with $x = 0.04$ (approximately) is the most common substrate for epitaxial growth of $Hg_{(1-x)}Cd_xTe$ used to manufacture infrared detector arrays. This application requires large area single crystals (typically 4cm x 6cm or slightly larger) possessing uniformly high crystal quality, high purity and excellent stoichiometry. Considerable progress has been made in meeting these requirements, although further improvements in yield are needed. This paper presents some results and observations relevant to obtaining high purity, low defect CdZnTe in a production environment. The lessons learned in producing substrates for infrared detector applications seem relevant to producing CdZnTe at higher Zn concentrations, x = 0.10 or 0.20, suitable for use in room temperature detectors of gamma rays and X-rays.

EXPERIMENT

The CdZnTe boules were grown by a modified vertical Bridgman [1,2] process with the material in a sealed, evacuated (fused) quartz ampoule. In most cases the melt and crystal were in a pyrolytic boron nitride (pBN) crucible which fit inside the ampoule. Some runs were performed without a crucible but with a pyrolytic carbon coating on the ampoule.

Starting materials were Johnson Matthey XTAL Grade Cd, Zn and Te. GDMS (glow discharge mass spectroscopy) analysis typically shows almost no detectable impurities with the exception of carbon, nitrogen and oxygen, which are not reliably measured by this technique. Metallic impurities of particular interest, Cu, Ni, Na and Li are undetected. For these elements, detection limits are no higher than 2 or 3 ppba (parts per billion, atomic); except the detection limits for Cu in Zn may be greater than 50 ppba since Cu and Zn are adjacent elements in the periodic table. Cd and Zn were cut and chemically etched to remove surface contamination and to trim the weight. Te was broken into large pieces and then cast into a slug under flowing hydrogen.

In most of the results reported here, the boules were 1.3 kg with a diameter of 55-57 mm. These are the smallest of the production boules, most of which are 3.5 kg with diameter 75-80 mm. Two different methods of charge preparation were used. In the *in situ* process, the Te slug, in the pBN crucible when used, was placed in the lower end of the ampoule. The Cd was placed in a quartz reservoir near the top of the ampoule. The Zn was loaded with either the Te or the

Cd. During the initial part of the growth process, reaction of the elements took place as the Cd (and Zn if also in the reservoir) vapor was transported to the molten Te. The temperature was raised to keep the CdZnTe-Te solution molten as the reaction proceeded. After a soak above the CdZnTe melting point, the temperatures were equilibrated at their growth values and crystal growth was carried out at a translation rate of 1.1 mm/h. The second method of charge preparation, *ex situ*, was similar except the vapor transport reaction was performed in a separate, carbon coated ampoule using a horizontal furnace configuration. An appropriate weight of the resulting polycrystalline CdZnTe material was then loaded into a growth ampoule.

Both compounding and growth were performed in the presence of excess Cd to provide a vapor pressure (roughly one atmosphere) sufficient to avoid excessive deviation from stoichiometry. The correct amount of excess Cd had been determined empirically through measurements of infrared transmission and through observation of second phase particles (precipitates) by infrared microscopy. These measurements were made on polished slices cut from the boules. When the Cd vapor pressure is correct, infrared transmission is at its theoretical value, >65%, across the 2.5-16 μm measured spectral range and precipitates, if detectable, are less than 10 μm in diameter and have a density no greater than 5×10^4 cm^{-2}. To avoid the possibility of excessive pressures, the Cd vapor pressure was limited by using a long ampoule with the top of the ampoule at a controlled temperature no greater than 825°C.

As an additional check on the use of Cd vapor pressure, and to resolve questions about the possible relationship between stoichiometry and crystal defects or yield, a few "boulette" experiments were run. The boulettes were grown in smaller, 15 mm diameter, carbon coated quartz ampoules. A cluster of five such ampoules was loaded into the furnace in place of a single 57 mm ampoule. Each of the small ampoules contained 100 g of *ex situ* compounded CdZnTe. Four of the five ampoules within each run contained either a small amount of extra Cd (up to 0.05 g) or Te (up to 0.1 g). The fifth ampoule contained only the CdZnTe. Each cluster of five ampoules was run through the same growth sequence as an *ex situ* 1.3 kg run. Although the temperature distribution would have been different for five boulettes than for a single large boule, all five boulettes within one run should have experienced very similar conditions except for the effects of the different deviations from stoichiometry.

Initial characterization of boules began by slicing off 3 cm from the conical, first-to-freeze (head) end and 1 cm from the last-to-freeze (tail) end. "Head" and "tail" slices were then cut and polished for preliminary evaluation by Fourier transform infrared spectrometry and infrared microscopy. GDMS samples were cut from the middle of each slice. In most cases the remaining center section was oriented and sliced on (111) planes. Representative slices were polished for evaluation including measurement of etch pit density using the Nakagawa etch [3].

Boulettes were subjected to an external visual inspection for grain size and twinning. Subsequently, each boulette was sliced in its entirety using a multi-blade slurry saw. The slices were made perpendicular to the grow direction with no attempt to obtain a particular crystal orientation. There were approximately 60 slices from each boulette and every tenth slice was polished for evaluation.

RESULTS

Figure 1 displays GDMS analyses for Cu in 35 boules. The data are grouped into three sets: *in situ* process in Furnace 2, *ex situ* process in Furnace 2 and *in situ* process in Furnace 3. Furnaces 2 and 3 were quite similar. For each boule the ppba Cu in the tail sample is plotted along with either one or two values for the head sample. If Cu was detected in the head sample, then that value is plotted. If Cu was undetected, zero and the detection limit are plotted. Although no Cu was detected in the starting elements, all boules had detectable concentrations of Cu. Cu levels in the tail samples were always higher than in the head as expected for slow solidification from head to tail.

For the head samples from *in situ* Furnace 2 boules, Cu was undetected in 6 of the 10 samples and the highest detected value was 8 ppba. The tail samples from these boules had Cu at values ranging from 10 to 65 ppba, with a mean of 35.2 ppba.

The *ex situ* Furnace 2 boules generally exhibited higher Cu content in both head and tail, with considerable variability. Only 2 head samples had undetected Cu. Detected values in head samples ranged up to 24 ppba. The mean of all head samples, calculated using the detection limit for the two samples with undetected Cu, was 11 ppba. The tail samples had a mean of 96 ppba, but the variance was considerable.

The *in situ* boules grown in Furnace 3 should have given results similar to those grown in Furnace 2. However, there were substantial differences. Excluding for the moment the last four boules from Furnace 3, Cu in the head samples was slightly higher while the tail samples had significantly higher Cu. Four of the nine head samples had undetected Cu and detected values ranged up to 15 ppba. The mean value for the corresponding tail samples was 93 ppba, about the same as for the *ex situ* boules although the variance is much smaller and the maximum concentration was only 150 ppba.

The observation that Cu content was similar for head samples from Furnaces 2 and 3 but that Furnace 3 showed much higher concentrations of Cu in the tail suggests that Furnace 3 contained a significant source of Cu. This conclusion was supported when the liner tube in Furnace 3 was replaced after the ninth boule, run number 3017. The subsequent four runs showed Cu values similar to those for *in situ* growths in Furnace 2. It should be noted that runs 3019 and 3021 were at a composition of $x = 0.20$ and run 3020 was at $x = 0.10$. Cu was undetected in the head samples of all three higher Zn boules. The detection limits for these three boules, ranging from 7 - 10 ppba, were somewhat higher than for the lower Zn compositions where detection limits ranged from 3 - 8 ppba. This suggests that the Zn is a source of Cu, but probably not the major source.

All of the *in situ* boules were grown in pBN crucibles. Five of the twelve *ex situ* boules were grown in pBN, with the other seven (the first two and the last five in Fig. 1) in carbon coated quartz. From Fig. 1 it does not appear that there was a significant difference in Cu content between the two sub-groups of *ex situ* boules. The higher Cu levels found in *ex situ* boules may have been introduced during the compounding process, which was performed in a different furnace. On the other hand, the great variability in Cu content among the *ex situ* boules suggests a more sporadic source such as exposure to contaminants during transfer of the material to the growth ampoule. Li and Na also were found to occur very sporadically. Ni was not detected in the *in situ* boules grown in Furnace 2 but was found in very small concentrations in about half the tail samples from the *ex situ* boules and the Furnace 3 boules. This may indicate that some furnaces contain a source of Ni.

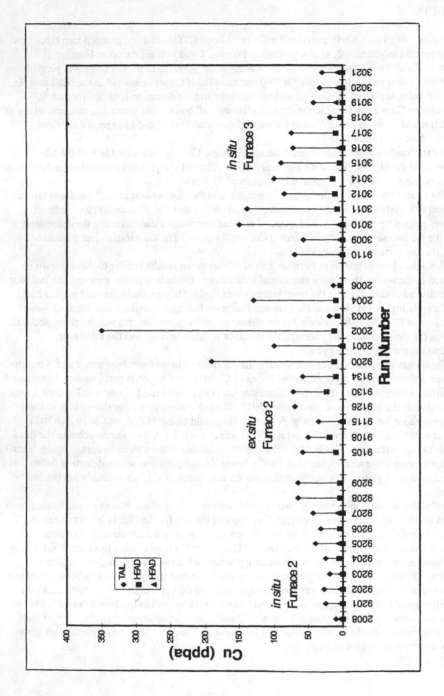

Figure 1. Glow discharge mass spectroscopy analysis of Cu in head and tail samples from CdZnTe boules.

One significant observed difference between growths in pBN and in carbon coated quartz is in dislocation density. Figure 2 shows the minimum and maximum average etch pit densities measured on the evaluation wafers from six boules grown in carbon coated quartz and four boules grown in pBN. Although the growths in quartz were done earlier than those in pBN, the growth processes were similar and the results are consistent with those from many other runs. Growth in carbon coated quartz typically resulted in etch pit densities of 1-2 x 10^5 cm^{-2}, while growth in pBN generally gave values in the low to mid x10^4 cm^{-2} range.

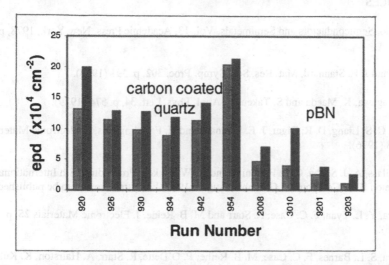

Figure 2. Minimum and maximum Nakagawa etch pit densities on (111) evaluation slices from CdZnTe boules grown in carbon coated quartz and pyrolytic boron nitride.

The main results of the boulette experiments were that a small excess of Cd is beneficial in maintaining stoichiometry. Boulettes which were grown without addition of extra Cd exhibited two characteristics of deviation from stoichiometry [4]: infrared transmission which dropped off at the longer wavelengths and a tendency to have a high density of precipitates [5]. Also, some ampoules which did not contain extra Cd exhibited noticeable sublimation of the CdZnTe to the cooler, top end of the ampoule. No undesirable effects of extra Cd were observed. Specifically, neither the growth of a large single crystal grain nor the occurrence of twins in that grain showed any dependence on the addition of Cd or Te or the absence of an addition.

CONCLUSIONS

Starting with the highest purity, commercially available elements, performing *in situ* compounding into a pBN crucible and adding excess Cd to provide an empirically determined optimum vapor pressure, CdZnTe crystals were obtained which meet the requirements for infrared substrates. Infrared transmission is high, etch pit and precipitate densities are low and purity is sufficiently good for these applications. Substrates grown by methods described here have been used successfully for epitaxy of device grade $Hg_{(1-x)}Cd_xTe$ by metalorganic chemical vapor deposition without the need for any impurity extraction process [6, 7]. These methods also are expected to be applicable to growth of CdZnTe for gamma ray detectors.

ACKNOWLEDGMENTS

C.L. Parfeniuk suggested the boulette experiments and P. Bodie performed polishing and detailed characterization of the boulette slices. Their contributions are greatly appreciated. The work reported here was supported in part by the Defense Advanced Research Projects Agency under Contract MDA972-91-C-0046.

REFERENCES

1. K.'Zanio, Semiconductors and Semimetals, Vol. 13, Academic Press, New York, 1978, pp. 11-14.

2. S. Sen and J. E. Stannard, Mat. Res. Soc. Symp. Proc. **302**, p. 391 (1993).

3. K. Nakagawa, K. Maeda and S. Takeuchi, Appl. Phys. Lett. **34**, p. 574 (1979).

4. S. Sen, C. S. Liang, D. R. Riger, J. E. Stannard and H. F. Arlinghaus, J. Electronic Materials **25**, p. 1188 (1996).

5. H. L. Glass, A. J. Socha, C. L. Parfeniuk and D. W Bakken, Proceedings 8th International Conference on II-VI Compounds, Grenoble, August 1997, J. Crystal Growth, to be published.

6. P. Mitra, Y. L. Tyan, F. C. Case, R. Starr and M. B. Reine, J. Electronic Materials **25**, p. 1328 (1996).

7. P. Mitra, S. L. Barnes, F. C. Case, M. B. Reine, P. O'Dette, R. Starr, A. Hairston, K. Kuhler, M. H. Weiler and B. L. Musicant, J. Electronic Materials **26**, p. 482 (1997).

DOPANTS IN HgCdTe

M.A. BERDING, A. SHER
Applied Physical Sciences Laboratory, SRI International, 333 Ravenswood Ave., Menlo Park, CA 94025, marcy@plato.sri.com

ABSTRACT

In this paper we discuss our *ab initio* calculations of native point defect and impurity densities in HgCdTe. Our calculations have explained the experimental finding in general, and in particular have explained the in-active incorporation of the group VII elements under mercury-deficient conditions; have shown that the group I elements have a large fraction of interstitial incorporation, thereby explaining their fast diffusion; and have described a microscopic mechanism for the amphoteric behavior of the group V elements. We discuss the trends found among the compounds in terms of the underlying bond strengths to understand why the various elements behave the way they do.

INTRODUCTION

$Hg_{1-x}Cd_xTe$ is a continuously soluble, pseudobinary alloy, with a band gap ranging from -0.3 eV for HgTe ($x=0$) to 1.6 eV for CdTe ($x=1$) [1]. The primary application of HgCdTe is for detectors in the long-wave infrared region of the spectrum. Unfortunately, the HgTe bond is weak, and consequently native point defect densities are high, with the primary defect being the cation vacancy [2].

A typical device structure in use today is a p-on-n double layer heterojunction, in which the infrared radiation incident through the substrate is absorbed in a lightly doped n-type base layer with $x=0.22$ and is confined in the junction region by a more heavily doped p-type cap layer with $x=0.3$. The material is typically grown on lattice-matched $Cd_{0.96}Zn_{0.04}Te$ substrates by liquid phase epitaxy (LPE) [3,4], molecular beam epitaxy (MBE) [5], or metal-organic chemical vapor deposition [6]. Although the n-type doping of the base layer with indium is pretty well controlled for both LPE [7,8] and MBE [9] growth, gas phase reactions of the tellurium and indium precursors has led to a renewed interest in the group VIIA elements for n-type doping [10]. Although cation vacancy densities can be high enough to be useful as the p-type dopant, the relatively high diffusivity of the vacancy and the short minority carrier lifetimes in vacancy-doped material have made vacancy doping undesirable. The group IA and IB elements behave as acceptors, but their use as p-type dopants is considered undesirable for most applications because of their relatively high diffusivity [11], which may lead to unstable doping profiles during subsequent processing. The group VA elements on the other hand are slow diffusers [12], in particular under mercury-rich conditions, but unfortunately are amphoteric [13], necessitating growth under mercury-rich conditions or post-growth activation anneals to produce p-type behavior [14].

APPROACH

The calculations of thermodynamic behavior of impurities and native defects in HgCdTe can be divided into two major parts: the first consists of the statistical theory from which the concentrations of the defects in various positions in the lattice are predicted; the second consists of all of the parameters that enter into the statistical theory.

The statistical theory we use is based on an extended quasichemical formalism which includes an arbitrary cluster size and overlap, and which includes the equilibration of the electronic subsystem simultaneously with the atomic system. In the theory, the real space lattice is divided into clusters. Here we choose clusters consisting of four lattice sites (two cation and two anion) and four tetrahedral interstitial sites. This is the minimum size cluster for calculating the defect complexes considered in this paper. An energy, a set of ionization states, a set of ionization-dependent degeneracies, and a chemical identity is associated with each cluster. The free energy of the system can be expressed in terms of the cluster-specific free energies, the configurational entropy, and the free energy of the electronic excitations. The equilibrium set of clusters is determined by minimizing the free energy, subject to a set of constraint equations. Preliminary details of the theory can be found in ref. [15].

One of the inputs to the statistical theory is the set of neutral cluster energies; these energies are calculated using the full-potential linear muffin-tin orbital method (FP-LMTO) within the local density approximation (LDA) [16]. Gradient corrections [17] to the LDA were added so that the vapor phase of mercury could be used in the calculations to establish the state of the material within the existence region [15]. Defect energies were calculated in the FP-LMTO using 32-lattice-site supercells, and 4-lattice-site cluster energies for the statistical theory were extracted by subtracting off the energy of a 28-lattice-site cell with no defect. Ionization energies were also calculated in the LDA and are cast in terms of one-electron excitations. No negative-U states were found. Vibrational excitations were also calculated, as described in ref. [15]. Electronic excitations are calculated using Fermi-Dirac statistics. The temperature- and x-dependent band gap were taken from experiment [18]. The density of states hole effective mass of 0.43 was used, and the conduction band density of states was fit to the intrinsic carrier concentrations using a linear dispersion for the band shape.

NATIVE POINT DEFECTS

We have previously reported on the properties of the native point defects in $Hg_{0.8}Cd_{0.2}Te$ [15]. More recent calculations have been performed that included larger supercells, calculation of the localized levels within the FP-LMTO, and consideration of various defect complexes. Our recent findings are in general agreement with results reported in ref. [15], and with experiment, and are summarized here.

For equilibration at all temperatures and pressures, we find that the cation vacancy is the dominant defect, in agreement with experiment [2], although our calculations indicate that the vacancy is singly ionized in $x=0.2$ material, rather than doubly ionized as deduced from electrical measurements in ref. [2]. Although it never dominates the carrier concentration, the anion antisite density is significant at low mercury partial pressures. The tellurium antisite is a donor, and the cation vacancy-tellurium antisite pair is a strongly bound complex (~ 1 eV), with a relatively high density at low mercury partial pressures. This complex is important both for the diffusion of the tellurium antisite and as a precursor to precipitation. If the cation vacancies are annihilated in a typical low-temperature mercury-saturated anneal before the tellurium antisites can equilibrate, nonequilibrium densities of the antisites may be frozen into the material, and may account for an uncontrolled recombination center. The defect densities at 500 °C as a function of mercury partial pressure throughout the existence region are shown in Fig. 1. As shown in Fig. 2, we predict that the material is anion-rich throughout the existence region.

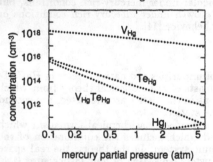

FIGURE 1: Native point defect densities as a function of mercury partial pressure throughout the existence region, at 500 °C.

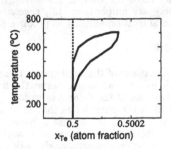

FIGURE 2: The equilibrium deviation from stoichiometry in $Hg_{0.8}Cd_{0.2}Te$.

N-TYPE DOPANTS

Group III Elements

Indium is the element most often used for n-type doping in both LPE [7,8] and MBE [9] growth, and is generally considered to be "well behaved." Our calculations reflect these findings. We predict that the indium incorporates nearly 100% on the cation sublattice where it behaves as a shallow donor. There is a significant binding of an indium substituting on the cation sublattice to a cation vacancy, and these complexes are present in the material, but account for less than 1% of the indium incorporation for all temperatures. The indium interstitial fraction is extremely small, and therefore the vacancies-indium complexes are probably the means by which indium diffuses. In material doped with 10^{15} cm^{-3} indium and subjected to a typical 250 °C, mercury-saturated anneal, the indium-vacancy complex density is only $\sim 10^7$ cm^{-3}, accounting for the observed stability of indium-doped devices.

Group VII Elements

Like indium, iodine has a low interstitial fraction and low incorporation fraction on the "wrong" sublattice, and in this sense iodine is a well-behaved n-type donor. Iodine substituting on the anion sublattice behaves as a donor and binds to the cation acceptor vacancy, resulting in a neutral complex. For high iodine densities and mercury-deficient conditions, this complex accounts for the majority of the iodine incorporation. In Fig. 3 we have plotted the iodine incorporation and native point defects for a fixed iodine concentration of 3×10^{19} cm^{-3}, as a function of the mercury partial pressure at 500 °C. We have also calculated the carrier concentrations at 77 K assuming the high-temperature defect structure is quenched into the crystal, and we compare this to some experimental results on bulk grown and annealed samples [19]. One can see that the theory is in very good agreement with experiment, showing a p-to-n type conversion as the mercury partial pressure is increased, with the p-type behavior at low pressures due to cation vacancies and the n-type behavior at higher mercury partial pressure due to the iodine. The discrepancy with experiment can be accounted for by a small shift in our predicted position in the existence region.

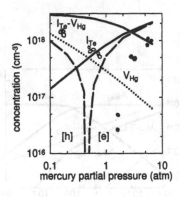

FIGURE 3: Iodine incorporation as a function of mercury partial pressure at 500 °C, compared to experimental results [19] for [h] (open circles) and [e] (closed circles).

P-TYPE DOPANTS

Group Ia and Ib Elements

Lithium and sodium are known fast diffusers [11] and are typically classified as "bothersome" impurities. In Fig. 4 we have plotted the lithium, sodium, and native point defect densities as a function of mercury partial pressure throughout the existence region at 250 °C, a typical temperature used for a mercury-saturated anneal to remove cation vacancies. Both lithium and sodium repel cation acceptor vacancies, and therefore have a negligible pairing with vacancies. The interstitial concentration is quite high under mercury-rich conditions, especially for sodium, and we expect both lithium and sodium to be interstitial diffusers. Furthermore, the interstitial fraction increases as the temperature decreases, and thus we expect these impurities to be very mobile under conditions of low-temperature, cation-rich anneals. Copper, silver, and gold all incorporate primarily on the cation sublattice, show negligible antisite incorporation and pair densities, and have a small interstitial density that increases in going from copper to silver to gold [20].

Group V Elements

We will discuss arsenic as the prototypical group V dopant; qualitatively similar behavior for phosphorus and antimony was found in our previous work [20]. Arsenic has been known to behave amphoterically in HgCdTe, with the desired p-type behavior under mercury-rich conditions, and n-type behavior under tellurium-rich conditions [13]. Our calculations predict this amphoteric behavior, as illustrated in Fig. 5. We find negligible incorporation of arsenic at interstitial sites, but do find incorporation on the cation sublattice that dominates the arsenic incorporation at low mercury partial pressures. The arsenic on the cation sublattice behaves as a donor, thus explaining the observed n-type behavior of arsenic-doping in mercury-deficient materials. Some of the arsenic on the cation sublattice are bound to cation vacancies, creating neutral complexes. These complexes are most likely the means by which arsenic diffuses in this material.

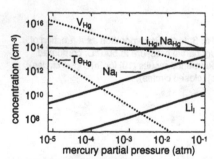

FIGURE 4: The lithium, sodium, and native point defect densities throughout the existence region at 250 °C for fixed impurity concentration of 10^{14} cm^{-3}.

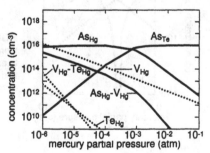

FIGURE 5: The arsenic and native point defect densities throughout the existence region at 220 °C for fixed arsenic concentration of 10^{16} cm^{-3}.

DISCUSSION

Only the group V elements have a significant incorporation on the "wrong," or antisite, sublattice, although the group VI antisite is predicted to be present, but never dominant. The formation of the group V and VI antisites can be attributed to the weakness of bonds that mercury makes with the group V elements and tellurium, and the relative strength of the group V-Te and the Te-Te bonds. Unlike the group V and VI elements, the group VII element iodine does not show significant incorporation on the cation sublattice. As more electrons are added to the valency in going from the V to the VI to the VII elements, the chemical difference with the group II element for which they are substituting become more significant – for example, the iodine has seven valence electrons compared to two for mercury – thus disfavoring incorporation on the cation sublattice in the tetrahedral environment. Furthermore, stable metal-iodine compounds such as HgI_2 are stable, whereas pure tellurium metal and arsenic-telluride compounds like to form, demonstrating the different chemistry of these elements.

We find no significant incorporation of the group I, II, or III elements on the anion sublattice. This can be attributed in part to the relative weakness of the bonds mercury and cadmium make with these elements, compared to those they make with tellurium. For the group I elements, the chemical differences with the tellurium are significant (one valence electron compared with six for tellurium), which also accounts for the lack of substitution of these elements on the anion sublattice. Hg-Hg bonds are very weak, as is evident from the low melting temperature of mercury. The strength of bonds that the group III elements make with tellurium is evidenced by the presence of well-xsybound phases, such as In_2Te_3. No significant incorporation of the group VII elements on the cation sublattice is found, attributable to lack of strong I-Te bonding, and relatively strong Hg-I bonds, as evident from the presence of a HgI_2 phase.

All of the impurities that incorporate into HgCdTe as donors (which includes the group III and VII elements, and tellurium and the group V elements incorporating on the cation sublattice) have a significant binding energy to the cation vacancy (greater than 1 eV). For indium and iodine which are introduced as intentional n-type dopants, this pairing is undesirable, although for the low doping densities used and under mercury-saturated conditions at low temperature the pair densities will be negligible and nearly 100% activation of the n-type dopants will be realized.

The group I elements have been shown to have the highest percent of interstitial substitution. This is in part due to the fact that these elements do not readily make sp^3 hybrids for bonding in the tetrahedral environment and therefore interstitial incorporation can compete effectively. The group III and V elements readily make sp^3 hybrids, so although the bond energies they make at substitutional sties may be small, they can still compete effectively against interstitial incorporation. Based on these arguments, one would expect iodine to have a significant interstitial incorporation, which we do not find in the present calculations.

Although the determination of the dominant diffusion mechanism for the impurities in HgCdTe involves more than an examination of the various defect densities, the defect densities can suggest the mechanism that dominates. With this caveat in mind, we will see how our results give a hint about how the various impurities will diffuse in the material.

We expect the group I elements to be interstitial diffusers. They all have relatively large interstitial fractions, and as acceptors, have very low pair densities with the cation vacancies. For lithium and sodium, we have shown that the interstitial fraction is highest for low-temperature, mercury-saturated conditions [21], accounting to the high mobility of these impurities under typical annealing conditions.

The group III element indium is expected to diffuse via cation vacancies, due to the large fraction of vacancy-indium pairs, and the negligible fraction of indium interstitials. Because the diffusion correlates with the presence of the cation vacancies, indium should be relatively stable after the vacancies are filled in a typical low-temperature, mercury-saturated anneal. The diffusion of group V and VII elements, along with the group VI tellurium self diffusion, are more complicated because of the exceedingly low anion vacancy and anion interstitial densities [15]. The group V and VI elements may diffuse by their antisites via the cation vacancy, which show a significant binding. We have demonstrated a correlation of the pressure dependence of the diffusion with the concentration of the arsenic antisite-cation vacancy pairs [21]. No specific means for iodine diffusion is suggested by our work.

CONCLUSIONS

We have given an overview of the behavior of a wide array of elements in HgCdTe, and have shown that theory is quite capable of predicting their fundamental properties. Accurately identifying the position of the material within the existence region in the theory is essential to correlating the theory with the experimental findings.

ACKNOWLEDGMENTS

This work was supported by the U.S. Air Force Wright Laboratories through a subcontract with Universal Technology Corporation, subcontract no. 97-S402-22-13-C1, and by DARPA through AFOSR contract no. F49620-95-C-0004. The work benefited from discussions with Mark van Schilfgaarde at SRI International.

REFERENCES

1. D. Long and J. L. Schmit, in Semimetals and Semimetals, Vol. 5, Edited by R. K. Willardson and A. C. Beer, Academic Press, NY, 1970, p. 175.
2. H. R. Vydyanath, J. Electrochem. Soc. 128, p. 2609 (1981).
3. B. Pelliciari, Prog. Crystal Growth and Charact. 29, p. 1 (1994).
4. M. H. Kalisher, P. E. Herning, and T. Tung, Prog. Crystal Growth and Charact. 29, p. 41 (1994).
5. See for example, J. M. Arias, Properties of Narrow Gap Cadmium-Based Compounds, EMIS Datareviews Series No. 10, Edited by P. Capper, INSPEC, 1994, p. 30.
6. See for example, A. J. C. Irvine, ibid., p. 24.
7. T. Tung, M. H. Kalisher, A. P. Stevens, and P. E. Herning, Mater. Res. Soc. Symp. 90, p. 321 (1987).
8. L. Colombo, G. H. Westphal, P. K. Liao, M. C. Chen, and H. F. Schaake, Proc. SPIE (USA) 1683, p. 33 (1992).
9. P. S. Wijewarnasuriya, M. D. Lange, S. Sivanathan, and J. P. Faurie, J. Electron. Mater. 24, p. 5 (1995).
10. P. Mitra, T. R. Schimert, F. C. Case, S. L. Barnes, M. B. Reine, R. Starr, M. H. Weiler, and M. Kestigian, J. Electron. Mater. 24, 1077 (1995).
11. P. Capper, in Properties of Narrow Gap Cadmium-Based Compounds, EMIS Datareviews Series No. 10, Edited by P. Capper, INSPEC, 1994, p. 158 and p. 163, and references within.
12. D. Chandra, M. W. Goodwin, M. C. Chen, and J. A. Dodge, J. Electron. Mater. 22, p. 1033 (1993).
13. H. R. Vydyanath, Semicond. Sci. Technol. 5, p. S213 (1990).
14. H. R. Vydyanath, L. S. Lichtmann, S. Sivanathan, P. S. Wijewarnasuriya and J. P. Faurie, J. Electron. Mater. 24, p. 625 (1995).
15. M. A. Berding, M. van Schilfgaarde, and A. Sher, Phys. Rev. B. 50, p. 1590 (1994).
16. O. K. Andersen, O. Jepsen, and D. Glotzel, Highlights of Condensed Matter Theory, Edited by F. Bassani et al. Amsterdam, The Netherlands: North Holland, 1985, p. 59.
17. Langreth and D. Mehl, Phys. Rev. B 28, p. 809 (1983).
18. G. L. Hansen, J. L. Schmit, and T. N. Casselman, J. Appl. Phys. 53, p. 7099 (1982).
19. H. R. Vydyanath and F. A. Kröger, J. Electron. Mater. 11, p. 111 (1982).
20. M. A. Berding, M. van Schilfgaarde, and A. Sher, J. Electron. Mater. 26, p. 625 (1997).
21. M. A. Berding, presented at The 1997 U. S. Workshop on the Physics and Chemistry of II-VI Materials, October 21-23, 1997, Santa Barbara, CA.

STUDY ON DEEP LEVEL TRAPS IN p-HgCdTe WITH DLTFS

S. KAWATA, I. SUGIYAMA, N. KAJIHARA and Y. MIYAMOTO

Fujitsu Laboratories Ltd., 10-1 Morinosato-Wakamiya, Atsugi, 243-0124, JAPAN.

skawata@flab.fujitsu.co.jp

ABSTRACT

We studied the characteristics of deep-level traps in p-type HgCdTe diodes using the Deep Level Transient Fourier Spectroscopy (DLTFS) method. For both holes and electrons, two types of traps were observed. The DLTFS signal intensity of one type of trap increased with the carrier density in the HgCdTe, while the other did not exhibit a monotonic increase. While measuring the stability of these traps during cooling cycles, the DLTFS signal intensity of the first group was almost constant while that of the latter fluctuated with every cooling cycle. Stable traps originated from Hg vacancies, unstable traps are attributed to vacancy-impurity complex defects.

INTRODUCTION

HgCdTe is used for high performance infrared detectors. It is well known that the responsivity and noise of HgCdTe photodiode devices depend on the minority carrier lifetime. The presence of deep-level traps in HgCdTe limits the minority carrier lifetime, because they act as carrier generation-recombination centers. Characterizing the deep-level traps in HgCdTe is important for improving the performance of HgCdTe photodiodes. Deep-level traps in HgCdTe have been reported by elsewhere [1-3], but the data needed to characterize the complete behavior of these traps in photodiodes is insufficient.

Deep Level Transient Spectroscopy (DLTS) [4] is the most popular technique for studying the characteristics of deep-level traps. However it is difficult to use this technique for measuring deep-level traps in HgCdTe, because the depletion layer capacitance of a typical HgCdTe photodiode is invariably very small and the DLTS method measures the differential capacitance of the depletion layer only at the two given times of t_1 and t_2. On the other hand, Deep Level Transient Fourier Spectroscopy (DLTFS) method [5] applies a numerical Fourier transformation to the transient spectra of the capacitance and consequently obtains the time constant and amplitude. This technique uses the entire transient spectra of the capacitance to calculate the Fourier coefficients, enabling a highly sensitive measurement.

In this paper, we study the characteristics of deep-level traps using the DLTFS method in samples having different carrier densities and note their instability through cooling cycles. We then identify the

347

origin of these traps.

EXPERIMENT

The samples used were p-type $Hg_{1-x}Cd_xTe$ (x = 0.29) doped with Hg vacancies grown by liquid phase epitaxy on CdZnTe wafers. We chose samples having carrier densities of 4.5, 5.1, 7.5 and 12 $\times$ 10^{15} cm^{-3} to evaluate the trap characteristics. The n$^+$ region of n$^+$-p diodes was formed by B$^+$ implantation. The measured carrier density of the n$^+$ region was on the order of 10^{18} cm^{-3}. The junction area was 100 $\times$ 100 μm^2.

We used a Bio-Rad DL8000 system for DLTFS measurements. Our measurement configuration is shown schematically in Fig. 1. The base reverse bias voltage to eliminate trapped carriers was 0.6 V. The voltage and width of the carrier induced pulse were 0.2 V and 5 ms for hole trap measurements and -0.2 V and 0.5 ms for electrons. The temperature range was from 100 to 20K. For evaluating trap stability through cooling cycles, we kept the samples at room temperature for an hour before each measurement.

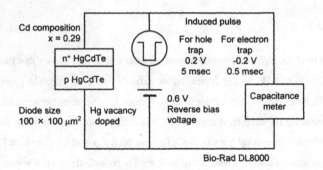

Fig. 1. Schematic diagram of DLTFS measurement system.

RESULTS and DISCUSSION

Trap measurement

We designated one of the DLTFS signals as a b_1 signal. These signals correspond to those signals that are obtained when we change the rate window in the DLTS measurement. The b_1 represents a first order Fourier sine coefficient, which is the largest in the transient spectrum [5]. Figure 2 shows the temperature dependence of the b_1 signal in samples having different carrier densities; Figures 2(a) and 2(b) show such

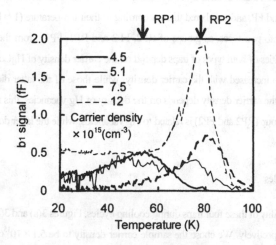

(a) Temperature dependence of b_1 signal from hole traps.

Reverse bias is 0.6 V, pulse voltage and width are 0.2 V and 5 ms, respectively.

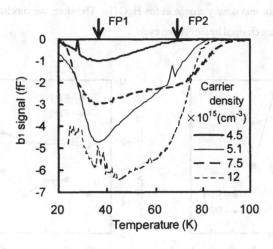

(b) Temperature dependence of b_1 signal from electron traps.

Reverse bias is 0.6 V, pulse voltage and width are -0.2 V and 0.5 ms, respectively.

Fig. 2. The b_1 signals of p-type HgCdTe with the carrier densities of 4.5, 5.1, 7.5 and 12×10^{15} cm^{-3}.

results for the hole and electron traps. In these figures, two traps are observable for both holes and electrons. We called these traps RP and FP, and numbered them according to their temperature (1 = low, 2 = high). We classified these traps into two types, as groups of RP2, FP2 and RP1, FP1. From these figures, we found that the relative densities of both types of traps depend on the carrier density of HgCdTe. The signal intensities of the first group increased with the carrier density, while those of the latter did not exhibit a monotonic increase. Since the carrier density depends on the density of Hg vacancies, this result indicates that the origin of the first group (RP2 and FP2) is related to Hg vacancies, while the latter dose not relate to it directly.

<u>Stability during cooling cycles</u>

Figure 3 shows the stability of these four traps during cooling cycles. Figures 3(a) and 3(b) show that of hole and electron traps, respectively. We chose the sample carrier density to be $5.1 \times 10^{15} cm^{-3}$ to observe the four traps simultaneously. We kept the sample at room temperature for an hour before each measurement. We found that the intensities of RP1 and FP1 fluctuated in every cooling cycle, while those of RP2 and FP2 remained almost constant. Some impurities seemed to readily diffuse to vacancy sites at room temperature to change the vacancy concentration [2]. These results indicate that b_1 signal intensity fluctuations are caused by the trap density change in the HgCdTe. Therefore, we conclude that the trap densities of RP1 and FP1 were changed by the cooling cycle.

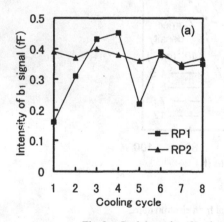 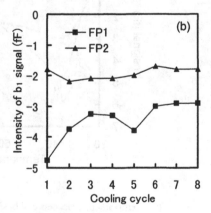

Fig. 3.   Peak intensity of b_1 signal in cooling cycle for

(a) hole traps and (b) electron traps.

The carrier density of the sample was $5.1 \times 10^{15} cm^{-3}$.

Level identification

Figure 4 summarizes the energy levels of the traps that we measured. The band-gap energy (E_g) of the samples was 220 meV at 80K. The energy levels of hole traps RP1 and RP2 are in the range of 32 to 50 meV and 80 to 90 meV above the valence band, respectively. Electron traps FP1 and FP2 are in the range of 18 to 44 meV and 56 to 80 meV below the conduction band, respectively. The energy levels of the stable traps agree with the values of the well-known Hg vacancy (0.4 and 0.7 E_g above the valence band). From these results, we conclude that the origin of these stable traps is the Hg vacancies, while the unstable traps are caused by another defect. Myles *et al.* [6] reported that the origin of the trap (in the range of 30 to 35 meV above the valence band) in $Hg_{0.7}Cd_{0.3}Te$ corresponds to the origin of the trap (in the range of 140 to 150 meV above the valence band) in CdTe, the latter of which is believed to occur due to a vacancy-impurity complex. In our case, the energy level of RP1 is between the deep level of Hg vacancies (0.4 E_g above the valence band) and the shallow acceptor level of Hg vacancies (15 meV above the valence band), and agrees with the value of complex defects level in $Hg_{0.7}Cd_{0.3}Te$. The trap density of RP1 did not exhibit a monotonic increase while increasing the carrier density, and it fluctuated in the cooling cycles. Therefore, we believe that the origin of this trap is due to vacancy-impurity complex defects. In the case of FP1, since the trap density fluctuated during the cooling cycles as did RP1, and the energy level is shallower than the Hg vacancy level (88 meV below the conduction band). We believe that this trap also arises from a vacancy-impurity complex. From these results, we believe that the unstable traps originated from the vacancy-impurity complex defects.

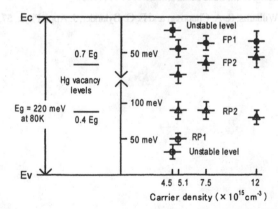

Fig. 4.   Energy levels of measured traps.

SUMMARY

In summary, we studied the characteristics of deep-level traps in p-type $Hg_{1-x}Cd_xTe$ ($x = 0.29$) diodes using the DLTFS method. For both holes and electrons, we observed two types of traps. We found that the relative densities of both types of traps depend on the carrier density of HgCdTe. We also found that one of these traps was unstable during cooling cycles. The energy levels of the stable traps agree with the well-known Hg vacancy levels in HgCdTe, and the densities of these traps increased with the carrier density in HgCdTe. On the other hand, the densities of the unstable traps did not exhibit a monotonic increase with the carrier density, and the energy levels of these traps correspond to vacancy-impurity complex defects. These results indicate that the stable traps originate from Hg vacancies, and that the unstable traps arise from complex defects. The cause of the density fluctuations of the unstable traps during the cooling cycles seems to be the diffusion of the impurity in HgCdTe at room temperature. We believe that these trap instabilities cause the fluctuation of the current-voltage characteristics of HgCdTe diodes.

REFERENCES

1. D. L. Polla and C. E. Jones, J. Appl. Phys., **52**, 5118 (1981).
2. C. E. Jones, K. James, J. Merz, R. Braunstein, M. Burd, M. Eetemadi, S. Hutton and J. Drumheller, J. Vac. Sci. Technol., **A3**, 131 (1985).
3. C. L. Littler, A. J. Syllaios and V. C. Lopes, Prog. Crystal Growth and Charact., **28**, 145 (1994).
4. D. V. Lang, J. Appl. Phys., **45**, 3023 (1974).
5. S. Weiss and R. Kassing, Solid-State Electronics, **31**, 1733 (1988).
6. C. W. Myles, P. F. Williams, R. A. Chapman and E. G. Bylander, J. Appl. Phys., **57**, 5279 (1985).

REACTIVE ION ETCHING (RIE) INDUCED
p- to n-TYPE CONVERSION IN EXTRINSICALLY DOPED
p-TYPE HgCdTe

C.A. Musca, E.P.G. Smith, J. F. Siliquini, J.M. Dell, J. Antoszewski, J. Piotrowski[*]
and L. Faraone
Department of Electrical and Electronic Engineering, The University of Western
Australia, Nedlands, WA, 6907, Australia, charlie@ee.uwa.edu.au
*Vigo System Ltd., HERY 23, 01-494, Warsaw, Poland

ABSTRACT

Mercury annealing of reactive ion etching (RIE) induced p- to n-type conversion in
extrinsically doped p-type epitaxial layers of HgCdTe (x=0.31) has been used to reconvert n-type
conversion sustained during RIE processing. For the RIE processing conditions used (400mT,
CH_4/H_2, 90 W) p- to n-type conversion was observed using laser beam induced current (LBIC)
measurements. After a sealed tube mercury anneal at 200°C for 17 hours, LBIC measurements
clearly indicated no n-type converted region remained. Subsequent Hall measurements confirmed
that the material consisted of a p-type layer, with electrical properties equivalent to that of the
initial as-grown wafer (N_A-N_D=2×10^{16} cm^{-3}, μ=350 cm^2.V^{-1}.s^{-1}).

INTRODUCTION

Reactive Ion Etching

The development of dry etching techniques is of great importance for the further
advancement of HgCdTe infrared device technology. Anisotropic dry etching, as opposed to wet
etching methods, provides opportunity for delineation of a higher density of active device
elements incorporating smaller features. However, HgCdTe has a relatively low damage threshold
resulting from weak Hg-Te bonds in the crystal lattice [1]. Consequently, dry etching techniques
such as ion milling and reactive ion etching (RIE) result in the conversion of p-type HgCdTe to n-
type [2-6]. The physical mechanism used to explain this type conversion in vacancy doped p-type
HgCdTe material is that during the etching process Hg atoms are liberated near the surface with
some proportion of these interstitial atoms diffusing into the underlying bulk material. These
interstitial Hg atoms annihilate acceptor vacancies, resulting in residual uncompensated donors
converting the region to n-type. The rapid diffusion of Hg interstitials in HgCdTe, even at room
temperature, has led to large conversion depths (over 100µm) in some experiments [2,5].
However, it has been suggested that cooling the substrate to 100K during ion milling and RIE can
limit the extent of p- to n-type conversion. Published work has also established that etching
induced type conversion is evident even in extrinsically doped p-type HgCdTe in which the Hg
vacancies have been previously filled using a mercury anneal [3,7]. This suggests that the
conversion mechanism extends beyond Hg interstitial atoms annihilating acceptor vacancies and
may point to an additional doping mechanism. Mercury annealing is a technique that is commonly
used to alter the electrical properties of HgCdTe to suit device requirements [8,9]. In this work

results are presented of a process whereby mercury annealing was used to eliminate the n-type converted regions induced by RIE processing of extrinsically doped p-type HgCdTe.

Laser Beam Induced Current

Laser beam induced current (LBIC) is a high resolution, non-destructive optical characterisation technique that has been used to spatially map and identify electrically active defects and active regions in HgCdTe materials and devices [10]. In the LBIC technique a low power laser is focused on a region of the semiconductor thus generating electron-hole pairs. If the material is uniform, the electron-hole pairs will recombine within a few diffusion lengths and no current will flow between two remote contacts. In the presence of any built-in electric fields, due to localised band-bending within a few diffusion lengths of the laser illuminated region, the

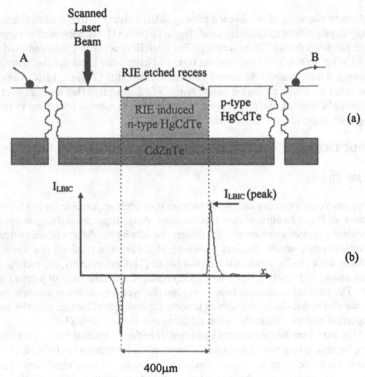

Figure 1 (a) Cross section schematic of LBIC configuration showing position of remote contacts (not to scale) (b) LBIC signal flowing between contacts A and B as a function of laser position for an RIE processed area on extrinsically doped p-type $Hg_{0.69}Cd_{0.31}Te$ measured at 80K.

photogenerated electron-hole pairs separate and induce a lateral photovoltage which causes a current to flow between the two remote contacts. Figure 1 shows a cross-section of the semiconductor under study, along with a typical LBIC current signal. By scanning the laser beam across the sample, a two-dimensional LBIC map can be obtained representing the spatial distribution of localised electrically active regions within the semiconductor. Furthermore, by considering the sign of the LBIC signal the direction of band-bending may be deduced. LBIC may be used both for analysing semiconductor wafers prior to any device processing as a means of detecting any electrically active defects, and for analysing the wafer at intermediate steps during processing, such as after mesa etching, to detect any process induced damage.

EXPERIMENT

The starting wafer used in this study was vapour phase epitaxially (VPE) grown arsenic doped p-type $Hg_{0.69}Cd_{0.31}Te$ on CdZnTe purchased from Vigo System Ltd. The p-type dopant was introduced during growth of the material before a standard low temperature near saturated mercury atmosphere anneal to fill Hg vacancies resulting in a p-type doping density of $N_A - N_D = 2 \times 10^{16}$ cm^{-3}. Note that for vacancy doped p-type HgCdTe, such an anneal would have type-converted the layer to n-type due to the presence of residual donors. Upon the commencement of processing two samples were etched in a 0.1% bromine methanol solution to a thickness of approximately 8μm. A 400μm×400μm square was patterned on each HgCdTe sample using a AZ4262 photoresist mask and low temperature photolithographic processing. Both samples were subjected to RIE processing in a Plasma Technology parallel plate reactor with the samples mounted on a water cooled cathode which was kept at a constant temperature of 16°C. The RIE processing conditions were: hydrogen flow rate=26sccm, hydrogen partial pressure=397mT; methane flow rate=5sccm, methane partial pressure: 20mT; rf power = 90W; dc bias=200V; processing time=30 minutes, resulting in 0.75μm of HgCdTe being etched away. After RIE processing for 30 minutes it is expected that the RIE processed area will be type converted all the way to the substrate [11].

Following RIE processing one sample was subjected to annealing in a mercury atmosphere. In HgCdTe the electrical properties are determined by both electrically active impurities and native defects, and annealing in a mercury atmosphere is a technique commonly used to adjust the electrical properties of HgCdTe to suit device requirements. This is due to the fact that mercury related defects, such as vacancies and interstitial atoms have a significant impact on the electrical properties of HgCdTe [8]. In addition, the rapid loss of mercury that occurs when HgCdTe is heated, due to weak mercury bonding, makes it necessary to add mercury to the ambient during any high temperature annealing process [8]. The sample was placed inside a quartz ampoule containing a separate mercury reservoir, followed by a sequence of evacuation, sealing, and annealing at a temperature of 200°C for 17 hours.

Laser beam induced current (LBIC) measurements were performed on the sample which had been RIE processed, and also on that which had been RIE processed and mercury annealed. The LBIC measurements were used to characterise the spatial extent of the n-type doping induced by RIE processing and to investigate the effect of mercury annealing on the RIE processed material.

355

A Scanning Laser Microscope purchased from Waterloo Scientific Inc. was used for LBIC measurements in this work. The system has a diode-pumped Nd:YLF laser that was operated at a wavelength of 1.047μm in cw mode, power 400 Wcm^{-2}, and focused to approximately a 3μm spot size. Figure 1(b) presents the LBIC line scan signal measured in the plane of the two remote contacts (A and B) as a function of laser beam position, for the 400μm×400μm square of HgCdTe exposed to RIE processing. The contacts were made to the edge of the wafer on each side of the processed area of the p-type sample using gold chloride and pressed indium, and are separated by approximately 8mm. The observed LBIC signal is typical of the bipolar characteristic exhibited by an n-on-p junction, indicating that the RIE processing has converted the p-type HgCdTe to n-type [12]. The magnitude and shape of the LBIC signal is dependent on several factors including, the distance from the laser illuminated area to the junction, junction geometry, distance between the two remote contacts, and material lifetime. If the laser beam generates electron-hole pairs further than a few diffusion lengths from the vertical or lateral

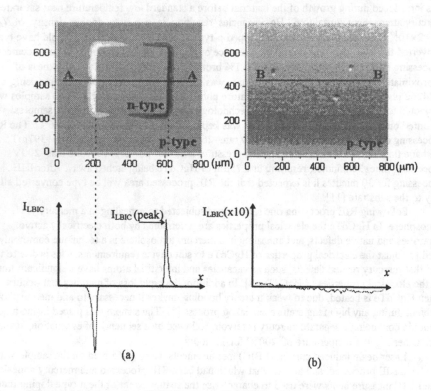

(a)   (b)

Figure 2 (a) False colour LBIC image and LBIC line scan through AA of RIE processed area of extrinsically doped p-type Hg$_{0.69}$Cd$_{0.31}$Te. The light and dark regions represent negative and positive signals, respectively, generated by the presence of a p-n junction. (b) False colour LBIC image and LBIC line scan through BB of mercury annealed RIE processed area of extrinsically doped p-type Hg$_{0.69}$Cd$_{0.31}$Te. All LBIC images were taken with the samples at 80K.

junction boundaries, little or no contribution will be made to the LBIC signal. In figure 1 the peaks of the LBIC signal correspond to the p-n junction boundary, which induces the greatest separation of charge and therefore magnitude of LBIC current. The distance between the peaks, 400μm, is that of the region exposed to RIE processing and represents the vertical boundaries of the p-n junction. Figure 1 also shows that the converted region must extend down to the substrate (junction depth>8μm) since no LBIC signal is generated very far beyond the vertical junctions. A shallow p-n junction would exhibit an LBIC signal with a linear spatial characteristic within the converted region because an LBIC signal would be generated by photogenerated carriers diffusing to the underlying p-n junction.

Figure 2 presents false colour contour maps of two dimensional LBIC scans and LBIC line scans on the RIE processed samples before and after mercury annealing. The bipolar behaviour of the spatial LBIC scan evident from the grey scale distribution and LBIC line scan AA in Figure 2a indicate that the square area exposed to RIE processing was converted from p-type to n-type, thus forming an n-on-p junction. Figure 2b presents a spatial LBIC scan and LBIC line scan BB on the RIE processed sample after a mercury anneal. Apart from a few localised electrically active defect clusters evident on the spatial image, and noise in the line scan signal, it is clear that the n-type region created after RIE processing and evident in Figure 2a, no longer exists. This result clearly indicates that mercury annealing can be used to successfully remove RIE induced n-type conversion on extrinsically doped p-type HgCdTe. It should be noted that the current scales in Figures 2(a) and (b) are not the same, with the current in Figure 2(b) being measured with an amplifier with gain one order of magnitude greater than that in Figure 2(a).

In order to confirm the above results, Hall and resistivity measurements were used to examine in greater detail the electrical properties of the annealed sample. Measurements were performed at room temperature on the mercury annealed RIE processed sample using a Van der Pauw structure and a variable magnetic field up to 12T. The results confirmed conclusively that the sample had the same electrical properties as the starting wafer, with N_A-N_D=2×10^{16} cm^{-3}, and μ = 350 cm^2.V^{-1}.s^{-1}.

CONCLUSIONS

In conclusion, it has been experimentally verified that the process of RIE induced p-to-n type conversion in extrinsically doped p-type Hg$_{0.69}$Cd$_{0.31}$Te can be readily reversed using mercury annealing. The sample upon which n-type conversion was successfully removed was subjected to a sealed tube annealing process in a mercury ambient at 200°C for 17 hours after RIE processing. A powerful optical characterisation tool, in the form of scanning laser microscopy utilising the LBIC phenomena, was used in the experimental investigation. Two dimensional mapping of the LBIC signal displayed the spatial extent of RIE induced p- to n-type conversion and subsequent removal of the converted region following mercury annealing. Hall measurements confirmed that the mercury anneal completely restored the RIE converted n-type material to p-type with electrical properties identical to those prior to any processing. This work demonstrates that anisotropically etched mesa structures in extrinsically doped p-type HgCdTe can be defined successfully using a sequence of RIE followed by a mercury anneal. However, the precise nature of the RIE induced p- to n-type conversion process in extrinsically doped p-type HgCdTe, and its subsequent removal by mercury annealing, has yet to be established.

ACKNOWLEDGMENTS

This work was supported by the Australian Research Council (ARC), and the Cooperative Research Centre for Broadband Telecommunications and Networking (CRC-BTN).

REFERENCES

1. G. P. Carey, D. J. Friedman, A. K. Wahi, C. K. Shih, and W. E. Spicer, J. Vac. Sci. Technol. A **6**, p. 2736 (1988).

2. E. Belas, J. Franc, A. Toth, P. Moravec, R. Grill, H. Sitter, and P. Hosch, Semicond. Sci. Technol. **8**, p. 1116 (1996).

3. L. O. Bubulac, W. E. Tennant, D. S. Lo, D. D. Edwall, J. C. Robinson, J. S. Chen, and G. Bostrup, J. Vac. Sci. Technol. A **5**, p. 3166 (1987).

4. L. O. Bubulac, W. E. Tennant, S. H. Shin, C. C. Wang, M. Lanir, E. R. Gertner, and E. D. Marshall, Japanese J. App. Phys. **19**, p. 495 (1979).

5. E. Belas, R. Grill, J. Franc, A. Toth, P. Hoschl, H. Sitter, P. Moravec, J. Cryst. Growth **159**, p. 1117, (1996).

6. G. Panin, P. Fernandez, and J. Piqueras, Semicond. Sci. Technol. **11**, p. 1354 (1996).

7. J.F. Siliquini, J. M. Dell, C. A. Musca, E. P. G. Smith, L. Faraone, and J. Piotrowski, Appl. Phys. Lett. in press, Jan (1998)

8. C. L. Jones, M. J. T. Quelch, P. Capper, and J. J. Gosney, J. Appl. Phys. **53**, p. 9080 (1982).

9. T. Sasaki, N. Oda, M. Kawano, S. Sone, T. Kanno, and M. Saga, J. Cryst. Growth **117**, p. 222 (1992).

10. J. Bajaj, W.E. Tennant, R. Zucca, and S.J. Irvine, Semicond. Sci. Technol. **8**, p. 872 (1993).

11. J. F. Siliquini, J. M. Dell, C. A. Musca and L. Faraone, Appl. Phys. Lett. 70(25) p. 3443 (1997)

12. J. Torkel Wallmark, Proc. IRE, **45**, p. 474 (1957).

DX-LIKE CENTERS AND PHOTOCONDUCTIVITY KINETICS
IN PbTe-BASED ALLOYS

B.A.AKIMOV, V.A.BOGOYAVLENSKIY, V.N.VASIL'KOV, L.I.RYABOVA
Moscow State University, Moscow 119899, Russia, mila@mig.phys.msu.su

ABSTRACT

Results of the study of photoconductivity (PC) kinetics, thermally stimulated currents (TSC) and electrothermal instabilities in PbTe(Ga), $Pb_{1-X}Mn_XTe(In)$ (0<X<0.11) single crystals and $Pb_{1-X-Y}Sn_XGe_YTe(In)$ (0.06<X<0.2; 0.06<Y<0.12) films are presented. The impurity states in these compounds exhibit a DX-like behavior. Reconstruction of the energy spectrum of solid solutions induced by the composition (X,Y) variation allows us to conclude that the metastable electronic states take pa:· ·n PC kinetics and accelerate the relaxation process when the metastable levels appear inside the energy gap.

INTRODUCTION

Doping of lead telluride and a number of its solid solutions with indium and gallium results in the formation of impurity centers which in some features are similar to DX-centers in wide gap semiconductors [1,2]. The Fermi level (FL) pinning and the effect of persistent PC at low temperatures are considered as the main properties of these alloys [3]. The energy spectrum, FL position and the PC parameters depend on the composition X, Y and the kind of doping impurity. In PbTe(Ga) the ground state of the impurity center responsible for FL pinning lies at 70 meV below the bottom of the conduction band. So the crystals have high resistance at low temperatures, the electron concentration appears to be close to the intrinsic values. Under continuous radiation of the sample by a heat source the persistent PC is observed at $T<T_C\sim80$ K. In PbTe(In) the pinned FL lies in the conduction band and moves down the energy scale in all studied solid solutions for both signs of the energy gap coefficient dE_G/dX [3,4]. The T_C value for In doped alloys is about 20 K, except $Pb_{1-X-Y}Sn_XGe_YTe(In)$, where T_C reaches 38 K [5]. Besides the ground state of the impurity, the metastable states have been observed in a number of experiments [1,3]. For PbTe(Ga) a pronounced peak, whose energy is shifted with respect to E_G by approximately 20 meV, is observed in PC spectra [6]. The spectral characteristics of the photoresponse in the long-wave part of the spectrum correlate with the presence or absence of excitation in the short-wavelength part region of the spectrum [7]. Study of the influence of metastable states on the character of PC kinetics, thermally stimulated currents and electrothermal instabilities in the alloys mentioned above is the subject of the present paper.

EXPERIMENT

Photoconductivity kinetics

Difficulties in measuring the kinetics of PC with large lifetimes of nonequilibrium charge carriers are associated with the background illumination. We used the previously devised low-temperature chamber to perform background-free measurements [3]. Pulsed illumination of the samples was performed with semiconductor LEDs and lasers. A miniature incandescent lamp was used in the experiments for irradiating the sample with white light. For PbTe(Ga) single crystals no persistent PC is observed under radiation of the crystals with pulse sources with a quanta energies both higher and lower than E_G if the radiation pulse duration is less than 100 μs and the power of the source less than 1 mW. The increase of the radiation intensity or pulse duration is followed by the appearance and gradual rise of the persistent PC response. The same tendency is observed under the radiation of the sample with a microlamp. Under continuous radiation with a higher lamp power (~100 mW) the photocurrent rises during several seconds to a level which corresponds to the resistivity value ~1 Ω cm, and then remains nearly constant.

For $Pb_{1-x}Mn_xTe(In)$ single crystals and $Pb_{1-x-y}Sn_xGe_yTe(In)$ films the view of the kinetic curves was found to change qualitatively under the composition X, Y variation accompanied by the energy spectrum reconstruction. For $Pb_{1-x}Mn_xTe(In)$ with X≤0.07 (T=4.2K) the persistent PC takes place after the light source is switched off. The subsequent increase of X value results in the appearance of a relatively fast relaxation region at the beginning of the kinetic process which is transformed into persistent PC at the relaxation tail. This process is illustrated in fig.1. The photoresponse is shown under pulses of irradiation from LED with the power 1 mW and wave length 1 μm in the regime of fixed voltage U=0.1 V. The pulse duration has been successively increased up to 3s. Dark current is equal to 0.03μA, samples dimensions are 3x1.6x0.6 mm. The quantity of the electrons participating in the fast relaxation process rises with X increase and at X>0.1 the persistent PC is not observed at all. It should be mentioned that the acceleration of the relaxation process is accompanied by the impurity energy activation E_A increase under the composition variation both for $Pb_{1-x}Mn_xTe(In)$ crystals and $Pb_{1-x-y}Sn_xGe_yTe(In)$ films.

Thermally stimulated currents

In order to register the TSC peaks the samples were cooled to the liquid helium temperature. The unequilibrium concentration of electrons was created by means of irradiation with a heat source during several seconds. Then the temperature was slowly increased (the heating rate was 5-9 K/s; U=1 V). An example of the TSC pictures is presented in fig.2.The most pronounced peaks of TSC were found for $Pb_{1-x-y}Sn_xGe_yTe(In)$ films with activation energies lower than 50 meV. For $Pb_{1-x}Mn_xTe(In)$ crystals TSC picture is observed only for the samples with the definite value of X~0.07-0.08 (E_A about 40 meV). The temperature T_M corresponding to the peak of TSC varies from 5 K up to 14 K. It should be noted that T_M is significantly lower than the values known for classical and wide gap semiconductors. However the dependence of T_M on the composition of the sample or thermal impurity activation energy E_A is not clearly pronounced. It may be caused by the existence of a number of variable parameters which determine the peak position. The main problem is connected with the redistribution of the electrons between the ground and metastable impurity states during the recombination and generation processes. That leads to the variation of the concentration of the trapping centers and relaxation rates during the experiments. For PbTe(Ga) single crystals the TSC structure appears to be more complicated: a set of peaks of different amplitude is observed at T lower than T_C.

Electrothermal instabilities

PbTe(Ga) may be taken as a model object for studying autovibrations and autowave processes. If the irradiation of the sample results in the appearance of unequilibrium charge carriers excited from the metastable electronic states and the applied electric field is enough to reach the heat breakdown, the periodic vibrations of the electric current and temperature will occur. The current increase is followed by the temperature rise. At some critical temperature the process of the relaxation of unequilibrium carriers into the ground states becomes dominating.

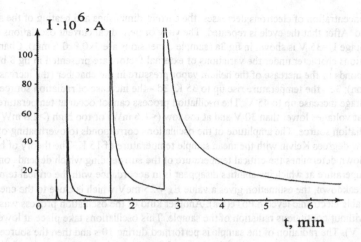

Fig.1. PC kinetics for $Pb_{1-X}Mn_XTe(In)$ (X=0.1) sample under a number of irradiation pulses.

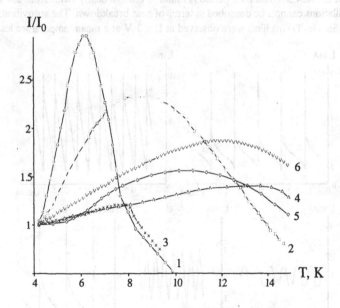

Fig.2. The view of TSC for $Pb_{1-X-Y}Sn_XGe_YTe(In)$ films with different activation energies E_A (curve 1 - E_A= 44 meV, 2 - 35 meV, 3 - 4 meV, 4 - 22 meV, 5 - 34 meV) and $Pb_{1-X}Mn_XTe(In)$ (X=0.08; E_A = 48 meV) single crystal (curve 6).

The concentration of electrons decreases, the current diminishes and heating of the sample is stopped. After that the cycle is repeated. The view of periodical current oscillations at T=4.2 K at the voltage U=35 V is shown in fig.3a (sample dimensions are 3x0.8x0.5 mm). Changes of the oscillations character under the variations of external factors are presented in fig.3 b,c,d,e. Fig.3b corresponds to the increase of the helium vapor pressure in the chamber (the increase of heat rejection); 3c - the temperature rise up to 35 K; 3d - the increase of radiation source power; 3e - the voltage increase up to 45 V. The oscillation process cannot occur at temperatures higher than 42 K, at voltages lower than 30 V and at too low (<1.6 mW) or too high (>2.8 mW) powers of the radiation source. The amplitude of the oscillations corresponds to overheating of the sample by a few degrees Kelvin with the mean sample temperature of 15 K. The theory of heat breakdown determines the critical temperature of the surrounding which depends on E_A value. If the temperature at which instabilities disappear is in accordance with the critical temperature of heat breakdown, the estimation gives a value $E_A \sim 25$ meV which is close to the energy of the metastable electronic level in PbTe(Ga). Another kind of the oscillation process was observed even without continuous radiation of the sample. This oscillations take place at lower voltages (U=24 V). The radiation of the sample is performed during 10 s and then the source is switched off. The monotonous current relaxation is transformed into the oscillatory process after the current reaches some critical value (~2 mA) and this state may exist for an indefinitely long time. Such oscillations take place near the mean temperature ~7 K, that is significantly lower than in the previous case. The latter kind of oscillations may be observed only in a rather narrow voltage interval 23.5-24.5 V and their period is found to change nearly twice from 2.8 s to 1.5 s. This type of oscillations cannot₁ be described in terms of heat breakdown. The autovibration processes in $Pb_{1-X-Y}Sn_XGe_YTe(In)$ films were observed at $U \leq 3$ V at a mean temperature less than T_M.

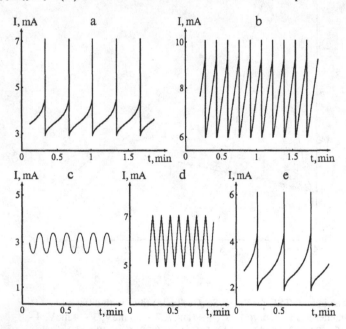

Fig.3. Thermoelectric instabilities (current vibrations) in PbTe(Ga) single crystals.

CONCLUSIONS

To explain the experimental data we use a scheme which is associated with the rearrangement of a deep impurity center in the configuration space. According to the scheme, the total energy of an electron plus the elastic energy of the lattice cell can be distorted with the capture and the release of an electron from an impurity center. The ground state E_2 (two electrons localized at the center) and the metastable state E_1 (one electron) correspond to parabolas in the configuration space. FL is pinned at the E_2 level minimum. The distinguishing feature of the investigated alloys is that the metastable levels cannot be associated with shallow states of the impurity. A barrier separates the metastable states from the conduction band. According to the numerical estimations [8] for PbTe(In) the minimal energy of one electron state exceeds Fermi energy by a value $\Delta \sim 40$ meV. If we assume that in PbTe(In)-based solid solutions the deformation potential and rigidity coefficient do not change significantly with composition variation, the E_1 level appears to enter the band gap when the condition $E_A > \Delta$ is satisfied. It is in this case when the relatively rapid relaxations appear in $Pb_{1-X}Mn_XTe(In)$ $(X \geq 0.085)$. The situation seems to be more complicated for $Pb_{1-X-Y}Sn_XGe_YTe(In)$ because of the phase transition. For this case we can only define the tendency to the acceleration of the relaxations with E_A increase. The TSC picture proves the existence of the metastable states within the energy gap. The peculiarities of PC kinetics and the autovibration processes observed in PbTe(Ga) show that the conduction band edge together with the metastable and ground state impurity levels can actually be considered as a 3-level system.

ACKNOWLEDGMENTS

This work was supported by INTAS - RFBR under Project N 95-1136 and RFBR under Project N 96-02-16275-a.

REFERENCES

1. D.R.Khokhlov and B.A.Volkov, Proc.23 Int.Conf.Phys.Semicond.,Berlin, Germany, World Sci. **4**, p.2941 (1996).

2. P.M.Mooney, J. Appl.Phys., **67**, R1 (1990).

3. B.A.Akimov, A.V.Dmitriev, D.R.Khokhlov, and L.I.Ryabova, Phys.Stat.Sol.(a) **137**, 9 (1993).

4. A.I.Lebedev and Kh.A.Abdullin, Phys.Stat.Sol.(a) **91**, 225 (1985).

5. V.F.Chishko, V.T.Hryapov, I.L.Kasatkin, V.V.Osipov, E.I.Slyn'ko, O.V.Smolin, and V.V.Tretinik, Infrared Phys. **3**, 197 (1992).

6. B.A.Akimov, A.V.Albul, V.Yu.Il'in, M.Yu.Nekrasov, and L.I.Ryabova, Semiconductors **29**, 1051 (1995).

7. A.I.Belogorokhov, E.I.Slyn'ko, and D.R.Khokhlov, Pis'ma Zh.Tekh.Fiz. **18**, 207 (1992).

8. V.I.Kaidanov, S.A.Rykov, M.A.Rykova, and O.V.Syuris, Fiz.Tekh.Poluprov., **24**, 144 (1990).

RAMAN AND PHOTO-LUMINESCENCE STUDIES ON INTRINSIC AND CR-DOPED ZNSE SINGLE CRYSTALS

BRAJESH K. RAI, BHASKAR S., H. D. BIST* AND R. S. KATIYAR
Department of Physics, University of Puerto Rico, Rio Piedras campus, San Juan, PR 00931.
*Permanent address: Physics Dept., IIT Kanpur, 208016.
K.-T.- CHEN, AND A. BURGER⁺
Center for Photonic Materials & Devices,⁺Dept of Physics, Fisk University,Nashville,TN

ABSTRACT

Single crystals of ZnSe, with varying amounts of Cr doping have been studied using Raman and photoluminescence(PL) spectroscopy. The Cr-doped samples show the existence of a coupling mechanism of longitudinal optical(LO) phonons of ZnSe with hole-plasmons. The dependence of intensity ratio of LO and transverse optical(TO) mode on temperature and excitation wavelength, has been attributed to the interaction of the field of LO phonons with the surface electric field in the depletion layer. The interaction of discreet phonons with the electronic continuum of conduction band in ZnSe is responsible for the shift of Raman peaks. The large electron capture cross-section of deep-level Cr^{2+} and Cr^{1+} impurities is inhibitive for the observation of band-to-band PL transition at ~2.7eV in ZnSe:Cr.

INTRODUCTION

ZnSe is a direct-band-gap semiconductor, with the room temperature bandgap of 2.67 eV. The development of efficient light-emitting p-n junctions [1] and blue ZnSe laser has been realized[2]. Highly conductive n-type materials of ZnSe, have been obtained.[3] However, barring Li and N dopants, the efforts to achieve low resistivity in p-type ZnSe have not met with any success[4]. In the case of p-type ZnSe, with Li dopant, a high concentration of free carriers is difficult to obtain due to the auto-compensation of acceptors.[5] Another problem, which hinders the efficient performance of ZnSe p-n junctions, is the presence of deep-level centers within the bandgap, providing nonradiative and radiative routes at longer wavelengths, and thus lowering the efficiency of the desired blue light emission[6]. A new class of materials for tunable solid state lasers based on Cr^{2+} doped II-VI compounds including Cr^{2+}:ZnSe has been investigated[7] .The study of Cr-doped ZnSe, which is known to contain deep-levels of Cr^{1+}/Cr^{2+},[8] should provide some insight into the behavior of deep-levels in ZnSe. In this paper we present the Raman and PL results of the intrinsic and Cr-doped crystals of ZnSe. The discreet-continuum interaction in intrinsic ZnSe and doped single crystals, the presence of depletion layer on the surface of Cr-doped samples and the existence of localized modes of Zn_{Cr}-Se, are, to the best of our knowledge, being reported for the first time.

EXPERIMENTAL

Intrinsic ZnSe crystals grown by Bridgman and seeded physical vapor transport(SPVT) methods (Eagle-Pitcher Co.), and pure CrSe powder (Alfa Co.) were used for doping the crystals. A one-zone furnace for diffusion doping was used. The annealing conditions were (800 °C, 2days), (900 °C, 5 days) and (1000 °C, 5.5days) for dopant concentration of 1.0, 1.8, 10.2 (x 10^{19} cm^{-3}) respectively. The Cr concentrations were determined by absorption measurements described elsewhere[7]. The Raman and PL measurements were performed using a JobinYvon T64000 spectrophotometer. Raman signals were detected using a liquid cooled charge coupled device [(CCD) at 140 K], while the PL spectra were accumulated using a photomultiplier tube (PMT) from Product for Research Inc. The excitation lines of 514.5, 488 and 457.9nm were obtained from the Coherent Innova 99 cw Ar+ laser. Details for recording the spectra and band analysis have been reported elsewhere . [8]

RESULTS AND DISCUSSIONS

Phonon-Plasmon Coupling

Typical spectra for ZnSe single crystals and three Cr-doped ZnSe at room temperature are shown in Fig.1. In each case, the 514.5 nm Ar-ion laser line was used for excitation. The (111) surface of ZnSe was kept perpendicular to the incident and scattered lights. While, in the case of ZnSe:Cr samples, the orientation of the surface was not defined. Both, intrinsic and doped, samples of ZnSe show essentially the same features in their Raman spectra. The two sharp and intense lines at ~250 and ~205 cm^{-1} have been identified as LO and TO phonon modes of ZnSe [10]. The weak and broad structure at ~150 cm^{-1} has been reported as 2TA, the first overtone of transverse-acoustical phonon.[11] In addition, a weak and broad combination mode 2TA+TO and the first overtone of LO phonon is observed in all the spectra at ~270 and ~500 cm^{-1}, respectively.[11]

Fig.1 Room temperature Raman spectra of (a)intrinsic and (b, c, d)Cr-doped ZnSe single crystals with Cr concentration of 1.0, 1.8, 10.2 (x 10^{19} cm^{-3}) respectively, using an excitation line of 514.5 nm. A Peak fit analysis of spectrum (d) shows additional four weaker bands, which are assigned to LO and TO modes of two types of Chromium ions Cr^{2+} and Cr^{+}. [9]

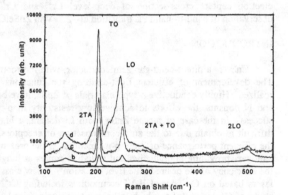

The peak position of the LO mode shows a downward shift as the doping concentration is increased; the peak position of the TO mode, however, remains unchanged. Effect of temperature on the peak positions of the LO and TO modes for the three samples is shown in the Fig. 2. A consistent downward shift of the LO mode with increased doping concentration is noticeable, at each temperature. The upward shift of the Raman modes with decreasing temperatures is a manifestation of anharmonicity and has long been studied in various semiconductors, including ZnSe [12]. Our data for the frequency shift in the temperature range of 70-290K show an excellent agreement with the theoretical calculations based on cubic anharmonicity to second order. However, for a wider range of temperature (70 to 550K), the cubic harmonic terms to second order are insufficient to fit the data and the inclusion of higher-order terms involving cubic and quartic anharmonicity in necessitated for a satisfactory fitting [12].

Two types of interactions of optical-phonons with free-carriers in the semiconductors have been known to occur[13] : (i) interaction of single-particle electron (hole) excitations with TO phonons through the deformation potential mechanism. (ii) interaction of LO phonons with collective excitations, i.e. plasmons, of electron (hole). Owing to the associated longitudinal electric fields, only LO phonons couple with the plasmons. The shifts in the peak position of LO phonons, in this process are much larger than the shifts of the TO modes, occurring as a result of its coupling with single particle excitations. The coupling of LO phonon with plasmons leads to two coupled modes denoted by $L^{\pm}(q)$. For zone-center optical phonons, i.e. $q \approx 0$, the frequency of the $L^{-}(q)$ mode is less than or equal to the TO phonon frequency, while the frequency of the $L^{+}(q)$ mode exceeds or equals the frequency of the LO mode [14]. We do not observe $q \approx 0$ plasmon-phonon coupled modes, L^{+} and L^{-}. A strong dependence of the LO mode, and a negligible dependence of the TO mode, on concentration of chromium impurity, establishes the predominance of hole-plasmon-phonon interaction mechanism for the shifts in the Raman modes. One-mode behavior of the plasmon-phonon coupling in n- and p- type ZnSe, doped with shallow donor and shallow acceptor impurities, have been reported [13]. L^{+} component of the plasmon-phonon coupled mode shows an upward shift in its peak position as the doping concentration is increased. The downward shift due to coupling of LO-phonon with hole-plasmon, was observed in p-type GaAs by Olego et al.; with doping concentration [15]. They associated this phenomenon to the wave-vector non-conservation by the ionized impurities. Considering the fact that deep-levels due to the ionized Cr^{+} and Cr^{2+} are formed in ZnSe:Cr system, we attribute the origin of LO-phonon-hole-plasmon coupled mode to the breakdown of the '$q = 0$' selection rule .

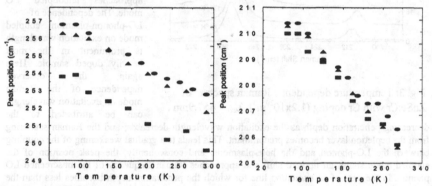

Fig. 2: Temperature dependence of LO and TO modes (λ_{ext} = 514.5 nm). The impurity concentrations N_A of the doped samples denoted by •, Δ and ■ are 1.0, 1.8, and 10.2 (x 10^{19} cm^{-3}) respectively.

Schottky Barrier and Depletion layer

The effect of temperature on the relative intensities of LO and TO modes of heavily doped ZnSe:Cr is shown in Fig. 3. The value of I_{LO}/I_{TO} increases with increasing temperature. The increase in the ratio of I_{LO}/I_{TO} comes about by a large(marginal) increase(decrease) in the total integrated intensity of the LO(TO) mode. We attribute the aforementioned behavior of the intensity of LO and TO modes and their ratio to the presence of a depletion layer and to the existence of an internal electric field associated with the surface space-charge layer. The surface electric field interacts with the electric field of LO mode and this results in a stronger electric-field-induced LO scattering intensity; however, the intensity of the TO mode remains unaffected. In addition, the depletion-layer leads to the observation of unscreened LO phonons. Using an excitation wavelength for which the penetration depth is less than or equal to the width of the depletion layer, only the unscreened LO phonons would be seen. The absorption coefficient of ZnSe single crystal, in the visible region, lies in the range of 10^{4} to10^{5} cm-^{16}.This value would give us a net penetration depth of the order of 1,000 to 10,000A^{0}. The value of net acceptor concentration in the range of $10^{16} - 10^{17}$ cm^{-3} would give a depletion layer width, of the same order as penetration depth. However, a strong auto-compensation of carriers in the presence of deep levels is expected to lower the value of net acceptor concentration significantly. The increase in the value of I_{LO}/I_{TO} with increasing excitation energy can be attributed to the decreasing value of the penetration depth. The interaction of the electric field of the LO phonon with the surface electric field would be stronger for lower excitation wavelength.

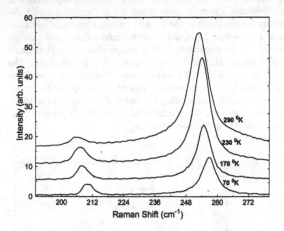

Fig. 3: Temperature dependent Raman spectra of ZnSe:Cr with Cr doping (1.2×10^{20} m^3), $\lambda_{ext} = 457.9$ nm .

The LO-phonon-plasmon coupled mode shows an upward shift with increasing excitation energy and approaches uncoupled LO mode. The dependence of LO-phonon-plasmon coupled mode on excitation wavelength is prominent in the most heavily doped sample. Here again, the observed dependence of the Raman mode on excitation wavelength can be attributed to the decreasing penetration depth as the excitation wavelength decreases; and the Raman scattering from the depletion layer becomes predominant. This leads to a gradual weakening of the coupling between the LO-phonon and the hole-plasmons, and consequently, the peak position of the coupled mode shows an upward shift and approaches Raman frequency of the unscreened LO phonon. In the case of an excitation line for which the penetration depth becomes less than the width of the depletion layer, unscreened LO phonon is expected to be observed. The absorption coefficient in ZnSe is known to increase with temperature in the visible region[16]. Thus, the penetration depth of the excitation lines will decrease at higher temperature, decreasing penetration depth will result in more contribution to the Raman scattering coming from within the depletion layer, and will lead to an increase in the ratio of I_{LO}/I_{TO} with 'Temperature', as evident from Fig. 3.

Photo-generation of free carriers in ZnSe

In the intrinsic ZnSe single crystal, an upward shift in the peak position of TO, LO and 2LO modes is observed with increasing excitation energy (Fig.4). The effect is more pronounced in the 2LO mode. In addition, an asymmetry to the low frequency side in the bands, is also more pronounced in the 2LO modes. The electronic continuum and discrete-phonon interaction leads to quantum mechanical interference. This quantum mechanical interference results in the development of asymmetry in the Raman LO modes. On excitation with the laser line of 514.5 nm, a weak anti-stokes PL band at ~ 2.61 eV (the band-to-band transition) is observed from intrinsic ZnSe. This band-to-band PL emission increases in intensity with increasing laser power. In addition, the efficiency of this transition increases with increasing excitation energy; and with the excitation energy of 2.7 eV a very strong PL band is observed at room temperature. The PL results indicate a process of photogeneration of electrons in the conduction band in the case of undoped ZnSe; which is inhibited in the case of Cr-doped ZnSe samples, due to the presence of deep levels of Cr^+ and Cr^{2+} impurities. The effect of the photogenerated electronic continuum in intrinsic ZnSe is manifested by the consistent upward shift in the frequencies of LO, 2LO, and most importantly of TO Raman modes with decreasing wavelength of excitation line. Therefore, we tend to attribute the observed dependence of the LO and TO modes on the excitation wavelength to the discreet-continuum interaction, which has equal effect on the shifting of the LO and TO modes. In addition, due to the discreet-continuum interaction, the halfwidth of the LO and TO modes show an increase with decreasing wavelength of the excitation line.

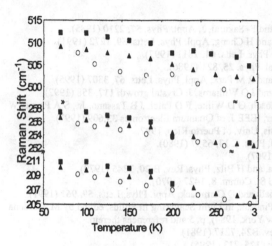

Fig. 4 Dependence of the of TO, LO, and 2LO modes in intrin sic ZnSe . ($\bullet$, Δ and) denote λ_{ext}=514.5, 488, and 457.9 nm respectively).

Photoluminescence

The bandgap of intrinsic ZnSe at room temperature is 2.67 eV and it increases with decreasing temperature. Using an excitation line of 2.71 eV from Ar^+ laser, a very intense luminescence band at 2.67 eV was observed at room temperature. In addition, at room temperature, we obtained weak anti-stokes band-to-band luminescence emission at 2.61 eV, using the excitation lines of 514.5 nm.. This band weakens with decreasing temperature and eventually disappears at a temperature of 110 K. Hence, the origin of the observed anti-stokes luminescence can be attributed to thermally assisted excitation of electrons from the valence to conduction band. In the PL spectra of Cr-doped samples two emission bands appear on the low energy side at 2.18 (strong) and 1.40 (weak) eV, at room temperature. With the help of previous ESR results[17], we can assign the origin of the two PL bands to the " $Cr^{2+} \leftrightarrow Cr^{1+}$ " charge-transfer process. This charge-transfer process gives rise to the 2.18-eV PL band, while the "$Cr^{1+} \to Cr^{2+}$" process is responsible for the origin of the 1.40-eV band. The weaker band-to-band transition in the doped samples can be attributed to the large electron-capture cross section of the deep-level chromium impurities.

CONCLUSIONS

The presence of deep levels of the Cr-impurity in ZnSe:Cr and the coupling of the LO-mode with the hole-plasmon showing a one-mode behavior is established. The doping of chromium has been found to create a surface depletion layer on the surface of ZnSe:Cr crystals. The localized modes of Zn_{Cr}-Se have been observed in heavily doped samples. A phenomenon of discreet-continuum interaction, has been observed in ZnSe crystals.

ACKNOWLEDGEMENTS

The authors at Fisk would like to acknowledge the help of Reed, Schaffers, Page and Payne of LLN Laboratory and Research Collaboration Program for HBCU's through the LLNL EPO. This work was supported in part by NSF (Grant NSF-OSR-(9452893)), NASA-NCCW-0088 and by NASA through the Center for Photonic Materials and Devices, Grant NCC8-133. HDB thanks CSIR' (India) for Emeritus Scientistship award.

REFERENCES

1. J- Nishizawa, K-Itoh, Y-Okuno, and F- Sakurai, J. Appl. Phys. **57**, 2210 (1985).
2. M A Haase, J Qiu, J M Depuydt, and H Cheng, Appl. Phys. Letts. **59**, 1272 (1991).
3. Z Zhu, H Mori, and T Yao, Appl. Phys. Letts. **61**, 2811 (1992).
4. T Yao and K Okada, Japan J. Appl. Phys. **25**, 821 (1986).
5. C Kothadaraman, G F Neumark, and R M Park, Appl. Phys. Letts. **67**, 3307 (1995).
6. J-Zheng, J W Allen, H M Yates, and J O Williams, J. Crystal growth **117**, 358 (1992).
7. R H Page, K I Schaffers, L D Deloach, G D Wilke, F D Patel, J B Tassano, Jr., S A Payne, W F Krupe, K T Chen, and ABurger, IEEE J. of Quantum Electronics **33**,609 (1997).
8. Brajesh Kumar Rai , Masters thesis, Univ. of Puerto Rico, 1997.
9. M Godlewski and M Kaminska, J. Phys. C **13**, 6537 (1980).
10. W Taylor, Phys. Letts **24A**, 556 (1967).
11. R L Schmidt, K Kunc, M Cardona, and H Bilz, Phys. Rev. B **20**, 3345 (1979).
12. J L LaCombe and J C Irwin, Solid St. Comm. **8**, 1427 (1970).
13. D J Olego, J Petruzzello, T Marshall, and D Cammack, Appl. Phys. Letts. **59**, 961 (1991).
14. G Abstreiter, M Cardona, and A Pinczuk, in Light Scattering in Solids, edited by M. Cardona and G. Guntherodt (Springer, New York, 1984), p. 5 and references therein.
15. D Olego and M Cadona, Phys. Rev. B24, 7217 (1981).
16. J M Pawlikowski, Thin Solid Films **125**, 213 (1995).
17. G Grebe, G Rousso and H-J Schulz, J. Phys. C **9**, 4511 (1976).

GROWTH AND CHARACTERIZATION OF PbSe AND $Pb_{1-x}Sn_xSe$ LAYERS ON Si (100)

H. K. SACHAR *, I. CHAO, X. M. FANG, P. J. MCCANN
School of Electrical and Computer Engineering, University of Oklahoma, Norman, OK 73019.
*hksachar@mailhost.ecn.ou.edu

ABSTRACT

Crack-free layers of PbSe were grown on Si (100) by a combination of liquid phase epitaxy (LPE) and molecular beam epitaxy (MBE) techniques. The PbSe layer was grown by LPE on Si (100) using a MBE-grown $PbSe/BaF_2/CaF_2$ buffer layer structure. $Pb_{1-x}Sn_xSe$ layers with tin contents in the liquid growth solution equal to 3%, 5%, 6%, 7%, and 10%, respectively, were also grown by LPE on Si (100) substrates using similar buffer layer structures. The LPE-grown PbSe and $Pb_{1-x}Sn_xSe$ layers were characterized by optical Nomarski microscopy, X-ray diffraction (XRD), Fourier transform infrared spectroscopy (FTIR), and scanning electron microscopy (SEM). Optical Nomarski characterization of the layers revealed their excellent surface morphologies and good growth solution wipe-offs. FTIR transmission experiments showed that the absorption edge of the $Pb_{1-x}Sn_xSe$ layers shifted to lower energies with increasing tin contents. The PbSe epilayers were also lifted-off from the Si substrate by dissolving the MBE-grown BaF_2 buffer layer. SEM micrographs of the cleaved edges revealed that the lifted-off layers formed structures suitable for laser fabrication.

INTRODUCTION

The study of IV-VI semiconductors such as PbSe and $Pb_{1-x}Sn_xSe$ has been motivated by their use as mid-infrared laser materials. IV-VI semiconductor band gap and lattice parameter can be adjusted by varying the composition thereby providing the versatility of spanning the 3 to 30 μm spectral range. As the band gap of IV-VI materials is also temperature sensitive, realization of diode lasers with temperature and current tunability is possible. Ternary $Pb_{1-x}Sn_xSe$ has a smaller band gap and larger refractive index than PbSe, so it can be used as the active layer in double heterostructure lasers. According to the band gap model suggested for the $Pb_{1-x}Sn_xSe$ alloy system, the energy gap decreases as the SnSe content is increased [1]. At a composition of $x = 0.15$ and a temperature equal to 4.2 K, the band gap energy becomes zero. Subsequently, the energy gap increases as the tin content increases.

Growth of IV-VI semiconductors on Si substrates is advantageous because silicon substrates are available in large dimensions and are also less expensive than conventionally used IV-VI substrates. High quality layers of PbSe, PbTe and $Pb_{1-x}Sn_xSe$ have been grown heteroepitaxially on Si (111) substrates using group IIa fluoride insulators as buffer layers [2-4]. The growth of these IV-VI layers on Si (111) substrates enables the realization of monolithic infrared detector arrays in which the infrared detection is performed in the IV-VI semiconductor layer and the signal processing is performed in the Si wafer. Device fabrication requirements of IV-VI lasers, however, favor growth on (100)-oriented silicon substrates since IV-VI lead salt materials cleave preferentially along their {100} planes and this allows formation of in-plane cleaved Fabry-Perot cavities. Even when IV-VI materials are grown on (100)-oriented Si substrates, cleavage problems persist due to the tendency of Si to cleave along the {111} planes. But this problem can be solved by lifting off the IV-VI epilayer from the Si substrate through selective etching of the MBE-grown BaF_2 buffer layer. With a minimum of thermally resistive IV-VI materials, IV-VI lasers fabricated in this manner should have continuous wave (cw) operating temperatures greater than 260 K, within the range of thermoelectric cooling modules [5]. This paper discusses the growth of PbSe and $Pb_{1-x}Sn_xSe$ layers by LPE on Si (100) by a combination of MBE and LPE growth techniques. Characterization of the layers by optical Nomarski microscopy, XRD, FTIR spectroscopy and SEM is discussed in detail in the following sections.

EXPERIMENT

$PbSe/BaF_2/CaF_2$ buffer layer structures were grown on p-type Si (100) by MBE in an Intevac Gen II MBE system. Prior to growth, the wafer was cleaned by the Shiraki method [6] and heated at 1000°C for 30 minutes in the growth chamber to remove the Shiraki grown oxide. After oxide desorption, a 200 Å thick CaF_2 layer was grown at a wafer temperature of 580°C. This was followed by the growth of a 3200 Å thick layer of BaF_2 at the same substrate temperature, annealing at 800°C for 3 minutes and growth of an additional 1500 Å of BaF_2 at 700°C. The final MBE layer was 1000 Å thick PbSe grown at 280°C. The 3-inch silicon wafer with the buffer layer structure grown on it was cleaved to obtain 1 x 1 cm^2 substrates which were used for subsequent LPE growths. All the substrates obtained from this wafer were designated as W113. The $PbSe/BaF_2/CaF_2$ buffer layer structure was grown by MBE on two other silicon wafers as well. The substrates obtained from these growths were designated as W222 and W245, respectively. On W222, the growth procedure used was similar to that described above except for the thickness of the CaF_2 and PbSe layers which were 400 Å and 2.2 µm, respectively. For W245, following the growth of a 570 Å thick CaF_2 layer, a 2000 Å thick BaF_2 layer was grown. Annealing was performed at 900°C for 3 minutes after which a 7000 Å thick BaF_2 layer was grown at 700°C. The thickness of the PbSe layer grown at 280°C was 8100 Å.

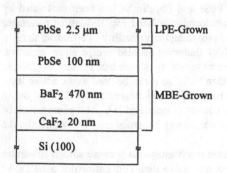

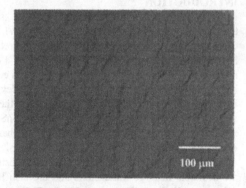

Figure 1. Schematic of the $PbSe/BaF_2/CaF_2$ structure grown on Si (100) by a combination of LPE and MBE.

Figure 2. Optical Nomarski micrograph showing the crack-free surface morphology of a PbSe layer (W113-H3) grown by LPE on Si (100).

The melt solution for LPE growth of PbSe was prepared by combining weighed amounts of Pb and PbSe according to relations that were derived using the molecular weights of the respective constituents [7]. The chalcogen concentration was chosen to be 0.2 wt. % based upon previously published phase equilibria data on PbSe [8]. The melt constituents were loaded into one of the wells of a graphite boat and the temperature of the furnace was increased to 650°C and maintained at this temperature for about an hour to allow for proper homogenization. The furnace temperature was then reduced to about 30 degrees above the expected nucleation temperature. The melt surface was observed with an optical microscope while reducing the furnace temperature at the rate of 2°C/minute. The temperature at which nucleation was observed on the melt surface was recorded. The furnace was then cooled back to room temperature and the substrate was now placed in the recess provided on the graphite slider. The furnace temperature was kept below 500°C in order to minimize the thermal stress to which the MBE-grown PbSe layer was subjected. A controlled cooling ramp of 2°C/minute was initiated and the graphite slider was pulled to position the substrate under the growth solution well at about 2 to 3 degrees above the measured nucleation temperature. After 80 degrees of cooling, the slider was pulled to position the substrate away from the melt thereby terminating the growth.

A schematic of the combined MBE/LPE grown structure is shown in Figure 1. $Pb_{1-x}Sn_xSe$ layers were grown by LPE on Si (100) substrates using the same procedure as outlined above. Based on previously published liquid-solid phase equilibria data for $(Pb_{1-x}Sn_x)_{1-z}Se_z$ [7], the chalcogen concentration, z, was chosen to be 0.25%. Percentage tin contents used for the growths were equal to 3%, 5%, 6%, 7%, and 10%, respectively. The LPE-grown PbSe and $Pb_{1-x}Sn_xSe$ layers were studied by optical Nomarski microscopy, high resolution X-ray diffraction using a Philips Materials Research XRD system, and Fourier transform infrared spectroscopy using a Bruker IR/98 spectrometer.

RESULTS AND DISCUSSION

Figure 2 is an optical Nomarski micrograph showing the surface morphology of a PbSe layer (W113-H3) grown by LPE on Si (100) using a $PbSe/BaF_2/CaF_2$ buffer layer structure. This layer was crack-free over a 8 x 8 mm^2 area and exhibited excellent growth wipe-off with no melt adhesions. Typical thicknesses for LPE-grown PbSe layers grown for 80° of cooling (40 minutes) were 2.5 μm as determined by Tencor step scan profiles. Full width half maxima (FWHM) values of about 200 arc-seconds obtained from X-ray diffraction studies of the LPE-grown PbSe layers indicated good crystalline quality of the layers. A symmetric (004) $\Omega/2\theta$ scan containing the silicon and PbSe peaks is shown in Figure 3.

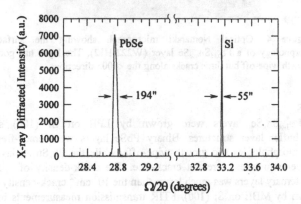

Figure 3. A symmetric (004) $\Omega/2\theta$ scan showing the PbSe and Si peaks and their respective full width half maxima values.

Unlike the LPE-grown PbSe layers, $Pb_{1-x}Sn_xSe$ layers grown by LPE on Si (100) substrates using similar $PbSe/BaF_2/CaF_2$ buffer layers were not crack-free. Figure 4 is an optical Nomarski micrograph of a $Pb_{1-x}Sn_xSe$ layer (W222-H12) grown using 5% tin in the liquid growth solution. As evident from the micrograph, the ternary layer had a good growth wipe-off but some cracks along the <100> directions. Crack-densities calculated over areas of 1.01 x 0.76 mm^2 of a Pb$_{0.95}$Sn$_{0.05}$Se layer (W222-H20) varied from 0 cracks/cm^2 to approximately 900 cracks/cm^2. An average crack density of approximately 150 cracks/cm^2 was measured over a 42 mm^2 area of this layer. This value is much lower than the 10^6 cm^{-2} crack-density observed for the MBE-grown PbSe layers on Si (100) [9]. The FTIR transmission spectrum for a Pb$_{1-x}$Sn$_x$Se layer grown using 7% tin in the liquid growth solution is shown in Figure 5. The thickness of this layer measured by Tencor step scan profile was 4.5 μm. Figure 6 is a plot of absorption edge energies versus percentage tin in the liquid growth solution for different Pb$_{1-x}$Sn$_x$Se layers. The expected decrease in absorption edge energies with increasing tin content is observed both for room temperature and low temperature transmission measurements. It is evident from the FTIR data that the 8-16 μm spectral range can be covered by these materials. Attempts made at lifting-off the PbSe layers from the Si substrate by dissolving the intermediate MBE-grown BaF$_2$ layer were successful and Fabry-Perot cavities with smooth parallel cleaves were obtained. A scanning electron micrograph of a cleaved edge is shown in Figure 7.

Figure 4. Optical Nomarski micrograph showing the surface morphology of a Pb$_{0.95}$Sn$_{0.05}$Se layer (W222-H12). The layer has good growth wipe-off but three cracks along the <100> directions.

CONCLUSIONS

PbSe and Pb$_{1-x}$Sn$_x$Se layers were grown by LPE on Si (100) substrates using PbSe/BaF$_2$/CaF$_2$ buffer layer structures. Binary PbSe layers were crack-free and exhibited excellent surface morphologies and growth wipe-offs, while the Pb$_x$Sn$_x$Se layers grown using similar buffer layer structures were not crack-free. The crack-density of approximately 150 cracks/cm^2 in the ternary layers was much lower than the 10^6 cm^{-2} crack-density observed for the PbSe layers grown by MBE on Si (100). FTIR transmission measurements indicated that the band gap energy of the ternary layers could be tailored by varying the tin content in the melt. A decrease in band gap energies was observed with a decrease in temperature and an increase in tin content. FTIR results indicated that PbSe and Pb$_{1-x}$Sn$_x$Se layers are suitable for forming the cladding and active layers of a double heterostructure laser. Efforts are being made to improve the epilayer lift-off procedures and the results will be reported in a future work.

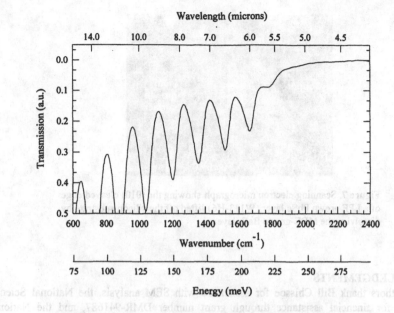

Figure 5. Room temperature FTIR transmission spectrum of a $Pb_{0.93}Sn_{0.07}Se$ layer (W222-H14) grown by LPE on Si (100). An absorption edge at about 6.4 µm and Fabry-Perot interference fringes are evident in the spectrum.

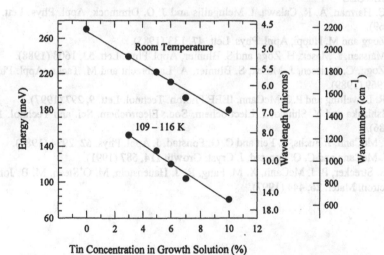

Figure 6. Absorption edge energy versus percentage tin for $Pb_{1-x}Sn_xSe$ layers with 0%, 3%, 5%, 6%, 7% and 10% tin (W113-H9, W222-H13, W222-H12, W113-H15, W222-H14, W245-H17, respectively) in the growth solution

Figure 7. Scanning electron micrograph showing the (010) cleaved edge of a LPE grown PbSe layer (W113-H6) lifted off the silicon substrate by dissolving the MBE-grown BaF₂ layer.

ACKNOWLEDGEMENTS

The authors thank Bill Chissoe for assistance with SEM analysis, the National Science Foundation for financial assistance through grant number DMR-941687, and the National Aeronautics and Space Association for financial assistance through grant number NGT-30308.

REFERENCES

1. T. C. Harman, A. R. Calawa, I. Melngailis and J. O. Dimmock, Appl. Phys. Lett. **14**, 333 (1969).
2. H. Zogg and M. Huppi, Appl. Phys. Lett. **47**, 133 (1985).
3. C. Maissen, J. Masek, H. Zogg and S. Blunier, Appl. Phys. Lett. **53**, 1608 (1988).
4. H. Zogg, C. Maissen, J. Masek, S. Blunier, A. Lambrecht and M. Tacke, Appl. Phys. Lett. **55**, 969 (1989).
5. K. R. Lewelling and P. J. McCann, IEEE Photon. Technol. Lett. **9**, 297 (1997).
6. A. Ishizaka and Y. Shiraki, J. Electrochem. Soc.: Electrochem. Sci. and Technol. **133**, 666 (1986).
7. P. J. McCann, J. Fuchs, Z. Feit and C. G. Fonstad, J. Appl. Phys. **62**, 2994 (1987).
8. P. J. McCann and C. G. Fonstad, J. Cryst. Growth **114**, 687 (1991).
9. B. S. Strecker, P. J. McCann, X. M. Fang, R. J. Hauenstein, M. O'Steen, M. B. Johnson, J. Electron. Mater. **26**, 444 (1997).

IN SITU SPECTROSCOPIC ELLIPSOMETRY FOR REAL TIME COMPOSITION CONTROL OF Hg₁₋ₓCdₓTe GROWN BY MOLECULAR BEAM EPITAXY

R. DAT,[*] F. AQARIDEN,[*,**] W.M. DUNCAN,[***] D. CHANDRA,[*] and H. D. SHIH[*]
* Raytheon TI Systems, Sensors and Infrared Laboratory, P.O Box. 655936, MS 150, Dallas, TX 75265, U.S.A
** Permanent address: Microphysics Laboratory, Department of Physics, M/C 273, University of Illinois at Chicago, 845 W Taylor #2236SES, Chicago, IL 60607, U. S. A.
*** Texas Instruments Incorporated, Components and Materials Research Center, P. O. Box 655936, MS 147, Dallas, TX 75265, U.S.A

ABSTRACT

Spectral ellipsometry (SE) was applied to in situ composition control of Hg₁₋ₓCdₓTe grown by molecular beam epitaxy (MBE), and the impact of surface topography of the Hg₁₋ₓCdₓTe layers on the accuracy of SE was investigated. Of particular importance is the presence of surface defects, such as voids in MBE- Hg₁₋ₓCdₓTe layers. While dislocations do not have any significant impact on the dielectric functions, the experimental data in this work show that MBE- Hg₁₋ₓCdₓTe samples having the same composition, but different void densities, have different effective dielectric functions.

INTRODUCTION

Accuracy of the composition control during epitaxial growth is always desirable. Spectral ellipsometry (SE) offers several advantages as an in situ, feedback sensor for molecular beam epitaxial (MBE) growth. The technique is non-intrusive and non-destructive. It offers reproducible growth and can provide real-time feedback when integrated with a fast data processor. It also exhibits high sensitivity to layer thickness and complex index of refraction (i.e., n-ik). Hence, it is ideal for sensing applications. This work is on the application of SE to composition control during MBE growth of Hg₁₋ₓCdₓTe on Cd₁₋ᵧZnᵧTe(211)B.

The issue of real-time computation with spectral ellipsometry is especially challenging since numerical solution methods depend not only upon the reflection equations, but also upon several other factors which strongly influence numerical solution. Some of the key factors include surface morphology of the starting substrate and the evolving thin film; library dielectric functions relevant at the deposition temperature; angle and plane of incidence at the substrate surface at the time the spectral data are acquired; and surface condition and bulk properties of the view ports. A poor surface morphology increases the scattering and depolarization of the incident radiation, degrades the signal-to-noise ratio of the spectral information and, consequently, leads to erroneous data to the feedback loop. In such a situation, the feedback becomes unstable and real-time control is lost. The second factor, library dielectric function, is

377

very sensitive to the quality of the MBE film. Defects such as voids and hillocks [1,2] need to be minimized in films that are specifically grown for the purpose of providing library functions. The last two points have been discussed in an earlier report [3]. Preliminary results show that MBE- $Hg_{1-x}Cd_xTe$ films with the same composition but different defect densities exhibit distinct spectral profiles. These results, as well as the influence of some of the other factors on real-time in situ SE of MBE- $Hg_{1-x}Cd_xTe$, are presented below.

EXPERIMENTAL

A schematic of the DCA Instruments MBE deposition chamber used for this work is shown in Figure 1. The chamber is fitted with ports for in situ sensors, a 3-inch manipulator with x, y, z movement, 4 standard Knudsen effusion cells, and a custom built Hg source. The manipulator is designed with piezoelectric elements placed at 120° intervals to suppress wobble during substrate rotation to better than ±0.05°. Control for the MBE growth process is provided

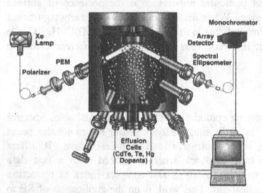

by a Siemens Simatic TI555 processor with individual DC power supplies. A vacuum better than $1x10^{-9}$ mb is maintained by employing both a cryopump and a 9-inch Hg diffusion pump. Single and multilayer epitaxial $Hg_{1-x}Cd_xTe$ films with various compositions were grown on 2x3 cm (211)B-oriented CdZnTe (4% Zn) substrates, using separate fluxes of Te, CdTe and Hg, and a nominal deposition temperature of 198° C. The growth rate was approximately 2.5 μm/hr. During $Hg_{1-x}Cd_xTe$ deposition, a substrate rotation of 3-10 rpm was used in order to achieve compositional uniformity of better than 0.001. Compositional control was implemented by establishing a feedback loop between the temperature controller of the CdTe cell and the output from the spectral ellipsometer.

Figure 1. Schematic of SE system as mounted on the DCA MBE chamber.

The spectral ellipsometer, attached to two opposing conflat flanges (and strain free quartz viewports), subtends an angle of approximately 74° with respect to the surface normal of the sample (see Figure 1 and [3]). The phase modulated spectral ellipsometer has been described previously [4-6]. This spectral ellipsometer is supported by multichannel detection and digital signal processing techniques to provide an analytical tool that is fast, cheap, and robust since it has no moving parts. The spectral ellipsometer employs a polarizer-modulator-sample-analyzer

(PMSA) configuration. The light passes through an analyzer after traversing the vacuum chamber and being reflected from the sample. The light is then taken off the MBE system and delivered via fiber optics to a grating monochromator and photodiode array detector. In its present form, the SE will typically scan 46 wavelengths in less than 0.5 sec to generate the ellipsometric parameters Ψ and Δ. The dielectric functions and thickness of the MBE- $Hg_{1-x}Cd_xTe$ layers are fitted to a model, where layer thickness, angle of incidence, complex index of refraction, n-ik, and wavelength are used as variables. In the case of $Hg_{1-x}Cd_xTe$, the effective medium approximation (EMA) model of Bruggeman has proved to be effective in determining the dielectric response [7], and Aspnes has discussed the general and specific forms of the effective medium theory for calculating the macroscopic dielectric response [8].

In this study, the layers are treated as a random heterogeneous mixture of $Hg_{1-x}Cd_xTe$. It is necessary here to interpolate ternary dielectric values by mixing two ternary compositions, one higher and one lower than the desired composition, with well known dielectric functions. The dielectric function of the investigated layer is obtained by interpolating between the two known dielectric library functions. In this work, for example, the calibrated library functions were obtained from MBE- $Hg_{1-x}Cd_xTe$ films with compositions of 0.203 and 0.367. These values were selected to cover film compositions in the LWIR and MWIR range of interest.

RESULTS AND DISCUSSION

Unlike the type of material properties and characteristics assumed in model calculations, real crystals frequently deviate from almost all of the ideal assumptions used in material system modeling. It is difficult to produce thin film crystals that satisfy stringent requirements, such as isotropic dielectric functions and surfaces that are terminated with smooth and planar boundaries.

Most often, the MBE- $Hg_{1-x}Cd_xTe$ films grown on CdZnTe substrates exhibit various surface irregularities such as voids, hillocks, grain boundaries, and other inhomogeneities which will affect the optical properties of the film. The dielectric response of such films will depend on the composition of the film and the density of the various irregularities and their interaction with the incident beam of polarized light from the ellipsometer. The observed change in the polarization state of the reflected beam is used to infer the optical properties of the film. The presence of defects on the films can, therefore, lead to apparent dielectric functions that are significantly different from their counterparts on a defect free film.

The effects of specific defects, such as voids and dislocations, on the optical properties of MBE- $Hg_{1-x}Cd_xTe$ layers grown in the DCA system are given in Table 1 and Figure 2. Table 1 gives the composition, etch pit density (EPD), and void density for four MBE- $Hg_{1-x}Cd_xTe$ layers. The void densities for DCA329 and DCA332, with compositions of approximately 0.18 and 0.20, were intentionally reduced by decreasing the deposition temperature by 3-4°C from that used for DCA318 and DCA319. However, the consequence of using a lower growth temperature is that the EPD increases for a fixed value of mercury flux. The real and imaginary parts of the complex index of refraction for samples DCA318 and DCA329, with the same composition, are plotted in Figure 2(a). Based on the data, it is obvious that the film with the higher void density exhibits a lower index of refraction. Similar results are observed for DCA319 and DCA332, as shown in Figure 2 (b). In both cases, the EPD does not appear to have any significant impact on the optical characteristics shown in Figure 2.

TABLE 1: EPD and void density for $Hg_{1-x}Cd_xTe$ films that were prepared for establishing library dielectric functions

Run #	Composition	EPD (cm^{-2})	Void Density (cm^{-2})
DCA318	0.185	6.8E5	9.0E4
DCA319	0.201	2.2E5	3.0E4
DCA329	0.182	3.3E5	1.2E4
DCA332	0.203	1.0E6	8.0E3

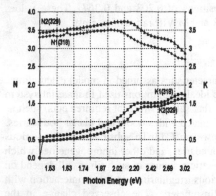

Figure 2 (a): Complex indices of refraction for two MBE $Hg_{1-x}Cd_xTe$ layers with same composition but different void densities.

Figure 2 (b): Another comparison of indices of refraction for two MBE $Hg_{1-x}Cd_xTe$ layers with same composition but different void densities.

These results show that the dielectric response, as probed by the ellipsometer, is intimately related to the void density of the MBE- $Hg_{1-x}Cd_xTe$ film. Hence, in order to use the ellipsometer as a feedback control tool to produce high quality MBE- $Hg_{1-x}Cd_xTe$ thin films in a reproducible manner, it is essential to acquire accurate dielectric library functions from samples grown under conditions yielding void concentrations similar to target samples.

The library dielectric functions used in this study are given in Figure 3. The films that produced those dielectric functions were grown in the DCA MBE chamber under conditions that yielded specular surfaces. Their compositions were determined by ex situ Fourier Transform Infrared (FTIR) measurements. To evaluate

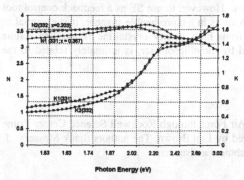

Figure 3. $Hg_{1-x}Cd_xTe$ library dielectric functions used in this study. Composition range 0.20-0.37.

the spectral ellipsometer as a viable, real-time, feedback tool to control the composition of MBE- $Hg_{1-x}Cd_xTe$ layers, a sequence of ten LWIR growth runs were performed. The composition of these ten growth runs, as acquired by SE, are plotted in Figure 4. For a target LWIR composition of 0.225, SE control for ten consecutive runs resulted in MBE- $Hg_{1-x}Cd_xTe$ films having an average composition of 0.2241, as determined by in situ SE, with a standard deviation of 0.0037. This type of control is approximately an order of magnitude better than the case without SE control. Analysis of the same ten LWIR samples by ex situ FTIR yielded an average composition of 0.2280, with a standard deviation of 0.0132. The reason for the difference between SE and FTIR data is not clear. However, it should be noted that for some of these samples, composition control was exercised at several intervals during the growth process by appropriate adjustments in the CdTe cell temperature. Consequently, the composition versus depth profiles for these films may vary depending on the degree of control at each interval. For a $Hg_{1-x}Cd_xTe$ sample with a non-uniform composition profile, ex situ FTIR evaluation will report on that segment of the layer having the lowest composition. The latter FTIR data will most likely be different from the recorded SE composition of the same sample.

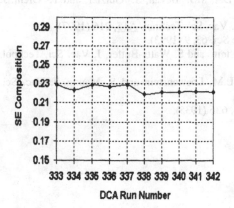

Figure 4. Real-time in-situ composition control of MBE $Hg_{1-x}Cd_xTe$ using SE for 10 consecutive runs.

CONCLUSION

Real time, in situ feedback control using SE is very effective in providing repeatable composition control of MBE- $Hg_{1-x}Cd_xTe$ layers. However, to use SE as a feedback control tool to produce high quality MBE- $Hg_{1-x}Cd_xTe$ thin films in a reproducible manner, it is essential to acquire accurate dielectric library functions from high quality crystalline films. The results indicate that the dielectric response, as probed by the ellipsometer, is intimately related to the surface features of the MBE- $Hg_{1-x}Cd_xTe$ film.

ACKNOWLEDGMENTS

The work was supported in part by the Air Force through Rockwell Science Center Prime Contract No. F33615-95-C-5424 and monitored by Lyn Brown. The authors thank Messers. J. Frazier and S.W. Gutzler for their excellent technical assistance.

REFERENCES

[1] M. Zandian, J.M. Arias, J. Bajaj, J.G. Pasko, L.O. Bubulac, and R.E. DeWames, J. Electronic Materials 24, 1207 (1995).
[2] D. Chandra, H.D. Shih, F. Aqariden, R. Dat, M.J. Bevan, S. Gutzler, and T. Orent, J. Electronic Materials, 1998 (in press).
[3] W.M. Duncan, M.J. Bevan, and H.D. Shih, J. Vac. Sci. Technol. A15, 216 (1997).
[4] W.M Duncan and S.A. Henck, Appl. Surface Sci. 63, 9 (1993).
[5] S.A. Henck, W.M. Duncan, L.M. Lowenstein, and S. Watts-Butler, J. Vac. Sci. Technol. A11, 1179 (1993).
[6] W.M. Duncan, S.A. Henck, J.W. Kuehe, L.M. Loewenstein and S. Maung, J. Vac. Sci. Technol. B12, 2779 (1994).
[7] D.A.G. Bruggeman, Ann. Phys. (Leipzig) 24, 636 (1935).
[8] D.E. Aspnes, Am. J. Phys. 50, 704 (1982).

NONMONOTONOUS BEHAVIOR OF TEMPERATURE-DEPENDENCE OF PLASMA FREQUENCY AND EFFECT OF A LOCAL INSTABILITY OF THE PbTe:In,Ga LATTICE

A.I.BELOGOROKHOV*, L.I.BELOGOROKHOVA**, D.R.KHOKHLOV**
* State Scientific & Research Institute of Rare Metals. Leninsky Prosp., 156-517, Moscow 117571, Russia. Fax: (095) 4387664. E-mail: abelog@glas.apc.org
**Moscow State University, Physics Department, Russia

ABSTRACT

The optical spectra of PbTe:In,Ga samples were studied in the Far Infrared region at temperatures T=11K-350K. Anomalous behavior of the temperature dependence of the plasma frequency at 65K-80K was observed for the first time. This is explained by means of a model of localized changes in the lead telluride lattice structure near impurity centers.

1.INTRODUCTION

There have recently been several studies of the influence of various added impurities on the electron spectra of ternary compounds of lead, tin, and tellurium. The alloys most investigated till now have been $Pb_{1-x}Sn_xTe:In,Ga,Cd$ [1]. A number of interesting results have been obtained with sample temperatures below 25K, mainly with single crystals, including anomalous photoconductivity, an insulator phase in the conductivity [2], a spontaneous ferroelectric voltage in the crystal [3], and Shubnikov-de Haas oscillations of the Hall coefficient [4]. The present investigation concerned the temperature dependencies of the reflectivity $R(\omega)$ and the transmittance $T(\omega)$ of lead telluride samples containing indium and gallium, in the infrared spectral region (ω = 10 - 5000 cm^{-1}) and temperatures T = 11K - 350K.

2. EXPERIMENTAL DETAILS

The lead-telluride samples were grown from the gas phase, and 0.005-1.0 wt.% impurities were added. The single crystals obtained being 0.5-1.0 mm thick. X-ray diffraction (XRD) was used to determine the structural quality of the samples such as the period and defect structure. X-ray diffraction measurements were performed using a powder Θ-2Θ diffractometer and a high resolution four-crystal diffractometer (Philips MRD). Infrared absorption and reflection measurements were performed with a Bruker model IFS-113v Fourier transform infrared (FTIR) spectrometer using a variable temperature sample cryostat equipped with KRS-5 and polythene (for far IR) windows. The crystals, mounted in a flow-helium cooled cryostat, were illuminated with a mercury lamp and a globar. The temperature was kept constant within ±0.1K. The wavenumber resolution was 0.1 cm^{-1}. The sample surfaces were chemically polished and then chemically etched. The carrier concentration and mobilities were found by means of the usual technique based on the Hall effect. The residual impurity concentration was monitored by X-ray spectrometry method (Camebax instruments and a Link elemental composition recorder) and with laser micro-mass spectrometry system LAMMA-1000 (Leibold-Heraeus).

3. RESULTS AND DISCUSSION

Figure 1 shows typical reflection spectra of *p*-type PbTe:<1.0 wt.%In> at various temperatures.

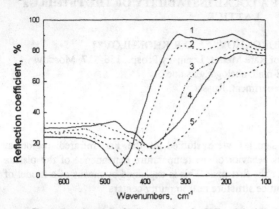

Fig. 1. Experimental reflection spectra of a PbTe:<0.4 wt.%> sample measured at temperatures, T(K): 1) 10; 2) 60; 3) 65; 4) 80; 5) 150.

There is a clear minimum (ω_p) in the wave number range ω=380-450cm^{-1}, corresponding to the plasmon-phonon modes. As the temperature rises from 11K to 60K, there is first a small shift of ω_p to longer wavelength, corresponding to the change in the carrier effective mass in that temperature range. Then, over a range ΔT=15K, the minimum, ω_p moves by $\Delta\omega$=70cm^{-1}. As the temperature increases further, above 100K, R(ω) reverts to "normal" behavior. So far as we know, this is the first observation of anomalous behavior of R(ω) at T=65K. The amount of anomalous deviation of R(ω), $\Delta\omega$, and the shift temperature T_i were dependent on the group III impurity content of the lead telluride. Other conditions being equal, $\Delta\omega$ increases with the impurity content. The dependence of T_i on N_{In} and N_{Ga} is less regular, with fluctuations in the range T_i=65-80K. The samples were studied with a temperature cycle 300K→10K→350K→300K several times repeated. It was noted that $\Delta\omega$ decreases when the crystal undergoes repeated rapid cooling. Samples raised to T=400K initially or in one cooling-heating cycle showed no anomalous properties subsequently. Figure 2 shows

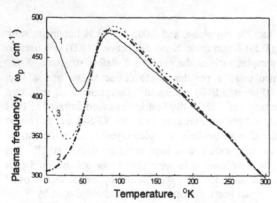

Fig. 2. Temperature dependences of the plasma frequency for PbTe samples containing In (wt.%): 1) 1.0; 2) 0.4; 3) 0.8.

the temperature dependencies of the plasma frequency ω_p. This parameter was calculated as proposed previously [5]. The increase of ω_p at low temperatures is explained by the contribution to the free carrier density from photocarriers due to radiation incident on the sample from the globar or Hg-lamp used to obtain the observed optical spectra. Nevertheless, the temperature at which ω_p begins to increase, and the photocarrier concentration, are considerably greater than the corresponding quantities found from electrical measurements. In order to exclude the effect of photocarriers on the further (at $T=60K$) sharp increase of ω_p, the reflection spectra $R(\omega)$ were recorded in a "negative flux" regime, whereby the cooled sample is put in the position of the radiation source, using the symmetry of the Michelson interferometer. The signal detector was a pyroelectric element. This also provided the radiation spectrum from the sample due to a temperature difference between the cooled sample and the instrument chamber. From the dependencies $R(\omega)$ thus obtained, we determined the values of ω_p, which in the

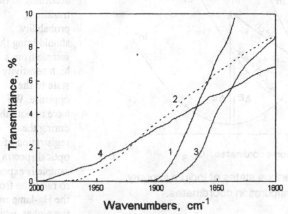

Fig. 3. Observed transmission spectra of a PbTe:<0.4 wt.% Ga> sample measured at T(K): 1) 10; 2) 73; 3) 80; 4) 150.

temperature range concerned were almost the same as those given by the usual experimental set-up. The photoexcited carriers thus have a negligible effect on the "anomalous" increase of ω_p at $T\sim65K$-70K. Figure 3 shows the transmission spectra of a PbTe:<0.4 wt.%Ga> sample at $T=10K$-350K in the PbTe fundamental absorption edge region. It is seen that up to 70K the edge moves to higher energies, corresponding to an increase band gap Eg with rising temperature, in full agreement with previous results [6]. On further heating up to 80K, the position of the absorption edge changes markedly by $\Delta\omega\sim170cm^{-1}$, after which E_g again increases smoothly, in full agreement with [6]. This thermal instability of E_g can be explained by the occurrence at $T\sim80K$ of a structural transition due to local lattice instability in the immediate neighborhood of defects. We will seek to explain this effect by the interaction of light with the electron subsystem in the presence of optical phonons. Here, it is necessary to bear in mind the dependence of the impurity state energies on the configuration coordinate $Q_{k,s}$ which determines the lattice distortion in the immediate neighborhood of the impurity atom. In the lead telluride lattice there is dissociation of the state In^0 (relative to the lattice) into a positive ion In^+ and a negative ion In^-, on account of the high static permittivity of PbTe[7]. The In^+ and In^- states are separated by a potential barrier, and the energies of these ionized states have minima below that of In^0 (Fig.4). The electron wave functions are highly localized at the impurity centers in lead telluride. When the amount of impurity in the sample increases, one can expect a stronger interaction between them by virtue of the deformational perturbation. That is, there is a change in the local structure of the lattice near an impurity center when the charge state of a self-localized perturbation changes. Electrons in the conduction band E_c of the semiconductor (with strong degeneracy at the Fermi level) are in equilibrium with those occupying the In- minimum. There is a certain probability of electron transitions through the potential barrier separating the In^+ and In^- states when they acquire an additional energy ΔE. The Fermi level E_F must then move toward the conduction band E_c, thus reducing the resistivity of the sample. It is seen from Fig. 4 that, when the additional energy ΔE is acquired, the In⁻ state becomes In⁰, and then, with its potential energy reduced by the electron-phonon interaction reaches the minimum of In⁺. The contribution from the two released electrons to the total electron density decreases the gap E_c-E_F. The process,

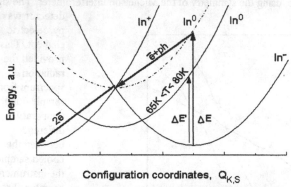

Fig. 4. Energies of various states of indium impurity in PbTe, in configuration coordinates.

after summation according to the transition probability, should bring the sample from a high resistivity state to the opposite. We have to note that during the registration of optical spectra the sample is exposed to radiation from the Hg-lamp or the globar, which is not monochromatic. The situation is therefore such that photons with various energies, interacting with the In⁻ state than convert it to In⁰ state. There are many such process. In consequence, even at the initial instant, there would have to be a large change in the free carrier density in the crystal at any temperature 300K>T>4.2K. This would further cause a sharp increase in the plasma density ω_p and a shift of the minimum reflectivity $R(\omega)$ in the plasma oscillation region. Nevertheless, as noted above, no such effect occurs. It is most likely that, as far as T~85K, the In⁰ state minimum energy in configuration coordinates is fairly high, as shown, for example, by the chain line in Fig. 4, and the energy ΔE needed to convert In⁻ into In⁰ (that is, to overcome the energy barrier between In⁺ and In⁻) is greater than the energy corresponding to the fundamental absorption range. Since the absorption cross section for interband transitions is greater than for In⁺→In⁰, the contribution of the latter at high temperatures is small. In order to account for the increased number of transitions In⁺→In⁰+ph→In⁻, where ph is the electron-phonon interaction, let us suppose that the In⁰ minimum energy at 65K<T<80K falls, for example, to the level shown by the continuous line in Fig. 4. In consequence, the number of In⁻ →In⁰ transitions increases. Next, since an estimate of the Huang-Rhys factor shows them to be greater than unity, there is every reason to postulate a strong interaction with the phonon modes, in this case optical ones. The further relaxation of the excited electron state is the result of lattice relaxation and the In⁰→In⁺ transition. The two electrons released make an additional contribution to the ω_p, which shifts the plasma minimum. This is what we in fact observed for $R(\omega)$ and $\omega_p(T)$ at 80K>T>65K. It is natural to consider the change in the energy of the In⁰ state relative to the immediate local lattice ion environment. The $Pb_{1-x}Sn_xTe$ solid solutions with 0.22<x<0.25 are known [8] to pass from the cubic to the rhombohedral phase at T_c~20-22K. However, if the sample contains a sufficient number of free carriers, there may be no phase transition, since they stabilize the cubic phase [9,10]. Lead telluride is regarded as a virtual ferroelectric. The addition of In or Ga favors the phase transition, for three reason: the Jahn-Teller instability of the cubic phase, the softening of the optical phonon when some of the lead atoms are replaced by the lighter and smaller indium (Ga) atoms, and the compensating effect of impurities, shown by a reduction in the free carrier density, which in turn stabilized the cubic phase. Thus, the addition of In or Ga causes instability of the PbTe lattice and also the formation of traps for free electrons, which increases the phase transition probability. On the other hand, the absence of the ferroelectric potential when the sample is illuminated [3] is regarded as indicating that the lattice returns to the cubic phase.

4. SUMMARY.

We can therefore suppose that lead telluride, which when pure is a virtual ferroelectric, if alloyed with group III impurities (In,Ga) and illuminated with strong, not necessarily monochromatic, radiation, has at $T\sim65K$ a local structural phase transition from the cubic to the rhombohedral phase, remaining in the latter down to liquid helium temperatures. It is precisely the lowering of the potential barrier between In^+ and In^- states, because of the local configuration change in the environment of an impurity atom, that reduces the free carrier density and causes the local phase transition. This conclusion is confirmed by results [11] concerning many-phonon recombination of carriers at frozen photoconduction centers in PbTe films. It was noted that the secondary center recombination cross section σ_2' is much less than the carrier trapping cross section σ_1'. This was explained by means of a model with a large deformation of the lattice - that is, the trapping of carriers with no change in the configuration of atoms near the defect is held to be impossible. It was further found [11] that illuminating the sample at $T\sim50K$ forms traps having a larger photoionization cross section, in agreement with our results. The presence of a strong electron-phonon interaction against the background of the local change in the lead telluride lattice structure near impurity centers, which is responsible for the results found here, may also affect the photoconductivity spectrum of the material at optical phonon frequencies [12]. Another possible consequence is the formation of local plasma filaments with a long lifetime ($\sim10^{-3}$s). These ideas need to be studied along with further experimental results.

ACKNOWLEDGMENTS
We acknowledge support of this work by the Russian Foundation for the Basic Research (RFBR), Grant No.95-02-04658, 96-02-18853 and by the INTAS-RFBR Grant No. 95-1136.

REFERENCES

1. C.M.Penchina,A.Klein,and K.Weiser, J.Phys.Soc.Jpn. **49S,A,** 783 (1980).
2. B.A.Akimov,N.B.Brandt,B.S.Kerner, et.al.,Solid State Commun. **43**, 31 (1982).
3. K.H.Herrmann,G.A.Kalyuzhnaya,K.-P.Mollmann, and M.Wendt, Phys.Status Solidi **A71**, K21 (1982).
4. K.Murase,S.Takaoka,T.Itoga,S.Ishida, Lecture Notes in Physics **177**, 55 (1983).
5. A.I.Belogorokhov,A.G.Belov,I.L.Petrovich, Opt.Spektrosk. **63**, 1293 (1987).
6. W.W.Anderson, IEEE J.Quantum Electron. **QE-13**, 532 (1977).
7. I.A.Drabkin, B.Ya.Moizhes, Sov.Phys.Semicond. **17**, 611 (1983).
8. K.Weiser,A.Klein,M.Ainhorn, Appl.Phys.Let. **34**, 607 (1979).
9. A.I.Belogorokhov,A.G.Belov,I.G.Neizvestnyi, et.al., Sov.Phys. JETP **65**, 490 (1987).
10.S.Takano,S.Hota,H.Kawamura, et.al., J. Phys.Soc.Jpn. **37**, 1007 (1974).
11.K.A.Baklanov,I.P.Krylov, JETP Let. **48**, 364 (1988).
12.A.I.Belogorokhov,I.I.Ivanchik,D.R.Khokhlov,S.Ponomarev, Brazilian J. of Physics 26, 308 (1996).

X-RAY CHARACTERIZATION OF PbTe/SnTe SUPERLATTICES

S.O. FERREIRA, E. ABRAMOF, P.H.O. RAPPL, A.Y. UETA, H. CLOSS, C. BOSCHETTI, P. MOTISUKE, and I.N. BANDEIRA.

Instituto Nacional de Pesquisas Espaciais INPE-LAS CP: 515 CEP: 12201-970 São José dos Campos, Brazil. E-mail: sukarno@las.inpe.br

Abstract

PbTe/SnTe superlattices have been proposed many years ago for use as base material for infrared detectors. However, many difficulties have prevented its use, mainly the ones related to obtaining low concentration SnTe. Recently we have shown that SnTe layers with relatively low hole concentration can be grown by molecular beam epitaxy at low temperature using stoichiometric charges. In this work we investigate the structural properties of PbTe/SnTe superlattices grown by molecular beam epitaxy on (111) BaF2 substrates. Sample characterization has been done by high resolution x-ray diffraction. Information on strain was obtained from reciprocal space maps of asymmetric Bragg reflections and used as input parameters for dynamical simulation of the diffraction spectra.

Introduction

The study of the PbTe-SnTe alloy system has attracted much interest due to their application for infrared detectors and also due to many interesting features as compared to III-V or II-VI semiconductors. The main properties are: i) the inversion of valence and conduction band symmetry between PbTe and SnTe, which, in principle, produces a zero gap compound for Sn composition around 35 % at T = 0 K; ii) very large dielectric constant and therefore high carrier mobility due to screening; iii) direct energy gap at the L-points of the Brillouin zone with many valleys; iv) a type II alignment of the band edges, with the valence band maximum of SnTe higher than the conduction band minimum of PbTe. The properties of PbTe/SnTe superlattices (SL's) grown on BaF$_2$(111) and on KCl(001) substrates have been studied in the last ten years.[1,2,3] Murase at all[1] have observed a superconducting transition at temperatures varying from 1.5 to 7 K, depending on PbTe layer thickness and sample substrate, and suggested that the phenomena are related with the presence of metal microprecipitates in the samples. Mironov at all[3] have explained the anomalous behavior of the conductivity in their samples by a structural phase transition undergone by the SnTe layer. Litvinov at all[4] have used the presence of two-dimensional interface states to explain the magnetic field dependence of Hall coefficient of short-period SL's. All work done until now, have been concentrated on the electrical characterization of PbTe/SnTe SL's. The structural quality has been almost forgotten, although many of the observed electrical properties and most of the open questions are related with structural details of the samples.

In this paper we have grown PbTe/SnTe superlattices on BaF$_2$(111) using solid source molecular beam epitaxy (MBE). Detailed structural characterization has been made by high resolution x-ray diffraction. Reciprocal space maps of asymmetric diffraction peaks, measured using a triple axis configuration, offer information about the sample strain state. Using this strain information and dynamical simulations of ω/2Θ scans the period of the superlattice and the thickness of the individual layers were obtained.

389

Sample Growth

The superlattices were grown in a RIBER 32P MBE system equipped with PbTe, SnTe and Te effusion cells. As substrates, we have used freshly cleaved $BaF_2(111)$ crystals, which were pre-heated at 500 °C for 15 min before growth. The growth temperature was 250 °C and the growth rate about 4 Å/s. Before growing the superlattices, PbTe and SnTe single layers were grown using stoichiometric sources. These samples have shown very good crystalline quality, with the rocking curve of the (222) diffraction line showing a full width at half maximum (FWHM) of about 100 arcsec, for 4 μm thick layers. The PbTe layers were p-type with hole concentration of 3×10^{17} cm^{-1} and the SnTe samples were also p-type with carrier concentration of 2×10^{19} cm^{-1}. Details about the growth and characterization of these layers are given elsewhere.[5,6]

The SL's were grown on the top of 4 μm thick PbTe buffer layers. The growth rates of PbTe and SnTe were checked by measuring the thickness of thick calibration samples and also by measuring reflection high energy electron diffraction (RHEED) intensity oscillations. Substrate rotation during the growth guarantees the thickness uniformity. The relatively low substrate temperature used (250 °C) was chosen to minimize interdiffusion and ensure abrupt interfaces.

In this work, we will focus attention in two kinds of SL's. In the first one the thickness of PbTe and SnTe layer is chosen in such a way that the average Sn content in the SL is under 10 % and the number of repetitions is restricted to less than 8. In this case, the in-plane lattice parameter of the whole structure is expected to be the same as the lattice parameter of the PbTe buffer. The sample is called to be pseudomorphic. The study of these structures gives precise information about the elastic constants of PbTe and SnTe.

The second one consists of 50 to 100 repetitions of PbTe and SnTe layers, with individual thickness ranging from 150 to 300 Å for PbTe and from 20 to 140 Å for SnTe. For SnTe layers thicker than this we have observed a rapid decrease in sample quality, as indicated by rocking curve FWHM. This fact is in agreement with observations of Fedorenko at all,[7] who have measured a critical thickness around 100 Å for SnTe on PbTe. Above this thickness, the strain is relieved by the introduction of misfit dislocations, which cause degradation in crystalline quality. Therefore, in these samples, the individual layers are strained, but, due to the high number of repetitions, the whole superlattice is expected to relax in relation to the buffer. In such case the SL is called free-standing having an in-plane lattice constant smaller than the lattice constant of the buffer layer.

High resolution x-ray characterization

The system used for structural analysis was Phylips X'Pert diffractometer in the triple axis configuration. This system employs a double crystal monochromator in the primary optics, which gives an axial divergence ($\Delta\omega$) of 12 arcsec and a wavelength dispersion of less than 3×10^{-5}. A Ge analyzer crystal in the secondary optics gives also a 12 arcsec resolution in the 2Θ direction ($\Delta\Theta$). Recently, a number of authors have described in detail the power of this technique for characterization of heterostructures.[8,9] Figure 1 shows a schematic representation of the reciprocal lattice points of a (111) oriented PbTe layer with x-ray incidence in the (11-2) azimuth. The shaded areas are not accessible since $\omega < 0$ or $2\Theta < \omega$. The horizontal and vertical coordinates give a measurement of the reciprocal lattice constant parallel (in-plane) and perpendicular (growth direction) to the sample surface, respectively. The area marked in the figure illustrates a map around the asymmetrical (224) diffraction peak. The map is obtained making a series of ω/2Θ scans (along the radius) at different ω settings or a series of ω scans (rocking curves, axial direction) at different 2Θ settings.

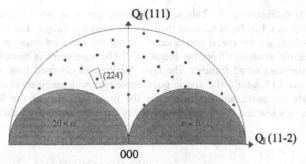

Figure 1- Schematic representation of the reciprocal lattice points of a (111) oriented PbTe sample, showing a map around the (224) reciprocal lattice point.

Results and discussion

Figure 2 shows the reciprocal space map around the (224) point for a totally pseudomorphic sample, representative of the first series. The sample consists nominally of a 4 μm PbTe buffer and 7 repetitions of PbTe (175 Å) and SnTe (17 Å). The x-ray intensity is shown in a gray scale, with the highest intensity corresponding to the darkest points. The map shows the PbTe buffer peak and two superlattice satellite peaks. The small three peaks, which appear in between, are interference fringes related to the total superlattice thickness. As can be seen, all peaks in the map have the same horizontal coordinate, indicating that all layers have the same in-plane lattice constant, as expected. Another important information, is that only the buffer peak has some deformation in the ω direction, what means that all defects are restricted to this layer.

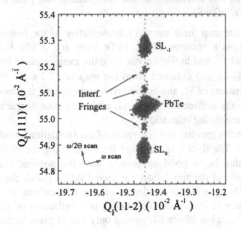

Figure 2 - Reciprocal Space map around the (224) asymmetric Brag diffraction point for a pseudomorphic SL.

In the Figure 3, we see a ω/2Θ scan, in a wide angle range, for the (222) diffraction of the same sample. This curve can be seen as a cut through the center of the map in Figure 2, but for a

391

symmetric (222) diffraction peak. This scan shows the peaks of the BaF₂ substrate and PbTe buffer and 8 SL peaks, besides of the total thickness interference fringes. The solid curve, with the experimental points, is a dynamical simulation assuming the strain information obtained from the map and the elastic constants from the literature.[10] The agreement is remarkably good in the position and intensities of all features. The layer thicknesses obtained from the simulation are 174 Å for PbTe and 17 Å for SnTe, also in good agreement with nominal values. The FWHM for all SL peaks are the same, and determined by the finite thickness of the SL, indicating that interdiffusion is very low and thickness reproducibility from layer to layer is also very good.

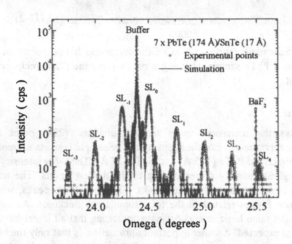

Figure 3 - X-ray data and simulation of the (222) diffraction peak for a pseudomorphic PbTe/SnTe SL

The (224) reciprocal space map for a sample representative of the free-standing SL's is shown in Figure 4. In this sample a nominal 225 Å PbTe layer and a 130 Å SnTe layer are repeated 50 times. It is clear, that SL and buffer do not have the same in-plane lattice parameter. The in-plane strain $((a_{||SL}- a_{||Buff})/a_{||Buf})$ obtained from the map is $- 7.3 \times 10^{-3}$. Where $a_{||SL}$ and $a_{||Buff}$ are the in-plane lattice constants of SL and buffer layer, respectively. It is interesting to note that, in this case, the FWHM in the ω direction of the SL peaks are even bigger than that of the PbTe buffer, due to defect creation during relaxation.

The (222) diffraction spectra (points) and the result of the simulation (solid curve) for this sample is shown in the Figure 5. The thickness obtained from the fitting for PbTe and SnTe are 232 and 132 Å, respectively, showing a good agreement with the nominal values. However, neither the relative peak intensities of the simulation nor their FWHM match the measured data. The larger experimental FWHM is a direct evidence of the dislocations introduced in the relaxation process. The difference in intensities, otherwise, is a reflection of the strain profile inside the sample. A simulation with the whole SL having only one in-plane lattice parameter, as expected from the free-standing assumption, does not give a good result, even if we use the strain extracted from the map. The best simulation, shown in Figure 5, was obtained assuming that the layers of the first 5 repetitions are completely pseudomorphic to the buffer, while the other have a linear strain profile with a mean value equal to that obtained from the map. Since this simulation can only be classified as reasonable, the real strain profile within the SL is much more complex than this one and shows the importance of a detailed structural characterization of these samples.

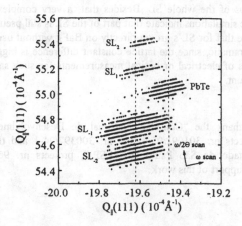

Figure 4- Reciprocal Space map around the (224) asymmetric Brag diffraction point for a free-standing SL.

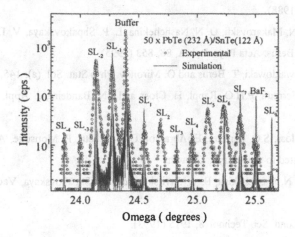

Figure 5- X-ray data and simulation of the (222) diffraction peak for a free-standing PbTe/SnTe SL

Conclusions

We have shown that x-ray reciprocal space mapping is an essential technique for characterization of PbTe/SnTe superlattices. For SL's where the total thickness is small enough, we have shown that they are completely pseudomorphic to the buffer. Dynamical simulations show that almost perfect samples, with very small interdiffusion, if any, can be grown by MBE. In these samples, the strain state is completely determined. On the other hand, for thicker SL's, the

393

map shows that they start to relax, having an in-plane lattice constant different from the buffer. In such samples, although the individual layers are under the critical thickness, defects are introduced due to the relaxation process of the whole SL. Besides that, a very complex strain profile is present within the SL and the simulations indicate that part of the SL is still pseudomorphic. From these results we can conclude that for SL's grown directly on BaF_2, without using a buffer layer, the situation is even more dramatic, since the lattice constant difference is bigger. Therefore, in the explanation of the results of electrical or optical measurements for such samples, this strain profile must be taken in account.

Acknowledgments

We would like to thank the "Conselho Nacional de Desenvolvimento Científico e Tecnológico (CNPq)", projects nr: 301091/95-1 and nr: 300397/94-1 and the "Fundação de Amparo à Pesquisa do Estado de São Paulo (FAPESP)", projects nr: 96/4546-0 and nr: 95/6219-4 for the financial support of this work.

References

[1] K. Murase, S. Ishida, S. Takaoka and T. Okumura. Surf. Sci, 170, 486 (1986).

[2] M. A. Tamor, H. Holloway, L. C. Davis, R. J. Baird and R. E. Chase. Superlatt. and Microstruct., 4, 493 (1988).

[3] O. A. Mironov, O. N. Makarovskii, O. N. Nashchekina, L. P. Shpakovskaya, V. I. Litvinov, M. Oszwaldovski and T. Berus. Acta Phys. Pol. A, 88, 853 (1995).

[4] V. Litvinov, M. Oszwaldowski, T. Berus and O. Mironov. Phys Stat. Sol. (a), 145, 503 (1994).

[5] E. Abramof, S.O. Ferreira, P.H.O. Rappl, H. Closs and I.N. Bandeira, J. Appl. Phys., 82(5), 2405 (1997).

[6] P.H.O. Rappl, H. Closs, S.O. Ferreira, E. Abramof, C. Boschetti, P. Motisuke, A.Y. Ueta and I.N. Bandeira, Submitted to the J. Cryst. Growth.

[7] A. I. Fedorenko, O. N. Nashchekina, B. A. Savitskii and L. P. Shpakovskaya, Vacuum, 43(12), 1191 (1992)

[8] P. F. Fewster, Semicond. Sci. Technol. 8, 1915 (1993).

[9] E. Koppensteiner, G. Bauer, H. Kibbel and E. Kasper, J. Appl. Phys., 76, 3489 (1994).

[10] G. Nimtz, in Numerical Data and Functional Relationships is Science and Technology, eds. K. Hellwege and ° Madelung, Landolt-Börstein, New Series Group III, Vol. 17 Part F p146 and p168 (Springer Verlag, Berlin, 1993).

Part VI

Interdiffusion in Quantum-Well Materials and IR Applications

Quantum Well Intermixing for Optoelectronic Applications

C. JAGADISH*, H.H. TAN*, S. YUAN* and M. GAL**
*Department of Electronic Materials Engineering, Research School of Physical Sciences and Engineering, Australian National University, Canberra, ACT 0200, Australia (cxj109@rsphysse.anu.edu.au)
**School of Physics, University of New South Wales, Sydney, NSW 2052, Australia

ABSTRACT

Ion implantation induced intermixing of GaAs-AlGaAs quantum well structures with H, O and As ions is investigated by low temperature photoluminescence. Large energy shifts are observed in all the cases, though recovery of photoluminescence intensities in proton case is more significant than others. Energy shifts are linear with proton dose and no saturation was observed even up to a dose of $\sim 5 \times 10^{16}$ cm^{-2}. Saturation in energy shifts are seen for both As and O ions at high doses. Energy shifts are also found to be dependent on Al composition of the barriers. Wavelength shifted quantum well lasers are fabricated with energy shifts of about 5 nm with no changes in threshold current using proton implantation. Anodic oxide induced intermixing of GaAs-AlGaAs quantum wells is demonstrated.

INTRODUCTION

Quantum Well Intermixing (QWI) has drawn considerable attention [1-3] in recent years due to its wide applicability for the fabrication of optoelectronic devices. QWI allows the modification of quantum well shape in selected regions which in turn modifies subband energies in the conduction band and the valence band [4]. This leads to changes in optical properties such as bandgap energy, absorption coefficient and refractive index [5-8]. QWI allows tuning of the laser wavelength due to the modification of bandgap energy and changes in absorption coefficient will allow us to fabricate non-absorbing mirrors for high power laser applications. QWI also leads to reduction in refractive index which could be used for fabrication of waveguides and improved optical confinement in ridge waveguide edge emitting lasers and thus leads to side-mode suppression. Changes in sub-band energies could be used to tune the wavelength of the quantum well infrared photodetectors. Monolithic integration of various opto-electronic devices such as lasers, modulators, waveguides, amplifiers can be achieved by selective area quantum well intermixing. Interdiffusion has been applied to the fabrication of high-reliability blue-shifted InGaAsP/InP, GaAs/AlGaAs and InGaAs/GaAs lasers [9-11]. QWI has been used to improve the performance of high power lasers, multiple wavelength lasers, low threshold current lasers and lasers with saturable absorbers [12-22].

A wide variety of techniques are used to create quantum well intermixing. Thermal interdiffusion is the simplest of all the techniques employed. However, thermal interdiffusion lacks spatial selectivity hence has found limited use in optoelectronic device fabrication. Other techniques could be classified into two categories, one is impurity induced intermixing and another is impurity free intermixing. Impurity induced intermixing has been extensively studied [23-29] and main impurities used for these studies are Zn and Si. However, impurity induced interdiffusion usually introduces substantial concentration of impurities in the active regions which is undesirable particularly for active devices such as lasers. Impurity free interdiffusion [30, 31] is achieved by using cap layers of SiO_2. Annealing of quantum wells with cap layers such as SiO_2 leads to creation of Gallium vacancies which enhance the atomic interdiffusion. In the GaAs/AlGaAs system, SiO_2 and Si_3N_4/SrF_2 are commonly used to enhance and suppress interdiffusion, respectively. However, SiO_2 reacts with Al when in direct contact with AlGaAs, thus generates Si which acts as an impurity source. Reproducibility has been found to be one of the main problems as interdiffusion has been found to be sensitive to the deposition conditions of SiO_2. Laser assisted interdiffused InGaAs/InGaAsP lasers have been fabricated and has been mainly used for this material system [18].

Ion implantation is widely used in microelectronics industry, but found limited use in optoelectronic device fabrication. This is mainly due to creation of defects during ion bombardment and difficulties associated with removal of defects during post-implant thermal processing. Despite these limitations, ion implantation has been used for creation of electrical and optical isolation [32] as well as fabrication of intermixed lasers [33].

Recently, we have also developed a novel intermixing technique, namely anodic oxide induced intermixing [34, 35]. This is based on creation of pulsed anodic oxides on GaAs substrates and subsequent thermal processing of the quantum well structures with anodic oxide cap layers. We have applied this technique for intermixing of quantum well structures based on GaAs/AlGaAs system and also to increase confinement on quantum wire structures grown on patterned substrates [36].

In this paper, we present results of systematic studies of variation of various implantation, anodic oxidation and annealing parameters on the intermixing of GaAs/AlGaAs quantum well structures. This material system is chosen for its simplicity due to close lattice matching which will avoid further complications of strain induced effects.

EXPERIMENTAL

Four quantum wells of nominal thicknesses 1.4 (QW#1), 2.3 (QW#2), 4.0 (QW#3) and 8.5 (QW#4)nm, were grown by low pressure metal organic chemical vapour deposition (MOCVD) on a semi-insulating (100) GaAs substrate for ion implantation study and on p+-GaAs for anodic oxide induced intermixing study. GaAs quantum wells are sandwiched between $Al_{0.54}Ga_{0.46}As$ barriers of 50 nm thick, with narrowest wells closer to the surface. All layers were undoped and grown at 750°C. The structure was then terminated by a GaAs capping layer to prevent oxidation of the Al rich layer and also to use it for the creation of anodic oxide of GaAs on the surface.

Ion implantation was carried with either 40keV protons, 280keV oxygen or 1MeV arsenic in the dose range of 1×10^{12} to 5×10^{16} cm^{-2} at various temperatures (from room temperature to 300°C). The ions' energies were chosen such that the peak of the atomic displacement distribution coincides for all ions and lies in the GaAs buffer at a depth of ~ 350 nm. Therefore, the entire QW structure was subjected to irradiation with less than a factor of two variation of nuclear energy deposition across the wells, as calculated by FASTRIM (a modified version of TRIM85-90) [37, 38].

Pulsed anodic oxidation technique developed by Zory and co-workers [39] has been employed in this study. The anodic oxidation was performed at room temperature in an electrolyte consisting of ethylene glycol : de-ionized water : phosphoric acid (40:20:1 by volume). The pulse width was 1msec, the period was 12 msec, the total pulsed anodization time was 4 minutes. The pulsed anodization current density was in the range of 40-160 mA/cm^2 as determined by the leading edge of the pulse. When the pulse was on, anodization took place and when pulse was off, the anodic oxide was etched slowly

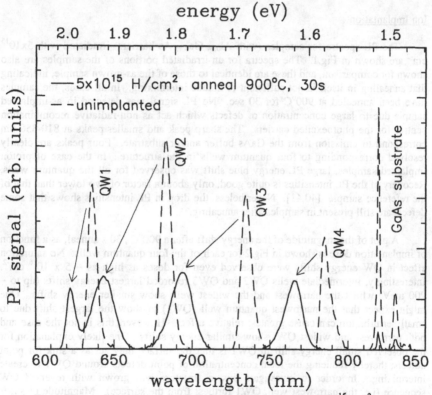

Fig.1 PL spectra of proton irradiated samples to a dose of 5×10^{15} cm^{-2} annealed at 900°C for 30 sec. The spectra for the unirradiated sample (annealed) is also shown.

by the electrolyte. Thermal glue was used to mask half of the sample and uncovered area was oxidized. After anodization, samples were rinsed in de-ionized water and thermal glue was removed by dipping in acetone and nitrogen blow dried.

Half of the samples were masked during irradiation / oxidation for direct comparison with unirradiated material and also to minimise the thickness/composition variation across the samples. After irradiation / oxidation, the samples (both the irradiated / oxidised and unirradiated/unoxidised halves) were annealed face down on a fresh pice of semi-insulating GaAs (to prevent excessive loss of arsenic from the surface) in a rapid thermal annealer (proximity method). The annealing temperatures and times were in the range of 850-1000°C and 10-120 sec, respectively. Low temperature photoluminescence (PL) measurements at ~12K were carried out with an Ar ion laser (514.5nm) as the excitation source and luminescence was detected with a silicon photodetector through a monochromator.

RESULTS AND DISCUSSION

Ion Implantation

The PL spectra for samples (with $Al_{0.54} Ga_{0.46} As$ barriers) irradiated with 5×10^{15} cm^{-2} are shown in Fig.1. The spectra for un-irradiated portions of the samples are also shown for comparison, and these are identical to those of the as-grown sample, indicating that annealing in itself is not sufficient to induce intermixing. In all cases, the samples have been annealed at 900°C for 30 sec. No PL signal was detected in as-implanted sample due to large concentration of defects which act as non-radiative recombination centres for the photoexcited carriers. The sharp peak and smaller peaks at 810-850 nm correspond to emission from the GaAs buffer and/or substrate. Four peaks are clearly resolved corresponding to four quantum wells in the structure. In the case of proton implanted samples, large PL energy blue shift was observed for all the quantum wells. Recovery of the PL intensities is quite good, only about a factor of two lower than that of the reference sample [40,41]. Nevertheless, the drop in PL intensities shows that some defects are still present in samples after annealing.

A plot of the magnitude of the energy shift after a 900°C, 30 s anneal, as a function of implantation dose is shown in Fig.2 for each of the four quantum wells. No saturation effect in QW energy shifts were observed even for doses as high as ~ 5 x 10^{16} cm^{-2}. Interestingly, intermediate wells QW2 and QW3 recorded largest energy shifts (up to ~ 200 meV) whilst the narrowest and the widest wells show smaller energy shifts. One might expect that the narrowest quantum well (QW1) to show the largest shift due to small perturbation can cause greatest relative effect. However, this is not the case and both narrowest and widest QWs show similar energy shifts. One likely explanation for the observed small energy shift in QW1 is that the surface may act as a sink for point defects, thereby reducing the local concentration of point defects around QW1 to create intermixing. In order to investigate this, a structure was grown with reversed QW sequence (i.e. the narrowest well, QW1 furthest from the surface). Magnitude of shifts are similar in both the cases for all quantum wells suggesting that the surface effects are not playing a major role. The small energy shift of narrow QWs could be due to the higher ground state energy level. PL from QW1 disappeared for doses higher than 5×10^{15} cm^{-2} which could be due to complete intermixing in this region. However, transmission electron microscopy studies revealed that the QW1 is still present. This suggests that QW1 has possibly became indirect, due to higher energy of the ground state in the Γ- band in the QW with respect to X-band in the barrier, similar to the type I to type II transitions reported in narrow quantum wells.

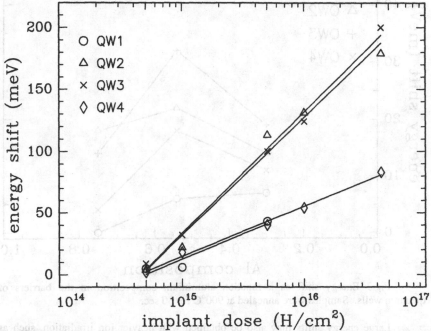

Fig. 2 Dose dependence of the magnitude of energy shift for the four quantum wells in samples annealed at 900°C for 30 sec.

A systematic study has been carried out to look at the barrier composition by growing quantum wells with Al$_{0.30}$Ga$_{0.70}$As and Al$_{0.75}$Ga$_{0.25}$As barriers. Fig. 3 shows a plot of the energy shift of four quantum wells versus Al composition of the barriers for a proton dose of 1×10^{15} cm^{-2}. All samples were annealed at 900°C for 30 sec.

Increase in the barrier Al composition from 0.3 to 0.54 is found to increase the energy shift of the quantum wells which is expected due to the higher concentration gradient across the well-barrier interface. However, any further increase in barrier composition from 0.54 to 0.75 led to reduction in the energy shifts. This is unexpected and most likely due to presence of large concentration of impurities such as carbon and oxygen in the barrier layers which are acting like pinning centres for the point defects thereby reducing the intermixing. This hypothesis needs further confirmation and experiments are underway to verify this.

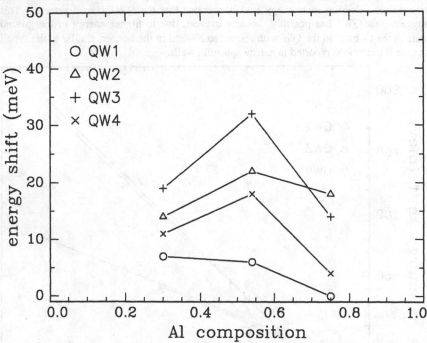

Fig.3 Energy shift as a function aluminium composition in the barriers of quantum wells. Samples were annealed at 900°C for 30 sec.

Large energy shifts may also be obtained with heavier ion irradiation, such as arsenic, but at significantly lower irradiation doses. From FASTRIM calculations, the collisional displacement density caused by arsenic ions is expected to be a factor of ~1000 higher than that of protons. Thus, the ion doses must be reduced by about 2-3 orders of magnitude to maximise optical recovery upon annealing as shown by the PL spectrum in Fig. 4 for an irradiation dose of 1×10^{12} cm^{-2} after 900°C, 30 sec anneal.

It can be seen from this figure that the energy shifts for all the quantum wells are significant and also recovery of PL intensity is reasonable for this dose. Above a certain dose (5×10^{12} cm^{-2}) the PL signal from QW1 disappears, similar to the case of proton irradiation. Energy shifts in excess of 200 meV are achieved at a dose of 1×10^{14} cm^{-2}. However, at this dose PL intensities are very weak. A major difference between this result and that of proton irradiation is that a saturation effect is observed in the case of

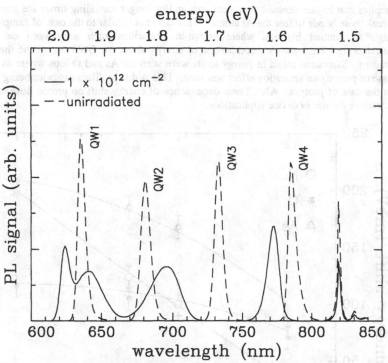

Fig. 4 Low temperature (12 K) PL spectrum of a sample irradiated with 1 MeV arsenic ions to a dose of 1×10^{12} cm^{-2} and annealed at 900°C for 30 s. The spectrum for the unirradiated samples is also shown for reference.

arsenic ions, especially in the two deeper QWs (closer to the end of range of the ions). Although the irradiation doses have been lowered significantly, it is known that the nature of defects caused by heavier ions could be quite different to those from light ions. The damage cascade by an individual ion track is expected to be denser and larger in size for heavier ions and, hence, increase the chance of agglomeration of the surrounding point defects into clusters and other extended defects by the overlap of the damage cascades. Unlike in the case of protons, increase in implantation temperature was found to marginally increase the energy shift. This further confirms that the nature of defects created by heavy and light ions are fundamentally quite different and intermixing is governed by the type and stability of the defects created during and after irradiation. A study of O ion implantation induced intermixing was carried out to see the effect of medium mass ion effects. At low doses, energy shifts were between for proton and arsenic cases but at higher doses saturation effects were seen. This could be due to implanted oxygen acting as an impurity and pinning defects and forming Al bonds to form complexes [42]. The formation of these complexes consumes point defects and hence less of them are available for intermixing. The thermal stability of complexes implies that higher annealing temperatures and/or longer annealing times are required to break these bonds to free the Al atoms. This effect is similar to the case of samples with high Al content barriers, where grown-in impurities such as oxygen can retard intermixing. Fig.5 shows a comparison of energy shifts for QW3 for all the three ions studied. Saturation effect in energy shifts were seen for As and O ions where as for the case of protons no saturation effect was seen. This is due to dilute cascades being formed in the case of protons. Also, linear dependence of energy shift on proton dose makes it attractive for use in device applications.

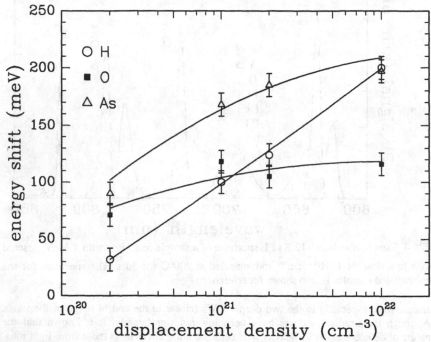

Fig.5 Comparison of the magnitude of energy shifts for H, O and As ions as a function of displacement density calculated by FASTRIM

Proton implantation (220 keV) was carried out in to a Graded-Index Separate-Confinement-Heterostructure (GRINSCH) laser structure with three 7 nm GaAs quantum wells and $Al_{0.30}Ga_{0.70}As$ barriers. Details of this structure are given elsewhere. The energy of the ions are chosen such that it corresponds to creating maximum displacements in the active region. To ensure direct comparison and minimise the effect of uniformity spread across the wafer, strips of about 2mm wide were masked adjacent to implanted areas. After irradiation, the samples (both the implanted and masked regions) were annealed simultaneously at 900°C for 30 sec in a rapid thermal annealer. After annealing, the samples were fabricated into 50 μm stripe width broad-area lasers by standard photolithography techniques. The lasers were cleaved into bars with each bar containing about 20 lasers spaced 300 μm apart and isolated from each other by a stripe which is not metallised.

The light output-current (L-I) curves of the implanted lasers are compared to that of the unimplanted lasers as shown in Fig.6 for a 400 μm long cavity. In all cases, the curves are essentially identical and the threshold currents of the lasers are found to be about 90 mA (Jth ~ 0.45 kA/cm^2). Hence, the current threshold chracteristics are not affected by the process of intermixing at these doses.

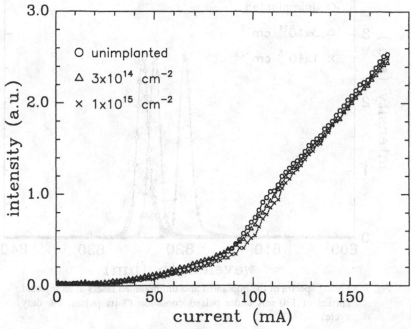

Fig. 6   L-I curves of the implanted and unimplanted lasers under pulsed
condition with 2 μs pulses of 1% duty cycle.

However, the emission spectra of the implanted lasers show a small but significant shift in the wavelengths as shown in Fig. 7. The spectra were collected under identical conditions at an injected current of 130 mA in all three cases. The lasing wavelengths of the implanted lasers are blue-shifted by 1 nm (2 meV) and 5nm (10 meV) in comparison with the unimplanted sample, for the lower and higher doses, respectively. Although the wavelength shift is small, these values are that would be expected for a 7 nm GaAs QW with Al$_{0.30}$Ga$_{0.70}$As   barriers at these doses. It should be noted that at lower implantation doses (< 3x10^{14} cm^{-2}), no measurable wavelength shift is observed, there is a threshold dose to induce wavelength shifting in this structure. The L-I and spectral characteristics measured at 30% duty cycle are also similar to those at 1% duty cycle.

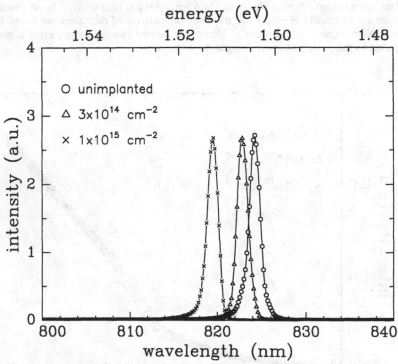

Fig. 7   Lasing spectra of the implanted and unimplanted lasers at an injected current of 130 mA under pulsed condition (2 μs pulses, 1% duty cycle).

These results are the first reported successful demonstration of wavelength shifting in GaAs-AlGaAs laser structures by ion implantation. The results also show that ion implantation is a very promising and simple technique of integrating lasers of different wavelengths for optical communication applications, particularly as WDM sources. To achieve wavelength shifts of more than 5 nm, higher doses are required. Our subsequent studies [33] have shown that indeed shifts upto 30-40 nm are possible by increasing the proton dose, but this led to increase in threshold current by a factor of 3 which is mainly due to increased concentration of residual non-radiative recombination centres at these high doses and are undesirable. Further work is underway to improve the characteristics of these lasers.

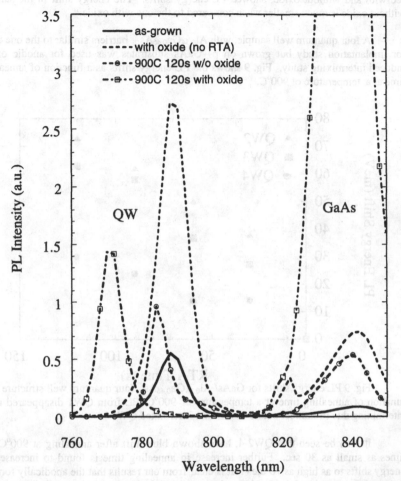

Fig. 8 PL spectra of GaAs/ Al$_{0.30}$Ga$_{0.70}$As single quantum well annealed at 900°C for 120 sec with and wihout anodic oxide. PL spectra of as-grown and anodized samples without annealing are also shown for comparison.

Anodic Oxidation

Fig. 8 shows the photoluminescence spectra of GaAs/ Al$_{0.30}$Ga$_{0.70}$As single quantum well (well thickness 3.2 nm) structure. After oxidation, PL Peak intensity has increased possibly due to reduced absorption by the surface oxide as well as surface passivation. However, there is no shift in the peak position suggesting that no intermixing took place after anodic oxidation. Annealing of the samples at 900°C for 120 sec with and without oxide, showed PL energy shifts. The energy shift in the sample without oxide is much smaller when compared to the one with oxide.

A four quantum well sample with $Al_{0.54}Ga_{0.46}As$ barriers similar to the one used for implantation study but grown on p+-GaAs substrates was used for anodic oxide induced intermixing study. Fig. 9 shows the PL energy shifts as a function of annealing time at a temperature of 900°C.

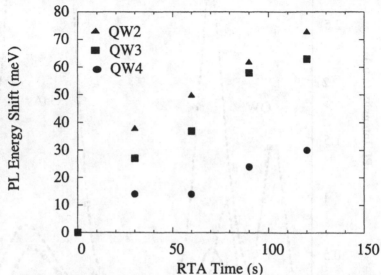

Fig. 9 PL energy shifts for GaAs/ $Al_{0.54}Ga_{0.46}As$ four quantum well structure as a function of annealing time at a temperature of 900°C. PL from QW1 disappeared after intermixing due to indirect nature of this thin intermixed QW.

It can be seen that QW2-4, have shown blue shift after annealing at 900°C for times as small as 30 sec. Further increase in annealing time is found to increase the energy shifts to as high as 70 meV. It is clear from our results that the anodically formed oxides are associated with the enhancement of group III interdiffusion. The interdiffusion results imply that the concentration of a native point defect, e.g. a group III vacancy or interstitial, is increased at elevated temperatures when annealing under an anodic oxide. For the impurity-free interdiffusion induced in GaAs by a deposited SiO_2 layer, it is often assumed that an increased Ga vacancy concentration enhances the interdiffusion. However, the chemical interaction between GaAs and a hydrated mixture of Ga and As oxides are presumably quite different than those between GaAs and SiO_2 [43]. Work is in progress to under the processes affecting the interdiffusion. Anodic oxide induced intermixing is a simple and novel way of creating quantum well intermixing and is a promising technique for the fabrication of photonic integrated devices.

CONCLUSIONS

We have demonstrated that implantation induced intermixing of quantum wells results in large energy shifts. Both lighter and heavier ions, create intermixing, though lighter ions are preferred due to linear dependence of the energy shift with ion dose as well as lack of saturation effect. Using proton implantation, wavelength shifting of quantum well lasers has been achieved. A simple and new technique of quantum well intermixing based on anodic oxides is developed and its potential for interdiffusion is demonstrated.

ACKNOWLEDGMENTS

This research is partially funded by the Australian Research Council and Research Grants Council of Hong Kong. Authors gratefully acknowledge fruitful discussions with Prof. Peter Zory, Dr. Craig Largent, Jason O (University of Florida, Gainesville), Prof. E. Herbert Li and Michael Chan (University of Hong Kong), Prof. Richard M. Cohen (University of Utah), Prod. David Cockayne, Dr. Jou Zin, D.Q. Cai (University of Sydney), Dr. L.V. Dao, M.B. Johnston, P.T. Burke (University of New South Wales), Dr. Yong Kim (Korea Institute of Science and Technology), Prof. Ian Mitchell, Dr. Richard Goldberg (University of Western Ontario), Dr. X.Q. Liu, Dr. Yong Chang (Shnaghai Institute of Technical Physics) and Professor James S. Williams (Australian National University).

REFERENCES

1. E.H. Li, Ed, Quantum Well Intermixing for Photonics, (SPIE Milestone Series, Bellingham, 1997).
2. D.G. Deppe and N. Holonyak, Jr., J. Appl. Phys. 64, R93 (1988) and references there in.
3. J.H. Marsh, Semicond. Sci. Technol. 8, 1136 (1993) and references there in.
4. E.H. Li, B.L. Weiss and K.S. Chan, Phys. Rev. B46, 15181 (1994).
5. S. Yuan, N. Frank, G. Bauer and M. Kriechbaum, Phys. Rev. B50, 5286 (1994).
6. E. Kapon, N.G. stoffel, E.A. Bobisz and R.Bhat, Appl. Phys. Lett. 52, 251 (1988).
7. A. Wakatsuki, H. Iwamura, Y. Suzuki, T. Miyazawa and O. Mikami, IEEE Photon. Technol. Lett. 3, 905 (1991).
8. T. Wolf, C.L. Shieh, R. Engelmann, K. Alavi and J. Mantz, Appl. Phys. Lett. 55, 1412 (1989).
9. J. P. Noel, D. Melville, T. Jones, F.R. Shepherd, C.J. Miner, N. Puetz, K. Fox, P.J. Poole, Y. Feng, E.S. Koteles, S. Charbonneau, R.D. Goldberg and I.V. Mitchell, Appl. Phys. Lett. 69, 3516 (1996).
10. Y. Nagai, K. Shigihara, S. Karakida, S. Karimoto, M. Otsubo and K. Ikeda, IEEE J. Quantum Electron. 31, 1364 (1996).
11. E.C. Larkins, W. Benz, I. Esquivias, W. Rothemund, M. Bauemler, S. Weisser, A. Schonfelder, J. Fleissner, W. Jantz, J. Rosenzweig and J.D. Ralston, IEEE Photon. Technol. Lett. 7, 16 (1995).

12. S. Burkner, J.D. Ralston, S. Weisser, J. Rosenzweig, E.C. Larkins, R.E. Sah and J. Fleisner, IEEE Photon. Technol. Lett. 7, 741 (1995).

13. D. Hofstetter, H.P. Zappe, J.E. Epler and P. Riel, Appl. Phys. Lett. 67, 1978 (1995).

14. B. S. Ooi, S.G. Ayling, A.C. Bryce and J.H. Marsh, IEEE Photon. Technol. Lett. 7, 944 (1995).

15. S.Y. Hu, M.G. Peters, D.B. Young, A.C. Gossard and L.a. Coldren, IEEE Photon. Technol. Lett. 7, 712 (1995).

16. N. Yamada and J.S. Harris, Jr., Appl. Phys. Lett. 60, 2463 (1992).

17. D.F. Welch, W. Streifer, R.L. Thornton and T. Paoli, Electron.Lett. 23, 525 (1987).

18. A.McKee, C.J. McLean, G. Lullo, A.C. Bryce, R.M. De La Rue, J.H. Marsh and C.C. Button, IEEE J. Quantum Electron. 33, 45 (1997).

19. A. Ramdane, P. Krauz, E.V. Rao, A. Hamoudi, A. Ougazzaden, D. Robein, A. Gloukhian and M. Carre, IEEE Photon. Technol. Lett. 7, 1016 (1995).

20. S.R. Andrew, J.H. Marsh, M.C. Holland and A.H. Ken, IEEE Photon. Technol. Lett. 4, 426 (1992).

21. S.F. Yu and E.H. Li, IEEE Photon. Technol. Lett. 8, 482 (1996).

22. A.S.W. Lee and E.H. Li, Appl. Phys. Lett. 69, 3581 (1996).

23. C. Vieu, Defect and Diffusion Forum, 119-120, 127 (1995).

24. W.D. Laidig, N. Holonyak, Jr., M.D. Camras, K. Hess, J.J. Coleman, P.D. Dapkus and J. Bardeen, Appl. Phys. Lett. 38, 776 (1981).

25. J.J. Coleman, P.D. Dapkus, C.G. Kirkpatrick, M.D. Camras and N. Holonyak, Jr., Appl. Phys. Lett. 40, 904 (1982).

26. Y. Hirayama, Y. Horikoshi and H. Okamoto, Jpn. J. Appl. Phys. 23, 1568 (1984).

27. S.T. Lee, G. Braunstein, P. Fellinger, K.B. Kahen and G. Rajeswaran, Appl. Phys. Lett. 53, 2531 (1988).

28. E.P. Zucker, A. Hashimoto, T. Fukunaga and N. Watanabe, Appl. Phys. Lett. 54, 564 (1989).

29. L.J. Guido, K.C. Hsieh, N. Holonyak, Jr., R.W. Kaliski, V. Eu M. Feng and R.D. Burnham, J. Appl. Phys. 61, 1329 (1987).

30. S. Burkner, M. Maier, E.C. Larkins, W. Rothemound, E.P. O'Reilly and J.D. Ralston, J. Electron. Mater. 24, 805 (1995).

31. I. Gontijo, T. Krauss, J.H. Marsh and R.M. De La Rue, IEEE J. Quantum Electron. 30, 1189 (1994).

32. S. J. Pearton, Mater. Sci. Rep. 4, 313 (1990).

33. H.H. Tan and C. Jagadish, Appl. Phys. Lett. 71, 2680 (1997).

34. S. Yuan, Y. Kim, C. Jagadish, P.T. Burke, M. Gal, J. Zou, D.Q. Cai, D.J.H. Cockayne and R.M. Cohen, Appl. Phys. Lett. 70, 1269 (1997).

35. S. Yuan, Y. Kim, H.H. Tan, C. Jagadish, P.T. Burke, L.V. dao, M. Gal, M.C.Y. Chan, E.H. Li, J. Zou, D.Q. Cai, D.J.H. Cockayne and R.M. Cohen, J. Appl. Phys. (in Press – 1998).

36. Y. Kim, S. Yuan, R. Leon, C. Jagadish, M. Gal, M. Johnston, M.R. Phillips, M. Stevens-Kalceff, J. Zou and D.J.H. Cockayne, J. Appl. Phys. 80, 5014 (1996).

37. M.J. Grove, D.A. Hudson, P.S. Zory, R.J. Dalby, C.M. Harding and A. Rosenberg, J. Appl. Phys. 76, 587 (1994).

38.   J.F. Ziegler, J.P. Biersack and U. Littmark, The Stopping and Range of Ions in Solids, vol.1, Pergamon, New York (1989).

39.   H.J. Hay, FASTRIM is a modified version of TRIM85-90 which takes into account the multilayer target (interafces) problems inherent with TRIM (Unpublished).

40.   H.H. Tan, J.S. Williams, C. Jagadish, P.T. Burke and M. Gal, Appl. Phys. Lett. 68, 2401 (1996).

41.   H.H. Tan, J.S. Williams, C. Jagadish, P.T. Burke and M. Gal, Mater. Res. Soc. Symp. Proc. 396, 823 (1996).

42.   H.H. Tan, Ph.D. Thesis, Australian National University, Canberra (1996).

43.   R.M. Cohen, Mater. Sci. & Eng. Rep. 20, no. 4-5 (1997).

THE EFFECT OF TENSILE STRAIN ON AlGaAs/GaAsP INTERDIFFUSED QUANTUM WELL LASER

K.S.CHAN* and MICHAEL C.Y.CHAN**
*City University of Hong Kong, Department of Physics and Materials Science, Kowloon, Hong Kong
**University of Hong Kong, Department of Electrical & Electronic Engineering, Pokfulam, Hong Kong

ABSTRACT

In this paper, we study the interdiffusion of tensile strained $GaAs_yP_{1-y}$ /$Al_{0.33}Ga_{0.67}As$ single QW structures with a well width of 60Å. Different P concentrations in the as-grown well are chosen to obtain different tensile strains in the QW. Interdiffusion induces changes in the tensile strains and confinement potentials, which consequently change the valence band structure and the optical gain.

INTRODUCTION

Several studies have reported that tensile strained QW lasers show improvement in performances including lower threshold currents, higher differential gains, and wider frequency modulation bandwidths as compared to the unstrained QW lasers [1,2]. In AlGaAs/GaAsP tensile strained QWs, the biaxial tensile strain shifts the light hole bands closer to the conduction band minimum than the heavy hole bands. Apart from tuning strain, interdiffusion of constituent atoms, the rates of which depend on lattice distortion, impurities, defects and the process temperatures, is a versatile technique to modify the bandstructure. Using this technique, the QW composition profiles, the confinement potentials and the optical properties can be modified as a result of the diffusion of constituent atoms. Three categories of intermixing of GaAsP/AlGaAs can occur for a quaternary based system: group-III only (GpIII) interdiffusion (Ga and Al), group-V only (GpV) interdiffusion (As and P), and both group-III and group-V (GpIII&V) sublattices diffusion. An interdiffusion of AlGaAs/GaAsP has been reported by Utpal et al with implantation of flourine [3].

This paper presents a theoretical study of the effects of interdiffusion in a GaAsP/AlGaAs single QW structure. The effects of interdiffusion on the strains, the splitting of the HH and LH subbands and the optical gains of the QW laser are calculated based on a multi-band k•p model including valance band-mixing. The optical gain spectra are calculated using the density matrix approach.

DIFFUSION

The group-III and group-V interdiffusion processes are modeled using the Fick's Law. The interdiffusion of the Al and Ga atoms is characterized by a diffusion length L_d^{III} , which is defined as $L_d^{III} = \sqrt{(D^{III}t)}$, where D^{III} is the diffusion coefficient of group-III atoms and t is the diffusion time. The interdiffusion of the As and P atoms is characterized in much the same way by the diffusion length L_d^V . The structure to be modeled consists of a layer of GaAsP sandwiched between two thick AlGaAs barriers. After intermixing, the Al and the P concentrations are described by

$$x_{Al}(z) = 1 - \frac{1-x_o}{2} [erf(\frac{L_z + 2z}{4L_d^{III}}) + erf(\frac{L_z - 2z}{4L_d^{III}})] \tag{1}$$

$$y_P(z) = \frac{y_o}{2} [erf(\frac{L_z + 2z}{4L_d^{V}}) + erf(\frac{L_z - 2z}{4L_d^{V}})] \tag{2}$$

where x_o is the as-grown Al content in the barrier, y_o is the as-grown P content in the well, L_z is the as-grown well width, z is the growth direction, and the QW is centered at z=0.

A QW structure with a 60Å thick GaAsP well layer sandwiched between 1000Å thick $Al_{0.33}Ga_{0.67}As$ barriers is considered in the present study. In our calculation, the P concentration in the as-grown quantum well, y_o, is set to be 0.08, 0.13 and 0.2. When y_o =0.08, heavy hole is the top valence subband; while y_o=0.2, light hole is the top valence subband. The heavy and light holes are degenerate at the zone centre when y_o=0.13.

THEORY

After interdiffusion, the alloy composition deviates from the square well profile leading to a non-uniform strain. The in-plane strain $\varepsilon(x,y)$, which is a function of compositions x and y, depends on the co-ordinate z along the <001> growth direction. The strain components, after interdiffusion, are given by:

$$\varepsilon_{xx} = \varepsilon_{yy} = \varepsilon(x,y) \tag{3a}$$
$$\varepsilon_{zz} = -2[c_{12}(x,y)/c_{11}(x,y)]\varepsilon(x,y) \tag{3b}$$
$$\varepsilon_{xy} = \varepsilon_{yz} = \varepsilon_{zx} = 0 \tag{3c}$$

where $\varepsilon(x,y)$ is negative for compressive strains, and $c_{ij}(x,y)$ are the composition-dependent elastic stiffness constants. The change in the bulk bandgap, $S_\perp(x, y)$, due to the biaxial strains is given by:

$$S_\perp(x,y) = -2a(x,y)[1-c_{12}(x,y)/c_{11}(x,y)]\varepsilon(x,y) \tag{4a}$$

where $a(x,y)$ is the hydrostatic deformation potential. The splitting energy, $S_{//}(x,y)$, between the HH and LH band edges induced by the shear strains is given by:

$$S_{//}(x, y) = -b(x, y)[1+2c_{12}(x, y)/c_{11}(x, y)]\varepsilon(x, y) \tag{4b}$$

where $b(x,y)$ is the shear deformation potential. The parameters a, b, c_{ij}, in the above equations are assumed to obey Vegard's law, so that their respective values depend directly on the compositional profiles across the QW. Including the coupling of the LH band to the spin-orbit split-off band, the valence band splitting at Γ for the HH band and for the LH band are given by:

$$S_{//HH}(x, y) = S_{//}(x, y) \tag{5a}$$

$$S_{//LH}(x,y) = -1/2[S_{//}(x,y) + \Delta_o(x,y)] + 1/2[9\{S_{//}(x,y)\}^2 + \{\Delta_o(x,y)\}^2 - 2S_{//}(x,y)\Delta_o(x,y)]^{1/2} \quad (5b)$$

respectively, where $\Delta_o(x,y)$ is the spin-orbit splitting. The QW confinement potential after the disordering process, $U_r(x,y)$, is obtained by modifying the as-grown potential profile by the non-uniform strain effects, and is given by:

$$U_r(x,y) = \Delta E_r(x,y) - S_{\perp r}(x,y) \pm S_{//r}(x,y) \quad (6)$$

where $S_{\perp r}(x,y) = Q_r S_{\perp}(x,y)$, the '+' and '-' signs are used for the HH and LH confinement potentials, respectively, and $S_{//c}(x,y) = 0$. Q_r (r = c , v) is the band offset ratio for the conduction and valence bands.

To calculate the electron and hole wave functions in QW, we use the effective mass approximation. For most III-V semiconductors, it is a good approximation that the conduction and valence bands are decoupled. A parabolic band model and the Luttinger-Kohn-Hamiltonian with strain components are used for the conduction and valence bands respectively. The electron states near the conduction subband edge are assumed to be s-like and nondegenerate (excluding spin), while the hole states near the valence subband edge are p-like and four-fold degenerate (including spin). The envelope function scheme is adopted to describe the slowly varying (spatially extended) part of the wavefunction. The wavefunctions of the electron and hole subbands at the zone centre can be calculated separately by solving the one-dimensional Schrodinger-like equation as follows:

$$-\frac{\hbar^2}{2}\frac{d}{dz}\left[\frac{1}{m^*_{\perp r}(z)}\frac{d\psi_{rl}(z)}{dz}\right] + U_r(z)\cdot\psi_{rl}(z) = E_{rl}\psi_{rl}(z) \quad (7)$$

where $\psi_{rl}(z)$ is the wavefunction of the lth subband for electrons(r=cl) or holes (r=vl), respectively; $m_{\perp r}^*(z)$ is the corresponding carrier effective mass in the z direction; E_{rl} is the subband-edge energy. Equation (7) is solved numerically using a finite difference method with the above confinement profile. For valence band sturcture, it is necessary to diagonalize the Luttinger-Kohn Hamiltonian with appropriate confinement potentials for heavy and light holes. The hole envelope functions depend on $k_{//}$ as a result of the mixing of the heavy and light hole bands. In this work, the effective Hamiltonian approach described in Chan [4] is used to solve the Luttinger-Kohn Hamiltonian to obtain the valence subband structure.

CONFINEMENT PROFILE

The electron cofinement potentials of interdiffused quantum wells with phosphorous y_o equal to 0.13 are plotted in figure 1 for both GpIII and GpV diffusion. Owing to the lack of space, results for holes, y_o=0.08 and 0.2 are not shown as the main features are very similar to those in figure 1. In GpIII diffusion, the aluminium concentration in the well increases with L_d as aluminium atoms diffuse into the well from the barriers. The confinement potentials have an error function profile as the distribution of Al is described by the error functions. There is a discontinuity in the potential at the well-barrier interface due to the discontinuity in strains of the AlGaAs and AlGaAsP layers. In the AlGaAs layer, the strain is negligible for all values of L_d, while the strain of the well layer depends on the Al and P concentration. For GpV diffusion , when the L_d is small, the electron confinement potential has an error function

415

profile inside the well. The well potential flattens when the L_d is increased. This is the result of the out-diffusion of P from the well. When L_d is small, the P concentrations near to the interface drop quickly while the concentration at the centre remains unchanged, which leads to a lower potential near to the interface.

STRAINS AND SUBBAND STRUCTURES

It is interesting to look at how the strain is affected by the interdiffusion process. Owing to the lack of space we only show the strain at the centre of the well as a function of L_d. For GpIII diffusion, we notice that the strain does not change very much with interdiffusion. It is because the change in Al concentration does not affect much the lattice constant of AlGaAsP, if the P concentration is kept constant. In GpIII interdiffusion the strain in the well changes substantially. The strains are decreased by about 40% when L_d is increased from 0Å to 25Å for these three values of y_o.

For tensile strained QWs, it is interesting to look at the splitting between the first heavy hole (HH1) and first light hole (LH1) subbands as it is an important factor affecting the laser performance. In figure 3, we have plotted the energy separation between the LH1 and HH1 subbands. The separation is positive when the light hole is above the heavy hole. In GpV interdiffusion, the separation increases with L_d which means that the LH1 subband is shifted up faster than the HH1 subband for all three values of y_o. This can be explained by the fact that the tensile strain is decreased by the increase in interdiffusion. In GpIII interdiffusion, the separation decreases with L_d when y_o=0.08. For y_o=0.13 and 0.2, the energy separation first decreases and then increases with L_d when L_d is greater than 20 and 15Å respectively. In fact for sufficiently large L_d, the LH1 is the lowest subband. This can be explained by the fact that the strains remain constant and become more important while the confinement potential becomes shallow as interdiffusion proceeds. For large L_d, the subband energy is mainly determined by the strain potential.

OPTICAL GAIN

The optical gain spectra are calculated by the density matrix approach and the transition between the pth conduction subband and the qth valence subband is given by

$$
\begin{aligned}
g_{pq}(E) = & \frac{2\pi q^2 \hbar}{(2\pi)^2 n\varepsilon_0 cm_0^2 L_w E} \cdot \int dk \left| \hat{e} \cdot P_{pq}(k_{//}) \right|^2 \rho\left(E_p^e(k_{//}) - E_q^h(k_{//}) - E\right) \\
& \times \left[f^e\left(E_p^e(k_{//})\right) - f^h\left(E_q^h(k_{//})\right) \right]
\end{aligned}
\tag{8}
$$

where q is the electric charge, n is the refractive index, ε_0 is the dielectric constant of the vacuum, c is the speed of light, L_w is the width of the QW, E is the photon energy, P_{pq} is the optical matrix element, $\hat{e}$ is a unit vector along the polarization direction of the optical electric field, and f^e and f^h are the Fermi distribution functions for electrons in the conduction and valence subbands respectively. To include the spectral broadening of each transition, the total gain in a single quantum well structure is convoluted with a lineshape function over all transition energies E' and is given by

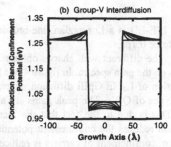

(a) Group-III interdiffusion

Conduction Band Confinement Potential (eV)

Growth Axis (Å)

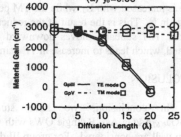

(b) Group-V interdiffusion

Conduction Band Confinement Potential (eV)

Growth Axis (Å)

Fig.1 Electron confinement profile of interdiffused QW. The diffusion length is from as-grown well to 25Å in step of 5Å.

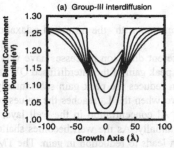

Lattice Mismatch-tensile strain (%)

$y_0 = 0.2$

$y_0 = 0.13$

$y_0 = 0.08$

GpIII ——
GpV - - - -

Diffusion Length (Å)

Fig.2 In-plane strain at well centre.

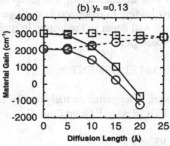

(a) $y_0 = 0.08$

Material Gain (cm^{-1})

GpIII —— TE mode ○
GpV - - - TM mode □

Diffusion Length (Å)

(b) $y_0 = 0.13$

Material Gain (cm^{-1})

Diffusion Length (Å)

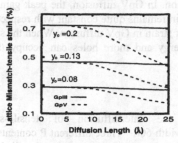

(a) Group-III interdiffusion

Subband energy spacing (eV)

$y_0 = 0.08$

$y_0 = 0.13$

$y_0 = 0.2$

Diffusion Length (Å)

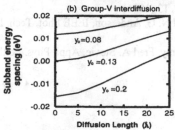

(b) Group-V interdiffusion

Subband energy spacing (eV)

$y_0 = 0.08$

$y_0 = 0.13$

$y_0 = 0.2$

Diffusion Length (Å)

Fig.3 Energy separation between ground state heavy hole and light hole.

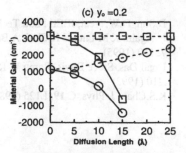

(c) $y_0 = 0.2$

Material Gain (cm^{-1})

Diffusion Length (Å)

Fig.4 The peak gain of interdiffused QW.

417

$$G(E) = \int dE' \sum_{p,q} g_{pq}(E') L(E'-E). \tag{9}$$

where L(E-E') is a Lorentzian line-broadening function with the intraband relaxation time taken to be 0.1ps.

The different well shapes obtained in different diffusion processes have very strong effects on the gain spectra. In figure 4, we plot the peak gain of the interdiffused QW lasers as a function of L_d. In GpIII diffusion, interdiffusion reduces the peak gain substantially in all three types of QWs. The peak gains are about zero when the L_d reaches the values of 15 to 20Å. This is due to the fact that the increase in Al concentration in the well layer due to diffusion reduces the confinement potential substantially. As the well becomes shallow, the quantum confinement of carriers is reduced which leads to reduction in gain. The TM mode peak gain is larger than that of TE mode in all three cases. This is due to the fact that the strain is not affected very much by GpIII diffusion. In GpV diffusion, the peak gain of TE mode is increased with L_d, while the TM peak gain remains quite constant with respect to the change in L_d. This is the result of the reduction of strain in GpV diffusion. When the strain is reduced, HH subband moves downward in energy and more holes can occupy the HH subband, which leads to increase in TE gain.

CONCLUSION

In conclusion, we have studied the interdiffusion of tensile-strained $Al_{0.33}Ga_{0.67}As/Ga_yAs_{1-y}P$ single QWs with well width 60Å. Three different P contents in as-grown well are considered. For group III diffusion, the depths of the potential wells become shallow leading to reduction in peak gain. For group V only diffusion, the increase in the potential well depth and reduction in strain result in an increase in TE peak gain and decrease in TM peak gain.

ACKNOWLEDGEMENT

This work is supported partially by City University of Hong Kong Strategic Grant and Hong Kong UGC Research Grant Council.

REFERENCES

[1]   D.Sun, D.W.Treat, IEEE Phot. Tech. Lett. **8,** 13 (1996)
[2]   F. Agahi, K M Lau, H K.Choi, A Baliga and N. G. Anderson, IEEE Phot. Tech. Lett. **7,** 140 (1995)
[3]   Uptal Das, Steve Davis, Ramu V. Ramaswamy, Fred A. Stevie, Appl. Phys. Lettl **60** ,210 (1992)
[4]   K.S.Chan, J. Phys. C **19,** L125 (1986)

Reduced Al-Ga interdiffusion in GaAs/AlGaAs multiple quantum well structure by introducing low hydrogen content SiNx capping layer for dielectric cap quantum well disordering

W.J. CHOI**, S.M. HAN*, S.I. SHAH, S.G. CHOI, D.H. WOO, S. LEE, H.J. KIM,
I.K. HAN, S.H. KIM, J.I. LEE, K.N. KANG, J. CHO*
Photonics Research Center, Korea Institute of Science and Technology, P.O.Box 131,
Cheongryang, Seoul 130-650, Korea
* Physics Dept., Kwangwoon Univ., 447-1 Wolgae-Dong, Nowon-Gu, Seoul 139-701,
Korea
** ECE Dept., Univerity of California, Santa Barbara, CA 93106,
  E-mail: wjchoi@xanadu.ece.ucsb.edu wjchoi@kistmail.kist.re.kr

ABSTRACT

Dielectric cap quantum well disordering (DCQWD) of GaAs/AlGaAs multiple quantum well (MQW) structure was carried out by using SiNx capping layer grown by plasma enhanced chemical vapor deposition (PECVD). By varying the NH_3 flow rate with fixed SiH_4 flow rate during the SiNx growth, the characteristics of the capping film were varied. There was an increase in the energy shift of quantum well photoluminescence (PL) peak after thermal treatment of the samples with rapid thermal annealing (RTA) as the NH_3 flow rate was increased, although the thickness of SiNx decreased. This is thought to be due to the increase of hydrogen content in SiNx film grown at higher NH_3 flow rate.

INTRODUCTION

Recently, DCQWD technique has been used to fabricate laser diodes (LDs) and photonic integrated circuits (PICs) [1-3], without any regrowth steps by disordering its structure selectively. Many dielectric films have been used to enhance QWD or to prevent QWD in DCQWD technique for the fabrication of LDs and PICs. SiO_2 has been widely used as a capping layer to promote QWD, and SiNx [1], SrF_2 [4] and WN_x [5] have been used to prevent QWD in the DCQWD technique. Since a SiO_2 capping layer induces a relatively larger blue shift than a SiNx capping layer in a GaAs/AlGaAs QW system, SiO_2 is generally used to promote QWD while SiNx is used as a mask to prevent QWD under the capped areas of QW [1].

The difference of the degree of QWD between SiO_2 capped QWD and SiNx capped QWD is thought to be a result of the difference in Ga out-diffusion from GaAs into SiO_2 capping layer and into SiNx capping layer during thermal annealing [6]. Although for the fabrication of PICs with DCQWD technique, SiNx capping layer has been proposed to be used as a mask to prevent QWD, it has been shown that SiNx capping layer can enhance QWD in GaAs/AlGaAs QW system [1,7,8]. It also has been shown that the degree of QWD caused by the SiNx capping layer is dependent on the film quality and/or the film growth process [7,8]. Therefore, if an optimum film growth condition could be found by varying the process condition, selective QWD can be achieved with the same dielectric capping material.

In this study, in order to investigate the dependence of DCQWD on the film growing condition, reactant gas ratio was varied during the growth of SiN_x capping layers on MQW substrate by PECVD method. The dependence of DCQWD on the quality of SiN_x capping layer was characterized by the hydrogen content in the capping layer.

EXPERIMENT

The GaAs/AlGaAs MQW structure used in this study was grown by MBE on semi-insulating GaAs substrate without any intentional doping. The schematic diagram of MQW structure is shown in Fig. 1. As one can see, MQW structure consists of four 7 nm thick GaAs wells with 10 nm thick $Al_{0.2}Ga_{0.8}As$ barrier.

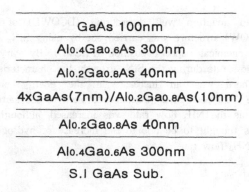

GaAs 100nm
$Al_{0.4}Ga_{0.6}As$ 300nm
$Al_{0.2}Ga_{0.8}As$ 40nm
4x GaAs(7nm)/$Al_{0.2}Ga_{0.8}As$(10nm)
$Al_{0.2}Ga_{0.8}As$ 40nm
$Al_{0.4}Ga_{0.6}As$ 300nm
S.I GaAs Sub.

Fig. 1 MQW substrate structure.

In order to find the dependence of DCQWD on the SiN_x film growth condition, reactant gas ratio was changed during SiN_x growth by PECVD method. NH_3 gas flow rates were changed from 0 sccm to 40 sccm at fixed SiH_4 gas flow rate of 20 sccm. During the growth, 30 W RF power was applied and the base pressure was kept at 0.9 Torr by adding N_2 gas. In this case, diluted SiH_4(5% in nitrogen) gas was used. The growth temperature and growth time were 300 °C and 20 min., respectively. After SiN_x growth on MQW sample, thermal treatment of the samples was accomplished by RTA at 950 °C for 30 sec.

Disordering of the MQW samples was observed by PL spectra at 9 K after removing the SiN_x capping layer. The characterization of SiN_x capping layers on the MQW sample was carried out using an ellipsometer, a surface profiler, Auger electron spectroscopic method and elastic recoil detection (ERD) method of Rutherford Back Scattering (RBS).

RESULTS AND DISCUSSION

As shown in Fig. 2, the thickness of SiN_x film decreases with the increase of the

NH₃ flow rate from 270 nm to 160 nm. The refractive index of SiN$_x$ film also decreases slightly with the increase of the NH₃ flow rate from 1.85 to 1.83. The refractive indices indicate that these SiN$_x$ films are N-rich films. The data from Auger electron spectroscopy also showed that the atomic percent of nitrogen of the SiN$_x$ film is almost 67 % for all SiN$_x$ films as seen in Fig. 3, although nitrogen content of stoichiometric Si₃N₄ is 57 %. However oxigen was not detected. As reported by Dun et al.'s [9], N₂ gas can be used as a nitrogen source of the SiN$_x$ film grown by PECVD method. Since N₂ gas was used not only as a carrier gas but also as a nitrogen source in our experimental condition, all SiN$_x$ films could exhibit nearly same composition. This N-rich SiN$_x$ might be the result from the lack of SiH₄ in our experimental condition.

However, although all SiN$_x$ films exhibit the same composition, the data from ERD reveals different characteristics in hydrogen content of the film. The hydrogen

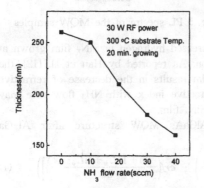

Fig. 2 Thickness of SiN film as a function of NH₃ flow rate

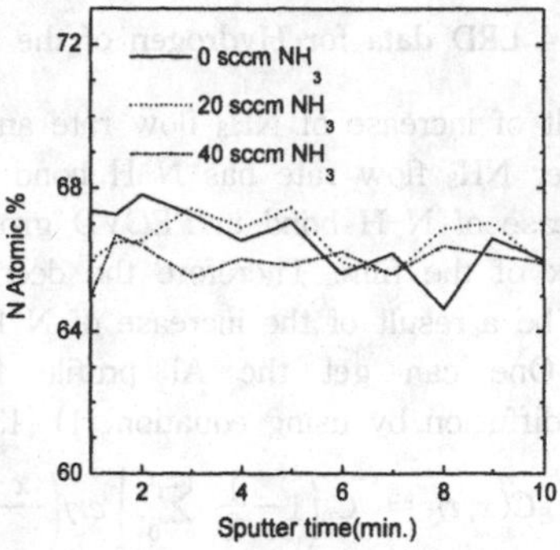

Fig. 3 Nitrogen atomic percent in SiN films grown at different NH₃ flow rate

content of the film increases with the increase of the NH₃ flow rate as shown in Fig. 2. This result is well coincident with the results of Dun et al.'s [9] who used SiH₄/N₂ gas mixture to lower hydrogen content in the SiN$_x$ film.

Fig. 5 shows PL spectra of MQW samples after RTA treatment at 9 K. As one can see, the amount of blue shift increases with NH₃ flow rate. In this case, one may consider the thickness of SiN$_x$ capping layer to compare the QWD to each other. It has been well known that the thicker capping layer causes larger QWD in DCQWD [8,10]. However, as shown in Fig. 2, the thickness of SiN$_x$ film decreases with NH₃ flow rate. This indicates that SiN$_x$ film grown at higher NH₃ flow rate causes larger QWD than that grown at lower NH₃ flow rate at the same thickness. Therefore one can conclude that SiN$_x$ film grown with higher NH₃ flow rate has larger vacancy density to cause larger QWD.

As shown in Fig. 4, SiN$_x$ film grown with higher NH₃ flow rate exhibited higher hydrogen content. As reported by Kapoor et al. [11], SiN$_x$ film with high hydrogen concentration exhibits large trap density. Therefore higher hydrogen content is thought to be a source of larger vacancy density of SiN$_x$ film, which induces larger QWD. Since SiH₄ was fixed during SiN$_x$ growing, the increase of hydrogen content in SiN$_x$ film is a

421

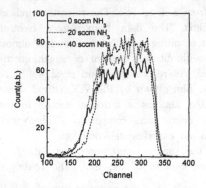 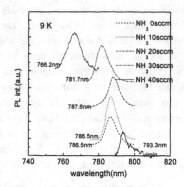

Fig. 4 ERD data for Hydrogen of the film　　Fig. 5 PL spectra of the MQW samples

result of increase of NH$_3$ flow rate and hence increased hydrogen in SiN$_x$ film grown at higher NH$_3$ flow rate has N–H bond configuration. As reported by Han et al. [12], the increase of N–H bond in PECVD grown SiN$_x$ film results in the decrease of refractive index of the film. Therefore the decrease of refractive index with NH$_3$ flow rate may also be a result of the increase of N–H bond in SiN$_x$ film.

One can get the Al profile for GaAs/AlGaAs MQW structure after Al–Ga interdiffusion by using equation (1) [13].

$$C(x,t) = C_0\left(1 - \frac{1}{2}\sum_{n=0}^{m-1}\left[erf\left(\frac{x - nL_w - nL_b}{2L_d}\right) + erf\left(\frac{(n+1)L_w + nL_b - x}{2L_d}\right)\right]\right), \quad (1)$$

where erf is the error function, C_o is the initial concentration of Al, t is the diffusion time, L_b is the barrier width, L_d ($=\sqrt{Dt}$) is the diffusion length and D is the diffusion coefficient. By using this equation one can calculate the Al concentration and the corresponding intermixed quantum well potential profile for arbitrary diffusion lengths. The Schrodinger equation was then solved numerically to get quantized energy levels of electrons and holes. With these calculations, one can get the energy shift as a function of diffusion length. By comparing the energy shift from PL spectra after QWD with the calculation result, one can get the diffusion length and corresponding diffusion coefficient.

Fig. 6 shows the calculated energy shift as a function of diffusion length for the GaAs/AlGaAs MQW structure described in Fig. 1. There is the difference of the 42 meV in QW transition energy shift between the sample with 0 sccm NH$_3$ flow rate and the sample with 40 sccm NH$_3$ flow rate as seen Fig. 6. Fig. 7 shows diffusion coefficient as a function of NH$_3$ flow rate. It is noteworthy that the diffusion coefficient due to SiN$_x$ grown at 40 sccm NH$_3$ flow rate is 5 times larger than that grown at 0 sccm NH$_3$ flow rate as seen in Fig. 7.

The difference of 42 meV in QW transition energy shift is large enough to use to spatially-selectively disorder MQW structure for the fabrication of optoelectronic devices monolithically. Moreover, this technique may reduce the problem caused by the difference of thermal expansion coefficient between different dielectric capping layers used for selective QWD. Therefore this technique is well suited for the fabrication of integrated optoelectronic devices.

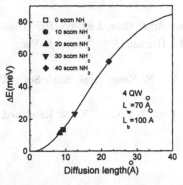

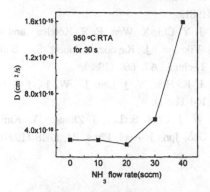

Fig. 6 Energy shift versus diffusion length   Fig. 7 Diffusion coefficient

CONCLUSIONS

The dependence of DCQWD on the characteristics of dielectric capping layer was studied using PECVD grown SiN_x film as the capping layer. The characteristics of the SiN_x capping layer were varied by changing the NH_3 flow rate during SiN_x deposition by PECVD. The higher NH_3 flow rate condition resulted in thinner SiN_x film, higher hydroge content. DCQWD was carried out by heat treatment of SiN_x capped MQW sample with RTA. The degree of QWD with SiN_x capping layer grown at higher NH_3 flow rate was larger than that with SiN_x film grown at lower NH_3 flow rate. This was explained by the increase of trap density in high hydrogen content SiN_x film grown at high NH_3 flow rate.

REFERENCES

[1] H. Ribot, K. W. Lee, R. J. Simes, R. H. Yan and L. A. Coldren, Appl. Phys. Lett. **55**, 672 (1989).

[2] T. Miyazawa, H. Iwamura, and M. Naganuma, IEEE Photon. Technol. Lett. **PTL-3**, 421 (1991).

[3] J. Beauvais, G. S. Ayling, and J. H. Marsh, IEEE Photon. Technol. Lett. **PTL-4**, 372 (1993).

[4] I. Gontijo, T. Krauss, J.H. Marsh and R. M. De La Rue, IEEE J. of Quantum Electron. **QE-30**, 1189 (1994).

[5] E. L. Allen, C. J. Pass, M. D. Deal, J. D. Plummer and V. F. K. Chia, Appl. Phys. Lett. **59**, 3252 (1991).

[6] M. Kuzuhara, T. Nozaki, and T. Kamejima, J. Appl. Phys. **66**, 5833 (1989).

[7] W. J. Choi, J. I. Lee, I. K. Han, K. N. Kang, Y. Kim, H. L. Park and K. Cho, J. Mat. Sci. Lett. **13**, 326 (1994).

[8] W. J. Choi, S. Lee, Y. Kim, S. K. Kim, J. I. Lee, K. N. Kang, N. Park, H. L. Park and K. Cho, J. Mat. Sci. Lett. **14**, 1433 (1995).

[9] H. Dun, P. Pan, F. R. White and R. W. Douse, J. Electrochem. Soc. **128**, 1555 (1981).

[10] J. Y. Chi, X. Wen, E. S. Koteles and B. Elman, Appl. Phys. Lett. **55**, 855 (1989).

[11] Vikram J. Kapoor, Robert S. Bailey and Herman J. Stein, J. Vac. Sci. Technol. **A1**, 660 (1983).

[12] I. K. Han, Y. J. Lee, J. W. Jo, J. I. Lee and K. N. Kang, Appl. Surf. Sci. **48/49**, 104 (1991).

[13] W. J. Choi, S. Lee, J. Zhang, Y. Kim, S. K. Kim, J. I. Lee, K. N. Kang and K. Cho, Jpn. J. Appl. Phys. **34**, L418 (1995).

THEORY OF CRITICAL LAYER THICKNESS OF NONCONSTANT QUANTUM-WELL WIDTH PRODUCED BY INTERDIFFUSION AND ITS OPTOELECTRONICS CONSEQUENCE

MICHAEL C.Y.CHAN[1] and E. HERBERT LI[1,2]
[1]University of Hong Kong, Department of Electrical & Electronic Engineering, Pokfulam, Hong Kong
[2]Harvard University, Division of Engineering & Applied Sciences, Pierce Hall 225, 29 Oxford Street, Cambridge MA 02138

ABSTRACT

In this paper, the concept of critical layer thickness is applied to the interdiffused quantum well (DFQW) structure. For the as-grown InGaAs/InP lattice matched quantum well, the interdiffusion process will induce in-plane strain into the DFQW forming a lattice mismatched system. The relation between the as-grown well width (L_z) and the diffusion length (L_d) for formation of dislocation is presented.

INTRODUCTION

Heteroepitaxy in lattice-mismatched systems is becoming more and more important for achieving high-performance electronic and optoelectronic devices. The fabrication of lattice mismatched semiconductor epitaxial layers in heterostructure is required to consider the critical layer thickness (CLT). When epitaxial layers of two different materials with their own lattice constants are grown on each others forming the strain layers QW material systems. The concept of CLT is essential for realising dislocation-free pseudomorphic QW materials. The CLT exists beyond which coherently strained pseudomorphic growth alters the growth with misfit dislocations. This restricts the design of device structures. In the theoretical calculations, Van der Merwe [1] calculated the CLT based on energy considerations in 1963. However, the first accepted model introduced by Matthews and Blakeslee [2] in 1974 is based on the mechanical equilibrium. Later, two models were further developed by several research groups [3, 4]. Both of the models are generally in agreement with experimental data. Recently, thermally induced composition intermixing in heterostructure quantum wells (QW) is becoming a popular choice in order to design multi-wavelength optical devices applications. [5] The intermixing process involves the interdiffusion of the constituent atoms of the heterostructure, the processing temperature, and time. During the process the as-grown square-QW compositional profile is modified to a graded profile thereby altering the confinement profile and subband structure in the interdiffused QW (DFQW). The CLT of the DFQW structures is not clearly defined by conventional methods. The conventional way to determine the CLT of a strained QW is to consider the abrupt interface while the DFQW has a graded compositional profiles of the interface. Therefore, it is difficult to define the QW width for DFQW. In this paper, the proposed theoretical model for tackling the CLT in DFQW is now considered. The theory is based on the equilibrium of two type of energies: the strain relief energy and the self-energy of misfit dislocation. These two energies can be calculated with the graded and continuous profile of the DFQW.

THEORETICAL MODEL

QW Interdiffusion

In our calculation, we consider the lattice matched $In_{0.53}Ga_{0.47}As/InP$ as-grown QW lattice matching to InP substrate. [6] After interdiffusion, group-III and group-V sublattices move across the interface of the heterostructure, so that the interdiffused QW structure may or may not be lattice matched to InP. Therefore, a strained-layer structure may result after interdiffusion. The constituent atoms interdiffusion has been modeled by an error function distribution. The group-III and group-V interdiffusion processes can be modeled by two different diffusion lengths. The diffusion of In and Ga atoms is characterized by a diffusion length, L_d^{III}, which is defined as $L_d^{III} = (D^{III}t)^{1/2}$, where D^{III} is the diffusion coefficient of group-III atoms and t is the diffusion time; the diffusion of As and P atoms is characterized by the diffusion length $L_d^{V} = (D^{V}t)^{1/2}$. The QW structure to be modeled consists of an as-grown InGaAs square well sandwiched between thick InP barriers. When the intermixing of QW occurs, the concentration of the diffusion atoms across the QW structure is assumed to have an error function distribution. The constituent atom compositional profiles can be represented as the group-III sublattice and group-V sublattice. In the group-III sublattices, the In concentration after interdiffusion is described by

$$x_{In}(z) = 1 - \frac{1-x}{2}[erf(\frac{L_z + 2z}{4L_d^{III}}) + erf(\frac{L_z - 2z}{4L_d^{III}})] \qquad (1a)$$

where L_z is the as-grown well width, z is the growth direction, and the QW is centered at z=0. In the group-V sublattices, the As concentration after diffusion is given by

$$y_{As}(z) = \frac{y}{2}[erf(\frac{L_z + 2z}{4L_d^{V}}) + erf(\frac{L_z - 2z}{4L_d^{V}})] \qquad (1b)$$

where x and y are the as-grown In and As concentration, respectively.

Critical Layer Thickness of DFQW

In the square QW epitaxial growth, the problem of lattice mismatch is due to a barrier-layer with different lattice constants of well-layer. This is generally accommodated by a combination of coherent strain and misfit dislocations. A lattice misfit parameter is defined as $f = |a_w - a_b|/a_w$, where a_w and a_b are the lattice constants of well and barrier, respectively. The CLT (h_c) for pseudomorphic epitaxy is derived by considering the thickness dependence of the strain energy and dislocation energy, and by minimizing the total energy. The CLT is obtained by [2]

$$h_c = \frac{b(1 - \gamma \cos^2 \Theta_{db})[\ln(\frac{h_c}{b}) + 1]}{8\pi(1 + \gamma)f \cos \lambda} \qquad (2)$$

where γ is Poisson's ratio, Θ_{ab} is the angle between the dislocation line and its Burgers vector (b), and λ is the angle between the slip direction and that line in the interface plane which is normal to the line of intersection between the slip plane and the interface. This model is only valid for calculating the CLT of the square-QW structure. It cannot be applied in the DFQW structure. In the DFQW structure, the thickness of well layer is not clearly defined because the abrupt interface is changed to the continuous distribution. The in-plane strain could not be uniformly induced in the QW layers.

The in-plane strain, $\varepsilon(x, y)$, across the well will vary with DFQW so that the strain effects are also z-dependent. Assuming that the growth direction z is along <001>, then for the biaxial components parallel to the interface of the strain components, after interdiffusion, are given by:

$$\varepsilon_{xx} = \varepsilon_{yy} = \varepsilon(x, y) = [a_w(x, y) - a_b]/a_w(x, y), \tag{3a}$$
$$\varepsilon_{zz} = -2[c_{12}(x, y)/c_{11}(x, y)]\varepsilon(x, y), \tag{3b}$$
$$\varepsilon_{xy} = \varepsilon_{yz} = \varepsilon_{zx} = 0, \tag{3c}$$

where $a_w(x, y)$ and a_b are the lattice constant of the well layer and the *InP* substrate materials, respectively. $\varepsilon(x, y)$ is defined to be negative for compressive strain, $c_{ij}(x, y)$ is an elastic stiffness constant and ε_{ij} is a strain component.

For the zincblende structure semiconductor, the strain energy density is given by [7]

$$U_{st} = \frac{1}{2}c_{11}\left(\varepsilon_{xx}^2 + \varepsilon_{yy}^2 + \varepsilon_{zz}^2\right) + c_{44}\left(\varepsilon_{yz}^2 + \varepsilon_{zx}^2 + \varepsilon_{xy}^2\right) + c_{12}\left(\varepsilon_{yy}\varepsilon_{zz} + \varepsilon_{zz}\varepsilon_{xx} + \varepsilon_{xx}\varepsilon_{yy}\right). \tag{4}$$

Substituting equations (3) into (4), we get

$$U_{st} = \left(c_{11}(x,y) + c_{12}(x,y) - \frac{2c_{12}^2(x,y)}{c_{11}(x,y)}\right)\varepsilon^2(x,y). \tag{5}$$

The dislocation energy density is calculated by elastic theory. For a 60° dislocation, the energy density is given by [7]

$$U_{dis} = \frac{\mu b^2}{4\pi(1-v)}\left(1 - v\cos^2\alpha\right)\ell n\left(\frac{B}{b}\right), \tag{6}$$

where μ is the shear modulus and is equal to $\frac{1}{2}(c_{11}-c_{12})$, v is Poisson's ratio and is equal to $c_{12}/(c_{11}+c_{12})$, B is the extent of the distortion produced by a dislocation and $\alpha=60°$ for the 60° dislocation. For the strain energy large than dislocation energy, it is energetically favourable to form dislocations and the mismatch is accommodated by a combination of coherent strain and misfit dislocations.

RESULTS AND DISCUSSIONS

In our calculation, an undoped $In_{0.53}Ga_{0.47}As$ single QW layer sandwiched between InP barriers is considered. The as-grown structure is lattice matched and the effects of interdiffusion on the QW structure are considered both for group-III and group-V sublattice intermixing. The parameters were generally determined by interpolating scheme between the binary parameters at room temperature and listed in the Table 1. [8]

427

Table 1 : Material parameters for InGaAsP DFQW at room temperature.

	Unit	$In_xGa_{1-x}As_yP_{1-y}$
a_w	Å	$5.6533(1-x)y + 6.0583xy + 5.4505(1-x)(1-y) + 5.8687x(1-y)$
C_{11}	$\times 10^{11}$ dynes/cm^2	$11.9(1-x)y + 8.329xy + 14.05(1-x)(1-y) + 10.11x(1-y)$
C_{12}	$\times 10^{11}$ dynes/cm^2	$5.38(1-x)y + 4.526xy + 6.203(1-x)(1-y) + 5.61x(1-y)$

Before calculating the dislocation in DFQW, the model is applied in the case of $In_xGa_{1-x}As$/GaAs square QW structure. A critical layer thickness of 104Å is found in the well layer for x=0.2. The value is agreed with the experimental results. [9]

For InGaAs/InP DFQW system of group-III interdiffusion, Ga atoms diffuse out to the InP barrier layer and In atoms diffuse into the well layer, and a thin and graded InGaP/InGaAs interface is formed. The distribution of the In and Ga atoms are described by the error function distribution, while the As and P concentration profiles do not change. The In compositional profiles are shown in Fig.1(a). In the early stages of the interdiffusion, the In atoms near the interface diffuse into the well, while Ga atoms diffuse into the barrier, but at the well centre the In concentration change slightly. As the interdiffusion process, the In concentration at the well center is changed from 0.53 to 0.85 as $L_d \rightarrow 120$Å. Thus, the intermixing of In and Ga atoms will result in a change of structural properties such as in-plane strain. The as-grown square well structure due to the intermixing of atoms will gradually change from an abrupt interface to a non-continuous profile.

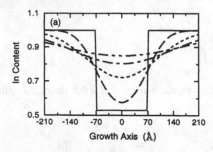
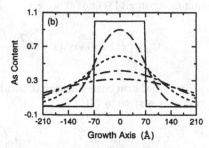

Fig.1 The (a) Indium and (b) Arsenic compositional profiles of InGaAs/InP DFQW with as-grown well width of 140 Å for various diffusion lengths. As-grown (————); Ld=30Å (— — — —); (· · · · · · ·); Ld=90Å (— · — · —); Ld=120Å (— · · — · ·).

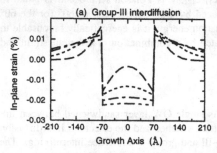
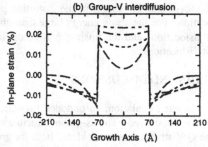

Fig.2 In-plane strain across the (a) group-III and (b) group-V interdiffused QW for various diffusion lengths. These lines are represented to the diffusion lengths same as figure 1.

The variation of the in-plane strain profile with interdiffusion across the structure is illustrated by Fig.2(a). Since the InP lattice constant is always larger than that of InGaP, tensile strain arises in the barrier near the interface, while the InGaAs well becomes compressively strained due to the increase in In content. Consequently the intermixing process results in a strained QW structure. At the interface of the heterostructure, the strain has its maximum value (compressive in the well and tensile in the barrier). The compressive strain in the well increases with interdiffusion. At the initial stage of interdiffusion, the strain in the well is smaller in the centre as compared with that at the interface. For extensive interdiffusion, such as $L_d = 120$Å, the well strain variation between the centre and the interface is reduced. On the other hand, the tensile strain in the barrier layers is a maximum at the interface and then decreases gradually to zero outward away from the interface. When the interdiffusion increases, the barrier strain induced at the interface decrease while that at the outer barrier increase.

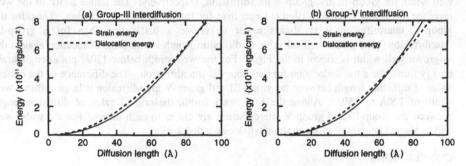

Fig.3 The dislocation energy and strain energy of InGaAs/InP DFQW
against the diffusion lengths for (a) group-III and (b) group-V interdiffusion.

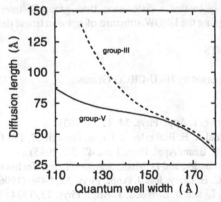

Fig.4 The critical diffusion length forming the dislocation against
the as-grown well width for group-III and group-V interdiffusion.

429

In the case of group-V interdiffusion, only As and P atoms will inter-diffuse with each other between the well-barrier interfaces. Since in the as-grown QW structure, an As atom in the well will move out and a P atom in the barrier will move in, the group-V interdiffusion will result in an InGaAsP/InAsP interface. The distribution of the As and P atoms are also described by the error function distribution in Fig.1(b). The in-plane strain profile is quite different from that of the group-III interdiffusion and is shown in Fig.2(b). Since the InAsP lattice constant is always larger than the InP substrate, the InAsP barrier layer becomes compressively strained due to an increase of the As content. In the well layer, the increase in P content will induce a tensile strain because of a larger lattice constant InGaAsP layer with respect to InP substrate.

The variation of the strain energy and dislocation energy with the diffusion length in the case of group-III and group-V interdiffusion is shown in Fig.3(a)-(b). The strain energy and dislocation energy are represented by the solid line and dash line respectively. The dislocation is generated when the strain energy is larger than the dislocation energy. For example, the dislocation will be created for a diffusion length greater than 86Å and 70Å for a 140Å as-grown well width for group-III and group-V interdiffusion, respectively. The lattice misfit of the well layer for the group-V interdiffusion is larger than for the group-III interdiffusion. Thus, for the group-V interdiffusion it is always easier to create a dislocation than for a group-III interdiffusion. The plot of critical value of diffusion length forming the dislocation against the as-grown well-width is shown in the Fig.4. For the well width below 120Å, not enough strain energy forms for a the dislocation for the group-III interdiffusion. The difference of the critical value of diffusion length between the group-III and group-V interdiffusion is large within a well width of 120Å to 150Å. Above the 150Å well width, the critical value of diffusion length between the group-III and group-V interdiffusion are closer to each others. For the wider well width, a small amount of in-plane strain will make a dislocation.

CONCLUSIONS

In this paper we present the critical layer thickness of lattice mismatched InGaAs/InP DFQW with different well widths and diffusion lengths. The results show that group-V interdiffusion more easily created a dislocation than group-III interdiffusion. Moreover, the model is useful for designing the DFQW structure of optoelectronic devices.

ACKNOWLEDGEMENTS

This work is supported by HKU-CRCG Grants.

REFERENCES

[1]   J. H. van der Merwe, J. Appl. Phys. **34**, 123 (1963)
[2]   J. W. Matthews and A. E. Blakeslee, J. Cryst. Growth **27**, 118 (1974)
[3]   R. People and J. C. Bean, Appl. Phys. Lett. **47**, 32 (1985)
[4]   S. C. Jain, M. Willander, and H. Maes, Semicond. Sci. Technol. **11**, 641 (1996)
[5]   E. H. Li and W. C. H. Choy, Jpn. J. Appl. Phys. **35**, L496 (1996)
[6]   J. Micallef, E. H. Li and B. L. Weiss, J. Appl. Phys. **73**, 7524 (1993)
[7]   C. Mailhiot and D. L. Smith, Solid State and Mat. Sci. **16**, 131 (1990)
[8]   S. Adachi, J. Appl. Phys. **53**, 8775 (1982)
[9]   T. G. Andersson, Z. G. Chen, V. D. Kulakovskii, A. Uddin, and J. T. Vallin, Appl. Phys. Lett. **51**, 752 (1987)

ACTIVE ANTI-GUIDE VERTICAL CAVITY SURFACE EMITTING LASERS WITH DIFFUSED QUANTUM WELLS STRUCTURE

S.F. YU, E. HERBERT LI, W.M.MAN
Department of Electrical & Electronic Engineering, University of Hong Kong, Pokfulam Road, Hong Kong, sfyu@hkueee.hku.hk

ABSTRACT

The enhancement of single transverse mode operation in vertical cavity surface emitting lasers by using interdiffused quantum wells is proposed and analyzed. It is observed that the influence of self-focusing (arising from carrier spatial hole burning and thermal lensing) on the profile of transverse modes can be minimized by introducing a step diffused quantum wells structure inside the core region of quantum-well active layer. Stable single-mode operation in vertical cavity surface emitting lasers can also be maintained.

INTRODUCTION

An ideal VCSEL should have single longitudinal mode operation, low threshold current and narrow output beam . However, at high power, VCSEL's exhibit multiple transverse modes operation, which is excited by the increase of refractive index arising from self-focusing effect (owing to carrier spatial hole burning and thermal lensing) [1]. This would deteriorate the performance of lasers and should be avoided. Thus, the goal of this paper is to utilize diffused quantum wells (DFQW's) structure of vertical cavity surface emitting lasers (VCSEL's) for the enhancement of high power single-mode operation.

The purpose of using step DFQW's structure can be explained as follows: The step-diffusion is defined selectively within the core region of the QW active layer. A non-uniform stepped refractive index profile is created, resulting in an anti-guiding structure for the transverse modes. Hence, DFQW's structure can compensate the influence of self-focusing and stabilize the profile of transverse modes. Also, the major advantages of utilizing DFQW's structure are that it is simple and compatible with existing fabrication technologies of lasers.

DESIGN AND ANALYSIS

Laser Structure

The schematic diagram of VCSEL with DFQW's structure is as in figure 1. We assume the circular metal contact of the laser has diameter of 10 µm on the p-side for current injection. There is an active layer sandwiched between two undoped spacer layers and two Bragg reflectors. The thickness of each undoped spacer layers is half-wavelength (~0.1 µm). The Bragg reflectors are formed by alternate layers of AlGaAs and AlAs, with quarter wavelength thickness dielectric layers on p-side and n-side respectively. The total number of layers on the p-side is 90 (~5.4 µm), while on the n-side is 36 (~2.1 µm). For the active layer, it consists of three GaAs/$Al_{0.3}Ga_{0.7}As$ QW having total thickness of half-wavelength (~0.1 µm).

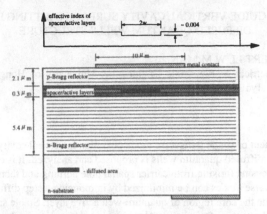

Figure 1 Schematic diagram of a VCSEL with step DFQW's structure

Introduction of DFQW's structure by impurity induced disordering

Step DFQW's structure is introduced along the active layer by compositional impurity induced disordering (IID) of QW. Interdiffusion is applied into the QW's active layer of laser. The implanted ions on the as-growth QW's layers are shielded by the circular mask. After implantation, we apply annealing to induce compositional disordering and to restore the impurities damage. Afterwards, the cladding region is resulted from the area covered by the mask, and the core region is resulted from the diffused area. Consequently, a small refractive index step is established between the core and cladding region. Thus, an anti-index guided structure is obtained. We then grow the spacer layer, p-type Bragg reflector and circular metal contact on the active layer to complete the structure of the device.

For easy penetration of impurities into the active layer, argon is proposed to be the impurity used. It is because argon has deep penetration power (>2 µm). Also, its influence on the electrical properties of p-Bragg reflector is less after thermal annealing [2].

Numerical Laser Model

We use k.p method to model the optical gain and refractive index of DFQW's under external carrier injections [3][4]. The QW's active layer under analysis consists of three GaAs-Al$_{0.3}$Ga$_{0.7}$As QW's with well width 100Å and barrier thickness 150Å. In the model, the extent of interdiffusion is controlled by the diffusion length of impurities, L$_d$. It is because the diffusion strength is increased with the magnitude of L$_d$.

The optical gain spectrum of QW material, G, with a particular set of λ and L$_d$ can be expressed as

$$G(\lambda) = a(\lambda)\log(N/N_o) \qquad (1)$$

with a(λ) being the gain coefficient and N$_0$ the carrier concentration at transparency.

Also, we can evaluate the change of refractive index, Δn$_r$, inside the QW's active layer by considering the variation of gain spectral through Kramers-Kronig dispersion relation[5]. It can be shown that at certain values of L$_d$ and λ$_0$, the relation between Δn$_r$ and N is

$$\Delta n_r = d(\lambda) \log(N / N_r) \qquad (2)$$

where $d(\lambda)$ and N_r are fitting parameters.

We can analyze the characteristics of transverse modes of VCSEL's by making some modifications on the recently developed model [6]. The Poisson equations for voltage and temperature can be solved by finite difference method self-consistently with the wave equation of optical field and rate equation of carrier concentration. Therefore, the 3-D distribution of voltage and temperature can be included into the model.

In solving the voltage equation, we assume the electrical conductivity of n-Bragg reflector, p-Bragg reflector, spacer/active layer and n-substrate to be $1.5 \text{cm}^{-1}\Omega^{-1}$, $7 \text{cm}^{-1}\Omega^{-1}$, $3 \text{cm}^{-1}\Omega^{-1}$ and $500 \text{cm}^{-1}\Omega^{-1}$, respectively. Also, several assumptions are made.
1) Charge density is zero.
2) Metal contacts are fixed at certain voltage level.
3) First derivative of voltage at the surface of device without any metal contact is zero.
4) Voltage across p-n junction, V_F, is related to carrier concentration by the following equation:

$$V_F = \frac{1}{q}\left[E_g + k_B T \cdot \log\left\{(\exp(N / N_C) - 1)(\exp(N / N_V) - 1)\right\}\right] \qquad (3)$$

where N is carrier concentration, E_g is bandgap energy of GaAs, N_C and N_V are effective conduction and valence edge density of states respectively and T is temperature.

For the Poisson equation of heat, it is assumed at 300°K, the thermal conductivity of n-Bragg reflector, p-Bragg reflector and spacer/active layer to be equal to $0.07 \text{Wcm}^{-1}\text{K}^{-1}$, and that of n-substrate to be $0.45 \text{Wcm}^{-1}\text{K}^{-1}$. Also, the boundary conditions are similar to that shown in [1]. Moreover, we assume the reasons for internal heat source are as follows:
1) There is joule heating in both p- and n- side Bragg reflectors and spacer layers.
2) There is non-radiative recombination spontaneous radiation inside the active layer.

In solving the wave equation, we assume the total absorption and scattering losses of the QW waveguide to be 20cm^{-1} for the case without interdiffusion and to be 30cm^{-1} when with diffusion. By varying the carrier concentration and temperature, the refractive index will change also. The change of refractive index, Δn, is given by

$$\Delta n = \Gamma \Delta n_r + \frac{\partial n}{\partial T} \Delta T \qquad (4)$$

where $\partial n / \partial T$ $(= 2 \times 10^{-4} \text{K}^{-1})$ is the variation of refractive index with temperature, $\Delta T = T - 300^\circ$K, and Γ $(= 0.5)$ is the longitudinal confinement factor taking into account the penetration of standing wave into the spacer layers.

Design Consideration on Diffusion Length and other parameters

Small value of L_d $(< 5 \text{Å})$ is preferred for VCSEL with step DFQW's structure. It is because small L_d will not reduce the optical gain and refractive index of the QW active layer significantly. However, large L_d may cause damage to the lattice structure of p- Bragg reflector owing to its high implantation energy $(> 5 \text{MeV})$.

433

Practically, L_d is set within a particular range. It is because small value of L_d is difficult to maintain, as L_d is affected by the variation of thickness and quality of p-Bragg reflector. In our investigation, we define an effective value, $< L_d >$, to be 3Å. Also, the operation wavelength, λ, is set to be 0.85μm.

For $L_d = 0$Å or 3Å, the values of a, N_0, background refractive index, d and N_r are shown in Table I. The values of other parameters can be found in references [6] and [7].

Table I Values of parameters

	$L_d = 0$ Å	$< L_d > = 3$ Å
a	1591cm^{-1}	1500cm^{-1}
N_0	$1.94 \times 10^{18}\text{cm}^{-3}$	$1.93 \times 10^{18}\text{cm}^{-3}$
Background refractive index	3.6270	3.6185
d	-0.0283	-0.0279
N_r	$2.06 \times 10^{18}\text{cm}^{-3}$	$2.04 \times 10^{18}\text{cm}^{-3}$

RESULTS

Curves of light power against current are plotted in fig. 2, with radius of diffusion area varying between 3 μm and 3.8 μm. The dotted and solid curves show the case of VCSEL with diffusion and without diffusion respectively. From the graph, it is observed that a kink is present in both cases with the excitation of first-order mode (LP_{11}). However, when diffusion is introduced and at w = 3 μm, the kink is seen to shift upward and to the right, resulting in a stable fundamental-mode (LP_{01}) operation at high power.

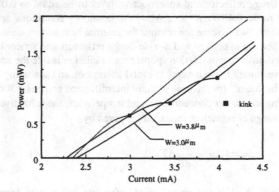

Figure 2 Light/current characteristics of VCSEL's without (dotted line) and with (solid line) a step DFQW structure

The variation of refractive index with change in radius is shown in Fig. 3. The output power is set to 1 mW. As seen from the plot, lasers without step-diffused QW's structure exhibit continuous increase of refractive index near the center of the core region. This can be explained by the self-focusing effect arising from spatial hole burning and thermal lensing. As a result, optical field with transverse modes is shifted towards the center region and the LP_{01} mode is excited. However, for lasers with step-diffusion introduced and at w = 3 μm, no increase of refractive index is observed, as it is counteracted by the step-diffused QW's structure. Thus, single-transverse-mode operation can be maintained.

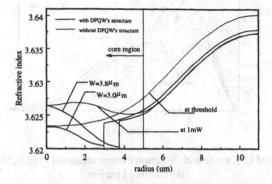

Figure 3 Transverse distribution of refractive index of VCSEL's without (dotted line) and with (solid line) a step DFQW structure at output power of 1 mW

The profile of normalized intensity of LP_{01} mode and LP_{11} mode is shown in Fig. 4. From the graph, it can be seen that for DFQW's lasers with $w \leq 3$ μm, a kink appears in both the LP_{01} and Lp_{11} modes. However, for DFQW's lasers with $w > 3$ μm and for lasers without diffusion, no kink is observed. The profile of the transverse modes is deformed so as to minimize the influence of self-focusing.

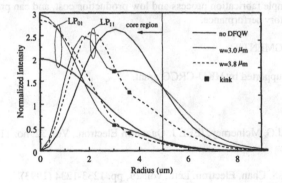

Figure 4 Profile of LP_{01} and LP_{11} modes for VCSEL's with and without DFQW's structure at threshold

A graph of normalized intensity of Lp_{11} mode against radius at different injection level is shown in fig. 5. It can be seen that for DFQW's lasers with $w = 3$ μm, the effect of self-focusing is less than that for lasers without diffusion. The beam-width of LP_{11} mode is reduced significantly by self-focusing effect for lasers without diffusion. Also, we can observe that the volume of LP_{11} mode inside the cladding region remains unchanged with increasing injection current. Thus, stable operation of LP_{01} mode is maintained.

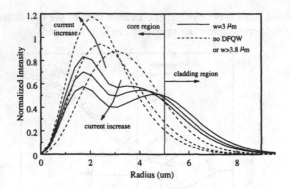

Figure 5 Variation of LP$_{11}$ mode with injection current increased from I$_{th}$ to 1.6 I$_{th}$ where I$_{th}$ is the threshold current

CONCLUSIONS

The enhancement of single transverse mode operation of VCSEL by using DFQW's structure is analyzed. Impurity induced disordering (IID) is used to introduce step DFQW's structure as it is simple and requires low cost. From our investigation, it is found that the optical loss in cladding region remains unchanged at high injection current. Thus stable LP$_{01}$ mode operation can be maintained. Also, the output power of stable LP$_{01}$ mode operation is doubled by using the proposed device. On the whole, VCSEL with DFQW's structure requires only simple fabrication process and low production cost, and can produce high yield rate and satisfactory performance.

ACKNOWLEDGMENTS

This work was supported by HKU-CRCG grant.

REFERENCES

1. Y.G. Zhao & J.G. McInerney, IEEE J. Quantum Electron., Vol. 32, no. 11, pp. 1950-1958 (1996).

2. E.H. Li and K.S. Chan, Electron. Lett., vol. 29, pp. 1233-1234 (1993).

3. E.H. Li, B.L. Weiss, K.S. Chan and J. Micallef, Appl. Phys. Lett. 62, pp.550-552 (1992).

4. C.H. Herny, R.A. Logan and K.A. Bertness, J. Appl. Phys. 52, pp. 4457-4461 (1981).

5. S.F. Yu, IEEE J. Quantum Electron., Vol. 32, no. 7, pp. 1168-1179 (1996).

6. S.F. Yu, C.W. Lo and E.H.Li, IEEE J. Quantum Electron, Vol. 33, no. 6, pp. 999-1009 (1997).

Modeling Interdiffusion in Superlattice Structures

Richard G. Gass and Howard E. Jackson

Department of Physics, University of Cincinnati, Cincinnati, OH 45220-0011

Compositional interdiffusion in $Al_{0.3}Ga_{0.7}As/GaAs$ superlattices induced by Si focused ion beam implantation and subsequent rapid thermal annealing is modeled using a set of diffusion equations which take into account the dynamics of the vacancy spatial profile. The inclusion of a new phenomenological term, which depends on the time derivative of the vacancy concentration spatial profile, provides good agreeement with experiment.

I. INTRODUCTION

The modification of optical properties of quantum well structures by post fabrication processing is of both intrinsic interest and of importance in the design, fabrication and integration of quantum well structures in optoelectronic applications. Interdiffusion and compositional intermixing affect the optical properties of quantum well structures and thus the understanding of intermixing is critical to the understanding of quantum well structures fabricated in this manner. In this paper we model compositional interdiffusion in $Al_{0.3}Ga_{0.7}As/GaAs$ superlattices induced by Si focused ion bean implantation and subsequent rapid thermal annealing and compare the result to data from several experimental groups.[1-4]

II. MODELING COMPOSITIONAL MIXING

Compositional interdiffusion in superlattices is enhanced by the presence of defects, importantly vacancies, in the superlattice.[1] Using the Al ion concentration as a marker for the compositional mixing, one standard model for compositional mixing in rapid thermal annealing (RTA) due to Kahen and Rajeswaran[3] assumes that diffusing ions and type III lattice-site atoms effect the diffusion of Al atoms solely through recombination with the vacancies. This assumption leads to the following set of coupled differential equations[3] governing the diffusion of the vacancies and Al atoms:

$$\frac{\partial C_v}{\partial t} = \frac{\partial}{\partial x}\left(\kappa_v \frac{\partial C_v}{\partial x}\right) - \frac{C_v - C_{v,eq}}{\tau}, \tag{1}$$

and

$$\frac{\partial C_{Al}}{\partial t} = \frac{\partial}{\partial x}\left(\kappa_{Al}\frac{\partial C_{Al}}{\partial x}\right) \tag{2}$$

where

$$\kappa_{Al}(x,t) = \kappa_{Al,eq}(C_v/C_{v,eq}). \tag{3}$$

This model is in good agreement with the data for lattice depths greater than the projected range (R_p), but does not reproduce the "pinch-off" effect seen in several experiments.[2,4] Experiments show that intermixing is small at the surface and increases to a maximum at around R_p and then falls off at greater depths. The maximum interdiffusion occurs *not* at the maximum of the vacancy concentration but, as Chen and Steckl[4] noted, at the maximum of the second derivative of the vacancy concentration. This led them to propose a modification of the Al diffusivity in which Eq. 3 is replaced by

$$\kappa_{Al}(x,t) = \kappa_{Al,eq}(C_v/C_{v,eq})exp\left(\alpha \frac{\partial^2 C_v(x,t)}{\partial x^2} - \phi\right) \tag{4}$$

where ϕ takes into account the effect of vacancy recombination and α is a phenomenologically determined constant.

Physically, Eq. 4 implies that interdiffusion during RTA is affected by the excited movement of vacancies due to the implantation induced gradient. Chen and Steckl[4] did not carry out detailed calculations using this model but suggested that the dependence on the second derivative of the vacancy concentration would lead to a "pinch-off" in the depth dependence of the intermixing.

We propose a different phenomenological equation for the Al diffusivity:

$$\kappa_{Al}(x,t) = \kappa_{Al,eq}(C_v(x,t)/C_{v,eq})exp\left(\beta\frac{\partial C_v}{\partial t}\right). \tag{5}$$

This allows us to actually calculate the effect of vacancy recombination. We have carried out extensive numerical calculations with this model and find very good agreement with experiment including the presence of a "pinch-off" region. Our calculation proceeds in several steps.

First, the initial vacancy concentration was fit to data from a TRIM calculation of a $1x10^{14}cm^{-2}$ implantation of 100Kev Si into GaAs.[5] The fit is shown in Fig 1.

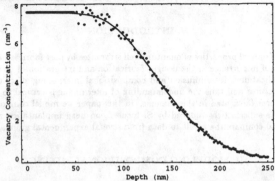

FIG. 1. Initial vacancy concentration. From a TRIM calculation (points); the solid line is a fit to the TRIM data.

Once the initial vacancy concentration is known, the time evolution of $C_v(x,t)$ can be calculated from Eq. 1. We use a finite difference method to solve the partial differential equations. For the purposes of this calculation we take the ends of the superlattice to be sinks for the vacancies, although the details of the intermixing are not materially effected by the boundary conditions. Following Kahen and Rajewaran[3] we take $\kappa_{Al,eq} = 0.1nm^2/s$, $\kappa_v = 1000nm^2/s$, $C_{v,eq} = (1/250)/nm^3$ and $\tau = 4s$. For the calculation of $\kappa_{Al}(x,t)$ we take $\beta = 1$

A "snap-shot" of the vacancy concentration $C_v(x,t)$ at a $t = 0.3$ seconds is shown in Fig 2.

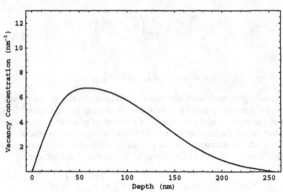

FIG. 2. A "snap-shot" of $C_v(x,t)$ at $t = 0.3s$

With $C_v(x,t)$ in hand one can calculate $\kappa_{Al}(x,t)$ and then $C_{Al}(x,t)$. A "snap-shot" of $\kappa_{Al}(x,t)$ is shown in Fig. 3.

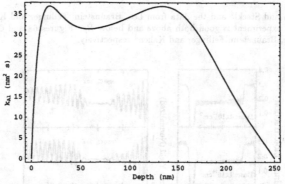

FIG. 3. A "snap-shot" of $\kappa_{Al}(x,t)$ shown at $t = 0.3$.

The results of compositional mixing are shown in Figs. 4 and 5. This calculation was run on a superlattice similar to the superlattice used by Chen and Steckl[4]. Figure 4 shows the results of compositional mixing using our model while Fig. 5 shows the results using model of Kahen, Rajewaran and Lee[1].

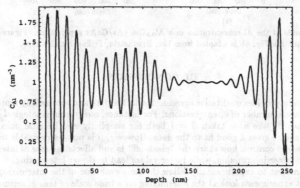

FIG. 4. Interdiffusion shown at $t = 0.4$. Note the appearance of a "pinch-off" region.

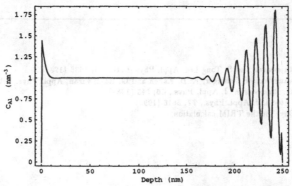

FIG. 5. Interdiffusion shown at $t = 0.4$ in the model of Kahen, Rajewaran and Lee.

Our calculations reproduce the "pinch-off" seen in the data and are in good qualitative agreement with the data from Chen and Steckl[4] and the data from Lee, Braunstein, Fellinger and Kahen.[2] Importantly the agreement with experiment is good both above and below R_p. Figures 6a and 6b are data of Chen and Steckl[4] and Lee, Braunstein, Fellinger and Kahen[2] respectively.

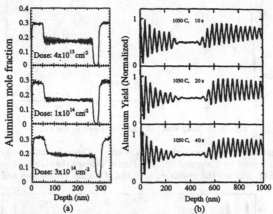

FIG. 6. SIMS profile of the Al concentration in a $Al_{0.3}Ga_{0.7}As/GaAs$ superlattice. Figure 6a is adapted from Chen and Steckl while the Fig. 6b is adapted from Lee, Braunstein, Fellinger and Kahen

III. CONCLUSIONS

Although our model provides qualitative agreement with the data and successfully reproduces the "pinch-off" region, there are a number of open questions. For instance, our phenomenological model contains an unspecified constant β. We have taken $\beta = 1$ both for simplicity and because numerical experiments indicate that this choice gives a good fit to the data. However, β is not yet derivable from the underlying physics. The value of β controls how sharp the "pinch-off" is and affects the rate of interdiffusion. Smaller values of β lead to faster intermixing, while larger values lead to slower intermixing.

It would be of interest to experimentally follow the time evolution of the intermixing. This is not easy to do. The current experiments look at the intermixing on a time scale of tens of seconds. As can be seen from figure 6b the intermixing is essentially complete after 10 seconds or less. Our calculations suggest that most of the intermixing occurs on an even shorter time scale, within the first one to two seconds.

[1] K. B. Kahen, G. Rajewaran and S.-Tong Lee, Appl. Phys. Lett., **53**, 1635 (1988)
[2] S.-Tong Lee, G. Braunstein, P. Fellinger, K. B. Kahen and G. Rajeswaran, Appl. Phys. Lett., **53**, 2531 (1988)
[3] K. B. Kahen and G. Rajeswaran, J. Appl. Phys., **66**, 545 (1989)
[4] P. Chen and A. J. Steckl, J. Appl. Phys., **77**, 5616 (1995)
[5] We thank Irving Chyr for the TRIM calculation.

CATION INTERDIFFUSION IN GaInP/GaAs SINGLE QUANTUM WELLS

Joseph Micallef, Andrea Brincat, and Wai-Chee Shiu*
Department of Microelectronics, University of Malta, Msida MSD 06, Malta
* Department of Mathematics, Hong Kong Baptist University, Waterloo Road, Hong Kong

ABSTRACT

The effects of cation interdiffusion in $Ga_{0.51}In_{0.49}P$/GaAs single quantum wells are investigated using an error function distribution to model the compositional profile after interdiffusion. Two interdiffusion conditions are considered: cation only interdiffusion, and dominant cation interdiffusion. For both conditions the fundamental absorption edge exhibits a red shift with interdiffusion, with a large strain build up taking place in the early stages of interdiffusion. In the case of cation only interdiffusion, an abrupt carrier confinement profile is maintained even after significant interdiffusion, with a well width equal to that of the as-grown quantum well. When the interdiffusion takes place on two sublattices, but with the cation interdiffusion dominant, the red shift saturates and then decreases. The model results are consistent with reported experimental results. The effects of the interdiffusion-induced strain on the carrier confinement profile can be of interest for device applications in this material system.

INTRODUCTION

The direct bandgap ternary semiconductor $Ga_{0.51}In_{0.49}P$ lattice matched to GaAs has attracted much attention for a variety of electronic and optical applications including high-speed heterojunction bipolar transistors [1], field effect transistors [2], tandem solar cells [3], light emitting diodes [4], and infrared intrasubband photodetectors [5]. GaInP is a potential alternative to AlGaAs since it is not easily oxidised and does not suffer from a high concentration of deep trap centres.

Quantum well (QW) intermixing involves the interdiffusion of constituent atoms across the well-barrier interfaces, and is of particular interest since it offers the possibility of continuous, controlled modification of the material composition [6]. This change in composition alters the confinement profile and subband edge structure in the QW, resulting in a modified effective bandgap of the QW structure. The extent of the intermixing, and thus of the modification of the subband edge structure, is a function of the QW intermixing technique parameters, such as the nature and concentration of the impurity species present, and the process temperature and time. Moreover the interdiffusion process can be localised to selected regions of the QW structure so that the optical properties of only the selected areas are modified by the interdiffusion process. QW intermixing can thus provide a useful tool for bandgap engineering.

Intermixing of lattice-matched AlGaAs/GaAs QW structures results in the interdiffusion of only group III atoms (Al, Ga) since there is no As concentration profile across the heterointerface. In contrast, the intermixing of lattice-matched $Ga_{0.51}In_{0.49}P$/GaAs QW structures can result in interdiffusion of both group III (Ga, In) and group V (P, As) atoms. Moreover, if the rates of interdiffusion on the two sublattices are not comparable, then a strained-layer structure will result. Reported experimental results of thermal interdiffusion of lattice-matched GaInP/GaAs QWs have been interpreted in terms of substantial interdiffusion of the group III atoms together with minor interdiffusion of group V atoms [7]. Results of Si-impurity-induced intermixing have also been interpreted in terms of dominant group III interdiffusion [8].

COMPUTATIONAL CONSIDERATIONS

An undoped GaAs single QW layer lattice-matched to $Ga_{0.51}In_{0.49}P$ barriers is considered here. The constituent atoms compositional profile after interdiffusion is modeled using an error function distribution [9]. The interdiffusion of Ga and In atoms is characterized by a diffusion length L_d, which is defined as $L_d = (Dt)^{1/2}$, where D is the diffusion coefficient and t is the diffusion time; the interdiffusion of P and As atoms is characterized by a different diffusion length L_d'. For cation only interdiffusion $L_d' = 0$, while for dominant cation interdiffusion, $L_d' \neq L_d$; in both cases a strained QW structure results.

The QW structure will be coherently strained after interdiffusion if the layer thickness is within the critical thickness limit [10]. Tetragonal deformation results in a biaxial strain parallel to the interface and a uniaxial strain perpendicular to the interface. A compressive (tensile) biaxial strain causes an increase (decrease) in the bandgap energy whilst the uniaxial strain lifts the degeneracy of the heavy hole (HH) and light hole (LH) subbands, so that the HH subband moves towards (away from) the conduction band and the LH subband moves away from (towards) the conduction band. After interdiffusion, the well confinement profile, the carrier effective mass and the strain effects vary across the QW structure. Under these conditions, the subband-edge structure is obtained using the envelope function scheme by introducing these variations in the appropriate Schrödinger equation, which is then solved numerically to obtain the subband energy levels, the interband transition energies and the envelope wavefunctions. Details of the calculations are reported in [11].

RESULTS AND DISCUSSION

The structure considered here is a 6 nm thick GaAs layer sandwiched between semi-infinite $Ga_{0.51}In_{0.49}P$ barriers. Parameter k is defined as $k = L_d'/L_d$, and results are presented for $0 \leq k < 1$. A conduction band offset $Q_C = 45\%$ [12] has been used in the calculations.

For $k = 0$ interdiffusion takes place on the group III sublattice only. In this case a graded compositional profile results for the Ga and In atoms, while the P and As atoms compositional profile remains abrupt, as shown in Fig. 1(a) for $L_d = 0.5$ nm. Ga diffuses into the barrier while In diffuses into the well forming an InGaAs/GaInP interface. Since the InGaAs lattice constant is always less than GaAs, the QW is under compressive strain, while a tensile strain arises in the barrier near the interface since the Ga concentration here is less than the lattice-matching composition of 0.51. The strain profile that is induced by the intermixing process is shown in Fig. 1(b). For $L_z = 6$ nm and $L_d = 0.5$ nm, no compressive strain results at the centre of the well, a 1.75% compressive strain in the well close to the interface, and a 1.75% tensile strain in the barrier close to the interface.

The abrupt change in the group V atoms compositional profiles produces an abrupt bandgap change at the interface from the InGaAs well to the GaInP barrier so that the carrier confinement profiles remain abrupt after interdiffusion, with a well width equal to that of the as-grown QW. The group III atoms graded compositional profile across the QW structure modifies the shape of the confinement profiles, while the effects of the strain distribution on this bandgap affects both the shape and separation of the conduction and valence bands, and the HH and LH potential wells no longer coincide, Fig. 1(c). In the well, the HH potential profile is shifted towards the electron (C) potential profile, while the LH potential profile is shifted away from the electron potential profile. In the barrier, near the interface, the tensile strain again separates the HH and LH potential profiles, with the HH potential profile shifted away from, and the LH potential profile shifted towards, the electron potential profile. As a result of these strain effects, the C and HH confinement profiles exhibit a double well profile at the bottom of the well, and a double barrier structure similar to a resonant tunnelling structure. In the case of the LH confinement profile, the compositional and strain separation effects result in an almost square confinement profile after intermixing. Numerical results show that eigen states can be supported in the HH double-welled bottom potential, while tunnelling enhancement is possible for the topmost states in both the C and HH wells. The latter result could be of interest for intrasubband photodetector applications.

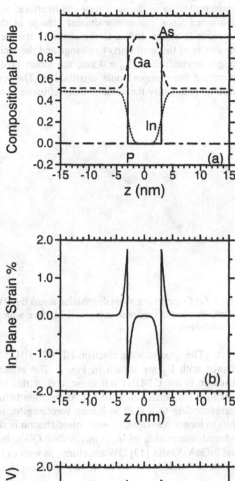

Fig. 1(a). Composition profile for L_z = 6 nm, L_d = 0.5 nm, with well centre at z = 0.

Fig. 1(b). In-plane strain for the interdiffused QW.

Fig. 1(c). Carrier confinement profiles for electron (C), heavy hole (HH) and light hole (LH), showing double-well feature at the well bottom.

The variation of strain inside the well with interdiffusion is shown in Fig. 2(a). In the initial stages of interdiffusion ($L_d \leq 0.5$ nm for the QW under consideration) a large compressive strain builds up in the well near the interface, while the centre of the well is still practically unstrained since the compositional change at the centre is still minimal. As interdiffusion proceeds In atoms diffuse to the centre of the well so that the group III atoms composition in the centre of the well starts to change and the well centre becomes compressively strained. For longer interdiffusion, $L_d = 4$ nm, the strain profile across the QW becomes fairly uniform, mirroring the compositional distribution. The tensile strain in the barrier near the interface remains practically the same as interdiffusion proceeds, Fig. 2(b).

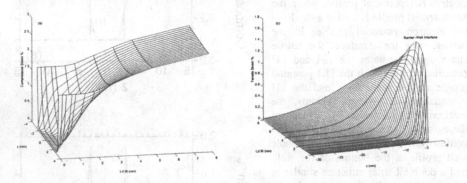

Fig. 2. (a) Compressive strain distribution across the QW with variation in interdiffusion; (b) tensile strain distribution in the barrier near the well interface with variation in interdiffusion. Group III atoms only interdiffusion.

The ground state electron-HH (C1-HH1) and electron-LH (C1-LH1) transition energy change with L_d are shown in Fig. 3. The effective QW bandgap energy of the intermixed structure is the C1-HH1 transition, and in this case the HH1 state occurs in the double-well potential, for small values of L_d. As the interdiffusion proceeds the bandgap energy decreases, corresponding to a shift to longer wavelengths, as evidenced in experimental results [7]. The shift to longer wavelengths with interdiffusion is similar to experimental results for Zn-diffusion induced intermixing of $In_{0.53}Ga_{0.47}As/InP$ QWs. In contrast, interdiffusion in AlGaAs/GaAs [9] and InGaAs/GaAs [13] QW structures, as well as InGaAs/InP induced by sulphur diffusion [14], results in bandedge shifts to shorter wavelengths.

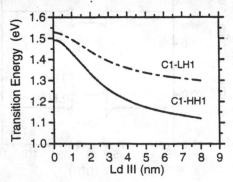

Fig. 3. Heavy-hole and light-hole ground state transition energy variation with L_d for group III atoms only interdiffusion.

For $0 < k < 1$, interdiffusion takes place predominantly on the group III sublattice, along with minor interdiffusion on the group V sublattice. Strain is again induced in the intermixed QW structure since $L_d' \neq L_d$, having a similar profile as for $k = 0$, with compressive strain in the well and tensile strain in the barrier close to the interface. For the case $k = 0.25$, $L_z = 6$ nm, and $L_d = 0.5$ nm, the strain at the well centre, in the well close to the interface, and in the barrier close to the interface, shows a decrease of about 40%, Fig. 4(a). Since interdiffusion now takes place on the two sublattices the carrier confinement profiles are no longer abrupt, while the double-well potential at the well bottom and the double barrier profile at the top of the well are much less pronounced, as shown in Fig. 4(b).

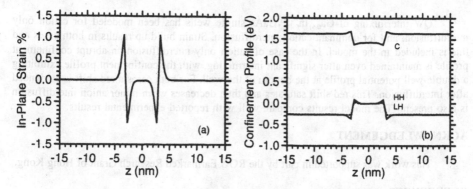

Fig.4. (a) In-plane strain and (b) carrier confinement profiles, for the interdiffused QW with $L_z = 6$ nm, $L_d = 0.5$ nm, $k = 0.25$.

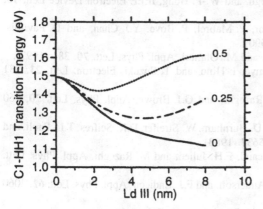

Fig. 5. Ground state transition energy shift with interdiffusion for different values of $k = L_d'/L_d$.

The variation of the ground state transition energy with interdiffusion, for different values of k, is shown in Fig. 5. The QW bandgap energy is the C1–HH1 transition, reflecting the compressive nature of the strain induced in the QW by the intermixing process. As already noted the bandgap energy decreases for $k = 0$ as the interdiffusion proceeds, corresponding to a large red shift of the effective bandgap of the interdiffused QW. When $k = 0.25$, a red shift in the effective bandgap again results. However, for long enough interdiffusion duration, the red shift saturates and even decreases. The results obtained from our model correspond to experimental results reported for disordering of $Ga_{0.51}In_{0.49}P/GaAs$ QWs by thermal annealing [7], showing the red shift saturating and decreasing with increasing duration of the QW disordering process. For k = 0.5, the red shift again saturates and then decreases, but at an earlier stage of the interdiffusion process, since the competing group V sublattice interdiffusion is now more pronounced. Thus for k < 1, the ground state transition energy variation (saturation and subsequent decrease of the red shift) provides an indication of how dominant the group III atoms interdiffusion is with respect to the group V interdiffusion.

CONCLUSIONS

QW intermixing of $Ga_{0.51}In_{0.49}P/GaAs$ single wells has been modeled for cation only interdiffusion, and for dominant cation interdiffusion. Strain build up results in both cases and this is included in the model. In the case of cation only interdiffusion an abrupt confinement profile is maintained even after significant intermixing, with the confinement profile exhibiting a double-well potential profile at the bottom of the well. For both cases a red shift is evidenced after interdiffusion. This red shift saturates and then decreases when minor anion interdiffusion is also present. The model results compare well with reported experimental results.

ACKNOWLEDGEMENT

This work was supported in part by the RGC Earmarked Research Grant of Hong Kong.

REFERENCES

1. J.-I. Song, C. Caneau, K.-B. Chough, and W.-P. Hong, IEEE Electron Device Lett. **15**, 10 (1994).
2. M. Razeghi, F. Omnes, M. Defour, P. Maurel, P. Bove, Y.J. Chan, and D. Pavlidis, Semicon. Sci. Technol. **5**, 274 (1990).
3. T. Takamoto, E. Ikeda, H. Kurita, and M. Ohmori, Appl. Phys. Lett. **70**, 381 (1997).
4. K. Koyabashi, S. Kawata, A. Gomyo, I. Hino, and T. Suzuki, Electron. Lett. **21**, 931 (1985).
5. C. Jelen, S. Slivken, J. Hoff, M. Razeghi, and G.J. Brown, Appl. Phys. Lett. **70**, 360 (1997).
6. M.D. Camras, N. Holonyak, Jr., R.D. Burnham, W. Streifer, D.R. Scifres, T.L. Paoli, and C. Lindström, J. Appl. Phys. **54**, 5637 (1983).
7. C. Francis, M.A. Bradley, P. Boucaud, F.H. Julien, and M. Razeghi, Appl. Phys. Lett. **62**, 178 (1993).
8. R.L. Thornton, F.A. Ponce, G.B. Anderson, and F.J. Endicott, Appl. Phys. Lett. **62**, 2060 (1993).
9. T.E. Schlesinger and T. Kuech, Appl. Phys. Lett. **49**, 519 (1986).
10. R.L. Thornton, D.P. Bour, D. Trent, F.A. Ponce, S.C. Tramontana, and F.J. Endicott, Appl. Phys. Lett. **65**, 2696 (1994).
11. J. Micallef, E.H. Li, and B.L. Weiss, J. Appl. Phys. **73**, 7524 (1993).
12. S.D. Gunapala, B.F. Levine, R.A. Logan, T. Tanbun-Ek, and D.A. Humphrey, Appl. Phys. Lett. **57**, 1802 (1990).
13. G.P. Khotiyal and P. Bhattacharya, J. Appl. Phys. **63**, 2760 (1988).
14. I.J. Pape, P. Li Kam Wa, J.P.R. David, P.A. Claxton, and P.N. Robson, Electron. Lett. **24**, 1217 (1988).

INTERDIFFUSION MECHANISMS IN GaAs/AlGaAs QUANTUM WELL HETEROSTRUCTURES INDUCED BY SiO$_2$ CAPPING AND ANNEALING

A.PÉPIN *, C.VIEU *, M. SCHNEIDER *, H.LAUNOIS *, E.V.K. RAO **
*Laboratoire de Microstructures et de Microélectronique (L2M/CNRS), 196 ave. Henri–Ravéra, 92225 Bagneux, FRANCE, anne.pepin@bagneux.cnet.fr
**Centre National d'Etudes des Télécommunications (CNET–Bagneux), 196 ave. Henri–Ravéra, 92225 Bagneux, FRANCE

ABSTRACT

We have investigated intermixing enhancement in GaAs/AlGaAs quantum well heterostructures achieved by SiO$_2$ capping obtained by rapid thermal chemical vapor deposition. Evidence of fast Ga pumping inside the SiO$_2$ layer during anneal and simultaneous generation of excess Ga vacancies under the SiO$_2$/GaAs interface is presented. A simple model involving the thermal stress arising from the difference in thermal expansion coefficients between SiO$_2$ and GaAs, is proposed to account for the abnormally fast Ga vacancy diffusion inside the heterostructure. A spatial control of the interdiffused areas can be achieved if a suitable stress field is imposed on the semiconductor surface by the capping layers. We show experimental evidence of this effect using a specific patterning of SiO$_2$/Si$_3$N$_4$ bilayers.

INTRODUCTION

For the past decade, extensive work has been dedicated to semiconductor quantum well (QW) intermixing as an avenue to monolithic optoelectronic integration [1]. Interdiffusion of well and barrier atoms creates a compositional disorder which results in a modification of the electronical and optical properties of QWs. This thermally activated phenomenon can be further enhanced by introducing impurities or additional defects into the heterostructure. When intermixing is stimulated only in selected areas of a wafer, permanent built–in lateral barriers, e.g. localized regions of larger effective bandgap, can be created inside the QW. Selective intermixing therefore represents an attractive post–growth process for the fabrication of a variety of integrated devices involving the confinement of photons and/or carriers. In addition, when defects are injected in QW heterostructures to stimulate layer disordering, investigation of interdiffusion becomes a unique mean to study the diffusion properties of the injected defects during annealing. QWs can thus be regarded as sensitive probes for defect diffusion.

Dielectric encapsulant layers, such as silicon dioxide and silicon nitride, have long been used for passivation in III–V semiconductor device fabrication technology. More recently, their utility as intermixing sources in specific III–V systems was demonstrated. In GaAs/AlGaAs QW heterostructures, SiO$_2$ encapsulation of samples was found to significantly enhance Ga–Al interdiffusion during high temperature annealing, while on the other hand, capping undoped GaAs/AlGaAs samples with Si$_3$N$_4$ proved to effectively inhibit interdiffusion [2,3]. For its relative simplicity of implementation, silicon dioxide capping and annealing has raised significant interest over the past few years. This distinctive property has been attributed to the special affinity of SiO$_2$ for gallium. During high temperature annealing, preferential absorption of Ga atoms by the SiO$_2$ capping layer occurs, which causes the generation of excess Ga vacancies (V$_{Ga}$) under the oxide/semiconductor interface. These additional point defects then diffuse to the QW heterointerfaces where they promote interdif-

447

fusion of group III species. Although SiO_2 encapsulation–induced intermixing was investigated by several groups and quickly led to the fabrication of novel built–in optoelectronic devices such as low–loss optical waveguides, modulators, and quantum well laser diodes [2,4], the basic understanding of the specific diffusion mechanism involved has suffered from a lack of reproducibility and discrepancies due to the wide variety of experimental conditions used (differences in oxide deposition techniques, annealing conditions, epitaxial layer composition, etc.). The main objective of this paper is to investigate the basic interdiffusion mechanisms of the SiO_2 capping technique.

EXPERIMENT

Undoped GaAs/AlGaAs QW heterostructures of different types were grown by molecular beam epitaxy on (001) semi–insulating GaAs substrates for this study. In order to prevent possible reduction of the SiO_2 by Al and resulting in–diffusion of silicon and/or oxygen inside the epitaxial layers as reported by various authors [3], all of our structures are capped with a final 5 nm–thick GaAs layer. The SiO_2 and Si_3N_4 layers used in this study were deposited by rapid thermal chemical vapor deposition (RTCVD). In previous studies [5], we have shown that SiO_2 RTCVD films give the highest quality interdiffusion compared to other deposition techniques. Conventional thermal anneals were carried out at either 850°C or 900°C in a As overpressure atmosphere. Samples were characterized before and after SiO_2 deposition, as well as after heat treatment. For patterning purposes, high resolution electron beam lithography and subsequent metal lift–off were used in association with reactive ion etching. The second RTCVD step was then performed in the case of SiO_2/Si_3N_4 arrays, and thermal annealing was carried out afterwards. A great care was taken in choosing effective characterization techniques. Our results were confronted through the use of the following complementary techniques: low temperature optical spectroscopy, photoluminescence (PL), photoluminescence excitation spectroscopy (PLE) and linear polarization anisotropy analysis, secondary ion mass spectroscopy (SIMS) and finally, cross–sectional transmission electron microscopy (XTEM).

RESULTS AND DISCUSSION

Intermixing under uniform SiO_2 capping layers

To appreciate better the kinetics of SiO_2 capping–induced intermixing, we studied more specifically the influence of anneal duration on a multiple quantum well (MQW) heterostructure coated with a 480 nm–thick layer of RTCVD SiO_2. The structure comprises a 0.45 μm–thick GaAs (5 nm)/ $Al_{0.33}Ga_{0.67}As$ (10 nm) MQW. Several samples were annealed at a fixed temperature of 850°C but for different durations varying from a few seconds (using either rapid thermal annealing or a push–pull procedure in a conventional furnace) up to 4 hours. Secondary ion mass spectroscopy (SIMS) was used to monitor the evolution of the Ga distribution inside the oxide layer as well as the changes in the Al depth profile within the heterostructure. Ga and Al profiles are shown on Figures 1 and 2, respectively. We see on Fig. 1 that even for the shortest anneals a significant quantity of Ga has been pumped inside the oxide cap as opposed to the as–deposited situation. It clearly appears that the total amount of Ga inside the SiO_2 capping layer is roughly the same after a push–pull, a 10 minute or a 4 hour anneal. Concurrently, the Al depth distribution within the heterostructure of Fig. 2 reveals a striking fact: for all anneal durations, a nearly uniform damping of the amplitude of Al oscillations throughout the thickness of the heterostructure is observed. In other words, a nearly uniform and depth–independent intermixing throughout the MQW is generated very early in the annealing stage. Longer anneal durations only uniformly increase the interdiffusion rate. Meanwhile (data shown elsewhere [5]) no diffusion of Si and/or O species inside the semiconductor is induced, i.e., the intermixing process is indeed impurity–free.

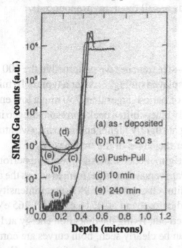

Figure 1: SIMS Ga concentration profiles obtained in the SiO_2 layer of several samples from strthe MQW structure F: a) after deposition and before annealing; b) after a 20 s RTA at 850°C; c) after a push–pull anneal at 850°C in a conventional furnace; d) after a 10 min. conventional anneal at 850°C; e) after a 4 h conventional anneal at 850°C.

Figure 2: SIMS Al concentration profiles obtained inside the same samples as on the left: a) after removal of the oxide layer before annealing; b) after a 10 min. anneal at 850°C and subsequent removal of the oxide layer ; c) after a 30 min. anneal at 850°C and subsequent removal of the oxide layer ; d) after a 4 h anneal at 850°C and subsequent removal of the oxide layer.

In order to explain this behavior, we propose a two stage diffusion model directly involving the thermal stress field generated inside the semiconductor due to the difference in thermal expansion coefficients between SiO_2 and GaAs. At high temperature, the SiO_2 film is under tension while the surface of the semiconductor is under compression. Because of the small distortions they induce in the crystal lattice, point defects such as vacancies will elastically interact with a macroscopic external stress field. This effect is at the origin of so–called Cottrell atmospheres [6] which take place in the highly stressed regions located around dislocations. It indeed seems quite intuitive that vacancies, which induce a slight tensile stress around their sites, would diffuse preferentially towards compressive regions where they can partially relax the imposed stress. By analogy, in our case, an elastic interaction will exist between the generated V_{Ga} and the external stress gradient. In the early stages of the anneal, V_{Ga} are injected into a compressive region which is thus in favor of their penetration. A driving force originating from the elastic interaction between the macroscopic thermal stress gradient and the V_{Ga} will thus quickly push the V_{Ga} into the depth of the semiconductor. In this transient phase, the interdiffusion rate is very small. Once the end of first phase is reached, a dynamic equilibrium develops between the transport due to the stress gradient, the diffusion due to the concentration gradient, and the presence of traps under the SiO_2/GaAs interface. A steady V_{Ga}

449

concentration, quasi–constant in depth is established. The vacancies are now free to diffuse around their sites and Ga/Al exchange is promoted. Intermixing develops with time until excess V_{Ga} are trapped by undetermined sites of either recombination or agglomeration, located most likely at the epilayer/substrate interface. The following results will show how such a simple model can help us to understand patterning experiments where stress fields are all the more pronounced.

Patterning of the SiO₂ layer at a nanometric scale

A schematic illustration of the arrays of nanometric–size trenches we patterned in the 500 nm–thick SiO_2 cap deposited on a single QW structure is displayed on Fig. 3. After a typical 10 minute anneal at 900°C, three PL peaks are observed in arrays of period larger than 750 nm: a low energy peak corresponding to non–intermixed regions, a strongly blue–shifted peak corresponding to inter-mixed regions, and a lower intensity third peak emitting at slightly higher energy which we attribu-ted to more strongly interdiffused regions located under the edges of the trenches [7].

In order to determine more precisely the spatial origin of each emission, we have performed spatially resolved PL experiments on several arrays. A transversal scan is performed across the array of interest at a fixed detection energy, and corresponding changes in the PL signal intensity are recorded. Two distinct PL emissions had been observed beforehand at 1.716 eV and 1.765 eV for this 8 μm period array. Two 15 μm–long scan was carried out, fixing the detection energy at 1.716 eV (Fig. 4(a)) first and 1.765 eV (Fig. 4(b)) next. As can be clearly seen, both curves are comple-mentary: low energy and high energy emissions originate from distinct regions of the array. Further-more, the two low energy emissions are indeed 8 μm apart. Moreover, the full width at half maxi-mum of the corresponding peaks is roughly 2 μm, which is the resolution of the experimental set–up. The actual width of the regions emitting at low energy is thus very likely much smaller than 2 μm. The low intensity emission at highest energy, corresponding to the bordering regions, could how-ever not be successfully located. These results thus unambiguously indicate that in arrays of well separated wires, the regions of the QWs located under the openings are preserved from the interdif-fusion, while the adjacent regions capped by the SiO_2 strips are strongly interdiffused lateral barri-ers. This behavior can be understood if we take into account the stress field distribution at annealing temperature, indicated in Fig. 3. We believe that excess V_{Ga} created under the SiO_2/GaAs interface will therefore be attracted towards the higher compressive regions located under the film edges

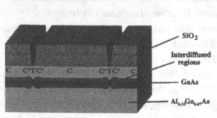

Figure 3: Schematic cross–sectional view of the quantum wire structure and stress field distribu-tion generated during anneal. C and T indicate compressive and tensile regions, respectively.

Figure 4: Spatially resolved PL spectra obtained by scanning an array of period 8 μm and trench width 120 nm a) detection energy was fixed at 1.716 eV b) detection was fixed at 1.765 eV.

while the adjacent strongly tensile regions located under the trenches will act as barriers to this vacancy diffusion. Such a stress–driven anisotropic diffusion can thus give rise to high lateral resolution of the intermixing process i.e. steep lateral potential modulation which would be impossible to obtain through a simple isotropic diffusion process. This unique property was indeed exploited for the fabrication of quantum wires which, for the first time using a selective interdiffusion technique, exhibited signatures of 1D confinement in PL, PLE and linear polarization analysis [7].

Selective interdiffusion under patterned SiO_2/Si_3N_4 layers

Fig. 5 is a striking example of selective interdiffusion obtained via combined SiO_2 and Si_3N_4 capping layers. The bright field (BF) image presented on Fig. 5(a) shows very clearly a 0.9 μm–wide trench patterned inside the SiO_2 film, further coated with a thin Si_3N_4 layer. The corresponding dark field (DF) image displayed on Fig. 5(b) reveals that the intermixed region is surprisingly located underneath the Si_3N_4–capped region, while the areas located under the adjacent $(SiO_2+Si_3N_4)$–capped regions have been protected from interdiffusion. We also notice an accumulation of defects immediately under the SiO_2. Such a behavior can also be explained by taking into account the stress field distribution at annealing temperature illustrated schematically in Fig. 6. Due to the higher stress of the nitride film, the semiconductor surface is under tension under the $(SiO_2+Si_3N_4)$–capped strips. V_{Ga} are thus generated inside a highly tensile region which is not in favor of their penetration. They quickly agglomerate, form the observed extended defects, and consequently, intermixing is inhibited underneath. However, a small amount of V_{Ga} generated near the edges of the SiO_2 strips are drained laterally towards the adjacent compressive regions located under the nitride, where they can effectively stimulate interdiffusion. Results obtained in the inverted configuration (patterned Si_3N_4 coated with SiO_2) also show good agreement with the stress field distribution [8].

CONCLUSIONS

The results we presented bring new light to the understanding of the mechanisms responsible for the enhancement of Ga–Al interdiffusion in SiO_2–capped GaAs/AlGaAs heterostructures during high temperature anneals. We have identified the main interdiffusion mechanism to be the quasi–instantaneous generation of excess Ga vacancies under the SiO_2/GaAs interface, resulting from the fast pumping of Ga atoms by the SiO_2 film during high temperature anneal, and their rapid diffusion inside the heterostructure where they can then promote Ga–Al exchanges at the heterointerfaces. To account for this unique property, we propose a qualitative interdiffusion model based on the fast V_{Ga} diffusion towards depth driven by the thermal stress field imposed on the semiconductor by the SiO_2 film. Our optimized RTCVD SiO_2 intermixing technique can be spatially localized on a length scale compatible with the lateral confinement of carriers into quantum wires, and double barrier wires are spontaneously fabricated with this process. Evidence of this stress–dependent vacancy diffusion could also be observed in selective intermixing experiments carried out through the combined use of Si_3N_4 and SiO_2 encapsulation layers, and brings additional support to this simple model. The high quality selective interdiffusion induced by SiO_2 capping make it an attractive alternative to widely–used ion implantion for various applications. Most particularly, SiO_2 encapsulation appears as a better candidate for the fabrication of such highly sought–after active devices as quantum wire lasers. Moreover, our work clearly demonstrates that point defects such as V_{Ga} can be "piloted" by the stress field imposed on the semiconductor by patterned capping layers. This

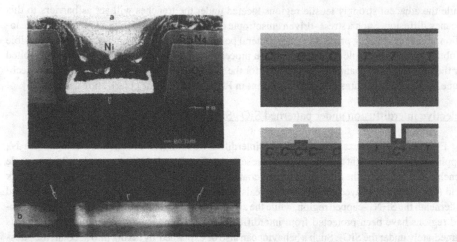

Figure 5: XTEM micrographs obtained after annealing on a 0.9 μm–wide trench opened in the SiO$_2$ layer and covered with Si$_3$N$_4$. (a) is a BF image. The horizontal arrow indicates the depth of the QW and the vertical one, the surface of the semiconductor. (b) is the corresponding DF (200) image. The edges of the SiO$_2$–capped region are indicated by tilted arrows and the surface of the semiconductor, by a vertical arrow.

Figure 6: Schematic representation of the SiO$_2$/Si$_3$N$_4$ arrays. The schematic stress distributions in the near surface region of the semiconductor induced by the dielectric layers is indicated, at high temperature (850°C). Letters C and T denote compressive and tensile states respectively. Symbol C+ and C$^-$ indicate highly compressive and lesser compressive regions.

effect can be generalized to others kind of point defects. Beyond these potential applications, SiO$_2$ capping–induced intermixing also turned out to be a unique experimental tool to investigate diffusion mechanisms and dynamics under stress in III–V compounds.

REFERENCES

1. Opt. Quantum Electron. **23** (1991) Special Issue on Quantum Well Mixing for Optoelectronics
2. D.G. Deppe, L.J. Guido, N. Holonyak Jr., J.J. Coleman and R. D. Burnham, Appl. Phys. Lett. **49**, 510 (1986)
3. L.J. Guido, N. Holonyak Jr., K.C. Hsieh, R.W. Kaliski, W.E. Plano, R.D. Burnham, R.L. Thornton, J.E. Epler and T.L. Paoli, J. Appl. Phys. **61**, 1372 (1987)
4. Y. Suzuki, H. Iwamura and O. Mikami, Appl. Phys. Lett. **56**, 19 (1989)
5. A. Pépin, Ph.D. Thesis, Université de Paris 6, (1995)
6. J. Philibert, <u>Diffusion et Transport de la Matière dans les Solides</u> (Les Editions de Physique, Les Ulis,1985)
7. A. Pépin, C. Vieu, M. Schneider, R. Planel, J. Bloch, G. Benassayag, H. Launois, J.Y. Marzin and Y. Nissim, Appl. Phys. Lett. **69**, 61 (1996)
8. A. Pépin, C. Vieu, M. Schneider, H. Launois, and Y.Nissim, J. Vac. Sci. Technol. B **15**(1), 142 (1997)

DISORDER-DELINEATED AlGaAs/GaAs QUANTUM-WELL PHASE MODULATOR

WALLACE C.H. CHOY, BERNARD L. WEISS
School of Electronic Engineering, Information Technology & Mathematics, University of Surrey, Guildford, Surrey, GU2 5XH, UK

ABSTRACT

Modeling is used to investigate waveguide phase modulators, with 0.5 μm and 1 μm quantum well active regions which are defined by implantation induced disordering. By controlling the extent of the interdiffusion in the lateral claddings, the refractive index difference between the core and claddings is used to provide single mode operation. The performance of the phase modulator is studied in terms of optical confinement, phase change per unit voltage per unit length, chirping property and absorption loss. Our result shows that the 0.5 μm one is a more efficient structure and its absorption loss can be reduced by increasing the applied field from 50 kV/cm to 100 kV/cm.

INTRODUCTION

Quantum well (QW) electro-optical phase modulators are of interest for a range of applications in optical communication and signal processing due to their large electro-optic effect [1,2]. However, most QW phase modulators are discrete devices which are integrated with other devices to form photonic integrated circuits (PICs). Interdiffusion occurs across the well-barrier interface and modifies the transition energies, and thus the optical properties, which results in a change of refractive index between the as-grown and interdiffused QW regions [3]. Conventional AlGaAs/GaAs QW modulators use the composition of the AlGaAs cladding layers to control the number of guided modes in the waveguide section of the modulator [4]. In this work the annealing time is used to control the extent of the interdiffusion, and therefore the refractive index change, to govern the modal characteristics of the waveguide modulator structure.

In this paper we model an electro-optic phase modulator with lateral confinement provided by implantation induced disordering (IID) in an $Al_{0.3}Ga_{0.7}As$/GaAs QW active region. This study addresses the effects of interdiffusion on the modulator performance and the waveguide characteristics of the modulator in terms of the thickness of the active region. The phase modulator is designed to be a single mode waveguide device to reduce dispersion and achieve good performance. The results demonstrate the effects of structural parameters of the device and interdiffusion conditions on the modulator characteristics and shows how a useful phase modulator can be designed using interdiffusion.

MODELING THE PHASE MODULATOR STRUCTURE

For the structure of the phase modulator, starting from the n^+-GaAs substrate, the layers are an n^+-type $Al_{0.3}Ga_{0.7}As$ lower cladding layer, a number of undoped quantum wells (QWs) consisting of 100 Å $Al_{0.3}Ga_{0.7}As$ barriers and 100 Å GaAs wells which serve as the active region of the device, and p^+-type $Al_{0.3}Ga_{0.7}As$ upper cladding with metal contacts on the GaAs substrate and the upper cladding layer. IID is used to provide the lateral confinement in the active region of the modulator.

In order to investigate an ion implanted phase modulator, a two-dimensional implantation induced defect profile of the QW active region after annealing and the corresponding refractive index profile are modeled. By solving the wave equation, the optical confinement of the

453

waveguide structure is determined from the refractive index profile. The electro-optic and electro-absorption properties of the QW structure are then calculated. Several modulator parameters are determined to characterize the performance of the device. Besides the breakdown voltage characteristics, the effects of the applied voltage across the p-i-n structure are determined by solving the Poisson's equation.

Annealing the implanted structure modifies the defect profile which is indicated by the diffusion length (L_d) [5] and results in a graded L_d profile between the waveguide core and cladding regions of the QW structure. Consequently, the electro-optic and electro-absorption properties of the modulator are determined by the sum of DFQWs with different extents of interdiffusion. Under reverse bias, the optical properties, including absorption coefficient and refractive index, of these DFQWs change to produce the modulation characteristics.

The application of an external applied voltage tilts the QWs, red shifts the transition energies and reduces the wavefunction overlap integrals (i.e. quantum confined Stark effect). The effective absorption coefficient, α_{eff}, is given by:

$$a_{eff} = \left. \int \underset{\substack{\text{the wells within} \\ \text{the active region}}}{a(x,y)j(x,y)j^*(x,y)dA} \middle/ \int \underset{\substack{\text{the entire range of} \\ \text{the guiding field}}}{j(x,y)j^*(x,y)dA} \right. \tag{1}$$

where $\varphi(x,y)$ is the guiding optical field, dA is a small but finite area normal to the optical field at (x,y) and $\alpha(x,y)$ is the absorption coefficient of the $Al_{0.3}Ga_{0.7}As/GaAs$ QW structure, since the extent of the interdiffusion in the QW region is inhomogeneous throughout the cross section of the phase modulation, i.e. the absorption is dependent on both x and y. Equation (1) shows that α_{eff} is determined by the fraction of the optical field intensity $\varphi(x,y)\varphi^*(x,y)$ within the wells of active region. The field-induced change of effective absorption coefficient ($\Delta\alpha_{eff}$) in the device is calculated using:

$$Da_{eff} = a_{eff}(F \neq 0) - a_{eff}(F = 0) \tag{2}$$

where the $\alpha_{eff}(F \neq 0)$ and $\alpha_{eff}(F = 0)$ are the effective absorption coefficients with and without an applied field respectively.

The TE refractive index and the change of refractive index are modeled using [5,6]. The waveguide effective refractive indices with and without an applied field are determined by solving Maxwell's equations using the refractive indices with the corresponding applied field which are then used to determine the change of the effective refractive index Δn_{eff}.

The important performance characteristics of the phase modulator are the phase change per unit modulation length per unit applied voltage, the chirp parameter, the optical confinement factor Γ, the absorption loss α_{loss} and the required bias voltage. The modulation efficiency of the phase modulator is measured from the phase change per unit modulation length per unit applied voltage, which is the normalized phase shift, $\Delta\theta_N$, and is given by:

$$Dq_N = (2pDn_{eff})/(V(applied)l_{op}) \tag{3}$$

where V(applied) is the applied voltage and λ_{op} is the operating wavelength. A high modulation efficiency requires a large $\Delta\theta_N$.

The static chirp parameter β_{mod} is given by:

$$b_{mod} = (4p \, Dn_{eff})/(l_{op} \, Da_{eff}) \tag{4}$$

where both $\Delta\alpha_{eff}$ and Δn_{eff} are functions of the applied voltage [7]. A useful phase modulator requires chirp parameter >10 [8].

The optical confinement factor Γ is determined using

$$G = \left. \int j(x,y) \, j^*(x,y)dA \middle/ \int j(x,y) \, j^*(x,y)dA \right. \tag{5}$$

the depletion region of   the entire cover range
the waveguide device   of a guiding field

where Γ parameter indicates the portion of the optical power which overlaps the depletion region of the p-i-n structure, which consists of a QW active region and part of the two (top and bottom) cladding layers, within the entire device structure. Therefore, an efficient modulator requires a large value of Γ. Moreover, a good phase modulator requires low α_{loss}, where α_{loss} is defined as α_{eff} for $F = 0$. The electrical bias required for modulation is also a crucial parameter which indicates the power consumption of the device, which is due to power dissipation in the load resistance of the drive circuit.

RESULTS AND DISCUSSIONS

The active region of the modulator consists of undoped multiple periods of 100 Å $Al_{0.3}Ga_{0.7}As$ barriers and 100 Å thick GaAs wells. The two structures studied here comprise 0.5 μm and 1.0 μm thick QW regions which contain 25 and 50 QW periods, respectively. The effects of the implantation induced defects on the confinement of the optical field in these two structures are studied. The performance of the modulator for applied fields (F) of 50 kV/cm and 100 kV/cm are also addressed. The implantation parameters, including the projected ion range and standard deviation and the lateral spread of the implanted ions around the mask for different implanted ion energies are taken from [9] and the Al and Ga interdiffusion coefficients are taken from [10].

The selection of λ_{op} is important for high modulation performance, including a large $\Delta\theta_N$, a large β_{mod}, i.e. a high refractive index change together with low absorption change for phase modulator, a low applied voltage and a low α_{loss}. The results show that 0.868 μm > λ_{op} > 0.86 μm and 0.878 μm > λ_{op} > 0.87 μm should be used for F = 50kV/cm and F = 100kV/cm respectively, to achieve $\Delta\theta_N \geq 0.59$ rad/Vmm of the existing device [8], for which acceptable values of α_{loss} (<500 cm^{-1}) can be obtained.

The refractive index profiles are obtained for the designed value of λ_{op} to ensure single mode operation. The transverse refractive index profiles of the waveguide structure are inhomogeneous and are highly dependent on the implanted ion concentration, the implanted ion energy, the annealing temperature and time. The ion dose modeled here is 2.5×10^{13} ion/cm^2 so that the lattice damage levels are minimized to reduce the damage induced waveguide loss and to retain the electro-optical properties of the material. The implanted ion energy is optimized to achieve maximum optical confinement. The optimized conditions and mode field for the 0.5 μm QW structure are an ion energy of 650 KeV and an $\Gamma > 0.75$. Figure 1 shows that the peak of the mode field coincides with the peak of the implantation induced defect profile at the center of the QW guiding layer.

455

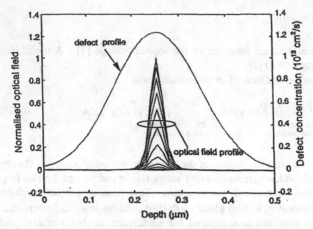

Figure 1. The defect profile and mode field profile of the phase modulator.

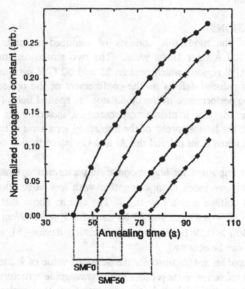

Figure 2. The variation of the normalized propagation constant with annealing time, showing the single mode range of operation for various applied field values.

The results show that using the selected ion energy and dose, a range of refractive index profiles can produce single mode modulation at the selected operation wavelengths. This is best illustrated by plotting the normalized propagation constant as a function of annealing time, see Figure 2, for 0.5 μm structures and λ_{op} = 0.863 μm, which shows the variation of the guiding properties of the QW waveguide, in terms of the normalized propagation constant, for different annealing times. The mode cutoff intervals SMF0 and SMF50 correspond to the ranges in which only Single Mode waveguides are formed for applied electric Fields of 0 and 50 kV/cm, respectively.

456

The applied electric field modifies the refractive index profile and thus the normalized propagation constant of the guided mode, as shown in Figure 2. In the case of 0.5 μm active region, λ_{op} = 0.863 μm and applied field F = 0, the interval (SMF0) covers the annealing time from 42 s to 62.5 s. However, when the applied field is increased to 50 kV/cm, the interval (SMF50) shifts to longer annealing times which range from 52 s to 74.5 s, outside which the waveguide will change from a single mode to multimode device at λ_{op} = 0.863 μm. To maintain single mode operation, only the overlapping region between the two mode cutoff intervals of the two selected applied fields should be used. For instance, for an operation with F = 0 and F = 50 kV/cm in the case of Figure 2, the overlapping region between SMF0 and SMF50, i.e. annealing time between 52 s to 64 s, should be used for a single mode phase modulator. Here, the annealing time at the mid-point of the overlapping region is used for the design of the ion implanted phase modulator. By plotting a similar diagram for the same 0.5 μm QW structure, single mode operation of phase modulation can also be obtained at λ_{op} = 0.873 μm with F = 0 and 100 kV/cm while the 1 μm QW structure can only operate with 0 and 50 kV/cm for single mode operation at λ_{op} = 0.863 μm.

Several parameters are used to characterize the performance characteristics of the modulator, see Table 1. A large phase change with $\Delta\theta_N$ greater than $\pi/2 \approx 1.57$ rad/Vmm can be obtained at the selected λ_{op}'s of 0.863 μm and 0.873 μm for F = 50 kV/cm and 100 kV/cm, respectively, in the 0.5 μm structure and 0.863 μm for F = 50 kV/cm in the 1 μm structure. Their α_{loss} values are also <500 cm^{-1}. For the 0.5 μm active region, this value reduces to 111.6 cm^{-1} when the applied field increases to 100 kV/cm because the λ_{op} can be further away from the unbiased heavy hole (HH) exciton peaks, as compared to that for F = 50 kV/cm.

Table 1. Modulation properties of the two cavity structures.

QWs thickness (μm)	0.5				1.0	
λ_{op} (μm)	0.863	0.867	0.873	0.877	0.863	0.867
bias field (kV/cm)	50	50	100	100	50	50
bias voltage (V)	2.54	2.54	5.06	5.06	5.04	5.04
depletion width (μm)	0.507	0.507	0.512	0.512	1.06	1.06
Γ	0.80	0.79	0.78	0.77	0.86	0.87
Δn_{eff} ($\times 10^{-3}$)	2.26	0.31	2.03	0.42	2.48	0.35
$\Delta\theta_N$ (rad/Vmm)	6.48	0.89	2.88	0.60	3.58	0.50
α_{loss} (cm^{-1})	302.7	174.6	111.6	85.2	332.3	195.6
$\Delta\alpha_{eff}$ (cm^{-1})	107.6	8.1	76	5.2	117.3	9.0
β_{mod}	3.1	5.5	3.8	11.6	3.1	5.6

For the 1 μm QW active region, the total phase modulation for 50 kV/cm is 18.1 rad/mm, which is larger than that for the 0.5 μm QW active region (16.4 rad/mm) and it is due to the greater optical confinement in the 1.0 μm QW active region, where twice the number of QWs contribute to the modulation mechanism. However, the operation of the 1.0 μm QW structure requires approximately twice the applied voltage as compared to that of 0.5 μm QW structure. This makes the $\Delta\theta_N$ of the 1.0 μm QW structure weaker than that of 0.5 μm QW structure, i.e. the 0.5 μm QW structure provides a more efficient modulation.

In both structures, β_{mod} is low with values of 3.1 and 3.8 for λ_{op} of 0.863 μm and 0.873 μm respectively. In order to increase β_{mod} a longer wavelength is used, such as for F = 50 kV/cm when λ_{op} is increased to 0.867 μm and β_{mod} increases to ~5.5. This improvement of β_{mod} is mainly due to the reduction of $\Delta\alpha_{eff}$, as shown in Table 1, but the penalty is a reduction of $\Delta\theta_N$. This case clearly shows the trade-off between $\Delta\theta_N$ and $\Delta\alpha_{eff}$. A larger β_{mod} can be obtained using F = 100 kV/cm when λ_{op} increases from 0.873 μm to 0.877 μm where β_{mod} increases to 11.6. The $\Delta\theta_N$ here is 0.6~34.4° rad/Vmm and this phase change is compatible with the exiting device [8].

CONCLUSION

Two waveguide type phase modulators with 0.5 μm and 1 μm QW active regions using masked ion implantation to produce lateral confinement have been investigated theoretically. These devices are designed for single mode operation at a wavelength of 0.867 μm for an applied field of 50 kV/cm and 0.877 μm for 100 kV/cm, where interdiffusion can be used to tune the propagating modes of the device. The design of a single mode phase modulator with strong optical confinement requires the peak ion concentration to be at the center of the guiding layer. Moreover, the annealing time is selected to be within the range required to achieve single mode operation for the two applied field values. Low implanted ion concentrations are also required to minimize lattice damage and retain the electro-optical properties of the material.

Comparing the modulation properties of 0.5 μm and 1 μm QW structures, the 1 μm structure can provide higher optical confinement and larger total phase change per unit length. The 0.5 μm structure can operate with either 0 and 50 kV/cm or with 0 and 100 kV/cm. The more important comparison is $\Delta\theta_N$ of the two structures. The 0.5 μm structure can provide more efficient modulation since its value of $\Delta\theta_N$ is approximately twice that of the 1.0 μm structure. β_{mod} and α_{loss} of the 0.5 μm structure can be further increased and reduced respectively, by increasing the applied field from 50 to 100 kV/cm since the longer λ_{op} can be used. It is important to note that a β_{mod} of 11.6, which is obtained for a field of 100 kV/cm, is large enough for a good phase modulator. The results are a guideline for the development of an optical phase modulator using lateral confinement provided by implantation induced disordering, which is also useful when these devices are used in photonic integrated circuits.

ACKNOWLEDGEMENT
W.C.H. Choy would like to thank the financial support of the Croucher Foundation.

REFERENCES
1. G.W. Yoffe, J. Brubach, W.C. Van der Vleuten. F. Karouta and J.H. Wolter, Trans. on Electron Dev., 40, p.2144 (1994).
2. S. Yoshida, Y. Tada, I. Kotaka and K. Wakita, Electron. Lett., 30, p.1795 (1994).
3. C. Vien, M. Schneider, D. Mailly, R. Planel, H. Launois, H.Y. Marzin and B. Descouts, J. Appl. Phys., 70, p.1444 (1991).
4. P.J. Bradley and G. Parry, Electron. Lett., 25, p.1349 (1989).
5. W.C.H. Choy and E.H. Li, IEEE J. of Quantum Electron., 33, p.382 (1997).
6. J. Micallef, E.H. Li and B.L. Weiss, Appl. Phys. Lett., 62, p.3164 (1993).
7. E.H. Li and W.C.H. Choy, IEEE Photon. Technol. Lett., 7, p.881, (1995).
8. T. Hausken, R.H. Yan, R.I. Simes and L.A. Coldren, Appl. Phys. Lett., 55, p.718 (1989).
9. J.F. Gibbons, W.S. Johnson and S.W. Mylroie, Projected range statistics: semiconductors and related materials 2nd ed. Stroudsburg, Pa: Dowden, Hutchinson & Ross; New York, 1975.
10. J. Cibert, P.M. Petroff, D.J. Werder, S.J. Pearton, A.C. Gossard and J.H. English, Appl. Phys. Lett., 49, p.223 (1986).

THERMAL INTERDIFFUSION IN InGaAs/GaAs STRAINED MULTIPLE QUANTUM WELL INFRARED PHOTODETECTOR

Alex S. W. Lee and E. Herbert Li

Department of Electrical and Electronic Engineering, University of Hong Kong, Pokfulam Road, Hong Kong

Gamani Karunasiri

Department of Electrical Engineering, National University of Singapore, Singapore 119260.

Abstract:

RTA at 850 °C for 5 and 10 s is carried out to study the effect of interdiffusion on the optical and electrical properties of strained InGaAs/GaAs quantum well infrared photodetector. Photoluminescence measurement at 4.5 K shows that no strain relaxation or misfit dislocation formation occurs throughout the annealing process. Absorption and responsivity peak wavelengths are red shifted continuously without appreciable degradation in absorption strength. The normal incident absorption, which is believed to be the result of band-mixing effects induced by the coupling between the conduction and valence and is usually forbidden in conventional polarization selection rule, is preserved after interdiffusion. Responsivity spectra of both 0° and 90° polarization are of compatible amplitude and the shape of the annealed spectra becomes narrower. Dark current of the annealed devices is not very sensitive to temperature variation and is found to be an order of magnitude larger than the as-grown one at 77K.

Introduction

Much progress has been made in bound-to-bound and bound-to-continuum[1] quantum well infrared photodetectors (QWIPs) after the first observation of intersubband absorption and a large dipole moment in AlGaAs/GaAs multiple quantum wells (MQW) has been reported.[2] With the development of strained layer QW and bandgap engineering, high quality pseudomorphic QW is achievable and it has been demonstrated that normal incident strained InGaAs/GaAs QWIP[3] is possible without grating coupling. However, the thermal stability of strained layers subjected to heat treatment is of prime importance and of great interest for optoelectronic device applications, especially for structure with higher In concentration. This is because highly strained heterostructure will result in smaller critical layer thickness[4] and will increase the risk of strain relaxation by the generation of misfit dislocation. Recently, fabrication of high-speed semiconductor lasers containing highly strained InGaAs/GaAs MQW in the active region has been reported using impurity-free interdiffusion by means of rapid thermal annealing (RTA).[5] Postgrowth tuning of AlGaAs/GaAs absorption peak[6] and QWIP detection wavelength have also been demonstrated.[7] In this communication, we report on the effect of dopant-enhanced layers interdiffusion on the performance of n-type strained $In_{0.3}Ga_{0.7}As$/GaAs QWIP annealed at 850 °C at different annealing times using RTA. Other than the continuous redshift of the detection wavelength, we also demonstrate that both the transverse magnetic (TM) and transverse electric (TE) infrared (IR) intersubband transitions are retained and that the responsivity performance of these annealed samples is compatible with the as-grown one by means of interdiffusion.

Results and discussions

The MQW structure was grown by molecular beam epitaxy on a (100) semi-insulating substrate. It consists of 50 periods of 40 Å wide as-grown $In_{0.3}Ga_{0.7}As$ well and 300 Å thick GaAs barrier. The Si doping is about 2×10^{18} cm^{-3} in the well. The MQW is sandwiched between a n^+ buffer (1 μm) and a cap layer (0.5 μm) as ohmic contact. It is designed to have only one bound state inside the well and the first excited state is in the continuum above the barrier. Before annealing, the samples were capped with approximately 250 nm thick electron-beam evaporated SiO_2 dielectric layer. RTA was carried out in a halogen lamp annealing system (AST SHS10) with double strip graphite heater under flowing nitrogen ambient. One of the samples was annealed for 5s at 850 °C while the other for 10s at the same temperature. Photoluminescence (PL) measurements were performed at 4.5 K after annealing using the 514.5 nm Argon laser at a power of 200 mW. Fig. 1 shows the PL spectra of the as-grown and interdiffused MQW. The PL peak shifts progressively to higher energy with anneal time from as-grown 1.316 eV to 1.319 eV and 1.323 eV, respectively. The blue shift of the bandgap energy indicates the intermixing of group III elements near the heterostructure interfaces. The PL peak intensity of the 5s annealed sample is increased by nearly one fold while the 10s annealed sample decreases by almost one fold in comparison with the as-grown intensity. The full width at half maximum (FWHM) PL line-width does not vary very much as compare to the as-grown sample; less than 4 meV difference for the 5s annealed sample and 1 meV difference for the 10s annealed sample. Since the well width of the sample currently under investigation is below the critical thickness for 30 % In concentration, the small variation in FWHM indicates that there is no strain relaxation or misfit dislocation formation during annealing and that there may even be a recovery of strain or an improvement in structural quality after RTA.[8, 9] Peaks were also observed at about 1.5 eV, which were red shifted with interdiffusion, in contrast to the PL peaks observed above. These peaks may due to the luminescence from GaAs either in the top cap layer or the bottom buffer layer.

Room temperature intersubband absorption measurement is taken using Nicolet Magna-IR 850 Fourier transform infrared spectrometer with a 45° polished multipass waveguide geometry. Effect of interdiffusion on the optical properties of annealed QWs is evidenced in Fig. 2. It shows the absorption spectra with 0° angle polarization, i.e., a mixture of TE and TM polarizations that contains a component of photon electric field along the growth direction as well as a component in the plane of the layers. The absorption peak of the as-grown sample originally at 10.2 μm is shifted to 10.5 μm after 5s annealing, and continuously red shifted to 11.2 μm in the subsequent 10s annealing. The red shift of the absorption peaks indicates both the bound state energy and the first excited state energy are being modified and/or the interdiffusion-induced changes in the depolarization shift,[6] which result in the postgrowth tuning of the absorption wavelength. But the first excited energy remains in the continuum under the different annealed conditions produced here, as can be seen from the high-energy tail and asymmetry of all the absorption spectra shown in Fig. 2, which are the characteristic features of bound-to-continuum intersubband transition in QWs.[1]

The absorption spectra of the annealed samples reduced in amplitude and broadened proportionally with increasing annealed time. This can possibly be attributed to the layers intermixing by RTA and to the modification in the QW profile. It is known that the solubility of Ga is very high in SiO_2 and so an increase in the concentration of group III vacancies is expected by the diffusion of Ga into the SiO_2 dielectric layer. This in turn will increase the dopant (Si) diffusion into the undoped GaAs barrier and converts it to a strongly n-typed material.[10] The Si diffusion across the heterointerfaces not only reduces the free carrier concentration but also enhances layers intermixing which results in the modification of the subband structure. Since absorption coefficient $\alpha(h\omega) \propto \rho_s$, the two dimension electron density in the well, the reduction in the number of carriers available to be excited by the incident IR radiation may render to a reduction in the absorbance. The change in the subband structure, which may result in smaller intersubband transition oscillator strength, together with impurity scattering may give rise to the broadening and decreasing in amplitude of the absorption spectra.

Mesa diodes (200 x 200 μm) were fabricated by standard lithography technique and 45° facet was polished at one end of the sample for responsivity measurement. The photocurrent was measured using grating monochromator and glowbar source with lock-in detection. Polarizer was inserted before the glowbar source in order to study the polarization dependence of the photoresponse. Figure 3 and 4 show the response spectra for 0° and 90° polarizations, respectively, as a function of wavelength at 25 K. For both polarization responsivity spectra, the peak positions were observed to be red shifted and independent of polarization. Note in both figures that there are a few peaks appear in the spectra with rather identical wavelength positions for both polarizations. Since the QW structure is designed to have the first excited state above the barrier, they are most probably due to intersubband transition from the bound state E_1 to other excited states in the continuum[11(15)] or the interaction between the first excited state E_2 and other states in the continuum.[12(16)] With the modification of QW structure by means of interdiffusion, the annealed spectra in Fig. 3 and 4 show that regardless of polarization, all these peaks are subdued except the designed main transition peak. For 0° polarization, the corresponding responsivity amplitudes 0.8, 0.79, and 0.77 A/W do not vary much for the as-grown and annealed detectors, as shown in Fig. 4(a), where all the spectra have almost identical rising edge. This is as expected since the MQW properties and its structure have not been substantially modified or deteriorated after interdiffusion, once the photoexcited carriers overcome the threshold barrier into the continuum states they are ready to be collected as photocurrent. The normal incident absorption, which is believed to be the result of band-mixing effects induced by the coupling between the conduction and valence[13(17)] and is usually forbidden in conventional polarization selection rule,[2] is preserved after interdiffusion. As shown in Fig. 4, the responsivity peaks in the as-grown spectrum due to the transition to other excited states in the continuum are subdued and the designed transition peak that is weak and lower in amplitude originally has become dominant and red shifted in the annealed spectra.

Leakage current is measured at 77 K using 4156A Parameter Analyzer and cold finger. The (I-V) characteristic is shown in Fig. 5 for the three devices. Note the asymmetry of the I-V curves between the two polarities. For the as-grown sample, leakage current is larger in reverse bias (i.e., mesa top negative) than in forward bias, which is attributed to inhomogeneity in material composition introduced during growth.[1] While for the annealed devices, the trend is just the opposite with leakage current larger at positive voltage. This is most probably due to the difference in diffusion rate of In and Ga species across the interfaces of the annealed QWs, which results in asymmetric barrier height[14(11)] seen by the thermal excited electrons, and to the re-distribution of dopant impurity as described in the previous section. These two factors together with the thinner 300 Å barrier not only explain the asymmetry[15(12)] I-V curves but also give rise to nearly an order of magnitude larger in leakage current than the as-grown one at 77 K. This is evidenced in Fig. 6. It shows that the annealed leakage current is not very sensitive to temperature variation from 25 to 90 K and remains almost constant for T < 50 K, if compare to the as-grown one. In this temperature range, the over all dark current increased nearly by one order of magnitude for the 5s annealed samples and by a factor of 8 for the 10s annealed sample, whereas the as-grown sample has increased by more than 5 orders of magnitude. Note also the leakage current of both the annealed devices is a few orders of magnitude larger than the as-grown one for T < 50 K. The huge increase in leakage current below this temperature is related to the defect-assisted tunneling mechanism as a result of the diffusion of Si and the group III constituent atoms in the heterostructure,[10] which introduced defects and dopant impurity into the barrier. At temperature larger than this where thermionic emission mechanism is dominant, all the leakage currents increase linearly to almost the same magnitude at T = 90K.

Conclusion

In conclusion, high In composition pseudomorphic interdiffused InGaAs/GaAs QWIP using dopant-enhanced vacancy interdiffusion has been demonstrated for its post-growth tunability. No strain relaxation and deterioration in the MQW structure are observed. The TE polarization infrared intersubband transition, as a result of the band-mixing effects, is preserved. Both 0° and 90° polarizations absorption peaks are red shifted with respect to the as-grown one without much degradation in absorption strength. Photoresponse peaks due to resonances in the continuum states are subdued after interdiffusion, which is most probably a consequence of the modification in subband structure. The annealed photoresponse spectra for 0° polarization are comparable to the as-grown device with a narrower FWHM, while the designed photoresponse peak becomes dominant for 90° polarization. Dark current of the annealed devices is about an order higher in amplitude than the as-grown one at 77 K. The I-V characteristic is less sensitive to the variation in temperature from 25-90 K where the overall dark current of both the annealed devices are varied in between a range of about one order in magnitude.

This work is support in part by the HKU-CRCG, RGC-Earmarked Research Grants and Academic Research Fund of National University of Singapore. The author would like to thank Prof. S. J. Chua for valuable suggestions, T. Mei, and Dr. S. J. Xu for technical assistance.

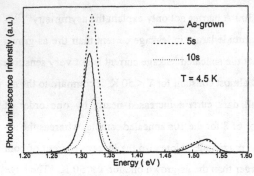

Fig. 1. Photoluminescence spectra of the as-grown, 5 s, and 10 s interdiffused InGaAs/GaAs MQW at T = 4.5 K.

Fig. 2. Absorption spectra of the as-grown, 5 s, and 10 s annealed samples at 300 K as a function of wavelength for 0° polarization.

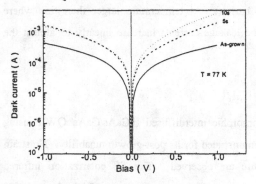

Fig. 3. I-V curve of the as-grown, 5 s, and 10 s annealed samples at 77 K as a function of bias.

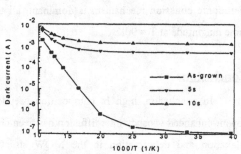

Fig. 4. Arrhenius plot of leakage current at 1 V bias as a function of the reciprocal of temperature in the range between 25-90 K.

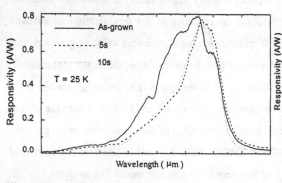

Fig. 5. Photoresponse spectra at 25 K of the as-grown, 5 s, and 10 s annealed samples bias at2.5 V, 1.05 V, and 1.65 V for 0° polarization as a function of wavelength.

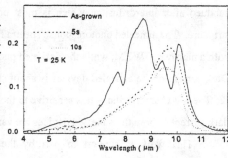

Fig. 6. 90° polarization photoresponse spe for the as-grown, 5 s, and 10 s annealed samples at 25 K and bias at 2.5 V, 1.05 V and 1.65 V as a function of wavelength.

Reference

1. B. F. Levine, A. Zussman, S. D. Gunapala, M. T. Asom, J. M. Kuo, and W. S. Hobson, J. Appl. Phys. Lett. **72**, 4429 (1992)

2. L. C. West, S. J. English, Appl. Phys. Lett. **46**, 1156 (1985).

3. R. P. G. Karunasiri, J. S. Park, J. Chen, and R. Shih, Appl. Phys. Lett. **67**, 2600 (1995).

4. J. W. Matthews and A. E. Blakeskee, J. Cryst. Growth **27**, 118 (1974).

5. S. Bürkner, J. D. Ralston, S. Weisser, J. Rosenzweig, E. C. Larkins, R. E. Sah, and J. Fleißner, IEEE Photon. Technol. Lett. **7**, 941 (1995).

6. J. D. Ralston, M. Ramsteiner, B. Discher, M. Maier, G. Brandt, P. Koidl, and D. J. As, J. Appl. Phys. **70**, 2195 (1991).

7. A. G. Steele, M. Buchanan, H. C. Liu, and Z. R. Wasilewski., J. Appl. Phys. **75**, 8234 (1994).

8. B. Elman, E. S. Koteles, P. Melman, C. Jagannath, C. A. Armiento, and M. Rothman, J. Appl. Phys. **68**, 1351 (1990).

9. S. Burker, M. Baeumler, J. Wanger, E. C. Larkins, W. Rothemund, and J. D. Ralston, J. Appl. Phys. **79**, 6818 (1996).

10. D. G. Deppe and N. Holonyak, Jr., "Atom diffusion and impurity-induced layer disordering in quantum well III-V semiconductor heterostructures," J. Appl. Phys. **64**, R93 (1988).

11. K. M. S. V. Bandara, B. F. Levine, and M. T. Asom, J. Appl. Phys. **74**, 346 (1993).

12. K. K Choi, M. Taysing-Lara, P. G. Newman, and W. Chang, "Wavelength tuning and absorption line shape of quantum well infrared photodetectors." Appl. Phys. Lett. **61**, 1781 (1992).

13. L. H. Peng and C. G. Fonstad, "Multiple coupling effects on electron quantum well intersubband transitions," J. Appl. Phys. **77**, 747 (1995).

14. A. S. W. Lee and E. H. Li, "Effects of interdiffusion of quantum well infrared photodetector," Appl. Phys. Lett. **69**, 3581 (1996).

15. H. C. Liu, Z.R. Wasilewski, M. Buchanan, and Hanyou Chu, Appl. Phys. Lett. **63**, 761 (1993).

ANALYSIS OF AlGaAs/GaAs MULTIPLE QUANTUM WELL DUAL WAVEGUIDES DEFINED BY ION IMPLANTATION INDUCED INTERMIXING

KAI-MING LO

32 Greenwood Ave. #7, Quincy, MA 02170, kmlo@xensei.com

ABSTRACT

A simple and accurate model is presented for the study of ion-implanted AlGaAs/GaAs multi-quantum well dual waveguides. The impurity induced disordering defined multi-quantum well dual waveguides are shown to have similar optical properties as conventional dielectric rib waveguides. They also provide a more flexible control over the waveguiding and coupling characteristics by changing parameters such as diffusion time, ion implant energy, mask width, and waveguide separation.

INTRODUCTION

The selective disordering or intermixing of III-V semiconductor quantum well (QW) structures has been widely employed to realize various optoelectronics devices [1]. This technique offers a planar technology to alter the bandgap energy, refractive index and other optical properties of the QW material. Impurity-induced disordering (IID) by ion implantation [2] can enhance the interdiffusion rate of the diffused quantum well and provides an accurate depth control of the disordered region.

In recent years, multi-QW waveguides and their applications have aroused much interest in the optoelectronics field. Optical directional couplers like power splitters, switches, wavelength and polarization (de)multiplexers are important components of optoelectronic integrated circuits. For optimum design of such devices, the propagation characteristics of the couplers have to be determined. By adopting a quasi-vector method based on the Galerkin's method which is applied to a mapped infinite domains [3], coupled structures are analyzed. While varying parameters such as ion implantation energy, operating wavelength, mask width and waveguide separation, both the waveguiding and coupling properties at different diffusion times are examined. In next section, the modelling and mathematical formulations for the dual waveguide structure are described. Numerical results are then presented. Finally, a conclusion will be given.

MODEL

Masked ion implantation technique is used to alter the band-gap and hence the refractive index of the as-grown square quantum well (SQW) material in selective regions. The non-implanted area has higher refractive indices than the implanted area, thus producing lateral confinement for light and a two-dimensional waveguide is formed [4].

The schematic of the waveguiding structure to be analyzed is shown in Fig. 1. It is composed of AlGaAs/GaAs multi-quantum well layers on a thick AlGaAs buffer grown on a GaAs substrate. Two identical masks are placed side by side on the top and the whole structure is exposed in air. Impurity Ga^+ ions are injected at high energy, followed by annealing at 950°C. Two parallel waveguides are fabricated as a consequence.

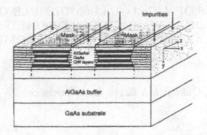

Figure 1: Schematic of the multi-quantum well dual waveguide defined by ion implantation.

Assuming the mask has an infinitely steep edge and is thick enough to avoid penetration, the as-implanted impurity concentration profile is given by

$$
N(x,y) = \frac{N_o}{\sqrt{2\pi}\Delta R_p} \exp\left[-\frac{(y-R_p)^2}{2\Delta R_p^2}\right] \times \frac{1}{2}\left\{\operatorname{erfc}\left(\frac{x+L_s}{\sqrt{2}\Delta R_{pL}}\right) - \operatorname{erfc}\left(\frac{x-L_s}{\sqrt{2}\Delta R_{pL}}\right)\right.
$$
$$
+ \operatorname{erfc}\left(\frac{x+L}{\sqrt{2}\Delta R_{pL}}\right) - \operatorname{erfc}\left(\frac{x-L}{\sqrt{2}\Delta R_{pL}}\right) + \operatorname{erfc}\left[\frac{x+(L_m+L_s)}{\sqrt{2}\Delta R_{pL}}\right]
$$
$$
\left. - \operatorname{erfc}\left[\frac{x-(L_m+L_s)}{\sqrt{2}\Delta R_{pL}}\right]\right\}
\tag{1}
$$

where N_o is the implanted dose of Ga, L_m is the mask width, $2L$ is the total width of the coupled waveguide, $L_s = s/2$ is half of the waveguide separation, R_p is the projected range, ΔR_p and ΔR_{pL} are the vertical and lateral standard deviation respectively.

The diffusion coefficient D_{imp} of impurity under annealing is assumed a constant. The impurity concentration profile $N(x,y,t)$ can be solved from the heat equation

$$
\frac{\partial N(x,y,t)}{\partial t} = D_{\text{imp}}\nabla^2 N(x,y,t) .
\tag{2}
$$

It is assumed that the interdiffusion coefficient D_{atom} of Al and Ga atoms depends on the local defect density only, and is described by the relation

$$
D_{\text{atom}}(x,y,t) = \alpha D_{\text{imp}} N(x,y,t)
\tag{3}
$$

where α is a constant determined from experimental data.

The square of the diffusion length $L_d^2(x,y)$ can be calculated by integrating $N(x,y,t)$ with respect to time,

$$
L_d^2(x,y) = \int_0^T D_{\text{atom}}(x,y,t)\,dt = \alpha D_{\text{imp}} \int_0^T N(x,y,t)\,dt
\tag{4}
$$

where T is the annealing time.

The concentration profile $C(x,y,t)$ of the Al atom is obtained numerically from

$$
\frac{\partial C(x,y,t)}{\partial t} = \nabla[D_{\text{atom}}(x,y,t)\nabla C(x,y,t)] .
\tag{5}
$$

Now that the diffusion length profile has much more grid points in the y direction than the number of QWs, the profile is re-gridded into regions equivalent to single QWs, and the mean value in each region is chosen. In this way each QW is matched with a specific diffusion length L_d. The Al concentration profile $w(y)$ of a single QW with well width L_z is given by

$$w(y) = w_o \left\{ 1 - \frac{1}{2} \left[\text{erf}\left(\frac{L_z + 2y}{4L_d} \right) + \text{erf}\left(\frac{L_z - 2y}{4L_d} \right) \right] \right\} \qquad (6)$$

where w_o is the initial Al concentration. A previously developed model [5] is adopted to find the refractive index profile $n_r(x, y)$ from the diffused QW structure.

The wave guiding properties of the dual waveguide are solved numerically using a quasi-vector solution which takes into account the polarization effects. For quasi-TE polarization, the wave equation to be solved is

$$\frac{\partial^2 E_x}{\partial x^2} + \frac{\partial^2 E_x}{\partial y^2} + (k^2 n_r^2 - \beta^2)E_x + 2\frac{\partial}{\partial x}\left(E_x \frac{\partial}{\partial x} \ln n_r \right) = 0 \qquad (7)$$

where E_x is the x component of the electric field, $k = 2\pi/\lambda$ is the free-space wavenumber, and β is the modal propagation constant. The wave equation is solved using a Galerkin's method [3] which transform the equation into an eigenvalue problem with eigenvalues β and eigenvectors E_x. The two largest eigenvalues β_e and β_o are for the even and odd supermodes of the dual waveguide. The shortest distance required for complete power transfer between the individual channel is given by the coupling length $L_c = \pi/(\beta_e - \beta_o)$. This length is commonly used in the measurement of the coupling phenomenon.

NUMERICAL RESULTS AND DISCUSSIONS

Dual channel waveguides with different geometric and optical parameters are investigated. All calculations are done for quasi-TE mode with several fixed parameters: the total lateral width $2L = 16\mu m$; the thickness of MQW layers is 0.6 μm, which consists of 30 periodic layers of 100 Å GaAs wells and 100 Å $Al_{0.3}Ga_{0.7}As$ barriers; this is followed by a 2.5 μm thick $Al_{0.3}Ga_{0.7}As$ buffer on top of a GaAs substrate; the implant dose $N_o = 1 \times 10^{12}$ cm^{-2}, and the implantation is performed at room temperature.

For each implant energy, there are three specific values associated with it as listed in Table I [6]. A typical Ga$^+$ ion impurity concentration profile is shown in Fig. 2.

Prior to diffusion, there is no guiding mode in the MQW structure due to a uniform lateral refractive index profile. As interdiffusion proceeds, the refractive indices of the implanted regions fall gradually while those of the masked regions remain the same except near the mask edges. Figure 3 shows the change of refractive index at two diffusion times.

Table I: Project range, its vertical and lateral deviation associated with different implant energies for Ga$^+$ implantation of AlGaAs/GaAs.

Implant energy (keV)	R_p (μm)	ΔR_p (μm)	ΔR_{pL} (μm)
600	0.2308	0.8550	0.1081
800	0.3111	0.1069	0.1383
1000	0.3916	0.1261	0.1665

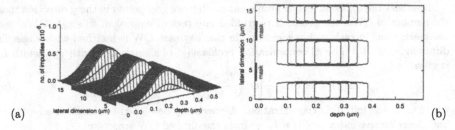

(a)   (b)

Figure 2: Impurity concentration profile. (a) 3-D plot; (b) contour plot, with implant energy of 600 keV, $L_m = 3\mu m$, and $s = 4\mu m$.

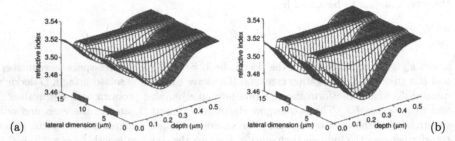

(a)   (b)

Figure 3: Refractive index profile changes with diffusion time. (a) t = 10 s; (b) t = 20 s, with implant energy of 600 keV, $L_m = 3\mu m$, $s = 4\mu m$, and $\lambda_{op} = 0.90\mu m$.

Waveguide structure produced at ion implant energy of 1000 keV with $L_m = 1.8\mu m$, and operating wavelength $\lambda_{op} = 0.90\mu m$ is taken as a typical case for illustrating the results. Considering the effect of waveguide separation, it is discovered that for all waveguide separations under investigation, diffusion time $t = 36$ s marks the beginning of guided modes and single-mode operation is still maintained up to $t = 100$ s.

It is evident that for small waveguide separation, there is a relatively large difference between the modal propagation constants of the even and odd supermodes. When the separation increases, their difference decreases and they both tend to that of the fundamental mode of a single guide. Since the coupling length is inversely related to this difference, the coupling length becomes longer as the waveguide separation increases. This is because the interaction between the guides naturally gets smaller as they are farther apart. Table II shows the calculated values of L_c in this case together with one at a shorter wavelength.

As diffusion proceeds (diffusion time $t > 36$ s), the coupling length rises. This may be explained by the fact that the difference in the refractive indices in the implanted and non-implanted regions gets larger, thus producing a stronger optical confinement effect. This implies less light can propagate into the adjacent channel. It can therefore be concluded that guiding improves while the extent of coupling decreases with diffusion time.

For shorter operating wavelength of 0.85 μm, guided modes will occur earlier but single-mode operation will last for a shorter time. Comparing the values of coupling length when guided modes have just begun, it is found that coupling length is smaller for longer wavelength, which suggests that the degree of coupling is greater for longer wavelengths.

Figure 4 shows the coupling length against diffusion time for different mask widths. In short, the wider the mask, the longer the coupling length. This is obvious as light is

Table II: Coupling length L_c at different diffusion time for different waveguide separation s with implant energy of 1000 keV and $L_m = 1.8\mu m$.

λ_{op} (μm)	Diffusion time (s)	Separation (μm)			
		1	2	3	4
	36	0.0051	0.1497	4.0120	8.6927
0.90	60	0.0095	0.4274	8.9305	27.1247
	100	0.0183	1.3407	74.5967	81.9981
	6	0.0053	0.1690	7.5794	9.1936
0.85	8	0.0084	0.3737	48.6159	59.5110
	10	0.0121	0.7119	77.0969	107.1463

confined better in a larger region under larger masks. Hence, the degree of coupling is smaller. Figure 5 shows the field contour plots for mask widths of 2.4 and 3.6 μm. It turns out that the diffusion time for guided modes to occur as well as the time duration for single-mode operation decreases with increasing mask width. It implies that to shorten the diffusion time, a wider mask is desirable; however, the undesirable multi-mode guiding will occur earlier.

Finally, the influence of implant energy on the coupling length is investigated. Lower implant energies are observed to shorten the coupling length. Figure 6 shows the coupling length against diffusion time for different implant energies. Higher implant energies will require longer diffusion time for guided modes to occur, and the time duration for single-mode operation is longer.

CONCLUSIONS

A model has been developed for studying multi-quantum well dual waveguide defined by the ion implantation technique. It enables a quasi-vector analysis of the guiding of optical wave. The effects of the implantation and geometric parameters on the guiding and coupling properties of the dual waveguide structures are analyzed. Knowledge of the the coupling length has been gathered to facilitate the design of IID defined integrated optoelectronic devices.

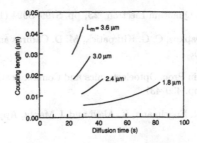

Figure 4: Coupling length L_c versus diffusion time for different mask widths L_m with implant energy of 800 keV, $s = 1\mu m$, and $\lambda_{op} = 0.90\mu m$.

471

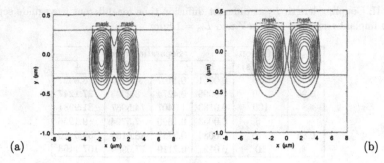

(a) (b)

Figure 5: Field contour of the even supermode for at (a) $L_m = 2.4\mu m$; (b) $L_m = 3.6\mu m$ with implant energy of 600 keV, $s = 1\mu m$, and $\lambda_{op} = 0.90\mu m$.

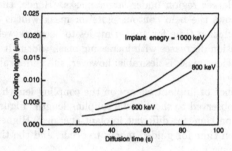

Figure 6: Coupling length L_c versus diffusion time for different implant energies with $L_m = 1.8\mu m$, $s = 1\mu m$, and $\lambda_{op} = 0.90\mu m$.

ACKNOWLEDGMENTS

The author would like to thank Mr. Steve Y. T. Wong of the University of Hong Kong for technical support.

REFERENCES

1. B. L. Weiss, Ed., Opt. Quantum Electron. **23**, pp. S799–S994 (1991).

2. J. J. Coleman, P. D. Dapkus, C. G. Kirkpatick, M. D. Camras and N. Holonyak, Appl. Phys. Lett. **40**, p. 904 (1982).

3. K. M. Lo and E. H. Li in Proc. Optoelectronics and Communications Conference (OECC '96, Chiba, Japan, 1996), pp. 430–431.

4. T. Wolf, C. L. Shieh, R. Engelmann, K. Alavi and J. Mantz, Appl. Phys. Lett. **55**, pp. 1412–1414 (1989).

5. E. H. Li, B. L. Weiss, K. S. Chan and J. Micallef, Appl. Phys. Lett. **62**, pp. 550–552 (1993).

6. F. Gibbons, W. S. Johnson and S.W. Mylroie, Range statistics in semiconductors, Academic, New York, 1975.

Part VII

Nonlinear Optical and OPO Materials

Second-Harmonic and Sum-Freqency Generation in CdGeAs$_2$

*Eiko Tanaka and **Kiyoshi Kato

*Department of Computer Science , Keio University,
Hiyoshi 3-14-1, Kohoku-ku, Yokohama, Japan
**Second Research Center, Japan Defense Agency
Ikejiri 1-2-24, Setagaya, Tokyo, Japan

Abstract

CdGeAs$_2$ has been found to be phase-matchable for type-2 SHG down to 2.833 μ m and type-2(b) SFG between the fundamental and second harmonic of a CO$_2$ laser down to 10.4406 μ m at 20.0°C. The Sellmeier's equations and the thermo-optic dispersion formula which reproduce well the phase-matching properties of SHG and direct FHG of the CO$_2$ lasers at 77K are presented together with the absolute value of the nonlinear optical constant.

Recent progress in the growth of CdGeAs$_2$[1],[2] has generated renewed interest in its potential for high power harmonic generation of a CO$_2$ laser and parametric oscillation in the mid to far IR when pumped by the solid state lasers. However, since the phase-matching conditions of this material thus far reported in the literature show large scatters from those predicted by the Sellmeier's equations of Kildal and Mikkelsen[3] and Bhar[4], we remeasured the phase-matching properties of this compound for type-2 SHG and SFG in the 2.65-5.30 μ m range and found that this material is 90° phase matchable for type-2 SHG of the 5.670 μ m and type-2(b) SFG between the fundamental and second harmonic of a CO$_2$ laser at 10.4406 μ m at 20.0°C. Here, we report some new experimental results on the IR frequency conversion in CdGeAs$_2$ along with the improved Sellmeier's equations and the thermo-optic dispersion formula which reproduce well the phase-matching conditions for SHG[5] and direct FHG[6] of the CO_2 laser observed at 100 and 77K. In addition, the absolute value of the nonlinear optical constant is presented.

The $CdGeAs_2$ crystals used in this experiment were supplied by the Eksma,Inc. in Lithuania and were fabricated at (θ =90° , ϕ =45°) and (θ =49.1° , ϕ =45°). The dimensions of these two crystals were $8 \times 8 \times 6$ mm³. The short and long cutoff wavelengths are 2.45 and 18.1 μ m.

Using a Ho:YLF laser pumped $AgGaSe_2$ OPO and the CO_2 laser as the pump source, we first measured the phase-matching angles for type-2 SHG in the short and long wavelength branches at 20.0℃. The experimental results were found to be in excellent agreement with the values calculated with the following Sellmeier's equations:

$$n_o^2 = 12.4008 + \frac{2.1603}{\lambda^2 - 2.0617} - 0.00133\lambda^2$$

$$n_e^2 = 13.0079 + \frac{3.2613}{\lambda^2 - 2.8382} - 0.00126\lambda^2$$

where λ is in micrometers. The shortest SHG and SFG wavelengths generated in this experiment were 2.833 and 3.4802μ m. Although the data points of Boyd et al[7] for type-1 and type-2 SHG of 10.5910μ m are ~2.5° larger than our measured and calculated values of θ_{eeo}=32.6° and θ_{oeo}=49.3° owing to the smaller birefringence of the BTLs crystals, our index formula reproduce well the experimental results of the·Stanford and MIT groups. In addition, this formula correctly reproduces the phase-matching conditions for type-2 DFG between the CO and CO_2 lasers presented by Kildal and Mikkelsen for the MIT crystal[3]. The comparison between theory and experiment is shown in Fig.1 together with the theoretical curves calculated with the index formulas of Kildal and MIkkelsen(3) and Bhar(4). Note that the birefringence of the current crystals is very similar to the value recently measured by Fisher et al[8](crystal 4Q).

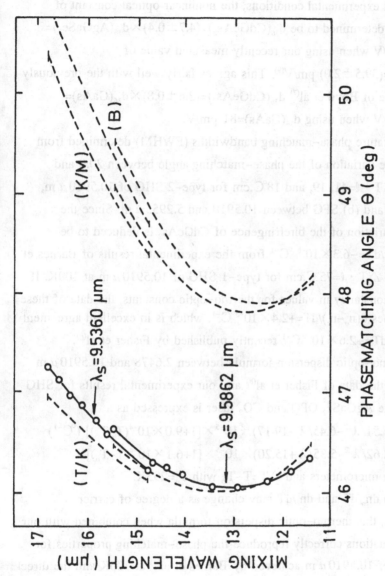

Fig.1 Phase-matching curves for type-2 DFG between the CO and CO_2 lasers in CdGeAs₂ at room temperature. The dashed lines (K/M), (B), and (T/K) are the theoretical curves calculated from the Sellmeier's equations of Kildal and Mikkelsen[3], Bhar[4], and the present authors. ○ : experimental points quoted from Ref.(3).

From direct comparison of SHG conversion efficiencies at $5.2955\,\mu$ m observed in the $49.1°$ cut CdGeAs$_2$ crystal and $55.5°$ cut AgGaSe$_2$ crystal under identical experimental conditions, the nonlinear optical constant of CdGeAs$_2$ was determined to be $d_{36}(CdGeAs_2)=(4.7\pm0.4)\times d_{36}(AgGaSe_2)=(186\pm16)$ pm/V when using our recently measured value of $d_{36}(AgGaSe_2)=(39.5\pm2.0)$ pm/V[9]. This agrees fairly well with the previously published value of Boyd et al[7] $d_{36}(CdGeAs_2)=(2.6\pm0.8)\times d_{36}(GaAs)=(212\pm63)$ pm/V when using $d_{36}(GaAs)=81$ pm/V.

The temperature phase-matching bandwidths (FWHM) determined from the temperature variation of the phase-matching angle between 20.0 and $120°C$ were $\triangle T \cdot \ell=41$, 19, and $18°C.cm$ for type-2 SHG of $10.5910\,\mu$ m, and type-2(a) and (b) SFG between 10.5910 and $5.2955\,\mu$ m. Since the temperature variation of the birefringence of CdGeAs$_2$ is deduced to be $d(n_e^1(\theta)-n_o^2)/dT=-6.3\times10^{-6}\ °C^{-1}$ from the experimental results of Barnes et al[5], we found $\triangle T \cdot \ell=75°C.cm$ for type-1 SHG of $10.5910\,\mu$ m at 100K. If we assume a constancy in values for thermo-optic constants, the data of these authors becomes $d(n_o-n_e)/dT=+2.4\times10^{-5}°C^{-1}$, which is in excellent agreement with $d(n_o-n_e)/dT=+2.6\times10^{-5}°C^{-1}$ recently published by Fisher et al[8].

The thermo-optic dispersion formula between 2.6478 and $10.5910\,\mu$ m deduced from the data of Fisher et al[8] and our experimental results for SHG and SFG of the AgGaSe$_2$ OPO and CO$_2$ laser is expressed as

$dn_o/dT=(26.51/\lambda^2-6.45/\lambda+19.17)\times10^{-5}\times[1+9.0\times10^{-4}(T-T_o)]\ (°C^{-1})$

$dn_e/dT=(22.62/\lambda^2-5.35/\lambda+15.20)\times10^{-5}\times[1+6.1\times10^{-4}(T-T_o)]$

where λ is in micrometers and $\triangle T=T-T_o$ with $T_o=293K$.
Although both dn_o/dT and dn_e/dT may change as a degree of carrier concentrations, this thermo-optic dispersion formula when combined with our Sellmeier's equations correctly reproduce the phase-matching properties for type-1 SHG of $10.5910\,\mu$ m achieved by Barnes et al[5] at 100K and the direct FHG of the CO$_2$ laser radiation of between 10.1946 and $10.5632\,\mu$ m measured by Kildal and Iseler[6] at 77K except in the vicinity of the absorption edge.

References

1) P.G.Schunemann, in Dig. Tech. Papers, CLEO'96, paper JTul4.

2) P.G.Schunemann, in Dig. Tech. Papers, CLEO'97, paper CThG5.

3) H.Kildal and J.C.Mikkelsen, Opt.Commun. 10, 306–309 (1974).
   There is a typographical error in the Sellmeier's equations: The first term
   of n_o^2 should read 3.4141.

4) G.C.Bhar, Appl.Opt. 15, 305–307 (1976).

5) N.P.Barnes, R.L.Eckhardt, D.J.Gettemy, and L.B.Edgett, IEEE J.Quantum
   Electron. QE–15, 1074–1076 (1979). The phase–matching angles of
   CdGeAs$_2$ presented in this paper were measured from x(=a). Thus, they
   should read θ =35.4° at 298K and θ =34.5° at 100K.

6) H.Kildal and W.Iseler, Phys.Rev.B. 19, 5218–5222 (1979).

7) G.D.Boyd, B.Buehler, F.G.Stortz, and J.H.Wernick, IEEE J.Quantum
   Electron. QE–8, 419–426 (1972).

8) D.W.Fisher, M.C.Ohmer, and J.E.McCrae, J.Appl.Phys. 81, 3579–3583
   (1997).

9) K.Kato, Appl.Opt. 36, 2506–2510 (1997).

Low Optical Loss Wafer Bonded GaAs Structures for Quasi-Phase-Matched Second Harmonic Generation

YewChung Sermon Wu, Robert S. Feigelson, Roger K. Route,
Dong Zheng, Leslie A. Gordon, Martin M. Fejer, Robert L. Byer
Center for Nonlinear Optical Materials, Stanford University, CA94305,
sermon@crystal.stanford.edu

ABSTRACT

A periodic GaAs wafer-bonded structure has been proposed for quasi-phase-matched (QPM) second harmonic generation (SHG). However, current bonding processes often lead to unacceptable optical losses and poor device performance. In this study, three sources of optical losses in wafer-bonded structures were investigated, (1) interfacial defects between the wafers, (2) bulk defects within the wafers, and (3) decomposition at the exposed outer surfaces. Surface losses due to incongruent evaporation were easily eliminated by repolishing the outer surfaces. However, to minimize the losses from interfacial and bulk defects, it was necessary to investigate the relationship between these defects and the bonding parameters. It was found that an increase in bonding temperature and/or time led to a decrease in interfacial defects, but an increase in bulk and surface defects. Through a trade-off process, optimized processing conditions were developed which permitted the preparation of bonded stacks containing over 50 (100)-oriented GaAs wafers, and about 40 layers of (110))-oriented GaAs wafers. Optical losses as low as 0.1-0.3% /interface (at 5.3 µm and 10.6 µm) were achieved.

INTRODUCTION

GaAs has a large nonlinear coefficient, and a high optical damage threshold. However, its use in nonlinear optical applications has been limited because GaAs cannot be birefringently phase-matched. To overcome this problem, a periodic GaAs (110) structure prepared by a high temperature wafer bonding process has been proposed for quasi-phase-matched (QPM) second harmonic generation (SHG)[1]. However, current bonding methodologies often lead to the introduction of defects which adversely affect optical transmission.

Three kinds of defects have been found in bonded GaAs wafers: (1) interfacial defects between wafers, (2) bulk defects within individual wafers and (3) surface defects at the exposed outer surfaces. Interfacial defects included voids and inclusions. Bulk and surface defects included antisite defects, dislocations, impurities, vacancies, interstitials, and precipitates.

EXPERIMENT

Three inch undoped semi-insulating GaAs wafers, either (100) or (110)-oriented, were used in this study. They were diced into 9 x 9 mm^2 dimensions, then solvent-cleaned in a class 100 clean room. Under a flowing gas mixture of H_2(60 cc/min.) and N_2(1l/min.), GaAs was bonded at 200 ~ 975°C for 0.5 ~19 hr with a compressive load ranging from 0 to 30 Kg. Finally, the effects of processing on bulk and surface defects, and the shrinkage of interfacial voids were analyzed using standard characterization techniques.

RESULTS

Individual GaAs wafers were used to study how the bonding process affected bulk and surface defects. Figure 1 shows the transmittance of single GaAs wafers heat-treated for 2 hr. at

Mat. Res. Soc. Symp. Proc. Vol. 484 © 1998 Materials Research Society

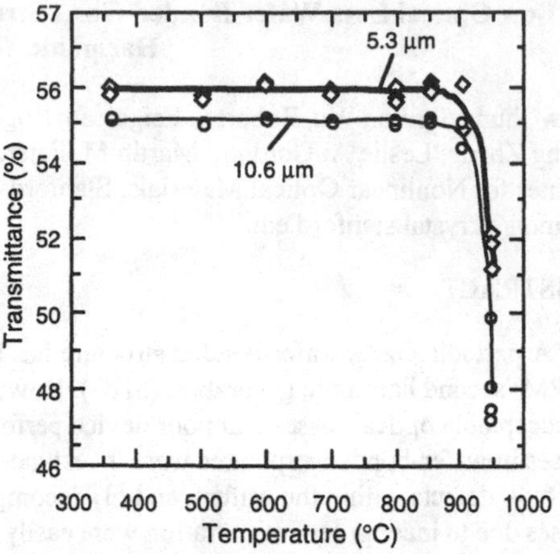

Fig. 1. Transmittance of single GaAs (100) wafers processed under a static 10 kg/cm^2 load for two hours at different temperatures ranging from 350 °C to 950 °C.

temperatures between 350°C and 950°C. When compared with unprocessed GaAs, the optical losses at the two wavelengths of interest for SHG were seen to increase when processing temperature exceeded 850 °C. The electrical properties also changed when the processing temperatures exceeded 850°C: the undoped semi-insulating wafers converted to p-type and their sheet resistance decreased at progressively higher temperatures. Most of the p-type conversion was on the surfaces (caused by arsenic depletion resulting from incongruent evaporation) and could be eliminated by repolishing.

In our interfacial defect study, artificial twin boundaries were created between GaAs wafer pairs as this is the configuration required for quasi-phase-matching. The investigation of interface void shrinkage followed the approach described in Refs. [2] in which artificial voids of various depths were created by bonding topographically-patterned wafers to unpatterned wafers in order to model the effect of random surface irregularities. Figure 2 shows IR transmission optical micrographs for two GaAs wafers containing 700 Å deep artificial voids that were bonded at 870 °C and 910 °C for 2 hr. The light features seen in the void area of Fig. 2(b) correspond to bonded areas which occurred through the nucleation and growth of islands by a mass transport process within the void space. The nature of the bonding in these regions was verified using high resolution transmission electron microscopy (HRTEM), as shown in Fig. 3. No other phases were found at the bonded interfaces, and only a few atomic layers were disrupted.

The "bonded fraction," as determined by estimating the light vs. dark areas on the IR transmission micrographs, was found to increase with temperature, consistent with the fact that mass transport increases with temperature[3]. We found that the detectable bonded fraction was a strong function of void depth. Bonding (or recrystallization) was not detected in 700 Å deep voids until the process temperatures reached 910 °C, as shown in Fig. 2. However, bonding was detected in 100 Å deep voids at process temperatures around 850 °C. These experiments showed that the bonded fraction in wafer-bonded GaAs depends strongly on the magnitude of the height of the surface irregularities at the wafer interfaces as well as on process temperatures.

Many different light features were observed in our study; most contained features with two morphologies: diamond/rhombohedral-geometry (with edges parallel to <100> directions) and dendrite-geometry (with dendrite arms in the <110> directions). As shown in Fig. 4, dendrite caps with <100> orientation were bounded by {111} planes and the (001) wafer plane, while the diamond features were bounded by {110} planes and the (001) wafer plane, again verified using transmission electron microscopy (TEM).

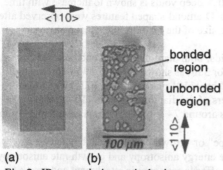

(a)   **(b)**

Fig. 2. IR transmission optical microscope images of interfacial voids (originally 700Å deep) on GaAs (100) wafers after bonding at temperatures of (a) 870 and (b) 910°C for 2hr.

Fig. 3. Cross-sectional high resolution TEM of (100)-oriented GaAs bonded area at 850 °C, showing intimate contact between the two GaAs plates with no intervening oxide layer.

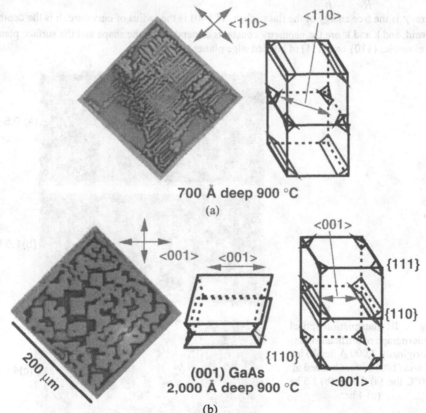

700 Å deep 900 °C

(a)

(001) GaAs
2,000 Å deep 900 °C

(b)

Fig. 4. Illustration of the (a) cap of dendrite light regions and (b) edges of the diamond.

In Fig. 5, the size of the bonded areas in 2,000 Å deep voids is shown to increase with time, as expected from the kinetic nature of the process. Diamond-shaped features were observed after annealing at 950°C for ≥ 0.5 hr. Regardless of the size of the features, their shape always remaineddiamond-like.

On the other hand, in 700 Å deep shallow voids, both diamond-shaped and dendrite-shaped features were observed after annealing at 900°C for 3 hr, as shown in Fig. 6. With increasing time, the stability of the diamond edges (bounded by {110} planes and the (001) wafer plane) broke down, and dendritic arms grew at the corners of the diamonds in the <110> directions. The average growth rate of the arms in length was around 7 - 8 μm/hr., which is much higher than that of the arms in width (0.4 - 0.6 μm/hr.).

It is important to note that the observed "shape" or "growth habit" of the bonded areas, diamond or dendrite, can result from both surface energy anisotropy and growth rate anisotropy, the latter occurring when a particular crystal face affords easy atomic attachment and grows rapidly. Growth rate anisotropies can result in shapes which exaggerate actual surface energy anisotropies[4].

From the Gibbs-Thomson relationship, the bulk driving force of a curved crystal is given by

$$\Delta G_\infty \approx k \frac{\gamma}{R} \approx 2k' \frac{\gamma}{h}$$

where γ is the free energy of the flat surface, R (<0) is the radius of curvature, h is the depth of the void, and k and k' are the geometry constants dependent on the shape and the surface planes (for example, {110} or {111}) of bonded edge planes).

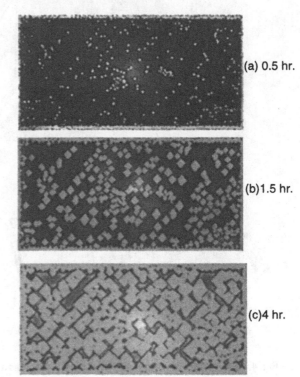

(a) 0.5 hr.

(b)1.5 hr.

(c)4 hr.

Fig. 5. IR transmission optical micrograph of artificial voids (originally 2,000 Å depth) in GaAs (100) wafers bonded at 950°C for: (a) 0.5 hr, (b) 1.5 hr, (c) 4 hr.

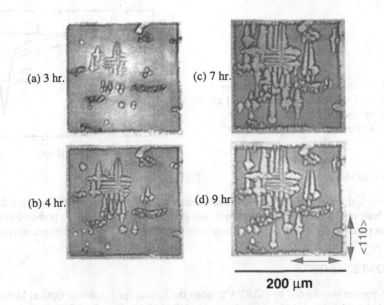

Fig. 6. IR transmission optical micrograph of artificial voids (originally 700 Å depth) in GaAs (001) wafers bonded at 900°C for: (a) 3 hr, (b) 4 hr, (c) 7 hr, and (d) 9 hr.

With a void depth h ≥ 2,000 Å, the driving force is small. According to the surface energy anisotropy, GaAs tends to grow into a low surface energy form. Regardless of the bonding temperatures (850-950 °C) and time, most the bonded areas were diamond-shaped (bounded by very flat {110} planes); therefore, we believe that the surface energies of {110} faces were lower than that of others[5]. On the other hand, when the void depth h ≤ 700 Å, the driving force of the fast grow {111} plane was large enough to break the constraint of the {110} planes, allowing dendrites to grow quickly in the <110> direction.

The interfacial, bulk and surface defect studies described above showed clearly that higher bonding temperatures and longer bonding times lead to an increase in the bonded fraction (i.e., a decrease in interfacial defects), but they also led to an increase in the density of surface and bulk defects. For this reason, a processing temperature of 850 °C was selected as the best temperature for the bonding of the multi-layer stacks necessary for practical nonlinear optical devices.

A 52 layer (100)-oriented stack, of 630 µm thick wafers, was bonded at 850 °C, for 0.5 hr (a somewhat shorter interval than used in the 2 layer studies). The average optical losses per layer were 0.9% at 5.3 µm, and 0.5% at 10.6 µm. Following these experiments, a stack of 39 (110)-oriented wafers, each 500 µm thick, was bonded under the same conditions. The average optical losses per layer were even lower, 0.1 - 0.3% at both 5.3 µm and 10.6 µm, as shown in Fig. 7. The reduced optical losses in the 39 layer stack may have come from the reduction of wafer thickness (shorter absorption pathlength - 630 µm compared with 500 µm), and/or the different properties (mechanical, physical, and chemical) of the different wafer surface orientations, (100) versus (110).

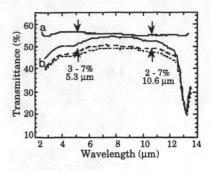

Fig. 7. Observed transmittance of (a) an unprocessed single GaAs wafer, and (b) a 39 layer stack of (110) bonded GaAs wafers at 3 different positions.

CONCLUSIONS

The optical losses achieved in these studies are considered low enough for practical devices, and we believe continued experiments with custom-fabricated GaAs wafers prepared to the thickness required for specific optical interactions are warranted. Such studies are currently under way.

ACKNOWLEDGMENTS

This project was funded by DARPA through the Center for Nonlinear Optical Materials at Stanford University, and by the Army Research Office. It has benefited from facilities and equipment made available to Stanford University by the NSF-MRSEC Program through the Center for Materials Research at Stanford University.

The authors would like to thank W. Tiller and D. Vanderwater for valuable discussions.

REFERENCES

1. L. A. Gordon, G. L. Woods, R. C. Eckardt, R. K. Route, R. S. Feigelson, M. M. Fejer, R. Byer, *Electronics Letters* (1993) **29**, no. 22, pp. 1942-4.
2. J. Rodel, A. M. Glaeser, *Journal of the American Ceramic Society* 73, pp. 592-601 (1990)
3. S. K. Ghandhi, *VLSI Fabrication Principles - Silicon and Gallium Arsenide, 2nd ed.* p.197, Wiley, New York (1994).
4. Y. Chiang, D. Birnie III, and W. D. Kingery, Physical Ceramics - Principle for Ceramic Science and Engineering, p.354, J. Wiley, New York (1997).
5. D. Vanderwater personal communications, 1997.

Phase-matched 2nd harmonic generation in asymmetric double-quantum wells

K. L. VODOPYANOV, C. C. PHILLIPS, I. VURGAFTMAN*, J. R. MEYER*

Solid State Group, Physics Department, Imperial College, London SW7 2BZ, UK
tel.+44-171-594 7589, k.vodopyanov@ic.ac.uk

* *Optical Sciences Division, Naval Research Laboratory, Washington DC 20375-5338, USA*

ABSTRACT

Phase-matched SHG, resonantly enhanced near $\lambda=8.6\mu m$, with ~1% efficiency and up to 30 W peak power has been observed in asymmetric double multi-quantum well structures. Waveguide-mode and 45^0 wedge multi-bounce geometries were used.

INTRODUCTION

Quantum well structures can be tailor-made to have giant 2nd and 3rd order intersubband optical nonlinearities[1-3], enhanced near subband resonances, which may be useful for optical frequency conversion, phase conjugation, optical bistability, etc.

Greatly enhanced nonlinearities in semiconductor multiple quantum wells (MQW's) look very attractive for the efficient generating new frequencies via frequency mixing and second harmonic generation (SHG). Nonetheless, SHG and difference frequency conversion efficiencies reported so far are rather small (<<1%)[4-7]. To improve on this, longer interaction lengths are needed, necessitating the matching of the phase velocities of the fundamental and second harmonic beams to maximise the energy transfer between them.

We report here on the efficient second harmonic generation of mid-infrared light, where a novel scheme[8] was used to "phase-match" the fundamental and second harmonic beams, i.e. to compensate for the natural dispersion of the waveguide bulk material which otherwise leads to dephasing within ~40μm.

EXPERIMENT

The structure studied, was grown by molecular beam epitaxy and consists of an active SHG region placed between two phase-matching (PM) regions. The active region incorporates 178 periods of repeated asymmetric InGaAs/InAlAs double quantum wells (27Å $In_{0.53}Ga_{0.47}As$ well - 18Å $In_{0.52}Al_{0.48}As$ barrier - 49Å $In_{0.53}Ga_{0.47}As$ well -100Å $In_{0.52}Al_{0.48}As$ barrier) with subband resonances detuned somewhat from the $\lambda=8-10$ μm pump wavelength and its second harmonic. The PM region consists of 139 periods of multiple QW (58Å $In_{0.53}Ga_{0.47}As$ well -100Å $In_{0.52}Al_{0.48}As$ barrier). The wider (49Å InGaAs) wells of the active region and the InGaAs wells in the PM region were n-doped with Si to a sheet carrier concentration 6.5×10^{11} cm^{-2}. The frequencies of the intersubband transition in the phase-matching QW's were chosen midway between the fundamental and SHG frequencies. The additional dispersion they produce (calculated from the Kramers-Kronig transform of the intersubband absorption line) compensates the normal material dispersion, yet they contribute minimal additional absorption of either beam. Detailed simulations predict that this approach can be used to maintain phase-matching over long propagation path, >> coherence length.

As a pump source we used a novel travelling-wave optical parametric generator (OPG) based on a nonlinear ZnGeP$_2$ crystal[9], which was tunable within the range $\lambda=4-10\mu m$. The OPG was pumped by single

487

100ps pulses (λ = 2.8 μm) from an actively mode-locked, Q-switched and cavity-dumped Cr:Er:YSGG laser-amplifier system (repetition rate 3 Hz). Single OPG pulses with energies of a few μJ and 90±15 ps pulse duration were focussed onto the samples (~100μm beam size) with peak radiation intensity up to 10^8 W/cm². Spectral linewidths were typically 10-20cm⁻¹, depending on output wavelength. These characteristics make the OPG system ideal to study subband non-linear optical effects in MQW structures.

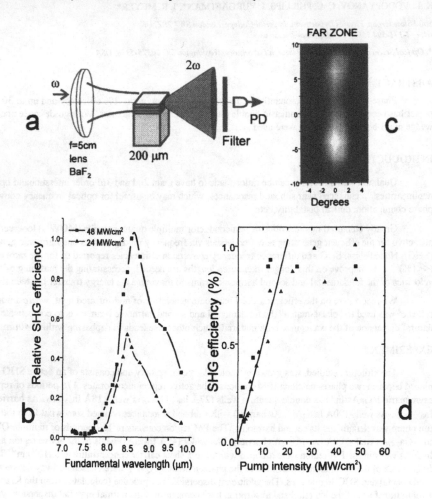

Fig.1. (a) Schematic of SHG using waveguide-mode geometry. (b) Second harmonic fundamental wavelength resonance curves. (c) Far zone SH beam profile. (d) SHG conversion efficiency as a function of input intensity.

The sample with a net epilayer thickness of ≈ 8 μm was cleaved to a cavity length of 200 μm. Two particular sample geometries were used: "true" waveguide (Fig.1a), where polarized IR pump radiation was coupled into the edge of the wafer with the electric field vector parallel to the growth direction, and a 45^0 waveguide geometry, where the edges were polished at 45^0 to the growth direction at each end, and the p-polarized light passed into the sample normal to these edges but at 45^0 to epitaxial layers. The light reflected 5 times internally at the epitaxial MQW layer before leaving the sample. In the former geometry, the phase-matching condition was achieved via "band structure engineering of the PM region to modify the dispersion[8]; while in the latter case it was achieved by adjusting the free path length within the InP substrate to make the second harmonics (SH) from consecutive bounces interfere constructively (although the PM region was still present in the SHG MQW structure, it did not play a significant role in that case because the path length was short).

A calibrated InSb (77K) photodiode with a 5.6μm cut-off was employed for SH detection. The second harmonic signal versus pump wavelength is shown in Fig.1b at two different pump intensities. Small variations of the pump intensity, due to linear absorption by the $ZnGeP_2$ crystal between 8-10μm, were taken into account by normalising the signal to the input power. The maximum SHG conversion efficiency is peaked at a fundamental wavelength of 8.6μm - close to the designed value of 8.5μm.

The far zone SH distribution, determined by XY scanning in the plane perpendicular to the beam, is shown in Fig.1c. The large divergence in the vertical direction corresponds to the diffraction limit ~λ/d, where λ is the SH wavelength (4.3 μm) and d is the waveguide thickness (8μm). The non-uniformity in this profile is thought to be caused by poor optical quality in the cleaved facet.

The SHG conversion efficiency (defined as the ratio of the SH intensity to the incoming pump beam intensity within the aperture of the waveguide) against pump intensity dependence is shown in Fig.1d. We used both attenuating filters and the "z-scan" method to vary the intensity of the pump. It is evident that the SHG efficiency is linearly dependent on the pump intensity below 15 MW/cm^2 and remains constant at ~0.9% for intensities above that value. The maximum SHG peak power generated at 4.3 μm is 5 W.

The almost 3-fild increase of the conversion efficiency as the sample length increased from 100 to 200μm indicates that good phase-matching has been achieved for a cavity length much greater than the bulk coherence length. While theoretical simulations of the multi-QW structure predict the loss of phase coherence and reduction in the SHG efficiency at high pump intensities, this effect is blurred under real experimental conditions by the Gaussian spacial and temporal profile of the pump pulses. The maximum theoretical conversion efficiency of ≈ 1.5% is in good agreement with the experimental results.

In the 45^0 waveguide structure, the SH is generated only while the light is passing through the MQW epilayer (we neglect bulk SHG in the InP substrate). To get maximum conversion efficiency the SH generated at each reflection must interfere constructively with the SH generated at previous reflections within the structure. Using a generalized phase matching condition, i.e.

$2k_\omega L = k_{2\omega} L + 2 m\pi$

where $k_{2\omega}$ and k_ω are the magnitudes of the wavevectors of the SH and fundamental beams, $L=L_1+L_2$ is the distance travelled by the light beam in the substrate between reflections at the epitaxial layer, and m is an integer. The change in angle of incidence of the pump light (normal to the polished edge) in the sample to bring an output beam from one maximum to the next was calculated to be ≈ 26^0. Such behaviour was observed, at a measured period of 30 ± 2^0.

The SHG conversion efficiency in a 45^0 waveguide with 5 double passes was found to increase linearly with pump intensity, and did not saturate up to the maximum intensity of 100 MW/cm^2. A maximum conversion efficiency of 0.2% was measured. The absence of saturation and lower SHG efficiency (compared to a true waveguide) may be explained by weaker coupling to intersubband transitions in this geometry i.e. the pump E-vector is not parallel to the growth axis. The maximum peak SHG power was over 30 W in this geometry, which is, despite the lower efficiency, larger than that for the true waveguide and is the result of better coupling of the pump into the waveguide. ·

As expected, the measured SH polarization in both geometries was the same as that of the pump (i.e. vertical in Fig.1), corresponding to the only non-zero nonlinear-optical tensor component d_{33}.

CONCLUSION

We have demonstrated for the first time phase-matched intersubband-based second harmonic generation in quantum well waveguides with relatively long interaction lengths. Resonant enhancement near the fundamental λ=8.6 μm was obtained with maximum conversion efficiency of 0.2% for the 45^0 wedge geometry and 0.9% (at 15 MW/cm^2) for the true waveguide geometry. This value for SHG conversion efficiency is, to our knowledge, the highest ever obtained for intersubband-base nonlinearities.

ACKMOWLEDGEMENTS

We would like to thank Geoff Hill from Dept. Elect. & Elect. Eng, Univ Sheffield for the samples processing. This work has been supported by the UK Engineering and Physical Sciences Research Council.

REFERENCES

[1] M. M. Fejer, S. J. B. Yoo, R. L. Byer, A. Harwit and J. S. Harris, Jr. , Phys. Rev. Lett. **62**, 1041 (1989)

[2] E. Rosencher, P. Bois, J. Nagle and S. Delaitre, Electron. Lett. **25**, 1063 (1989)

[3] P. Boucaud, F. H. Julien, D. D. Yang, J-M. Lourtioz, Appl. Phys. Lett. **57**, 215 (1990)

[4] P.Boucard, F.H.Julien, D.D.Yang, J-M. Lourtioz, E.Rosencher, and P.Bois, Opt. Lett. **16**, 199 (1991)

[5] Z. Chen, M.Li, D.Cui, H.Lu, and G.Yang, Appl. Phys. Lett. **62**, 1502 (1993)

[6] C.Sirtori, F.Capasso, J. Faist, L.N.Pfeiffer, K.W.West, Appl. Phys. Lett. **65**, 445 (1994)

[7] H.C. Chui, G.L.Woods, M.M.Fejer, E.L.Martinet, J.S.Harris, Jr., Appl. Phys. Lett. **66**, 265 (1995)

[8] I. Vurgaftman, J R Meyer and L R Ram-Mohan , IEEE J. Quant. Electron. **32**, 1334 (1996).

[9] K .L .Vodopyanov. JOSA B **10**, 1723 (1993); K.L.Vodopyanov, V. Chazapis, Opt. Communs., **135**, 98 (1997)

DARK SOLITON FORMATION FOR LIGHT-INDUCED WAVEGUIDING IN PHOTOREFRACTIVE InP:Fe

M. Chauvet*, S.A. Hawkins*, G.J. Salamo*, M. Segev**, D.F. Bliss***, and G. Bryant***.
* Department of Physics, University of Arkansas, Fayetteville, Arkansas 72701,501-575-5931; fax: 501-575-4580; salamo@comp.uark.edu
**Department of Electrical Engineering and, Advanced Center for Photonics and Optoelectronic Materials,Princeton University Princeton, New Jersey 08544, 609-258-1765; fax: 609-258-4454; segev@EE.Princeton.EDU
***U.S. Air Force, Rome Laboratory, Hanscom Air Force Base, Massachusetts 01731
617-377-4841; fax: 617-377-7812; bliss@maxwell.rl.plh.af.mil

ABSTRACT

We present experimental evidence for the observation of steady-state dark photorefractive screening solitons trapped in bulk InP:Fe.

INTRODUCTION

Spatial solitons in photorefractive materials have been the subject of recent interest.[1-6] Compared with Kerr spatial[7] solitons, the most distinctive features of spatial solitons are that they are observed at low light intensities and that trapping occurs in both transverse dimensions. Until now, photorefractive spatial solitons have been observed in the tungsten bronze ferroelectric oxides and in the non-ferroelectric sillenite oxides. Several reasons have led us to carry out similar experiments in the photorefractive semiconductor crystal InP:Fe.[8,9] First, this material is sensitive in the range of the near-infrared wavelengths used in optical telecommunications. Second, the photorefractive effect in semiconductors has the advantage of a faster response time than that observed in either the tungsten bronze or the sillenite crystal materials. Third, the possibility of monolithic integration with other optoelectronic components (lasers, detectors) is also attractive.

In this paper we will discuss self-trapping of optical beams in a semiconductor, using the photorefractive effect. The self-trapping of a bright soliton is observed at steady state with a low light intensity (50mW/cm²) and with a moderate d.c. applied field (9kV/cm) in bulk InP:Fe. The output diameter is observed to be trapped at the input value after propagation of both 5 and 10mm. Self-trapping of a beam with a dark notch, or dark soliton, is observed at steady state using only a 6kV/cm field.

RESULTS

To observe the self-trapping effects, we focus a continuous beam at 1.3μm with a 70-mn focal-length cylindrical or spherical lens onto the entrance face of a InP:Fe crystal. A beam profile system and an imaging lens are used to analyze the beam shape. The light beam is linearly polarized along the (110) direction and propagates along the (110) direction. Absorption is 0.15 cm^{-1} at 1.3μm. A d.c. field (E_0) is applied between the (001) faces. The temperature of the crystal is stabilized at 297K.

The resonance behavior of the two-wave-mixing (TWM) gain in InP:Fe has been characterized, both experimentally and theoretically.[10] In this model electrons are dominantly thermally excited from Fe^{2+} to the conduction band, whereas holes are dominantly optically exicted from Fe^{3+} to the valence band. The resonance in the gain as a function of intensity occurs at an intensity I_{res}, for which the optical excitation of holes is equal to the thermal excitation of electrons. In addition, the relative phase between the intensity spatial pattern and the space-charge-field pattern is also intensity and temperature dependent. The resonant intensity plays a critical role in the self-trapping, both bright and dark soliton, effects reported here.

The index change in InP which results in a trapped beam is very different in structure than that found in the ferroelectric oxides. In the oxides the index change is a simple one, where the index is lowered outside the beam, resulting in waveguiding. In InP, however, the focusing or waveguiding effect is induced by an increase of the refractive index in the center of the beam (for a negative applied field), while a strong decrease (10^{-4}) of the index appears on the right side of the beam. This amazingly strong decrease of the index implies that a large photorefractive space charge field is present in this region. We calculate the magnitude of this space charge field to be about 50 kV/cm when the magnitude of the applied field is only E_0=5kV/cm. Despite the small value of the electro-optic coefficient r_{41}, it is clearly possible to induce large index changes. Physically, the large index change is due to the large space charge field which can be created because of both the intensity temperature resonance and the low value of the dielectric constant associated with InP.

For a positive applied field the refractive index decreases where the beam is present, causing defocusing while producing an increase in index on the left side of the beam. For both positive and negative applied fields, the strong index produced on the side of the 1.3μm beam can be utilized to guide a second beam, as we have demonstrated in previous experiments. The point is that the observed photoinduced refractive index change profiles in InP:Fe have no center of symmetry. This indicates that, unlike two-dimensional solitons in (BTO) or (SBN), the 2D self-trapped beam in InP is not symmetrical in the direction parallel to the applied field. As a result, our experimental observations cannot be explained by the existing theory of photorefractive solitons or the existing theory for photorefractive focusing, defocusing, and deflection. Previous theories of photorefractive focusing, defocusing, deflection, and solitons include a single carrier and do not predict a resonant enhancement. A new model is clearly required, and several important clues now exist and will be discussed.

Fig. 1 shows the apparatus used for producing a dark notch in the laser beam. As seen in

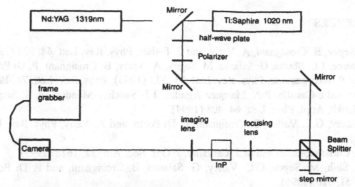

FIG.1. Experimental Apparatus

in the figure, a mirror is used to introduce a 180° phase shift between two halves of the beam. As a result of the phase shift, a dark notch develops as the beam propagates towards the crystal. The FWHM of the notch at the crystal entrance face is 12µm and 36µm at the exit face of the 1cm long crystal. This is shown in Fig. 2 where we plot the input waveform or the input intensity versus the horizontal size. In Fig.3 we plot the output waveform for various intensities. When a 6kV/cm field is applied across the crystal, the size of the output notch is reduced by an amount that depends on the incident intensity. For an incident intensity just below resonance, the notch has a FWHM of 12µm, indicating that the notch is trapped at 12µm.

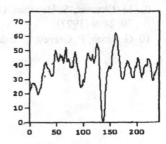

FIG.2. Input Waveform

For higher intensities the notch is greater than 12µm. These results are also shown in Fig.3.

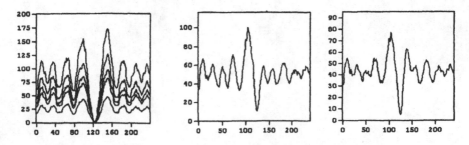

FIG. 3. Output intensity profile without field(left) and with field (center at an intensity below resonance and right at an intensity above resonance).

CONCLUSIONS

We have demonstrated dark soliton formation in InP:Fe.

REFERENCES

1. M. Segev, B. Crosignani, A. Yariv, and B. Fisher, Phys. Rev. Lett. **68**, 923 (1992).
2. G. Duree, J.L. Shultz, G. Salamo, M. Segev, A. Yariv, B. Crosignani, P. Di Porto, E. Sharp, and R.R. Neurgaonkar, Phys. Rev. Lett. **71**, 533 (1993); Phys. Rev. Lett. **74**, 1978 (1995).
3. M.D. Iturbe-Castillo, P.A. Marquez Aguilar, J.J. Sanchez-Mondragon, S. Stepanov, and V. Vysloukh, Appl. Phys. Lett. **64**, 408 (1994).
4. M. Segev, G.C. Valley, B. Crosignani, P. Di Porto, and A. Yariv, Phys. Rev. Lett. **74**, 1978 (1995).
5. D.N. Christodoulides and M.J. Carvalho, J. Opt. Soc. Am. **12**, 1628 (1995).
6. M.F. Shih, M. Segev, G.C. Valley, G. Salamo, B. Crosignani, and P. Di Porto, Electron. Lett. **31**, 826 (1995).
7. J.S. Aitchison, Opt. Lett. **15**, 471 (1990).
8. M. Chauvet, S. Hawkins, G. Salamo, M. Segev, D. Bliss and G. Bryant, Opt. Lett. **21**, 1333 (1996).
9. M. Chauvet, S. Hawkins, G. Salalmo, M. Segev, D. Bliss and G. Bryant, Appl. Phys. Lett. **70**, 2499 (1997).
10. G. Picoli, P. Gravey, C. Ozkul, and V. Vieux, J. Appl. Phys. **66**, 3798 (1989).

ZnGeP$_2$ AND ITS RELATION TO OTHER DEFECT
SEMICONDUCTORS

A.W. VERE, L.L. TAYLOR, P.C. SMITH, C.J.FLYNN, M.K.SAKER, AND J.JONES
Defence Evaluation and Research Agency (DERA), St Andrews Road, Malvern,
Worcestershire, WR14 3PS, UK   awvere@dera.gov.uk

ABSTRACT

The paper discusses progress in the development of ZnGeP$_2$ (ZGP) for optical parametric
oscillator (OPO) applications and draws parallels with other semiconductors with volatile
components, in which the presence of lattice defects gives rise to non-stoichiometry. In
particular, attention is drawn to the microprecipitation which accompanies deviation from
stoichiometry. In other materials this has been shown to result in spatial non-uniformity in the
density of point defects.

INTRODUCTION

A^{II}-B^{IV}-C$^V_2$ compounds, especially ZnGeP$_2$ and CdGeAs$_2$, have been studied for over
twenty years [1-6]. Until recently however, synthesis of the compounds and single-crystal
growth have been difficult. Problems have included high phosphorus or arsenic pressure at the
melting point (1027°C for ZnGeP$_2$; 660°C for CdGeAs$_2$), mosaic cracking during cooling
from the melting point and the presence of optical absorption centres in the OPO pump band.

Optical absorption is now the major choke-point in the development of these materials for
practical applications such as OPOs and second-harmonic generators. Although substantial
improvement has been obtained in the last few years [7][8], the detailed nature of the
absorption mechanism remains unclear. There is general agreement that native defects play a
controlling role, but vacancy species such as V$_{Zn}$ [9] and the antisite defect Zn$_{Ge}$ [10] have both
been suggested as the dominant centre.

In other semiconductor materials such as GaAs, ZnSe, CdTe and CdHgTe, early research
based on photoluminesecence, ESR and ENDOR, also showed the presence of high
concentrations of native defects. Subsequent studies however, revealed the controlling
influence of relatively low concentrations of residual deep-level impurities [11][12].
Moreover, the spatial distribution and concentration of other native and impurity defects within
the crystal was found to vary according to the dislocation and elastic strain distributions in the
ingot and the thermal history of the material [13 -16]. In this paper we report progress in the
single-crystal growth of ZGP and review the possible nature of the optical absorption and
scattering centres in the light of our observations on other semiconductor materials.

Mat. Res. Soc. Symp. Proc. Vol. 484 © 1998 Materials Research Society

CRYSTAL GROWTH

Synthesis of ZGP

Early attempts to synthesise ZGP by heating the constituent elements to the melting point inevitably failed, due to premature reaction of zinc and phosphorus to form zinc phosphides,

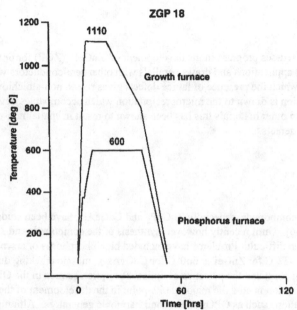

Fig. 1. Thermal cycle used in the synthesis of MCP Wafertechnology
polycrystalline ZGP by horizontal vapour transport

which formed a solid skin over part of the melt, entrapping free phosphorus and resulting in catastrophic explosions. 'Two-temperature' processes were the developed [17], in which phosphorus is pre-reacted with zinc in a low temperature zone operated at 400-600°C or transported into a zinc/germanium alloy in a high temperature (1000-1100°C) reaction zone. In the process developed by Smith and Grant [18], the temperature/time profile shown in Figure 1 is used to produce polycrystalline ZGP with a columnar grain structure with grain diameters in the 2-5mm range. Phosphorus transport processes such as this one and that used by and Bliss et al [19] are capable of producing large (700-1000gm) charges. As shown in Table 1, this can result in the minimisation of contamination by reaction vessel constituents. Other researchers prefer to synthesise smaller charges (50-200gms) aimed at achieving closer control of stoichiometry and compositional uniformity in the ingot.

TABLE 1 Glow Discharge Mass Spectrometric Analysis of MCP Wafertechnology
ZGP polycrystalline starting material of nominally stoichiometric
composition .
(A. Mykytiuk and P Semeniuk, NRC Institute for Environmental Chemistry)

GLOW DISCHARGE MASS SPECTROMETRIC REPORT - ppb (atomic)			
B	20	Cr	2
C	100	Mn	200
N	20	Fe	280
O	1500	Ni	1
Na	20	Zn	Major
Mg	2	Ge	Major
Al	16	Zr	8
Si	10	Mo	1
P	Major	In	10
Ti	2	Sn	20
V	0.7		

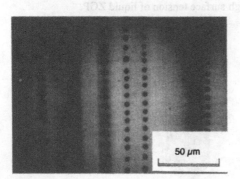

Fig. 2. Infrared transmission micrograph of 'Lineage
structure' formation in nominally stoichiometric ZGP

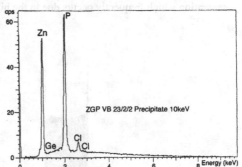

Fig. 3. EDAX analysis of the precipitates shown
in figure 2.

Stoichiometry is governed by the choice of cold zone temperature and the composition of the
starting charge. Nominally stoichoimetric charges, with excess phosphorus added to
compensate for losses to the vapour space, yield ingots slightly deficient in phosphorus and
zinc. Analysis of one starting charge [20] gave a composition of $Zn_{0.99} Ge_{1.01} P_{1.98}$. Despite
this, the presence of a precipitate-decorated lineage structure (Figure 2) containing Zn_xP_y
second-phase (figure 3), in the last-to-freeze parts of ingots grown from this material, indicates
that solidification occurred on the zinc/phosphorus rich side of the phase field.

Single-crystal Growth

Although attempts have been made to produce large single-crystals by liquid encapsulated Czochralski growth, using a B_2O_3 liquid seal to suppress loss of zinc and phosphorus from the melt, the high thermal strain resulting from the need to maintain a steep thermal gradient to control crystal growth($\Delta T > 15°C\ cm^{-1}$), inevitably led to mosaic cracking of the materials during cooling from the melt [21]. For this reason, Bridgman or Gradient Freeze (GF) growth techniques have emerged as the preferred route to large crack-free single crystals. In horizontal Bridgman (HB) or GF techniques, transparent furnace technology [22] has led to significant progress in maintaining crystal growth control under the very low thermal gradient conditions ($<5°C cm^{-1}$) required to suppress vapour transport. In vertical Bridgman (VB) processes, a steeper gradient (~10-$15°C cm^{-1}$) can be imposed, without significant vapour transport, by careful control of the temperature of the vapour-space above the melt. In this process the solidifying melt is totally enclosed by the crucible material and cooling stresses must be minimised using a flexible walled crucible material (usually boron-nitride) and careful choice seed orientation to ensure minimum resolved shear stress on the {112} cleavage plane [23]. In both horizontal and vertical configurations, the very shallow temperature gradient leads to significant difficulty in control of the liquid/solid interface. In horizontal systems, nucleation on the crucible wall ahead of the growing interface must be removed by local heating and good contact must be maintained between melt and seed, usually by tilting the charge to prevent the separation which frequently occurs due to the high surface tension of liquid ZGP.

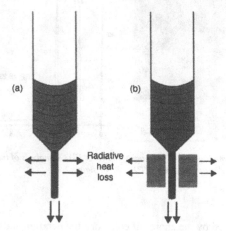

Fig. 4. Comparison of heat flow and melt isotherms in a) conventional Bridgman ampoule and b) seed-holder surrounded by refractory 'bobbin'.

We have recently found that single-crystal growth in the vertical Bridgman system is improved by the insertion of a ceramic jacket (bobbin) around the seed holder. (Figure 4) Finite element modelling [24] of the changes in thermal field induced by the bobbin show that, whilst these are small in magnitude, they result in the stabilisation of a convex liquid/solid interface. The effect of insertion of the ceramic bobbin on the axial temperature profile in the vertical

Bridgman furnace is shown in Figure 5. Using the combination of correct seed orientation, flexible-walled crucible and the insulated seed holder, it is now possible to grow crack-free single crystals 20mm in diameter and up to 70mm long.

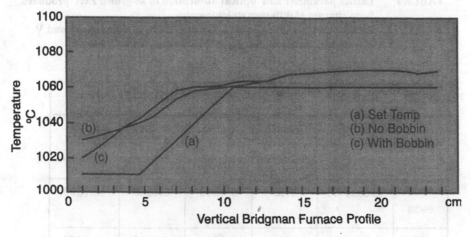

Fig. 5. Temperature profile in the vertical Bridgman process
(a) set temperature on furnace wall
(b) on centre line of ampoule, in the absence of the ceramic bobbin
(c) as position (b), but with ceramic bobbin

DEFECT STRUCTURE

Large volume, crack-free single crystals provide a necessary starting point for the study of optical loss in the material. Although this is predominantly attributable to absorption at midgap levels associated with native defects, adverse growth conditions can result in spatial non-uniformity on a macroscopic scale. This takes the form of growth striations, dislocation arrays and second phase precipitation.

Figure 6 shows an optical micrograph taken from the shoulder region of a vertical Bridgman ingot [24]. The original melt/seed interface is delineated by small voids associated with gas entrapment. Oscillatory growth then occurs, producing growth striae with an initial period of approximately 25µm, gradually extending to 120-150µm over the first 3-5mm of growth. After the first 5mm, striae are no longer detectable and the material transmits uniformly over the whole cross-section of the ingot. Despite this apparent uniformity, spot measurements of lattice parameter at different locations across the boule diameter show significant variations. Such variations have hampered attempts to monitor stoichiometry by studying systematic changes in lattice parameter as a function of starting-charge composition. Table 2 shows the results of a collaborative exercise in which optical absorption and lattice parameter of vertical-Bridgman grown ZGP, produced at IAO Tomsk, were measured at DERA Malvern. Three starting-charge compositions were used to give ingots which were nominally stoichiometric, Zn-rich and Ge-rich. The lattice parameter results obtained using the Fewster method [25] compare well with those of the JCPDS data base, the c-axis being slightly shorter than the determinations made by Ray et al [26].

TABLE 2   Lattice parameter and optical absorption in as-grown ZGP produced from charges of differing stoichiometry.
(Single -crystals supplied by A Gribenyukov, G Verozubova and V Korotkova, Institute of Atmospheric Optics, Tomsk, Russia

Sample	Composition	Pre-annealed Absorbance and Lattice parameter [1]			
	(melt)	α (cm^{-1}) $(\lambda=2.05\mu m)$	a (100) $\pm$ 0.0005 Å	b (010) $\pm$ 0.0005 Å	c (001) $\pm$ 0.0003 Å
IAO 89/3/2	Excess Ge	0.45	5.4659	5.4660	10.7090
IAO 89/3/4	Excess Ge	0.43	5.4661	5.4657	10.7079
IAO 91/2/2	Stoichio-metric	0.33	5.4654	5.4661	10.7091
IAO 91/2/4	Stoichio-metric	0.46	5.4674	5.4661	10.7092
IAO 93/3/2	Excess Zn	0.62	5.4668	5.4658	10.7088
IAO 93/3/4	Excess Zn	0.51	5.4658	5.4659	10.7091

(1)   X-ray source Cu k_α (λ = 1.54056Å) a (400); θ_B = 34.303° ; c(008); θ_B = 35.106°

        Ray *et al* Phys Stat Sol <u>35</u> (1969):197   a = 5.466 $\pm$ 0.001Å c = 10.722 $\pm$ 0.002Å

        JCPDS Data base (33-1471)   a = 5.467(2)Å   c = 10.715Å

Fig.6. IR transmission micrograph of the first-to-freeze region of a vertical Bridgman ingot. The original, slightly concave seed/melt interface, decorated by small gas bubbles, is shown at the extreme bottom. Growth striae of increasing periodicity occur during the first few millimeters of growth.

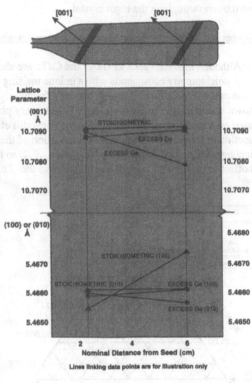

Fig. 7. Variation of a) (100) and b) {100} lattice parameters as a function of measurement position and stoichiometry in ZGP ingots grown from material of differing starting-charge compositions.

Figure 7 shows the location of the (001) axis slices used in the determinations and the (001) and (100) lattice parameters obtained. For the (001) lattice parameter, both positions in the boule have similar lattice parameters for the excess-zinc and stoichiometric charges but the Ge-rich ingot shows a change of 1 part in 10^3 between the two positions. In the (100) measurement, the stoichiometric material shows the maximum change (1 part in 2×10^3). More

501

data are required before it is possible to determine whether the variation is attributable to internal strain, micro-precipitation or variations in native defect or impurity species concentrations. The data do however indicate the need for caution when attributing the results of electrical, photoluminescence or ESR measurements to a single defect species of uniform distribution throughout the ingot crystal.

Comparison with other defect semiconductor materials

Although materials such as GaAs and CdTe are shown on elementary phase diagrams as fixed stoichiometry compounds with a unique melting point, closer inspection reveals a narrow phase range (typically <0.5 atomic percent), frequently exhibiting a retrograde solidus as shown schematically in the inset to the ZGP ternary phase diagram of figure 8. This diagram is based on a thermodynamic calculation by Voevodin et al. [27] of the phase field positions as a function of temperature and shows the deviation of the maximum melting point towards a zinc/germanium rich composition $Zn_{1.002}Ge_{1.008}P_{1.99}$ so that the congruent melting composition is offset to the Ge-rich side of stoichiometry in the ZnP_2 - Ge pseudo-binary section.

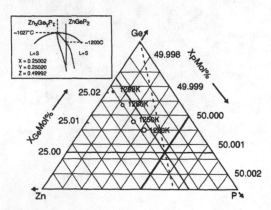

Fig. 8. Enlargement of the area of the Zn-Ge-P ternary phase diagram around the stoichiometric ZGP composition. The circles show isothermal sections through the single-phase region at different temperatures (V.G. Voevodin and O.V.Voevodina [27]). The inset shows schematically the corresponding T-X section on the zinc phosphide / zinc germanium pseudo-binary diagram.

Equivalent calculations for GaAs [28], indicate that the congruent composition lies to the As-rich side. In CdTe [29] the calculated position lies towards the Te-rich side of stoichiometry. The retrograde solidus on the As-rich side of the phase-field in GaAs means that growth from a nominally stoichiometric melt will result in precipitation of arsenic [13] during cooling. Similarly, in CdTe, tellurium precipitation occurs on cooling. A similar effect has also been observed in the precipitation of In in InP [30]. In CdHgTe, Williams and Vere [15] have shown that the morphology and distribution of precipitates depends on the dislocation distribution and density in the material, which in turn depends on the thermal history of the growth process and any post-growth annealing cycle. Figure 9 is an optical micrograph of a {112} section of ZGP after etching for 5 mins in 3HCl : 1HNO$_3$ at 25°C. It shows Zn_xP_y precipitates of 1μm to 10μm side-length. They frequently occur in pairs, presumably located on opposite sides of a dislocation loop, as has been observed previously in GaAs and CdTe. The precipitate edges lie along {001} directions. Work is now in progress to study their distribution and determine the effect of melt stoichiometry and cooling rate in order to minimise precipitation in material designed for OPO devices.

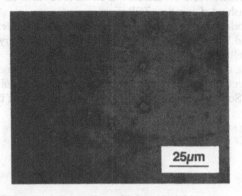

Fig. 9. Optical micrograph of ZGP after etching in 3HCl:1HNO3 for 5 minutes at 25C. The tetragonal features have been identified as Zn/P precipitates using SEM/EDAX analysis.

CONCLUSIONS

1. ZGP can be successfully synthesised by transport of phosphorus into an equimolar alloy of zinc and germanium. Careful control of the starting materials yields total impurity concentrations (excluding oxygen) of less than 1ppm in 750-1000 gm ingots. The resultant material is, however, germanium rich and this deficiency, together with provision for loss of zinc and phosphorus vapour space, must be compensated for in calculating the required charge-weight for single-phase, single-crystal growth.

2. In the VB process, single-crystal growth is enhanced by partial insulation of the seed holder to prevent radial heat loss which results in a concave liquid-solid interface.

3. The requirement to operate in low temperature gradients to suppress vapour transport of the volatile elements, zinc and germanium, can result in growth striations.

4. In common with other defect semiconductors such as GaAs, InP, CdTe and CdHgTe, ZGP exhibits a retrograde solidus on the ZnP_2 side of the single-phase field in the pseudo-binary ZnP_2 - Ge phase diagram. Initial observations indicate that the precipitation is not spatially uniform in concentration. Similar observations have been made for other materials such as GaAs and CdTe and is attributable to dislocations acting as preferred sites for nucleation of the second phase.

ACKNOWLEDGEMENTS

It is a pleasure to acknowledge the contribution made by other researchers in the preparation of this paper. Firstly we would like to thank Chris Reeves for the SEM analysis of the Zn_xP_y precipitates, Andrew Keir for discussions on the X-ray analysis and Audrey Parish for preparation and polishing of countless ZGP slices.

From other laboratories, we must acknowledge the part played by Alexander Gribenyukov, Galina Verozubova and Valentina Korotkova in the provision of ZGP crystals for analysis and the finite-element modelling of the thermal fields in Bridgman ingots.

Finally but not least, our thanks to Peter Schunemann (Lockheed Sanders), Ilya Zwieback (Inrad Co) and Mel Ohmer of Wright Patterson-Materials Labs for on-going discussion and stimulation and to Jim Solomon, also of ML/WPAFB for the analysis of the ZGP starting charge.

REFERENCES

1. K. Masumoto, S. Isomura and W. Goto, J. Phys. Chem. Sol., **27**, p1939 (1966).

2. E. Beuhler, J. Wernick and T. Wiley, J. Electron. Mater. **2** (3) p.445 (1973)

3. K. L. Vodopyamov, V. G. Voevodin, A. I. Gribenyukov and L. A. Kulevskii, Sov. J. Quantum Electron., **17**, p1159 (1987).

4. R. S. Feigelson and R. K. Route, J. Cryst. Growth, **49**, p261 (1980).

5. C. C. Xing and K. J. Bachmann, J. Cryst. Growth, **147**, p35 (1995).

6. P. G. Schunemann and T. M. Pollak, OSA Proceedings on Advanced Solid State Lasers, **10**, (Eds. G. Dubé and L. Chase), Hilton Head South Carolina, March 18-20 (1991).

7. G. A. Verozubova, A. I. Gribenyukov, V. V. Korotkova and M. P. Ruzaikin, Proc. Internat Workshop on Stoichiometry in Compound Semiconductors, Bad Suhl, Germany 1995 (unpublished)

8. P. G. Schunemann, P. A. Budni, L. Pomeranz, M.G. Knights, T.M. Pollak and E.P. Chicklis in Advanced Solid State Lasers, edited by C.R. Pollock and W.R. Bosenberg OSA Top. Proc. **10** pp253-255 (1997)

9. C. E. Halliburton, G. J. Edwards, P. P. Scripsick, M. P. Rakowsky, P. G. Schunemann and T. M. Pollak, Appl. Phys. Lett., **66** (20), p2670 (1995).

10. N. Dietz, I. Tsveybak, W. Ruderman, G. Wood and K. J. Bachmann, Appl. Phys. Lett., **65** (22), p2759 (1994).

12. NuoFu Chen, Hongjia He, Yutian Wang and Lanying Lin, J. Cryst. Growth, **173**, p325 (1997).

13. A. G. Cullis, P. D. Augustus, D. J. Stirland, J. Appl. Phys., **51**, p2556 (1980).

14. J. L. Pautrat, N. Magnea and J. P. Faurie, J. Appl. Phys., **53** (12), p8668 (1982).

15. D. J. Williams and A. W. Vere, J. Vac. Sci. Technol., **A4** (4), p2184 (1986).

16. A. W. Vere, Crystal Growth, Principles and Progress, Plenum Press, London & New York (1987) pp29-50

17. A. L. Gentile and D. M. Stafsudd, Mater. Res. Bull., **9**, p105 (1974).

18. M. Smith and I. R. Grant, MCP Wafer Technology Final Report UK Government DRA Contract MAL 16/2310, July 1986.

19. D. F. Bliss, M. Harris, J. Horrigan, M. M. Higgins, A. F. Armington and J. A. Adamski, J. Cryst. Growth, **137**, p145 (1994).

20. A. Mykytiuk and P. Semeniuk, NRC Institute for Environmental Chemistry, Canada (unpublished).

21. H. M. Hobgood, T. Henningsen, R. N. Thomas, R. H. Hopkin, M. C. Ohmer, W. C. Mitchel. D. W. Fischer, S. M. Hegde and F. K. Hopkins, J. Appl. Phys., **73** (8), p4030 (1993).

22. Peter G. Schunemann and Thomas M. Pollak, U.S. Patent No. 5 611 856 (18 March 1997)

23. A. W. Vere, L. L. Taylor, P. C. Smith, C. J. Flynn, M. K. Saker and J. A. C. Terry, IOP proceedings of the 11th International Conference on Ternary and Multinary Components, Salford, UK (1997) - In press.

24. A. Gribenyukov, G. Verozubova and V. Korotkova - private communication.

25. P. F. Fewster and N. L. Andrew, J. Appl. Cryst., **28**, p 451 (1995).

26. B. Ray, A. J. Payne and G. F. Burrell, Phys. Stat. Sol., **35**, p197 (1969).

27. V.G. Voevodin and O.V. Voevodina, J. Russ. Phys. **36** (10) p924 (1993)

28. D. T. J. Hurle, Materials Science Forum, **196-201**, p179 (1995).

29. D. de Nobel, Philips Research Rev. **14** (4), p361 (1959).

30. G. T. Brown, B. Cockayne and W. R. MacEwan, J. Electron. Mater., **12**, p93 (1983).

GROWTH OF NLO CHALCOPYRITE MATERIALS BY OMVPE

M.L. TIMMONS*, K.J. BACHMANN**
*Research Triangle Institute, Research Triangle Park, NC 27709, mlt@es.rti.org
**Material Science and Engineering, North Carolina State University, Raleigh, NC 27695

ABSTRACT

This paper describes the application of organometallic vapor phase epitaxy to the growth of $II-IV-V_2$ chalcopyrite materials that have high figures of merit for nonlinear optical (NLO) applications. $ZnGeAs_2$, although not a particularly interesting NLO material, is used as a model for the growth of $ZnGeP_2$, which is. Both compounds, as well as others, have been successfully grown by vapor phase on III-V substrates that provide a close lattice match. Doping studies using Group III and VI elements have been undertaken to control the p-type conductivity found in both compounds. Except for the possible case of indium, the results of these experiments are less than encouraging. Minority-carrier lifetimes of 150 ns have been measured in $ZnGeAs_2$.

The results of this work are used to make projections about the growth of $CdGeAs_2$. $CdGeAs_2$ is promising for the next generation of NLO mid-infrared materials if an absorption band that occurs at about 5 μm can be reduced. The growth projections suggest that this compound will be difficult to grow epitaxially and has no III-V substrate that provides a close lattice match. Mixing $CdGeAs_2$ with other $II-IV-V_2$ materials may offer solutions to the substrate problem. The defect properties of $CdGeAs_2$ have not, to our knowledge, been studied.

INTRODUCTION

As late as 1960 little was known about nonlinear properties of matter, partly because of the absence of intense, coherent light sources. In 1961, an experiment in which the wavelength of a ruby laser was doubled by a quartz crystal provided the first confirmation of nonlinear theories. Since then, applications using nonlinear effects have multiplied.

Examination of the nonlinear optical (NLO) properties of chalcopyrite crystals began in 1972. A summary of early crystal growth results and material properties of both the $I-III-VI_2$ and $II-IV-V_2$, chalcopyrite analogs of II-VI and III-V materials, respectively, as well as a chapter on nonlinear effects can be found in Shay and Wernick [1]. Preparation of these crystals generally used bulk growth or liquid phase techniques. Equilibrium thermodynamics control both of these techniques. Non-equilibrium approaches, molecular beam epitaxy (MBE) or organometallic vapor phase epitaxy (OMVPE), for example, offer advantages for growing crystals that might otherwise be unstable or metastable and for growing thin-layered structures where composition and thickness control are essential for device performance.

Crystals must satisfy four criteria to be useful in nonlinear optics[2]:

1. They must have adequate and uniform nonlinearity.
2. They must be adequately transparent.
3. They must have adequate, uniform birefringence for phasematching.
4. Crystal faces must be relatively immune to damage by intense optical radiation.

Major shortcomings of chalcopyrites in NLO are high absorption losses, nonuniform crystallinity, and relatively small-sized crystals. We believe that OMVPE may address these shortcomings. One advantage of epitaxial growth by OMVPE is that material properties can be very uniform. We expect the uniformity of nonlinearity and birefringence to be excellent. Optical losses (reduction in transparency) result from impurities and defects found in crystals. Again, epitaxial growth techniques such as OMVPE are capable of producing material that contains fewer impurities than host substrates. Non-equilibrium control of the growth environment may provide a means to alter the densities of native point defects that are difficult to control in bulk-grown crystals. OMVPE allows rapid, flexible changes of gas-phase reactant concentrations and resulting solid compositions.

We have grown several of the II-IV-V_2 compounds for applications other than nonlinear optics as well as $ZnGeP_2$, which is of keen interest for NLO. A summary of the compounds grown and their applications is given in Table I.

$ZnSiAs_2$ provides an example that illustrates the advantages of OMVPE growth of these materials. In earlier work using traditional vapor phase epitaxy (open tube with elemental zinc source), $K\alpha_1$ and $K\alpha_2$ lines of copper x-rays failed to resolve for $ZnSiAs_2$ growth on Ge. They do resolve for OMVPE growth (using dimethylzinc) on Ge and GaAs substrates, suggesting improved crystalline quality[3]. Other workers using OMVPE to grow $ZnSiAs_2$ have subsequently reported similar resolution of the $K\alpha$ radiation[4]. Bulk $ZnGeP_2$, grown at North Carolina State University, illustrates a second advantage of OMVPE; Figure 1 shows mass spectral analyses of the bulk-grown material and an epitaxial sample of $ZnGeP_2$ grown on GaP. The OMVPE sample contains fewer impurities, and we would expect this sample to have better optical properties.

Two of the alloys listed in Table I, $Zn(Ge_xSi_{1-x})As_2$ and $ZnGe(As_xP_{1-x})_2$, demonstrate the flexibility of mixed-crystal vapor phase growth. By mixing Ge and Si ($x \approx 0.72$), a perfect lattice match with GaAs substrates was obtained, and the band gap of $ZnGeAs_2$, nominally 1.15 eV, increased to 1.225 eV in a $ZnGe(As_{0.9}P_{0.1})_2$ alloy that also lattice matches GaAs. While we were fascinated by the possibility of forming the alloy $Zn(Ge_xSi_{1-x})(As_yP_{1-y})_2$, we have not attempted this growth. This compound is equivalent to a III-V quaternary mixed crystal.

To summarize, OMVPE has the potential to make contributions to growth of the II-IV-V_2 chalcopyrite compounds that are of interest for NLO applications and to the understanding of point-defect chemistry that controls conductivity and unwanted infrared absorption. However, these are not easy compounds to grow, particularly when compared to a material such as GaAs.

Table I.   Chalcopyrite Materials Grown by OMVPE at Research Triangle Institute.

Application	Material Grown	Substrate Used
Photovoltaics	$ZnSiAs_2$	GaAs, Ge
	$ZnGeAs_2$	GaAs, Ge
Photocathodes	$CdSiAs_2$	InP
	$ZnGeAs_2$	GaAs
	$ZnGeP_2$	GaP
	$Zn(Ge_xSi_{1-x})As_2$	GaAs
	$ZnGe(As_xP_{1-x})_2$	GaAs
Nonlinear Optics	$ZnGeP_2$	GaP

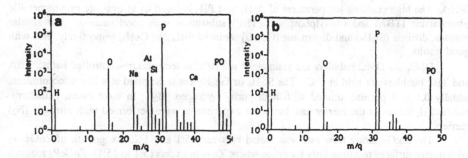

Figure 1.   Mass spectral analysis of (a) bulk-grown and (b) epitaxial ZnGeAs₂. The epitaxial
sample was grown on a GaP substrate by OMVPE.

Substantial technical obstacles must be overcome if the materials are to be useful in device applications. Most vexing and difficult to solve may be the heavy p-type conductivity that most exhibit because of native defects. The nature of these defects must be studied to devise ways to minimize their occurrence. The properties important to NLO, birefringence and nonlinearity, of the OMVPE-grown material should exhibit good uniformity, and the transparency may be superior to other growth techniques. Finally, OMVPE offers growth over large areas. For example, there are production OMVPE systems now that can process more than 1000 cm² of substrate material per growth run. Whether such large systems can be employed for the chalcopyrites remains to be proven although, on a smaller scale, we have grown high quality $ZnGeAs_2$ on single, 2-inch-diameter GaAs substrates and 4.5×4.5 cm² Ge.

The paper will address growth of $ZnGeP_2$ and $CdGeAs_2$ since these compounds are of the most interest for NLO of the II-IV-V₂ materials. $CdGeAs_2$, in fact, has the highest figure of merit of any compound semiconductor except for Te[2]. In the sections that follow, we use a rather extensive experience with $ZnGeAs_2$ and somewhat less with $ZnGeP_2$ and the other compounds listed in Table I to show general trends in OMVPE growth of these materials including extrapolations to $CdGeAs_2$, which has not grown been grown epitaxially, to our knowledge, by either OMVPE or MBE.

EXPERIMENTAL

Growth Details

The essential characteristics for OMVPE are that we must deal with elements that have widely different vapor pressures and that source gases have substantially different thermal decomposition rates. Zn, Cd, As, and P have high vapor pressures compared to Ge and Si, and the thermal decomposition of dimethylzinc (DMZ) or dimethylcadmium (DMC) begins at markedly lower temperatures than silane (SiH_4), germane (GeH_4), arsine (AsH_3), or phosphine (PH_3). The high cracking temperatures of AsH_3 and PH_3 induced us to evaluate organometallic t-butylarsine (TBA) and t-butylphosphine (TBP) substitutes with good success. For similar reasons, disilane (Si_2H_6) and digermane (Ge_2H_6) replaced SiH_4 and GeH_4, respectively, also with good results.

DMZ and DMC bubblers are maintained at -10°C in temperature-controlled baths. TBA and TBP bubblers are held at 0°C. The Si_2H_6 or Ge_2H_6 gas is supplied at a low concentration, usually 0.1 to 1 percent, diluted with high purity hydrogen (H_2). In most cases, palladium-purified H_2 serves as the carrier gas, but some experiments were performed with nitrogen (N_2) carriers.

The Zn:Ge molar flow rate was varied between 15:1 and 1:20 for growth of $ZnGeAs_2$ with mirror surfaces resulting only for ratios where Zn is in excess (5:1 to 15:1) $ZnGeP_2$ requires almost the same ratios for growth at 545°C to produce mirror surfaces. For both materials, the (Zn+Ge):As or (Zn+Ge):P ratios are about 1:25 or 1:50, respectively.

Group III elements have been used in an attempt to produce n-type conductivity, or at least to control the p-type conductivity. Group III sources are trimethylgallium (TMG), from a bubbler held at -10°C, trimethylaluminum (TMA), held at 20°C, or trimethylindium (TMI), held at 30°C. Boron from a dilute diborane (B_2H_6) source was also studied. Group III elements substitute on the cation sublattice in the chalcopyrite crystal, and if these elements selectively occupy Zn sites, n-type conductivity should result. If there is Ge-site preference or no site preference, then the elements would likely behave as acceptors or amphoterically, respectively. Gaseous sources of hydrogen selenide (H_2Se) and diethyltelluride (DET), both diluted to about 50 ppm with H_2, provide Group VI elements, Se and Te, that substitute on the anion sublattice. This substitution should produce n-type conductivity if the Se or Te concentration exceeds the defect concentration that causes p-type behavior.

Growth pressures ranged from near-atmospheric to about 100 Torr. Attempts to grow at pressures less than about 250 Torr make it difficult to keep Zn and Cd on the surface long enough to bond prior to desorption. Most of the work with $ZnGeAs_2$ was done at 600-700 Torr, and $ZnGeP_2$ was grown at atmospheric pressure.

Growth temperatures are controlled, in large measure, by the nature of the source gases that contribute to the cation sublattice. If temperatures too low, insufficient amounts of the Si- or Ge-containing gases decompose, and heavy deposits of a Zn- or Cd-rich phase (Zn_3As_2, for example) result. If too high, the Zn or Cd desorbs from the surface before bonding, and Ge- or Si-rich phases grow. $ZnGeAs_2$ is usually grown at 550-600°C, but this depends on the configuration of the reactor. $ZnGeP_2$ is grown at about 550°C.

Several different reactor configurations have been examined. These range from horizontal to a stagnation-flow vertical geometry. While $ZnGeAs_2$ was grown with all reactor geometries, most successful were vertical geometries.

Substrate Details

Since OMVPE is an epitaxial process that must produce exceedingly uniform material properties and excellent optical properties to make successful NLO devices, the proper choice of substrate is essential. The lattice mismatch between $ZnGeAs_2$ and GaAs or Ge is about 0.34 percent with the epilayer in compression. The lattice mismatch between $ZnGeP_2$ and GaP is less than 0.3 percent, again with the epilayer in compression. In both of these cases, high quality epitaxy has resulted. The lattice parameter of $CdGeAs_2$ is 5.943 Å[1]. The III-V material with the closest lattice match to this is InP (5.8688 Å), yielding a mismatch of about 1.3 percent (compressive). This amount of lattice mismatch is too large to expect the kind of high quality crystallinity needed for NLO devices. None of the II-VI substrates commonly available improve the lattice mismatch. So, this will remain a problem area that must be addressed.

Substrate preparation depends on the substrate being used. For GaAs, epi-ready wafers are available and require little preparation prior to insertion into the growth apparatus. When preparation is required, ultrasonic solvent cleaning followed by brief etching to remove surface contamination and damage is satisfactory. We generally use a 5 H_2SO_4:H_2O_2:H_2O or a 2 NH_4OH:H_2O_2:10 H_2O solution for GaAs surface etching. GaP substrates are etched with dilute Br:CH_3OH solutions (0.5%) after solvent cleaning. Ge is etched with NH_4OH:H_2O_2: 150 H_2O, and InP is etched with 5 H_2SO_4:H_2O_2:H_2O. These etches are brief, usually 1-3 minutes, with the etchant maintained at room temperature.

Characterization Details

X-ray diffraction is the most important tool that we use to characterize the chalcopyrites. Either single-crystal 2Θ scans or double-crystal rocking curves provide data about lattice parameters. Diffraction is useful to indicate non-stoichiometric amounts Ge in $ZnGeAs_2$ or $ZnGeP_2$ by using the split between the peak from the epitaxial layers and substrates as a measure. The full width at half maximum (FWHM) of rocking curves gives a qualitative measure of the crystalline uniformity and quality of the epilayer.

Strong emissions by photoluminescence (PL) give a qualitative measurement of the optical quality of $ZnGeAs_2$, and radio-frequency photoconductivity decay (RFPCD) provides a more direct measurement of minority-carrier lifetime with results that are surprising for $ZnGeAs_2$. Minority-carrier lifetime is an excellent probe of material quality since this lifetime is extremely sensitive to both chemical and structural defects.

Electron microprobe (EMP) is useful for determining chemical composition, and secondary ion mass spectroscopy (SIMS) provides profiles with adequate depth resolution and composition data.

Measurement of carrier concentration is a major concern because of the presence of native defects, thought to be mainly Zn or Cd vacancies or antisites, that produce heavy p-type conductivity. To study these defects and devise schemes to eliminate them, reliable carrier concentration data must be available. Hall-effect measurement is the standard means of determining carrier concentration (and mobility), but we have some reservation about this technique, at least for $ZnGeAs_2$ and $ZnGeP_2$. Our experience in attempting to dope $ZnGeAs_2$ with Group III and VI elements and getting essentially no variation in results with Hall-effect measurements led us to speculate that Zn diffusion into semi-insulating GaAs forms a thin doped surface region that has a carrier concentration of 10^{19} cm^{-3}. This observation is supported by results from an electrochemical etcher/profiler and SIMS. With this profiler, we see substantially different carrier concentrations that seem to reflect changes that were made with different experiments. The electrochemical profiler forms a low-leakage Schottky barrier between the electrolyte and the semiconductor surface. A capacitance-voltage measurement then determines the carrier concentration. The sample is etched and the process is repeated. Unfortunately, the Schottky barrier is not always good, and even with this technique there is still some uncertainty.

With electrochemical profiling, n-type conductivity has actually been measured for $ZnGeAs_2$, but annealing experiments lead us to conclude that this behavior results from interstitial Zn, which behaves as a donor and is highly mobile. Interstitial Zn would be consistent with the excess Zn used in the growth.

We have not had access to electron-nuclear double resonance (ENDOR) that has recently been used to verify that the dominant point defect in bulk $ZnGeP_2$ is the Zn vacancy[5]. It would be interesting to compare this kind of sample with one grown in a Zn-rich environment, such as the epitaxial materials described in this work, since the number of Zn vacancies should diminish and the number of Zn_{Ge} antisites should increase.

RESULTS

Growth Results

The characteristics of the sources used for the OMVPE growth of chalcopyrites produce narrow growth "windows". An example is given in Table II for $ZnGeAs_2$. With the sources and conditions used for the growths shown in Table II, single-phase $ZnGeAs_2$ results only with temperatures from about 560°C to 580°C[6]. GaAs, for comparison, can be grown from about 500°C to about 800°C. We see almost identical behavior for $ZnSiAs_2$ (at higher temperatures), as have other workers[4], and for $ZnGeP_2$ (single-phase growth range of 530°C to 570°C). If different source gases are used or different growth conditions, the window shifts. For example, switching to TBA and Ge_2H_6 sources allows $ZnGeAs_2$ growth at temperature less than 500°C

If we extrapolate this window concept to Cd-bearing compounds such as $CdGeAs_2$, potential difficulties arise. The vapor pressure of Cd at 550°C is roughly an order of magnitude greater than that of Zn[7]. We believe that this will make the $CdGeAs_2$ growth window even more restricted. This belief is supported by our experience with $CdSiAs_2$. While $ZnSiAs_2$ is relatively easy to grow, $CdSiAs_2$ is extremely difficult, and our ability to grow this compound reproducibly has been poor to date.

The crystalline quality of $ZnGeAs_2$ (and $ZnGeP_2$) is quite good. The double-crystal x-ray rocking curve for $ZnGeAs_2$, shown in Figure 2, has a FWHM almost as narrow as the GaAs substrate, using the (004) and (008) reflections of GaAs and $ZnGeAs_2$, respectively. The spacing between the two peaks is important because it shows if excess Ge has incorporated into the crystal. A spacing of about 1400 to 1420 s is the maximum that we have observed and suggests stoichiometry. Ge is soluble in both $ZnGeAs_2$ and $ZnGeP_2$, and excess Ge in the chalcopyrite reduces the splitting.

Table II. Phase Relationships for Zn-Ge-As Materials.

Growth Temperature (C°)	Phase Grown
500	Zn_3As_2
550	$Zn_3As_2 + ZnGeAs_2$
575	$ZnGeAs_2$
600	Ge + $ZnGeAs_2$ (trace)

Notes: Zn:Ge = 5:1;   (Zn+Ge):As = 1:25;   Pressure = 760 Torr
Sources: DMZ, GeH_4, AsH_3

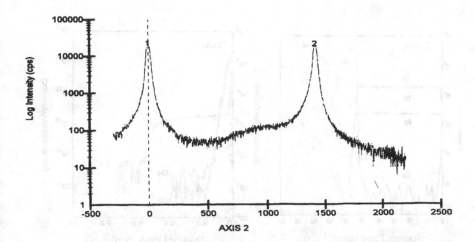

Figure 2.   Double-crystal rocking curve of 0.5-μm-thick ZnGeAs₂ layer grown on a GaAs
              substrate. Splitting between peaks is 1418 s (0.394°), indicative of stoichiometric
              ZnGeAs₂. FWHM of peaks are comparable. Substrate diffraction is marked as 1.

Diffusion into substrates is a problem, but one that depends upon the Zn:Ge ratio. This point is illustrated in Figure 3. In Fig. 3a, a SIMS profile of a ZnGeAs₂ layer is shown when the Zn:Ge ratio during growth was 5:1. There does not appear to be diffusion into either the GaAs or the epilayer. However, when the Zn:Ge ratio increases to 15:1, producing a substantial concentration of mobile, interstitial Zn, a SIMS profile of the GaAs, shown in Fig. 3b with the ZnGeAs₂ layer removed by selective etching, yields a 0.5-μm-thick layer doped with Zn above 10^{19} cm^{-3}. This clearly shows that care must be used when using Hall-effect data from these materials, and growth conditions must be taken into consideration.

Optical Results

Films of ZnGeAs₂ have been examined by PL. An example of a PL scan is shown in Figure 4. The excitation comes from a 488 nm line of an Ar ion laser. The peak emission occurs at 1242 nm (1 eV). The emission energy is about 150 meV less than the reported 1.15-eV bandedge[6]. Since this sample gave stoichiometric peak splitting, the radiative transition being measured is not a band to band transition. No transition assignment has been attempted. Had the sample been a nonstoichiometric $(ZnGeAs_2)_x(Ge)_{1-x}$ alloy, we would have expected a lower band gap. The FWHM of the sample is about 20 nm (about 16 meV).

Double heterojunctions (DH) of a ZnGeAs₂ layer surrounded by GaAs layers, which passivate the ZnGeAs₂ surfaces and confine minority carriers, were fabricated. Using RFPCD [see Ref. 8], estimates of the minority-carrier lifetime were made. Since another potential application for ZnGeAs₂ is in the field of photovoltaics (minority-carrier devices), the results

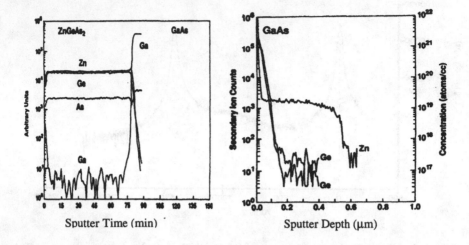

Figure 3.   SIMS profiles of a) ZnGeAs₂ on GaAs grown with Zn:Ge ratio of 5:1 and
b) profile of GaAs substrate after selective removal of ZnGeAs₂ grown with
Zn:Ge ratio of 15:1. Zn penetration of substrate is much greater in latter sample.

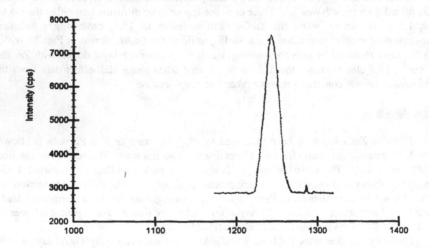

Figure 4.   Photolumnescence from ZnGeAs₂ layer. Peak emission is at 1242 nm (1 eV).
FWHM is 20 nm. Characteristics indicate good optical quality of the material.

came as a very pleasant surprise. The minority-carrier lifetimes that were measured ranged between 30 and 150 ns. These values are comparable to those found in very high quality III-V materials. Minority-carrier lifetime is sensitive to both chemical and structural defects. But the defects in $ZnGeAs_2$ seem to make little impact on the minority-carrier lifetime and result in very little minority-carrier trapping, suggesting as one possibility that capture cross-sections are small. Behavior in the infrared has not been measured.

As a final comment, it is worth noting that $ZnGeAs_2$ has been fabricated into a photocathode for spin-polarized electron emission. The process was a standard photochathode one; the surface was treated with a few monolayers of cesium to produce band bending and negative electron affinity. Minority-carrier electrons can then be emitted from the surface with optical excitation. The efficiency of this cathode was not particularly good, but none of the steps involved were rigorously optimized and the proof of concept for the material was established.

Doping Results

We cannot overemphasize the importance to being able to control the conductivity if the chalcopyrites are to be useful for devices, NLO or otherwise. As indicated above, two approaches were tried to effect this control. The first is to use Group III elements in hopes that there is preferential Zn-site occupation (Ge-site occupation produces p-type behavior). The Group III element becomes a donor and the concentration of Zn vacancies may be reduced. The second approach is to use Group VI elements to occupy As site, again producing n-type behavior but failing to address the Zn vacancy/antisite problem. The second approach is a "brute force" one that requires the solid solubility of the Group VI element to be greater than the concentration of defects. Results of both approaches are shown in Table III. The data in Table III show that of the six elements investigated only In seems to make a difference to the carrier concentrations. Indium compensation of p-CdSiAs$_2$ has been reported previously[9], supporting our observations here. Even with the reservations we have about both measurement techniques, we do see these data as some evidence that defect densities can be controlled. It would be interesting to examine material that has been co-doped with both In and Te or Se.

Another doping approach which we have not examined is to use a Group II element such as Mg or Be (both have good OMVPE sources). Since absorption bands that are detrimental for NLO are the result of native defects (vacancies and/or antisites), small cations (Be or Mg) may affect point-defect chemistries (occupy vacancies) so that absorption losses are reduced. This

Table III. Results of Doping $ZnGeAs_2$ with Group III and Group VI Elements

Element (Source)	Carrier Flow Rates Used (cm^3/min)	Carrier Concentration (cm^{-3})
B (100 ppm B_2H_6)	10-100	$p \approx 10^{19}$- 10^{20}
Ga (TMG at 0°C)	10-50	$p \approx 10^{19}$
Al (TMA at 20°C)	10-50	$p \approx 10^{19}$
In (TMI at 30°C)	10-20	$p \approx 10^{16}$- 10^{17}
Te (50 ppm DET)	10-50	$p \approx 10^{19}$
Se (50 ppm H_2Se)	10-100	$p \approx 10^{19}$

idea is as yet unprovem, and it is not intuitively obvious to us that a smaller cation would selectively substitute on Zn sites.

We have actually produced epitaxial layers of ZnGeAs₂ that show n-type conductivity (in parts of the layers) by electrochemical profiling (Figure 5). However, we believe that accumulation of interstitial Zn, which would behave as a donor and would be mobile, is responsible for the conductivity. In subsequent annealling experiments, done without a Zn overpressure, the n-type regions diminished and finally disappeared, and only a p-type profile remains.

Based on our experience with the Zn-bearing compounds, we suspect that conductivity in Cd-bearing compounds may be even more difficult to control, but we have no experimental evidence to back this up.

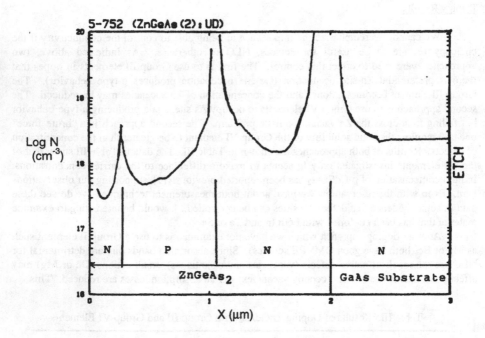

Figure 5.   Electrochemical profile of 2-μm-thick ZnGeAs₂ layer grown on n-type GaAs substrate. The ZnGeAs₂ layer shows regions with both n- and p-type conductivity. N-type regions disappeared after annealing. Peaks are an artifact of profiling through a depletion region. Layer thickness determined by SEM.

CONCLUSIONS

We have gained a good deal of experience growing II-IV-V₂ chalcopyrite materials by OMVPE. ZnGeAs₂ provides a useful model for the growth of ZnGeP₂, which holds considerable interest for NLO applications. Both of these compounds are challenging to grow because growth windows are small, but both have been grown with some degree of reproducibility. ZnGeAs$_2$, the easier of the two to grow, has been used to study the heavy p-type conductivity that occurs naturally in both materials. Unfortunately, we have not been able to control the point defects that produce this conductivity, but we do have some indication that an element such as In may substitute preferentially on a Zn site, thereby decreasing the number of vacancies as well as behaving as a donor. N-type conductivity has been observed in ZnGeAs$_2$, but this is likely the result of interstitial Zn that rapidly disperses during annealling, converting the conductivity to a uniform p type.

Recent work with bulk ZnGeP$_2$ has shown that the dominant defect is a Zn vacancy, but this may not to be the case in an OMVPE growth regime that uses excess Zn. We have not determined the dominant defect in ZnGeAs$_2$ or ZnGeP$_2$ grown by OMVPE.

The quality of the material grown by OMVPE is very good, determined from x-ray data, for growth on both GaAs and Ge substrates for ZnGeAs$_2$ and GaP for ZnGeP$_2$. ZnGeAs$_2$ has been grown on 2-inch-diameter GaAs substrate, and growth on larger substrates is conceivable. Minority-carrier lifetimes as long as 150 ns have been measured in ZnGeAs$_2$, and, consistent with the lifetime, strong photoluminescence has been observed.

Extrapolating this work to CdGeAs$_2$, the other II-IV-V$_2$ compound of interest for NLO, suggests that the growth will be more challenging. The higher vapor pressure of Cd compared to Zn is a concern. Growth windows will be smaller than for the Zn compounds. The conductivity control may also be more difficult to manage. We do not know if the infrared absorption bands that lower NLO efficiencies can be reduced or eliminated by an epitaxial process if the point defects causing the absroption are present in epitaxial material as well.

CdGeAs$_2$ is the analog of the III-V material Ga$_{0.5}$In$_{0.5}$As and has no substrates that provide a close lattice match. The lattice match with InP is the closest, but even here the mismatch is more than 1 percent, which is a large amount. A new approach to solve this problems might be to form compounds such as (Cd$_x$Zn$_{1-x}$)GeAs$_2$, Cd(Ge$_x$Si$_{1-x}$)As$_2$, or CdGe(As$_x$P$_{1-x}$)$_2$. There are compositions of each of these compounds that lattice match at least one of the commonly used III-V substrates. We have produced prototypes for these compounds with the growth of Zn(GeSi)As$_2$ and ZnGe(AsP)$_2$, both lattice matched to GaAs. The NLO properties of mixed compounds such as these have not been studied.

In summary, then, OMVPE offers a great deal of flexibility and advantages not shared by bulk growth. However, for the chalcopyrite class of compounds, growth problems are going to be challenging, perhaps no more so than for bulk growth.

ACKNOWLEDGEMENTS

The authors are pleased to acknowledge Dr. R. K. Ahrenkiel and Mr. B. Keyes of the National Renewable Energy Laboratory, Golden, CO, for the transient photoluminscence measurements. We are also pleased to recognize Dr. C. M. Sinclair of the Continuous Electron Beam Acceleration Facility, Hampton, VA, who fabricated the ZnGeAs$_2$ photocathodes. We also are pleased to acknowledge Dr. G.S. Solomon of Stanford University for his contributions to the growth of ZnGeAs$_2$ while he was employed at Research Triangle Institute. Finally, much early work on ZnGeP$_2$ was done by Dr. G.C. Xing while a graduate student at North Carolina State University, Raleigh, NC.

REFERENCES

1. J.L. Shay and J.H. Wernick, Ternary Chalcopyrite Semiconductors: Growth, Electron Properties, and Applications, Pergamon Press, Oxford, 1975.

2. R.L. Byer in Nonlinear Optics, edited by P.G. Harper an B.S. Wherrett (Academic Press, New York, 1977), p. 127.

3. J.E. Andrews, M.A. Littlejohn, G. Igwe, and R.T. Pickett, J. Crystal Growth 56, p. 1 (1982)

4. N. Achargui, B. Benachenhou, A. Foucaran, G. Bougnot, P. Coulon, J.P. Laurenti, and J. Camassel, J. Crystal Growth 107, p. 410 (1991).

5. L.E. Halliburton, G.J. Edwards, M.P. Scripsick, M.H. Rakowsky, P.G. Schunemann, and T.M. Pollack, Appl. Phys. Lett. 66(20), p. 2670 (1995).

6. G.S. Solomon, M.L. Timmons, and J.B. Posthill, J. Appl. Phys. 65(5), p. 1952 (1989)

7. Handbook of the Physicochemical Properties of the Elements, edited by G.V. Samsonov (Plenum, New York, 1968), p. 254.

8. R.K. Ahrenkiel and D.H. Levi, Proceedings of the 24th IEEE Photovoltaic Specialists Conference (First World Conference on Photovoltaic Energy Conversion), IEEE Press, Piscataway, NJ, 1995, p. 1368

9. V. Yu. Rud', Yu. V. Rud', and M. Serginov, Phys. Status Solidi A 121, p. K171 (1990).

LASER DAMAGE STUDIES OF SILVER GALLIUM SULFIDE SINGLE CRYSTALS

Warren Ruderman*, John Maffetone*, David E. Zelman**, and Derrick M. Poirier***
* INRAD, Inc., Northvale, NJ 07647
** Materials Directorate, Air Force Research Laboratory, WPAFB, OH 45433
*** Physical Electronics Laboratories, Eden Prairie, MN 55344

ABSTRACT

Although $AgGaS_2$ single crystals have desirable optical and nonlinear properties, the low laser damage threshold has limited their application for tunable infrared laser sources. Crystals have been studied by ESCA and Auger techniques and the surface of polished crystals has been found to be silver deficient, while the bulk composition is closer to stoichiometry.

INTRODUCTION

$AgGaS_2$ single crystals have exceptional fundamental optical properties with a nonlinear optical susceptibility d_{36}= 12-29pm/V^{1-3} and a transmission range of 0.45 to 13μm. No other nonlinear crystal that is transparent in this region can be phase-matched with a 1.06μm pump. Such a combination of properties makes $AgGaS_2$ a very attractive candidate for a variety of nonlinear optical devices, such as an OPO operating from 2.5 to 11mm with a 1.06μm as a pump. $AgGaS_2$ has a lower thermal conductivity (K $\approx$ 0.014–0.015 W×cm^{-1}×K^{-1})[4] than $ZnGeP_2$ (K $\approx$ 0.35–0.36 W×cm^{-1}×K^{-1})[4]. Low thermal conductivity results in more stringent conditions on the optical absorption coefficient, α, of $AgGaS_2$; a value for α of 0.001 to 0.003 cm^{-1} is desirable in order to minimize thermal lensing at high average powers.

$AgGaS_2$ crystals experience laser damage thresholds far below the values that are needed for useful OPO systems. Laser damage occurs at the surface of the crystal, rather than in the bulk. This property, which we do not believe to be intrinsic, limits the usefulness of $AgGaS_2$.

Fundamentally, the surface laser damage should be linked to the chemical and phase composition of the crystal surface and to its defect structure. Mechanical damage, adsorption of common contaminants such as carbon, oxygen and sodium, deviation from stoichiometry, presence of scattering centers, or surface instability in the presence of the high electric fields of the pump laser could contribute to the observed surface laser damage.

Due to its retrograde homogeneity region, all melt-grown $AgGaS_2$ single crystals appear "milky" due to dispersed second phase precipitates[5-7]. An isothermal annealing in an excess of Ag_2S has been commonly used to reduce the precipitates[5-7]. The annealed crystals are, however, not completely defect-free. Even the best samples exhibit a certain degree of residual optical absorption and scattering[5].

The impressive potential for $AgGaS_2$ nonlinear devices will not be realized until the surface laser damage threshold is substantially increased.

EXPERIMENT

Single Crystal Growth

Synthesis

The two-temperature vapor transport synthesis technique was used to synthesize pure $AgGaS_2$ for feedstock for crystal growth. The essence of this method is control of the speed of the synthesis reactions and composition of the condensed phase by the pressure of the volatile component. A sealed ampoule, loaded with Ag, Ga and S, is placed in a two-zone furnace with silver and gallium in the "hot" zone and sulfur in the "cold" zone. Sulfur vapor reacts with the molten Ag+Ga charge, while the sulfur pressure is determined by the temperature in the "cold" zone so that it does not exceed the working limits of the quartz ampoule.

The starting composition of the charge and temperature of the process can be varied to produce an as-synthesized ingot of any composition, including stoichiometric (50 mol% of Ag_2S + 50 mol% of Ga_2S_3) and "congruent" (49 mol% of Ag_2S+51 mol% of Ga_2S_3). High purity elements were used: 5N silver, 7N gallium and 5N5 sulfur.

Crystal Growth

Crystals were grown by the Horizontal Gradient Freeze technique (HGF) in a semi-transparent gold furnace which permits full visual observation during seeding and growth.

The temperature of the furnace is regulated with an accuracy of $\pm 0.1°C$ by a ramp-and-soak temperature controller and two S-type thermocouples (for temperature control and monitoring) were brought in direct contact with the ampoule's ends.

Crystal growth runs resulted in general in an ingot having a typical "milky" appearance and containing dispersed precipitates of a second phase.

Annealing

Crystals were annealed in a two temperature process to dissolve precipitants. The $AgGaS_2$ crystal and either Ga_2S_3 or Ag_2S nutrient in a 1 mol% excess over stoichiometry were loaded into separate vitreous carbon boats. The crystal was placed in the hot zone of the seeded quartz ampoule, and the nutrient was placed in the cold zone. The temperature of the hot zone was about 900°C and the temperature of the cold zone was varied from 600 to 800°C. The duration of the annealing was 15 days. The crystals were clear after annealing.

Sample Preparation and Characterization

After orientation with x-rays, crystal samples were cut on an i.d. diamond saw. Grinding was carried out to minimize subsurface damage. Crystal surfaces were polished to $\lambda/10$ using a superpolishing process developed at INRAD. The surface roughness of the polished samples was measured with a Chapman MP-2000 non-contact laser profilometer. Typical roughness values were in the range of 2 to 4 Å rms. A double crystal x-ray spectrometer was used to evaluate the perfection of the $AgGaS_2$ crystals using Cu-Kα radiation. The full width at half maximum of the rocking curve was about 190 seconds and the curve was symmetrical.

Laser Damage

Several samples of INRAD grown polished $AgGaS_2$ crystals and one commercial crystal polished at INRAD were subjected to laser damage tests at INRAD with 1.064mm, 18 nsec pulses. Damage was found only on the surfaces, no bulk damage was observed. In all cases the output surface damaged more extensively than the input face. The surface laser damage threshold was found to be ~ 25 Mw/cm^2.

Figure 1 shows scanning electron micrographs of a laser damage site on the exit surface of a polished $AgGaS_2$ crystal (Sample 1 described below) as acquired in a Auger electron spectrometer.

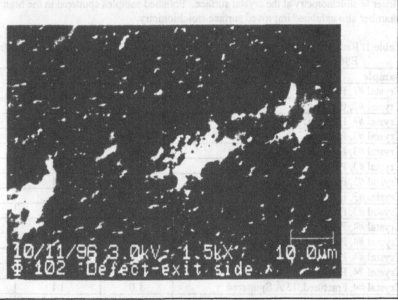

Figure 1. Scanning electron micrograph of laser exit damage site as acquired by Auger electron spectrometer.

ESCA and Auger Analysis

The following four samples were prepared for the analyses, which were carried out at Physical Electronics Laboratories with their Quantum 2000 system:

Sample 1. 5x5x5mm cube cut for Type II OPO, $\phi=0°$, $\theta=48.4°$, polished surface, commercial crystal.

Sample 2. Single crystal of $AgGaS_2$, with one surface cleaved at INRAD, as-grown (not annealed) at INRAD.

Sample 3.   Single crystal of AgGaS$_2$, as-grown (not annealed) at INRAD, polished.
Sample 4.   Single crystal of AgGaS$_2$, grown and annealed at INRAD, polished.

RESULTS

The cleaved surface of Sample 2 was analyzed and then the sample was fractured and immediately put into the high vacuum chamber and re-analyzed. The polished surfaces of Samples 3 and 4 were first analyzed, then material was sputtered off the crystal surface while in the vacuum chamber and then analyzed as a function of the amount of material sputtered away. In addition, a piece of each sample was broken off and the fractured surface was analyzed immediately.

Table 1 gives the ratio of the components in the various crystal samples as determined by ESCA analysis. All the polished samples show a large Ag deficiency at the surface. The same samples when fractured and analyzed immediately show elemental composition much closer to stoichiometry at the crystal surface. Polished samples sputtered in the high vacuum chamber also exhibited improved surface stoichiometry.

Table 1. Ratios of Ag, Ga and S at the surface of various AgGaS$_2$ crystals as determined by ESCA. Sputter depths are in SiO$_2$ equivalents.

Sample	Ag	Ga	S
Crystal #1, Polished	1.0	2.7	2.4
Crystal #2, Cleaved	1.0	2.3	2.0
Crystal #3, Fractured	1.0	1.4	1.8
Crystal #3, Polished	1.0	3.1	2.3
Crystal #3, Polished, 480Å Sputtered	1.0	1.6	1.7
Crystal #3, Polished, 1000Å Sputtered	1.0	1.2	1.3
Crystal #3, Fractured	1.0	1.1	1.6
Crystal #3, Fractured, 15Å Sputtered	1.0	1.1	1.6
Crystal #3, Fractured, 120Å Sputtered	1.0	0.9	1.2
Crystal #4, Polished	1.0	2.0	2.5
Crystal #4, Polished, 30Å Sputtered	1.0	1.3	1.7
Crystal #4, Polished, 450Å Sputtered	1.0	1.8	2.1
Crystal #4, Fractured	1.0	1.1	1.7
Crystal #4, Fractured, 15 Å Sputtered	1.0	1.1	1.6
Crystal #4, Fractured, 120Å Sputtered	1.0	0.9	1.1

Polished samples only reached constant stoichiometry after 9 minutes of sputtering (or 270 Å). This gives some measure of the thickness of the disturbed layer on a polished sample.

Figure 2 presents a summary of the surface composition of all four samples as received and after sputtering, cleaving and fracture. The large variation from stoichiometry at the polished surface and the convergence to stoichiometry after treatment, such as sputtering or fracture, is clearly seen.

One important caveat must be considered when looking at these profiles. The relative sputtering rates of the three elements are likely to differ, and although a steady state composition may be reached, it will not reflect the exact bulk composition. However, it is clear that the surface of the polished samples is very different from that of the fractured surfaces. In particular, the relative amount of Ag seems to be low on all polished surfaces. The Ag concentration appears to be lower near the surface of the annealed/polished sample than on the as-grown/polished sample.

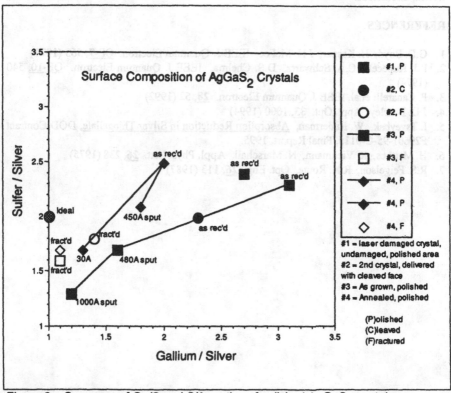

Figure 2. Summary of Ga/S and S/Ag ratios of polished AgGaS$_2$ crystals as received and after sputtering, cleaving, and fracture.

CONCLUSIONS

Surface ESCA and Auger analyses of superpolished AgGaS$_2$ single crystals showed that the atomic composition of polished surfaces differs radically from stoichiometry; the surface is highly deficient in Ag. Polished surfaces that were sputtered for different times in the vacuum chamber and analyzed were found to more closely conform to stoichiometry. The non-stoichiometric surface layer is about 250 Å deep. Fractured surfaces closely approximated stoichiometry. The low laser damage threshold at the surface is not surprising given the high deviation from stoichiometry of the surface.

ACKNOWLEDGEMENTS

This work was supported by a Phase I SBIR grant from the Materials Directorate, Air Force Research Laboratory, Wright Patterson AFB, OH. The authors acknowledge their appreciation to Dr. Ilya Zwieback, Manager, Crystal Growth Department of INRAD for fruitful discussions.

REFERENCES

1. G.D. Boyd, H. Kasper, J.H. McFee, IEEE J. Quantum Electron. QE-7, 563 (1971)
2. P.J. Kupecek, C.A. Schwartz, D.S. Chelma, IEEE J. Quantum Electron. QE-10, 540 (1974)
3. P. Canarelli et.al, IEEE J. Quantum Electron. 28, 52 (1992)
4. J.D. Beasley, Appl. Opt. 33, 1000 (1994)
5. I. Tsveybak, W. Ruderman, Absorption Reduction in Silver Thiogallate, DOD Contract F29601-95-C-0115, Final Report, 1995.
6. H. Matthes, R. Viehmann, N. Marschall, Appl. Phys. Lett. 26, 238 (1975)
7. R.S. Feigelson, R.K. Route, Opt. Eng. 26, 113 (1987)

Atomistic Calculations of Dopant Binding Energies in ZnGeP$_2$

Ravindra Pandey*, Melvin C. Ohmer**, A. Costales*** and J. M. Recio***

* Michigan Technological University, Houghton, MI, pandey@mtu.edu
** Air Force Research Laboratory, Wright Patterson AFB, OH,
                                              ohmermc@ml.wpafb.af.mil
***Universidad de Oviedo, Oviedo, Spain, mateo@carbono.quimica.uniovi.es

ABSTRACT

Atomistic model has been applied to study various cation dopants, namely Cu, Ag, B, Al, Ga and In in ZnGeP$_2$. The pairwise interatomic potential terms representing the interaction of dopants with the host lattice ions are derived using first principle methods. Defect calculations based on Mott-Littleton methodology predict small binding energies for Cu and Ag substituting Zn in the lattice which are in agreement with the available experimental data. The group III dopants (i.e. B, Al, Ga and In) at the Ge site are predicted to have large binding energies for a hole except B which shows a distinct behavior. This may be due to large mismatch in atomic sizes of B and Ge. At the Zn site, the calculated binding energies of the group III dopants place donor levels in the middle of the band gap.

INTRODUCTION

Zinc germanium phosphide (ZnGeP$_2$) is the NLO material for mid IR lasers allowing high-power tunability in the spectral region of 2-5 μm.[1-2] An absorption band around 1-2 μm is, however, found to affect the usefulness of this material. This band is attributed to photoionization of a deep native acceptor center associated with the zinc vacancy.[3-5] Experimental efforts involving selective doping of the material are underway to reduce the concentration of this acceptor-complex in the lattice. Earlier work on the properties of dopants in ZnGeP$_2$ have been limited to few photoluminescence and Hall effect studies. Averkieva et al. have reported[6] photoluminescence spectra and hole concentrations for crystals melt doped with Cu, Ga, In, Se, Fe, Cd and Si. The room temperature hole concentration for these samples were reported to vary from a low of 2.8x10^{10}/cm^3 for Si, nominally the value seen in undoped crystals, to a high of 10^{17}/cm^3 for In. Based on a photoluminescence study[7], it is reported that Cu diffusion can increase the hole concentration by a factor of 10,000 in ZnGeP$_2$ by a proper choice of the diffusion regime. Grigoreva et al., however, have reported[8] Hall effect data on samples doped with Au, Cu, Se, Ga and In and stated that the first two dopants were inactive and the rest were acceptors. The binding energy of Se was reported to be 0.40 eV and that of Ga to be 0.30 eV. In a more recent review paper, Rud[9] has indicated that Au, Cu, Ga, In, Se and Pt are acceptors with activation energies in eV of respectively 0.50, 0.30, 0.06 for heavy doping, 0.03 for heavy doping and ≈0.4 for light doping, 0.40 and 0.50. These results will be compared with our calculated values later in the paper.

We have initiated an extensive theoretical study[5] on $ZnGeP_2$ and, in this paper, we will report the results of atomistic calculations involving cation dopants, namely Cu, Ag, B, Al, Ga and In. The choice of dopants is expected to indicate the ion-size effect on the binding energies for the groups I and III dopants which lie left and right columns of the host Zn ion in the periodic table. Our approach is based on the pair-wise description of the lattice which has been very successful in yielding reliable defect energies in a wide variety of materials including sixfold- and fourfold-coordinated structures.[10-11]

INTERATOMIC POTENTIALS

In our approach, the lattice is considered to consist of ions interacting via a combination of electrostatics and interatomic potentials, while polarization is included by means of the shell model.[12] The total energy of a crystal is then taken to be a sum of two-body interactions in the lattice. The pairwise interaction term consists of the long-range Coulombic part and the short-range repulsive part which is given by an analytical expression of the Born-Mayer form :

$$V_{SR}(R_{ij}) = A_{ij} \exp(-R_{ij}/\rho_{ij}), \tag{1}$$

where A and ρ are the parameters yet to be determined.

We apply a parameter-free procedure to model interatomic interactions of dopant ions with the host P^{3-} anion. The general strategy has been described in detail previously[10] and consist of two steps : (i) generation of quantum-mechanical crystalline ionic electronic densities (IEDs) for the ions involved in the interaction, and (ii) application of the electron gas formalism[13] to derive the corresponding crystal consistent interatomic potential (CCIP). To obtain the quantum-mechanical description of P^{3-} embedded in the $ZnGeP_2$ environment, we have used the ab-initio Perturbed Ion (aiPI) model.[14] The minimization of the total crystal energy required by the Hartree-Fock approach provides a set of crystal-like atomic wavefunctions that respond self-consistently to the nearly exact crystal potential. In our case the aiPI computation has been performed at the experimental value of the lattice parameters of $ZnGeP_2$ using the Slater type orbitals given by Clementi and Roetti.[15] It is interesting to remark that the essential electron gas assumption of the total crystal density being a superposition of the individual IEDs is better satisfied by the aiPI crystal-like orbitals than by solutions employed in earlier electron gas calculations. This behavior is mainly due to the inclusion in the representation of the aiPI crystal potential of a projection operator that enforces the ion-lattice orthogonality.

For calculations based on the electron gas theory, we have used the following functional forms to evaluate the pairwise interatomic energy : Thomas-Fermi for kinetic, Lee-Lee-Parr for exchange and Wigner (with the new fitting by Clementi) for correlation. The coulombic contributions has been exactly evaluated by means of analytical algorithms recently developed.[16,17] To compute the rest of the contributions, we have used a spheriodal(ϕ, λ, μ) coordinate system. The angular integration over ϕ is trivially 2π, since the electron density is cylindrically symmetric along the internuclear axis. In the cases of λ and μ, we have used a 60x60 Gauss-Legendre quadrature. The total volume of integration contains at least 99.9998% of the electron density. For each set of interatomic potential representing interaction between dopant and the P ion, the interatomic energy is evaluated at separations starting from 2 bohr to 12 bohr at intervals of 0.1 bohr.

The calculated potential energy surface for all the dopants considered here is shown in Fig. 1. The effect of ion-size is clearly illustrated here; higher the atomic number of a dopant, stronger its short-range interatomic interaction with P in the lattice. The energy surface is then fitted to the analytical expression (Eq. 1) whose parameters (i.e. A and ρ) are listed in Table I. For the host lattice, the interatomic potential parameters (i.e. Zn-P, Ge-P, P-P, Zn-Zn, Ge-Ge and Zn-Ge) along with the shell-model parameters Y_P and k_P parameters are obtained by the empirical fitting using the program GULP.[19] Both dopant and host lattice cations are considered as rigid ions in the lattice. The derived interatomic potential set (Table I) for the host lattice reproduces the calculated crystal properties such as structure, elastic and dielectric constants very well.[20]

DOPANT ENERGETICS

The nature of a cation dopant in $ZnGeP_2$ depends on the host atom which it is replacing in the lattice. In the present case, the group III dopants (i.e. B, Al, Ga and In) would either act as a donor substituting Zn or act as an acceptor when they replace Ge. The group I cations (i.e. Cu and Ag) would always be acceptors substituting either Zn or Ge in the lattice. The binding energy of an electron or a hole to the dopant can then be written as :

$$\Delta E = E_{dop+iso} - (E_{iso} + E_{dop}) \tag{2}$$

where $E_{dop+iso}$ refers to energy of the dopant complex, E_{iso} is energy of an isolated hole or electron and E_{dop} is energy of the dopant in the lattice. Calculations to obtain energies for Eq. 2 include, for example, Cu at the Zn site (E_{dop}), an isolated hole localized over As ions (E_c) and Cu next to the localized hole (E_{dop+c}) in the lattice. Note that a direct comparison can be made between ΔE and the acceptor binding energy obtained from the Hall-effect measurements.

Total energies of defects including both native and dopants have been calculated using the Mott-Littleton methodology[21,22] in which the lattice is divided into a series of different regions around the defect by concentric spherical boundaries. Immediately surrounding the defect is region 1 in which all ions are treated explicitly and fully optimized. Beyond this is region 2a in which all atoms are still explicitly considered, but the relaxation effects are much smaller and can be treated more approximately. The displacements in region 2a can be calculated based on the force due to region 1. Beyond region 2a is region 2b in which the relaxation energy is treated yet more approximately and responds only to the total charge on the defect position at the defect center. Hence this component can be summed to infinity through the use of lattice summation methods analogous to the Ewald sum for the first inverse power of distance. In the present work, region 1 contains approximately 350 atoms which was found to be sufficient to converge the absolute defect energy to better than 0.01 eV.

The calculated binding energies (ΔE) for various dopants are given in Table II. At the Zn site in $ZnGeP_2$, Cu and Ag acting as acceptors are predicted to have the binding energy of 0.25 and 0.17 eV respectively. Comparison of the calculated value with the experimental value of 0.30 eV for Cu shows a very good agreement. The group III dopants, on the other hand, have a donor-like nature at the Zn site. Their electron binding energies vary from 0.36 to 1.50 eV with a significant change in energy going from B to Al. This may be due to a large mismatch between the atomic sizes of B and Zn in the lattice (Table II). For Al, Ga and In, calculations yield large binding energies for the electron placing the donor levels in the middle of the gap in $ZnGeP_2$. At the Ge site where the group III dopants act as acceptors, the results find a distinct behavior by B which forms a deeper

TABLE I. Interatomic potential parameters (given in Eq.   ) both for dopants (obtained by electron-gas methods) and the host lattice (obtained by emprical fitting method).

		A_{ij} (eV)	ρ_{ij} (Å)
electron-gas :	$Cu^{1+} - P^{3-}$	1562.56	0.3694
	$Ag^{1+} - P^{3-}$	2170.20	0.3514
	$B^{3+} - P^{3-}$	2566.54	0.2808
	$Al^{3+} - P^{3-}$	2310.8	0.3233
	$Ga^{3+} - P^{3-}$	2567.12	0.3474
	$In^{3+} - P^{3-}$	2879.71	0.3501
empirical :	$Zn^{2+} - P^{3-}$	2776.92	0.2946
	$Ge^{4+} - P^{3-}$	1030.93	0.4659
	$Zn^{2+} - Zn^{2+}$	2776.92	0.2946
	$Ge^{4+} - Ge^{4+}$	646.280	0.1996
	$Zn^{2+} - Ge^{4+}$	23944.3	0.3406
	$P^{3-} - P^{3-}$	1422.15	0.4932

Fig. 1 : Short-range interatomic potentials (Eq. 1) for various dopants in ZnGeP$_2$.

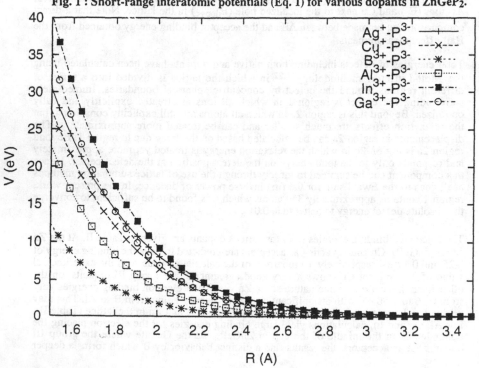

center as compared to other ions. This difference can again be understood in terms of the large lattice relaxations arising due to difference in covalent radii of B and Ge which tend to stabilize the hole more strongly. The group I dopants (i.e. Cu and Ag) at the Ge site have a very large hole binding energy due mainly to electrostatic interactions, for example, between $[Cu_{Ge}]^{...}$ and a hole.

In summary, we have derived the interatomic potentials for various cation dopants in $ZnGeP_2$ and have calculated their binding energies to a hole or an electron depending on the substitution site in the lattice. The calculations based on atomistic description of the lattice predict that the group-III dopants would bind the hole more strongly than the group-I dopants. Furthermore, the results find that dopants acting as donors introduce midgap states in $ZnGeP_2$.

ACKNOWLEDGMENTS :

This work is supported by the AFOSR Contract F49620-96-1-0319. AC and JMR acknowledges the support from Spanish DGICYT, Project No. PB96-0559.

TABLE II: Binding energies for various cation dopants in $ZnGeP_2$.

Dopant	Nature	Binding energy[*] (eV)	R_{dopant}[**] (Å)
$[Cu_{Zn}]^{.}$	acceptor	0.25	1.35
$[Ag_{Zn}]^{.}$	acceptor	0.17	1.53
$[B_{Zn}]^{'}$	donor	0.36	0.88
$[Al_{Zn}]^{'}$	donor	1.05	1.26
$[Ga_{Zn}]^{'}$	donor	1.40	1.26
$[In_{Zn}]^{'}$	donor	1.49	1.44
$[Cu_{Ge}]^{...}$	acceptor	2.2	1.35
$[Ag_{Ge}]^{...}$	acceptor	2.0	1.53
$[B_{Ge}]^{.}$	acceptor	0.83	0.88
$[Al_{Ge}]^{.}$	acceptor	0.61	1.26
$[Ga_{Ge}]^{.}$	acceptor	0.57	1.26
$[In_{Ge}]^{.}$	acceptor	0.56	1.44

[*]For Cu, Ga and In as acceptors, the experimental values[8-9] are 0.30, 0. 31 and 0.40 eV respectively.
[**]The covalent radii of the host ions : R_{Zn} : 1.31Å, R_{Ge} : 1.22Å.

REFERENCES

[1] P. G. Schunemann, P. A. Budni, M. G. Knights, T. M. Pollok, E. P. Chicklis, and C. L. Marquardt, Advanced Solid State Lasers and compact Blue-Green Laser Technical Digest (Optical Society of America, Washington, DC, 1993).

[2] Yu. V. Rud, Fiz. Techn. Poluprovodn, 28 633 (1994).

[3] M. H. Rakowsky, W. K. Kuhn, W. J. Lauderdale, L. E. Halliburton, G. J. Edwards, M. P. Scripsick, P. G. Schunemann, T. M. Pollak, M. C. Ohmer and F. K. Hopkins, Appl. Phys. Lett. 64, 1615 (1994).

[4] L. E. Halliburton, G. J. Edwards, M. P. Scripsick, M. H. Rakowsky, P. G. Schunemann, and T. M. Pollak, Appl. Phys. Lett. 66, 2670 (1995).

[5] P. Zapol, R. Pandey, M. Ohmer, and J. D. Gale, J. Appl. Phys. 79, 671 (1996).

6. G. K. Averkieva, V. S. Grigoreva, I. A. Maltseva, V. D. Prochukan, Yu. V. Rud, Phys. Stat. Sol (a) 39, 453 (1977).

[7] V. G. Voevodin, A. I. Gribenyukov, A. . Morozov and V. S. Morozov, Izve Vys. Ucheb. Zav., Fiz 2, 64, 1985.

[8] V. S. Grigoreva, V. D. Prochukhan, Yu. V. Rud, A. A. Yakovenko, Phys. Stat. Sol. (a) 17, K69 (1973).

[9] Physical Phenomena in Ternary Compounds and Devices by Yu. V. Rud (in Russian), Machine Translation-NAIC-ID(RS)T-0699-94, Distribution limited, available from DTIC.

[10] J. H. Harding and A. M. Stoneham, J. Phys. C15, 4649 (1982).

[11] P. Zapol, R. Pandey and J. D. Gale, 1997, J. Condens. Matter, in press, 1997.

[12] B. G. Dick and A. W. Overhauser, Phys. Rev. 112, 90 (1958).

[13] E. Francisco, J. M. Recio, M. A. Blanco, A. Martin Pendas and L. Pueyo, Phys. Rev. B 51, 2703 (1995).

[14] R. G. Gorden and Y. S. Kim, J. Chem. Phys. 56, 3122 (1972).

[15] V. Luana and L. Pueyo, Phys. Rev. B 41, 3800 (1990).

[16] E. Clementi and C. Roetti, At. Nucl. Data Tables 14, 177 (1974).

[17] E. Francisco, J. M. Recio, M. A. Blanco and A. Martin Pendas, Phys. Rev. B 51, 11289 (1995).

[18] A. Martin Pendas and E. Francisco, Phys. Rev. A43, 3384 (1991).

[19] J. D. Gale, Phil. Mag. B73, 3 (1996); J. D. Gale, JCS Faraday Trans 93, 629 (1997).

[20] A detailed comparison can be obtained from the authors : pandey@mtu.edu

[21] A. B. Lidiard and M. J. Norgett in Computational Solid State Physics, edited by F. Herman (Plenum, New York, 1972), p. 385.

[22] C. R. A. Catlow and W. C. Mackrodt, *Computer Simulations of Solids* (Springer, Berlin, 1982).

PHOTORESPONSE STUDIES OF THE POLARIZATION DEPENDENCE OF THE CdGeAs$_2$ BAND EDGE

G.J. BROWN *, M.C. OHMER *, and P.G. SCHUNEMANN **
*Air Force Research Laboratory, Materials & Manufacturing Directorate, AFRL/MLPO, Wright-Patterson AFB, OH 45433-7707
** Sanders, a Lockheed Martin Co., Nashua, NH 03061-2035

ABSTRACT

Mid-infrared photoresponse studies were performed on an oriented p-type cadmium germanium arsenide uniaxial crystal. The effects of optical polarization alignment, parallel and perpendicular to the c-axis of the crystal, were studied, as well as the effects of the transport electric-field direction. The measured optical band edge was 0.578 eV at 10 K for all polarization and bias configurations. This band gap energy is in good agreement with absorption and photoluminescence results for this sample. However, the photoresponse spectrum measured with unpolarized light at 10 K showed a much lower onset at 0.50 eV. This difference in the low temperature activation energy of the photoresponse is attributed to deep native defect levels near the band edge. These deep levels at times obscure the true band edge and can cause under estimates of the band gap energy. These results can explain the wide disparity in the reported CdGeAs$_2$ band gap in the literature. In addition, the intensity of the photoresponse was found to be only slightly dependent on the optical polarization direction, but strongly dependent on the bias electric-field direction. The largest photoreponse was observed when the optical polarization was parallel to the c-axis and the bias electric field was perpendicular to the c-axis. The bias electric-field direction also had a significant effect upon the temperature dependence of the peak photoresponse intensity. The temperature dependence of the CdGeAs$_2$ energy band gap was determined by empirical fitting.

INTRODUCTION

CdGeAs$_2$ has the largest reported second order non-linear susceptibility [1] of any birefringent compound semiconductor, and it is transparent in the infrared from 2.4 to 18 microns. In principle, it is possibly, the ultimate choice for frequency doubling CO$_2$ laser wavelengths ranging from 9 to 11 microns. As CdGeAs$_2$ is a pseudo-III-V chalcopyrite, it has a much larger thermal conductivity than the state-of-the-art material AgGaSe$_2$, a pseudo-II-VI chalcopyrite. As it has a smaller band gap, its nonlinear optical coefficient is much larger than that of AgGaSe$_2$. [2] The thermal conductivity and the conversion efficiency of CdGeAs$_2$ are respectively, 6 times larger [3] and 50 times larger [1] than those of AgGaSe$_2$. This larger thermal conductivity allows higher laser powers to be transmitted through the crystal without damage. Additionally, the study of chalcopyrites in general promises to add significantly to the science base of compound semiconductors. As a result, the study of this material is both very interesting from a fundamental science and from an application point of view.

At present, the performance of CdGeAs$_2$ at room temperature is limited by extrinsic absorption due to native defects, although cooling to 77K can largely overcome the limiting mechanism of free carrier absorption. The as-grown material is typically p-type with a free carrier concentration in the 10^{16} to 10^{17} cm^{-3} range corresponding to resistivities of 10 to 0.1 ohm-cm at room temperature. [4]   Ongoing research on this material is directed toward understanding the nature of these defects and developing approaches to eliminate them.   Mid-infrared photoresponse studies were performed on an oriented p-type cadmium germanium arsenide uniaxial crystal as a function of temperature.   The effects of optical polarization alignment, parallel and perpendicular to the c-axis of the crystal, were studied, as well as the effects of the transport electric-field direction. The results of these studies are discussed below.

EXPERIMENT

The sample was cut from a CdGeAs$_2$ single crystal boule grown by a horizontal gradient freeze technique. [5] The sample was 8mm x 7.3mm x 1mm in size with the c-axis parallel to a long dimension and lying in the plane of the sample surface. The sample was contacted on two parallel sides along the 1 mm thick edges by gold wires solder bonded with indium strips.

The photoresponse measurements were made using a Bio-Rad FTS-20V Fourier Transform spectrometer with a wavelength range from 2 to 40 microns. The sample was mounted on the coldfinger of a closed cycle refrigerator allowing the sample temperature to be varied from 10 K to 290 K. The sample was mounted in series with a load resisitor and biased at 9 volts. Electrical contacts to the sample were arranged such that the bias electric field could be oriented along the c-axis or perpendicular to the c-axis. The optical polarization was selected using a KRS5 wire grid polarizer in front of the sample. The polarizer was kept at zero degrees and the sample was rotated 90° to change the orientation of the optical polarization with respect ot the c-axis. This means of changing the polarization orientation was chosen because the infrared beam of the spectrometer is not completely randomly polarized.

RESULTS

Initially, a baseline photoresponse spectrum was collected without the wire grid polarizer. The bias electric field was applied perpendicular to the c-axis. This spectrum showed a strong onset in the photoresponse at 0.50 eV. Several other spectra were also collected to check if the photoresponse varied with time in the infrared beam. The spectrum remained very repeatable at 10 K. Next the infrared polarizer was placed in front of the sample and a new set of spectra were collected. For this data the optical polarization was aligned parallel the c-axis. With the polarizer, the photoresponse band edge changed dramatically and now showed an onset at 0.573 eV. This higher band gap energy is in good agreement with absorption [4] and photoluminescence [6] results for this sample.

A comparison of the photoresponse spectra with and without the polarizer is shown in Figure 1. The difference in the low temperature onset of the photoresponse is attributed to deep native defect levels near the band edge. It is not clear how the polarization affects the sensitivity of these native defects, but it is conceivable that the defects are preferentially oriented in the anisotropic crystal structure of a chalcopyrite. These results may explain the wide disparity in the reported CdGeAs$_2$ band gap in the Russian literature. Also worth noting is the large increase

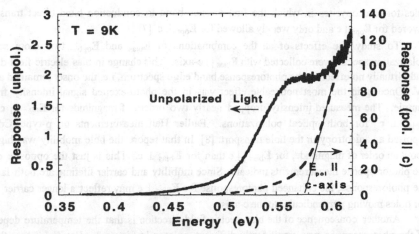

Figure 1. Comparison of the photoresponse band edge for randomly polarized light and light polarized parallel to the c-axis of CdGeAs$_2$.

in the signal intensity with the polarizer in place. The elimination of the band edge defect response appears to have improved the overall photodetection in this material.

Next the sample was rotated 90° so the optical polarization would be perpendicular to the c-axis. The bias electric field across the sample was kept perpendicular to the c-axis. The resulting photoresponse spectrum is compared to the previous spectrum taken with $E_{opt} \parallel$ c-axis in Figure 2. A few meV shift in the photoresponse onset to higher energy is observed. But the most notable difference at low temperature is the reduction in the signal intensity by nearly a factor of two. Clearly, the intrinsic band to band, from the highest valence band, photo-excitation favors the infrared polarization parallel to the c-axis. This agrees with the polarization selection

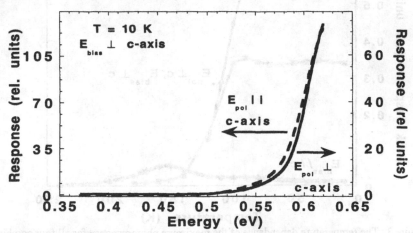

Figure 2. Comparison of the photoresponse band edge spectra for two different infrared optical polarizations, parallel and perpendicular to the c-axis of the crystal.

rules for chalcopyrites in which the first valence band to conduction band direct transition is favored for $E_{opt} \parallel c$ and only weakly allowed for $E_{opt} \perp c$. [7]

To study the effects of all the combinations of E_{bias} and E_{opt}, two more series of polarization spectra were collected with $E_{bias} \parallel$ c-axis. This change in bias electric field direction had virtually no effect on the photoresponse band edge spectrum, i.e. the onset remained at 0.579 eV. Once again the most noticeable effect was in the photo-excited signal intensity from the sample. The measured intensity dropped nearly two orders of magnitude for $E_{bias} \parallel c$ versus $E_{bias} \perp c$ for both optical polarizations. Earlier Hall measurements on p-type CdGeAs2 reported an anisotropy in the hole transport. [8] In that report, the hole mobility was higher, by about an order of magnitude, for $E_{bias} \parallel c$ than for $E_{bias} \perp c$. This is just the opposite of what the photoresponse measurements indicate. Since mobility and carrier lifetime are both factors in the photoresponse, the enhanced photoresponse for $E_{bias} \perp c$ may reflect a longer carrier lifetime for holes moving perpendicular to the c-axis.

Another consequence of the bias electric field direction is that the temperature dependence of the photoresponse was significantly different, as seen in Figure 3. For $E_{bias} \perp c$, the peak response drops rapidly between 90 K and 140 K. Since the band gap absorption is relatively insensitive to temperature changes, this decrease has to be related to the hole transport and lifetime. The thermal activation of a hole recombination center would decrease the measured response. For $E_{bias} \parallel c$, the peak response shows a very different trend with the signal gradually increasing with temperature up to 150K and then slowly decreasing. It is interesting that the temperature trends for the two transport directions appear to be "out-of-phase" with one another, with the peak for $E_{bias} \parallel c$ occurring at the minimum for $E_{bias} \perp c$. No immediate explanation is available for this unexpected behavior.

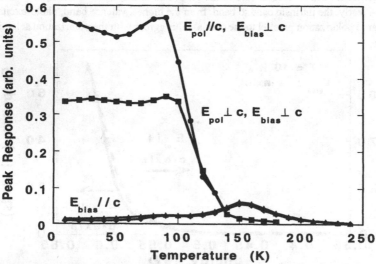

Figure 3. The temperature dependence of the maximum photoresponse for all four combinations of optical polarization and applied electric field.

As Fig. 3 indicates, photoresponse spectra for all four polarization and electric field combinations were taken from 10 K up to a maximum of 240 K, above 240 K the signal was no longer detectable. From these spectra, the energy gap (E_g) at different temperatures could be extracted using the Moss rule, i.e. the long wavlength edge of the intrinsic photoconductivity was extrapolated by a straight line to zero onset. Two such linear extrapolations are contained in Figure 1. In this process not all the spectra could be used since there were temperaature ranges where extrinsic processes would mask the true band edge. This extrinsic features did not always occur in the same temperature ranges for the different run conditions, so a composite of the band gap energy versus temperature could still be constructed.

The temperature dependent shift of these extrapolated band gap energies is shown in Figure 4. The data points were fit with the standard Varshni equation for empirically determining the temperature dependence of semiconductor band gaps. [9] The results of this fit produced : $E_g (T) = 0.580 - [3.6 \times 10^{-4} T^2 / (T + 160)]$ eV. The fitted values for α (3.6×10^{-4}) and β (160) are similar to those reported for other III-V semiconductors. There was one other reported linear fit of E_g (T) using data from temerature depèndent Hall measurements. [10] This linear fit yielded $E_g (T) = 0.673 - 3.5 \times 10^{-4} T$, which indicates a very similar temperature dependence to that determined from our photoresponse spectra.

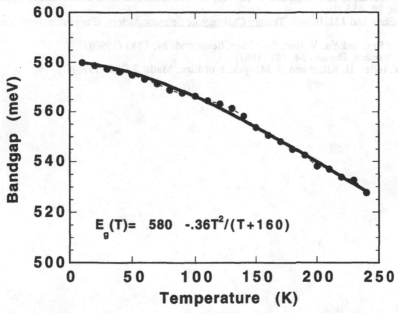

Figure 4. Temperature dependent shift of the CdGeAs$_2$ band gap energy. Data points are extrapolated from the measured photoresponse onset. The solid line is an empirical fit using the standard equation developed by Varshni [9]. The best fit results are shown for E_g (T).

CONCLUSIONS

The anisotropic nature of the crystalline and energy band structure of CdGeAs$_2$ leads to orientational dependent electrical and optical properties. We have studied these orientational effects on the intrinsic photoconductivity with respect to the crystal's optical axis (c-axis). The band edge photoresponse was strongest for an optical polarization parallel to the c-axis and an applied electric field perpendicular to this axis. Multiple sets of photoresponse spectra from 10 to 240 K, in 10 K increments, allowed a complete set of band gap energy versus temperature data to be constructed. A standard empirical equation was fit to this data to determine Eg (T). The large number of data points used allowed an excellent fit to to the data.

REFERENCES

1. V.G. Dmitriev, G. G. Gurzadyan, and D. N. Nikogosyan, Handbook of Nonlinear Optical Crystals, (Springer-Verlag, Berlin, 1991).
2. A. G. Jackson, M. C. Ohmer, S. R. Leclair, Infrared & Technology **38**, 233 (1997).
3. Y. V. Rud, Physical Phenomena in Ternary Compounds and Devices, (Fan, Taskent, 1994). (in Russian).
4. D. W. Fischer and M. C. Ohmer , J. E. McCrae, J. Appl. Phys. **81**, 3579 (1997).
5. P. G. Schunemann and T. M. Pollack, J. Cryst. Growth **174**, 272 (1997).
6. J. E. McCrae, R. L. Hengehold, Y. K. Yeo, M. C. Ohmer and P. G. Schunemann, Appl. Phys. Lett **70**, 455 (1997).
7. J. L. Shay and J.H. Herick, Ternary Chalcopyrite Semiconductors, (Pergamon, New York, 1975).
8. V. Yu. Rud and Yu. V. Rud, Sov. Phys. Semicond. **24**, 1352 (1990).
9. Y. P. Varshni, Physica **34**, 149 (1967).
9. G. W. Iseler, H. Kildal and N. Menyuk, J. of Elec. Matls. **7**, 737 (1978).

REFRACTIVE INDEX MEASUREMENTS OF BARIUM TITANATE FROM .4 TO 5.0 MICRONS AND IMPLICATIONS FOR PERIODICALLY POLED FREQUENCY CONVERSION DEVICES

DAVID E. ZELMON*, DAVID L. SMALL*, and PETER SCHUNEMANN**
*Materials Directorate, Air Force Research Laboratory, Wright-Patterson AFB, OH 45433-7707
**Sanders, A Lockheed-Martin Company, Nashua, NH 03061-0868

ABSTRACT

Barium titanate has recently been suggested as a possible candidate for use in periodically poled optical frequency conversion devices. We report refractive index measurements which are critical to periodically poled device designers and present quasi phase matching loci which can be used to design an optical parametric oscillator in periodically poled barium titanate

INTRODUCTION

Tunable frequency conversion devices, especially those with output in the infra red spectral region, have a wide variety of applications including high resolution spectroscopy, pollution detection, communications, petrochemical exploration, and medicine. Optical sum and difference frequency generators and parametric oscillators have been fabricated using many different device configurations and materials. However, because the performance of these devices depends on birefringent phase matching, they have been limited by inefficient energy transfer from the pump to the output beams and, in some cases, inability to perform certain frequency conversion processes due to lack of a phase matching direction. To circumvent the problem of birefringent phase matching, device designers have begun to take advantage of the concept of quasi phase matching to develop efficient frequency conversion devices[1-4]. Most of these quasi phase matched optical devices have been fabricated from periodically poled lithium niobate but recently, barium titanate has been suggested as a possible candidate for use in frequency conversion devices made by periodic poling. Its advantages over lithium niobate include transmission further into the infra red spectral region and a lower coercive field. Precise measurements of the refractive indices of this material as a function wavelength are necessary to predict the grating period required for any given frequency conversion process. We report measurements of the refractive index of barium titanate from 0.4 to 5.0 microns and discuss a sample grating structure for an optical parametric oscillator pumped at 1.064 microns.

Mat. Res. Soc. Symp. Proc. Vol. 484 © 1998 Materials Research Society

EXPERIMENT

The refractive indices of barium titanate were measured from 0.4 to 5 microns using the minimum deviation method on a prism fabricated from a boule grown from a top seeded solution. The prism was oriented and polished using standard techniques. The apex angle was measured optically and was 40.015 ±.0015 degrees.

A Gaertner L-124 precision spectrometer was used for the minimum deviation measurements. The spectrometer was calibrated by measuring indices of a calcium fluoride prism several times and determining the standard deviation. Our data were within 2×10^{-5} of the published values[12]. An Oriel mercury-xenon lamp source was coupled to a Digikrom L 240 monochromator to permit selection of wavelength at which the index was to be measured. The near IR measurements were made using an Electrophysics hand held IR viewer. For measurements from 2.5 to 5 microns the quartz optics of the spectrometer were replaced by a single ZnSe collimating lens. An infrared imaging camera (Cincinnati Electronics IRRIS 160 LN) was used in place of the imaging optics to detect the refracted beam. Five separate runs were made throughout the entire spectrum and the standard deviation of the data at any wavelength was less than 2×10^{-4}. The temperature at which the measurements were taken was 21^0 C.

RESULTS

The refractive indices as a function of wavelength are shown in figure 1.

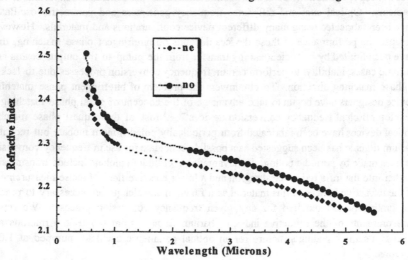

Figure 1. Ordinary and extraordinary refractive indices of barium titanate as a function of wavelength. Dotted lines represent the Sellmeir fit to the data.

The experimental error was less than .0002 over the entire transmission range of the material. The data were fit to a Sellmeir equation of the form[13]

$$n^2 = A + B \lambda^2 / (\lambda^2 - C) + D \lambda^2$$

using a modified Levenburg-Marquardt algorithm. The difference between the data and the predicted refractive index at any wavelength was less than 2×10^{-4} which is within the error in the experiment. The Sellmeir coefficients are shown in Table I for both ordinary and extraordinary refractive indices and the dotted lines through the data points in figure 1 represent fits to the Sellmeir equation. The Sellmeir coefficients derived from the data were used in all subsequent calculations.

Coeff	Ne	No
A	3.0584	3.02479
B	2.27326	2.14062
C	.074090	.067007
D	-.02428	-.02169

Table I. Sellmeier Coefficients for Barium Titanate

The combination of birefringence and dispersion characteristics of BaTiO$_3$ make it unsuitable for conventional phase matched frequency conversion. Extensive calculations for frequency doubling of 1.06 µ and optical parametric oscillation utilizing a 1.06 µ pump showed that birefringent phase matching is impossible for these processes. However, success of periodic poling in lithium niobate has led to the suggestion that parametric oscillation might be possible for BaTiO$_3$ albeit with a different grating structure. In order to take advantage of the largest nonlinear coefficient of BaTiO$_3$, d$_{15}$, we considered the case for which the propagation direction was along the y-axis, the polarizations of the pump wave and the signal wave were along the x-axis and the polarization of the idler wave was along the z-axis (Type II quasi phase matching). We then calculated the grating periods required for a type II optical parametric oscillator pumped at 1.06 µ. The results are shown in figure 2 in which the output signal and idler wavelengths of an optical parametric oscillator are plotted versus the grating period.

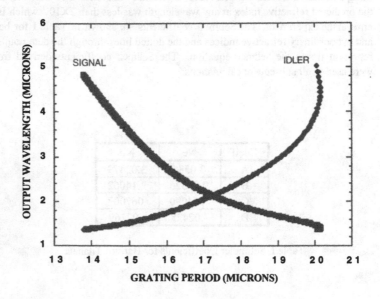

Figure 2. Output wavelengths vs. grating period for a type II quasi phase matched optical parametric oscillator.

The calculations indicate that parametric oscillation is possible using a multigrating structure and wavelengths of both the signal and idler waves can be varied between 1.35 μ and 5.0 μ using structures with grating periods ranging from 14 to 20 μ.

CONCLUSION

We have presented the most extensive and accurate refractive index data on barium titanate. Calculations based on this data show that birefringent phase matching for optical processes pumped at 1.06 μ are not possible. However, using quasi phase matching, we have shown that parametric oscillation is possible with output wavelengths between 1.35 μ and 5.0 μ using a series of gratings with periods varying from 14 to 20 μ.

REFERENCES

1. J. A. Armstrong, N. Bloembergen, J. Ducuing, , and P. S. Pershan, Phys. Rev., **127**, p. 1918 (1962)

2. M. M. Fejer, G. A. Mangel, D. H. Jundt, and R. L. Byer, IEEE J. Quant. Elec., **28**, p. 2631 (1992)

3. L. E. Myers and W. R. Bosenberg, IEEE J. Quant. Elec., **33**, p. 1663 (1997)

4. J. Webjorn, S. Siala, D. W. Nam, R. G. Waarts, and R. J. Lang, IEEE J. Quant. Elec., **33**, p. 1673 (1997)

5. W. R. Bosenberg, A. Drobshoff, J.I. Alexander, L.E. Myers, and R. L. Byer, Optics Lett., **21**, p. 713 (1996)

6. M. L. Bortz, M. A. Arbore, and M. M. Fejer, Optics Lett., **20**, p. 49 (1995)

7. L. E. Myers, R. C Eckardt, M. M. Fejer, R. L. Byer, W. R. Bosenberg, and J. W. Pierce, J. Opt. Soc. Am. B, **12**, p. 2102 (1995)

8. L. E. Myers, R. C Eckardt, M. M. Fejer, R. L. Byer, and W. R. Bosenberg, Opt. Lett., **21**, p. 591 (1996)

9. L. E. Myers, G. D. Miller, R. C. Eckardt, M. M. Fejer, R. L. Byer, and W. R. Bosenberg, Opt. Lett., **20**, p. 52 (1995)

10. A. Balakrishnan, S. Sanders, S. DeMars, J. Webjörn, D. W. Nam, R. J. Lang, D. G. Mehuys, R. G. Waarts, and D. F. Welch, Opt. Lett., **21**,952 (1996)

11. M. Yamada, N. Nada, M. Saitoh, and K. Watanabe, Appl. Phys. Lett., **62**, p. 435 (1993)

12. Irving Malitson, Appl. Opt., **2**, 1103 (1963)

13. Max Born and Emil Wolf, <u>Principles of Optics, 6th ed.,</u> , Pergamon Press, 1980, p. 94

Polarized Raman Scattering Study of ZnGeP$_2$ Single Crystals

Spirit Tlali*, Howard E. Jackson*, M. C. Ohmer**, P. G. Schunemann***
and T. M. Pollak***

* Department of Physics, University of Cincinnati, Cincinnati OH 45221-0011
** Wright Laboratory, Wright-Patterson AFB, OH 45433
*** Lockheed Sanders, Nashua, NH 03061

Abstract

Raman scattering experiments on high quality ZnGeP$_2$ single crystals grown by the seeded horizontal dynamic gradient technique have been carried out. Polarized Raman spectra were obtained in the backscattering geometry at both room and low temperatures for several crystal orientations and compared with group theoretical predictions. Raman spectra from as-grown and annealed samples display distinctive differences which were explored by utilizing two different excitation wavelengths: 514.3 nm and 632.8 nm; the observed differences are attributed to a surface interdiffusion effect.

Introduction

In this work high quality ZnGeP$_2$ single crystal samples were investigated using polarized Raman scattering. The orientation and polarization dependent anisotropic electrical and optical properties of ZnGeP$_2$ make it an excellent candidate for parametric infrared frequency shifting of radiation and for second harmonic generation/sum frequency mixing.[1] ZnGeP$_2$ has positive birefringence which makes it appropriate for phase matching from 1 to 11 micrometers even at room temperature, and is extremely transparent in the mid-infrared.[2] In addition, it has excellent mechanical and thermal properties particularly its large thermal conductivity (35 times larger than AgGaSe$_2$, for instance), which implies that thermal gradients and subsequent thermal lensing during a high-average power operation of a optical parametric oscillator (OPO) can be avoided.[3]

In the past the use of ZnGeP$_2$ based devices has been limited, to a large extent, by lack of large high quality single crystals[4] and a strong optical absorption extending from 0.7 μm to 2.25 μm.[4,5] Recent improvements in crystal growth procedures and post growth anneals have led to a major progress in the study of

ZnGeP$_2$ material.[6] These efforts have culminated in several recent demonstrations of 2.05 μm pumped type-I ZnGeP$_2$ optical parametric oscillators (OPOs).[2] Schunemann et al.[2] reported a 26% overall conversion efficiency and 37% slope efficiency at an average continuous power output of 0.6W. Additional improvements in the performance of ZnGeP$_2$ based OPOs can be expected if the material's extrinsic absorption near the pump wavelength at 2.05 μm is further reduced.

In spite of the great improvement in the growth and processing techniques, which may be expected to result in more perfect single crystals, no experimental studies using Raman scattering and/or infrared techniques on the phonon behavior of ZnGeP$_2$ single crystals as a result of thermal annealing process have been recently reported. In the work reported here, we investigate Raman scattering from ZnGeP$_2$ single crystals and explore the effects of thermal annealing.[7] Two excitation wavelengths were used for the acquisition of the Raman spectra, one above the bandgap, 514.5 nm, and the other below, 632.8 nm. In the following sections experimental Raman spectra obtained using two different excitation wavelengths will also be presented and discussed.

Experimental results

A. As-growth sample results with 514.5 nm excitation

Figure 1 shows a typical unpolarized Raman scattering spectrum acquired at room temperature from an as-grown sample excited $\bar{z}$ using the 514.5 nm line of

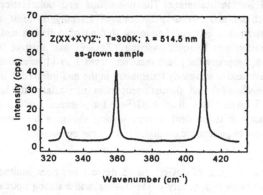

Figure 1: Raman spectrum of as-grown sample using 514.5 nm excitation.

the argon ion laser. For the data displayed in Fig. 1, the wavevector of the incident light was parallel to the c-axis with polarization along the x-axis, and the backscattered light thus had its polarization in the xy-plane (reflected in the notation z(xx+xy)z). In this geometry only longitudinal optic (LO)modes should be observed; transverse optic (TO) modes are not allowed. The narrow Raman lines observed in the spectra are indicative of a single crystal response and affirm that we are investigating high quality, well-ordered, single crystal sample. The full width at half maximum (FWHM) of the prominent peaks in this spectrum, estimated using a Lorentzian line fit, was less than 3 cm^{-1}, most which was an instrumental contribution. The spectral resolution available for this case was less than 1 cm^{-1}.

In order to identify vibrational modes with definite atomic displacements and symmetries, we have obtained a number of spectra in different geometries. For our purposes here, we wish, however, to explore the role of *annealing* and the changes which might then be observed in the Raman response. Annealing was attempted to try to reduce the near-band-edge losses that effect the performance of the material as an optical parametric oscillator. One might reasonably expect that annealing might sharpen the observed Raman linewidths, for instance, reflecting increased crystal order.

B. Annealed sample results with 514.5 nm and 632.8 nm excitation

The sample was thermally annealed at 500C for 300 hours in ZnP$_2$ powder; the near band edge absorption was reduced by annealing by a factor of two. The Raman results using 514.5 nm excitation just as above, however, were unexpected. We display in Fig. 2 a spectrum of the annealed sample where there is clear evidence of additional Raman response on the high energy side of the line at ~ 410

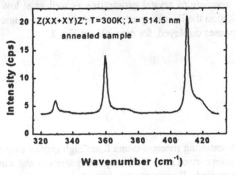

Figure 2: Raman spectrum from an annealed sample
using 514.5 nm excitation Compare to *Figure1*.

cm^{-1} as well as an additional response at ~ 370 cm^{-1}. These additional responses are not expected from group theoretical arguments, nor can they be attributed to other possibilities such as second order Raman scattering. In addition, these responses persist down to low temperatures. Changing the excitation wavelength, however, provides an interesting insight.

In Fig. 3 we show a Raman spectrum from an annealed sample in the same

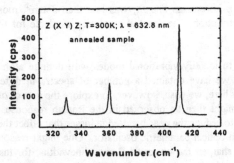

Figure 3: Raman spectra of annealed sample using 632.8 nm excitation.

orientation as the sample of Figs. 1,2, but excited with a wavelength of 632.8 nm. This spectrum is obtained by accepting only one of the scattered polarizations and should be sensitive to the response at ~ 410 cm^{-1}. One sees immediately that the response on the high energy side of this line is absent. We have pursued these differences in a variety of crystal geometries as well as at low temperatures. For 632.8 nm excitation the Raman spectra always indicate a single crystal without the additional responses displayed, for example, in Fig. 2.

Conclusions

Raman scattering measurements from high quality single crystal samples of ZnGeP$_2$ have been carried out. Both as-grown samples and samples that had been annealed were studied. Raman spectra from as-grown samples using an excitation wavelength of 514.5 nm indicated high quality single crystals, but data from annealed crystals with this sample excitation wavelength displayed unexpected additional features. Raman spectra from this same samples obtained using 632.8 nm excitation, a wavelength that is above the bandgap and thus able to penetrate into the bulk of the sample, do not display the additional features. We attribute these features to a change of the surface composition by interdiffusion and surface order during the extended annealing in the ZnP_2 powder.

References

1. See, for instance, P. A. Bundi, K. Ezzo, P. G. Schunemann, S. Minnigh, J. C. McCarthy, and T. M. Pollak, in *OSA Proceedings on Advanced Solid State Lasers*, ed. G. Dube and L. L. Chase, **10**, 335 (1993).

2. See, for instance, P. G. Schunemann, P. A. Budni, M. G. Knights, T. M. Pollak, E. P. Chicklis, and C. L. Marquardt, in *OSA Proceedings on Advanced Solid State Lasers*, ed. A. A. Pinto and T. Y. Fan, **15**, 166 (1993).

3. C. L. Marquardt, D. G. Cooper, P. A. Budni, M. G. Knights, K. L. Schepler, R. DeDomenico, and G. C. Catella, Appl. Opt. **33**, 3192 (1994).

4. M. G. Roelofs, J. Appl. Phys., **65**, 4976 (1989).

5. Y. M. Andreev et al., Sov. J. Quant, Elect. **14**, 1022 (1984).

6. P. G. Shunemann and T. M. Pollak, in *OSA Proceedings on Advanced Solid State Lasers*, ed. G. Dube and L. L. Chase, **10**, 332 (1991).

7. For a complete version of these experiments, including low temperature results and a symmetry analysis of the Raman response, see S. Tlali, H. E. Jackson. M. C. Ohmer, P. G. Schunemann, and T. M. Pollak, to be published.

ELECTRON-NUCLEAR DOUBLE RESONANCE STUDY OF THE ZINC VACANCY IN ZINC GERMANIUM PHOSPHIDE (ZnGeP2)

K. T. STEVENS*, S. D. SETZLER*, L. E. HALLIBURTON*, N. C. FERNELIUS**, P. G. SCHUNEMANN***, AND T. M. POLLAK***
*Department of Physics, West Virginia University, Morgantown, WV 26506
**Air Force Research Laboratory, WPAFB, Dayton, OH 45433
***Sanders, A Lockheed Martin Company, Nashua, NH 03061

ABSTRACT

As-grown crystals of ZnGeP$_2$ are highly compensated and contain significant concentrations of donors and acceptors. The dominant acceptor in ZnGeP$_2$ is believed to be the zinc vacancy. This center is paramagnetic in its normal singly ionized state, and gives rise to an electron paramagnetic resonance (EPR) signal characterized by a resolved primary hyperfine interaction with two equivalent phosphorus nuclei adjacent to the vacancy. The present investigation has focused on electron-nuclear double resonance (ENDOR) measurements of additional hyperfine interactions which are not resolved in the regular EPR spectra. Principal values and principal axes directions for four additional phosphorus nuclei are determined from the ENDOR angular dependence. These parameters support the zinc-vacancy assignment for the acceptor and they provide an experimental check of wave functions generated in future computational modeling efforts.

INTRODUCTION

The identification and characterization of point defects continues to be a primary focus of research in zinc germanium phosphide (ZnGeP$_2$). Much of this interest arises because of the detrimental effect that the point defects can have on the performance of nonlinear optical devices which employ this material. It has been demonstrated that an optical parametric oscillator (OPO) based on ZnGeP$_2$ crystals can be tuned across large regions of the mid-infrared [1,2]. The performance, however, of such an OPO is affected by an unwanted optical absorption band in the crystal which overlaps the desirable 2-µm pump region and thus limits the maximum pump intensity that can be used. This broad defect-related absorption band extends from 0.7 to 2.5 µm and must be eliminated, or greatly reduced, if ZnGeP$_2$-based OPOs are to reach their expected potential. A recent study has suggested that this near-edge absorption in ZnGeP$_2$ is associated with the presence of the zinc-vacancy acceptor [3].

A very intense electron paramagnetic resonance (EPR) spectrum is present in every ZnGeP$_2$ crystal. Initially, it was suggested [4] that this spectrum could be due to either a singly ionized zinc vacancy (V_{Zn}^-) or a singly ionized zinc on a germanium site (Zn_{Ge}^-). A subsequent investigation using electron-nuclear double resonance (ENDOR) provided lattice-distortion evidence which supported the zinc vacancy assignment [5]. Because of the importance of this acceptor in the extrinsic optical absorption and the need to further substantiate its model, we have extended the ENDOR investigation to include several additional weak phosphorus hyperfine interactions.

EXPERIMENT

The ZnGeP$_2$ crystals used in the present investigation were grown by the horizontal gradient freeze technique at Sanders, a Lockheed Martin Company (Nashua, NH). A typical sample had approximate dimensions of 3 x 3 x 3 mm³ with the faces perpendicular to the high symmetry directions. The EPR data were taken on a Bruker ESP-300 spectrometer operating at 9.45 GHz with 100-kHz static field modulation. An Oxford Instruments ESR-900 helium-gas-flow system maintained the sample temperature at values between 8 and 30 K. A Varian E-500 digital gaussmeter was used to measure the resonant values of magnetic field and a Hewlett Packard 5340A counter was used to measure the microwave frequency. A small MgO:Cr^{3+} crystal was used to correct for the difference in magnetic field between the ZnGeP₂ sample and the gaussmeter probe. The Bruker ESP 300 spectrometer also was used for the ENDOR measurements. In these double resonance experiments, the rf field was frequency modulated at 12.5 kHz with the depth of modulation varying from 50 to 100 kHz. The ENDOR coil was helical and was directly attached to the Oxford Instruments glassware extending through the Bruker ENDOR cylindrical cavity.

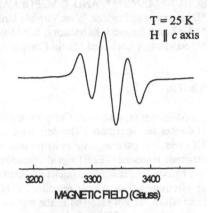

Figure 1. EPR spectrum of the dominant acceptor in ZnGeP₂.

RESULTS

The dominant singly ionized acceptor in ZnGeP₂ is paramagnetic (S = ½) and is easily seen without photoexcitation at temperatures below 50 K [4]. Its c-axis EPR spectrum is shown in Fig. 1. The unpaired spin is shared equally by two phosphorus nuclei (I = 1/2, 100% abundant), and this gives rise to triplets (1:2:1 line intensity ratios) in the EPR spectra. As previously described by Rakowsky et al. [4], the angular dependence of the EPR spectrum can be explained in terms of four crystallographically equivalent orientations of the defect. One of these sites is illustrated on the left in Fig. 2, where the two phosphorus ions labeled P_A are central to the defect. The principal values of the g matrix are 2.002, 2.021, and 2.074 and the corresponding principal axes for the particular site in Fig. 2 are the [011], [$\bar{1}$00], and [0$\bar{1}$1] directions, respectively [4].

An ENDOR study of the resolved hyperfine interactions (i.e., with the two primary phosphorus nuclei) has been reported [5]. This work revealed that the acceptor has significant lattice distortion, and it was argued that this provided strong evidence in favor of the zinc-vacancy model for the dominant acceptor in ZnGeP₂. In Fig. 2, the parameter φ represents the angle between the interphosphorus axis (heavy dashed line) and the basal plane of the crystal, i.e., the (001) plane. The initial ENDOR analysis [5] gave a value of 37.8° for this angle, which is considerably different from the value of 44.5° for the undistorted lattice. Such a large lattice distortion would not be expected for a Zn_{Ge} center because of its regular tetrahedral bonding.

It was recently suggested [6], and it now has been verified in the present investigation, that the sample was slightly out of the c-a plane when the ENDOR angular data reported in Figure 3 of Reference 5 was taken. An analysis of in-plane ENDOR data (both from the c-a plane and the a-a plane) shows that the two EPR-resolved phosphorus hyperfine interactions, are equivalent, within experimental error. The previous study had concluded that the unique principal values ($A_{1,z}$ and $A_{2,z}$) for these two primary phosphorus interactions differed by about 2.4%. The revised values for the spin-Hamiltonian parameters are given in Table I. These more recent results

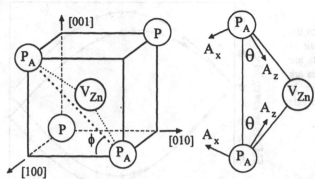

Figure 2. Zinc-vacancy model for the dominant acceptor in ZnGeP$_2$. The P$_A$ phosphorus ions sharing the unpaired spin are shown on the left. The hyperfine principal axes for these two nuclei are on the right.

do not affect the earlier arguments made in favor of the zinc-vacancy model, however, since the parameter ϕ changes only slightly from 37.8° to 37.9°.

In the present paper, we have extended the ENDOR technique to study additional phosphorus hyperfine interactions not resolved in the EPR spectra. These weaker interactions have ENDOR lines below 20 MHz, while the stronger hyperfine from the primary P$_A$ pair of phosphorus nuclei had ENDOR lines between 40 and 75 MHz. Figure 3 shows the lower frequency ENDOR spectrum when the magnetic field is parallel to the c axis. This spectrum contains many lines, which indicates that the wave function for the acceptor overlaps a large number of neighboring nuclei. In the remainder of this paper, attention is focused on the two lines located near 17 MHz and 14 MHz in Fig. 3. For convenience, they are labeled the P$_B$ pair and the P$_C$ pair of phosphorus nuclei, respectively. Because each of these ENDOR lines has a sufficiently large hyperfine parameter A, they and their companion lines are separated by $2\nu_N$ and centered on A/2. In our ENDOR experiments, the free nuclear-resonance frequency for phosphorus (ν_N) was approximately 5.74 MHz. This places the two companion lines near 2.5 MHz and 5.5 MHz, but these are regions where the spectrometer has less sensitivity and where large numbers of ENDOR lines overlap. Lower frequency companion lines have been observed for the 17 and 14 MHz lines when the magnetic field is parallel to the c axis, thus demonstrating that the responsible nuclei are phosphorus, but it proved impossible to follow the angular dependence of these low-frequency companions. The angular dependence of the 17 MHz and the 14 MHz ENDOR lines were easy to follow, however. Figure 4 shows the observed angular variation associated with the

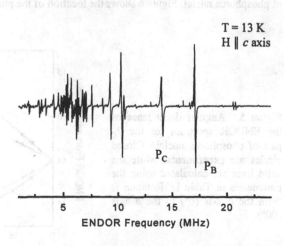

Figure 3. Low-frequency ENDOR spectrum for the acceptor.

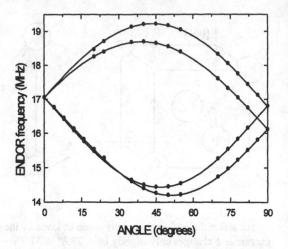

Figure 4. Angular dependence of the ENDOR spectrum for the P_B pair of phosphorus nuclei. Data points are experimental while the solid lines are calculated using the parameters in Table I. Rotation is from the c axis (0°) to the a axis (90°).

17 MHz interaction (P_B pair) as the magnetic field is rotated from the c axis to the a axis. Similar angular data is shown in Fig. 5 for the 14 MHz interaction (P_C pair). Data for both interactions also were taken with the magnetic field along the [110] direction.

The experimental data in Figs. 4 and 5 have been fit to the following spin-Hamiltonian with S = 1/2 and I = 1/2.

$$H = \beta S \cdot g \cdot B + S \cdot A \cdot I - g_N \beta_N B \cdot I \qquad (1)$$

The first term is the electron Zeeman, the second term is the hyperfine interaction with one phosphorus, and the third term is the phosphorus nuclear Zeeman. Values for the g matrix were taken from Reference 4. A least-squares fitting program repeatedly diagonalized the 4 x 4 Hamiltonian matrix to obtain the set of parameters which best fit the experimental data. These results are presented in Table I. Five parameters were included in the fitting procedure for each of the two pairs of phosphorus nuclei. Figure 6 shows the location of the phosphorus ions we have assigned to the

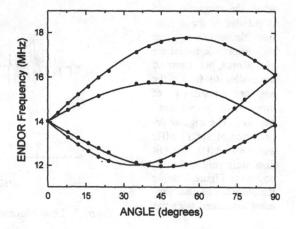

Figure 5. Angular dependence of the ENDOR spectrum for the P_C pair of phosphorus nuclei. Closed circles are experimental while the solid lines are calculated using the parameters in Table I. Rotation is from the c axis (0°) to the a axis (90°).

552

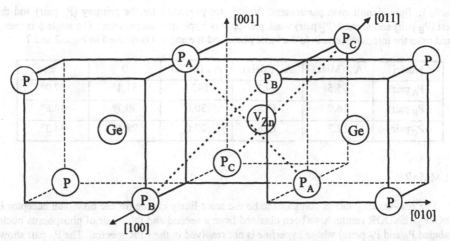

Figure 6. Zinc-vacancy model for the dominant acceptor in ZnGeP$_2$ showing the P$_A$, P$_B$, and P$_C$ pairs of phosphorus nuclei. The zinc vacancy is in the rear center "cube," while the left and right front "cubes" have germanium ions at their centers.

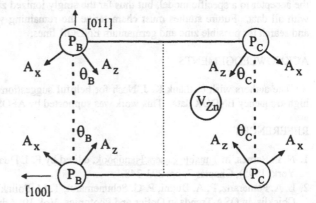

Figure 7. Hyperfine principal axes directions for the P$_B$ and P$_C$ pairs of phosphorus nuclei.

P$_B$ and P$_C$ pairs. There are three connected "cubes" illustrated in Fig. 6; the left and right "cubes" in the front are GeP$_4$ units while the back "cube" in the center contains the zinc vacancy. The P$_A$ pair is located between the P$_B$ and P$_C$ pairs. The P$_A$ and P$_C$ pairs are adjacent to the zinc vacancy, but on opposite sides. The P$_B$ pair is not adjacent to the vacancy. Figure 7 represents the (01$\overline{1}$) plane, containing the P$_B$ and P$_C$ pairs of phosphorus nuclei, and describes the principal axes directions. We have assigned the P$_C$ pair of phosphorus to be adjacent to the vacancy because they appear to have significant lattice distortion (a value of 34.3° for ϕ compared to 44.5° for the unperturbed lattice). This is similar to the result for the P$_A$ pair and is expected for all nuclei neighboring the vacancy. In contrast, the value of ϕ for the remaining pair is 46.6°, which suggests only a small amount of lattice distortion. Thus, the P$_B$ pair is assigned to nuclei away from the vacancy.

Table I. Spin-Hamiltonian parameters. Values are presented for the primary (P_A pair) and the first (P_B pair) and second (P_C pair) weak phosphorus hyperfine interactions. The angle ϕ is measured from the internuclear axis to the basal plane and the angle θ is defined in Figs. 2 and 7.

Nuclei	A_x (MHz)	A_y (MHz)	A_z (MHz)	θ	ϕ
P_A pair	95.5	99.3	143.5	31.4°	37.9°
P_B pair	16.9	17.5	30.0	30.7°	46.6°
P_C pair	12.3	12.6	27.0	28.6°	34.3°

SUMMARY

The zinc-vacancy model continues to be the most likely choice for the dominant acceptor in ZnGeP$_2$. ENDOR results have been obtained from a second and third pair of phosphorus nuclei (labeled P_B and P_C pairs) whose hyperfine is not resolved in the EPR spectra. The P_C pair shows significant lattice distortion and the P_B pair does not, although they have similar hyperfine interactions with the unpaired spin, i.e., their principal values are comparable. This is understandable if the P_C pair is adjacent to the zinc vacancy and the P_B pair is more distant from the vacancy, e.g., part of the neighboring GeP$_4$ units. It is difficult to obtain absolute evidence on which to assign the acceptor to a specific model, but thus far the singly ionized zinc vacancy (V_{Zn}^-) is consistent with all data. Future studies must characterize the remaining weak phosphorus ENDOR lines and search for possible zinc and germanium ENDOR lines.

ACKNOWLEDGEMENTS

The authors wish to thank K. J. Nash for helpful suggestions concerning the analysis of the high-frequency ENDOR data. This work was supported by AFOSR Contract F49620-96-1-0452.

REFERENCES

1. N. P. Barnes, in Tunable Lasers Handbook, edited by F. J. Duarte (Academic Press, New York, 1995), Chapter 7, pp. 293-348.
2. L. A. Pomeranz, P. A. Budni, P. G. Schunemann, T. M. Pollak, P. A. Ketteridge, and E. P. Chicklis, in OSA Trends in Optics and Photonics, Vol. 10, Advanced Solid State Lasers, edited by C. R. Pollock and W. R. Bosenberg, (Optical Society of America, Washington, DC 1997), pp. 259-261.
3. S. D. Setzler, L. E. Halliburton, N. C. Giles, P. G. Schunemann, and T. M. Pollak, in Infrared Applications of Semiconductors - Materials, Processing, and Devices, edited by M. O. Manasreh, T. H. Myers, and F. H. Julien (Mater. Res. Soc. Proc. 450, Pittsburgh, PA, 1997), pp. 327-332.
4. M. H. Rakowsky, W. K. Kuhn, W. J. Lauderdale, L. E. Halliburton, G. J. Edwards, M. P. Scripsick, P. G. Schunemann, T. M. Pollak, M. C. Ohmer, and F. K. Hopkins, Appl. Phys. Lett. 64, 1615 (1994).
5. L. E. Halliburton, G. J. Edwards, M. P. Scripsick, M. H. Rakowsky, P. G. Schunemann, and T. M. Pollak, Appl. Phys. Lett. 66, 2670 (1995).
6. K. J. Nash (private communication).

INFLUENCE OF QUANTUM CONFINEMENT ON THE PHOTOEMISSION FROM NONLINEAR OPTICAL MATERIALS

KAMAKHYA P. GHATAK,[*] P.K. BOSE[**] AND GAUTAM MAJUMDER[**]
Department of Electronic Science, University of Calcutta,
University College of Science and Technology,
92, Acharya Prafulla Chandra Road, Calcutta-700 009,
INDIA.
[**] Department of Mechanical Engineering,
Faculty of Engineering and Technology,
Jadavpur University, Calcutta-700 032, INDIA.

ABSTRACT

In this paper we have studied the photoemission from quantum wells (QWs), quantum well wires (QWWs) and quantum dots (QDs) of nonlinear optical materials, respectively on the basis of a newly derived electron dispersion law considering all types of anisotropies within the framework of k.p. formalism. It is found, taking $CdGeAs_2$, GaAs and InAs, as exmaples, that the photoemission increase with increasing photon energy in a ladder like manner and also exhibits oscillatory dependences with changing electron concentration with film thickness respectively for all types of quantum confinement. The photoemission current density is greatest in QDs and least in QWWs. In addition, the theoretical results are in agreement with the experimental observation as reported elsewhere.

Introduction

With the advent of Molecular beam Epitaxy, Fine line lithography and other experimental techniques, low-dimensional structures having quantum confinements of one, two and three dimensions such as QWs, QWWs, and QDs have in the last few years attracted much attention not only for their potential in uncovering new phenomena in material science and technology but also for their interesting device applications [1,5]. In this paper we shall study the photoemission from QWs, QWWs, and QDs of quantum confineed nonlinear optical materials on the basis of a newly derived electron dispersion law considering the anisotropic crystal potential, anisotropic effective electron masses and the spin-orbit splitting parameters of the valence bands respectively. We shall investigate the doping, thickness and the incident photon energy dependences of such photoemission from such quantum confined compounds taking $CdGeAs_2$, GaAs and InAs as examples.

Theoretical Background

The form of $\bar{k}.\bar{p}.$ matrix for nonlinear optical mateirals can be written as

$$H = \begin{bmatrix} H_1 & H_2 \\ H_2^+ & H_1 \end{bmatrix} \tag{1a}$$

where

$$H_1 = \begin{bmatrix} E_g & P_{\shortparallel}k_z & 0 & 0 \\ P_{\shortparallel}k_z & -(\delta+\frac{1}{3}\Delta_{\shortparallel}) & (\sqrt{2}/3)\,\Delta_{\perp} & 0 \\ 0 & \frac{\sqrt{2}}{3}\Delta_{\perp} & -\frac{2}{3}\Delta_{\shortparallel} & 0 \\ 0 & 0 & 0 & 0 \end{bmatrix}$$

$$H_2 = \begin{bmatrix} 0 & 0 & f_- & f_+ \\ 0 & 0 & 0 & 0 \\ f_- & 0 & 0 & 0 \\ f_+ & 0 & 0 & 0 \end{bmatrix}$$

in which E_g is the bandgap, $P_{\shortparallel}$ and $P_{\perp}$ are the momentum matrix elements ll and $\perp$ to the C-axis respectively, δ is the crystal field splitting parameter, $\Delta_{\shortparallel}$ and $\Delta_{\perp}$ are the spin orbit splitting parameters $\shortparallel$ and $\perp$ to the C-axis respectively $f_{\pm} = (P_{\perp}/\sqrt{2})(k_x \pm ik_y)$. The diagonolization of the above matrix leads to the dispersion relation of the conduction electrons in bulk specimens of non-linear optical materials as

$$U(\epsilon) = k_s^2 + V(\epsilon)\,k_z^2 \tag{1b}$$

where $U(\epsilon) = [c(\epsilon)/A(\epsilon)]$, $C(\epsilon) = \epsilon(\epsilon+Eg)\,[(\epsilon+E_g+\Delta_{\shortparallel})$

$$+ \delta\,(\epsilon+E_g + \frac{2}{3}\Delta_{\shortparallel}) + \frac{2}{9}(\Delta_{\shortparallel}^2 - \Delta_{\perp}^2)),$$

$A(\epsilon) = \hbar^2 E_g(\epsilon+\Delta_{\perp})\,[2m_{\perp}^*\,(E_g + \frac{2}{3}\Delta_{\perp})]^{-1}\,[\delta(\epsilon+E_g+\frac{1}{3}\Delta_{\shortparallel})$

$$+ (\epsilon+E_g)\,(\epsilon+E_g+\frac{2}{3}\Delta_{\shortparallel}) + \frac{1}{9}(\Delta_{\shortparallel}^2 - \Delta_{\perp}^2)],$$

$B(\epsilon) = \hbar^2 E_g(E_g + \Delta_{\shortparallel})\,[2m_{\shortparallel}^*\,(E_g + \frac{2}{3}\Delta_{\shortparallel})]^{-1}[(\epsilon+E_g)$

$(\epsilon + E_g + \frac{2}{3}\Delta_{\shortparallel})]$, $V(\epsilon) = [B(\epsilon)/A(\epsilon)]$,

$m_{\shortparallel}$ and $m_{\perp}$ are the effective electron masses at the edge of the conduction band, $\shortparallel$ and $\perp$ to the direction of C-axis respectively.

The modified electron energy spectra in QWs, QWWs and QDs can respectively be written as

$$U(\epsilon) = (n_x\pi/d_x)^2 + k_y^2 + V(\epsilon) k_z^2 \tag{2}$$

$$U(\epsilon) = (n_x\pi/d_x)^2 + (n_y\pi/d_y)^2 + V(\epsilon) k_z^2 \tag{3}$$

$$U(\epsilon) = (\pi n_x/d_x)^2 + (\pi n_y/d_y)^2 + V(\epsilon) (\pi n_z/d_z)^2 \tag{4}$$

where n_x, n_y and n_z are size quantum numbers along x, y and z directions respectively, d_x, d_y, d_z are the thickness along x, y and z axis respectively. The electron concentration in QWs, QWWs and WDs can, respectively, by written as

$$n_0 = (2\pi)^{-1} \sum_{n_x=1}^{n_{max}} (A_1 + B_1) \tag{5}$$

$$n_0 = (\pi)^{-1} \sum_{n_x=1}^{n_{xmax}} \sum_{n_y=1}^{n_{ymax}} (A_2 + B_2) \tag{6}$$

$$n_0 = (2/d_x d_y d_z) \sum_{n_x=1}^{n_{xmax}} \sum_{n_y=1}^{n_{ymax}} \sum_{n_z=1}^{n_{zmax}} F_{-1}(\eta) \tag{7}$$

where $A_1 = (U(\epsilon_F) - (n_x\pi/d_x)^2)/\sqrt{V(\epsilon_F)}$, $B_i = \sum_{r=1}^{S} \nabla_r [A_i]$,

i=1 and 2, $\nabla_r = 2(k_B T)^{2r}(1-2^{1-2r})\xi (2r)d^{2r}/d\epsilon_F^{2r}$, r is the set of integers, ϵ_F is the respective Fermi energy,

$A_2 = ((U(\epsilon_F) - (\pi n_x/d_x)^2 - (\pi n_y/d_y)^2)^{1/2}$, $\eta = (k_B T)^{-1}$

(ϵ_F-E'), E' is the root of (4) and $F_j(\eta)$ is the Fermi-Dirac integral of order j.

Thus the photoemission from QWs, QWWs and QDs can respectively, be expressed as

$$J_{2D} = (ew_0/2d_y) \sum_{n_{ymin}}^{n_{ymax}} (vn_0') \tag{8}$$

$$J_{1D} = (w_0 e/\pi^2\hbar d_y) \sum_{n_x=1}^{n_{xmax}} \sum_{n_{ymin}}^{n_{ymax}} (\overline{E}/n_y) (A_2+B_2) \tag{9}$$

$$\text{and } J_{3D} = (2ew_0/\pi\hbar d_x d_y)\sum_{n_{x=1}}^{n_{xmax}} \sum_{n_{y=1}}^{n_{ymax}} \sum_{n_{zmin}}^{n_{zmax}} (\rho/n_z)F_{-1}(\eta) \qquad (10)$$

where v is the electron velocity and n_0' is the electron concentration per subband in QWs, $h\nu + \epsilon_{nymin} \geq W$, $h\nu$ is the energy of the incident photon, W is electron affinity, $\overline{E}$ is obtained from (2) by putting $n_y = 0$ and $\epsilon = \overline{E}$, ρ is obtained by putting $n_x = n_y = 0$ and $\epsilon = \rho$ in (3), ω_0 is the probability of photon absorption. Under the conditions $\delta = 0$, $\Delta_{\shortparallel} = \Delta_{\perp} = \Delta$ and $m^*_{\shortparallel} = m^*_{\perp} = m^*$, (1) assumes the form [3]

$$\hbar^2 k^2/2m^* = \gamma(E), \qquad (11)$$

where $\gamma(E) = E(E+E_g)(E+E_g+\Delta)(E_g+2/3\Delta)[E_g(E_g+\Delta)(E+E_g+2/3\Delta)]^{-1}$

Which is known as three-band Kane model and must be used as such for studying the electronic properties of n-InAs where $\Delta \sim$ Eg.

For bulk specimens of degenerate wide bandgap matéirals, the expressions of the photoemission in quantum confined sturctures as given by (8) to (10) can be simplified as

$$J = (4\pi ew_0 m^* k_B^2 T^2/h^3)F_1[(h\nu - \phi_0)/k_B T] \qquad (12)$$

where ϕ_0 is the work function.

Under the condition of non-degeneracy together with the condition $\omega_0 = 1$, (12) assumes the well-known form [5] as

$$J = (4\pi em^* k_B^2 T^2/h^3) \exp((h\nu - \phi_0)/k_B T) \qquad (13)$$

Results and Discussion

Using the appropriate equations together with the appropriate physical parameters [6] we have plotted the photoemissions from quantum confined structures of CdGeAs$_2$, GaAs and InAs, we have plotted the photoemission from these materials with various physical variables in Figs.1 to 4 where the circular plots exhibit the experimental results [6].

The influence of quantum confinement is immediately apparent from the Figs. since the photoemission depends strongly on the thickness of the quantum confined structures in contrast with bulk specimens of the same compound. The appearance of the humps of the above figs. is due to the redistribution of the electrons among the quantized energy levels when the size quantum number corresponding to the highest occupied level changes from one fixed value to the other. The photoemission due to quantum confinement varies

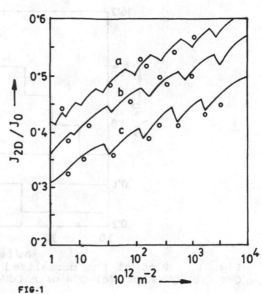

Fig. 1 Plot of the normalized photoemission as a function of n_o in QWs of (a) $CdGeAs_2$ (b) GaAs; (c) InAs; lattice matched to InP. The circular plot exhibits the experimental results. (d_x = 40 nm).

FIG·1

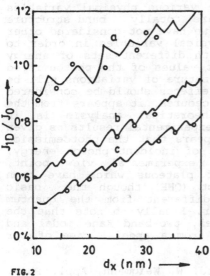

FIG.2

Fig. 2 Plot of the normalized photoemission versus d_x in QWWs of all the cases of Fig. 1. The circular plots exhibit the experimental results

(d_y = 60 nm, n_o = $(10^9$ $m^{-Y})$)

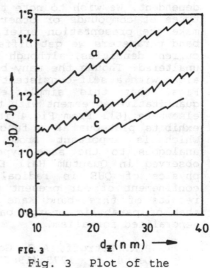

FIG. 3

Fig. 3 Plot of the normalized photoemission versus d_z in QDs of all the cases of Fig. 1 (d_x = d_y 40 nm, n_o = $(10^{14}$ $m^{-3})$

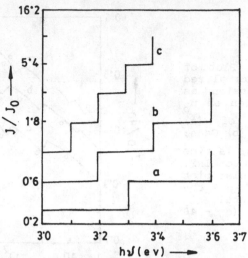

Fig. 4   Plot of the normalized photoemission versus h in QWs, (b) QWWs and (c) QDs of n-CdGeAs$_2$

with different manners with the various physical variables and the rates of variations are totally band-structure dependent. We wish to note that we have not considered other types of compounds or other physical variables in order to make the presentation brief. With different sets of energy band parameters we get different values of the photoemitted current densities, although the nature of variations will be unaltered. Though the many-body effects should be considered along with a self-consistent procedures, it appears from the Figs. that this simplified theoretical analysis is in quantitative agreement with the experimental results as given elsewhere [6]. From Fig.4 it appears that the photoemission exhibits plateous as a function of incident photon energy, which is important from the experimental view point, analogous to the same type of plateous which have been observed in Quantum Hall Effect (QHE) though the basic physics of QHE is radically different from the quantum confinement of our present paper. Finally to note that the results of three-band Kane model, two-band Kane model and that of parabolic energy bands from a special case of our generalized formalism.

1.   P.K. Petroff, A.C. Gossard, W. Weigmann, Appl. Phys. Letts. 49, 1275 (1986).
2.   K.P. Ghatak and B. Mitra, Int. J. Electron. 72, 541 (1992).
3.   K.P. Ghatak and S.N. Biswas, J. Vac. Sci. Tech. 7B, 104 (1989).
4.   K.P. Ghatak and M. Mitra, Phys. Scripta 46, 182 (1992).
5.   K.P. Ghatak, M. Mondal and S.N. Biswas, J. Appl. Phys. 68, 3032 (1990).
6.   L.M. Brandt and A. Aronov, Jour. of Exp. and Theo. Phys. 160, 719 (1997).

EXTRA-WIDE TUNING RANGE MID-IR OPTICAL PARAMETRIC
GENERATOR PUMPED BY Er-LASER PULSES

K.L.VODOPYANOV

Solid State Group, Physics Department, Imperial College, Prince Consort Road, London SW7 2BZ, UK. Tel. 0171-594 7589, Fax.0171-594 7580. e-mail k.vodopyanov@ic.ac.uk

ABSTRACT

Infrared pulses, continuously tunable in the 3-19 μm range, and with up to 1MW peak power, have been achieved using single-stage frequency conversion in a travelling-wave optical parametric generator (OPG) based on $ZnGeP2$, CdSe and GaSe crystals. The OPG was pumped by 100ps pulses from an actively mode-locked, Q-switched and cavity dumped 2.8 μm Cr,Er:YSGG laser. The quantum conversion efficiency reached 17%.

INTRODUCTION

Generation of coherent, continuously tunable mid-infrared radiation is of vital importance for remote sensing, molecular and solid state laser spectroscopy, isotope separation, biology and medicine. Nonlinear optical frequency conversion is an attractive way to achieve broad tunability in the middle infrared range of frequencies. As a general rule, nonlinear optical figure of merit of crystals dramatically increases, as their bandgap decreases (roughly as one over the 9-th power the bandgap of the material), consequently the nonlinear optical figure of merit of "infrared" crystals can be very high (e.g. $ZnGeP_2$ with its bandgap of 1.67eV has the figure of merit, which is 95 times bigger than in $LiNbO_3$ with its bandgap of 3.1 eV).

As it was shown recently, short Er-laser λ=2.8μm pulses can be successfully used for pumping travelling-wave optical parametric generators (OPG) based on highly nonlinear $ZnGeP_2$ and GaSe crystals - to achieve high peak power mid-infrared radiation with extra-wide tunability1. The principle of operation is based on the high gain ($>10^{10}$) amplification of quantum noise in a nonlinear crystal pumped by intense short light pulses. The main advantage of such devices is that the travelling wave configuration makes it possible to have the OPG output in the form of a single intense burst of radiation.

EXPERIMENT

We used in this work three types of highly nonlinear crystals, namely $ZnGeP_2$, GaSe and CdSe. $ZnGeP_2$ is remarkable for having one of the largest nonlinear-optical coefficients, among all the nonlinear-optical crystals in practical use, however its transparency is limited to 10μm. As seen from the Table 1, GaSe crystal has an extreme transparency which starts from the visible (0.65-18μm) and high second order nonlinearity (d_{NL}= 5.44x10^{-11}pV/m = 0.72 d_{NL} ($ZnGeP_2$)). However GaSe is a soft layered crystal and its maximum length is limited to 1-2 cm. As for CdSe - despite its nonlinear optical coefficient of d_{31} =18 pm/V being 3 and 4 times smaller than that of GaSe and $ZnGeP_2$ respectively, it has certain advantages, namely wide transparency range (0.75-

25μm), extremely low optical losses (<0.01 cm^{-1} over the 1-10 μm region), and availability of long (>5cm) crystals.

Table 1. Linear and nonlinear optical properties of ZnGeP$_2$, CdSe and GaSe

Crystal	ZnGeP$_2$	CdSe	GaSe
Transparency range (μm)	0.74-12	0.75-25	0.65-19
n$_o$ (near λ=3 μm)	3.13	2.455	2.85
n$_e$	3.17	3.475	2.46
Nonlinearity d$_{eff}$ (10^{-12} m/V)	88	18	54.4
Nonlinear figure of merit, d^2/n^3 (10^{-24} m^2/V^2)	248	22	128

Here we report travelling-wave (resonator-free) optical parametric generation based on ZnGeP$_2$, GaSe and CdSe.

As a pumping source we used single pulses of an actively mode-locked, Q-switched and cavity dumped Cr3$^+$,Er^{3+}:YSGG laser[2] (using the Er^{3+} ion 4I$_{11/2}$ - 4I$_{13/2}$ transition, λ=2.797μm) operating at a repetition rate of 3 Hz. The laser pulses had 100 ps duration and an energy of 2-3 mJ in the TEM$_{oo}$ mode.

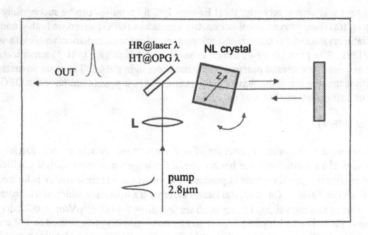

Fig. 1. Double-pass travelling -wave optical parametric generator (pump @ 2.8μm)

We used a two-pass travelling-wave OPG setup (Fig.1). A dichroic beamsplitting mirror (DM) which reflects the laser beam and transmits the OPG signal and idler output. The back reflecting gold mirror was located about 3-4 cm from the end of the nonlinear optical crystal.

The laser beam was astigmatically (resulting in an elliptical beam-shape) focused into the crystal in such a manner that the pump laser intensity was approximately 1 GW/cm^2 on the first pass through the crystal and 4 GW/cm^2 on the second pass (stronger focusing on the second pass compensates for the absorption and Fresnel reflection losses). Thus the first pass served as a superluminescent seed, while the second as a parametric amplifier. The advantages of the two-pass (as compared to a single-pass) scheme are the absence of the walk-off of the output beam when the crystal is angular-tuned (which is quite important for our applications in the pump-probe spectroscopic measurements), suppression of the off-axis parametric generation, and improved spectral and spacial quality of the output pulses

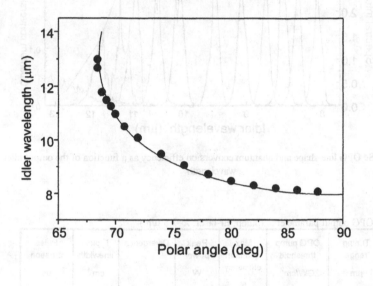

Fig.2. CdSe angular tuning curve (idler wave). Solid line - calculated.

The experimental angular tuning curve (CdSe crystal) is shown in Fig.2 and manifest an excellent agreement with the calculated tuning curve (solid line) based on published dispersion relations.

Fig. 3 shows spectral lineshape and quantum conversion efficiency as a function of the idler wavelength; the linewidths were typically 10 cm^{-1} for both signal and idler waves, except for in the vicinity of the "turning point" near $\lambda=13\mu m$ (Fig.2), where the spectrum becomes much broader, typically ~ 100cm⁻¹. The OPG divergence in our experiment was measured to be ~1.5x(diffraction limit), which enabled us to routinely focus the idler beam into a 10λ spot with $>10^9$ W/cm² peak power density.

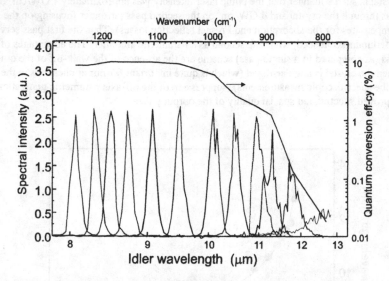

Fig.3. CdSe OPG line-shape and quantum conversion efficiency as a function of the output idler wavelength

Table 2. OPG output parameters   (pump: Er-laser: $\lambda=2.8\ \mu m$)

Crystal	Tuning range μm	OPG pump threshold GW/cm²	Max. quantum efficiency %	Peak power W	Divergence	Typical linewidth cm⁻¹	Pulse duration ps
ZnGeP₂	3.9-10	0.09 (L=4 cm)	17%	~10⁶	~1.5 diffr	10	75-90
CdSe	8-13 3.57-4.3	0.47 (L=5 cm)	10%	~10⁶	~1.5 diffr	10	75-90
GaSe	3.3-19	1.1 (L=1.4cm)	5%	~0.3x 10⁶	~1.5 diffr	20-30	75-90

The other OPG parameters are summarised in the Table 2.

CONCLUSION

Travelling-wave OPG was achieved in $ZnGeP_2$, GaSe, and CdSe with up to 3-19 μm tunability range, ~1MW peak output power and typical quantum conversion efficiency of 10%. The comparison of three crystals shows that:

- $ZnGeP_2$, which has the largest nonlinear-optical coefficient, has the smallest OPG pump threshold (90 MW/cm^2)

- GaSe OPG has the largest tunability range of 3-19 μm but higher pump threshold

- CdSe OPG is superior in the sense of efficiency and the linewidth to $ZnGeP_2$ or GaSe, in the spectral range of 8-12 μm

ACKNOWLEDGMENTS

I would like to thank C.C.Phillips for his interest to this work, L A Kulevskii and A.V. Lukashev for valuable discussions and help in testing the crystals, and G. B. Serapiglia for help in performing the experiment. This work has been supported by the UK Engineering and Physical Sciences Research Council.

REFERENCES

1. K. L. Vodopyanov and V. Chazapis, Opt. Commun. **135**, 98 (1997).
2. K. L. Vodopyanov, J. Opt. Soc. Am. B **10**, 1723 (1993).

Operational Characteristics of GaSe Crystals for Mid-IR and Far-IR Applications

N. C. Fernelius+, F.K. Hopkins+, N.B. Singh*, D. Suhre*, M. Marable**, R. H. Hopkins*, R. Meyer**, and P. Mui**
+Materials Directorate, Air Force Research Laboratory, Wright-Patterson AFB, OH 45433
*Northrop Grumman Corporation, STC-ESSD, 1350 Beulah Road, Pittsburgh, PA 15253
** Northrop Grumman Corporation, ESID, Rolling Meadows, IL 60008

ABSTRACT

GaSe has a number of attractive properties for nonlinear optical applications including large birefringence for ease in phase matching. Its biggest drawback is its mechanical properties. GaSe has a strong tendency to cleave along the <100> plane which has made it difficult to grow and fabricate. We have developed a method to modify GaSe by structurally strengthening the material by doping. We have synthesized large boules of GaSe reacted mixtures and grown centimeter size single crystals by the Bridgman technique. Depending on the dopant and crystal quality, SHG measurements indicate a d_{eff} of 51 to 76 pm/V. SHG power levels were theoretically calculated and appear to be in good agreement with the experimental data. The measured performance of crystals for the fourth harmonic generation and laser damage threshold are also reported in this paper. The damage threshold was greater than 2.8 J/cm^2 and 85 KW/cm^2 at the surface of the crystal.

BACKGROUND

There is a great need for optical wavelength conversion devices for a number of applications including electro-optic countermeasures. The particular interest is in high-average-power devices spanning the atmospheric windows of 3-5 μm (Mid IR) and 8-12 μm (Far-IR) range; tunable devices in these wavelength ranges are also highly desirable. Since most suitable high-average-power lasers lie outside the 2-12 μm range, their output must be shifted to these ranges using nonlinear devices fabricated from single crystal materials. To perform effectively, these materials must possess a unique combination of optical, thermal, and mechanical properties. None of the currently available materials have all the characteristics needed for high-average-power operation. GaSe has some interesting properties for nonlinear frequency conversion applications of infrared laser light either as second harmonic generation, sum or difference frequency generation, or optical parametric oscillation. GaSe's most outstanding feature is its broad transparency from 0.7 to 18 μm with no intermediate absorption features. It has a rather large nonlinear coefficient, moderately high thermal conductivity perpendicular to the optic axis, and an above average damage threshold. These latter properties facilitate high power operation. This makes GaSe a very useful nonlinear material for the 2-18 μm range and is a good candidate for use between the 10 and 18 μm. GaSe has a strong tendency to cleave along the <100> plane which makes them very difficult to grow and fabricate. We have synthesized large at batches of GaSe reacted mixtures and grown centimeter size doped single crystals by the vertical Bridgman technique. With proper growth conditions and slow cooling we have grown cm size crystals with good optical transparency and reduced cracking. These results along with demonstrated capability of growing device-sized single crystals and high second harmonic generation (SHG) "d" values for modified GaSe crystals are reported in this paper.

EXPERIMENTAL METHOD

Gallium and selenium (6-9's pure) were further purified remove residual impurities. The stoichiometric mixture was prepared by mixing elemental Ga and Se. The temperature of the mixture was raised in steps to 1050°C and maintained for many hours to ensure complete mixing.

The GaSe phase was examined by X-ray diffraction powder pattern which confirmed the GaSe phase.

Crystal growth was carried out by the capillary seeded Bridgman method. The details of this growth method is described in reference [1-3]. We used a pyrolysed quartz tube for the ampoule. The quartz tube was well cleaned and annealed at 1000°C to remove the residual impurities. The temperature gradient was 30 K/cm and the growth speed was 2 cm/day during the growth of pure and doped GaSe. The ampoule size were between 15 to 23 mm in diameter.

Crystals of cm size were fabricated by cutting with a string saw and polishing with diamond powder. The surfaces of the slabs were polished to examine the transparency and evaluate second harmonic (SHG) and fourth harmonic generation characteristics.

A CO_2 laser beam was quadrupled using GaSe as the final output doubler crystal. It consists of a Lumonics Model 203 CO_2 laser tuned to the 9P(22) transition at 9.569 μm. The laser operated at about 0.5 Hz, and the pulses were sent through a series of attenuators consisting of five CaF_2 plates, each with an optical density of about 0.2 at 9.6 μm. The plates were moved in and out of the laser beam, so that the laser intensity could be varied without altering the characteristics of the laser pulse, as would occur by adjusting the laser voltage.

RESULTS AND DISCUSSION

In our studies single crystal growth of undoped GaSe crystals always resulted in cleaved crystals. These crystals were not suitable for testing. Undoped crystals have a tendency of cleaving in the form of thin plates. Due to the softness of crystals, bending and deformation was also very common. In-doped crystals had better mechanical property and fabricability. These crystals had less tendency to cleave than pure GaSe. When we doped GaSe crystals with In, we were not able to increase the concentration of dopant in the matrix due to the low distribution coefficient in GaSe (<0.2). However we were able to grow cm size crystals and tested the SHG performance of these crystals. In-doped crystals did not show precipitates in the matrix.

To further improve the GaSe fabricability we doped the crystals with silver. We introduced this in the form of compound to maintain the stoichiometry of GaSe. We observed that the Ag doped GaSe was much improved mechanically and showed even better fabricability than in-doped GaSe crystals.

The conversion efficiency was measured in a manner similar to that described in reference 4. A CO_2 laser operating at 9.25 μm was used as the SHG pump source, which was operated at 1 kHz, could produce pulses having a FWHM pulse width of 20 nsec, and the pulse energy could be varied up to a mJ. The energy was focused using a ZnSe lens with a focal length of 28 cm, and the GaSe samples were placed at the focus, and the angular alignment was adjusted for maximum SHG output. This resulted in an incidence angle of about 42°, which corresponds to an internal angle of about 14°.

The optical power into the crystal was measured with a power meter, along with the transmitted primary power. The optical faces of the GaSe samples were uncoated, and about 44% of the primary power was transmitted through the crystal, as predicted by the Fresnel equations for these angles of incidence and transmission. The pulse energy inside the crystal could therefore be accurately scaled by dividing the input power by the repetition rate and scaling by the reflection loss at the first surface.

The SHG efficiency was measured using a sapphire plate to block the 9.25 μm beam, and the fundamental power was also measured without the sapphire plate. After scaling by the reflection losses of the sapphire plate, the ratio of these powers gave the SHG efficiency. A "d" value of 51pm/V was determined by this method. The "d" coefficient for silver doped compound was determined in a manner similar to that described above. The source was a pulsed CO_2 waveguide laser with a pulse repetition rate of 15-100 kHz and a pulse width of 11 ns (FWHM) reduce in the fundamental. The output second harmonic power was measured after the crystal, using a sapphire plate to block the fundamental. The measurement of the SHG efficiency and nonlinear *d* coefficient were carried out using the GaSe AF-39 crystal. The measured value for the "d" coefficient was 76 pm/V for this crystal which is much higher than 51 measured for earlier GaSe crystals. This higher value of "d" increases the conversion efficiency (d^2/n^3) by an order of magnitude compared to commercially available materials such as Tl_3AsSe_3 and $AgGaSe_2$.

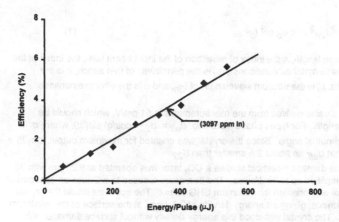

Figure 1. Measure SHG performance of 9-mm doped GaSe sample.

For the fourth harmonic conversion efficiency a Tl_3AsSe_3 (TAS) harmonic generator was used to convert part of the energy at 9.569 μm to 4.785 μm, and an aperture with a diameter of 7 mm was used to define the beam size of the frequency doubled beam. This beam was directed onto a 6 mm long sample of GaSe doped with 7184 ppm of indium, which formed the final output doubler that converted part of the energy at 4.875 μm into 2.392 μm. The wavelengths were separated using sapphire and pyrex plates, which could also be moved in and out of the beam path. The signal energy was measured with a Gentec ED-100 pyroelectric joulemeter having a flat response over wavelength, so that the signals at the first, second, and fourth harmonic frequencies could be directly compared. The TAS crystal was first oriented for maximum second harmonic output, and the GaSe crystal was next adjusted for maximum fourth harmonic output. The first harmonic beam was absorbed by the sapphire plate, leaving the second and fourth harmonics. The pyrex plate was then used to absorb the second harmonic beam, leaving only the fourth harmonic. Since the beams were not focused and the interaction lengths were fairly long, the angular alignment of both crystals was quite critical.

The silver doped GaSe crystal faces were rotated for maximum output. The c-axis was rotated approximately 30° to the beam direction, with ordinary polarization of the incidence beam. This orientation corresponds to Type I phase matching, and the internal polar phase matching angle of 10° is approximately correct for this wavelength conversion. The crystal was also rotated about the azimuthal angle for maximum output. We observed that heavily Ag-doped crystals had slightly different phase matching angle (smaller) than pure or In-doped crystals. Further study is underway to determine the exact phase matching angle for Ag-doped crystals.

The GaSe crystal faces were not coated, and the Fresnel losses were taken into account using indices of refraction calculated from the Sellmeier equations [5-10]. The reflection losses from the sapphire and pyrex plates were also accounted for, so that both the conversion efficiency

569

of the GaSe from the second to the fourth harmonic along with the second harmonic energy inside the crystal could be determined. The results are shown in Fig. 2, where the conversion efficiency from 4.785 to 2.392 μm is plotted as a function of the 4.785 μJ energy inside the crystal. A straight line was then projected through the data, which corresponds to the prediction of plane wave harmonic generation without depletion, with a nonlinear d value of 51 pm/V.

Since the laser beam was of a large diameter, both focusing and double refraction do not have to be taken into account, and the intensities of the second and fourth harmonics are given by [5]

$$I_{4n} = 8 p d^2 L^2 I_{2n}^2 / (c e_o n^3 I_o^2)$$ (1)

where L is the interaction length, n the index of refraction of the input beam (also the index of the output beam under phase matched conditions), e_o is the permittivity of free space, c is the vacuum velocity of light, I_o is the vacuum wavelength of I_{2n}, and d is the effective nonlinear coefficient in m/V.

The effective d value derived from the measurements is 51 pm/V, which should be independent of wavelength. For type I phase matching, $d_{eff} = -d_{22} \cos(q) \sin(3f)$, where q is the polar and f the azimuthal angle. Since the crystal was oriented for maximum output, $sin(3f) = \pm 1$, and $q = 10°$, so that d_{eff} is about 2% smaller than d_{22}.

For the damage threshold studies a CO_2 laser was operated at 9.25 μm with 20 ns pulses that were amplified up to 1mJ/pulse. The beam was focused to a Gaussian spot size of 150 μm and the crystal was oriented for maximum SHG output. The spot size inside the crystal was same as at the surface, giving an energy density of 2.8 J/cm² at the surface at the maximum energy of 1 mJ/pulse. The crystal withstood this energy density without surface damage. We increased the power up to the limit laser could provide, which was 30 watts at a rep rate of 30 KHz. This represents 85 KW/cm² power at the surface of the crystal.

SUMMARY

We have grown cm size GaSe crystals doped with indium and silver. The doped crystals showed "d" values of 51 and 76 pm/V which are very high compared to commercially available crystals for the mid-IR and far-IR wavelength region. We could not damage the crystal up to the power limit our laser could provide, which was 2.8 J/cm² and 85 KW/cm² at the surface of the crystal.

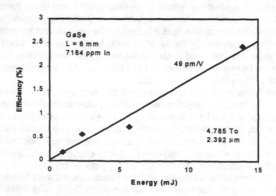

Figure 2 Measured FHG performance of 6-mm doped GaSe sample

ACKNOWLEDGMENTS

Authors are grateful to Materials Directorate, Air Force Research Laboratory, Wright-Patterson AFB, for financial assistance. Technical help of Mr. R. P. Storrick, A. Stewart, and R. Hamacher is sincerely appreciated.

REFERENCES

1. N. B. Singh, T. Henningsen, V. Balakrishna, D. R. Suhre, N. C. Fernelius, F. K. Hopkins and D. E. Zelmon, J.Crystal Growth, **163**, p.398 (1996).
2. N. B. Singh, R. Narayanan, A.X. Zhao, V. Balakrishna, R.H. Hopkins, D. R. Suhre, N. C. Fernelius, F. K. Hopkins, and D. E. Zelmon, Materials Science and Engineering, (In Press).
3. N. B. Singh, T.Henningsen, R. H. Hopkins, K. C. Yoo, R. Mazelsky, Z. Kun, Progress in Crystal Growth and Characterization, **20**, p.175 (1990).
4. D. R. Suhre, N. B. Singh, V. Balakrishna, N. C. Fernelius, and F. K. Hopkins, Optics Letters, **22**, p. 775 (1997).
5. B. Abdullaev, K. R. Allakhverdiev, L. A. Kulevskii, A. M. Prokhorov, E. Yu. Salaev, A. D. Saval'ev, and V. V. Smirnov, Sov. J. Quantum Electron. **5,** P.665 (1975).
6. A. Yariv, *Introduction to Optical Electronics*, (Holt. Rinehart, and Winston, New York, 1971), p. 190.
7. K. L. Vodopyanov and V. G. Voevodin, Opt. Commun. **114,** P.333 (1995).
8. G. B. Abdullev, K. R. Allakhverdiev, M. E. Karasev. V. I. Konov, L. A. Kulevskii, N. B. Mustafaev, P. P. Pashinin, A. M. Prokhorov, Y. M. Starodumov, and N. I. Chapliev, Sov, J. Quantum Electron. **19,** P.492 (1989).
9. J. L. Oudar, Ph. J. Kupecek, and D. S. Chemla, Opt. Commun. **29,** P.119 (1979).
10. A.V.Vanyakin, Quantum Electronics (Moscow),**27(2)**, 137 (1997).

ACKNOWLEDGEMENTS

REFERENCES

FREQUENCY DOUBLING OF CW AND PULSED CO_2 LASERS USING DIFFUSION-BONDED, QUASI-PHASE-MATCHED GAAS STACKS

MELVIN C. OHMER *, SHEKHAR GUHA* , RONALD E. PERRIN*, LAURA S. REA*, PHIL WON YU*, AND AYUB FATHIMULLA **
*USAF Research Laboratory, Materials Directorate Wright Patterson AFB, OH 45433
**AlliedSignal Aerospace Co., Microelectronics & Technology Ctr, Columbia MD, 21045

ABSTRACT

We describe the fabrication, characterization and frequency doubling properties of first order stacks of diffusion bonded 2" diameter GaAs wafers. Near IR imaging through the stacks indicated that excellent bonding was obtained over 40%-70% of the central area of the wafers. A power spectrum analysis of the spectral noise (due to interface reflections) appearing in the transmission data is shown to be a quantitative diagnostic tool useful for determining interface quality and for accurately estimating the thickness of the bonded layers in a stack. The conversion efficiency for a four layer stack at 10.6 microns was found to be 0.03% or 3 mJ for a 10 mJ pulse with a 100 nanosecond pulse length. The corresponding efficiency for cw SHG was found to be 0.0002%.

INTRODUCTION

In 1962, Armstrong et. al.[1] suggested that is would be possible to use a stack of wafers of a semiconductor such as GaAs with proper crystal orientations and thicknesses to efficiently convert the wavelength of a pump laser via the nonlinear processes of second harmonic generation or optical parametric oscillation. The high value of the second order nonlinear optical susceptibility, the laser damage threshold and the thermal conductivity of GaAs make it a suitable candidate for frequency conversion of multiwatt lasers. The wafer thickness should be an odd multiple of the coherence length of the pump wavelength which for GaAs is in the range of 103-107 microns. [2,3] The orientation should be such that the E vector will be parallel to the [1,1,1] direction in the crystal. Most recent work has utilized a stack of the form (100),(110), -(110), (110), -(110).......(100) where the (100) wafers are inactive end caps. The efficiency increases as the square of the number of wafers in the stack and it is inversely dependent on the stack order. In 1996 Szilsagi et. al. and Thompson et. al. successfully demonstrated the viability of this approach. Szilsagi et. al. used a 3rd order stack of 5 (110) wafers aligned at the Brewster angle to eliminate Fresnel reflection losses. Thompson et. al. used a 1st order stack of nine (111) wafers also aligned at the Brewster angle. Note that to attain conversion efficiencies above 50%, ninety or more 100 micron thick wafers are required. As the 1st order stack wafer thickness is nominally 1/5 the thickness of a conventional wafer, the wafers are quite fragile and free standing wafers stacks did not seem to be practical. Additionally, tolerance to error for large n is almost[5] non-existent. However, in 1993 Gordon et. al. proposed diffusion-bonding GaAs wafer stacks as an approach to overcome some of these limitations. In principle, this approach could make quasi-phase matched GaAs stacks competitive with and potentially a replacement for the two state-of-the-art nonlinear optical crystals, $ZnGeP_2$ and $AgGeSe_2$ and the emerging materials, $CdGeAs_2$ and $AgGaTe_2$.[6]

Pragmatically, this approach deserves a serious assessment and the results of one such assessment will be described in this paper. The fabrication process for the wafer stacks, the optical properties of these stacks, and their 10.6 micron pump frequency doubling efficiencies will be discussed. Stacks consisting of 3, 4, 5, and 7 (110) oriented wafers capped on each end with a nonfunctional (100) wafer have been fabricated and tested. The thickness of the (110) oriented wafers was equivalent to one coherence length at 10 microns taken as 103 microns in thickness, that is, a first order quasi-phase matched stack for optimum efficiency.

EXPERIMENTAL PROCEDURES AND RESULTS

The GaAs wafer bonding process is as described below. The wafers must be double side polished. As they are only about one fifth as thick as a standard wafer, they are extremely fragile. Additionally, they are (110) oriented wafers where the standard wafer is oriented (100). The steps in the bonding process are: 1) Inspect as received double side polished wafers and select wafers having smooth surfaces and equal thicknesses; 2) Clean and surface treat the wafers (proprietary process developed at MTC for bonding Si-Si and Si-to-III-V); 3) Contact the GaAs wafers to form a stack and inspect via IR transmission for particulates, voids, etc.; 4) Place the stack in a PBN fixture and place it in a "hot press" and apply pressure at 300° C; 5) Inspect the stack via IR transmission for bonds or voids; 6) Anneal in a furnace at 850° C for one hour to increase the bond strength; 7) A final inspection of the stack via IR transmission. The PBN fixture had a 54 mm inner diameter cup and 52 mm outer diameter cap. This process has been successfully used to produce perfect bonds between two 3" diameter GaAs wafers as determined by near IR imaging.

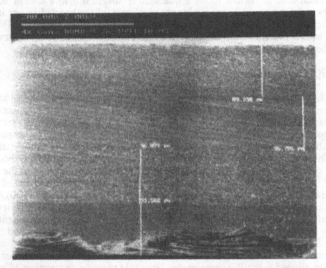

Figure 1. Vertical TEM Of A Well Bonded 4-Layer GaAs Stack. Thicknesses Shown In Microns Are 89.3, 96.8, 96.1, and 91.6. A 200 Micron Bar Is Shown For Scaling.

A vertical TEM of a well bonded 4-layer GaAs stack is shown in Figure 1 along with a 200 micron bar for scaling. The wafer thicknesses shown in microns are respectively from top to bottom 89.3, 96.8, 96.1, and 91.6. Note that thicknesses vary from 89.3 to 96.8 microns in this stack. This corresponds to about an 8% variation in thickness. If one takes 103-107 microns as the coherence length for GaAs, then all of the wafers are running about 6.5-10 % too thin. A vertical TEM of a poorly bonded 5-layer GaAs stack is shown in Figure 2 along with a 500 micron bar for scaling. Delaminations and slip regions are visible. Interfaces like those depicted in Figure 2 with air gaps will be far from perfectly transmissive and they will produce large undesirable reflections. In fact, each wafer interface is a source of some reflection which as we shall see later can be used as a measure of the quality of the interfaces in a stack. In this stack, there are three wafers of thickness 97.2 microns and two of 100.0 microns. Again all wafers are on average 4.4-8% too thin. The large air gap is 3.8 microns wide.

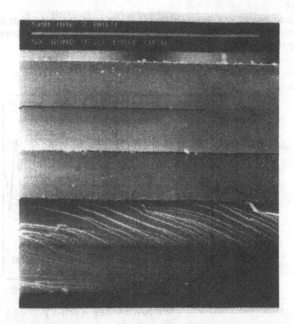

Figure 2. Vertical TEM Of A Poorly Bonded 5-Layer GaAs Stack. Delaminations And Slip Regions Are Visible. A 500 Micron Bar Is Shown For Scale.

The near IR transmission of 2" wafers showing large well bonded central regions is shown in Figure 3. The well bonded region is basically equal to the well contacted region obtained in the contacting step. The left stack consists of 5 thin wafers plus 1 standard wafer, and the right stack consists of 7 thin wafers plus 2 standard wafers. As a .5 x .5 cm² cross-sectional area is adequate for most laser applications, these wafers can be diced to provide a number of useful parts. This techniques was used to produce the four "identical samples" identified in Table 2. Spectral infrared transmission data was acquired on a Digilab FTS-20E Fourier-transform spectrometer in the range of 2.5-12 microns at a resolution of 1 cm⁻¹ at room temperature, nominally 19° C. The spectral infrared transmission data for these stacks is typically very noisy due to coherent interference of the transmitted light with that reflected from the interfaces in the stack. In Figure 4, the data shown is for one of four samples cut from the same bonded wafer stack. A periodic pattern of 7 lines in two groups of lines, one with 3 lines and one with 4 lines is apparent for this sample. It will be shown that the analysis of this noise can be useful for determining the interface quality and for accurately determining the thicknesses of the bonded

Figure 3. Near IR Transmission Of 2" Wafers Showing Large Well Bonded Central Regions. At Left, 5 Thin Plus One Standard Wafer, And At Right, 7 Thin Plus Two Standard Wafers.

layers in the stack. Note that an uncoated solid GaAs crystal will transmit about 51% at 1000 cm⁻¹ or 10 microns, and that this sample with four interfaces has a transmission of 38%. However, this is still a useful value.

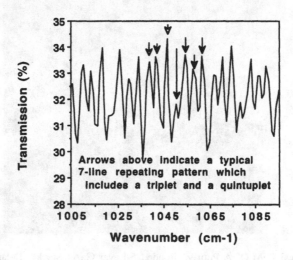

Figure 4. Transmission Near 10 Microns For A 5-Layer Stack, Three Thin (110)s And Two Thick (100)s.

Figure 5 shows the results of a power spectrum analysis of the spectral transmission for sample GaAs5-X1. Power is plotted vertically in arbitrary units, and interestingly, frequency is plotted in centimeters, a unit of length. Standard analysis provides a direct correlation between the frequency position and the physical thickness. For the case of GaAs it is, $t=10,000f/n=1515.15f$ where t is in microns and n is the index of refraction of GaAs in the IR. This spectrum is basically a Fourier transform of the transmission spectra. A library program in MatLab, called SPECTRUM, was used to perform the analysis. Sample GaAs5-X1, is one of four "identical samples", that is, they were cut from the same wafer stack. The stack consisted of five wafers in the order, (100), (110), -(110), (110), (100), where the (110) wafers were nominally 100 microns in thickness. Figure 6 shows a power spectrum analysis of the spectral transmission for sample GaAs5-X4 for comparison to GaAs5-X1. Notice that the spectra for these two samples cut from the same wafer stack are quite different. For sample X1, the largest peak is near the frequency 0.25, while for X4, the largest peak is near the frequency 0.30, and that the relative intensity of the line in both spectra near 0.15 is quite different. The power spectrum differences and similarities and their implications for the four "identical samples", X1, X2, X3, and X4 will be discussed in detail.

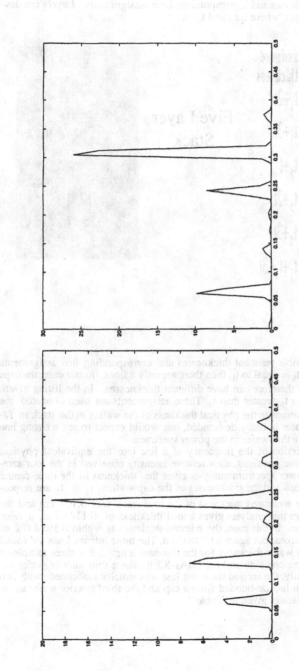

Figure 5. (left) & Figure 6. (right). Power Spectrum Analysis Of Spectral Transmission. For Respectively, Sample GaAs5-X1 And GaAs5-X4. Line positions 1-8 at 0.0674, 0.1354, 0.1736, 0.2047, 0.3076, 0.4442, and 0.066.

Table 1. Possible Resonant Thicknesses and Corresponding Line Assignments. Layers are designated from left to right t_1 through t_5 where t_2, t_3, and $t_4 = t$.

line #	assigned thickness	line#	assigned thickness
L1	t	L7	t_1+2t
L2	2t	L8	t_1+3t
L3	3t	L9	t_5+t
L4	t_1	L10	t_5+2t
L5	t_5	L11	t_5+3t
L6	t_1+t	L12	t_1+t_5+3t

Five Layer Stack

Table 1 identifies the possible resonant thicknesses and corresponding line assignments used in the analysis. Note that if t_5 is equal to t_1, then there are only 8 lines. In our case, the caps are repolished after the anneal and therefore can have different thicknesses. In the listing given, the assigned thickness increases for t_5 greater than t_1. These assignments are used to related the power spectrum analysis peak positions to the physical thickness of the wafers in the stack in Table 2. For example if one thin wafer is badly de-bonded, one would expect to see a strong line corresponding to the thickness of a thin wafer in the power spectrum.

Table 2 provides a conversion of the frequency of a line into the equivalent physical thickness and a summary of the line positions, their relative intensity observed in the four samples studied. The results of the power spectrum analysis gives the thickness of the three central wafers, t, as 102.7+ or - 0.6 microns and the thicknesses of the cap wafers, t_1 & t_5 are respectively 374 and 466 microns. The measured thickness of the total stack was 0.107 cm and the sum of the wafer thicknesses from this analysis gives a total thickness of 0.1148 cm, a value within 0.7% of the measured value. In this case, the relevant thickness is within 0.3%-4.0% of the coherence length, 103-107 microns, but again a bit too thin. The most intense lines are found to usually correspond to a single wafer thickness for the resonant length. For three samples it was a capping layer de-bonding on one side and for GaAs-X2 it was a thin interior wafer de-bonding on both sides. Additionally, the second strongest line was usually associated with two well bonded interior wafers which had de-bonded from a cap and the third interior wafer as inferred from the relative intensity listings given in Table 2.

Table 2. Summary of the line positions and their relative intensities observed in power spectra and sample transmission at ten microns.

Line #	Position Microns	Relative -	Proposed ID -	GaAs5-X1 Relative	GaAs5-X2 Intensity	GaAs5-X3	GaAs5-X4
1	102	1.00	t	3	1	3	2
2	205	2.01	2t	2	2	2	4
?	263	2.58	Ghost?	7	-	-	7
3	310	3.04	3t	6	-	-	6
4	374	3.57	t1	1	-	1	3
5	466	4.56	t5	5	3	4	1
9	566	5.54	t5+t	4	5	5	5
10	673	6.59	t5+2t	8	4	6	-
Trans	at 10	microns		32%	22%	28%	38%

A CO_2 TEA laser with a 100 ns pulse length and an energy per pulse of 10 mJ and a 5 W cw CO_2 laser both operating at 10.6 microns were used as pump sources to evaluate the SHG conversion efficiency of stacks containing four (110) wafers similiar in cross-section to the sample shown in Figure 1. A 10mJ pulse produced 3 microjoules of output for .03% conversion efficiency. Two stacks used in tandem produces 3.7 times more power than a single stack. For the 5W cw CO_2 laser, 4.8W incident produced 10 microwatts of output for .0002% efficiency.

These conversion levels attained using just a four layer stack, are certainly useful to provide researchers with easy access to laser wavelengths they might require for testing optical devices such as IR detectors or for other research projects. However for high power, to be competitive with $AgGaSe_2$ or $ZnGeP_2$, a stack of 80 to 90 wafers would be required in a stack fabricated with a high degree of perfection. At present, GaAs stacks do not appear to be competitive with these other approaches for obtaining high power IR laser sources via SHG.

SUMMARY

GaAs first order stacks perform adequately to be useful as providers of low power IR laser emission via SHG that is otherwise not readily available. A 10 mJ pulse with a 100 ns pulse length produced 3 microjoules of output for .03% conversion efficiency. Two stacks used in tandem produces 3.7 times more power than a single stack. The demonstration of cw frequency conversion is also reported. For the 5 W cw CO_2 laser, 4.8W incident produced 10 microwatts of output for .0002% efficiency. The first power spectrum analysis of transmission data has proved to be a very useful non-destructive method to assess the quality of the bonds in the stack.

REFERENCES

1. J. A. Armstrong, N. Bloembergen, J. Ducuing, and P. S. Pershan, Phys. Rev. **127**,1918(1962).
2. A. Szilsagyi, A. Jhordvik, and H. Schlossberg, J. Appl. Phys. **47**,2025(1976).
3. D. E. Thompson, J. D. McMullen, and D. B. Anderson, Appl. Phys. Lett.,**29**,113(1976).
4. L. Gordon, G. L. Woods, R. C. Eckardt , R. R. Route, R. S. Feigelson, M. M. Fejer, and R. L. Byer, Electronic Letters,**29**,1942(1993).
5. Martin M. Fejer, G. A. Magel, Dieter H. Jundt, and Robert Byer, IEEE J. of Quantum Elec.,**28**, 2631(1992).
6. A. G. Jackson, M. C. Ohmer, S. R. Leclair, Infrared & Technology,**38**, 233(1997).

CHARACTERIZATION OF CdGeAs$_2$ USING CAPACITANCE METHODS

S.R. SMITH*, A.O. EVWARAYE**, and M.C. OHMER
Materials & Manufacturing Directorate, Air Force Research Laboratory, MLPO, Wright-Patterson Air Force Base, OH 45377-7707
*University of Dayton Research Institute, 300 College Park, Dayton, OH 45469-0178
**University of Dayton, Department of Physics, 300 College Park, Dayton, OH 45469-2314

ABSTRACT

Thermal Admittance Spectroscopy(TAS) has been used to detect energy levels in the bandgap of CdGeAs$_2$ specimens. Capacitance-Voltage(CV) measurements were used to determine the net free carrier concentration of the specimens as well as the conductivity type. All specimens were found to be p-type. CV measurements determined that the free carrier densities ranged from 1.2×10^{17} cm^{-3} to 8×10^{18} cm^{-3}. Usually one peak (but in some cases two) was observed in the thermal admittance spectra. One peak present in two samples indicates an acceptor with a thermal activation energy of E$_V$+(0.10-0.13) eV which corresponds closely to the value of 0.10-0.12 eV found from Hall effect measurements on these specimens. The additional peak observed could correspond to a second deeper acceptor at E$_V$+0.346 eV, however, the energy could not be accurately determined because the peak was not fully resolved. Evidence for the existence of two native acceptors from electron paramagnetic resonance has recently been reported which tends to support a two acceptor model.

INTRODUCTION

CdGeAs$_2$ is a ternary chalcopyrite nonlinear optical material which is potentially useful for frequency conversion in the infrared because it has the largest nonlinear optical coefficient of any known phase-matchable compound (235 pm/V).[1] It also has a wide transparency range (2.4-18 μm). The primary application of CdGeAs$_2$ is for frequency doubling the output of CO$_2$ lasers. Until recently large single crystals were not available; however, Schunemann has succeeded in overcoming the growth problems by using a seeded horizontal gradient-freeze growth technique with a temperature gradient of only a few degrees per centimeter and a very slow cooling rate.[2]

CdGeAs$_2$ is a narrow band gap semiconductor. The generally accepted band gap value is 0.57 eV at room temperature.[3] The band gap increases to 0.65 eV at 77 °K.[4] The principal source of charge carriers is acceptor-like defects. These acceptors are thought to originate primarily from native defects such as cation vacancies and cation disorder. These defects have been characterized by Hall effect[5] and photoluminescence(PL),[6] by infrared absorption techniques, and by electron paramagnetic resonance(EPR).[7] We have applied Thermal Admittance Spectroscopy (TAS) to determine the energies of shallow electronic levels. In addition, these measurements have yielded information about the band structure of the material.

Mat. Res. Soc. Symp. Proc. Vol. 484 © 1998 Materials Research Society

EXPERIMENTAL

The technique of thermal admittance spectroscopy involves monitoring the response of a Schottky diode as a function of frequency and temperature. The details of this experiment have been adequately presented elsewhere, hence they will be omitted in this discussion.[8]

The specimens used in these experiments were obtained from Lockheed-Sanders Corp. They were received with both sides polished, and no further processing was done. Al Schottky contacts were fabricated on one side and In Ohmic contacts were applied to the other. C-V measurements were made at 1 MHz with an AC amplitude of 20 mV. The quality of the Schottky diodes was monitored using the linearity of the $1/C^2$ vs V curve. TAS measurements were made at frequencies ranging from 100 kHz to 100 Hz, and from 5 °K to 300 °K with an AC amplitude of 50 mV. The results of C-V and TAS measurements were compared to previously reported temperature dependent Hall effect measurements.

RESULTS AND DISCUSSION

The net doping concentration of the specimens was determined by C-V measurements. Four specimens were examined. The results are summarized in Table I. The net free carrier concentrations ranged from 1×10^{17} cm^{-3} to 8×10^{18} cm^{-3}.

Table I Net free carrier concentration as determined by C-V measurements and activation energy of the shallow acceptor as determined by TAS for four CdGeAs$_2$ specimens.

	5600 (4Q)	5601 (4O)	5603 (4N)	5604 (2G)
N_A-N_D (cm^{-3})	1.2×10^{17}	7.8×10^{17}	8.4×10^{18}	---
E_A (eV)	0.10	0.111	0.004-0.005	---

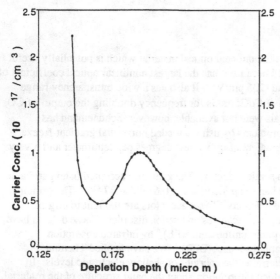

Figure 1. C-V concentration profile of specimen 5600.

In most cases, the CV profile was similar to that shown in figure 1, where a peak in the carrier concentration indicates a possible damage layer beneath the surface of the material. Many of the specimens showed visible polishing damage, so this is not unreasonable. This layer did not affect the CV measurement, nor did it affect the TAS measurement, except to possibly present another electrically active level within the bandgap.

Specimens 5600 and 5601 yielded good values for the activation energy of the shallow acceptor, but the concentration of the acceptor in 5600 was so low that the peak was not very well resolved, hence the error is larger.

Specimen 5603 exhibited hopping conduction at very low temperatures, thus it was not possible to obtain an energy for the ground state of the shallow acceptor. It was not possible to obtain data on specimen 5604.

Optical microscopy revealed numerous scratches on most of the specimens, although surface conductivity related to the scratches was not large enough to corrupt the performance of the Schottky diodes fabricated on these surfaces. Specimen 5601 was the smoothest surface, having only a few random scratches. Specimen 5603 was very badly scratched, however, it was still possible to obtain TAS data from the diodes on this material. It is probable that the peak seen in the C-V profile of figure 1 is polishing damage that was concentrated below the surface, and is not related to the scratches seen optically.

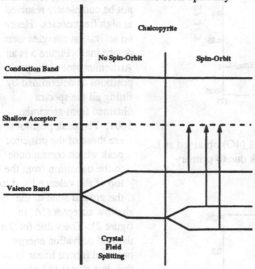

Figure 2 Schematic of CdGeAs₂ band structure, including non-degenerate sub valence bands.

TAS was used to determine the energy of thermally activated transitions within the temperature range 5 °K to 350 °K. Since this material has a relatively narrow bandgap, these transitions could include transitions from non-degenerate subvalence bands. Figure 2 is a schematic of the band structure of ordered CdGeAs₂.[9] This figure includes subbands made non-degenerate by order induced splitting. Transitions from the subbands would be detected by TAS as peaks at higher temperatures. While peaks were seen in some high temperature TAS spectra, they could not be resolved at higher frequencies, so a thermal activation energy could not be determined. Nevertheless, we believe that these peaks indicate that TAS measurements can be used to determine the valence band splitting in CdGeAs₂.

Figure 3 presents the primary TAS peak for specimen 5601 (4O) acquired at a frequency of 1 MHz. Fitting of the peak was accomplished using a commercially available spectroscopic fitting program. The activation energy of the transition responsible for this peak was obtained from the Arrhenius plot of $1/kT$ vs $\ln(\omega/T^2)$, where T is the fitted peak temperature, and ω is the measurement frequency. The energies obtained for two of the four specimens used in these experiments are tabulated in table I. Two of the specimens (5603 and 5604) were heavily doped and did not present peaks for the shallow acceptor. One of the specimens, 5604 exhibited a peak at very low temperatures (5-10 °K) indicative of hopping conduction. An energy was estimated from two peaks to be about 4-5 meV. Only two peaks were resolved, because the lowest temperature peaks were below the temperature capability of our system. A shoulder is evident on the low-temperature side of the primary peak.

Figure 4 shows the TAS spectrum obtained at 60 kHz, and the fit to the spectrum. Resolution of the shallowest (lowest T) peak is only marginal. The degradation of the resolution of the shallower peaks prevented the determination of an accurate thermal activation energy. The identity of the shallow acceptor which appears as a small peak in figure 4 is not known. It could not be resolved in all of the spectra to a degree that would permit an accurate determination of energy. It may, however, be a small transition from one of the split-off valence bands to the top of the valence band. Further work is necessary to confirm this.

Higher temperature TAS measurements indicated that even deeper transitions were present by bumps on the quadratic increase of the conductance. However, these peaks could not be completely resolved at high frequencies. Hence no activation energies were determined. Figure 5 is an Arrhenius plot of the peak positions as determined by fitting all the spectra obtained from specimen 5601. The peak positions are those of the principal peak which corresponds to the transition from the top of the valence band to the ground state of the shallow acceptor ('A' in figure 2). The value for the thermal activation energy obtained from a linear fit to this data was 0.111 eV. This value agrees with the value obtained by temperature dependent Hall effect measurements.

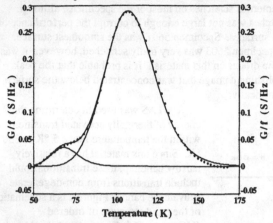

Figure 3. TAS spectrum for specimen 5601 (4O) obtained at 1 MHz measurement frequency showing peak due to primary acceptor.

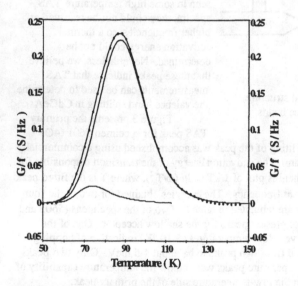

Figure 4. TAS spectrum for specimen 5601 (4O) obtained at 60 kHz measurement frequency showing position of fitted peaks.

CONCLUSIONS

Thermal Admittance Spectroscopy has been used to determine the energy of the principal acceptor in CdGeAs$_2$ crystals. The energy obtained agrees well with that obtained via such methods as temperature

584

dependent Hall effect. Furthermore, TAS reveals peaks possibly attributable to transitions from non-degenerate sub valence bands to the shallow acceptor. Even deeper levels were detected, but incomplete data prevented the accurate resolution of the peaks by the fitting routine.

ACKNOWLEGEMENTS

The authors would like to acknowledge the technical contributions of Mr. Gerald Landis and Mr. Robert V. Bertke in the preparation of the specimens. One of us (SRS) was supported by Air Force contract no. F33615-96-C-5445.

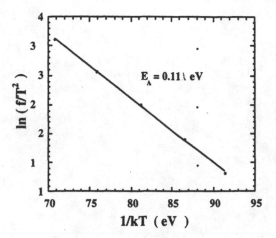

$E_A = 0.11 \text{ eV}$

Figure 5. Arrhenius plot of the fitted peak positions for specimen 5601 (4O). The activation energy is that of the shallow acceptor in $CdGeAs_2$.

REFERENCES

1. V.G. Dmitriev, G.G. Gwzadyan, D.N. Nikogosyan, Handbook of Nonlinear Optical Crystals, Springer-Verlag, Berlin, 1991
2. P.G. Schunemann and T.M. Pollack, J. Cryst. Growth 174, p.272 (1997)
3. R. Madelon, E. Paumier, and A. Hairie, Phys. Stat. Sol. (b) 165, p.435 (1991)
4. L.P. Achimchenko, V.S. Ivanov, and A. S. Boschchevskii, Sov. Phys. Semicond. 7, p.309 (1973)
5. J.E. McCrae, R.L. Hengehold, and Y.K. Yeo, Appl. Phys. Lett. 70, p.455 (1997)
6. D.W. Fischer, M.C. Ohmer, and J.E. McCrae, J. Appl. Phys. 81, p.3579 (1997)
7. L.E. Halliburton, G.J. Edwards, P.G. Schunemann, and T.M. Pollack, J. Appl. Phys. 77, p.445 (1995)
8. A.O. Evwaraye, S.R. Smith, and W.C. Mitchel, J. Appl. Phys. 75, p.3472 (1994)
9. J.L. Shay and J.H. Wernick Ternary Chalcopyrite Semiconductors: Growth, Electronic Properties and Applications, Pergamon, New York, 1975

Part VIII

Related Contributions

Chemical Bonding on GaAs (001) Surfaces Passivated Using SeS₂

Jingxi Sun[*], Dong Ju Seo[**], W. L. O'Brien[***], F. J. Himpsel[**], T. F. Kuech[*]
[*]Department of Chemical Engineering
[**]Department of Physics
[***]Synchrotron Radiation Center
University of Wisconsin-Madison
Madison, WI53706

ABSTRACT

Selenium disulfide has been demonstrated to be an effective passivant for GaAs (001) surfaces. This chemical treatment can be more robust and effective in reducing surface-states-based Fermi level pinning than other analogous chemical treatments. We have studied SeS₂-passivated surfaces, formed by treatment of GaAs in SeS₂:CS₂ solution, with synchrotron radiation photoemission spectroscopy. The SeS₂-treated surface consists of a chemically stratified structure of several atomic layers thickness. The As-based sulfides and selenides appear to reside on the outermost surface with the Ga-based compounds adjacent to the bulk GaAs substrate. The motion of the Fermi level within the band gap was monitored during controlled annealing conditions allowing for the specific chemical moieties responsible for the reduction in surface charge to be identified. As-based species are removed at low annealing conditions with little motion of the Fermi level. GaSe-based species, formed on the surface, are clearly shown to be associated with the unpinning of the Fermi level.

INTRODUCTION

A stable Se-based passivation layer can be produced on the GaAs (001) surface by contact with a SeS₂ solution [1,2,3,4]. This treatment appears to be very effective in depositing Se-based species on the GaAs surface and in improving the electronic properties through the formation of a thermally and chemically stable passivating layer [4]. As in other cases of chemical-based surface passivation, the underlying passivating mechanism of the SeS₂-treated GaAs (001) surface still is not clearly understood. Knowledge of the chemical bonding at the treated surfaces and the corresponding electronic properties has been incomplete and is presently addressed in this study.

The GaAs (001) surfaces passivated by SeS₂, using wet chemical techniques, have been studied with synchrotron radiation photoemission spectroscopy. Core-level spectra exhibiting predominantly either surface or bulk features have been obtained using the tunability of synchrotron radiation source. These results reveal a rich surface-derived substructure that is due to changes in the near surface chemical bonding environment. The core-level shifts induced by surface Fermi level movements have been identified. A structural layer model is proposed based on the analysis of the spectroscopic results.

EXPERIMENT

Si-doped n-type GaAs (001) wafers with a carrier density of 1×10^{18} cm^{-3} were used this study. The native oxide on the surface was initially removed with HCl:H₂O (1:1) solution and blown dry with N₂ gas. The samples were subsequently dipped into a saturated solution of SeS₂ in CS₂ for 40 seconds. Any excess SeS₂ on the surface was removed by repeated CS₂ rinsing. The above treatment was performed at room ambient. Samples prepared using above described

SeS$_2$-treatment will be referred to as the 'as-treated' samples. The photoemission experiments were carried out at the Synchrotron Radiation Center (SRC) of the University of Wisconsin-Madison. Monochromatic light of different photon energies was used to control the emitted electron kinetic energy to enable the depth profiling. The photoelectrons were analyzed with a double pass cylindrical mirror electron kinetic energy analyzer (CMA). Kinetic energies (KE) of 40 eV and 300 eV, corresponding to photoelectron escape depths of 0.5nm and 1.5 nm, were utilized to measure the surface-sensitive and bulk-sensitive spectra, respectively. A clean GaAs (001) surface, as a reference for subsequent measurements, was obtained by thermal annealing up to 600°C in the photoemission chamber under ultra-high vacuum (UHV). The Fermi level position on the sample holder was determined relative to a reference sample of Ta foil that was cleaned by Argon sputtering and thermal annealing. In order to investigate the nature of chemical bonding on the surface, the 'as-treated' samples were annealed in the photoemission chamber under UHV at increasing temperatures. The sample temperature during thermal annealing was measured by an optical pyrometer.

RESULTS

The changes of photoemission peak intensity ratios between the different elements were used to deduce the in-depth elemental composition resulting from the SeS$_2$-treatment. The atomic concentration ratios can be approximately determined from the photoemission peak intensity ratio using corrections for both photoionization cross section and the transmission function of the spectrometer. The atomic concentration ratio is given by

$$\frac{C_i}{C_j} = \frac{I_i}{I_j} \times \frac{\sigma_j}{\sigma_i} \times \frac{T_j}{T_i}$$ (1)

where C_i is the atomic concentration, I_i is the intensity of the core-level photoemission peak, σ_i is the photoionization cross section, T_i is the transmission function of the spectrometer for a specific i component. Therefore, a homogeneous distribution of the i component over the probing depth is assumed. In our case, this is not quite true, as discussed in the layer structure model below. Therefore, the resulting concentration should only be taken as qualitative guide. The relationship

$$T_i \propto \left[\frac{E_{kinetic}}{E_{pass}} \right]^{-1}$$ (2)

is used, as previously reported [5], where $E_{kinetic}$ is the kinetic energy of photoelectrons, E_{pass} is the pass energy of the spectrometer. Table1 presents the atomic concentration ratios calculated for both surface-sensitive and bulk-sensitive measurements from the 'as-treated' surface. An increase was seen in the As-to-Ga atomic concentration ratio for the surface-sensitive measurement with respect to the bulk-sensitive measurement. The Se-to-S atomic concentration ratio for the bulk-sensitive measurement is substantially larger than that for the surface-sensitive measurement. These results indicate that the surface composition changes in depth.

To further study surface chemistry of the 'as-treated' surface, Ga 3d, As 3d, Se 3d and S 2p core-level photoemission spectra were measured individually from both the 'as-treated' and thermally annealed surface. The combined energy resolution was less than 0.2 eV. Spectral deconvolution was performed on the measured photoemission spectra using a spectral synthesis approach. The spectral lineshape was simulated by a suitable combination of two single lineshapes possessing a Gaussian broadening function. The spin-orbit coupling constants for As 3d, Ga 3d, and Se 3d were taken as 0.72, 0.45, and 0.86 eV, respectively. The spin-orbit coupling constant for S 2p was taken as 1.15 eV. The experimental data are represented by square dots and the solid line by the sum of the simulated components of identical lineshape.

Table1. The atomic concentration ratios of As-to-Ga, Ga-to-Se and Se-to-S for the SeS$_2$-treated GaAs (001) surface. Kinetic energies (KE) of 40 and 300 eV, corresponding photoelectron escape depth of 0.5nm (surface-sensitive) and 1.5 nm(bulk-sensitive), were utilized.

Escape Depth	As/Ga	Ga/Se	Se/S
0.5 nm	2.50	1.85	1.30
1.5 nm	1.85	3.66	1.70

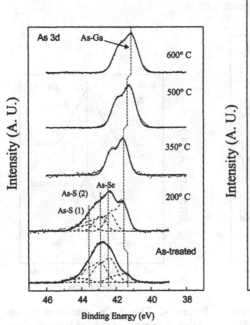

Figure 1. The As 3d core-level photo-emission spectra from the 'as-treated' GaAs (001) surface and the changes induced in the core levels due to thermal annealing. (Photon energy = 88 eV in all cases.)

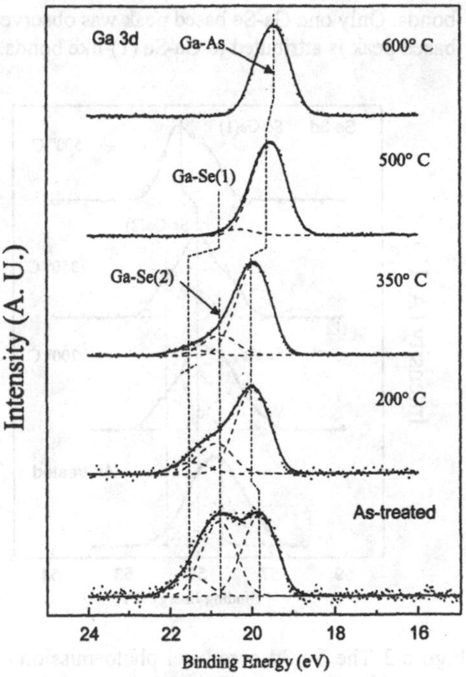

Figure 2. The Ga 3d core-level photo-mission spectra from the 'as-treated' GaAs (001) surface and the changes due to thermal annealing. (Photon energy = 65 eV in all cases.)

The As 3d photoemission spectra from the 'as-treated' GaAs (001) surface and the changes due to thermal annealing is shown in Figure 1. For the 'as-treated' surface, there are three shoulders beside the main As-Ga based peak. The first two peaks, which have chemical shifts from the As-Ga based peak of 2.5 eV and 1.43 eV respectively, are ascribed to As-S based peaks and labeled as As-S (1) and As-S (2) respectively. A third peak having a chemical shift of 0.86 eV is assigned to the As-Se peak. The chemical shifts of these peaks, from the As-Ga reference peak, are very close to previously reported values [6][8]. After thermal annealing at 350°C, the As 3d core level spectra can be fitted well with one peak. It is assumed that this peak is attributed to As-Ga bonds. The Ga 3d spectra from the 'as-treated' surface and the changes due to thermal annealing are shown in Figure 2. For the 'as-treated' surface, a clear shoulder peak is observed at 1.0 eV higher binding energy than the main Ga-As peak. There is also a smaller shoulder at 1.7 eV higher binding energy from Ga-As peak. From the S 2p spectra presented later, there are no observed Ga-S based peaks. These peaks are therefore ascribed to Ga-Se bonds and labeled as Ga-Se (1) and Ga-Se (2) respectively. Ga-Se (2) bonds are assumed to be Ga_2Se_3-like, and Ga-Se (1) bonds are assumed to be the bonding states corresponding to $GaSe_x$ (x>1.5)-like bonds. In a comparison to the Ga_2Se_3-like bonds, more charge transfer from Ga to Se per unit atom cause a bigger chemical shift for $GaSe_x$ (x>1.5)-like bonds. Only one Ga-Se based peak was observed after thermal annealing at 500°C. This Ga-Se based peak is attributed to Ga-Se (1)-like bonds.

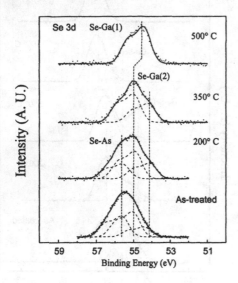

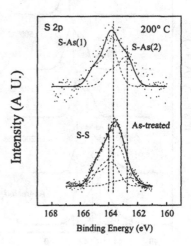

Figure 3 The Se 3d core-level photoemission spectra from the 'as-treated' GaAs (001) surface and the changes due to thermal annealing. (Photon energy = 100 eV in all cases.)

Figure 4. The S 2p core-level photoemission spectra of the 'as-treated' GaAs (001) surface and the changes due to the thermal annealing. (Photon energy = 209 eV in both cases.)

Figure 3 presents the Se 3d photoemission spectra from the 'as-treated' surface and the changes due to thermal annealing. Three peaks are observed for the 'as-treated' surface. The peak with highest binding energy is assigned to the Se-As bonds. The other two peaks are ascribed to the Se-Ga bonds. The amount of charge transfer from Ga to Se is considered to be larger than that from As to Se since the electronegativity difference between Ga and Se is larger than that between As and Se. Therefore, these peaks correspond to Se-As, Se-Ga (1) and Se-Ga (2) bonds. The binding energies for these peaks are very close to the previously reported values [6]. Figure 4 shows the S 2p photoemission spectra from the 'as-treated' surface and the changes due to thermal annealing. The spectrum from the 'as-treated' surface is composed of two peaks of lower binding energy ascribed to S-As and the highest kinetic energy peak ascribed to S-S bonds. The binding energy for the S-S peak (164 eV) is very close to the previously reported value [7]. The binding energy differences between S-S peak and the other two peaks are 0.81 eV and 1.26 eV, respectively. These peaks are not attributed to S-Ga based bonds since the magnitude of the chemical shifts from S-S reference peak for these other two peaks are smaller than expected for S-Ga bonds [8]. The S 2p spectra became undetectable after thermal annealing at 350°C.

The difference between the Ga 3d core-level binding energy and valence band maximum in bulk GaAs is 18.75±0.03 eV [9]. Therefore, the following equation is used to determine the position of surface Fermi level:

$$E_F = E_{Ga3d} - 18.75eV \qquad (3)$$

where E_F is the surface Fermi level position relative to the valence band maximum and E_{Ga3d} is the observed Ga 3d core level binding energy for Ga-As bonds. Figure 5 presents the position of surface Fermi level for the 'as-treated' and thermally annealed surfaces. The shift of the surface Fermi level within band gap was monitored under controlled thermal annealing conditions allowing for the determination of the specific chemical entities responsible for the movement of the surface Fermi level. The breaking of Ga-Se (2) based bonds by thermal annealing at 500°C is correlated with the dramatic shift of the surface Fermi level toward the valence band, indicating that the chemical bonding, corresponding to Ga-Se (2)-based bonds, is most likely responsible for the elimination of midgap states and the measured surface Fermi level position on the 'as-treated' surface.

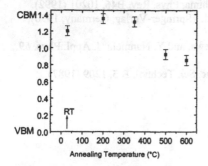

Figure 5. Position of the surface Fermi level relative to the band edge for the 'as-treated' and thermally annealed n- GaAs (001) surface.

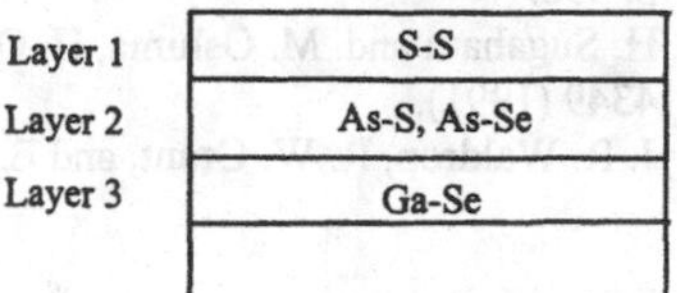

Figure 6. Structural model for the SeS₂-treated n-GaAs (001) surface. The layer 3 (Ga-Se) provides the surface electronic passivation.

CONCLUSIONS

These experimental results obtained from the SeS_2-treated GaAs (001) surface can be used to develop a consistent picture of the surface chemical structure and electronic properties underlying the passivation mechanism of the 'as-treated' surface. A structural layer model is proposed based on these results. A schematic diagram of the proposed structural layer model is shown in Figure 6. The surface is terminated with S-S bonds (layer 1). The uppermost surface region is As-rich (layer 2), while the adjacent region is As-deficient (layer 3). The As atoms are bonded to both S atoms and Se atoms in the As-rich region. The As-deficient region consists of Ga-based selenides. The thin Ga-based selenide layer below the surface provides the requisite chemical bonding that improves the electronic properties of the surface, while the S atoms adsorbed on the surface and the adjacent As-rich region may serve to protect the underlying passivating layer from the atmosphere or other chemical environments.

ACKNOWLEGMENTS

This work was supported by NSF under Award No. DMR-9632527. It is based upon research conducted at the Synchrotron Radiation Center, University of Wisconsin-Madison, which is supported by the NSF under Award No. DMR-9531009.

REFERENCES

1. B. A. Kuruvilla, S. V. Ghaisas, A. Datta, S. Banerjee, S. K. Kulkarni. J. Appl. Phys. **73**, 384 (1993).
2. S. Nozaki, S. Tamura and K. Takahashi. J. Vac. Sci.Technol. B **13**, 297 (1995).
3. B. A. Kuruvilla, A. Datta, G. S. Shekhawat, A. K. Sharma, P. D. Vyas, R. P. Gupta, and S. K. Kulkarni, Appl. Phys. Lett. **69**, 415 (1996).
4. B. A. Kuruvilla, A. Datta, G. S. Shekhawat, A. K. Sharma, P. D.Vyas, R. P .Gupta, and S. K. Kulkarni, J. Appl. Phys. **80**, 6274 (1996).
5. J. Barth, F. Gerken, C. Kunz, Nucl. Instrum. Method. **208**, 797 (1983).
6. T. Scimeca, Y. Watanabe, R. Berrigan, and M. Oshima, Phys. Rev. B**46**, 10201 (1992).
7. Stefan Hafner, *Photoelectron Spectroscopy*, 2nd ed. (Springer-Verlag, Germany, 1996), p. 456.
8. H. Sugahara and M. Oshima, H. Oigawa, H. Shigewa, and Y. Nannichi J. Appl. Phys. **69**, 4349 (1991).
9. J. R. Waldrop, R. W. Grant, and E. A. Kraut, J. Vac. Sci. Technol. B **5**, 1209 (1987).

Er-RELATED EMISSION IN IMPURITIES (NITROGEN, OXYGEN) IMPLANTED Al$_{0.7}$Ga$_{0.3}$As

S. Uekusa*, M. Wakutani*, M. Saito*, and M. Kumagai**
* Department of Electrical Engineering, School of Science and Technology, Meiji University, 1-1-1 Higashi-mita, Kawasaki, Kanagawa, 214 Japan
** Kanagawa High-technology Foundation, 3-2-1 Sakado, Takatsu-ku, Kawasaki, Kanagawa, 213 Japan.

ABSTRACT

Erbium (Er) was co-implanted with (i) nitrogen (N), (ii) oxygen (O) and (iii) N and O into Al$_{0.7}$Ga$_{0.3}$As substrates. Compared with the Al$_{0.7}$Ga$_{0.3}$As:Er sample, the 20 K Er^{3+}-related 1.54 μm integrated photoluminescence (PL) intensity from the Al$_{0.7}$Ga$_{0.3}$As:Er,N and Al$_{0.7}$Ga$_{0.3}$As:Er,N,O samples were enhanced approximately fifteen and ten times more, respectively. Thermal quenching of Er^{3+}-related emission from Al$_{0.7}$Ga$_{0.3}$As:Er,N,O was smaller than that of Er^{3+}-related emission from Al$_{0.7}$Ga$_{0.3}$As:Er,N. The 20 K Er^{3+}-related 0.98 μm PL which was radiated by transition from the second excited state ($^{4}I_{11/2}$) to the ground state ($^{4}I_{15/2}$) was observed. The 0.98 μm PL intensity from Al$_{0.7}$Ga$_{0.3}$As:Er,N,O generally decreased with increasing O dosage from 1×10^{13} cm^{-2} to 1×10^{15} cm^{-2}. These results suggest the formation of different complexes composed of Er and the impurities (N,O). This leads to the generation of complex related traps in the band-gap of Al$_{0.7}$Ga$_{0.3}$As as a result of the co-implantation of the impurities. It was found that the trap level of the Er-N complex center lay between 2.05 eV and 1.26 eV, and that of the Er-N-O complex center lay between 1.26 eV and 0.82 eV.

INTRODUCTION

The optical behavior of rare-earth (RE)-doped III - V compound semiconductors has been the focus of much attention because of its sharp and temperature-stable infrared emission due to intra-4f-shell transition of RE [1] [2] [3]. In particular, Er-doped semiconductors exhibit various sharp emissions around 1.54 μm, which coincides with the wavelength of minimum loss in silica-based optical fibers, due to transitions between the crystal field split states $^{4}I_{13/2}$ and $^{4}I_{15/2}$ of Er^{3+}. Therefore, Er doped semiconductors would be of great utility for light emitting diodes, optical lasers and optical amplifiers in optical fiber communication. However, the observed emissions have been weak and difficult to observe at room temperature in narrow band gap semiconductors. Thus, it is important for Er-doped semiconductors to improve upon the rapid thermal quenching properties of the Er^{3+}-related emissions and the low efficiency of energy transition from the host semiconductor to intra-4f-shell of Er^{3+} ions. Numerous studies have been carried out in order to solve these problems. Co-doping of O, C, N, and F with Er into float-zone (FZ) Si has resulted in a significant enhancement of Er^{3+}-related emission [4]. Recently, it has been reported that the co-doping of O into Er-doped GaAs by metalorganic chemical vapor deposition (MOCVD) has a remarkable effect on their luminescence characteristics [5]. Hence we are extremely interested in the influence of impurities on Er^{3+}-related emission in Er-doped semiconductors, and we have already reported the effect of N on Er^{3+}-related emissions in GaP:Er and Si:Er [6] [7].

In this work, in order to investigate the effect of N and O on Er^{3+}-related emissions in Al$_{0.7}$Ga$_{0.3}$As:Er, an ion-implantation of (i) N, (ii) O and (iii) N and O into Al$_{0.7}$Ga$_{0.3}$As:Er was performed, and their PL properties were systematically studied as a function of the impurity dose. Furthermore, selected samples were examined by variable-temperature PL measurements to understand their thermal quenching properties. The expected position of the trap

595

levels related to the Er center is discussed in light of these results and the PL measurement of Er^{3+}-related 0.98 μm emissions from the second excited state of Er^{3+}

EXPERIMENT

Er ion-implantation was carried out at an elevated substrate temperature of 350 ℃ at an ion energy of 1 MeV with a dose of 1×10^{13} cm^{-2}. The substrate was a liquid-phase-epitaxialloy grown un-doped (110) $Al_{0.7}Ga_{0.3}As$ layer on GaAs. From the calculation of the TRIM (TRansport Ions in Materials), the project range (Rp) and straggling ($\triangle Rp$) for the Er implant profile were estimated to be 205.7 nm and 58.4 nm, respectively. Either N or O ion was implanted into the $Al_{0.7}Ga_{0.3}As$:Er with a dose of between 1×10^{13} cm^{-2} and 1×10^{15} cm^{-2} at the energy of 120 keV and 130 keV, respectively. The Rp for the implanted N and O ions were almost the same as that for the Er ions. Following the ion-implantation, these samples were annealed at high temperatures, from 750 ℃ to 900 ℃, for 10 minutes using the proximity cap method in a N_2 atmosphere. Photoluminescence (PL) measurements were carried out using a 1-m-focal length double-grating monochromator and a liquid-nitrogen-cooled germanium p-i-n photodiode. Samples were excited with 488.0 nm line of an Ar ion laser with a power of 10 mW.

RESULT AND DISCUSSION

In all the samples annealed at above 900 ℃, the Er^{3+}-related emission was weak due to surface oxidation and the desorption of arsenic atoms. From these observations, we found that the best annealing temperature for Er^{3+}-related emissions to radiate was 850 ℃. Following the ion-implantation, all the samples, mentioned in this article were annealed at 850 ℃ for 10 minutes.

The effect of N co-doping on the Er^{3+}-related emission is shown in Fig. 1, which shows the 20 K Er^{3+}-related 1.5 μm PL spectra obtained from the $Al_{0.7}Ga_{0.3}As$:Er,N samples implanted with

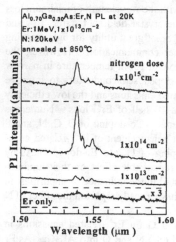

FIG. 1. The nitrogen dose dependence of Er-related PL spectra annealed at 850 ℃.

Er at a dose of 1×10^{13} cm^{-2} and implanted with N at different doses. The broken lines shown in the figure indicate the zero luminescence level for each spectrum. Two weak Er^{3+}-related peaks (1538.0, 1583.0 nm) were observed for the sample implanted with only Er. The N co-doping samples exhibited a different emission structure (1538.0, 1543.1, 1546.2 and 1549.6 nm) as compared with the sample implanted with only Er. Above a N dose of 1×10^{14} cm^{-2}, the Er^{3+}-related emission showed a sharp enhancement. The integrated PL intensity of the Er^{3+}-related emissions from the sample co-implanted with N at a dose of 1×10^{14} cm^{-2} was fifteen times stronger than that from the sample implanted with only Er. An improvement of Er^{3+} luminescence could be realized by the introduction of N impurity. The optimum N dose to obtain the maximum PL intensity was 1×10^{14} cm^{-2}. In AlGaAs, it is known that nitrogen atoms occupy the As site, and then the isoelectronic traps N_{As} are formed below the conduction band [8]. The nitrogen N_{As} is a very localized potential well which can trap an electron created by the photo-excitation of the host materials. The resulting coulomb field attracts a hole, and the two trapped carriers form an exciton bound to the N_{As}. Taking into consideration the difference in spectra between $Al_{0.7}Ga_{0.3}As$:Er and $Al_{0.7}Ga_{0.3}As$:Er,N, it is suggested that N creates isoelectronic traps N_{As} and new Er^{3+}-related radiative center (Er-N complex center), and that an efficient energy-transfer to Er^{3+} ions occurs through the recombination of the exciton bound to the N_{As}.

The effect of O co-doping on the Er^{3+}-related emission is shown in Fig. 2, which shows the 20 K Er^{3+}-related 1.5 µm PL spectra obtained from the sample implanted with Er at a dose of 1×10^{13} cm^{-2} and implanted with O at different doses. As shown in the figure, the sample co-implanted with O at a dose of 1×10^{13} cm^{-2} showed an emission structure similar to the sample implanted with only Er. The integrated PL intensity of the Er^{3+}-related emission from the sample co-implanted with O at

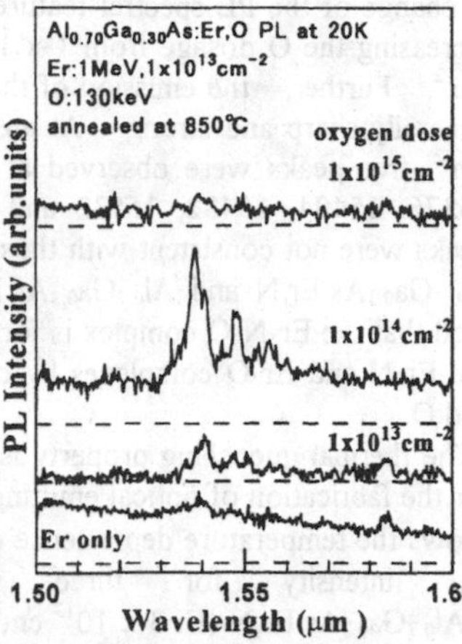

FIG. 2. The oxygen dose dependence of Er-related PL spectra annealed at 850 °C.

a dose of 1×10^{14} cm^{-2} was five times stronger than that from the sample implanted with only Er. The sample showed a different emission structure as compared with $Al_{0.7}Ga_{0.3}As$:Er,N in Fig. 1 since three main peaks became apparent at the wavelengths of 1538.0, 1548.0, and 1555.0 nm. At an O dose of 1×10^{15} cm^{-2}, the Er^{3+}-related emission was not enhanced. It appears that an O dose of 1×10^{15} cm^{-2} is too much for an Er dose of 1×10^{13} cm^{-2} to give rise to the optical activation of Er atoms in the host. As a result, we found that the introduction of oxygen atoms in $Al_{0.7}Ga_{0.3}As$:Er significantly changed the Er^{3+}-related optical properties, where the optimum O dose to obtain the maximum PL intensity was 1×10^{14} cm^{-2}. Since the differences are observed between $Al_{0.7}Ga_{0.3}As$:Er, $Al_{0.7}Ga_{0.3}As$:Er,O, and $Al_{0.7}Ga_{0.3}As$:Er,N in the spectral feature, it is concluded that a new type of Er-O complex is formed by co-doping of O and annealing at

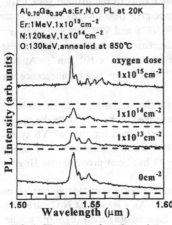

FIG. 3. The oxygen dose dependence of Er-related PL spectra in AlGaAs :Er,N,O annealed at 850 °C.

850 °C. Oxygen atoms have recently attracted increasing attention as co-dopants in Er-doped semiconductors [5] [9] [10]. This is because co-doping of O into an Er-doped semiconductor forms some types of Er-O radiative complex centers such as an Er-2O complex center in MOCVD-grown GaAs:Er,O and enhances the intra-4f-shell emission of Er^{3+} ions [11]. Hence we believe that O co-implanted into Er-doped AlGaAs also forms some type of Er-O radiative complex, but the efficiency of the Er^{3+}-related emission of this center is lower than that of the Er-N complex center.

The effect of co-doping of both N and O on the Er^{3+}-related emission is shown in Fig.3, which shows the 20K Er^{3+}-related 1.5 µm PL spectra obtained from the sample implanted with Er at a dose of 1×10^{13} cm^{-2}, N at a dose of 1×10^{14} cm^{-2} and O at different doses. The PL intensity decreased slightly with O implantation.

A change of the PL spectral feature was observed with increasing the O dosage from 1 x 10^{13} cm^{-2} to 1 x 10^{15} cm^{-2}. Further, the emission of the main peak became generally sharp and strong. At an O dose of 1 x 10^{15} cm^{-2}, five peaks were observed at the wavelengths of 15376, 15404, 15482, 15532 and 15578 nm. These peaks were not consistent with those of Al$_{0.7}$Ga$_{0.3}$As:Er, Al$_{0.7}$Ga$_{0.3}$As:Er,N and Al$_{0.7}$Ga$_{0.3}$As:Er,O. It is considered that the Er-N-O complex is formed differently from the Er-N and Er-O complexes by co-doping of both N and O.

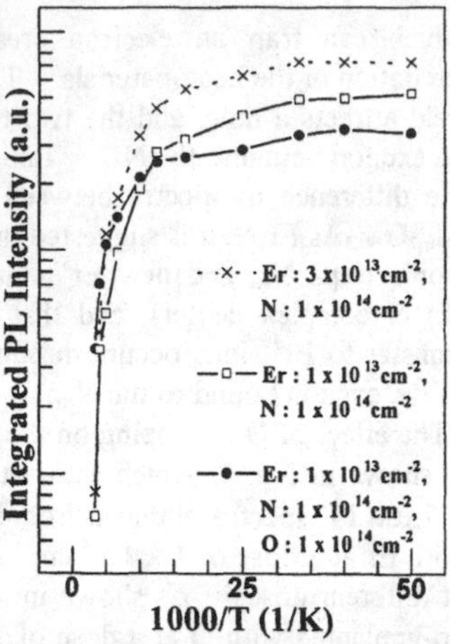

The thermal quenching property is an important factor for the fabrication of optical emitting devices. Figure 4 shows the temperature dependence of integrated 1.5 μm PL intensity for three selected samples { Al$_{0.7}$Ga$_{0.3}$As:Er,N (Er:3 x 10^{13} cm^{-2}, N:1 x 10^{14} cm^{-2}), Al$_{0.7}$Ga$_{0.3}$As:Er,N(Er:1 x 10^{13} cm^{-2}, N:1 x 10^{14} cm^{-2}) and Al$_{0.7}$Ga$_{0.3}$As:Er,N,O (Er:1 x 10^{13} cm^{-2}, N:1 x 10^{14} cm^{-2},

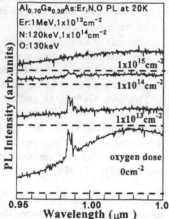

FIG. 4. The temperature dependence of Er-related integrated PL intensity in selected samples annealed at 850 ℃.

O: 1 x 10^{14} cm^{-2})}. In Al$_{0.7}$Ga$_{0.3}$As:Er,N, the PL integrated intensity increased with increasing the Er dosage from 1 x 10^{13} cm^{-2} to 3 x 10^{13} cm^{-2}. However, these thermal quenching properties showed a very similar tendency. It may follow from this that an increase in the number of optically active Er atoms in the host causes the enhancement of the PL intensity, and these emissions are due to the same Er radiative center. In the Al$_{0.7}$Ga$_{0.3}$As:Er,N and Al$_{0.7}$Ga$_{0.3}$As:Er,N,O samples, the integrated PL intensity at RT was one-eightieth and one-tenth the value of that at 20K, respectively. Therefore, it was found that the N and O co-doping into Al$_{0.7}$Ga$_{0.3}$As:Er had a significant effect on the thermal quenching property.

Figure 5 shows the 20 K Er^{3+}-related 0.98 μm PL spectra from Al$_{0.7}$Ga$_{0.3}$As:Er,N,O samples as a function of O doses, which is radiated by transition from the second state (^{4}I$_{11/2}$) to the ground state (^{4}I$_{15/2}$) in Er^{3+} 4f-shell. Three main peaks were observed at the wavelengths of 9849, 9876 and 9893 nm. The ^{4}I$_{11/2}$→ ^{4}I$_{15/2}$ luminescence decreased dramatically with increasing the O dosage from zero to 1 x 10^{15} cm^{-2}. At an O dose of 1 x 10^{15} cm^{-2}, the ^{4}I$_{11/2}$→^{4}I$_{15/2}$ luminescence was absent.

The mechanisms for the excitation of an Er center by host-excitation are assumed to be similar to the mechanism experimentally determined for the InP:Yb system [1] [12]. The model for the mechanism of energy transfer for GaAs:Er,O has been proposed by Hogg et al. [13].

FIG. 5. The oxygen dose dependence of Er-related 0.98 μm PL spectra in AlGaAs:Er,N,O annealed at 850 ℃.

Based on their model, our experimental results on Al$_{0.7}$Ga$_{0.3}$As:Er,N and Al$_{0.7}$Ga$_{0.3}$As:Er,N,O are discussed as follows. Figures 6(a) and 6(b) show a schematic diagram detailing the possible excitation processes. The diagram shows the energy transfer mecha-

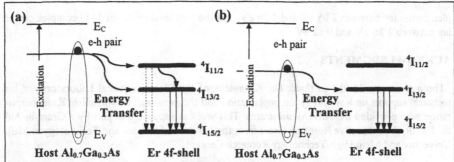

FIG. 6. Schematic diagram showing the energy transfer mechanism for an Er-related trap level with bound exciton energy (a) higher and (b) lower than the $^4I_{11/2}$ level. Straight arrows pointing up (down) indicate excitation (emission). Intra-4f-shell emission from $^4I_{11/2} \rightarrow {}^4I_{15/2}$ ($^4I_{13/2} \rightarrow {}^4I_{15/2}$) is marked by dashed (solid) arrows. The curved arrow indicates non-radiative energy transfer.

nism for an Er-related trap level with energy (a) higher and (b) lower than the $^4I_{11/2}$ level. Only the $^4I_{13/2}$ level is excited in (b). The expected position of the Er-related trap is discussed in light of the thermal quenching properties and measurement of the $^4I_{11/2} \rightarrow {}^4I_{15/2}$ (0.98 μm) luminescence of both $Al_{0.7}Ga_{0.3}As$:Er,N and $Al_{0.7}Ga_{0.3}As$:Er,N,O. In $Al_{0.7}Ga_{0.3}As$:Er,N, it is considered that the energy transfer mechanism is consistent with the mechanism shown Fig. 6(a). Therefore, the trap level of the Er-N complex center lies between 2.05 eV (energy of $Al_{0.7}Ga_{0.3}As$ band-gap) and 1.26 eV (energy of the $^4I_{11/2} \rightarrow {}^4I_{15/2}$ luminescence). In $Al_{0.7}Ga_{0.3}As$:Er,N,O, it is found that the energy transfer mechanism generally turns from the mechanism as shown in Fig. 6(a) to that of Fig. 6(b) with increasing the O dosage from zero to 1×10^{15} cm^{-2}. Above an O dose of 1×10^{15} cm^{-2}, the trap level of the Er-N-O complex center lies between 1.26 eV (energy of the $^4I_{11/2} \rightarrow {}^4I_{15/2}$ luminescence) and 0.82 eV (energy of the $^4I_{13/2} \rightarrow {}^4I_{15/2}$ luminescence) because of the absence of the $^4I_{11/2} \rightarrow {}^4I_{15/2}$ luminescence.

According to these results, the difference of the thermal quenching properties between $Al_{0.7}Ga_{0.3}As$:Er,N and $Al_{0.7}Ga_{0.3}As$:Er,N,O can also be explained by the difference in the amount of energy back-transfer (from the trap level to host) due to the increasing temperature. The energy back-transfer generally decreases as the trap level becomes deeper. Therefore, we found that the thermal quenching properties were improved because of the formation of the deep Er-related trap by the N and O co-doping.

CONCLUSION

The influence of the impurities (either nitrogen or oxygen, and both of them) on Er^{3+}-related emission in $Al_{0.7}Ga_{0.3}As$:Er were studied using the co-implantation method. We observed Er^{3+}-related 1.54 μm emission at room temperature from $Al_{0.7}Ga_{0.3}As$:Er,N and $Al_{0.7}Ga_{0.3}As$:Er,N,O samples. The enhancement of the 1.54 μm emission at low temperatures was observed from $Al_{0.7}Ga_{0.3}As$:Er,N, $Al_{0.7}Ga_{0.3}As$:Er,O and $Al_{0.7}Ga_{0.3}As$:Er,N,O. We found that the impurity atoms (N,O) formed three radiative complex centers (Er-N, Er-O and Er-N-O) with Er ions, and the centers affected the Er^{3+}-related PL spectra and the 1.54 μm emission efficiency. The Er-N-O complex center has the maximum influence on the improvement of the thermal quenching property of the Er^{3+}-related emission among the three complexes. By investigating the result of the thermal quenching properties and the $^4I_{11/2} \rightarrow {}^4I_{15/2}$ luminescence of both $Al_{0.7}Ga_{0.3}As$:Er,N and $Al_{0.7}Ga_{0.3}As$:Er,N,O, we found that the trap level of the Er-N complex center lay between 2.05 eV and 1.26 eV, and the trap level of the Er-N-O complex center lay between 1.26 eV and 0.82 eV.

ACKNOWLEDGMENTS

The authors would like to thank Dr. Kanayama of the Electrotechnical Laboratory for his technical support on low-energy ion implantation and the personnel of Mitsubishi Kasei corporation who provided the AlGaAs substrates. This work is supported in part by a Grant-in-Aid for Faculty Collaborative Research Grant from the Institute of Science and Technology at Meiji University and Meiji High-Technology Research Center.

REFERENCES

1. H. Katsumata, S. Uekusa, A. Majima and M. Kumagai, J. Appl. Phys. **77,** 1881 (1995).

2. H. Katsumata, S. Uekusa and H. Sai, J. Appl. Phys. **80,** 2383 (1996).

3. S. Uekusa, T. Ohshima, A. Majima, and M. Kumagai, Proc. 1994 International Workshop on Electroluminescence, Beijing 279-283 (1994).

4. J. Michel, J.L. Benton, R.F. Ferrante, D.C. Jacobson, D.J. Eaglesham, E.A. Fitzgerald, Y.H. Xie, J.M. Poate and L.C. Kimerling, J. Appl. Phys. **70,** 2672 (1991).

5. K. Takahei and A. Taguchi, J. Appl. Phys. **74,** 1979 (1993).

6. S. Uekusa, K. Shimazu, A. Majima and K. Yabuta, Nucl. Instr. and Meth. B **106,** 477 (1995).

7. S. Uekusa, K. Y. Yano, K. Fukaya and M. Kumagai, Nucl. Instr. and Meth. B **127/128,** 541-544 (1997).

8. S. Gonda, Y. Makita, S. Mukai, T. Tsurushima and H. Tanoue, Appl. Phys. Lett. **29,** 196 (1976).

9. P.N. Favennec, H. L'Harridon, D. Mountonnet, M. Salvi and M. Gaunneau, Jpn. J. Appl. Phys. **29**(4), 524 (1990).

10. J.E. Colon, D.W. Elsaesser, Y.K. Yeo, R.L. Hengehold and G.S. Porenke, Mater. Res. Soc. Symp. Proc. **301** 257 (1993).

11. K. Takahei, A. Taguchi, Y. Horikoshi and J. Nakata, J. Appl. Phys. **76,** 4332 (1994).

12. A. Taguchi, K. Takahei, and Y. Horikoshi, J. Appl. Phys. **76,** 7288 (1994).

13. R.A. Hogg, K. Takahei, and A. Taguchi, J. Appl. Phys. **79,** 8682 (1996).

PHOTOACOUSTIC STUDY OF THE EFFECT OF 0.9 eV LIGHT ILLUMINATION IN SEMI-INSULATING GaAs

Atsuhiko Fukuyama*, Yoshito Akashi*, Kenji Yoshino**, Kouji Maeda**, and Tetsuo Ikari**
* Department of Materials Science, Miyazaki University and
** Department of Electrical and Electronic Engineering, Miyazaki University
1-1 Gakuen kibanadai-nishi, Miyazaki 889-21, Japan

ABSTRACT

The effect of the secondary light illumination of $h\nu = 0.9$ eV on the photoquenched and the enhanced states in semi-insulating GaAs are investigated by using piezoelectric photoacoustic (PPA) measurements at 80 K. It is found that the secondary light causes an optical recovery from EL2* to EL2^0 and this is in agreement with the result reported by using infrared optical absorption measurements. We observed a broad peak around 0.9 eV after the secondary light illumination for the first time. The most important finding is that the PPA spectra after the secondary light illumination on the quenched and the enhanced states are the same in the whole photon energy region. We concluded that the secondary light of $h\nu = 0.9$ eV induces both an optical recovery and generation of metastable state of the EL6 level. The difference of the transformation rates of these two processes explains well the observed complex natures of the PPA signal under the secondary light illumination.

INTRODUCTION

Deep lying defect level EL2 is known to be a dominant donor to accomplish a semi-insulating (SI) nature of GaAs. EL2 transforms to its metastable state by the light illumination at low temperature, so-called photoquenching effect [1]. One of the properties about this metastable state, EL2*, is that the initial state before photoquenching can be obtained after annealing the sample above 130 K. The existence of an optical recovery from EL2* to the initial state, EL2^0, was observed in SI samples by Manasreh and Fischer [2]. They reported that photons around 0.9 eV and/or ranging from 1.4 to 1.51 eV are able to restore EL2^0 partially.

In our previous papers [3, 4], we reported by using piezoelectric photoacoustic (PPA) measurements that when the quenching light of 1.12 eV illuminates on sample at 80 K, there are two kinds of photo-induced states of the PPA signal in SI GaAs. One is the quenched state generated by a short period illumination, which is due to photoquenching effect of the EL2 level. The other is the enhanced state attained by a prolonged illumination. The presence of deep donor EL6 and its metastable state, EL6^m, were proposed to explain the enhanced state.

In this paper, we report a study on effects of the secondary light illumination of $h\nu = 0.9$ eV on two photo-induced states from a nonradiative recombination point of view. The spectral and the time dependent PPA measurements were carried out after the secondary light illumination for 30 min. Optical recovery from EL2* to EL2^0 was observed by the secondary light illumination on both the photoquenched and the enhanced states. We observed a broad peak around 0.9 eV in the PPA spectrum after the secondary light illumination for the first time. The most important finding is that the PPA spectra after the secondary light illumination on the quenched and the enhanced states are the same in the whole photon energy region. We concluded that the secondary light of $h\nu = 0.9$ eV induces both an optical recovery and generation of metastable state of the EL6 level. The difference of the transformation rates of these two processes explains well the observed complex natures of the PPA signal under the secondary light illumination.

EXPERIMENTAL PROCEDURES AND RESULTS

The sample was prepared from carbon concentration controlled SI GaAs wafer grown by the liquid encapsulated Czochralski (LEC) method. The total EL2 and carbon concentration of the sample are 1.1 and $(1.2 - 1.4) \times 10^{16}$ cm^{-3}, respectively. Since this wafer was thermally treated by three-stage annealing method [5], a minimum amount of irrelevant intrinsic defects is expected. Details of the experimental setups for PPA measurements are reported elsewhere [6].

The typical time dependence of the PPA signal during the quenching light illumination of 1.12 eV is shown in Fig. 1 [4]. The PPA signal shows the complex feature specified by three states. A decrease of the PPA signal is observed for a short period to 2.5 min due to photoquenching of EL2 (Q-state). The PPA signal next increases through a local minimum and to a saturation level for more than 10 min (E-state). If the temperature of the sample is kept at 80 K, both two photo-induced states are quite stable. N-state can be obtained by heating the quenched and/or enhanced sample above 130 K and subsequent cooling down to 80 K.

The typical PPA spectra of N-, Q-, and E-states are shown in Fig. 2 [4]. A hump up to the band gap energy that corresponds to the photoionization cross-section spectrum of EL2 is observed in N-state PPA spectrum shown by the solid curve. A vanishing of this hump is achieved by the quenching light illumination of 1.12 eV for a short period as shown by the broken curve (Q-state). E-state produced by the quenching light illumination for a long time shows a wide band over the range from 0.6 to 1.4 eV as shown by the dotted curve. We refer this wide band to E-band.

A sequence of measuring an effect of the secondary light illumination for Q-state is shown in Fig. 3(a). After the sample was cooled down to 80 K in the dark, the quenching light of 1.12 eV was illuminated on sample for 2.5 min to change to Q-state and was cut off [curve 1 in Fig. 3(a)]. Next, the secondary light of hν = 0.9 eV was illuminated on sample for 30 min [curve 2 in Fig. 3(a)]. At this time, both the probe and the secondary light were set at the photon energy of 0.9 eV.

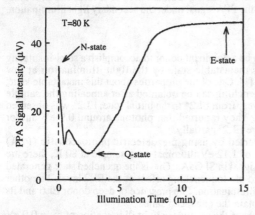

FIG.1 Typical time dependence of the PPA signal at 1.12 eV during the quenching light illumination.

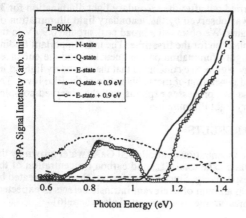

FIG.2 The typical PPA spectra of N-, Q- and E-states at 80 K. The PPA spectra after the secondary light illumination of 0.9 eV on Q- and E-states for 30 min are also shown in the figure. A broad peak around 0.9 eV and a hump above 1.1 eV up to the band gap energy were observed.

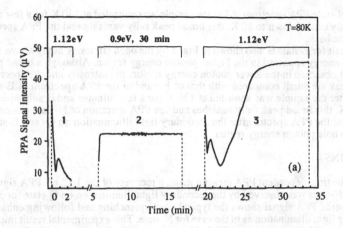

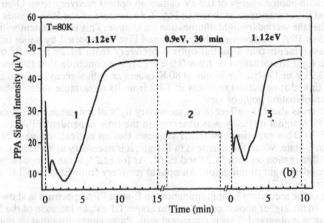

FIG. 3   The sequence of measuring the effect of the secondary light
illumination of 0.9 eV on (a) Q-state and (b) E-state at 80 K.

Finally, the time dependent PPA signal at 1.12 eV was measured again with illuminating the
quenching light. The result is shown in Fig. 3(a) by curve 3. Recovery of the 1.12 eV PPA signal
intensity to a value of N-state and following photoquenching and enhancement were observed.

A sequence of measuring an effect of the secondary light illumination of 0.9 eV on E-state is
shown in Fig. 3(b). As in the case for Q-state, the quenching light of 1.12 eV was first illuminated
for 15 min to change the sample to E-state and was cut off [curve 1 in Fig. 3(b)]. The time dependence
of the PPA signal intensity was measured after the secondary light illumination for 30 min [curve
3 in Fig. 3(b)]. Recovery of the 1.12 eV PPA signal was observed as well as in the case for Q-state.

The effect of the secondary light illumination on the PPA spectra were also measured. As in
the case for the time dependent PPA measurements, the sample was first cooled down to 80 K in
the dark and underwent to Q- or E-state by the quenching light illumination of 1.12 eV. After the
secondary light of 0.9 eV illuminates on sample for 30 min, the PPA spectrum was measured as a
function of photon energy of the probe light ranging from 0.6 to 1.45 eV. The result for Q-state is
shown in Fig. 1 by the open triangles. A broad peak around 0.9 eV and a hump above 1.1 eV up to
the band gap energy appear. It seems that the PPA signals in the higher photon energy region
above 1.1 eV almost recover to a value of N-state. However, a broad peak around 0.9 eV is not

603

observed in N-state PPA spectrum. After the sample was annealed at 130 K for a few minutes and is subsequently cooled down to 80 K, this broad peak fully vanishes and the PPA spectrum of N-state can come back.

The result for E-state is also shown in Fig. 3 by the open circles. A hump above 1.1 eV up to the band gap energy appeared in the higher photon energy region. Although a broad peak around 0.9 eV is still observed in the lower photon energy region, the intensity and the spectral width of this broad peak are small compared with that of E-band in the PPA spectrum of E-state. It was found that after the sample was annealed at 130 K for a few minutes and is subsequently cooled down to 80 K, this broad peak fully vanishes and the PPA spectrum of N-state can come back. It was found that the PPA spectra after the secondary light illumination on Q- and E-states are the same in the whole photon energy region.

DISCUSSIONS

From the time dependent PPA measurement, a recovery of the 1.12 eV PPA signal intensity to a value of N-state was observed by the secondary light illumination on Q-state for 30 min at 80 K. The recovered PPA signal shows the typical photoquenching and following enhancement by the quenching light illumination as in the case for N-state. This experimental result implies that the secondary light with photon energy of 0.9 eV causes an optical recovery from Q-state to N-state. From the spectral PPA measurement shown in Fig. 2, an increase of the PPA signals above 1.1 eV was observed after the secondary light illumination on Q-state. This is consistent with the result of the time dependent PPA measurement. Manasreh and Fischer reported by using infrared optical absorption measurements that a partial optical recovery from EL2* to EL2^0 occurs by the monochromatic light illumination of hv = 0.9 eV [2]. We conclude that the secondary light illumination of 0.9 eV on Q-state for 30 min at 80 K causes an optical recovery from EL2* to EL2^0. This is the first time that an optical recovery of EL2 from its metastable state is observed from a nonradiative recombination point of view.

We next discuss about the effect of the secondary light illumination on E-state. A recovery of the 1.12 eV PPA signal intensity was observed in the time dependent PPA measurement as shown in Fig. 3(b). This experimental result suggests that an optical recovery of N-state also occurs even from E-state. When the sample is in E-state, it is necessary to consider the effect of the secondary light illumination on both EL2* and EL6^m. As for EL2*, it can be optically recovered to EL2^0 by the secondary light illumination. An optical recovery from EL6^m to EL6$^+$ may occur as well as in the case for EL2*.

The effect of the secondary light illumination on E-state PPA spectrum is shown in Fig. 3 by the open circles. In the higher photon energy region above 1.1 eV, an increase of the PPA signal to a value of N-state was observed. This is consistent with above suggestion that an optical recovery of EL2 also occurs even from E-state. However, the observed broad peak around 0.9 eV does not appear in the PPA spectrum of N-state. If the optical recovery of both EL2 and EL6 are fully accomplished, the sample should show the N-state PPA spectrum. After the sample is heated to 130 K and is subsequently cooled down to 80 K, this broad peak vanishes and sample shows N-state PPA spectrum that only a hump appears. These experimental results cannot be explained by simply supposing that the secondary light of 0.9 eV causes an optical recovery from Q-state to N-state as reported by Manasreh and Fischer [2].

The most important finding is that the PPA spectra after the secondary light illumination on Q- and E-states are the same in the whole photon energy region as shown in Fig. 3. It can be considered that the states generated by the secondary light illumination of 0.9 eV on Q- and E-states for 30 min at 80 K are identical. Although the intensity and the spectral width of a broad peak are small compared with that of E-band in E-state PPA spectrum, the origin of E-band and this broad peak seems to be the same origin. We hereafter refer the broad peak around 0.9 eV in the PPA spectra after the secondary light illumination to A-peak.

To investigate a change of the PPA spectrum during the secondary light illumination, the PPA spectra were measured one by one by illuminating the secondary light on Q- or E-state for short period. It was found that the overall PPA signals of E-band decrease monotonously and remain A-peak with increasing the secondary light illumination time. In the higher photon energy region above 1.1 eV, the PPA spectrum first decreases to a value of Q-state and becomes increase to that of N-state. As for Q-state, the PPA signals above 1.1 eV monotonously increased with increasing the illumination time. At the same time, an increase of the PPA signals around 0.9 eV was observed. As a result, the PPA spectra after the secondary light illumination on Q- and E-states for 30 min are coincident in the whole photon energy region. It was also found that even if the secondary light further illuminates on this state for a long time, further change in the PPA spectra is not observed. We then conclude that the origins of E-band and A-peak are the same. We have already reported that the electron transition from VB and/or from the compensated carbon acceptor to EL6^m causes E-band [4]. Therefore, A-peak is also caused by the electron nonradiative transitions involving EL6^m.

We also reported that the metastable transition from EL6$^+$ to EL6^m effectively occur in the photon energy region from 1 to 1.3 eV. To explain the present experimental results, we propose here that the photon energy of 0.9 eV can introduce both the optical generation and extinction of EL6^m. When the sample is in Q-state, all of the EL6 levels exist as EL6$^+$ due to compensate the carbon acceptor and therefore no EL6^m exist. If the sample is in E-state, all of the EL6 level already transformed to EL6^m. When the secondary light of 0.9 eV illuminates on E-state, E-band partially decreases due to optical recovery from EL6^m to EL6$^+$. As a result, only a broad peak around 0.9 eV remains. As discussed above, since the secondary light illumination also introduces an optical recovery of EL2^0, the PPA signals in the higher energy region above 1.1 eV increase. The difference of the transformation rates of these two processes of the generation and extinction of EL6^m explains well the observed complex features of the PPA signal under the secondary light illumination.

It should be noted that the present model based on one assumption that when the metastable transition from EL6$^+$ to EL6^m occurs, the EL2 level should not necessarily exist as EL2*. That is, there is no correlation between the metastable generation mechanisms of EL2* and EL6^m. We have reported in our previous paper that E-state cannot be achieved without undergoing the photoquenched state [4]. In order to check the availability of the former assumption, we investigated an effect of the secondary light illumination of 0.9 eV on N-state. When the sample is in N-state, the EL2 and EL6 levels exist as normal EL2^0 and ionized EL6$^+$, respectively. If the metastable transition from EL6$^+$ to EL6^m does not need EL2*, the secondary light illumination should increase the PPA signal around 0.9 eV. The result is shown in Fig. 4 by the open circles. Small but a certain broad peak around 0.9 eV appears. This broad peak is quite stable unless the temperature of the sample increases above 130 K as well as the thermal behavior of A-peak. The PPA signals above 1.0 eV were not influenced by the secondary light illumination. This suggests that the secondary light illumination does not change the situation of the EL2 level. We then conclude that the metastable transition from EL6$^+$ to EL6^m occurs even from N-state and that there is no correlation between the metastable transition mechanisms EL6^m and EL2*. No matter what the EL2 level is in any situation of normal or metastable, the secondary light illumination of 0.9 eV causes the metastable transition from EL6$^+$ to EL6^m; the PPA signal around 0.9 eV increases.

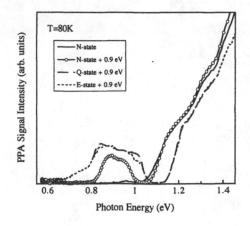

FIG.4   The PPA spectrum after the secondary light illumination of 0.9 eV on N-state for 30 min.

CONCLUSIONS

The effect of the secondary light illumination of $h\nu = 0.9$ eV on two photo-induced states was investigated by using PPA technique. We found that the secondary light illumination for 30 min at 80 K causes an optical recovery from EL2* to EL2⁰. We observed for the first time the broad peak around 0.9 eV after the secondary light illumination on Q- and E-states for 30 min, which cannot be observed in N-state spectrum. The most important finding is that the PPA spectra after the secondary light illumination on Q- and E-states are the same in the whole photon energy region. We concluded that the secondary light illumination induces both an optical generation and extinction of metastable state of the EL6 level. The difference of the transformation rates of these two processes of the generation and extinction of EL6ᵐ explains well the observed complex features of the PPA signal under the secondary light illumination. It was also found that EL6ᵐ can achieve without undergoing photoquenching of EL2. There is no correlation between the metastable generation mechanisms of EL2* and EL6ᵐ.

ACKNOWLEDGMENTS

The authors acknowledge to Dr. Yohei Otoki of Hitachi Cable Co., Ltd. for supplying the GaAs wafers and to Komatsu Electronic Metals Co., Ltd. for the financial supports.

REFERENCES

1. D. W. Fischer, Appl. Phys. Lett. **50**, 1751 (1987).

2. M. O. Manasreh and D. W. Fischer, Phys. Rev. **B40**, 11756 (1989).

3. A. Fukuyama, Y. Morooka, Y. Akashi, K. Yoshino, K. Maeda, and T. Ikari, Mat. Res. Soc. Symp. Proc, **442**, 459 (1997).

4. A. Fukuyama, Y. Morooka, Y. Akashi, K. Yoshino, K. Maeda, and T. Ikari, J. Appl. Phys. **81** (1997) 7567.

5. Y. Otoki, M. Nakamori, R. Nakazono and S. Kuma, *Proc. 4th Conf. Semi-Insulating III-V Materials*, Hakone, 1986 (Ohm-sha, Tokyo, 1986) p.285.

6. T. Ikari, S. Shigetomi, Y. Koga, H. Nishimura, H. Yayama and A. Tomokiyo, Phys. Rev. **B37**, 886 (1988).

PHOTO-ASSISTED RESONANT TUNNELING THROUGH LOCALIZED STATES IN AlAs/GaAs DOUBLE-BARRIER STRUCTURE WITH UNDOPED SPACER LAYERS

H.Y. Chu, K.-S. Lee, H.-H. Park, and E.-H. Lee
Basic Research Laboratory, Electronics and Telecommunications Research Institute, Taejon 305-600, Korea ; hychu@idea.etri.re.kr

ABSTRACT

We report on the evidence of photo-assisted resonant tunneling through localized states in AlAs/GaAs double-barrier structures (DBSs) with undoped GaAs spacers. In the photocurrent measurement, additional peaks were observed at voltages lower than that of resonance and were enhanced with the laser power. This behavior was more pronounced as the thickness of spacer layers in the DBS increased. These results are attributed to the resonant tunneling of electrons through the localized states, they are induced in the neighboring barriers, by the photoexcited carriers in spacers. We discuss the localization effect of photoexcited carriers on the resonant tunneling.

I. INTRODUCTION

Interband and intersubband transitions in resonant tunneling diode (RTD) structures have received increased attention due to its application to ultrafast optoelectronic devices [1-3]. The effects of local space charges on the resonant tunneling and the intersubband transitions have been investigated to optimize the performance of IR detector or modulator, etc [4]. Drexler et al. observed THz oscillations associated with the absorption and the stimulated emission between subbands in quantum wells [5]. For optoelectronic devices utilizing interband transitions, several works considered optical switching mechanisms and demonstrated its application as optically-controlled oscillator and modulator, etc [3,6]. Previously, we reported that the localization of photoexcited carriers outside the quantum well of RTD changed the resonant-tunneling condition [7,8]. In this paper, we report clear observations of resonant tunneling through localized states in the nearby barriers.

II. EXPERIMENT

The RTD structures studied in this work were grown on n-type GaAs substrate by molecular beam epitaxy. Table I shows the thickness and alloy composition of the samples. The mesa diodes with diameters of $7 \sim 30$ μm were fabricated by using chemically-assisted ion-beam etching with chlorine. AuGe/Ni/Au was used as n-type contacts at the top surface of the mesa and the bottom of the substrate. The photocurrent characteristics of RTDs were measured during the illumination of a tunable cw Ti-Sapphire laser. When the top surface was illuminated, the current-voltage (I-V) curve of the device was dependent on the bias directions [7], because more than 60% of the light was absorbed at the contact and the top spacer. In order to remove the influence of the scan direction of bias voltage on the I-V curve, we illuminated the laser on the sidewall of the diode. The laser was transferred and focused into a spot of 15-μm diameter by a tapered optical fiber. The I-V and capacitance-voltage (C-V) properties were measured at room temperature using HP4145B parameter analyzer and SI1260 impedance analyzer, respectively. A 5 mV root-mean-square ac voltage of 10 KHz was applied for the C-V measurement.

Table I. Structures of resonant tunneling diodes

Name	Material	N-type doping(cm^{-3})	Thickness($\overset{\circ}{A}$)/ Al composition		
			Diode A	Diode B	Diode C
Contact	GaAs	2x10^{18}	5,000	5,000	5,000
Spacer 4	GaAs	4x10^{18}	500	3,000	1,000
Spacer 3	GaAs	UD	25	100	200
Barrier	Al$_x$Ga$_{1-x}$As	UD	48, x=0.8	40, x=1	50, x=1
Well	GaAs	UD	45	45	45
Barrier	Al$_x$Ga$_{1-x}$As	UD	48, x=0.8	40, x=1	40, x=1
Spacer 2	GaAs	UD	25	100	100
Spacer 1	GaAs	4x10^{18}	500	3,000	1,000
Buffer	GaAs	2x10^{18}	10,000	10,000	10,000

*The UD layers are undoped

III. RESULTS AND DISCUSSION

Figure 1(a) ~ (c) show the I-V characteristics of diodes A, B, and C, respectively. The solid and the dashed lines represent the results obtained in the dark and under the illumination of laser with a 790-nm wavelength and about 0.15 mW power. Under the illumination of laser, the resonance voltage decreases. As the photocurrent reduces the series resistance in spacer layers, along with the fact that the photoexcited electrons and holes are piled up at the accumulation and the depletion region, respectively, the band bending is induced around the DBS. Therefore, the

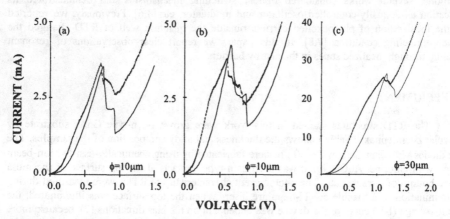

Figure 1. I-V characteristics of (a) diode A, (b) diode B, and (c) diode C. The solid and the dashed lines represent the results measured in the dark and under the illumination of laser with a 790-nm wavelength and about 0.15 mW power.

condition for the resonant tunneling is satisfied at a lower voltage [7,8].

On the other hand, an abnormal peak is observed at a voltage below the resonance condition. This peak is more pronounced for a diode with either the thicker spacers or the narrower mesa, indicating that the effect of the spacers layer thickness on the abnormal current peak is significant. The current also increases abruptly at voltages above the resonance condition due to the thermal emission. The increase in the non-resonant tunneling current in diode A is higher than that in diodes B and C, because the former diode has the lower AlGaAs barriers than the others, and the thermal emission in the former is higher than that of the others. It is noted that diode C shows the resonance peak at the highest voltage because of the largest voltage drop in the intrinsic region of the diode which corresponds to the thickest undoped spacers. For all the tested devices, the I-V curve in the opposite bias direction showed the similar characteristics as those in Fig. 1.

Figure 2 shows the I-V characteristics of diode B of 7-μm-diameter mesa for various laser power at 830-nm wavelength. Circles, squares, triangles, and diamonds represent I-V characteristics in the dark, 0.4 mW, 0.7 mW, 1.6 mW illumination, respectively. As the laser power increases, the abnormal photocurrent is enhanced and clearly reveals a negative-resistance behavior. When the photo-excited electrons and holes are piled up nearby barriers, the band bending across the undoped spacers causes the localization of carriers in the accumulation layer. The signature of the resonant tunneling between these localized states and the ground state in the quantum well is evident in the I-V curve. As these localized states are located very close to each other in energy, their contributions in the current response are difficult to observe at the low laser power. As the laser power is increased, the band bending is enhanced with accumulated carriers, and consequently the resonant tunneling through the localized states can be distinguished from that through the continuous states in accumulation layer, as shown in Fig 2. It is also emphasized that the low-power laser illumination on the sidewall of the device gives rise to the

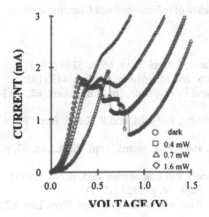 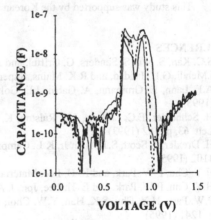

Figure 2. I-V characteristics of diode B for the various laser power at 830-nm wavelength.

Figure 3. C-V characteristics of diode A in the dark (solid line) and under the illumination of a 0.15 mW laser with wavelength of 830 nm (dashed line) and 770 nm (bold line).

more sensitive current response than that on the top-surface.

Figure 3 shows the C-V characteristics of diode A of 7-μm-diameter mesa. The solid, the dashed and the bold lines represent the results in the dark, and under the illumination of a 0.15 mW laser with wavelength of 830 nm and 770 nm, respectively. In the dark, a capacitance peak is observed at the bias near the resonance condition due to electron accumulation in the quantum well [9]. However, under the laser illumination, the periodic oscillations are revealed in the C-V curves from about 0.3 V to the resonant voltage. These oscillations are interpreted to be originating from the tunneling of electrons through the quasi-continuous states closely located in the accumulation region, although it is not observed in current measurement.

Photoassisted tunneling may cause some problems in the application of RTD as optical devices, because the photoassisted tunneling can increase the non-resonant tunneling current and decrease the peak-to-valley current ratio. But, these behaviors can provide a possibility of a new class of optically controlled multi-logic devices which utilize the property that two stable states are on the load line in the dark and more than three stable states intersect the load line under illumination.

IV. SUMMARY

We observed the photo-assisted resonant tunneling though the localized states in the accumulation region of RTDs. The current response was found to be more sensitive to the incident laser power when illuminated on the sidewall of the device. The present results provides a possibility of new optoelectronic devices based on the multiple resonant tunneling which can be controlled by a low power laser.

ACKNOWLEDGMENTS

This study was supported by the Korean Ministry of Information and Communications.

REFERENCES

1. S.C. Kan, S. Wu, S. Sanders, G. Griffel, and A. Yariv, J. Appl. Phys. **69**, p. 3384 (1991).
2. I.Mehdi, G.I. Haddad, and R.K. Mains, Superlattices and Microstructures **5**, p. 443 (1989)
3. A.F. Lann, E. Grumann, A. Gabai, J.E. Golub, and P. England, App. Phys. Lett. **62**, p.13 (1993)
4. H. Schneider, E.C. Larkins, J.D. Ralston, K. Schwartz, F. Fuchs, and P. Koidi, Appl. Phys. Lett. **63**, p. 782 (1993)
5. H. Drexler, J. Scott, S.J. Allen Jr, K.L. Campman, and A.C. Gossard, Appl. Phys. Lett. **67**, p. 4102 (1995)
6. H.Y. Chu, P.W. Park, and E.-H. Lee, Material Science and Engineering **B35**, p. 446 (1995)
7. H.Y. Chu, P.W. Park. and E.-H. Lee, Jpn. J. Appl. Phys. **34**, p. 1355 (1995)
8. P.W. Park, H.Y. Chu, S.K. Han, Y.W. Choi, G.O. Kim, and E.-H. Lee, Appl. Phys. Lett. **67**, p. 1241 (1995)
9. J. Jo, H.S. Li, Y.W. Chen, and K. L. Wang, Appl. Phys. Lett. **64**, p. 2276 (1994).

Oxygen-related Defects in $In_{0.5}(Al_xGa_{1-x})_{0.5}P$ Grown by MOVPE

[†]J.G. Cederberg, [†]B. Bieg, [††]J.-W. Huang, [††]S.A. Stockman, [††]M.J. Peanasky, [†]T.F. Kuech

[†] University of Wisconsin – Madison, Department of Chemical Engineering,
1415 Engineering Dr., Madison, WI 53706

[††] Hewlett-Packard Company, Optoelectronics Division,
370 West Trimble Road, San Jose, CA, 95131

ABSTRACT

Oxygen related defects in Al-containing semiconductors have been determined to degrade luminescence efficiency and reduce free carrier lifetime, affecting the performance of light emitting diodes and laser diodes. We have used the oxygen doping source, diethylaluminum ethoxide, $(C_2H_5)_2AlOC_2H_5$, to intentionally incorporate oxygen-related defects during growth of $In_{0.5}(Al_xGa_{1-x})_{0.5}P$ by Metalorganic Vapor Phase Epitaxy (MOVPE). Our investigations have identified several defects which are 'intrinsic' or present in non-intentionally oxygen-doped n-type $In_{0.5}(Al_xGa_{1-x})_{0.5}P$ as well as those due to oxygen, which introduces defect states near the middle of the conduction band. Deep level transient spectroscopy and photoluminescence data obtained for these defects over a range of composition, are presented illustrating the trends in defect structure with alloy composition. The impact of oxygen contamination on the visible emission spectrum is presented and discussed in terms of the defect structure.

INTRODUCTION

Ternary and quaternary alloys have become increasingly important in the fabrication of electronic and optoelectronic devices. The $In_{0.5}(Al_xGa_{1-x})_{0.5}P$ alloy system, which is lattice matched to GaAs, is becoming more important in a broad spectrum of applications. This material has been used to fabricate laser diodes [1] and light emitting diodes [2] emitting in the visible spectral region. $Al_{0.5}In_{0.5}P$ is also used to reduce hole leakage currents in heterojunction bipolar transistors (HBT) owing to its large valance band offset to GaAs [3].

Unfortunately, some of the problems which exist within the $Al_xGa_{1-x}As$ system are present in $In_{0.5}(Al_xGa_{1-x})_{0.5}P$. It is well established that oxygen is an unintentional impurity in $Al_xGa_{1-x}As$ grown by any epitaxy technique [4]. In MOVPE growth of Al containing alloys, oxygen is incorporated at the growth front by forming a strong covalent bond with Al. Oxygen introduces non-radiative recombination pathways and reduces free carrier lifetimes, thereby reducing luminescence intensity. It also forms deep states in the band gap that can trap electrons and holes [5]. We have utilized the molecular dopant source diethylaluminum ethoxide $(C_2H_5OAl(C_2H_5)_2$ or DEAlO) to incorporate oxygen-related defects into $In_{0.5}(Al_xGa_{1-x})_{0.5}P$. DEAlO incorporates oxygen in the form of an Al-O complex, providing a controlled technique for incorporating oxygen into the growing crystal. DEAlO has been used previously to effectively introduce oxygen into other binary and ternary compound semiconductors [6,7,8].

This study utilized a combination of Deep Level Transient Spectroscopy (DLTS), capacitance voltage (C-V), and secondary ion mass spectroscopy (SIMS) measurements to determine the concentration and electronic structure of defects both nominally present in these materials and introduced by oxygen in $In_{0.5}(Al_xGa_{1-x})_{0.5}P$.

EXPERIMENT

$In_{0.5}(Al_xGa_{1-x})_{0.5}P$ layers, 1 μm thick, were grown on GaAs substrates. The background doping was 4×10^{17} cm^{-3} provided by Te from diethyltelluride. DEAlO was used to intentionally incorporate oxygen. The additional amount of Al incorporated into the growing crystal, due to the Al in the DEAlO source, is comparable to the oxygen concentration and hence has a negligible effect on the overall composition of the quaternary. Backside Ohmic contacts were fabricated by depositing AuGe(12%)/Ni/Ti/Au multilayer by electron beam evaporation on the GaAs substrate. Schottky contacts were formed by depositing 20 nm Cr followed by a thick layer of Au through a shadow mask. DLTS measurements were made at a reverse bias of -2 V. The diodes were pulsed towards a forward voltage with a 0.5 V, 1 ms pulse. These pulsing conditions were chosen to insure that the measurement volume was well separated from the heterointerface (metal-$In_{0.5}(Al_xGa_{1-x})_{0.5}P$). The 454 nm line of an Ar laser, at a power density of 0.125 W/cm^2, was used to excite the photoluminescence (PL). The emission spectra produced were analyzed with a 1 m monochromator and detected with a photomultiplier tube sensitive from 200 to 900 nm. Secondary Ion Mass Spectroscopy (SIMS) was performed on selected samples to quantify the amount of oxygen present.

Results

Measurements on Al$_{0.5}$In$_{0.5}$P

Figure 1 presents the effect of oxygen on the DLTS spectrum of $Al_{0.5}In_{0.5}P$. Sample 1 was grown without intentional oxygen doping and was determined to be nominally oxygen-free by SIMS measurements. Two spectral peaks were measured in Sample 1 with a peak being located at ~170 K, designated as E1, and the second peak found above a measurement temperature of 390 K and is designated as E3. Defects similar to E1 have been previously observed in $Al_{0.5}In_{0.5}P$ alloys containing Si and Se shallow donors and have been attributed to DX-type centers [9,10]. The behavior of defect E3 under a wide variety of DLTS measurement conditions is complex and is currently under

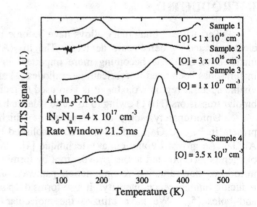

Figure 1: DLTS spectra of $Al_{0.5}In_{0.5}P$ doped with increasing amounts of oxygen. The main oxygen-related defect is centered ~350 K.

investigation. An additional peak appears at a measurement temperature of ~ 350 K in samples grown with intentional oxygen doping and will be identified as E2. This peak, E2, increases with the concentration of oxygen, as shown in Figure 2. The data of Figure 2 also indicates that the E1 defect is independent of the oxygen concentration, supporting its previous identification as a donor-'intrinsic' defect complex. The presence of oxygen has a very drastic effect on the sample luminescence as shown in Figure 3. Samples 1 and Sample A have oxygen concentrations below the SIMS detection limit of 1×10^{16} cm^{-3}, but the emission spectrum of Sample 1 is much lower than that observed for Sample A. While both samples were grown without any intentional oxygen introduction, it is believed that TMIn source used to grow Sample 1 possessed a higher oxygen level than in Sample A. These PL measurements indicate that that an oxygen concentration over 3.5×10^{17} cm^{-3} effectively quenches all band edge luminescence.

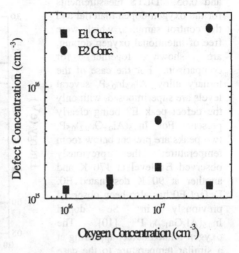

Figure 2. Variation of defect concentration, as determined by DLTS, as a function of the oxygen concentration in $Al_{0.5}In_{0.5}P$.

Measurements on $In_{0.5}(Al_xGa_{1-x})_{0.5}P$

Several defects-derived energy levels are present in the DLTS spectra of nominally oxygen-free $In_{0.5}(Al_xGa_{1-x})_{0.5}P$. Figure 4 presents several scans for $In_{0.5}(Al_xGa_{1-x})_{0.5}P$ with x = 1, 0.9, 0.8,

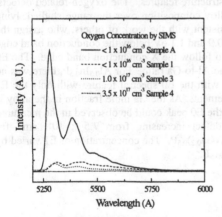

Figure 3. Visible PL spectra of $Al_{0.5}In_{0.5}P$ doped with increasing amounts of oxygen. A small amount of oxygen can drastically affect the optical quality of the epilayer.

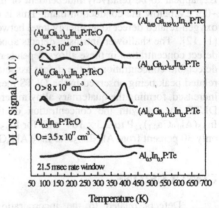

Figure 4. DLTS spectra for $In_{0.5}(Al_xGa_{1-x})_{0.5}P$ with and without intentionally incorporated oxygen . The presence of several 'intrinsic' defects at low and high measurement temperatures are also observed.

and 0.65. DLTS measurements for the oxygen-doped material and the control samples, which were free of intentional oxygen doping, are shown together for comparison. For the case of the ternary alloy, $Al_{0.5}In_{0.5}P$, several levels are superimposed, with only the defect peak E2 being clearly present. For $In_{0.5}(Al_{0.8}Ga_{0.2})_{0.5}P$, two peaks are present below room temperature; the previously observed E1 level at 170 K and another at 90 K designated E0. Defect E0 has also been observed previously in Se doped $In_{0.5}(Al_xGa_{1-x})_{0.5}P$ [10]. The oxygen-related defect appearing at a similar temperature to the case of $Al_{0.5}In_{0.5}P$. $In_{0.5}(Al_{0.9}Ga_{0.1})_{0.5}P$ was also investigated, but the magnitude of the E2 defect peak is not as great as in the other compositions and is superimposed with E3 defect peak. The trend in the defect activation energies as a function of Al composition for the defects found in $In_{0.5}(Al_xGa_{1-x})_{0.5}P$

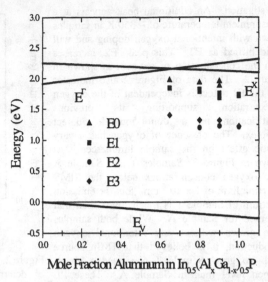

Figure 5. Variation of the defect energy position within the band gap of $In_{0.5}(Al_xGa_{1-x})_{0.5}P$ for $0.65 < x < 1.0$. The defects that are attributed to oxygen incorporation are represented by open symbols. The other defects are 'intrinsic' to the alloy system and not the result of intentional oxygen incorporation.

is shown in Figure 5 along with the major band structure features. The oxygen-related defects, E2, appear to be relatively independent of the alloy composition over the range studied, lying ~0.9 eV below the conduction band. This is consistent with the work of others, who assign the oxygen-related defect to an energy level between 0.9 and 1.1 eV below the conduction band edge [11,12]. The shallower 'intrinsic' defects appear to follow the X conduction band edge. The E0 defect concentration in these alloys depends on the Al-to-Ga ratio. For $Al_{0.5}In_{0.5}P$, there was no detectable concentration of the defect associated with the E0 spectral feature with only the E1 related peak being observed in the DLTS measurements. As the Ga mole fraction in the alloy is increased, forming the quaternary alloy materials the E0 peak could be observed in the measured DLTS spectra with the concentration of this defect increasing from 7.5 x 10^{15} cm^{-3} for $In_{0.5}(Al_{0.9}Ga_{0.1})_{0.5}P$ to 2.3 x 10^{16} cm^{-3} for $In_{0.5}(Al_{0.8}Ga_{0.2})_{0.5}P$. The concentration of E1 varied by only 30 percent from $Al_{0.5}In_{0.5}P$ to $In_{0.5}(Al_{0.8}Ga_{0.2})_{0.5}P$.

CONCLUSIONS

Defects related to the incorporation of oxygen into $In_{0.5}(Al_xGa_{1-x})_{0.5}P$ have been identified. The activation energy of these oxygen-related defects is ~0.9 eV below the conduction band edge, near the midgap for this alloy system. Several 'intrinsic' defects were also determined over the alloy system and appear in both the nominally non-oxygen-doped as well as intentionally oxygen-doped films. Defect E1 was assigned to the previously determined

DX-type center, while E0, the oxygen-related defect, becomes increasingly prominent with the increase of Ga in the alloy system. A fourth defect, E3, appears at high measurement temperatures and exhibits an emission behavior that is dependent on both thermal history and biasing of the sample. The intentional oxygen doping of the materials has identified the specific defect due to oxygen incorporation. Other defects are dominant in non-oxygen doped materials that could impact the performance of optical devices.

REFERENCES

1.   D.P. Bour, in *Quantum Well Lasers*, (P.S. Zory, Jr. ed.), Academic Press, p.427, (1993).
2   H. Sugawara, K. Itaya, M. Ishikawa, Jap. J. Appl. Phys., **31**, (1992), 2446.
3.   H.K. Yow, P.A. Houston, C.C. Button, T.-W. Lee, J.S. Roberts, J. Appl. Phys., **15**, (1994), 8135.
4.   T.F. Kuech, R. Potemski, F. Cardone, G. Scilla, J. Electronic Materials, **21**, (1992), 341.
5 .   M. Skowronski, in *Deep Centers in Semiconductors: A State-of-the-Art Approach*, (S. T. Pantelides ed.) , 2nd. ed., Gordon and Breach Science Publishers, Philadelphia, p. 401, (1992).
6.   J.W. Huang, D.F. Gaines, T.F. Kuech, R.M. Potemski, F. Cardone, J. Electronic Materials, **23**, (1994), 659.
7.   J.W. Huang, J.M. Ryan, K.L. Bray, T.F. Kuech, J. Electronic Materials, **24**, (1995), 1539.
8.   J.G. Cederberg, K.L. Bray, T.F. Kuech, J. Appl. Phys., **82**, (1997), 2263.
9.   S. Nojima, H. Tanaka, H. Asahi, J. Appl. Phys., **59**, (1986), 3489.
10.   M.O. Watanabe, Y. Ohba, J. Appl. Phys., **60**, (1986), 1032.
11.   M. Suzuki, K. Itaya, Y. Nishikawa, H. Sugawara, M. Okajima, J. Crystal Growth, **133**, (1993), 303.
12.   M. Kondo, N. Okada, K. Domen, K. Sugiura, C. Anayama, T. Tanahashi, J. Electronic Materials, **23**, (1994), 355.

DISLOCATIONS AND TRAPS IN MBE GROWN LATTICE MISMATCHED
p- InGaAs/GaAs LAYERS ON GaAs SUBSTRATES

A. Y. Du *, M. F. Li *, T. C. Chong *, Z. Zhang **

*Centre for Optoelectronics, Department of Electrical Engineering, National University of Singapore, Singapore 119260.

** Beijing Laboratory of Electron Microscopy, Chinese Academy of Sciences, Beijing P.O. Box 2724, Beijing 100080, P. R. China.

Abstract:

Dislocations and traps in MBE grown p-InGaAs/GaAs lattice-mismatched heterostructures are investigated by Cross-section Transmission Electron Microscopy (XTEM), Deep Level Transient Spectroscopy (DLTS) and Photo-luminescence (PL). The misfit dislocations and the threading dislocations observed by XTEM in different samples with different In mole fractions and different InGaAs layer thickness generally satisfy the Dodson-Tsao's plastic flow critical layer thickness curve. The threading dislocations in bulk layers introduce three hole trap levels H1, H2 and H5 with DLTS activation energies of 0.32 eV, 0.40 eV, 0.88 eV, respectively, and one electron trap E1 with DLTS activation energy of 0.54 eV. The misfit dislocations in relaxed InGaAs/GaAs interface induce a hole trap level H4 with DLTS activation energy between the range of 0.67-0.73 eV. All dislocation induced traps are non-radiative recombination centers which greatly degrade the optical property of the InGaAs/GaAs layers.

I. Introduction

The lattice mismatched $In_xGa_{1-x}As$/GaAs system has a considerable potential for the fabrication of heterojunction devices such as high electron mobility transistors (HEMTs)[1], heterojunction bipolar transistors (HBTs)[2] and strain quantum well lasers[3]. When the lattice mismatch is accommodated only by the coherent elastic strain, the structure is termed pseudomorphic. When the mismatch is accommodated by both the elastic strain and formation of misfit dislocations, it is named relaxed structure. The boundary between these two structures is not clear. The crossover is estimated by the concept of critical thickness[4]. In the relaxed structure, dislocations introduce deep traps which frequently act as recombination centers or scattering sites. In this paper, we report some new hole traps associated with dislocations in p-InGaAs/GaAs system. Combining the cross-section transmission electron microscopy (XTEM) and deep-level transient spectroscopy (DLTS) results, we identify that three hole traps and one electron trap H1, H2, H5 and E1 are associated with threading dislocations, and one hole trap H4 is associated with misfit dislocations. We also observed that the dislocations cause degradation in the optical properties of InGaAs and GaAs materials.

II. Sample Preparation

The detailed sample structures are shown in Fig. 1. Fig. 2 shows the In mole fraction x and thickness of $In_xGa_{1-x}As$ layers in the four samples grown at a substrate temperature of 550 °C and the curve for the $In_xGa_{1-x}As$ layer critical thickness, according to the Dodson-Tsao's plastic flow model[5]. The cap p-GaAs layer in sample #1, #2, #3 is used to stop Be out-diffusion in p^+ layer. The samples were grown using a RIBER MBE 32P molecular-beam epitaxy (MBE) system.

III Results and Discussion:

A. Investigation of structure defects and dislocation distribution by XTEM

The main structural defects are misfit dislocations and threading dislocations. Most misfit dislocations in the InGaAs/GaAs system are $60°$ dislocations with four $\frac{1}{2}\langle 110 \rangle$ Burgers vector inclined at $45°$ to the (001) interface plane[6]. In the InGaAs/GaAs system (epilayer under compression), any $60°$ dislocation at interface has the extra half plane lying in the GaAs substrate. We define the misfit dislocation as the $60°$ dislocations at the interface. Other dislocations are defined as threading dislocations. The XTEM pictures for samples #1-4 are depicted in Fig.3 a-d.

Sample #1: We performed XTEM on four specimens of sample #1 and found no dislocations in them. Since the InGaAs layer in our sample #1 is much thinner than the critical thickness (see Fig.2), misfit dislocations should not be found in the sample. No defects (dislocation, stacking faults) were observed in the four specimens by TEM.

Sample #2 : Owing to the GaAs/InGaAs/GaAs sandwich structure of sample #2, there are two misfit dislocation types in this system: the dislocation dipole and the single-dislocation[6,7]. The dislocation dipole is a pair of parallel $60°$ dislocations with Burgers vectors of opposite sign, lying in the same{111} glide plane, one at each interface. The single-dislocation which is not part of a dipole only appears at lower interface. Fig. 3(b) shows the two types of misfit dislocations, one is the dislocation dipole, marked as D, and the other is the single-dislocation, marked as S. In Fig.3(b), we also note that there is no dislocation in the cap GaAs layer, however there are some threading dislocations in the lower GaAs layer.

Sample #3: Fig. 3(c) is the weak beam dark field image. It shows that in the InGaAs layer there is no stacking faults or micro-twins which are the typical defects in 3D growth epitaxy layer. Therefore, the InGaAs layer is grown with 2D growth mode.Fig. 3(c) also shows that there is a very high density of threading dislocations in the InGaAs layer. Correspondingly, the quality of the cap GaAs layer, which has many defects, is poor. The stacking faults & threading dislocations are also found in the cap GaAs layer in the XTEM picture. However, the threading dislocation density in the cap GaAs layer is much lower than that in the InGaAs layer.The density of threading dislocations in the InGaAs layer is so high that the reacting dislocations cannot go into the lower GaAs substrate as in sample #2. The lower GaAs layer is dislocation free for sample #3.

sample #4 In this sample, no dislocation can be observed in the InGaAs layer except at the interface. The misfit dislocations are located in the interface, similar to the single dislocations in sample #2. The segments of interaction dislocations incline towards the GaAs substrate to form threading dislocations. In this sample, there are some threading dislocations in the GaAs buffer layer which are indicated by arrows in the picture. These threading dislocations propagate into GaAs layer as deep as 0.9μm below the InGaAs/GaAs interface.

Sample #1	#2	#3	#4
p⁺ GaAs (200 Å)	p⁺ GaAs (200 Å)	p⁺ GaAs (200 Å)	p⁺ In$_{0.05}$Ga$_{0.95}$As (200 Å)
p GaAs (1 μm)	p GaAs (2000 Å)	p GaAs (1 μm)	p In$_{0.05}$Ga$_{0.95}$As (2 μm)
p In$_{0.05}$Ga$_{0.95}$As (500 Å)	p In$_{0.12}$Ga$_{0.88}$As (500 Å)	p In$_{0.24}$Ga$_{0.76}$As (650 Å)	
p GaAs (0.1 μm)	p GaAs (0.1 μm)	p GaAs (0.1 μm)	p GaAs (0.1 μm)
GaAs (undoped 200 Å)	GaAs (undoped 200 Å)	GaAs (undoped 200 Å)	GaAs (undoped 200 Å)
n⁺ GaAs (buffer 0.5 μm)	n⁺ GaAs (buffer 0.5 μm)	n⁺ GaAs (buffer 0.5 μm)	n⁺ GaAs (buffer 0.5 μm)
n⁺ Sub GaAs	n⁺ Sub GaAs	n⁺ Sub GaAs	n⁺ Sub GaAs

Fig. 1 The structures of four samples: (100) n⁺
GaAs : Si substrates. n⁺ GaAs buffer layers,Si
doping 2x 10^{18}cm⁻³. p-GaAs or p-InGaAs
layers ,Be doping 1x10^{17}cm⁻³. p⁺ GaAs or
InGaAs cap layers, Be doping 2x10^{18}cm⁻³.

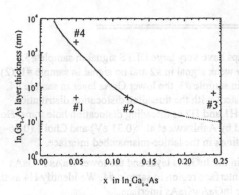

Fig.2 Four In$_x$Ga$_{1-x}$As samples (+) and the
Dodson-Tsao critical thickness curve
(solid line) at 550^{0}C growth temperature.

Fig.3 XTEM micrographs of #1-4 samples

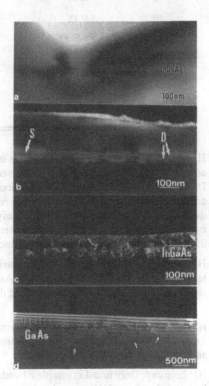

B.   Investigation of traps by DLTS measurements.

The p-n junction diodes used in DLTS measurements are made from the same sample structures studied by XTEM. All diodes have the same area (0.1257 mm^2). By using different bias and filling pulse voltage, there is different active DLTS region[8,9] and therefore we can distinguish different traps located in different layers. By this method, we can correlate the relationship between different DLTS trap signals and the corresponding structure defects found in XTEM studies.

Fig. 4 shows some typical DLTS spectra of samples #1- #4 . Six hole traps H1, H2, H3, H4, H5, H1' and two electron traps E1, E2 are found in these samples. The trap activation energies are calculated from the Arrhenius plots of the DLTS signals and are listed in Table I.

Table I. Hole and Electron traps characteristics

Trap Type	DLTS activation energy	Identification
H1'	0.20 eV	
H1	0.32 eV	threading dislocation
H2	0.38~0.40 eV	threading dislocation
H4	0.67~0.73 eV	misfit dislocation
H5	0.88 eV	threading dislocation
E1	0.54 eV	threading dislocation
E2		

H1 and H2 hole traps : (1)H1 and H2 hole traps have very large DLTS signal in sample #3, comparatively weak signal in sample #4, very weak signal in #2 and no signal in sample #1. (2) DLTS signal comes from the InGaAs region in sample #3, the lower GaAs layer in sample #4, and sample #2. These DLTS results are consistent with the threading dislocation distribution detected by XTEM and therefore we identify H1 and H2 as threading dislocation hole traps. Hole traps similar to H1 and H2 have been reported by Ashizawa et.al.[10](0.31 eV) and Choi[11] (0.16-0.40 eV) that they are associated with dislocations in the lattice-mismatched interface.

H4 hole trap :   H4 DLTS peak only appears in the InGaAs layer and the lower GaAs P-GaAs layer in sample #2, and in the InGaAs/GaAs interface region in sample #4. We identify H4 as the single and dipole misfit dislocations along the InGaAs/GaAs interface.

H5 hole trap : H5 DLTS signal mainly comes from the cap GaAs layer in sample #3, and from lower GasAs layer in sample #4. Comparing with the XTEM information, we identify H5 as another kind of threading dislocation trap.

H1' hole trap: Only found in sample #1 under large forward bias injection. Since sample #1 is free of dislocations, H1'is not related to any dislocations.The origin of H1' trap is not clear yet.

E1 electron trap : E1 DLTS signal was only detected in sample #3 under a very large forward bias injection. We tentatively assign this E1 electron trap to be threading dislocation trap, since

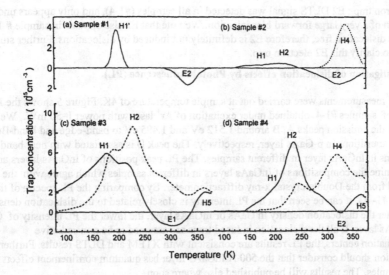

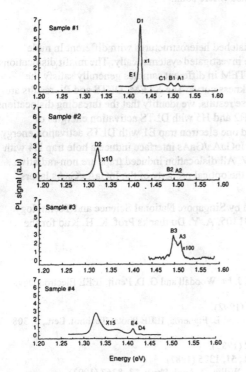

Fig.4 DLTS spectra of samples #1-4 with a rate window 200 s[-1]. The solide curves are measured at small reverse bias with large foward bias injection condition. The dotted curves are the typical DLTS spectra without foward bias injection.

Fig. 5 PL spectra at 4K for samples #1-4 under excitation of 488 nm Ar[+] laser with power 50 mW.

only sample #3 has high density of threading dislocations .Electron trap similar to E1 has been assigned as a dislocation trap by many authors[12-14].

E2 electron trap : E2 DLTS signal was detected in all samples (#1-4), and only appears under the condition of a very large forward bias injection. We note that E2 also appears in sample # 1 which is dislocation free, therefore E2 is definitely not induced by dislocations. Further study is needed to clarify this E2 electron trap.

C. Investigation of dislocation effects by Photoluminescence (PL).

All PL measurements were carried out at sample temperature of 4K. Figure 5 shows the PL spectra of samples #1-4, obtained under excitation of Ar^+ laser with power I=50 mW. We assigned the emission peaks A, B around 1.512 eV and 1.495 eV to band-edge and band-Be acceptor transitions in p-GaAs layer, respectively. The peak D is associated with near band-band transitions in InGaAs layer in different samples. The PL peak positions of InGaAs layers are used to determine the compositions of InGaAs layers in different samples which agree with the results obtained from the Double-crystal x-ray diffractionmeter. By comparing the PL spectra of the samples #1- #3, it can be seen that the PL intensity is closely related to the dislocation density. The higher the dislocation density in GaAs or InGaAs layer, the lower the PL intensity of GaAs or InGaAs layer. If all the dislocation induced traps are assumed to be non-radiative recombination centers, the PL results are consistent with XTEM and DLTS results. Further PL explanation should consider that the 500 Å InGaAs layer has quantum confinement effects in the #1-3 samples. The results will be published else where soon.

IV. Conclusions

MBE grown p-$In_xGa_{1-x}As$/GaAs lattice-mismatched heterostructures with different In mole fractions x and different layer thicknesses are investigated systematically. The misfit dislocations and the threading dislocations observed by XTEM in different samples generally satisfy the Dodson-Tsao's plastic flow critical layer thickness curve. The XTEM, DLTS and PL results are consistent with each other. By comparing these results, we identify that the threading dislocations in bulk layers introduce three hole traps H1, H2 and H5 with DLTS activation energies of 0.32eV, 0.40eV and 0.88eV, respectively, and one electron trap E1 with DLTS activation energy of 0.54eV. The misfit dislocations in relaxed InGaAs/GaAs interface induce a hole trap H4 with DLTS activation energy between 0.67-0.73eV. All dislocation induced traps are non-radiative recombination centers which greatly degrade the optical property of the InGaAs/GaAs layers.

Acknowledgment This work was supported by Singapore National Science and Technology Board RIC-university research project No. 681305. A. Y. Du thanks Prof. K. H. Kuo for the arrangement of TEM work at BLEM.

References
1. J. J. Rosenberg, M. Benlamri, P. D. Kirchner, J. M. Woodall and G. D. Pettit, IEEE Electron Dev. Lett. **EDL-6**, 491 (1985).
2. H.Ito and J.S. H. Jun, Electron. Lett., **28**, 655 (1992).
3. R. J. Fu, C. S. Hong, E. Y. Chan, D. J. Booher and L. Figueroa, IEEE Photo. Technol. Lett., **3**, 308 (1991).
4. R. People and J. C. Bean, Appl. Lett., **47**, 322 (1985).
5. B.W. Dodson and J. Y. Tsao, Appl. Phys. Lett., **51**, 1325 (1987).
6. T. J. Gosling, R. Bullough, S. C. Jain and J. R. Willis, J. Appl. Phys., **73**, 8267 (1993).
7. J. Zou, D. J. H. Cockayne and B. F. Usher, J. Appl. Phys., **73**, 619 (1993).

8. M. F. Li and C. T. Sah, IEEE Trans. Electron Devices, **ED-29**, 306(1982).

9. M. F. Li, *modern Semiconductor Quantum Physics*, (World Scientific, Singapore, 1994) chap.3.

10. Y. Ashizawa, S. Akbar, W. J. Schaff. L. F. Eastman E. A. Fitzgerald and D. G. Ast, J. Appl. Phys., **64**, 4065 (1988).

11.   Y. W. Choi, K. Xie, H. M. Kim and C. R. Wie, J. Electron. Mater., **20**, 545 (1991).

12.   A. C. Irvine, L. K. Howard and D. W. Palmer, Materials Science Forum, **83-8**, 1291 (1992).

13.   W. R. Buchwald, J. H. Zhao M. Harmatz and E. H. Poindexter, Solid-St. Electron., **36**, 1077 (1993).

14.   Y. Uchida, H. Kakibayashi and S. Goto, J. Appl. Phys., **74**, 6720 (1993).

STUDY OF THE HOMOGENEITY OF
Fe-DOPED SEMIINSULATING InP WAFERS

J. JIMENEZ*, R. FORNARI**, M. CURTI**, E. de la PUENTE*, M. AVELLA*,

L. F. SANZ*, M. A. GONZALEZ*, A. ALVAREZ*

* Física de la Materia Condensada, ETS Ingenieros Industriales, 47011 Valladolid, Spain

** MASPEC CNR Institute, via Chiavari 18A, 43100 Parma, Italy.

ABSTRACT

The homogeneity of semiinsulating Fe-doped InP wafers is studied using mapping techniques, Scanning Photocurrent (SPC) and Scanning Photoluminescence (SPL). These techniques allow to map with a micrometric spatial resolution the distribution of electrically active levels, in particular substitutional iron levels, Fe^{2+} and Fe^{3+}. The correlation between both measurements allows to obtain information about the local compensation in terms of the $[Fe^{3+}+Fe^{2+}]/[Fe^{2+}]$ ratio. Samples thermally treated were studied in order to analyse the consequences of the annealing on the homogeneity.

INTRODUCTION

Semiinsulating InP is produced by doping with iron, which introduces a deep acceptor level that compensates the residual shallow donors [1]. In LEC (Liquid Encapsulated Czochralski) InP crystals, iron is added to the starting charge [2]; owing to the low segregation coefficient of Fe in InP ($\cong 0.0015$) high iron amounts in the melt are necessary to obtain semiinsulating material and a gradient in the axial Fe distribution is always present, with detrimental consequences for the fabrication yield of these substrates. The electrical properties are determined by the presence of iron that is a substitutional impurity, (Fe_{In}). It can be in two different charge states, either neutral, Fe^{3+}, or ionised, Fe^{2+}, when it captures an electron from a shallow donor. The relative concentrations of both charge states depend on the total iron amount, $[Fe]_{tot}$, and the net donor concentration, $N_D-N_A \cong [Fe^{2+}]$. The assessment of the iron distribution is essential to improve the crystal quality in view of optoelectronics and high frequency device applications. It has been argued that the net shallow donors are incorporated almost homogeneously on a macroscopic scale [1, 3].

Typically, both microscopic and macroscopic non-uniformities are present in semiinsulating Fe-doped crystals [4]. The list of inhomogeneities includes growth striations, microprecipitates, dislocation atmospheres, microdefects and grain boundaries [4,5]. Iron plays a major role in this context, since is responsible for the high resistivity and its irregular distribution results in non-homogeneous electrical properties. A successful way to improve the substrates consists of appropriate thermal treatments, to produce more uniform material with better electrical properties [6,7]. On the other hand, it is important to produce semiinsulating material with low iron concentration in order to minimise the effects of the iron exodiffusion during the different steps of the device fabrication.

Scanning photoluminescence (SPL) [4] and Scanning Photocurrent (SPC) [7,8] are powerful techniques for studying the homogeneity of Fe-doped InP wafers at both microscopic and macroscopic scale, since they combine micrometric probe beams and the possibility of long scanning distances (cms); on the other hand they probe electronic transitions related to iron levels, which allows to map electrically active iron. The iron distribution can be assesed in terms of the contrast of both maps, PL and PC.

We present herein a study of the homogeneity of Fe-doped InP substrates combining these techniques. The results are discussed in terms of electrically active iron and net donor concentration distributions. Homogeneity maps are presented and discussed, considering the influence of thermal treatments.

EXPERIMENTAL AND SAMPLES

Photoluminescence (PL) measurements were carried out at room temperature. The excitation was done with the visible lines of an Ar+ laser focused onto the sample surface by means of a microscope objective, which results in laser beam diameters at the focal plane of ~1 μm. Intrinsic luminescence was measured. The monochromator was settled at λ_{max} of the Band to Band PL spectrum and the intensity at λ_{max} was recorded scanning the laser beam across the sample surface.

Photocurrent (PC) measurements were carried out at liquid nitrogen temperature, in order to minimise the influence of the dark current. The excitation is provided by a cw:YAG laser operating at 1.32 μm wavelength through a microscope objective. Transparent electrodes were evaporated on both faces of the wafers. A rough estimation gives a beam size of ~5 μm. Contrarily to PL measurements, where the probe depth is about 100 nm, the laser beam of the PC measurements crosses the wafer, probing 700 μm (wafer thickness).

In both systems the samples were mounted on a high precision motorised X-Y stage. Scanning photoluminescence (SPL) and Scanning Photocurrent (SPC) experimental set-up have been described elsewhere [7,8,9].

Several Fe-doped LEC (Liquid Encapsulated Czochralski) wafers were studied. All of them were semiinsulating with resistivities above 10^6 Ωcm, which should mean that [Fe] >> (N_D-N_A). Samples as-grown and annealed under different conditions were studied. All the annealings were carried out in ultrapure phosphorous atmosphere (1bar pressure). for more details see ref.7. Three different thermal treatments are considered, i.e. wafer annealing, ingot annealing and ingot annealing followed by wafer annealing. Table I summarises the different treatments, as well as the resistivity and mobility data.

	Electrical Properties						Annealing Conditions			
	Before Annealing			After Annealing						
Treatment (Sample)	Resist. (Ω cm)	Mobility (cm²/Vs)	Hall Conc. (cm³)	Resist. (W cm)	Mobility (cm²/Vs)	Hall Conc. (cm³)	Type (Waf./Ing.)	Temper. (°C)	Duration (h)	Cool rate (°C/min)
NF1 (159-53)	7.27 x 10⁶	1.500	5.72 x 10⁸	7.3 x 10⁷	1.900	4.5 x 10⁷	W	900	50	1
NF7 (13B4)	2.88 x 10⁷	1.597	1.36 x 10⁸	6.04 x 10⁷	2.344	4.41 x 10⁷	I	900	50	6
NF7+NF13 (13B4)							I+ (W)	900+ (900+1000)	50+ (50+80)	6+ (60+120)

Table I : Samples, electrical properties and annealing conditions.

RESULTS AND DISCUSSION

A reliable interpretation of the PL and PC maps requires knowing the main photoelectronic transitions involved in both phenomena. Intrinsic PL is excited by e-h generation; its intensity is limited by the competition between Band to Band recombination and trapping by other levels. In Fe-doped InP neutral iron, Fe^{3+}, traps one electron being converted to Fe^{2+}, while Fe^{2+} comes back to neutral iron, Fe^{3+}, trapping a hole. Both capture processes limit the intrinsic luminescence emission. In other words, the intensity of intrinsic PL is reduced by the presence of iron. The higher the iron concentration the lower the intrinsic photoluminescence intensity [4].

While the photoluminescence was excited with visible light, the photocurrent is excited with below band gap light, 1.32 μm wavelength. This light photogenerates electrons and holes directly from iron levels in the bulk; contrarily to what can happen in PL, the surface effects are minimised. Fe^{3+} is converted to Fe^{2+} and viceversa by optical excitation. Simultaneously, Fe^{3+} captures electrons and Fe^{2+} holes. The photocurrent intensity is a complex function of the free carrier generation rate and lifetime limited by these electronic transitions. The optical cross section for holes is at least one order of magnitude larger than the optical cross section for electrons at this wavelength [10]. Assuming that trapping by other levels is negligible, it is possible to analyse the photocurrent in terms of $[Fe^{3+}]$ and $[Fe^{2+}]$ and their optical and capture cross sections (for more details see ref. 8). In a first approximation the higher the iron concentration the higher the PC intensity; furthermore, for a predetermined constant total iron concentration the photocurrent increases with increasing $[Fe^{3+}]/[Fe^{2+}]$ ratios.

PL and PC maps obtained on an ingot-annealed sample are shown in fig.1. Following the above considerations an anticorrelation between PL and PC intensities was expected. However, such a relation is not always straightforward and regions where PL and PC appear correlated are also found. Some of these regions are indicated by **a** (anticorrelated) and **c** (correlated) in the maps of fig.1.

An accurate calculation of the photocurrent and photoluminescence intensities, in terms of the electronic transitions at iron levels [8], allows to understand the different behaviours that can be observed. These calculations show that PL decreases for increasing $[Fe]_{tot}$, while PC increases with $[Fe]_{tot}$. However, the $[Fe^{3+}]/[Fe^{2+}]$ ratio is found to be determinant for the behaviour of both PL and PC intensities. For an equivalent total iron amount, the PL and PC reach higher intensities

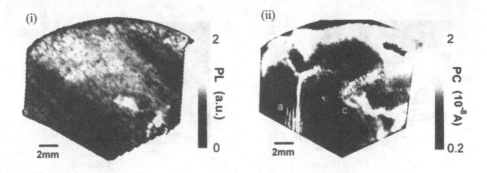

Figure 1.- PL map (i) and PC map (ii) of an ingot annealed sample (treatment NF7) showing correlated (c) and anticorrelated (a) PL-PC areas

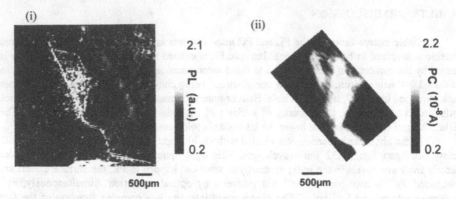

(i)

2.1

PL (a.u.)

0.2

(ii)

2.2

PC (10⁻⁸ A)

0.2

500μm   500μm

Figure 2.- High resolution PL map (i) and PC map (ii) from a grain boundary of the ingot annealed sample (treatment NF7)

for increasing $[Fe^{3+}]/[Fe^{2+}]$ ratio. These data allow to understand the complex correlations between PC and PL contrasts in the corresponding maps.

For example, in the map of fig.1, we have selected several regions in order to illustrate the correlation behaviour between PL and PC maps.

i) Regions with high PL and high PC. In principle, this observation should correspond to heavily compensated regions ($[Fe^{3+}]>>[Fe^{2+}]$) (high PC) with low $[Fe]_{tot}$ (high PL). This region is located near a twin boundary, which getters impurities, leaving the surrounding areas free of iron and donors, which should explain the above mentioned PL-PC correlation, a detail of this region is shown in fig.2, where, the above mentioned PL-PC correlation between both is observed. Probably, this gettering effect is strongly enhanced by the thermal treatment undergone by this specimen, that was submitted to treatment NF7, see table I. Similar effects are found around dislocations [7], fig.3. Apparently, donors diffuse faster than iron, which should account for the enhancement of the photocurrent in regions of low $[Fe]_{tot}$, which is unambiguously related to the increase of the $[Fe^{3+}]/[Fe^{2+}]$ ratio, according to the above discussion.

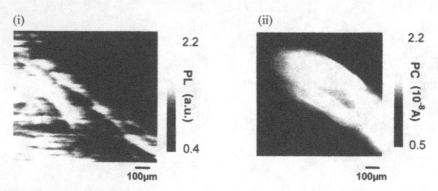

(i)

2.2

PL (a.u.)

0.4

(ii)

2.2

PC (10⁻⁸ A)

0.5

100μm   100μm

Figure 3.- High resolution PL map (i) and PC map (ii) of dislocations of the ingot annealed sample (treatment NF7)

(i)  (ii)

200 μm   200 μm

Figure 4.- Nomarski micrographs after DSL etching of (i) an as-grown reference sample and (ii) an ingot anneald sample. It can be observed how the Cottrell atmospheres spread out after thermal treatment.

ii) High PC and low PL, which should correspond to both high $[Fe]_{tot}$ and high $[Fe^{3+}]/[Fe^{2+}]$. See fig.1. These regions are probably due to iron accumulation. They appear usaually at the wafer rim.

iii) High PL and low PC corresponds to low $[Fe]_{tot}$ and low $[Fe^{3+}]/[Fe^{2+}]$, which is a region with poor compensation (small $[Fe]_{tot}$ / $[Fe^{2+}]$); however this situation is not expected in high resistivity Fe-doped InP, even after thermal treatment; which is consistent with the fact that donors are removed more efficiently than iron during the thermal treatment [11], making the above situation highly improbable.

iv) Low PL and low PC. This situation has to be considered only comparatively to other regions. It seems rather probable that the ratio $[Fe^{3+}]/[Fe^{2+}]$ is determining this behaviour.

These results suggest that the PC and PL fluctuations are related to both shallow donor and iron inhomogeneities. The net donor distribution, which is usually believed to be uniform [1,3,8], is probably such only at a macroscopic scale in untreated material. The non-uniformity of $[Fe^{2+}]$ is found to be pronounced around the crystal defects, dislocation, twin boundaries after thermal treatments. This probably depends on the different gettering action that the defects exercise towards iron and donors, respectively. Wafer annealing results in a better homogeneity as large free surface are available as sinks for both diffusing species, so that after polishing the bulk remains uniform. In ingot annealed material, where free surfaces are far away, the main role of sink for donors and iron is played by extended defects (but with different effectiveness), and this enhances the Cottrell atmospheres around them (see for example the twin boundary of sample NF7, fig.2 and the dilocation of fig.3). This hypothesis is furhter supported by chemical etching studies: the atmospheres, which are revealed by DSL photoectching [5,7] as grooves around dislocations, produce relatively shallow holes in wafer annealed material, while they give broader and deeper grooves in ingot annealed material, fig.4.

CONCLUSIONS

In summary SPC and SPL are powerful techniques for studying the homogeneity of Fe-doped InP wafers. A correlation between both mapping techniques allows information about the iron distribution, regarding not only the total iron amount but also the $[Fe^{3+}]/[Fe^{2+}]$

distribution on the microscale. Thermal treatments evidence improved electrical properties, but the homogeneity of the material is only improved by wafer annealing, while ingot annealed material enhances local inhomogeneity around crystal defects.

REFERENCES

1. G. Müller; Phys. Scripta **35**, 201 (1991)

2. C.R. Zeisse, G.A. Antypas, C. Hopkins; J. Cryst. Growth **64**, 217 (1983)

3. W. Meier, H. Ch. Alt, Th. Vetter, J. Völkl and A; Winnacker, Semicon. Sci. Technol. **6**, 297 (1991)

4. J.Y.Longere, K. Schohe, S. Krawczyk, R. Coquille, H. L'Haridon and P.N. Favennec, J. Appl. Phys. **68**, 755 (1990)

5. R.Fornari, C.Frigeri, J.L. Weyher, S.K. Krawczyk, F. Krafft and G. Mignoni, Proc. of 7th Semi-insulating III-V Materials Conference, Ed. C. Miner, W. Ford and E. Weber, IOP Bristol 1993, p. 39

6. R. Fornari, E. Gilioli, A. Sentiri, G. Mignoni, M. Avella, J. Jimenez, A. Alvarez, M.A. Gonzalez; Mater. Sci. Eng. B **44** (1997) 233.

7. M. Avella, J. Jimenez, A. Alvarez, R. Fornari, E. Gilioli, A. Sentiri; J.Appl. Phys. **82**, 3832 (1997)

8. A. Alvarez, M. Avella, J. Jimenez, M.A. Gonzalez and R. Fornari, Semicond. Sci. Technol. **11**, 941 (1996)

9. L.F.Sanz, M.Avella, J.Jimenez, M.A. Gonzalez, R. Fornari; 19th Int. Conf. on Defects in Semiconductors, Aveiro (Portugal) July 21-25, 1997. (to be published)

10. T. Takanohashi and K. Nakajima, J. Appl. Phys. **65**, 3933 (1989)

11. A. Zappettini, R. Fornari, R. Capelletti. Mat. Sci. Eng. B **45**, 147 (1997)

RELAXED In$_x$Ga$_{1-x}$As GRADED BUFFERS
GROWN WITH ORGANOMETALLIC VAPOR PHASE EPITAXY ON GaAs

M.T. BULSARA, C. LEITZ, and E.A. FITZGERALD
Department of Materials Science and Engineering, Massachusetts Institute of Technology,
Cambridge, MA, 02139.

ABSTRACT

In$_x$Ga$_{1-x}$As structures with compositionally graded buffers were grown with organometallic vapor phase epitaxy (OMVPE) on GaAs substrates and characterized with plan-view and cross-sectional transmission electron microscopy (PV-TEM and X-TEM), atomic force microscopy (AFM), and x-ray diffraction (XRD). The results show that surface roughness experiences a maximum at growth temperatures where phase separation occurs in In$_x$Ga$_{1-x}$As. The strain fields from misfit dislocations induce this phase separation in the <110> directions. At growth temperatures above and below this temperature, the surface roughness decreases significantly; however, only growth temperatures above this regime ensure nearly complete relaxed graded buffers with the most uniform composition caps. With the optimum growth temperature for grading In$_x$Ga$_{1-x}$As determined to be 700 °C, it was possible to produce In$_{0.33}$Ga$_{0.67}$As diodes on GaAs with threading dislocation densities $< 8.5 \times 10^6$/cm^2.

INTRODUCTION

Compositionally graded buffers are implemented in lattice mismatched heteroepitaxy to maintain a low threading dislocation density and achieve a completely relaxed growth template. OMVPE is a well-established growth technique which is capable of growth rates that are significantly greater than molecular beam epitaxy (MBE) growth rates, and therefore OMVPE is more practical growth tool for fabricating graded buffers. The ability to grow In$_x$Ga$_{1-x}$As graded buffers with OMVPE would facilitate the manufacture of commercial lattice-mismatched devices, including 1.3 μm wavelength emitting lasers on GaAs [1].

EXPERIMENT

In this study we examine In$_x$Ga$_{1-x}$As graded buffers on GaAs substrates with atmospheric OMVPE at different growth temperatures. The samples were grown in a Thomas Swan atmospheric research reactor on n$^+$ GaAs substrates from AXT. The PV-TEM and X-TEM characterization was done with a JEOL 2000FX. The XRD was performed with a Bede D^3 triple axis diffractometer. The AFM experiments were conducted with a Digital Instruments D3000 Nanoscope.

To explore graded In$_x$Ga$_{1-x}$As relaxation, the initial samples were graded to $x_{In} \cong 0.06$ ($\approx$ 0.4 % mismatch). Such a small amount of mismatch should produce excellent relaxed layers independent of most growth parameters.

Undoped In$_x$Ga$_{1-x}$As graded buffers with nominal final indium concentration of x_{In}=0.06 were grown at temperatures between 500-700 °C. In addition, x_{In}=0.15 and x_{In}=0.33 graded buffers were grown at the 700 °C and a x_{In}=0.33 graded buffer was grown at 550 °C. All growths were performed with a 5000 sccm H$_2$ carrier flow and 134 sccm AsH$_3$ flow. The TMG flow was fixed at 0.030 sccm throughout the graded buffer growth sequence. Compositional grading was accomplished by stepping the TMI flow rate by approximately 0.005 sccm up to a final flow of 0.031 sccm for the x_{In}=0.06 graded buffer, 0.077 sccm for the x_{In}=0.15 graded buffer, and 0.163 sccm for the x_{In}=0.33 graded buffer. Sufficient vent times were incorporated

631

after each change in TMI flow setting to ensure the expected composition, during which time the sample was kept at the growth temperature. All samples, except the structure with the x_{In}=0.33 graded buffer grown at 700 °C, had an undoped 1 μm uniform cap. The sample with an x_{In}=0.33 graded buffer at 700 °C had a 2 μm cap which incorporated a PIN diode structure.

RESULTS

A visual inspection of the surface morphology reveals a strong dependence on growth temperature, a surprising result for a such a low lattice mismatch. AFM surface topology data taken on 10 μm x 10 μm areas of each sample, including the x_{In}=0.15 and x_{In}=0.33 structures is depicted in figure 1. The data show that the rms roughness for the nominal x_{In}=0.06 sample grown at 550 °C has a significantly greater rms roughness value (52 nm) than the other structures which have an rms roughness value of about 10 nm. In fact, despite the low mismatch, the sample grown at 550 °C is not specular. In addition, the surface roughness for the x_{In}=0.33 structure grown at 700 °C was less than the surface roughness of x_{In}=0.06 structures grown at all lower temperatures. Most notably, the x_{In}=0.33 structure grown at 550 °C has a surface roughness 20X that of the structure grown at 700 °C, showing the greater disparity in surface roughness with increasing indium content.

A key criterion for applications is that there must be a great amount of strain relief and a low threading dislocation density. To determine the degree of strain relaxation and the indium composition, glancing exit (224) reciprocal space maps were conducted with triple axis XRD. Since the x_{In}=0.06 structures were of low mismatch and relatively thin (2 μm) the effect of epilayer tilt was expected to be negligible, and thus no glancing incidence (224) or (004) reciprocal space maps were acquired to extract this effect. X-TEM was used to measure the thickness and in combination with the final composition, the grading rate. Table I shows the growth temperature, composition, and grading rate for the nominal x_{In}=0.06 structures. The table shows that the indium composition steadily increased with decreasing temperature (with the exception of the structure grown at 600 °C), which is due to the lower cracking temperature for TMI. In addition, the growth rate decreased with decreasing temperature, which in turn provided

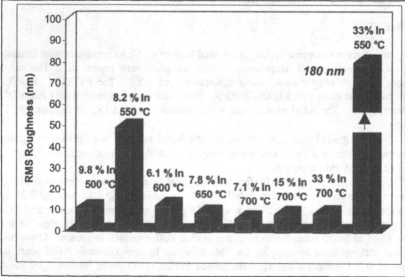

Figure 1- RMS roughness for $In_xGa_{1-x}As$ structures grown at different temperatures

for a higher grading rate at lower temperature. It should be noted that there is a small error ($\approx$ 300 Å) in the measurement of the graded buffer thickness due to the calibration of the TEM and the tilting of the TEM specimens.

Table 1 shows the residual strain in each of the nominal x_{In}=0.06 structures as a function temperature. The structure grown at 500 °C had a noticeably greater residual strain left in the structure, and there is no general trend among the other samples. However, since the compositions and grading rates differed, the efficiency of the graded buffers at relieving strain was compared after calculating the equilibrium plastic strain rate (strain/thickness) and the overall equilibrium plastic strain [2]. The equilibrium plastic strain rate is given by

$$C_\delta(h) = C_f + \frac{3D\left(1 - \dfrac{v}{4}\right)\ln\left(\dfrac{2\pi b C_\delta}{e}\right)}{2Yh^2} \qquad (1)$$

where C_f is the mismatch introduction rate (misfit/thickness), Y is the Young's Modulus, h is the film height and $D = Gb/[2\pi(1-v)]$ with G, the shear modulus; v, Poisson's ratio; and b, the magnitude of the Burgers vector (60˚ dislocations are assumed). The expression for the overall plastic deformation in a graded buffer is:

$$\delta_{eq}(h) = C_\delta(h)h \qquad (2)$$

The percentage of the equilibrium strain relieved (i.e., percent relaxation) is also listed in table 1. All the samples showed a similar degree of relaxation ($\approx$ 80-85%). At such a low mismatch it is difficult to distinguish the most effective growth conditions for strain relief. We would expect the disparities in strain relief to be more pronounced at higher indium compositions. In general, higher growth temperatures would allow for more efficient strain relief.

The X-TEM and XRD data exhibit great differences in microstructure between the x_{In}=0.06 sample grown at 550 ˚C and the same structure grown at 700 °C, in agreement with the drastic difference in surface morphology. Figure 2 shows the X-TEM micrographs of the two structures. Both structures have threading dislocation densities below the X-TEM limit (< 10^8/cm^2). Thus, the very poor surface morphology of the 550 °C sample is not due to a very high defect density in the top In$_x$Ga$_{1-x}$As layer. The uniform cap layer of the structure grown at 550 ˚C does show additional semi-circular regions in the top of the film. These features are not dislocations, as the contrast is weak, and are believed to be variations in strain from neighboring regions which have undergone phase separation during growth. A (224) glancing exit reciprocal space map of this structure shows a significantly greater spread in the 2θ direction for the uniform cap than any of the other samples grown at different temperatures (the FWHM data for the XRD peaks from the uniform caps are listed in table I. Note that the sharpest peak in the 2θ direction is from the sample grown at 700 °C). This spread is indicative of a spread in lattice constant or indium composition, which is consistent with phase separation [3,4]. In addition, the spread in the ω direction (FWHM data also tabulated in table I) for the cap in the structure grown at 550 °C was less than that of the other x_{In}=0.06 samples, creating a circular projection of the (224) spot in reciprocal spot, as opposed to the typical elliptical spot for a relaxed

Table I

Composition, Grading Rates, Relaxation, and X-Ray Data for x_{In}=0.06 for Structures Grown at Different Temperatures

Growth Temp (˚C)	% Indium in Graded Buffer	Grading Rate (%In/μm)	Residual ε (x 10⁻³)	% of Equilibrium ε Relieved	2θ FWHM for In$_x$Ga$_{1-x}$As Cap Layer	ω FWHM for In$_x$Ga$_{1-x}$As Cap Layer
500	9.81	15.32	-2.15	79	233	505
550	8.21	11.90	-1.59	85	283	356
600	6.06	8.81	-1.41	84	151	512
650	7.76	8.21	-1.28	89	144	415
700	7.14	7.51	-1.59	79	109	460

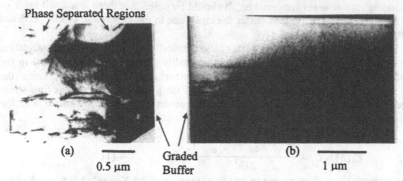

Phase Separated Regions

(a)

0.5 μm

Graded
Buffer

(b)

1 μm

Figure 2- X-TEM micrographs for x_{In}=0.06 $In_xGa_{1-x}As$ structures grown at 550 °C (a) and 700 °C (b). The $In_xGa_{1-x}As$ structure grown at 550 °C shows contrast from phase separated regions.

heterostructure. Figure 3 shows a planview TEM image of the x_{In}=0.06 $In_xGa_{1-x}As$ structure grown at 550 °C showing striations under a g=<220> diffraction condition in <110> directions (the same direction as the dislocations in the graded buffer) which is attributed to compositional variations due to phase separation. These striations disappear under the other g=<220> condition and g=<400> conditions.

The data suggests that there is a correlation between the dislocation strain fields from the dislocations in the graded buffer and regions of phase separation, which in turn cause the drastic surface roughening in the temperature regime where phase separation is favored. In the low temperature growth regime (500 °C), the phase separation and surface roughening are kinetically

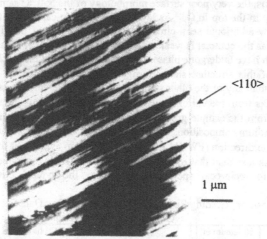

<110>

1 μm

Figure 3- PV-TEM micrograph of x_{In}=0.06 structure grown at 550 °C (g=<220>). The striations are in the direction of the cross-hatched pattern in a <110> direction. PV-TEM of the same structure grown at 700 °C did not show these features.

limited, as the atoms do not have the mobility to attach to sites which cause the long range variations which are seen at higher growth temperatures. In the high temperature growth regime (> 600 °C), thermodynamics dictate the growth conditions, as the growth occurs above the spinodal and atoms have the mobility to attach to favorable lattice sites. At the growth temperatures within the spinodal where the depositing atoms do not have kinetic limitations, either the strain fields from the graded buffer or the surface cross-hatch from the graded buffer encourage phase separation [5]. This phase separation occurs along the <110> directions since the misfit dislocations, their strain fields, and their cross-hatch surface lie along the <110> directions. As a consequence, the $In_xGa_{1-x}As$ layers roughen as surface energy is created and as strain energy is relieved. Note that this roughening (i.e., very pronounced cross-hatch pattern), produced by both phase separation and dislocation strain fields, is much more severe than the roughening that occurs in systems which lack phase instability. We note here that phase separation at only x_{In}=0.06 is not expected. Our observations suggest that the dislocation strain fields are large enough to encourage phase separation.

Although the X-TEM image of the x_{In}=0.06 structure grown at 550 °C did not show threading dislocations, it has been shown in the Si_xGe_{1-x} materials system that a rough surface with increasing lattice mismatch will eventually lead to a high threading dislocation density even in structures with slow grading rates [6]. Grading to greater indium compositions (x_{In}=0.33) at 550 °C produces very rough surfaces that eventually lead to high threading dislocation densities (ρ=2.0x10^8/cm^2). Although surface roughness can be decreased by growing at lower temperature, this can not be achieved without compromising the relaxation of the graded buffer. This implies that the only window for growth of $In_xGa_{1-x}As$ graded buffers which provides both relaxation and good surface morphology is at high temperature. It is important to note that MBE can not attain such high growth temperatures due to limited arsenic overpressure.

With surface roughness and relaxation conditions determined to be optimum at a growth temperature of 700 °C, an x_{In}=0.33 device structure was grown. Figure 4 is an X-TEM micrograph of this structure with no threading dislocations within the X-TEM limit. No threading dislocations could be observed in PV-TEM, showing that the threading dislocation density in this structure is < 8.5 x 10^6/cm^2 given this size of viewable area in TEM. (224) glancing incidence and glancing exit $\theta/2\theta$ double axis XRD scans showed the composition to be x_{In}=0.33 and the structure was 99.39% relaxed (the same structure grown at 550 °C was only 88% relaxed with the indium composition determined to be x_{In}=0.27). Examining the cap layer for the x_{In}=0.33 structure grown at 700 °C with X-TEM, there are additional striations layered in the <001> growth direction. The x-ray and TEM data do not the support the conclusion that this is phase separation or ordering. The cause of these striations is still under investigation. It is important to note that the surface does not roughen for this structure since the phenomenon occurs on the growth plane. A more detailed description of these features will be reported at a later date.

CONCLUSIONS

We have demonstrated that growth of $In_xGa_{1-x}As$ graded buffers is sensitive to the growth temperature regime where phase separation occurs in $In_xGa_{1-x}As$. In the temperature regime where phase separation occurs, the surface morphology of the structures degrades and rapidly leads to poor quality layers, even at relatively low mismatch. Only growth at higher temperatures produces relaxed layers with sufficient relaxation, good surface morphology, and low threading dislocation densities. The low threading dislocation densities are sufficient for the fabrication of electronic and optoelectronic devices.

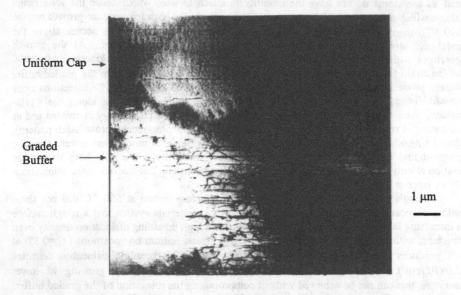

Uniform Cap →

Graded
Buffer →

1 μm
———

Figure 4- X-TEM micrograph of x_{In}=0.33 $In_xGa_{1-x}As$ structure grown at 700 °C. The threading density for this structure is less than x $10^6/cm^2$

ACKNOWLEDGMENTS

We would like to thank Prof. S. Mahajan at Carnegie Mellon University for helpful discussions. Funding for this work was provided by AFOSR SBIR Phase II contract DSI-500 with Discovery Semiconductors. This work made use of MRSEC Shared Facilities supported by the National Science Foundation under Award Number DMR-9400334.

REFERENCES

[1] T. Uchida, H. Kurakake, H. Soda, and S. Yamazaki, J. Elec. Mat., 12 (4), 581 (1996)
[2] E.A. Fitzgerald, Y.-H. Xie, D. Monroe, P.J. Silverman, J.M. Kuo, A.R. Kortan, F.A. Thiel, and B.E. Weir, J. Vac. Sci. Tech. B, 10 (4), 1807 (1992)
[3] K.C. Hsieh and K.Y. Cheng, Mat. Res. Soc. Symp. Proc. Vol. 379, 145 (1995)
[4] S.N.G. Chu, S. Nakahara, K.E. Strege, and W.D. Johnston, Jr., J. Appl. Phys., 57 (10), 4610 (1985)
[5] M. Bulsara, V. Yang, A. Thilderkvist, K. Häusler, K. Eberl, and E.A. Fitzgerald, to be published in January 1, 1998 issue of J. Appl. Phys.
[6] S. Samavedam and E.A. Fitzgerald, J. Appl. Phys., 81 (7), 3108, 1997

γ-RAY IRRADIATION EFFECT ON THE INTERSUBBAND TRANSITION IN InGaAs/AlGaAs MULTIPLE QUANTUM WELLS

M. O. Manasreh[*], J. R. Chavez[*], W. T. Kemp[*], K. Hoenshel[*], and M. Missous[**].
[*]Air Force Research Lab, 3550 Aberdeen Ave, SE, Kirtland AFB, NM 87117-5776.
[**]Center for Electronic Materials, Department of Electrical Engineering and Electronics, UMIST, Manchester, U.K.

ABSTRACT

Intersubband transitions in n-type InGaAs/AlGaAs multiple quantum wells were studied as a function of 1.0 to 5.0 MRad γ-ray irradiation dose using the optical absorption technique. The spectra were recorded at both 295 and 77K. The results show that the total integrated area of the intersubband transition is decreased as the irradiation dose is increased. This could be explained as follows: The secondary electrons generated from the γ-ray irradiation cause lattice damages where traps and point defects are created. Some of the electrons in the quantum wells are trapped by these defects causing the two dimensional electron gas (2DEG) density to decrease. The reduction of the 2DEG density thus leads to the reduction of the total integrated area of the intersubband transitions.

INTRODUCTION

Irradiation effects in III-V semiconductors have been studied extensively in the past several years [see for example Refs. 1-3]. On the other hand, irradiation effects on electronic and optoelectronic devices have been scarcely studied [see for example Refs. 4-6]. However, irradiation effects on the intersubband transitions in GaAs-related quantum wells and interband transitions in InGaSb-related superlattices (as well as GaN ternary compounds) have not been studied or reported in the open literature.

It is well known that point defects such as vacancy, antisite, and interstitial are formed due to the displacement of atoms from their regular sites in semiconductors during irradiation with high energy fast moving particles. More complex defects made of a combination of the above point defects are also formed. The introduction of these defects should produce energy levels in the band gap that could be detected by optical and electrical techniques. The introduction of these defects which may undergo radiative or nonradiative transitions affects the properties of the semiconductor materials and consequently the interband and intersubband transitions in quantum structures based on these semiconductors.

637

Part of the irradiation studies should include the production rates of the irradiation-induced defects and their thermal annealing properties. In addition, future work should focus on the study of these transitions during irradiation at low temperatures. This is because it is well known that many irradiation-induced defects are generated during irradiation at low temperatures and then annealed out by raising the temperature to 300 K.

In this paper, we present the latest results on intersubband transitions in n-type InGaAs/AlGaAs multiple quantum well samples irradiated with different γ-ray doses. The integrated area under the intersubband transition was measured, using the optical absorption technique, as a function of the irradiation dose.

EXPERIMENT AND RESULTS

Several samples were cut from an $In_{0.07}Ga_{0.93}As/Al_{0.4}Ga_{0.6}As$ multiple quantum wells grown on semi-insulating GaAs wafer by using a molecular beam epitaxial technique. The InGaAs quantum well was 75 Å thick and Si-doped ($[Si]=3 \times 10^{18}$ cm^{-3}) in the middle 55 Å. The barrier thickness was 100 Å and the total number of quantum wells is 50. The 10 Å thickness was left on both sides of the well doped region to avoid the dopant diffusion into the barrier. This is quite desirable since we try to prevent the modulation doping and DX center problems. The intersubband transition was measured at the Brewster's angle of GaAs (73°) using a Fourier-Transform infrared spectrometer. Five samples were irradiated at room temperature using 1.0 - 5.0 MRad γ-ray.

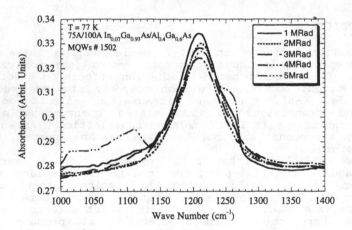

Fig. 1. *The optical absorption spectra measured at T = 77 K of five samples irradiated with γ-ray. The samples were irradiated at room temperature with 1.0 to 5.0 MRad.*

The optical absorption spectra of the intersubband transitions of five samples are shown in Fig. 1. The samples were irradiated with γ-ray using 1.0 to 5.0 MRad. The irradiation was performed at room temperature, while the spectra reported in this figure were measured at temperature (T) = 77K. It is clear from this figure that the absorbance of the spectra is reduced as the irradiation dose is increased. For the sample that was irradiated with 5 MRad, we noted that an extra band is appeared around 1100 cm^{-1}. An extra shoulder at 1250 cm^{-1} was also observed in the spectrum of the latter sample.

The total integrated area under the intersubband transition spectra, presented in Fig. 1, was extracted for the five samples and plotted as a function of the γ-ray dose as shown in Fig. 2. The integrated area of the intersubband transition in the control sample (un-irradiated sample) was also shown as the point at 0.0 MRad. This figure clearly demonstrates that the total integrated area is decreased as the irradiation dose is increased above 1.0 MRad. The intersubband transitions do not seem to be affected in samples irradiated with doses smaller than 1.0 MRad. The integrated area seems to decrease approximately linearly as a function of the irradiation dose above 1.0 Mrad. The magnitude of the vertical error bars was obtained from the standard deviation of the spectra data points and from the inhomogeneity of the wafer.

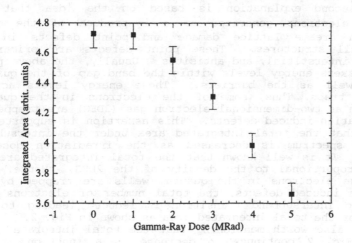

Fig. 2. *The total integrated area of the intersubband transitions in the five samples were plotted as a function of the γ-ray dose. The data point at 0.0 MRad is for the control sample (unirradiated sample).*

DISCUSSION

The present results indicate that there is a degradation in the intersubband transitions in the InGaAs/AlGaAs multiple quantum wells when the samples are exposed to 1.5 MeV γ-ray radiation with doses higher than 1.0 MRad. The reduction in the absorbance as well as the total integrated area of the intersubband transition as the irradiation dose is increased may have two possible explanations. First, the γ-ray perhaps does not produce extensive damage to the multiple quantum well structures. However, the secondary electrons produced as a by-product of γ-ray radiation could cause some damage to the interfaces leading to some sort of degradation to the intersubband transitions. This explanation require additional confirmation using a structural studies such as transmission electron microscopy technique. This explanation is supported by The appearance of the shoulder in the spectrum of the 5.0 MRad irradiated sample. It may be that the high dose irradiation causes significant damage to the interfaces that the energy levels in some of the quantum wells are shifted leading to some variation in the energy peak position of the intersubband transitions. The damage to the interfaces might be extensive in some of the quantum wells stack such that the electronic energy levels are shifted dramatically due to the fact that a small band appeared at the lower energy side of the spectrum as shown in Fig. 1, (5.0 Mrad irradiated sample).

The second explanation is based on the idea that the secondary electrons generated as a by-product of the γ-ray irradiation create lattice damage and point defects in the quantum well structures. These point defects are primarily, vacancies, interstitial, and antisites. Usually, the above point defects possess energy levels within the band gap of the quantum wells as well as the barriers. These energy levels act as electronic traps. Thus, some of the electrons in the quantum wells known as two-dimensional electron gas (2DEG) are trapped by the irradiation-induced defects. This assertion is supported by the fact that the total integrated area under the intersubband transition spectrum is decreased as the irradiation dose is increased. It is well known that the total integrated area is directly proportional to the density of the 2DEG. Hence, when some of the electrons in the quantum wells are trapped by the irradiation induced defects, the total number of electrons that undergo the intersubband transition is reduced leading to the reduction of the total integrated area as shown in Fig. 2.

It is also worth mentioning that the total integrated area shown in Fig. 2 continues to decrease as a function of the irradiation dose without reaching a saturation plateau up to the maximum irradiation dose reached in this study. This indicates that the production of the irradiation-induced electronic traps continues to increase as a function of irradiation does.

CONCLUSIONS

We have shown in this article the γ-ray irradiation effects on the intersubband transitions in InGaAs/AlGaAs multiple quantum well structure grown on semi-insulating GaAs wafer. The results show that the intersubband transitions exhibit some degradation as the irradiation dose is increased. This degradation is inferred from the reduction of the total integrated area under the intersubband transitions as a function of irradiation dose. The reduction of the total integrated area was explained as being due to the production of irradiation-induced electronic traps in the quantum wells and the barriers. Some of the electrons in the quantum wells were proposed to be trapped by these defects causing the reduction in the total integrated areas.

ACKNOWLEDGMENTS -- This work was partially supported by the Air Force Office of Scientific Research.

REFERENCES

1. M. O. Manasreh, D. W. Fischer, and B. C. Covington, Phys. Rev. **B 37**, 6567 (1988).
2. M. O. Manasreh and D. W. Fischer, Appl. Phys. Lett. **53**, 2429 (1988).
3. M. O. Manasreh, D. W. Fischer, and G. Matous, Appl. Phys. Lett. **63**, 3038 (1993).
4. V. D. Akhmetov et al., Infrared Phys. Technology **36**, 837 (1995).
5. N. C. Das et al., IEEE Electron Device Lett. **14**, 40 (1993).
6. N. C. Das et al., International J. of Electronics **73**, 1201 (1992).

1.95 μm COMPRESSIVELY STRAINED INGAAS/INGAASP QUANTUM WELL DFB LASER WITH LOW THRESHOLD

Jie Dong*, Akinori Ubukata and Koh Matsumoto
Tsukuba Lab., Nippon Sanso Corp., 10 Ohkubo, Tsukuba, Ibaraki 300-26, Japan
*e-mail: jii_don@sanso.co.jp

Abstract

In this study, we demonstrate the growth of highly compressively strained InGaAs/InGaAsP quantum well structures with large well thickness by low pressure metalorganic chemical vapor deposition for extending the emission wavelength of lasers. By comparing the photoluminescence characteristics of quantum wells grown at different temperatures, it is clarified that a relatively high quality quantum well layer emitting at 2.0 μm can be obtained at a growth temperature of 650°C. 1.95-μm-wavelength InGaAs/InGaAsP highly compressively strained quantum well DFB laser for laser spectroscopy monitors was also fabricated. Double quantum-well DFB laser operating at 1.95 μm exhibits threshold currents as low as 6 mA and 6.2 mW maximum output powers. 2.04-μm-wavelength DFB laser is also described.

Introduction

Compressively strained InGaAs/InGaAsP quantum well material is efficacious for near 2 μm lasing emission by increasing the strain and the well thickness [1-4]. In the wavelength range of 1.2 μm to 2 μm, there are many potential applications such as remote sensing, pollutant detection and molecular spectroscopy. Laser spectroscopy is a powerful tool in those applications such as high temperature furnaces and reactive gases environments, where chemically-based monitors may not function well. Recently, various efforts have been made in those new research areas, which undoubtedly expand the applications of the InGaAsP material semiconductor lasers. We have proposed an in-situ, real time and high sensitive moisture monitor with the laser absorption spectroscopy method [5]. In addition, there are also strong absorption spectra of gases such as CH_4, CO_2, CO, NH_3, HF, HCL and HBr within this wavelength range.

Therefore, semiconductor lasers covering the wavelength range 1.2 μm~2.0 μm become more and more important. However, most of the research efforts on the InGaAsP material lasers have been made with regard to 1.3 μm and 1.55 μm devices for the optical communications. Lasers with the unconventional wavelengths, especially near 2.0 μm, are not well explored. In this paper, first, we demonstrate the crystal growth results of single quantum well (SQW) structures with high strains and thick layers grown by low pressure metalorganic chemical vapor deposition (LP-MOCVD). Photoluminescence (PL) measurements of the SQW structures were conducted to determine the strained quantum well structure and the growth conditions. Second, we present the fabrication and characteristics of quantum well DFB lasers operating near 2 μm.

Growth of strained quantum wells by MOCVD

To extend the emission wavelength of InGaAs/InGaAsP lasers, highly compressive strains and thick layers in the quantum wells are required. In our experiments, the epitaxial growth conditions of SQW structures with highly compressive strains were studied by LP-MOCVD to obtain emission wavelength as long as possible. Epitaxial growth of SQW structures was used to evaluate epitaxial layer quality and to optimize the growth conditions. A LP-MOCVD system with a vertical reactor was used. The growth pressure was 76 torr. The epitaxial growths were carried out at three different temperatures of 580°C, 620°C and 650°C . Trimethylindium (TMI) and triethylgallium (TEG) for group III sources and phosphine (PH_3) and arsine (AsH_3) for group V sources were used. Compressively strained $In_xGa_{1-x}As/InGaAsP$ SQW structures were grown on a (100) InP substrate without growth interruption at each interface. The V to III ratios in the strained $In_xGa_{1-x}As/InGaAsP$ well and the confinement layer growth were approximately 340. The optical qualities of the SQWs were evaluated by the intensity and the full width at half maximum (FWHM) of the conventional photoluminescence spectrum.

The structure was a compressively strained $In_xGa_{1-x}As/InGaAsP$ SQW, in which the indium composition was changed from x=0.53 to 0.88. The indium composition was inferred from the thick layer mismatch The well thickness was varied from 2 nm to 20 nm. Lattice matched InGaAsP with a total thickness of 400 nm and a band gap wavelength of 1.25 μm was used as the separate confinement heterostructure (SCH).

Figure 1 shows the PL emission wavelengths of the SQWs measured at room temperature. The solid lines indicate the calculated emission wavelengths between the n=1 electron-heavy hole transitions. Table I lists the parameters and values used in our calculation. It can be seen, within a relatively wide growth temperature region, the measured PL emission wavelengths of the strained SQWs are in good agreement with our calculation results. With a strain of +2.0% and a well thickness of about 10 nm, an emission near 2.0 μm can be obtained.

Table I Parameters and values used in calculation

	GaAs	InP	InAs	GaP
a(eV)	-8.0	-6.6	-6.0	-8.6
b(eV)	-1.7	-2.0	-1.8	-1.5
C_{11}^*	11.88	10.22	8.329	14.12
C_{12}^*	5.38	5.76	4.526	6.253

* C_{11} & C_{12} ($\times 10^{11}$ dyne/cm²)

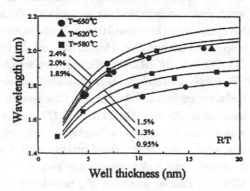

Fig. 1 PL emission wavelengths of SQWs with various strains and well widths.

Figure 2 shows the room temperature PL intensities of SQWs with a strain of +2.0% and various well thickness. Within the useful well thickness region, the SQWs grown at 650°C exhibit the strongest PL intensity compared with those grown at 580°C and 620°C. It is not clear, at present, why the SQWs grown at 620°C exhibit weaker PL intensities than those grown at 580°C. When the quantum well exceeds a certain thickness, the PL intensity decreases dramatically. From the result shown in Fig. 2, the epitaxial layer thickness up to 11.5 nm is available for a near 2.0 μm wavelength SQW structure in the case of +2.0% strain.

From the 77K PL linewidth, which is shown in Fig. 3, it can be seen that almost no broadening of the PL linewidth was observed when the strain is approximately +1%. With the increase of the strain, remarkable broadening happens. Narrowest PL linewidth of 10 meV is obtained with the well thickness of 7.5 nm in the case of +2.0% strain, which corresponds to a wavelength of 1.9 μm. This value of PL linewidth was comparable with those observed in conventional quantum well structures without strain or with slight strain. Increase of the well thickness led to a wider PL linewidth, which almost doubled when the well thickness is 11.5 nm. Further increase of the strain, for example +2.4%, extremely wide PL linewidth is observed.

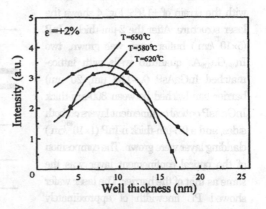

Fig.2 PL intensities of SQWs with strain of +2.0%. The measurements were carried out at RT.

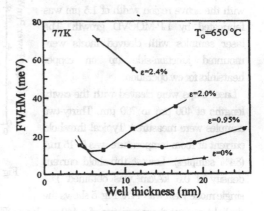

Fig. 3 PL linewidths of SQWs measured at 77K.

Laser fabrication and characteristics

According to the above results, we have fabricated double quantum-well DFB lasers with the strain of +1.9%. Fig. 4 shows the laser structure. After the 2-μm-thick p-InP $(5\times10^{17}/cm^3)$ buffer layer was grown, two $In_{0.81}Ga_{0.19}As$ quantum wells with lattice-matched InGaAsP (λ_g=1.36 μm, 20 nm) barrier sandwiched between 300-nm-thick InGaAsP optical confinement layers on both sides, and a 0.5-μm-thick n-InP $(1\times10^{18}/cm^3)$ cladding layer were grown. The composition of the optical confinement layer was the same as that of the barrier. The laser wafer showed PL linewidth of approximately 40 meV at room temperature. After the n-InP cladding was etched, first order corrugation with the pitch and the coupling factor of 298.9 nm and 87 cm^{-1} was formed by holographic lithography and wet chemical etching. A buried heterostructure with the active region width of 1.5 μm was fabricated by LP-MOCVD growth. The laser samples with cleaved facets were mounted junction-side up on copper heatsinks for cw operation.

Laser chips were cleaved with the cavity lengths of 400 μm to 760 μm. Thirty-two samples were measured. Typical threshold current at room temperature was 5~15 mA (84% samples). Lowest threshold current density of 0.6 kA/cm² was obtained The single mode yield was 44%. Fig. 5 shows the

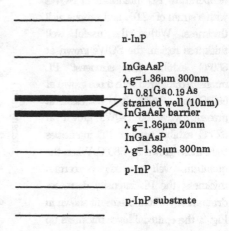

Fig. 4 Strained quantum well laser structure.

n-InP

InGaAsP
λg=1.36μm 300nm
In 0.81 Ga 0.19 As
strained well (10nm)
InGaAsP barrier
λg=1.36μm 20nm
InGaAsP
λg=1.36μm 300nm

p-InP

p-InP substrate

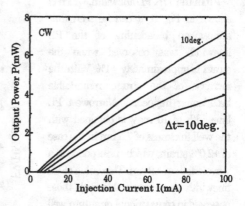

Fig. 5 Light-current characteristics of 1.95-μm-wavelength DFB laser.

cw light-current characteristics of a 440-μm-long DFB laser with κL=3.8. At 20°C, the threshold current is 6 mA and the maximum output is 6.2 mW. The maximum value of the external differential quantum efficiency is 14% and it decreases to 10% at the maximum output power (17I$_{th}$). In the temperature region of 10~40°C, the characteristic temperature T$_0$ is 34 K.

646

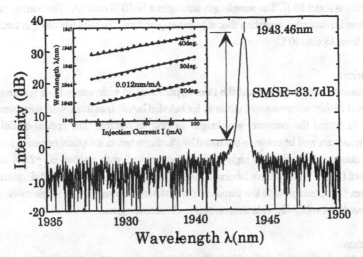

Fig. 6 Emission wavelength of DFB laser. Inset shows the tuning characteristics

Figure 6 represents the lasing emission wavelength with the injection current of 50 mA(3.4 mW). The side mode suppression ratio (SMSR) is 33.7dB. Inset shows the wavelength variation with respect to operation temperature and injection current. Single mode operation is obtained over relatively wide temperature and injection current regions. A wavelength tuning rate of 0.012 nm/mA in the same order as those of the conventional 1.3 μm and 1.55 μm lasers is observed. The total wavelength tuning range is 3.5 nm from 20~40°C. Generally, the absorption linewidths of molecules in low pressure are approximately one gigahertz, that approximately corresponds to 1 mA injection current variation in this DFB laser.

Another longer wavelength DFB laser was also fabricated by increase the quantum well thickness. Fig. 7 shows a typical light-current characteristics of the DFB laser with the quantum well thickness of 11 nm (+1.9% 2 pairs). The single longitudinal mode spectrum is also shown in the inset. The cavity length is 410 μm with cleaved facets. The wavelength is 2.043 μm with SMSR of 26 dB. At 20°C, the threshold current is 6 mA and the linear

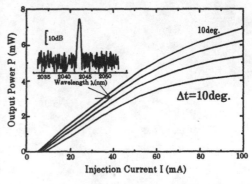

Fig. 7 Light-current characteristics of 2.04μm-wavelength DFB laser.

quantum efficiency is about 16%. In the temperature region from 10°C to 40°C, the characteristic temperature is about 59 K. The wavelength tuning rate is 0.016 nm/mA. The tuning rate at low temperature is relatively lower than that at high temperature. The total wavelength tuning range is 5.5 nm from 15°C to 40°C.

Conclusion

In conclusion, in order to expand the laser applications to wide range laser spectroscopy, we have grown highly compressively strained InGaAs/InGaAsP quantum well structures by LP-MOCVD to extend the emission wavelength as long as possible. The optical qualities of the epitaxial quantum well layers were evaluated by PL characteristics. Optical emission near 2.0 μm has been obtained in a relatively high quality quantum well with a strain of +2.0% and a well thickness of 11 nm. We also have fabricated near 2 μm strained InGaAs/InGaAsP quantum-well DFB lasers. Cw operation with low threshold current was obtained. Stable single mode operation was achieved over relatively wide temperature and current regions.

Reference

[1] J. S. Major Jr., J. S. Osinski and D. F. Welch, Electron. Lett. 29, p. 2112 (1993).
[2] M. Ochiai, H. Temkin, S. Forouhar and R. A. Logan, IEEE Photon. Tech. Lett. 7, p. 825 (1995).
[3] M. Oishi, M. Yamamoto and K. Kasaya, IEEE Photon. Tech. Lett. 9, p. 431 (1997).
[4] J.Dong, Akinori Ubukata and Koh Matsumoto, Electron. Lett., 33, p. 1090 (1997).
[5] Y. Ishihara, H. Masusaki, S-Q Wu, K. Matsumoto and T. Kimijima, Tech. Dig. International Symposium on Semiconductor Manufacturing, Austin,TX, p. 218 (1995).

MONOLITHIC 1.55μm SURFACE EMITTING LASER STRUCTURE WITH In$_{0.53}$Al$_{0.14}$Ga$_{0.33}$As/In$_{0.52}$Al$_{0.48}$As DISTRIBUTED BRAGG REFLECTOR AND SINGLE CAVITY ACTIVE LAYER GROWN BY MEATALORGANIC CHEMICAL VAPOR DEPOSITION METHOD

J.-H. Baek, B. Lee, W. S. Han*, J. M. Smith, B. S. Jeong** and E.-H. Lee,
Electronics and Telecommunications Research Institute,
Yusong P.O. BOX 106,Taejon 305-600, Republic of Korea
*Chungnam National University, Dept. of Physics, Seoul, Republic of Korea
**Yonsei University, Dept. of Physics, Seoul, Republic of Korea

ABSTRACT

Vertical cavity surface emitting laser (VCSEL) structure designed at 1.55 μm was grown by low pressure metalorganic chemical vapor deposition method. The VCSEL structure contains top and bottom distributed Bragg reflector (DBR) and single cavity active layer. The DBR was grown with In$_{0.53}$Al$_{0.14}$Ga$_{0.33}$As and In$_{0.52}$Al$_{0.48}$As quarter lambda wavelength layer, alternatively. The growth temperature was as high as 750°C to prevent ordering and phase separation of the In$_{0.52}$Al$_{0.48}$As layer. The In$_{0.52}$Al$_{0.48}$As buffer layer was subsequently grown on the InP buffer layer in order to make an abrupt uniform interface. Unity reflectance was achieved at the center of 1.55 μm with 35.5 pairs undoped DBR layer. The reflectance spectrum of undoped DBR showed a wide flat-band region (greater than 50 nm) where the reflectivity was more than 99.5 %. The center wavelength of DBR was previously determined by an in-situ laser reflectometry technique during the growth of the whole structure. An infrared laser operating at 1.53 μm, which was the design wavelength of DBR layer was used as an in-situ measuring tool. The In$_{0.53}$Ga$_{0.47}$As multiple quantum well was used as a cavity layer. The reflectance spectrum of VCSEL structure, which included a single cavity active layer, showed excellent square shaped stop band and also showed an absorption region at the center of the flat band.

INTRODUCTION

The vertical cavity surface emitting lasers(VCSEL) which are operated in the spectral regions of 780 ~ 1300 nm using GaAs, AlGaAs and In$_x$Ga$_{1-x}$(x<0.3)As materials system have been demonstrated in the past few years.[1] All the structures mentioned above were grown monolithically on the lattice matched or mismatched substrate. The study is recently focused at extension to longer wavelength (~1.55 μm) for the long distance fiber-optic communications due to its low loss in the fiber transmission. Many reports have been published for the DBR structure designed at 1.55 μm using AlAsSb/GaAsSb[2], AlAsSb/InGaAsSb[3], and InGaAsP/InP[4] heterostructures. But each material system has some problem such as relatively low contrast in refractive indices and technical difficulties in growth itself. Hence the fusion technique was proposed to overcome these problem by making high reflection mirrors and quantum well independently and fusing them each other. D. I. Babic et al. demonstrated the continuous room temperature operated 1.55 μm VCSEL with AlGaAs/GaAs distributed Bragg reflector(DBR) and InGaAsP/InP quantum well using two fusion bonding technique.[5] The future work of the fusion technique is the enhancement of the reliability and the longer life time. An InAlGaAs/InAlAs heterostructure lattice matched to InP is the attractive heterostructure for the

1.55 μm DBR due to the relatively large contrast of refractive indices as compare with InGaAsP/InP system and the structure showed a high quality of structural and optical properties. The heterostructure of $In_{0.53}Al_{0.14}Ga_{0.33}As/In_{0.52}Al_{0.48}As$ ($\Delta n \sim 0.35$) also has an applicable advantage as compared with the other materials mentioned above. The VCSEL grown with this heterostructure can be fabricated as an uncooled laser continuously operated at room temperature because the structure has an excellent electron confinement. We report the monolithically grown VCSEL structure with this material system. A highly reflecting (>99.9%) InAlGaAs/InAlAs distributed Bragg reflector (DBR), lattice matched to InP designed at 1.55 μm, were grown by low pressure metal-organic chemical vapor deposition. The cavity layer including InGaAs/InAlAs multiple quantum well emitting at a wavelength of 1.55 μm was embedded with a thickness of one lambda between top and bottom DBR. However, a previously reported $In_{0.52}Al_{0.48}As$ epitaxial layer grown at optimal growth temperature[6] showed anomalous optical behavior due to the ordering or phase separation.[7] To prevent the ordering or phase separation of InAlAs layer in this work, the growth temperature of all the layers were grown at high temperature. The laser reflectometry technique has been successfully applied as an in-situ growth monitoring tool for MOCVD grown AlAs/GaAs Bragg reflectors by measuring the oscillatory signals of buffer layers at the beginning of a run.[8] A diode laser operating at 1.53μm was used so that quarter wavelength optical thickness of each layer could be directly determined in a real time without previous measurement of thickness.

EXPERIMENT

The entire structure was grown in a vertical flow, low pressure (20 torr) MOCVD chamber. Trimethylindium (TMIn), Trimethylaluminum (TMAl), and Trimethylgallium (TEGa) were used as a metal-organic precursor. 100% pure gas of arsine (AsH_3) and phosphine (PH_3) were used as hydride sources. The growth temperature of the multi-layer stacks was 750 °C and the V/III ratio was about 80. The layers were grown on a (100) Fe-doped InP substrate rotating at a speed of 1400 rpm. Figure 1 shows a structural diagram of the VCSEL structure.

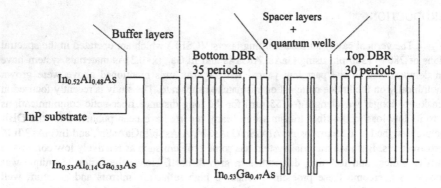

Figure 1. The schematic diagram of the $In_{0.52}Al_{0.48}As/In_{0.53}Al_{0.14}Ga_{0.33}As/$ $In_{0.53}Ga_{0.47}As$ VCSEL structure grown on InP substrate.

The abrupt interface between InP substrate and subsequently grown multistack layers including InGaAs and InAlGaAs was achieved by embedding an $In_{0.52}Al_{0.48}As$ buffer layer whose growth properties are stable at the high temperature. The VCSEL structure was subsequently grown on the two buffer layers. The entire structure consists of a 35 period bottom DBR, a λ thick cavity which includes two $In_{0.53}Al_{0.14}Ga_{0.33}As$ spacer layers and 9 nm/8 nm thick $In_{0.53}Ga_{0.47}As/In_{0.52}Al_{0.48}As$ multiple quantum wells and a 30 period top DBR. The thicknesses of the quarter-wavelength layers in the DBR are 128 nm for $In_{0.52}Al_{0.48}As$ and 116 nm for $In_{0.53}Al_{0.14}Ga_{0.33}As$. The total structure is about 17 μm thick, needing 6 hours single epitaxial run. We found that such a thick structure requires a precise lattice-matching control of the quarter-wavelength layers stack to within 5×10^{-4}. During the growth of $In_{0.53}Al_{0.14}Ga_{0.33}As$ and $In_{0.52}Al_{0.48}As$ buffer layers, the growth rate of each layer was determined by in-situ laser reflectometry. We used two different wavelength laser beams: One was 5 mW He-Ne laser operating at 0.633 μm and the other was 2 mW diode laser operating at 1.53 μm, and they were detected by Si and Ge photovoltaic detectors, respectively. Two laser beams were simultaneously entered and exited the reactor chamber, at an angle of 71° with respect to the substrate surface normal, through optical ports, which remained clear during the epitaxial growth. Only the S-polarized laser beam (TE mode) was injected to the growth chamber through polarizing filter. The period of a complete oscillatory signal, T_p, is given by $\lambda_i/2n_{eff}G$, where λ_i is the wavelength of the laser i, n_{eff} is the effective index of refraction of the layer, which takes into account the growth temperature and 71° incidence angle, and G is the epitaxial growth rate. We determined the quarter-wavelength optical thickness ($\lambda/4$) of the DBR by using a diode laser operating at 1.53 μm. The optical thickness of each layer ($\lambda/4$) is given by $(t_g \cdot T_p / 2)$, where t_g is the growth rate and T_p is the period of the oscillatory reflectance signal by 1.53 μm diode laser. The reflection beam intensity of He-Ne laser during the entire growth gave an information of the surface state whether the cross hatch pattern occurs or not. The n-type and p-type DBR structures were also grown by adding a Si_2H_6 and DEZn as a dopant sources, respectively.

RESULTS AND DISCUSSION

Epitaxial growth

A compromise between the refractive index step (Δn) and absorption must be found in the choice of quaternary alloy composition : The $In_{0.53}Al_{0.14}Ga_{0.33}As$ wavelength gap(λ = 1.48 μm) was chosen to minimize the absorption while keeping the Δn large enough. The $In_{0.52}Al_{0.48}As/In_{0.53}Al_{0.14}Ga_{0.33}As$ buffer layers were grown prior to growth of the DBR layer to determine the growth rate by in-situ laser reflectometry. In order to obtain the good surface morphology of the $In_{0.52}Al_{0.48}As/In_{0.53}Al_{0.14}Ga_{0.33}As$ at a temperature of 750 °C, a high quality InP buffer layer is essential. An InP buffer layer with good surface morphology was achieved by a temperature ramping technique. The initial growth temperature was 500 °C and then the temperature was slowly increased to 750 °C at a rate of 0.4 °C/second. As soon as the temperature reached to 750 °C, the $In_{0.52}Al_{0.48}As$ buffer layer was subsequently grown after a PH_3 purge. The $In_{0.53}Al_{0.14}Ga_{0.33}As$ buffer layer was grown on the $In_{0.52}Al_{0.48}As$ buffer layer without any hydride purge. Then the quarter-wavelength DBR layer was grown after growth of ternary and quaternary buffer layers also without any interruption at each interface. We confirmed the band edge of the $In_{0.53}Al_{0.14}Ga_{0.33}As$(1.48 μm) layer by the room temperature photoluminescence (PL). The active layer consists of 9 quantum wells of $In_{0.53}Ga_{0.47}As$ of 9 nm thickness with barrier layers of $In_{0.52}Al_{0.48}As$ of 8 nm thickness. PL measurement showed 1.55

μm emission peak of the active layer, which was grown on the InP substrate. We also have grown n-type and p-type DBR structure. The flow rates of the alkyl sources for the lattice matching condition of the n-type DBR were the same as the undoped DBR, which indicated that the Si_2H_6 did not affect the growth condition. In case of p-type DBR, however, the flow rates of the alkyl sources for the lattice matching condition were quite different from the undoped DBR. We found that the dopant source, Zn, cracked from metalorganic source reacted with In rather than Al or Ga. The mole fraction of a TMAl or TEGa had to be reduced to maintain the lattice matching condition in the p-type DBR. The reflectivity of both type of DBR 35 periods showed more than 99.5 %.

In-situ laser reflectance signal of the VCSEL

Figure 2 shows the in-situ oscillatory signal of the reflection intensity of the entire structure. $In_{0.52}Al_{0.48}As/In_{0.53}Al_{0.14}Ga_{0.33}As$ buffer layers and subsequently grown DBR layers and cavity layer as a function of the growth time simultaneously monitored by 0.633 μm and 1.5 μm lasers. The reflection intensity is in arbitrary units and has been controlled with a bias voltage applied to the photodetector. We previously reported the detailed technique of the double beam laser reflectometric method.[9]

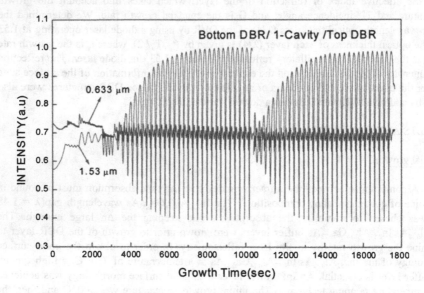

Figure 2. In-situ oscillatory reflectance signal of the VCSEL structure : $In_{0.52}Al_{0.48}As/In_{0.53}Al_{0.14}Ga_{0.33}As/In_{0.52}Al_{0.48}As$ buffer layers and subsequently grown bottom DBR, cavity layer and top DBR on InP substrate as a function of the growth time by two lasers.

During growth of the $In_{0.53}Al_{0.14}Ga_{0.33}As$ and $In_{0.52}Al_{0.48}As$ buffer layers, the period of the oscillatory signal (T) for each laser is determined. Since the wavelength of the diode laser is 1.53 μm, we can determine directly the center wavelength of the DBR as a $1.53 \times \alpha$ μm where α means the correction factor originated from the deviation angle of the laser beam path with a normal direction and from the difference of refractive indices of the multi-stack layers between growth temperature and room temperature. We found the correction factor was 1.024. The growth time of an each layer for DBR structure can be determined as $T_1/2 \times \alpha$ and $T_2/2 \times \alpha$, where T_1 and T_2 are oscillation periods for the $In_{0.53}Al_{0.14}Ga_{0.33}As$ and $In_{0.52}Al_{0.48}As$ buffer layer, respectively. We ignored the difference of the correction factor between $In_{0.53}Al_{0.14}Ga_{0.33}As$ and $In_{0.52}Al_{0.48}As$. As is shown in Figure 2, the reflection intensity monitored by 1.53 μm reached plateau after the growth of 25 periods of DBR layer, indicating that the reflectivity of the DBR reached to the unity.

Reflectance spectrum of the VCSEL

In the previous section, we had determined the center wavelength of the DBR by the product of the correction factor and the laser wavelength. But it was very difficult to have the reproducible center wavelength within 10 nm because the designed wavelength is so large that the slight deviation of the period determined by in-situ measurement shifted the center wavelength easily. Figure 3 showed the reflectance spectrum of the VCSEL structure. The reflectivity has been calibrated using a gold mirror.

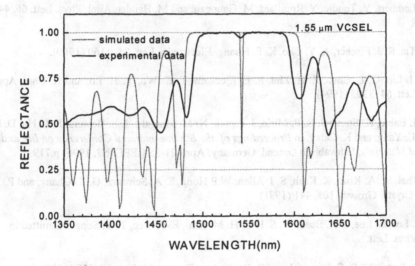

Figure 3. Reflectance spectrum of the 1.55 VCSEL structure. Solid line indicates the experimental data and dotted line indicates the simulated calculation.

The reflectance spectrum of VCSEL showed a wide flat-band region (greater than 50 nm) where the reflectivity was more than 99.5 %. As is shown in Figure 3, the width of the spectrum stop band of the simulated curve is larger than that of the experimental curve indicates that the difference of refractive indices of the mluti-stack layers is smaller than calculated values. We found the deep region in the flat band due to the cavity absorption was well controlled.

CONCLUSIONS

The VCSEL structure designed at 1.55 μm was grown by low pressure metalorganic chemical vapor deposition method. The entire structure consists of top and bottom DBR and single cavity active layer. The DBR was grown with $In_{0.53}Al_{0.14}Ga_{0.33}As$ and $In_{0.52}Al_{0.48}As$ quarter lambda wavelength layer, alternatively. The $In_{0.52}Al_{0.48}As$ buffer layer subsequently grown on the InP buffer layer produced an abrupt uniform interface. Unity reflectance was achieved at the center of 1.55 μm with 35 pairs undoped DBR layer. The reflectance spectrum of undoped VCSEL structure showed a wide flat-band region (greater than 50 nm) where the reflectivity was more than 99.5 %. The center wavelength of DBR could be previously determined by an in-situ laser reflectometry technique during the growth of the whole structure. The reflectance spectrum of VCSEL structure, which contained a single cavity active layer, showed excellent square shaped stop band and also showed an exact position of deep region at center of the band.

REFERENCES

1. Z.-H. Zhu, F. E. Ejeckam, Y. Quian, J. Zhang, Z. Zhang, G. L. Christenson, and Y. H. Lo, IEEE J. of Selected Topics in Quantum Electronics, V. 3, N. 3, 927 (1997).

2. B. Lambert, Y. Toudic, Y. Rouillard, M. Gauneau, and M. Baudet, Appl. Phys. Lett. 66, 442 (1995).

3. K. Tai, R. J. Fischer, A. Y. Cho, K. F. Huang, Electronics Lett. 25, 1160 (1989).

4. C. H. Lin, C. L. Chua, Z. H. Zhu, F. E. Ejeckam, T. C. Wu, Y. H. Lo, and R. Bhat, Appl. Phys. Lett. 64, 3395 (1994).

5. D. I. Babic, K. Streubel, R. P. Mirin, J. Piprek, N. M. Margalit, J. E. Bowers, E. L. Hu, D. E. Mars, L. Yang, and K. Carey, in *Proceedings of the 8th International Conference on InP and Related Materials*, Schwabisch Gmund, Germany, April 21-25 (IEEE, NJ, 1996) p.719

6. R. Bhat, M. A. Koza, K. Kash, S. J. Allen, W.P Hong, S. A. Schwarz, G. K. Chang, and P. Lin, J. Crystal Growth, 108, 441 (1991).

7. J. H. Lee, B. Lee, J.-H. Baek, W. S. Han, H. M. Kim, E.-H. Lee, T. Y. Sung, submitted in Appl. Phys. Lett.

8. N. C. Frateschi, S. G. Hummel, and P. D. Dapkus, Electronics Lett. 27, 157 (1991).

9. J.-H. Baek, B. Lee, S. W. Choi, J. H. Lee, and E.-H. Lee, Appl. Phys. Lett. 68, 2355 (1996)

EFFECT OF GEOMETRIC FACTORS ON POLARIZATION PROPERTIES OF VERTICAL-CAVITY SURFACE-EMITTING LASERS WITH TILTED PILLAR STRUCTURES

MIN SOO PARK*, BYUNG TAE AHN*, HYE YONG CHU**, BYUENG-SU YOO**, AND HYO-HOON PARK**
*Department of Materials Science and Engineering, Korea Advanced Institute of Science and Technology, 373-1 Koosung-dong, Yusong-gu, Taejon 305-701, Korea, btahn@cais.kaist.ac.kr
**Electronics and Telecommunications Research Institute, Yusong P. O. Box 106, Taejon 305-600, Korea

ABSTRACT

We have studied the effect of geometric factors on the polarization properties of vertical-cavity surface-emitting lasers with tilted pillar structures. Laser pillars with circular, square, diamond, and rectangular shapes were formed using reactive ion beam etching, by tilting the substrate with an angle of $15° \sim 30°$ toward the [110] or [1$\bar{1}$0] direction. We measured the polarization characteristics for the devices of $10 \sim 20$ μm size. We observed that an effective geometric factor on the polarization selectivity in tilted pillar structures is an asymmetric shape of vertical-cavity rather than an anisotropic shape of device area in planar direction.

INTRODUCTION

Vertical-cavity surface-emitting lasers (VCSELs) are considered promising light sources for applications in optical parallel processing, optical communications, and optical interconnections[1]. The control of polarization for VCSELs is the most critical issue for polarization sensitive optical systems. In VCSEL structure, the nearly degenerate orthogonal polarization states of fundamental mode are generally observed at and above the threshold. Due to the absence of the selectivity of polarization state, however, unstable polarization switching occurs during increasing current and may result in an excess intensity noise by hopping between polarization states. Several attempts have been made to control the polarization of VCSELs[2-5]. A birefringent metal/dielectric polarizer on the top distributed Bragg reflector (DBR)[2], an anisotropic transverse cavity geometry[3, 4], and metal interlaced gratings on the top DBR[5] were introduced. We also reported a polarization control method by tilted etching of laser post for index-guided structure. Using this structure, we could select a single dominant polarization state with a perpendicular electric field to the tilted direction of the laser pillar[6].

In this work, we investigate geometric effect on the polarization state of VCSELs with tilted pillar structure of circular, square, diamond, and rectangular shapes. From the results of diamond and rectangular shaped devices, we could expect that an effective geometric factor on the polarization selectivity in tilted pillar structures is an asymmetric shape of vertical-cavity rather than an anisotropic shape of device area in planar direction.

EXPERIMENT

For this work, we used a periodic gain InGaAs/GaAs structure with two-wavelength-thick

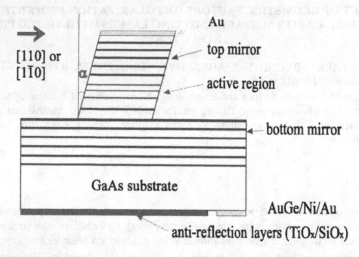

Figure 1. Schematic diagram of a tilted pillar VCSEL.

cavity. The epitaxial structure was grown on a 2°-off-(001) GaAs substrate toward <011> by metalorganic chemical vapor deposition. The top and bottom DBR mirrors consisted of 16 and 23.5 periods of AlAs/GaAs quarter wave stacks with $Al_{0.33}Ga_{0.67}As$ grading layers. The details for epitaxial structure were described in the previous reports[7-9]. We used four patterns of circular, square, diamond, and rectangular shapes for pillar structure. The sides of square and rectangular patterns were aligned along the [110] or [1$\bar{1}$0] direction and the longer side of rectangular pattern was along the [1$\bar{1}$0] direction. In rectangular pattern, the shape of overlapped region between the top contact and active region is designed to be square when the rectangular pattern is etched with $\alpha = 20°$ toward [1$\bar{1}$0] direction. Diamond pattern is a 45°-rotated shape of the square pattern. Tilted laser pillars were formed using reactive ion beam etching (RIBE) with chlorine, by tilting the substrate toward the [110] or [1$\bar{1}$0] direction against the ion beam direction. A Au(3000 Å)/photoresist-1/Al(500 Å)/photoresist-2 multilayer was used as mask layer for RIBE. The laser pillars were etched through the active region, as illustrated in Figure 1. Prior to the deposition of n-type metal layer, the backside of the wafer was first polished and antireflection (AR) layers of TiOx/SiOx were deposited. N-type contact was formed by depositing AuGe/Ni/Au layers on the edge of the substrate where the AR layers were removed in advance. Figures 2a and 2b show the scanning electron micrographs of circular and square shaped VCSELs etched by tilting toward the [110] direction with $\alpha = 15°$.

We examined various device geometries changing the tilted angle α between 15° and 30° and changing the device size d between 10 μm and 20 μm where d is defined as the diameter of circular patterns or the length of a side of square and diamond patterns or a shorter side of rectangular patterns on designed mask. The micrographs illustrate the tilted laser pillars. Two orthogonal polarization states of VCSELs were simultaneously measured using a polarized beam splitter at room temperature without a heat sink.

(a)   (b)

← [110]   ← [110]

Figure 2. Scanning electron micrographs of (a) circular and (b) square shaped VCSELs with d = 10 μm tilted toward [110] direction.

RESULTS AND DISCUSSIONS

Figure 3 shows the light output power versus current (L-I) characteristics measured under continuous wave (CW) operation for (a) circular, (b) square, (c) diamond, and (d) rectangular shaped devices of d = 10 μm and α = 30°. The dotted and solid lines represent the devices tilted along the [110] and [1$\bar{1}$0] direction, respectively. The polarization directions indicated by ⊥ and // represent the direction of electric fields perpendicular or parallel to the tilted direction of laser pillars, respectively. Figures 3 show that the light polarized perpendicularly to the tilted direction is predominantly emitted. As device size increased, devices showing reversely polarized or switching behavior increased. For the devices of d ≥ 18 μm, devices emitting unpolarized light were more than the devices emitting polarized light.

Figure 4 shows that the data of the polarization suppression ratio (PSR) for circular shaped devices with various sizes. The polarization suppression ratio was decided in terms of the ratio of the light output power of a dominant polarized state to that of a minor state at 1.5 ~ 2 times threshold current. For [1$\bar{1}$0] tilted direction, PSR seems to decrease overall as device size increases. However, the devices tilted along the [110] direction show no remarkable dependence of device size on PSR within a range of d = 10 μm and 20 μm.

Figure 5 shows the average PSR for various tilt angles and device shapes. It was averaged for devices between d = 10 μm and 20 μm. One standard deviation was set as error range. The PSR is about 15 dB for the devices tilted along the [1$\bar{1}$0] direction and about 10 dB for the devices tilted along the [110] direction. The average PSR is nearly same regardless of the device size and tilt angle within the error range.

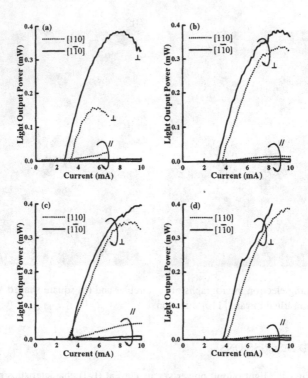

Figure 3. L-I characteristics of (a) circular, (b) square, (c) diamond, and (d) rectangular shaped
devices of d = 10 μm and α = 30°. The dotted and solid lines represent the tilted
direction.

From the results of Figures 3-5, we discuss the geometric effects on the polarization
selectivity. In our previous work[6], the selectivity of polarization in the tilted pillar structures
was interpreted by a difference in optical losses for two polarization directions, which is arisen
from asymmetry in vertical direction of the tilted cavity. We expect a higher optical loss can
be induced at the tilted side of the pillar, compared to the vertical side located parallel to the
tilted direction. The dominant polarization state could be determined along the directions
having the lower optical loss. In other work[3], it was reported that anisotropic cavity in
planar direction, like dumbbell or rhombus shape, also provides an effect on the selection of
one dominant polarization state.

In our tilted pillar structure, the devices made by circular, square, and rectangular patterns
have anisotropic geometry in planar direction of cavity, considering the overlapped area in the
vertical resonance of light. In the tilted pillar structures with diamond pattern, however, the
overlapped area maintains the same isotropic geometry regardless of tilted direction and
degree. Nevertheless, the diamond shaped devices show a selectivity of polarization, as seen
in Figure 5. Furthermore, the rectangular devices which should provide more anisotropic

658

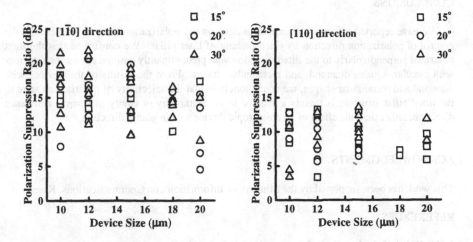

Figure 4. Polarization suppression ratio of circular devices versus device sizes for various tilt angles.

effect does not show a strong selectivity comparing to other circular and square devices, as seen in Figure 5. These results indicate that the selectivity of polarization in tilted structures is more strongly affected by the vertical asymmetry in the tilted cavity rather than an anisotropy of device area in planar direction.

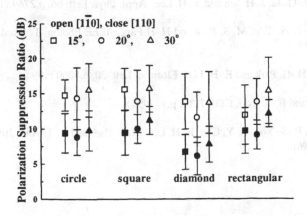

Figure 5. Average polarization suppression ratio for various tilt angles and device shapes.

CONCLUSIONS

We have reported effects of geometric factors on the polarization state of VCSELs in the control of polarization direction by tilted etching of laser pillar. We confirmed that the light polarized perpendicularly to the tilted direction was predominantly emitted for various devices with circular, square, diamond, and rectangular shapes. From the results of the devices with diamond and rectangular shapes, we could conclude that the selectivity of polarization state in the tilted pillar structure is mainly attributed to an asymmetry of cavity in vertical resonance direction, rather than the effect of an anisotropic device area in planar direction.

ACKNOWLEDGMENTS

This work has been supported by the Ministry of Information and Communications, Korea.

REFERENCES

1. Y. H. Lee, J. L. Jewell, A. Scherer, S. L. McCall, J. P. Harbison and L. T. Florze, Electron. Lett. **25**, p.1377 (1989).

2. T. Mukaihara, N. Ohnoki, Y. Hayashi, N. Hotori, F. Koyama and K. Iga, IEEE J. Selected Topics in Quantum Electron. **1**, p.667 (1995).

3. K. D. Choquette and R. E. Leibenguth, IEEE Photon. Technol. Lett. **6**, p.40 (1994).

4. T. Yoshikawa, H. Kosaka, K. Kurihara, M. Kajita, Y. Sugimoto and K. Kasahara, Appl. Phys. Lett. **66**, p.908 (1995).

5. J.-H. Ser, Y.-G. Ju, J.-H. Sin and Y. H. Lee, Appl. Phys. Lett. **66**, p.2769 (1995).

6. H. Y. Chu, B.-S. Yoo, M. S. Park and H.-H Park, IEEE Photon. Technol. Lett. **9**, p.1066 (1997).

7. B.-S. Yoo, H.-H. Park and E.-H. Lee, Electron. Lett. **30**, p.1060 (1994).

8. H.-H. Park and B.-S. Yoo, ETRI J. **17**, p.1 (1995).

9. H.-H. Park, B.-S. Yoo, H. Y. Chu, E.-H. Lee, M. S. Park and B. T. Ahn, Jpn. J. Appl. Phys. **35**, p1378 (1996).

HIGH RESOLUTION X-RAY REFLECTOMETRY AND DIFFRACTION OF CaF₂/Si(111) STRUCTURES GROWN BY MOLECULAR BEAM EPITAXY

E. ABRAMOF, S.O. FERREIRA, P.H.O RAPPL, A.Y. UETA, C. BOSCHETTI, H.CLOSS, P. MOTISUKE, and I.N. BANDEIRA

Instituto Nacional de Pesquisas Espaciais - INPE, Laboratório Associado de Sensores e Materiais - LAS, CP 515, 12201-970 São José dos Campos - SP, Brazil. abramof@las.inpe.br

ABSTRACT

CaF₂ layers were grown by molecular beam epitaxy on differently prepared Si(111) substrates. X-ray reflectivity spectra were measured and fitted. From the fitting process, the thickness of the CaF₂ layer was precisely (within 1 Å) determined and the CaF₂/Si interface roughness was also obtained. This roughness was used as an evaluation parameter for the quality of the layers. The CaF₂/Si sample from which the intentional oxide was desorpted at 800°C inside the growth chamber exhibited the most clear x-ray reflectivity spectrum with very well resolved interference fringes. The epitaxial relations of the CaF₂/Si samples grown at temperatures between 250 and 700°C were determined from x-ray diffraction analysis.

INTRODUCTION

Epitaxy of stacked BaF₂/CaF₂ buffer layers on silicon has received much attention in order to obtain the monolithic integration of lead salt detector arrays with silicon read-out circuits[1-2]. Due to problems with the BaF₂ layer during wet-processing techniques, the epitaxial growth of lead salt layers directly on CaF₂/Si has been tried. Recently, the fabrication of infrared sensor arrays of PbSnSe grown on Si with only a thin (~3 nm) CaF₂ buffer layer has been reported[3].

The large difference in thermal expansion coefficients between the CaF₂ (also lead salts) and Si builds a tensile strain during cooling after growth. Normally, this strain is relieved through the gliding of dislocations which ultimately limits the infrared detector performance. The residual strain in CaF₂/Si structures has been mainly investigated by x-ray diffraction and Rutherford back scattering channeling[4-5], and the dislocation dynamics by atomic force microscopy[6].

In this work, the structural properties of CaF₂ grown on Si(111) are investigated by grazing incidence x-ray reflectivity analysis. For this purpose, CaF₂ layers were grown by molecular beam epitaxy (MBE) on differently prepared Si(111) substrates. By fitting the measured reflectivity spectra, the CaF₂ layer thickness was precisely determined and the roughness of the CaF₂/Si interface was obtained. Combining the reflectivity analysis with reflection high energy electron diffraction (RHEED) observations, the influence of the Si preparation processes on the quality of the CaF₂ layers was evaluated. The epitaxial relations of the CaF₂/Si samples grown at temperatures between 250 and 700°C were determined by scanning the asymmetrical (224) Bragg reflections at two opposite azimuthal directions.

EXPERIMENTAL

The CaF₂/Si samples were grown in a Riber 32P MBE system equipped with proper CaF₂ and BaF₂ effusion cells. The MBE system is dedicated for the growth of IV-VI compounds

(PbTe-SnTe) on Si substrates using IIa fluoride buffer layers. The temperature of the CaF_2 effusion cell was 1250°C which leads to a growth rate of approximately 1 nm/min at the substrate temperatures used (250 to 700°C). The growth chamber had a background pressure below 10^{-10} Torr during the fluoride growth. Reflection high energy electron diffraction (RHEED) of 10 keV was used to monitor *in situ* the growth.

A high resolution x-ray materials research diffractometer (X'Pert Philips) was used for the x-ray analysis with $CuK\alpha$ radiation. The configuration for the grazing incidence x-ray reflectivity measurements was: x-ray tube in line focus; Soller slit, 1/32° divergence slit and an attenuator in the primary optics; parallel beam collimator, 0.1 mm anti scatter slit and flat crystal graphite monochromator as secondary optics. Before starting measuring, the sample height and the goniometer zero points ($2\Theta=\omega=0$) was precisely adjusted. The reflectivity spectra were recorded with an $\omega/2\Theta$ scan between $\omega=0.1°$ to $\omega=2°$. The attenuator is used for angles lower than a pre-defined angle to avoid the detector damage due to a very intense incident radiation. The attenuator factor is automatically computed allowing a dynamical scale of eight orders of magnitude during measurement. The high resolution x-ray diffraction measurements were done with the x-ray tube in point focus, a Ge(220) four crystal monochromator in primary optics and an open detector.

RESULTS AND DISCUSSION

A) X-ray Reflectivity

In order to investigate the influence of the Si substrate preparation processes on the quality of MBE grown CaF_2/Si structures, CaF_2 layers with approximately 10 nm were grown at 700°C on differently prepared Si(111) surfaces.

The Si wafers for the growth of the CaF_2 layers were prepared through four different processes, namely: (a) degreasing + thermal remove of native oxide inside growth chamber (10^{-10} Torr) at 900°C; (b) degreasing + chemical remove of native oxide immediately before load-in; (c) degreasing + chemical remove of native oxide + chemical growth of intentional oxide + thermal remove of intentional oxide inside growth chamber at 800°C; (d) degreasing + chemical remove of native oxide + chemical growth of intentional oxide + chemical remove of intentional oxide immediately before load-in.

The chemical processing steps cited above were performed in an exhaustion hood and are described below: <u>degreasing</u> by boiling in triclorethylene and methanol followed by immersion for 15 minutes in a solution of $NH_4-OH:H_2O_2:H_2O$ (1:1:4) at 75°C; <u>chemical remove of oxides</u> by dipping the wafers in a buffered fluoride acid, $NH_4F:HF$ (3:1), at room temperature for a few seconds; <u>intentional oxide growth</u> by immersing the wafers for 10 minutes at 85°C in a solution of $HCl: H_2O_2:H_2O$ (3:1:1). After each step the wafers were rinsed in H_2O DI and dried in N_2.

Figure 1 shows the x-ray reflectivity spectra for the CaF_2 layers on Si wafers prepared by the four methods described above. Each spectrum was normalized and shifted in order to allow a better comparison between them. As expected, the critical angle (Θ_C) for all samples was approximately the same (0.22°), indicating that, if the sample is well aligned, the instrument angle off-set can be neglected.

The CaF_2/Si samples in which the oxides were removed thermally inside the growth chamber (curves (a) and (c)) showed clear spectra with very well resolved interference fringes. On the other hand, for the samples which the oxides were removed chemically before load-in, the reflectivity spectra (curves (b) and (d)) showed damped interference fringes indicating a more rough interface between the CaF_2 and Si.

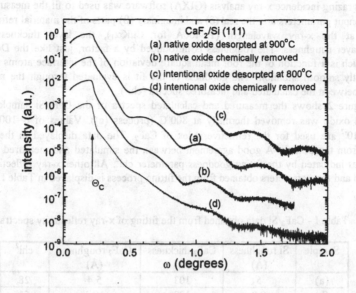

Fig. 1- X-ray grazing incidence reflectivity spectra of 10 nm CaF$_2$ layers grown by MBE on Si(111) wafers prepared by four different processes.

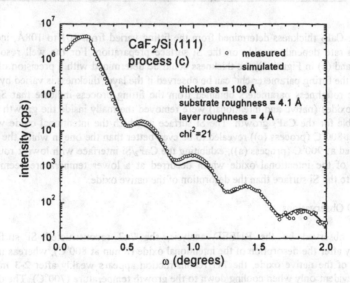

Fig. 2- Measured and calculated x-ray reflectivity spectra of CaF$_2$ grown on a Si surface from which the intentional oxide was removed inside the growth chamber at 800°C.

A grazing incidence x-ray analysis (GIXA) software was used to fit the measured spectra. The main input parameters are: the real (δ) and imaginary (β) parts of the material refractive index $n=1-\delta-i\beta$ at the x-ray wavelength (1.5406Å for $CuK\alpha_1$), the layer thickness, and the substrate/layer roughness. The roughness is introduced by a factor, just like the Debye-Waller factor, which is a function of the root mean square deviation of the interface atoms (in Å) from the perfectly smooth situation. The goodness of the fit is evaluated through the mean square deviation between the calculated and measured spectra (chi^2).

Figure 2 shows the measured and calculated spectra of the CaF_2/Si sample which the intentional oxide was removed thermally at 800°C (process (c)). Values of $\delta=100.5\times10^{-7}$ and $\beta=3.707\times10^{-7}$ are used for the refractive index of CaF_2. The data displayed in the graph was obtained from the best fit. A good agreement between the simulated and measured spectrum is observed as indicated by the fitting goodness parameter chi^2. All other x-ray reflectivity spectra were fitted and the parameters obtained from the fitting process is displayed in Table I.

Table I - CaF_2/Si data obtained from the fitting of x-ray reflectivity spectra

Sample	Si roughness (Å)	CaF_2 thickness (Å)	CaF_2 roughness (Å)	chi^2
(a)	5	103	5.4	28
(b)	9.9	102	10.1	51
(c)	4.1	108	4	21
(d)	10.8	101	10	128

The CaF_2 thickness determined from the fitting varied from 101 to 108Å, indicating that the growth rate depends weakly on the Si surface preparation. For the well resolved spectra (curve (a) and (c) in Figure 1), the thickness could be determined with a precision of 1Å, i.e., an increase in the fitting parameter chi^2 can be observed if the layer thickness is varied by 1Å.

The roughness parameters obtained from the fitting process indicate that Si surfaces in which the oxides (native and intentional) were removed thermally inside the growth chamber are more suitable for the CaF_2 growth. The Si surface in which the intentional oxide was removed thermally at 800°C (process (c)) revealed to be even better than the one in which the native oxide was removed at 900°C (process (a)), exhibiting the CaF_2/Si interface with lowest roughness. The desorption of the intentional oxide which occurred at a lower temperature seems to be less aggressive to the Si surface than the desorption of the native oxide.

B) RHEED Observations

As observed by the RHEED analysis, the 7x7 reconstructed Si surface appears immediately after the desorption of the intentional oxide (8 min at 800°C), whereas in case of the desorption of the native oxide, the 7x7 reconstruction appears weakly after 2-3 min at 900°C, becoming evident only when cooling down to the growth temperature (700°C). The desorption of the intentional oxide occurred in a more controllable and reproducible manner than the desorption of the native one.

The RHEED patterns of the Si surfaces where the oxides were chemically removed before load-in were not reconstructed even at growth temperature. They showed a bit elongated points lying on a semi-circle which is characteristic of an atomically flat surface.

As the CaF_2 layer deposition starts, the Si reconstruction disappears and the RHEED pattern changes to a streaky one. This pattern of long streaks remains until the end of the CaF_2 growth.

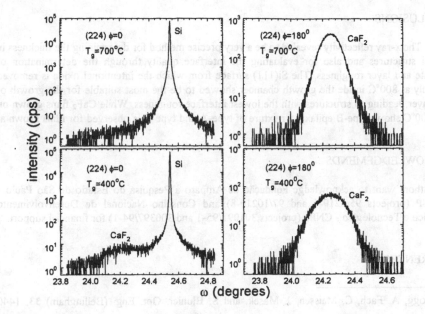

Fig. 3 - $\omega/2\Theta$ scans of (224) Bragg reflection of CaF_2/Si samples grown at 700°C and 400°C for two opposite azimuthal directions (ϕ=0 and ϕ=180°). A type-B epitaxy is observed for the sample grown at 700°C whereas a mixture of type-A and type-B epitaxy is found for the sample grown at 400°C.

C) Epitaxial Relations

The epitaxial relation of the CaF_2 layers grown on Si at substrate temperatures between 250 and 700°C was determined by scanning the (224)CaF_2 and (224)Si asymmetrical Bragg reflections at two opposite azimuthal directions (ϕ=0 and ϕ=180°). Figure 3 shows the $\omega/2\Theta$ scans of two samples with thickness of approximately 500 nm grown at 700°C and at 400°C for both azimuths.

The results indicate that for the sample grown at 700°C a clearly type-B epitaxy occurs with the crystalline orientations of the layer rotated by 180° in relation to the substrate. For the sample grown at 400°C, a mixture of type-A and type-B epitaxy was observed as indicated by the shoulder observed near the (224)Si peak. In case of a pure type-A epitaxial relation, (224)CaF_2 and (224)Si peaks should appear when scanning at the same azimuthal direction and no diffraction should occur when scanning at an 180° rotated azimuth. A pure type-A epitaxy was not observed even when growing at temperatures lower than 400°C. The samples grown at these lower temperatures were policrystalline, as indicated by the rings observed in the RHEED pattern during the CaF_2 growth. Our results are in perfect agreement with the ones published by Cho et al.[7], where he found a transition between type-A and type-B epitaxy around 400°C. It was possible to observe the type-B epitaxial relation even for the thin CaF_2 layers (10 nm) grown at 700°C. The crystalline quality of the samples grown at 700°C was much better than for the samples grown at lower temperatures as can be observed from the width of the $(224)CaF_2$ peaks in Fig. 3.

CONCLUSIONS

The x-ray reflectivity revealed to be a very precise method for determining the thickness in CaF_2/Si structures and also for evaluating the interface quality through the determination of substrate and layer roughness. The Si(111) surface from which the intentional oxide is removed thermally at 800°C inside the growth chamber showed to be the most suitable for the growth of CaF_2 layer, leading to structures with the lowest interface roughness. While CaF_2 films grown on Si at 700°C shows type-B epitaxy, a mixture of type-A and type-B is observed for films grown at 400°C.

ACKNOWLEDGEMENDS

The authors want to acknowledge Fundação de Amparo à Pesquisa do Estado de São Paulo - FAPESP (projects 95/6219-4 and 97/10210-8) and Conselho Nacional de Desenvolvimento Científico e Tecnológico - CNPq (projects 301091/95-1 and 300397/94-1) for financial support.

REFERENCES

[1] H. Zogg, A. Fach, C. Maissen, J. Masek, and S. Blunier. Opt. Eng. (Bellingham) **33**, 1440 (1994).

[2] C. Boschetti, P.H.O. Rappl, A.Y. Ueta, and I.N. Bandeira. Infrared Phys. **34**, 281 (1993).

[3] H. Zogg, A. Fach, J. John, J. Masek, P. Müller, C. Paglino, and W. Buttler. J. Electron. Materials **25**, 1366 (1996).

[4] S. Hashimoto, J.-L. Peng, W.M. Gibson, L.J. Schowalter, and R.W. Fathauer. Appl. Phys. Lett. **47**, 1071 (1985)

[5] H. Zogg, S. Blunier, A. Fach, C. Maissen, P. Müller, S. Teodoropol, V. Meyr, G. Kostorz, A. Dommann, and T. Richmond. Phys. Rev. B **50**, 10801 (1994).

[6] S. Blunier, H. Zogg, C. Maissen, A.N. Tiwari, R.M. Overney, H. Haefke, P.A. Buffat, and G. Kostorz. Phys. Rev. Lett. **68**, 3599 (1992)

[7] C.C. Cho, H.Y. Liu, B.E. Gnade, and T.S. Kim. J. Vac. Sci. Technol. A **10**, 769 (1992)

OPTICAL ABSORPTION IN $Hg_{1-x}Cd_xTe$

VAIDYA NATHAN
Air Force Research Laboratory/VSSS, 3550 Aberdeen Ave SE, Kirtland AFB, NM
87117-5776

ABSTRACT

The theory of optical absorption due to interband transitions in direct-gap
semiconductors is revisited. A new analytical expression for linear absorption
coefficient in narrow-gap semiconductors is obtained by including the nonparabolic
band structure due to Keldysh and Burstein-Moss shift. Numerical results are
obtained for $Hg_{1-x}Cd_xTe$ for several values of x and temperature, and compared with
recent experimental data. The agreement is found to be good.

INTRODUCTION

$Hg_{1-x}Cd_xTe$ is a material widely used for modern infrared detectors. The detector's
photo-responsivity is directly dependent on it's optical absorption coefficient. Besides, if
the incident radiation is of very short duration and the absorbed energy cannot be
efficiently conducted away, it can give rise to temperature increase in the material and
degrade it. Hence it is important to thoroughly study the optical absorption in this
material. However, to date there have been very few studies of this subject[1-7]. There is
only one published work[2] that predicts absorption coefficients for energies greater than the
bandgap, and this was based on the band structure model of Kane[8]. However, the
formulae obtained in Ref. 2 are cumbersome and not in readily useable form. In this paper
we present theoretical calculation of the intrinsic absorption spectrum of $Hg_{1-x}Cd_xTe$
employing a nonparabolic band structure due to Keldysh[9], taking into account the
Burstein-Moss shift. The theoretical predictions are compared with recent experimental
data[7] for several values of x and temperature. The agreement is found to be good.

ABSORPTION COEFFICIENT CALCULATION

Following Anderson[2] we start with the following equation for the linear absorption
coefficient of electromagnetic radiation of angular frequency ω in a solid with direct
energy gap E_g

$$\alpha(\omega) = \frac{\varepsilon_\infty^{0.5}}{c} \int W(k) \left[\frac{1}{1 + \exp[(E_v(k) - F)/k_BT]} - \frac{1}{1 + \exp[(E_c(k) - F)/k_BT]} \right] \frac{2d^3k}{(2\pi)^3} \quad (1)$$

where

$$W(k) = \frac{2\pi}{\hbar} |H_{vc}|^2 \delta[E_c(k) - E_v(k) - \hbar\omega]. \quad (2)$$

667

In Eqs. (1) and (2) c is the velocity of light in vacuum, $\hbar$ is Planck's constant divided by 2π, ε_∞ is the high frequency dielectric constant, H_{vc} is the interaction Hamiltonian, k_B is Boltzmann's constant, F is the Fermi energy, $\mathbf{k}$ is the wave vector, and T is the temperature. In order to evaluate the integral in Eq. (1) one needs detailed knowledge of $E_v(\mathbf{k})$ and $E_c(\mathbf{k})$. For this, we use the model described by the following equations due to Keldysh [9]

$$E_v(\mathbf{k}) = -\frac{E_g}{2}\left(1 + \frac{\hbar^2 k^2}{\mu^* E_g}\right)^{0.5}, \tag{3a}$$

$$E_c(\mathbf{k}) = \frac{E_g}{2}\left(1 + \frac{\hbar^2 k^2}{\mu^* E_g}\right)^{0.5}, \tag{3b}$$

in a system where the zero of energy is taken to be at the center of the fundamental energy gap, and μ^* is the reduced effective mass of the valence and conduction bands. After some mathematical simplifications we obtain

$$\alpha(\omega) = \frac{2q^2}{c\hbar^2}\left(\frac{\mu^* E_g}{\varepsilon_\infty}\right)^{1/2}\left[\left\{\frac{\hbar\omega}{E_g}\right\}^2 - 1\right]^{1/2} BM, \tag{4}$$

where BM is the Burstein-Moss shift given by

$$BM = \left\{1 + \exp\left[\frac{-\hbar\omega - 2F}{2k_B T}\right]\right\}^{-1} - \left\{1 + \exp\left[\frac{\hbar\omega - 2F}{2k_B T}\right]\right\}^{-1} \tag{5}$$

Eqs. (4) and (5) represent closed-form, compact formulae to calculate the absorption coefficient of narrow, direct-gap semiconductors, in contrast to the cumbersome formulae obtained by Anderson[2]. These equations were used to calculate the absorption coefficient of $Hg_{1-x}Cd_x Te$ for x values of 0.265, 0.33 and 0.443 at 300K and 4.2K as a function of photon energy. This was done separately for the light hole and heavy hole bands, and the results were added to obtain the effective absorption coefficient. The composition and temperature dependence of the energy gap was obtained from the equation [10]

$$E_g = \left[-304 + \frac{0.63T^2(1-2x)}{11+T} + 1858x + 54x^2\right] meV. \tag{6}$$

The effective masses of the electron and hole were obtained from $\mathbf{k}.\mathbf{p}$ theory[11]. According to this method the electron effective mass m_e^* is given by

668

$$\frac{m}{m_e^*} = 1 + \frac{8mP^2\pi^2}{3h^2}\left[\frac{2}{E_g} + \frac{1}{E_g + \Delta}\right],\tag{7}$$

where m is the free electron mass, P is the interband momentum matrix element, and Δ is the spin-orbit splitting. For $Hg_{1-x}Cd_xTe$, P is given[12] by $8mP^2\pi^2 / h^2 \approx 19eV$, and Δ is equal to 1 eV. The light-hole effective mass is equal to the electron effective mass. The heavy-hole effective mass is insensitive to composition and temperature. Most of the magneto-optical experiments are consistent with a heavy-hole effective mass between 0.38m and 0.53m and a value of 0.443m appears to be the best value to date[13,14], and is used in this calculation. A value of 13 was used for the optical dielectric constant of HgCdTe [15]

RESULTS

We used Eqs. (4) and (5) to calculate the intrinsic absorption spectrum, due to interband transitions, of $Hg_{1-x}Cd_xTe$ for x values of 0.265, 0.33 and 0.443 at temperatures of 4.2 and 300 K. In figures 1-3 we compare the theoretical spectra with the experimental spectra[7], and find good agreement for energies above the band gap, where the theory is valid. The absorption below the band-gap is due to impurities, defects and free carriers, with which we are not concerned here. The good agreement between theory and experment gives confidence in the accuracy of our analytical formula which is compact and easy to use.

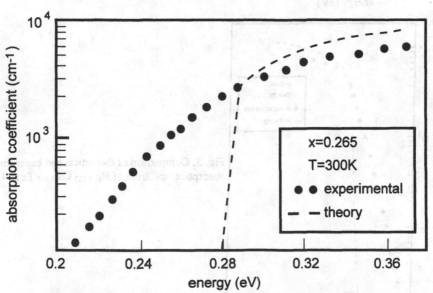

Fig. 1. Comparison of theoretical and experimental absorption spectrum of $Hg_{0.735}Cd_{0.265}Te$ at 300 K.

$$\alpha = \frac{2\pi e^2}{n c} \cdots \left| \frac{2}{3} \frac{\sin \theta}{\theta} \frac{P_{cv}}{E_g} \cdots \frac{J_{cv}}{\cdots \lambda \lambda} \right|^2 \cdots$$

where m is the free electron mass, $\cdots$

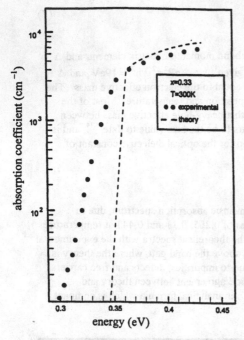

Fig. 2. Comparison of theoretical and experimental absorption spectrum of Hg $_{0.67}$ Cd $_{0.33}$ Te at 300 K.

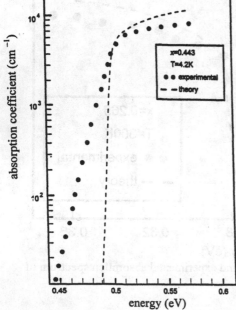

Fig. 3. Comparison of theoretical and experimental absorption spectrum of Hg $_{0.557}$ Cd $_{0.443}$ Te at 4.2 K.

REFERENCES

1. M. W. Scott, J. Appl. Phys. **40**, 4077 (1969).

2. W. W. Anderson, Infrared Phys. **20**, 363 (1980).

3. J. A. Mroczkowski, D. A. Nelson, R. Murosako, and P. H. Zimmerman, J. Vacuum. Sci. Technol. **A1**, 1756 (1983).

4. K. P. Mollman and H. Kissel, Semicond. Sci. Technol. **6**, 1167 (1991).

5. J. Chu, Z. Mi, and D. Tang, J. Appl. Phys. **71**, 3955 (1992).

6. K. H. Hermann, M. Happ, H. Kissel, K. P. Mollman, J. W. Tomm, C. R. Becker, M. M. Kraus, S. Yuan, and G. Landwehr, J. Appl. Phys. **73**, 3486 (1993).

7. B. Li, J. H. Chu, Y. Chang, Y. S. Gui, and D. Y. Tang, Infrared Phys. Technol. **37**, 525 (1996).
8. E. O. Kane, J. Phys. Chem. Solids **1**, 249 (1957).

9. L.V. Keldysh, Soviet Phys.-J. Exper. Theor. Phys. **6**, 763 (1958).

10. M. H. Weiler, in *Semiconductors and Semimetals*, eited by R. K. Willardson and A. C. Beer, (Academic Press, New York 1981) Vol 16, p.180.

11. E. O. Kane, J. Phys. Chem. Solids **2**, 181 (1960).

12 Y. Guldner, C. Rigaux, A. Mycielski and Y. Couder, Phys. Stat. Sol. **(b) 82**, 149 (1977).

13. R. W. Miles, in *Properties of Mercury Cadmium Telluride*, edited by J. Brice and P. Capper, (INSPEC, Inst. Electrical Engineers, New York 1987) p.116.

14. M. H. Weiler, R. L. Agarwal and B. Lax, Phys. Rev.**B16**, 3603 (1977).

15. J. Baars and F. Sorger, Solid State Commun. **10**, 875 (1972).

THE BURSTEIN-MOSS SHIFT IN QUANTUM DOTS OF III-V, II-VI AND IV-VI SEMICONDUCTORS UNDER PARALLEL MAGNETIC FIELD

KAMAKHYA P. GHATAK,[*] P.K. BOSE[**] AND GAUTAM MAJUMDER[**]

[*]Department of Electronic Science, University of Calcutta, University College of Science and Technology, 92, Acharya Prafulla Chandra Road, Calcutta-700 009, INDIA.
[**]Department of Mechanical Engineering, Faculty of Engineering and Technology, Jadavpur University, Calcutta-700 032, INDIA.

ABSTRACT

In this paper, we have investigated the Burstein-Moss shift (BMS) in QDs of III-V, II-VI and IV-VI semiconductors in the presence of a parallel magnetic field on the basis of newly formulated carrier disperson laws. It is found, taking QDs of InSb, Cds and CdTe as examples that the BMS increases with increasing doping and decreasing film thickness in ladder like manners. The numerical values of the BMS in QDs are much greater than that of their corresponding values for bulk specimens. The theoretical results as presented here are in agreement with the experimental observations as reported in literature.

Introduction

With the advent of fine line lithography, molecular beam epitaxy and other fabrication techniques, low dimensional structures having quantum confinement in three dimensions such as magneto size quantized systems and quantum dots (QDs) have in the last few years attracted much attention not only for their potential in uncovering new phenomena in material science but also for their interesting device applications [1]. In the former case the motion of the carriers are quantized in the three perpendicular directions in the wave-vector space and the carriers can only move due to broadening. In quantum dots, the freedom of motion of free carriers is not allowed and the density-of-states function is changed from Heaviside step function to Dirac's delta function [3].

For non-parabolic materials the absorption edge lay at much shorter wave length when it was very n-type than if it was intrinsic and the explanation of this effect is well-known [4]. Though, the Burstein-Moss shift (BMS), has been studied for various materials under different physical conditions, nevertheless it appears that the same shift in quantum confined structures has relatively been less studied [5-6]. It appears from the literature that the BMS in QDs of III-V, II-VI and IV-VI semiconductors for the more interesting case which occurs from the presence of a parallel magnetic field on the basis of newly formulated carrier dispersion laws. These materials find extensive applications in the fields of materials science and technology[7]. In this paper we shall study the doping and thickness dependences of BMS taking QDs InSb, Cds and CdTe as examples in the presence a parallel magnetic field.

Theoretical Background

The carrier energy spectra in bulk specimens of III-V, II-VI and IV-VI semiconductors can, respectively, be expressed [8]

$$E = a_{c,v} \, p^2 - b_{c,v} \, p^4 \tag{1}$$

$$E = A_{c,v} \, k_s^2 + B_{c,v} \, k_z^2 \pm C_{c,v} \, k_s \tag{2}$$

$$E = \phi_{c,v} - \alpha (\phi_{c,v})^2 + \alpha a_{2cv} \, p_y^2 / \phi_{c,v} - \alpha a'_{2v,c} \phi_{c,v} + \alpha a_{2c,v} a'_{2v,c} p_y^4 \tag{3}$$

where $a_{c,v} = \hbar^2 / 2m^*_{c,v}$, $b_{c,v} = \alpha a^2_{c,v}$, $\alpha = 1/E_g$.

$A_{c,v} = \hbar^2 / 2m_{\perp c,v}$, $B_{c,v} = \hbar^2 / 2m_{\parallel c,v}$, $C_{c,v}$ is the splitting constants of conduction or valence bands,

$\phi_{c,v} = a_{1\,c,v} p_x^2 + a_{2\,c,v} p_y^2 + a_{3\,c,v} p_z^2$, $a_{i\,c,v} = (2m_{i\,c,v})^{-1}$, $i = 1, 2$ and 3, $a'_{2v,c} = (2m'_{2v,c})^{-1}$ and the other notations have been defined in [8].

In the presence of a parallel magentic field B, along y-direction the modified carrier dispersion laws in QDs of the said materials can, respectively, be written as

$$E_{1\,c,v,\xi} = t_{1\,c,v} - \alpha \, t_{1\,c,v}^2 + (e^2 B^2 / 2m^*_{c,v}) <z^2>$$
$$-\alpha (\hbar e B / m^*_{c,v})^2 (L\pi/b)^2 <z^2> - 2\alpha (e^2 B^2 / 2m^*_{c,v}) \cdot$$
$$\cdot t_{1\,c,v} <z^2> - \alpha (e^2 B^2 / 2m^*_{c,v})^2 <z^4> \tag{4}$$

$$E_{2\,c,v,\xi} = (\hbar^2 / 2m^*_{1\,c,v}) [(L\pi/b)^2 + (R\pi/c)^2]$$
$$+ \hbar^2 / 2m^*_{2\,c,v} (n\pi/a)^2 + (e^2 B^2 / 2m^*_{2\,c,v}) <x^2> \tag{5}$$

and

$$E_{3\,c,v,\xi} = t_{2\,c,v} - \alpha t_{2\,c,v}^2 + \alpha a_{2\,c,v} \hbar^2 (R\pi/c)^2 t_{2\,c,v}$$
$$-\alpha \, a'_2 \, \hbar^2 (R\pi/c)^2 t_{2\,c,v} + \alpha a_{2\,c,v} a'_{2\,v,c} (\hbar R\pi/c)^4$$
$$+ a_{3\,c,v} e^2 B^2 <x^2> \pm 2\alpha a_{1\,c,v} a_{3\,c,v} \hbar^2 e^2 B^2$$
$$-\alpha [2a_{1c,v} a_{3c,v} (\hbar L\pi/b)^2 + a_{2\,c,v} a_{3\,c,v} (\hbar R\pi/c)^2$$

$+6 \ a_{3 \ c,v}^2 (\hbar n\pi/a)^2] e^2 B^2 <x^2> - \alpha a'_{2v,c} \ a_{3c,v} (\hbar R\pi/c)^2 .$

$. e^2 B^2 <x^2> - \alpha a_{3 \ c,v}^2 \ e^4 B^4 (x^4) \hfill (6)$

where $t_{1 \ c,v} = (\hbar^2 \pi^2 / 2m_{c,v}^*) \ [(n/a)^2 + (L/b)^2 +$

$(R/c)^2], \ <z^2> = a^2 (1/3 - 1/2\pi^2 n^2), \ \xi = L, n, R$

$<z^4> = a^4 (1/5 - 1/\pi^2 n^2 + 3/2\pi^4 N^4)$

$<x^2> = b^2 (1/3 - 1/2L^2\pi^2), \ t_{2 \ c,v} = \hbar^2 \pi^2 \ x$

$[a_{1 \ c,v}(L/b)^2 + a_{2 \ c,v}(R/c)^2 + a_{3 \ c,v}(n/a)^2],$

$<x^4> = b^4 (1/5 - 1/L^2\pi^2 + 3/2L^4\pi^4)$

The BMS can, in general, be written for the present case as

$\theta = E_g + E_F + E_{i,v,1} \hfill (7)$

where $i=1,2$ and 3 and E_F is the Fermi energy which can be related to the respective electron statistics as

$n_O = (2/abc) \sum_{n,L,r} F_{-1}(x_i) \hfill (8)$

where $x_i = (k_B T)^{-1} (E_F - E_{i,c,\xi})$, k_B is Boltzmann constant, T is temperature and $F_z(x_i)$ is the one parameter Fermi-Dirac integral of order z [9].

Results and Discussion

Using the appropriate equations together with the material constants for InSb, Cds and CdTe as given in [10] we have plotted the normalized BMS in and QDs of the said compounds versus n_o and a as shown in Figs. 1 and 2 respectively in the presence of the parallel magnetic field B in which the circular plots exhibit the experimental datas as given elsewhere [11]. The BMS oscillates both with n_o and a respectively. It may be noted that the 3D quantization in QDs leads to the discrete energy levels which produces very large changes. Under such quantization, there remains no free carrier states and consequently the crossing of the Fermi level by size qunatized subbands under 3D-quantization would have much more greater impact on the redistribution of the carriers as compared to the same for QWs. Besides presence of parallel magnetic field enhances the BMS.

In this paper we have first formulated the simple expressions of the electron statistics and the BMS for III-V, II-VI materials by formulating the new carrier dispersion laws in QDs of the said materials under parallel magentic

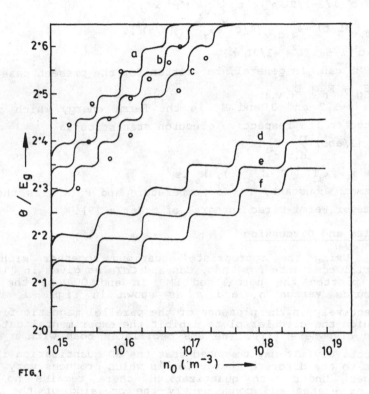

FIG.1

Fig. 1. Plot a,b and c exhibit the dependance of θ/E_g Vs n_o in QDs of InSb, CdS and CdTe in the presence of a parallel magnetic field. The plots d, e and f are for B=0. The circular plots exhibit the experimental datas as given in [10]

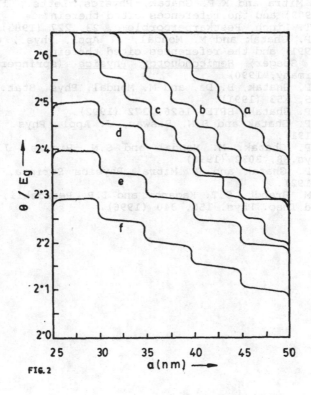

Fig. 2. Plots of the normalised BMS Vs a for all the cases of fig.1 (b=c=30nm)

field without any approximation of band constants. Besides the influnece of energy band models on the BMS of quantum confined materials can also be assessed from our work together with the experimental datas is significant. Finally it may be noted that the basic purpose of the present work is not only to investigate the BMS but also to derive the electron statistics since the formulation of the various material properties and the transport coefficients are based on the temperature dependent electron statistics in such quantum confined structures.

[1]   B. Mitra and K.P. Ghatak, Physics, Letts., 137A, 413 (1989) and the references cited therein.
[2]   M.T. Linch, Festkarperporbleme, 23, 227 (1983).
[3]   K.P. Ghatak and M. Mondal, J. Appl. Phys., 69, 1666 (1991) and the references cited therein.
[4]   K. Seeger, Semiconductor Physics (springer-Verlag, Germany, 1990)
[5]   K.P. Ghatak, B. De. and M. Mondal, Phys. Stat. So. (b) 165, K53 (1991).
[6]   K.P. Ghatak, SPIE, 1626, 372 (1992).
[7]   K.P. Ghatak and S.N. Biswas, J. Appl. Phys. 70, 4309 (1191).
[8]   K.P. Ghatak, M. Mondal and S.N. Biswas, J. Apppl. Phys.68, 3032 (1990).
[9]   K.P. Ghatak and B. Mitra, Physics Scripta, 46, 182 (1192).
[10]   L.M. Brandt, S.T. Kaganov and I.P. Petroskii, J. Exp. and Theo. Phys. 150, 340 (1996).

ON THE MOSS-BURSTEIN SHIFT IN QUANTUM CONFINED OPTOELECTRONIC TERNARY AND QUATERNARY MATERIALS

KAMAKHYA P. GHATAK,[*] P.K. BOSE[**] AND GAUTAM MAJUMDER[**]
Department of Electronic Science, University of Calcutta,
University College of Science and Technology,
92, Acharya Prafulla Chandra Road, Calcutta-700 009, INDIA.
[**] Department of Mechanical Engineering,
Faculty of Engineering and Technology, Jadavpur University,
Calcutta-700 032, INDIA.

ABSTRACT

In this paper we have studied the Burstein-Moss shift in quantum wires and dots of ternary and quaternary materials on the basis of a newly formulated electron dispersion law which occurs as a consequence of heavy doping. It is found taking $Hg_{1-x}Cd_xTe$ and $In_{1-x}Ga_xAs_yP_{1-y}$ lattice matched to InP as examples that the Burstein-Moss shift exhibits oscillatory dependences for quantum wires and dots of the said materials with respect to doping and film thickness respectively. Besides, the numerical values of the same shift is greatest in quantum dots and least in quantum wires. In addition, the theroretical analysis is in agreement with the experimental datas as given elsewhere.

Introduction

With the advent of fine line lithography, molecular beam epitaxy and other fabrication techniques, low dimensional structures having quantum confinement in two and three dimensions such as quantum wires (QWs) and quantum dots (QDs) have in the last few years attracted much attention not oly for their potential in uncovering new phenomena in material scinece but also for their interesting device applications [1]. In quantum wells (QWs), the motion of the carriers are quantized in the two perpendicular directions in the wave-vector space and the carriers can only move in the single free direction [2]. In quantum dots, the freedom-of motion of free carriers is not allowed and the density-of-states function is changed from Heaviside step function to Dirac's delta function [3].

For non-parabolic materials the absorption edge lay at much shorter wave length when it was very n-type than if it was intrinsic and explanation of this effect is well-known [4]. Though, the Burstein-Moss shift (BMS), has been studied for various materials under different physical conditions, nevertheless it appears that the same shift in quantum confined structures has relatively been less studied [5-6].

It appears from the literature that the BMS in QWs and QDs of ternary and quaternary materials has yet to be investigated on the basis of a newly formulated electron dipsersion law which occurs as a consequence of heavy doping. It would, therefore, be of much interest to study the BMS in QWs and QDs of ternary and quaternary materials.

We shall take $Hg_{1-x}Cd_xTe$ and $In_{1-x}Ga_xAs_yP_{1-y}$ lattice matched to InP as examples of ternary and quaternary materials. The compound $Hg_{1-x}Cd_xTe$ is an important optoelectronic compound, because by varying the alloy compostion its band gap can be adjusted to cover the spectral range from $0.8\mu m$ to $30\mu m$. $Hg_{1-x}Cd_xTe$ finds extensive applications in infrared detector materials and photovoltaic detector arrays in 8-12μm wave bands[7]. The quanternary alloys are widely used for the fabrication of heterojunction lasers, avalanche photodiodes, light emitting diodes and new types of integrated optical devices. We shall investigate the doping and film thickness dependences of the BMS in QWs and QDs of the aforementioned materials.

Theoretical Background

The electron energy spectrum in ternary and quaternary materials, in accordance with Kane[8], can be expressed as

$$E = a_o k^2 - b_o k^4 \qquad (1)$$

where E is the energy as measured from the edge of the unperturbed conduction band in the vertically upward direction, $a_o = \hbar^2/2m_o^*$, $\hbar = h/2\pi$, h is Planck's constant, m_o^* is the effective electron mass at the edge of the unperturbed conduction band

$$b_o = (1 - (m_o^*)(m_o)^{-1})^2 a_o^2 \left[3E_g^2 + 4\bar{\Delta}E_g + 2\bar{\Delta}^2\right] \left[E_g(E_g + \bar{\Delta})(2\bar{\Delta} + 3E_g)\right]^{-1}$$

in which m_o is the free electron mass, E_g is the unperturbed band gap and $\bar{\Delta}$ is the spin-orbit splitting constant. The modified electron energy spectrum due to heavy doping can be expressed, extending the method of D.K. Roy [9] as

$$\hbar^2 k^2/2m_o^* = \gamma_o(E)$$

where $\gamma_o(E) = G - \sqrt{H - IE}$, $G = a_o A/2B$, $A = a_o - \alpha b_o + \beta b_o$,

$\alpha = 2b_o/a_o$, $\beta = 4/a^2$, $a = (\pi\epsilon_s)^{1/2} (2\pi/6)^{1/6} (N^{-1/3}(\hbar^2/m_o e^2)$

ϵ_s is the material permittivity, $B = b_o - D\alpha\beta$, $D = C_o + D_o$,

$C_o = (4\pi N_i A_s/Nra^2) - (Ni/4\pi^2 N\Omega)(b_o\pi/a)(4\pi A_s)^2(a_o)^{-3}$,

Ni are the number of impurity atoms per N atoms of the crystal, (rystal, $H = H_o a_o^2$, $H_o = ((A^2/4B^2) - DB^{-1})$, $I = I_o a_o^2$, and $I_o = B^{-1}$ Thus the modified electron dispersion law in QWs can be written as

$$(\gamma_o(E) \, 2m_o^*/\hbar^2) = (n_x \pi/d_x)^2 + (\pi n_y/d_y)^2 + k_z^2 \qquad (2)$$

where n_x and n_y are the size quantum numbers along x and y directions respectively and d_x and d_y are the widths of QW along the said directions. The use of (2) leads to the expression of the electron concentration per unit length as

$$n_o = (2(2m_o^*)^{1/2}/\pi\hbar) \sum_{n_x, n_y} (C_1 + C_2) \qquad (3)$$

where $C_1 = \left[\gamma_o(E_{F1}) - (\hbar^2\pi^2/2m_o^*) \left((n_o/d_x)^2 + (n_y/d_y)^2 \right) \right]^{1/2}$,

$C_2 = \sum_{r=1}^{S} I_1[C_1]$, r is the set of positive integers whose upper limit is S, E_{F1} is the Fermi energy in the case of QW,

$I_1 = 2(k_B T)^{2r}(1-2^{1-2r})\xi(2r) \, d^{2r}/dE_{F1}^{2r})$, k_B is Boltzmann constant, T is temperature and $\xi(2r)$ is the zeta function of order 2r [11]. The heavy hole spectrum can be written in accordance with the warped energy band model as [12].

$$E(\hbar^2 k^2 \gamma_1/2m_o) - [(\hbar^2/m_o)\left[k^4 \gamma_2^2 - 3(\gamma_3^2 - \gamma_2^2)(k_x^2 k_y^2 + k_y^2 k_z^2 + k_z^2 k_x^2)^{1/2}\right] \qquad (4)$$

where m_o is the free electron mas and γ_1, γ_2 and γ_3 are the well-known Luttinger constants [12]. Therefore tha BMs in QWs of the said materials is given by

$$B_{1D} = E_{F1} + E_g + G_1 \qquad (5)$$

Where $G_1 = (\hbar^2\gamma_1/2m_o)(a_1+a_2) - (\hbar^2/m_o)\left[\gamma_2^2(a_1+a_2)^2\right.$

$\left. + 3(\gamma_3^2 - \gamma_2^2)(\pi^2/d_x d_y) + 3(\gamma_3^2 - \gamma_2^2) a_1 a_2\right]^{1/2}$,

$a_1 = \pi^2(d_x^{-2} + d_y^{-2})$ and $a_2 = \left[2m_o^* \hbar^{-2} \gamma_o(E_{F1}) - a_1\right]$

The BMs in QDs of ternary and quaternary compounds can also be written as

$$B_{3D} = E_g + G_2 + E_{F3} \qquad (6a)$$

where $G_2 = (\hbar^2\gamma_1 p_1/2m_o) - ((\hbar^2/m_o)\gamma_2^2 p_1^2 + 3(\gamma_3^2 - \gamma_2^2)p_2)^{1/2}$

$p_1 = \pi^2(d_x^{-2} + d_y^{-2} + d_z^{-2})$, d_z is the thickness of the QD along z-direction, $p_2 = \pi^4(d_x^{-2} d_y^{-2} + d_y^{-2} d_z^{-2} + d_z^{-2} d_x^{-2})$, E_{F3} is the Fermi energy is QD case which can be related to the electron concentration per unit volume as

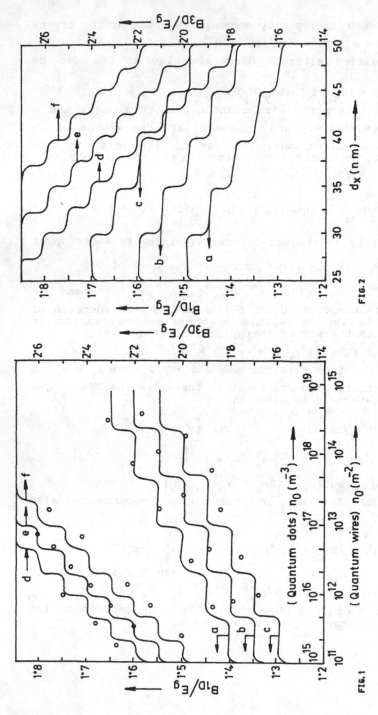

Fig. 1 Plots a, b and c exhibit the dependence of B_{1D}/E_g versus n_0 in QWs of $Hg_{1-x}Cd_xTe$, and $In_{1-x}Ga_xP_{1-y}$ lattice mathced to InP respectively. The plots d, e and f exhibit the dependence of B_{3D}/E_g versus n_0 in QDs of the said materials. The circular plots exhibit the experimental datas as given in [14] ($d_x=d=d_z=20nm$)

Fig. 2 Plots of normalized BMS versus d_x for all the cases of Fig.1 ($d=d_z=40nm$, $n_{1D}=10^{14} m^{-1}$, $n_{3D}=10^{23} m^{-3}$)

$$n_o = (2/d_x d_y d_z) \sum_{n_x, n_y, n_z} F_{-1} (g) \qquad (6b)$$

in which n_z is the size quantum number along z-direction, $g = (k_B T)^{-1} . (E_{F3} - E')$, E' can be obtained from (2) when E=E' and $k_z = n_z \pi / d_z$ and $F_j (g)$ is the one-parameter Fermi-Dirac integral of order j [13].

Results and Discussion

Using the appropirate equations together with the material constants for $Hg_{1-x} Cd_x Te$ and $In_{1-x} Ga_x As_y P_{1-y}$ lattice matched to InP as given in [10] we have plotted the normalized BMs in QWs and QDs of the said compounds versus n_o and d_x as shown in Figs. 1 and 2 respectively in which the circular plots exhibit the experimental datas given elsewhere [14].

The BMs oscillates both with n_o and d_x for both QWs and QDs respectively. It may be noted that the 3D quantization in QDs leads to the discrete eergy levels which produces very large changes. Under such quantization, there remains no free carrier states and consequently the crossing of the Fermi level by size quantized subbands under 3D-quantization would have much more greater impact on the redistribuion of the carriers as compared to the same for QWs. It is basically this impact which results in the increased numerical values of the BMs in QDs as compared to QWs.

In this paper we have first formulated the simple experssions of the electron statistics and the BMS for ternary and quaternary materials by considering the complex energy models without any approximation of band constants. Besides the influence of energy band models on the BMS of quantum confined materials can also be assessed from our work together with the fact that the agreement between the simplified analysis and the experimental datas is significant. Finally it may be noted that the basic purpose of the present work is not only to investigate the BMS but also to derive the electron statistics since the formulation of the various material properties and the transport coefficients are based on the temperature dependent electron statistics in such quantum confined structures.

[1]   B. Mitra and K.P. Ghatak, Physics, Letts., 137A, 413 (1989) and the references cited therein.
[2]   M.T. Linch, Festkarperprobleme, 23, 227 (1983).
[3]   K.P. Ghatak and M. Mondal, J. Appl. Phys., 69, 1666, (1991) and the references cited therein.
[4]   K. Seeger, Semiconductor Physics (springer-Verlag,

Germany, 1990.)

[5]   K.P. Ghatak, B.De and M. Mondal, Phys. Stat. Sol.(b) 165, K53 (1991).

[6]   K.P. Ghatak, SPIE, 1626, 372 (1992).

[7]   K.P. Ghatak, S.N. Biswas, J. Appl. Phys. 70, 4309 (1991).

[8]   G.E. Stillman, C.M. Wolfe and J.C. Dimmock, Semiconductor and Semimetal 12, 169 (1977).

[9]   D.K. Roy, Quantum Mechanical Tunnelling and Its Applications, (World Scientific, Singapore, 1988).

[10]   K.P. Ghatak and B. Mitra, Fizika, 21, 363 (1989).
        K.P. Ghatak, M. Mondal and S.N. Biswas, J. Appl. Physc. 68, 3032 (1990).

[11]   M. Abramowitz and I.A. Stegun, Handbook of Mathematical Functions (Dover, New York, 1965).

[12]   I.M. Tsidilkovskii, G.I. Karus and N.G. Shelushinna, Adv. in Phys. 34, 43 (1985).

[13]   K.P. Ghatak and B. Mitra, Physica Scripta, 46, 182 (1992).

[14]   L.M. Brandt, S.T. Kaganov and I.P. Petroskii, J. Exp. and Theo. Phys. 150, 340 (1996).

AUTHOR INDEX

SUBJECT INDEX

mechanical cracks, 241
medical imaging, 285
mercury cadmium telluride, 153
metallic
    meshes, 183
    photonic bandgap, 183
metalorganic
    chemical vapor deposition (MOCVD),
        13, 19, 643, 649
    vapor-phase epitaxy (MOVPE), 233
metastable transition, 601
mid(-)
    IR, 3, 89, 95, 101, 143, 531, 561
        detectors, 135
        lasers, 11, 19, 83
        MQW laser, 117
        (3–6 μm) lasers, 19
    wave IR, 205
molecular-beam epitaxy (MBE), 329, 371,
    377, 389, 617, 661
MOVPE, 233
multiplexer readout, 267
multistage injection lasers, 19

native point defect, 341
near-room temperature, 221
negative luminescent, 153
    sources, 143
Nomarski microscopy, 63
nonlinear optical, 567
    (NLO) material, 507, 525, 555, 581
    susceptibility, 531
nonradiative recombination, 601
nonstoichiometry, 495
nuclear medicine imaging, 267

optical
    absorption, 495, 667
    gain, 413
    losses, 481
    parametric
        generator (OPG), 561
        oscillation, 573
        oscillator (OPO), 495, 537
        recovery, 601
    optically pumped, 95
    laser, 107
organometallic
    chemical vapor deposition (OMCVD),
        507
    vapor-phase epitaxy (OMVPE), 37, 631
oxygen, 611

passivant, 589
patterning of SiO_2/Si_3N_4 epilayers, 447
PbSe, 371
$Pb_{1-x}Sn_x$ Se
PbTe/SnTe superlattices, 389
periodically poled, 537
phase modulator, 453

phase-matched, 487
phonons, 383
photoacoustic, 601
photo-assisted resonant tunneling, 607
photoconductivity, 301
    kinetics, 359
photoemission, 555
    current density, 555
photoluminescence, 37, 459, 595
    spectroscopy (PL), 365
photon
    energy, 555
    recycling, 191
photoreflectance (PR), 57
photorefractive semiconductor, 491
photoresponse, 129, 531
piezoelectric, 601
    heterostructure, 171
p-i-n devices, 273
p-InGaAs/GaAs, 617
pipes, 309
point defects, 495, 549
polarization, 531
    properties, 655
polyimide, 183
population inversion, 177
p-type dopant, 37
pulse height analysis, 253

quantum(-)
    cascade, 165
    confinement, 555
    dots, 555, 673, 679
    efficiencies, 89
    well, 95, 425, 441, 453, 467, 555, 643
        intermixing, 397, 441
        IR photodetector (QWIP), 205,
            459
    wires, 555
quasi-phase-matched (QPM), 481, 573
quaternary materials, 679

radiation detector, 253, 273
Raman scattering, 543
    spectroscopy, 365
reactive ion etching (RIE), 353
reciprocal space maps, 389
refractive index, 537
resistivity, 253
resonant tunneling diode (RTD), 607
responsivity spectra, 459
room-temperature radiation detectors, 291

scanning
    photocurrent (SPC), 625
    photoluminescence (SPL), 625
second-harmonic generation (SHG), 475,
    481, 487, 567, 573
selenium disulfide, 589